Seed Development: OMICS Technologies Toward Improvement of Seed Quality and Crop Yield

Ganesh K. Agrawal · Randeep Rakwal
Editors

Seed Development: OMICS Technologies Toward Improvement of Seed Quality and Crop Yield

OMICS in Seed Biology

Springer

Editors
Dr. Ganesh K. Agrawal
Research Laboratory for Biotechnology and Biochemistry (RLABB)
Lalitpur
Nepal

Prof. Randeep Rakwal
Graduate General Education Courses (GGEC) Program, Graduate School of Life and Environmental Sciences
University of Tsukuba
Tsukuba
Japan

ISBN 978-94-007-4748-7 ISBN 978-94-007-4749-4 (eBook)
DOI 10.1007/978-94-007-4749-4
Springer Dordrecht Heidelberg London New York

Library of Congress Control Number: 2012949569

Printed on acid-free paper

Springer is part of Springer Science+Business Media (www.springer.com)

This book is dedicated to Dominique Job and Claudette Job in recognition of their contribution to the field of seed research and development

The secret of joy in work is contained in one word—Excellence. To know how to do something well—is to enjoy it
(Pearl Sydenstricker Buck)

Foreword

Dominique grew up partly in France and partly in several African countries. Having been trained in Physics and Mathematics, he obtained his Master's degree in Physics at the University of Dakar, Senegal. Back in France, he joined the laboratory of Plant Biochemistry at the *Centre National de la Recherche Scientifique* (National Center for Scientific Research; CNRS)/Marseilles University (Prof. Jacques Ricard) to prepare a thesis during which he studied the physicochemical and enzymatic properties of various hemeproteins from plants. He was recruited at CNRS as a Research Associate in the year 1970. It was then that he met with Claudette (Research Engineer at CNRS). Claudette was working in the same laboratory and wrote a thesis on the structure of these hemeproteins, notably through establishment of their amino acid composition. After their wedding, they spent most of their scientific careers at CNRS, working together in different laboratories in Marseilles (years 1970–1992) and Lyon (since 1992).

Early works of Dominique focused on studying the oxidation mechanisms of auxin, specifically indole-3-acetic acid (IAA), by several isoperoxidases isolated from horseradish and turnip toots. Thanks to the purification to homogeneity, from hundred kilograms of turnip roots, of such enzymes, these studies relied on the use of rapid kinetic techniques (stopped-flow; T-Jump relaxation) and of spectroscopic approaches at very low temperature to considerably slow biochemical reactions and then be able to follow step-by-step, like in a movie, the reaction cycle of the enzymes during catalysis. A general mechanism was proposed to explain the mode of action of IAA as a regulator of plant growth. This work was the subject of his doctoral thesis in 1975. In particular, he clarified the possibility of oxidative degradation of auxin by peroxidases, supporting their role as auxin oxidases. Until 1978, he was during a postdoctoral fellowship at the Department of Chemistry (Professor Brian Dunford), University of Alberta, Canada; at the University of Newcastle upon Tyne, Radiation and Biophysical Chemistry Laboratory (Dr. Peter Jones), UK; and then his return to Marseilles, involved in the study of the physicochemical properties of plant peroxidases, notably toward deciphering their reactivity toward hydrogen peroxide (H_2O_2) and diverse aromatic substrates. This work contributed to understanding the mechanism of action of these hemeproteins in peroxidation reactions. The oxidation rates of a series of aromatic substrates proved to follow

Hammett relationship, allowing proposing a general mechanism in which the substrate gives an electron to the enzyme and simultaneously loses a proton. Furthermore, the second order rate constants measured for aromatic substrate oxidation approached the diffusion-controlled limit for bimolecular reactions. These results are regularly cited 30 years after their publication, to explain the features of electron transfer occurring at the heme iron in peroxidation reactions. Another highlight was the characterization by ESR spectroscopy at very low temperature of a new oxidation state of peroxidases in their reaction with H_2O_2, that is a compound which they named Compound Y, and whose formation proved to precede that of Compound I, which was previously classically assumed to be the first intermediate in peroxidation reactions.

In the year 1980, he decided to reorient his scientific activity and initiated, with Claudette, a study of the mechanisms of transcription in plants. In particular, they succeeded in purifying to homogeneity the three classes of nuclear DNA-dependent RNA polymerases from wheat germ, although these enzymes are very low abundant, extremely fragile and possess complex structure, each being composed of more than a dozen subunits. Major findings of this research were the elucidation of the mechanisms contributing to the processivity of transcription RNA chain elongation, the factors influencing the balance between abortive and productive initiation, the mechanism of action of α-amanitin (a specific inhibitor of class II DNA-dependent RNA polymerases) and the influence of the sequence and conformation of the DNA template (e.g., right-handed DNA or B-DNA and left-handed DNA or Z-DNA) on the velocity and fidelity of transcription. Part of this work was done in collaboration with the Department of Molecular Biology of the Max-Planck Institut für Biophysikalische Chemie/Institute for Biophysical Chemistry (Dr. Tom Jovin), Gottingen, Germany.

In the year 1992, Dominique and Claudette wished to move to the UMR041 CNRS/Rhône-Poulenc, a public-private joint laboratory (Prof. Roland Douce), Lyon, France, now the UMR5240 CNRS/Bayer CropScience. At the request of the Directorate of Research and Development of Rhône-Poulenc, they created a small research team on the biology of seed development, particularly on germination and initial seedling growth. The idea was to gain information on the seed system, in the context of the rapid development of commercial seed treatments to increase crop yields. Highlights of this research were the discovery of a new class of biotinylated proteins (SBP65, Seed Biotinylated Protein of 65 kDa), which belongs to the family of LEA (Late Embryogenesis Abundant) proteins and that is specifically expressed in seeds, as well as a mechanism of structural reorganization of the 11S globulin, a main storage protein present in seeds of dicotyledonous plants.

By that time, Dominique continued to be interested in enzymology and was involved, with several colleagues of the CNRS/Rhône-Poulenc joint laboratory, in several studies dealing with the characterization of the enzymatic pathways that in plants are responsible for essential amino acids (sulfur amino acids, branch chain amino acids) and vitamin (biotin) metabolisms.

More recently, in 1999, realizing the invaluable potential afforded by the knowledge of the first genome sequence of a plant, namely that of *Arabidopsis*, Dominique and Claudette decided to broaden the scope of their research toward the global elucidation of the germination process by a systematic study of the proteome and transcriptome of this model plant. Their publication in 2001 in Plant Physiology on the *Arabidopsis* seed proteome was probably one of the leading publications in plant proteomics combining two-dimensional gel electrophoresis and mass spectrometry. The development of reference protein maps then allowed approaching several questions important in seed biology, as to the role of gibberellins in seed germination, the influence of abscisic acid on seed dormancy, and the role of protein oxidation (carbonylation) in plants and seeds during development. In addition, an important discovery dealt with the role during germination of messenger RNAs that have been stored in the dry mature seeds during their maturation on the mother plant. This work showed that germination was insensitive to α-amanitin and therefore can occur in the absence of *de novo* transcription[1]. This feature, being originally documented in *Arabidopsis*, has now been extended by other authors to other species, including rice.

Dominique and Claudette were also always willing to translate their work from model to crop seeds. To this end, they developed programs in close collaboration with industry and the seed sector. The benefits of this research have included the development of biochemical markers for monitoring priming (an invigoration treatment of low-vigor seedlots widely used in the seed industry) and patents have been filed.

In parallel to these research activities Dominique was successively deputy head and head of the joint laboratory CNRS/Rhône-Poulenc then CNRS/Bayer CropScience. Since 2000, he was also heavily involved in collective tasks. Thus, he has been for several years the scientific coordinator of the Genoplante programs, the French program in plant genomics. Soon, he was willing to establish cooperation with other plant genomics programs. In 2000, he set up together with Jens Freitag the first joint GABI-Genoplante programs (GABI is the German plant genomics program funded by the BMBF). A few years later, with the enthusiastic help of Pablo Vera, they set up trilateral programs in plant genomics, gathering together France, Germany, and Spain, with the idea of strengthening public-private partnership. In particular, these actions resulted in the creation of the ERA-NET PG (European Research Area in Plant Genomics; an instrument of the EU to promote joint research in Europe), which brought together 16 European member states and Canada. However, the trilateral projects between France, Germany, and Spain continued to exist, giving rise to the Plant-KBBE program, in which now Portugal and Canada are partners. Altogether, these activities have enabled the selection and the funding of hundreds of plant genomics projects and helped strengthening this discipline in France and Europe, which is important in the context of the development of novel uses of plant products and the necessity to increase crop yields in a more sustainable agriculture. These initiatives also allowed creating a series of international conferences on plant genomics, the Plant GEMs (Plant Genomics European Meetings), of

[1] Rajjou, L, Duval M, Gallardo K, Catusse J, Bally J, Job C, Job D (2012) Seed Germination and Vigor. Annu Rev Plant Biol 63:507–533.

which Dominique organized in Lyon the third meeting (22–25 September 2004), which brought together 650 participants. None of these actions could have been developed without the constant support of Research Ministries and Funding Agencies. Yet, they demonstrated the power of combining bottom-up and top-down approaches.

More recently, Randeep Rakwal, Ganesh Kumar Agrawal, and Dominique have been active toward the establishment of an International Plant Proteomics Organization (INPPO)[2] to which have now joined many colleagues around the world (http://www.inppo.com/). The common goal is to establish the complete proteome of several plants, models and crops, as in being done in HUPO for the human proteome (http://www.hupo.org/). It is anticipated that this will help toward establishing sustainable agriculture in the context of an increasingly growing world population.

Dominique is a member of the Editorial Board of Molecular & Cellular Proteomics, Seed Science Research, and Associate Editor of Frontiers in Plant Proteomics. Together with Paul A. Haynes and Michel Zivy, he was the guest editor of a Special issue of PROTEOMICS dedicated to plant proteomics, which was published in May 2011. He is consulting professor at AgroParisTech and associate member of the French Academy of Agriculture.

Among the many successful collaborations that Dominique and Claudette have developed over the years, none have proved more successful and rich, both with respect to results and the work as a team, than those with their Ph.D. students and postdoctoral fellows: Jacques Dietrich, Laure de Mercoyrol, Yves Corda, Alain Kersulec, Manuel Duval, Bertrand Gakière, Karine Gallardo, Loïc Rajjou, Lucie Miché, Georgia Tanou, Julie Catusse, Rafika Yacoubi, and Julia Bally.

Still today, Dominique and Claudette continue their scientific journey toward '*nothing less than the excellence in their chosen fields of study*'. Last but not the least, their contributions to seed biology research and development will be remembered for many, many years to come.

[2] Agrawal GK, Job D, Zivy M, Agrawal VP, Bradshaw R, Dunn MJ, Haynes PA, van Wijk KJ, Kikuchi S, Renaut J, Weckwerth W, Rakwal R (2011) Time to articulate a vision for the future of plant proteomics—a global perspective. An initiative for establishing the international plant proteomics organization (INPPO). Proteomics 11:1559–1568.

Preface

The beginning of the twenty-first century is certainly a great time for plant biology research. The beginning of the new millennium has placed an ever growing amount of sophisticated technologies at the disposal of the modern scientist to the benefit of all. The research performed with these technologies has the potential to provide the answers to important and complex biological questions and problems (especially those relating to crop plants and the human food supply) within reach.

Of those technologies, transcriptomics, proteomics, and metabolomics are increasingly being utilized for the near complete understanding of plant biology including seed development. These technologies are also loosely called the twenty-first century omics technologies, which have revolutionized the way research was being performed in the past in the field of seed developmental biology. Since the year 2000 these technologies have been largely optimized for model and nonmodel plants. As far as seed development is concerned, these technologies have already generated huge amount of data and continue to do so with astonishing pace. Those data have been organized and integrated in an efficient and confident manner using systems biology approaches. Generated data are being exploited for seed and yield improvement by combing these technologies with breeding and other classical molecular approaches. Indeed, when one looks at the progress achieved to date in the field of seed development and its overall impact on biological research, it is clear just how essential these technologies are to our understanding of the physiology and biology of any organism. It would not be far from the truth to say that the "power of technologies" (and indeed the omic sciences as a whole) is one of the driving forces of the twenty-first century seed biology research. The principles of good science are as true in this age of omics still hold true today and the disciplined scientist must keep these principles in mind to avoid rushing blindly into the field (intentionally or unintentionally) without first obtaining a thorough understanding of its fundamental principles.

When one looks at the impressive progress of above mentioned technologies in seed development as well as its immense importance in biological sciences as a whole, it is clear that there was a need for a textbook of the subject to translate/disseminate the knowledge acquired by leading experts in the field to the wider scientific community for some time. This was the impetus for the book you are currently reading. Though we knew that such a project would be a formidable challenge, we also knew that it would bring us the opportunity to work closely with the leading experts of the field. What we did not fully appreciate when we started was how much of a truly unparalleled experience it would be to work with each and every one of the contributors of this book, whom we genuinely thank for being part of this ambitious endeavour.

This book is composed of seven sections including appendix in the following order: introduction to seed development and omics technologies/transcriptomics/proteomics/metabolomics/towards systems biology/and discovery-driven seed and yield improvement. There are 26 chapters which between them provide excellent coverage of almost all the studies conducted to date on seed development toward improvement of seed quality and crop yield. Each chapter also provides basic knowledge tightly associated with that particular topic. Seed physiology and developmental patterns have been discussed for most of the crops that are being widely utilized as research material; one might think of redundancy but we believe that those detailed descriptions are necessary for observed/recorded subtle differences, if any, associated with different seed species and to avoid any confusion especially for students. More than 1,000 references serve as a great resource for the academic and nonacademic communities. We hope this book will be beneficial in scope and practical knowledge to you the readers, whose response will be the final judge on the validity of the work. Moreover, the editing and organization of book have been done in a way that the book and its contents can also be used as text book for all level of students. This book is also dedicated to Dominique Job and Claudette Job in recognition of their contribution to the field of seed research and development (Foreword).

We also wish to thank our colleagues and collaborators around the world with whom we have struggled to do "good science", forming new partnerships and friendships in the process. Though during our long journey in science many persons had an effect on us, but Masami Yonekura (Ibaraki University, Japan), Shigeru Tamogami (Akita Prefectural University, Japan), Akihiro Kubo (National Institute of Environmental Sciences, Japan), Nam-Soo Jwa (Sejong University, South Korea), Oksoo Han and Kyoungwon Cho (Chonnam National University, South Korea), Shoshi Kikuchi (National Institute of Agrobiological Sciences, Japan), Yu Sam Kim and Hyung Wook Nam (Yonsei University, South Korea), Kyu Young Kang and Sun Tae Kim (Gyeongsang National University and Pusan National University, South Korea), and Oliver A.H. Jones (University of Cambridge, UK) deserve both mention and appreciation. We would also like to acknowledge two young people, Abhijit Sarkar (Banaras Hindu University, India and Administrative Officer at International Plant Proteomics Organization—INPPO, www.inppo.com) and Raj Agrawal (Computer Programmer and Webpage Administrator at INPPO) for their constant support in our scientific achievements. We would especially like to thank Prof. Vishwanath Prasad Agrawal (RLABB, Kathmandu, Nepal) for his directions and guidance in our research (this is especially true for Ganesh who started his research under Prof. Vishwanath's watchful eyes).

It is without doubt that the support of the Editorial Team (Scientific, Technical, Medical, and Scholarly Division) at Springer (The Netherlands), especially the Editorial Director, Jacco Flipsen and the Publishing Assistant, Ineke Ravesloot, was instrumental in bringing this book out to light. We greatly appreciate their professional support and patience with our queries and correspondence.

Finally, to this long list of supporters, we must add our thanks for the personal sacrifices by our families, especially our wives [Mrs. Nitu Agrawal for Ganesh Kumar Agrawal; Mrs. Junko Shibato Rakwal for Randeep Rakwal] and children [Mr. Dakshit Agrawal (son) and Ms. Divya Agrawal (daughter) for Ganesh Kumar Agrawal; Mr. Aryan Shibato Rakwal (son) for Randeep Rakwal], who have allowed us to devote time meant for them into efforts for completing this project. Randeep's wife Junko Shibato has also contributed greatly with the technical aspects of the book at both laboratory and home. Our parents [Mrs. Savitri Devi (mother) and Late Sri Gajanand Madwari (father) for Ganesh Kumar Agrawal; Mrs. Meera Rakwal (mother) and Brigadier (retd.) Om Parkash Rakwal (father) for Randeep Rakwal], who brought us into this world and taught and inspired us to contribute to the society and do our job well and in the right way under any circumstances, also deserve special mention.

To you the reader we also extend our thanks and appreciation. We hope this work will be useful to you.

Ganesh Kumar Agrawal and Randeep Rakwal

Contents

Editors Biographies

Ganesh Kumar Agrawal Ph.D. is the Associate Director of RLABB, a non-profit research organization focusing on biotechnology and biochemistry, in Kathmandu, Nepal. He is also a multidisciplinary scientist working on the global problem of food security, safety, and human nutrition using proteomics and its integration with other high-throughput and targeted omics techniques. Dr. Agrawal has been working on different aspects of plant proteomics since the year 2000 when proteomics, a high-throughput genome-enabled technology, was beginning to be utilized in plants to comprehend our understanding on plant biology at the system level. Along with Randeep, he has edited a comprehensive book entitled "Plant Proteomics: Technologies, Strategies, and Applications" (John Wiley & Sons, NY, USA, 2008). He received his M.Sc. in Physical Chemistry in 1989 from Nepal (Tribhuvan University, Chemistry Department, Kathmandu) and was awarded Ph.D. in Applied Biological Chemistry (Biochemistry & Biotechnology) in 1998 from Tokyo University of Agriculture and Technology, Tokyo, Japan. He did extensive post-doctoral research work in Japan (Genomics and Proteomics and Metabolomics-related biological questions) and the U.S.A. (Proteomics), before returning to Nepal for promotion of education and research for the younger generation. He continues to work closely with collaborators in India, South Korea, and Japan on various projects related to use of omics technologies in plants. He is also an initiator of the global initiative on plant proteomics, International Plant Proteomics Organization, INPPO (www.inppo.com).

Randeep Rakwal Ph.D. is a Professor of Graduate General Education Courses (GGEC) program at the University of Tsukuba (Tsukuba, Japan). Dr. Rakwal did B.Sc. (Hons.) Botany (Delhi University, 1989) and M.Sc. Agriculture (Plant Pathology) from G. B. Pant University of Agriculture & Technology (1992) in India. He completed Ph.D. in Biochemistry and Biotechnology (Tokyo University of Agriculture and Technology, Japan, with

Monbusho Scholarship from Government of Japan) in 1997. He has researched on environmental stress biology focusing on jasmonic acid, ozone, radiation, and other biotic abiotic stresses in rice model using "omics" approaches. Working with DNA-microarray based transcriptomics, proteomics and targeted metabolomics with close collaborators and experts in their respective fields, especially in Japan, South Korea, Nepal and India. In Japan, he also contributes to research in human health and mental stress with colleagues at Showa University School of Medicine (Department of Anatomy), Tokyo and Toho University (Laboratory of Neuroscience), Chiba, in Japan. He is also one of the initiators of INPPO.

Contributors

Aurélie Baldy Institut Jean-Pierre Bourgin, Unité Mixte de Recherche 1318, INRA-AgroParisTech, team "Physiology of Seed Germination", INRA Centre de Versailles-Grignon, Route de St-Cyr (RD10), 78026, Versailles Cedex, France

Vagner A. Benedito West Virginia University, Morgantown, WV, USA

Julie Catusse CNRS/Bayer CropScience Joint Laboratory (UMR5240), Bayer CropScience,14-20 rue Pierre Baizet, 69263, Lyon cedex, France

Quintiles – Navigating the new health, Parc d'innovation, rue Jean Dominique Cassini, 67404, illkirch cedex, France

Sreekala Chellamma Dow AgroSciences LLC, 9330 Zionsville Road, Indianapolis, IN 46268, USA

Gwendal Cueff Institut Jean-Pierre Bourgin, Unité Mixte de Recherche 1318, INRA-AgroParisTech, team "Physiology of Seed Germination", INRA Centre de Versailles-Grignon, Route de St-Cyr (RD10), 78026, Versailles Cedex, France

AgroParisTech, Unité de Formation et de Recherche de Physiologie Végétale, 16 rue Claude Bernard, 75231, Paris Cedex 05, France

Catherine Damerval Université Paris-Sud, UMR 0320/UMR 8120 Génétique Végétale, 91190, Gif sur Yvette, France

Howard Davies James Hutton Institute, Invergowrie, DD2 5DA, Dundee, UK
e-mail: howard.davies@hutton.ac.uk

Robert C. Day Department of Biochemistry, University of Otago, Dunedin 9054, New Zealand

Martine Devic Seed Development Group, Laboratoire Genome et Developement des Plantes, IRD-UPVD-CNRS UMR5096, IRD Centre de Montpellier, 911 avenue Agropolis BP5045, 34032, Montpellier cedex 1, France

Sangeeta Dhaubhadel Southern Crop Protection and Food Research Centre, Agriculture and Agri-Food Canada, N5V 4T3, London, Ontario, Canada
e-mail: sangeeta.dhaubhadel@agr.gc.ca

Frédéric Domergue Laboratoire de Biogenèse Membranaire, CNRS UMR 5200, Université Bordeaux Segalen, INRA Aquitaine BP 81, Villenave d'Ornon, France

Stéphane Dussert IRD, UMR DIADE, Diversity, Adaptation and Development of Cultivated Crops, BP 64501, 34394, Montpellier, France

Matthias Erb Root-Herbivore Interactions Group, Max Planck Institute for Chemical Ecology, Hans-Knoell-Str. 807745 Jena, Germany

Aaron Fait Jacob Blaustein Institutes for Desert Research, French Associates Institute for Agriculture and Biotechnology of Drylands Midreshet Ben-Gurion, Ben-Gurion University of the Negev French Associates, 84990 Negev, Israel
e-mail: fait@bgu.ac.il

Marc Galland Institut Jean-Pierre Bourgin, Unité Mixte de Recherche 1318, INRA-AgroParisTech, team "Physiology of Seed Germination", INRA Centre de Versailles-Grignon, Route de St-Cyr (RD10), 78026, Versailles Cedex, France
e-mail: marc.galland@versailles.inra.fr

AgroParisTech, Unité de Formation et de Recherche de Physiologie Végétale, 16 rue Claude Bernard, 75231, Paris Cedex 05, France

Karine Gallardo INRA (National Institute for Agronomic Research), UMR1347 Agroécologie, BP 86510, 21065 Dijon, France
e-mail: Karine.Gallardo@dijon.inra.fr

Neeru Gandotra Department of Molecular, Cellular & Developmental Biology, Yale University, P. O. Box 203104, New Haven, CT 06520-8104, USA

Eva Grafahrend-Belau Leibniz-Institute of Plant Genetics and Crop Plant Research (IPK), Correns str.3, Gatersleben, Germany

Abdelali Hannoufa Agriculture and Agri-Food Canada, Southern Crop Protection and Food Research Centre, 1391 Sandford Street, N5V 4T3, London, Ontario, Canada
e-mail: Abdelali.Hannoufa@agr.gc.ca

Rowan P. Herridge Department of Biochemistry, University of Otago, Dunedin 9054, New Zealand

Claudette Job CNRS/Bayer CropScience Joint Laboratory (UMR5240), Bayer CropScience, 14-20 rue Pierre Baizet, 69263, Lyon cedex, France

Dominique Job CNRS/Bayer CropScience Joint Laboratory (UMR5240), Bayer CropScience, 14-20 rue Pierre Baizet, 69263, Lyon cedex, France
e-mail: dominique.job@bayer.com

AgroParisTech, Unité de Formation et de Recherche de Physiologie Végétale, 16 rue Claude Bernard, 75231, Paris Cedex 05, France

Unité Mixte de Recherche 5240, CNRS-Université Claude Bernard Lyon-Institut National des Sciences Appliquées-Bayer Crop Science, Bayer Crop Science, Centre de Recherche de La Dargoire, 14-20 rue Pierre Baizet, 69263, Lyon Cedex 9, France

Thierry Joët IRD, UMR DIADE, Diversity, Adaptation and Development of Cultivated Crops, BP 64501, 34394, Montpellier, France
e-mail: thierry.joet@ird.fr

Mark L. Johnston USDA, Agricultural Research Service, Plant Genetics Research Unit, University of Missouri, Columbia, MO 65211, USA

Björn H. Junker Leibniz-Institute of Plant Genetics and Crop Plant Research (IPK), Correns str.3, Gatersleben, Germany
e-mail: junker@ipk-gatersleben.de

Kameswara Rao Kottapalli Center for Biotechnology and Genomics, Texas Tech University and Cropping Systems Research Laboratory, USDA-ARS, 3810 4th Street, Lubbock, TX 79415, USA

Pratibha Kottapalli Center for Biotechnology and Genomics, Texas Tech University and Cropping Systems Research Laboratory, USDA-ARS, 3810 4th Street, Lubbock, TX 79415, USA

Christian Krach Leibniz-Institute of Plant Genetics and Crop Plant Research (IPK), Correns str. 3, Gatersleben, Germany

Christine Le Signor INRA (National Institute for Agronomic Research), UMR 1347 Agroécologie, BP 86510, 21065 Dijon, France

René Lessire Laboratoire de Biogenèse Membranaire, CNRS UMR 5200 - Université Bordeaux Segalen, INRA Aquitaine BP 81 - Villenave d'Ornon - France
e-mail: rene.lessire@biomemb.u-bordeaux2.fr

Tie Liu Department of Molecular, Cellular & Developmental Biology, Yale University, P. O. Box 203104, New Haven, CT 06520-8104, USA

Department of Plant Biology, Carnegie Institute of Washington, 260 Panama Street, Stanford, CA 94305, USA

Imen Lounifi Institut Jean-Pierre Bourgin, Unité Mixte de Recherche 1318, INRA-AgroParisTech, team "Physiology of Seed Germination", INRA Centre de Versailles-Grignon, Route de St-Cyr (RD10), 78026, Versailles Cedex, France

AgroParisTech, Unité de Formation et de Recherche de Physiologie Végétale, 16 rue Claude Bernard, 75231, Paris Cedex 05, France

Richard C. Macknight Department of Biochemistry, University of Otago, Dunedin 9054, New Zealand
e-mail: richard.macknight@otago.ac.nz

Domenica Manicacci Université Paris-Sud, UMR 0320/UMR 8120 Génétique Végétale, 91190, Gif sur Yvette, France

Frédéric Marsolais Southern Crop Protection and Food Research Centre, Agriculture and Agri-Food Canada, N5V 4T3, London, Ontario, Canada

Guillaume Marti School of Pharmaceutical Sciences, EPGL, University of Geneva, University of Lausanne, Quai Ernest-Ansermet 30, 1211 Geneva, Switzerland
e-mail: guillaume.marti@unige.ch

Fumio Matsuda RIKEN Plant Science Center, Tsurumi-ku, 230–0045, Yokohama, Japan

Valérie Méchin Institut Jean-Pierre Bourgin, INRA, UMR 1318 INRA-AgroParisTech, 78000, Versailles, France

Ján A. Miernyk USDA, Agricultural Research Service, Plant Genetics Research Unit, University of Missouri, Columbia, MO 65211, USA
e-mail: MiernykJ@missouri.edu

Department of Biochemistry, University of Missouri, Columbia, MO 65211, USA

Interdisciplinary Plant Group, University of Missouri, Columbia, MO 65211, USA

Halima Morin Institut Jean-Pierre Bourgin, Unité Mixte de Recherche 1318, INRA-AgroParisTech, team "Physiology of Seed Germination", INRA Centre de Versailles-Grignon, Route de St-Cyr (RD10), 78026, Versailles Cedex, France

Timothy Nelson Department of Molecular, Cellular & Developmental Biology, Yale University, P. O. Box 203104, New Haven, CT 06520-8104, USA
e-mail: timothy.nelson@yale.edu

Mélanie Noguero INRA (National Institute for Agronomic Research), UMR 1347 Agroécologie, BP 86510, 21065 Dijon, France

Brendan O'Leary Department of Biology, Queen's University, K7L 3N6, Kingston, ON, Canada

Rodomiro Ortiz Swedish University of Agricultural Sciences, Department of Plant Breeding and Biotechnology, Sundsvagen 14, Box 101, SE 23053 Alnarp, Sweden
e-mail: rodomiro.ortiz@slu.se

Paxton Payton Center for Biotechnology and Genomics, Texas Tech University and Cropping Systems Research Laboratory, USDA-ARS, 3810 4th Street, Lubbock, TX 79415, USA
e-mail: paxton.payton@ars.usda.gov

Bhinu Pillai Dow AgroSciences LLC, 9330 Zionsville Road, Indianapolis, IN 46268, USA

William C. Plaxton Department of Biology, Queen's University, K7L 3N6, Kingston, ON, Canada
e-mail: plaxton@queensu.ca

Hart Poskar Leibniz-Institute of Plant Genetics and Crop Plant Research (IPK), Correns str.3, Gatersleben, Germany

Jean-Louis Prioul Institut de Biologie des Plantes, Université Paris-Sud, Bât. 630, UMR 8618, CNRS, 91405, Orsay, Cedex, France
e-mail: jean-louis.prioul@u-psud.fr

Loïc Rajjou Institut Jean-Pierre Bourgin, Unité Mixte de Recherche 1318, INRA-AgroParisTech, team "Physiology of Seed Germination", INRA Centre de Versailles-Grignon, Route de St-Cyr (RD10), 78026, Versailles Cedex, France
e-mail: loic.rajjou@agroparistech.fr

AgroParisTech, Unité de Formation et de Recherche de Physiologie Végétale, 16 rue Claude Bernard, 75231, Paris Cedex 05, France

Ombretta Repetto INRA (National Institute for Agronomic Research), UMR 102 Genetics and Ecophysiology of Grain Legumes, BP 86510, 21065 Dijon, France
e-mail: ombrettarepetto@gmail.com

Thomas Roscoe Seed Development Group, Laboratoire Genome et Developement des Plantes, IRD-UPVD-CNRS UMR5096, IRD Centre de Montpellier, 911 avenue Agropolis BP5045, 34032, Montpellier cedex 1, France

Serge Rudaz School of Pharmaceutical Sciences, EPGL, University of Geneva, University of Lausanne, Quai Ernest-Ansermet 30, 1211 Geneva, Switzerland

Paolo A. Sabelli School of Plant Sciences, University of Arizona, 303 Forbes Hall, Tucson, AZ 85721, USA
e-mail: psabelli@ag.arizona.edu

Kazuki Saito RIKEN Plant Science Center, Tsurumi-ku, 230–0045, Yokohama, Japan

Kai Schallau Leibniz-Institute of Plant Genetics and Crop Plant Research (IPK), Correns str.3, Gatersleben, Germany

Falk Schreiber Leibniz-Institute of Plant Genetics and Crop Plant Research (IPK), Correns str.3, 06466 Gatersleben, Germany
e-mail: schreibe@ipk-gatersleben.de

Clayton School of Information Technology, Monash University, Victoria 3800, Australia

Louise Shepherd James Hutton Institute, Invergowrie, DD2 5DA, Dundee, UK

Weibing Shi Department of Plant Pathology and Microbiology, Texas A & M University, 2123 TAMU, College Station, TX 77843, USA

Institute for Plant Genomics and Biotechnology, Texas A & M University, 2123 TAMU, College Station, TX 77843, USA

Penelope M. C. Smith School of Biological Sciences, University of Sydney, Sydney, Australia
e-mail: penny.smith@sydney.edu.au

Rentao Song Shanghai Key Laboratory of Bio-Energy Crops, School of Life Sciences, Shanghai University, No. 333 Nanchen Road, Shanghai, P. R. China, 200444, China
e-mail: rentaosong@staff.shu.edu.cn

Lena Y. C. Soo School of Biological Sciences, University of Sydney, Sydney, Australia

Charles N. Stewart Department of Plant Science, University of Tennessee, Knoxville, TN 37996, USA

Ryan D. Syrenne Department of Plant Pathology and Microbiology, Texas A & M University, 2123 TAMU, College Station, TX 77843, USA

Institute for Plant Genomics and Biotechnology, Texas A & M University, 2123 TAMU, College Station, TX 77843, USA

S. Lori Tausta Department of Molecular, Cellular & Developmental Biology, Yale University, P. O. Box 203104, New Haven, CT 06520-8104, USA

Richard Thompson INRA (National Institute for Agronomic Research), UMR 1347 Agroécologie, BP 86510, 21065 Dijon, France
e-mail: thompson@dijon.inra.fr

David Toubiana Jacob Blaustein Institutes for Desert Research, French Associates Institute for Agriculture and Biotechnology of Drylands Midreshet Ben-Gurion, Ben-Gurion University of the Negev French Associates, 84990 Negev, Israel

Ted Turlings Laboratory for Fundamental and Applied Research in Chemical Ecology, University of Neuchâtel, Neuchâtel, Switzerland

Michael K. Udvardi Plant Biology Division, The Samuel Roberts Noble Foundation, Ardmore, OK, USA
e-mail: mudvardi@noble.org

Jérôme Verdier Plant Biology Division, The Samuel Roberts Noble Foundation, Ardmore, OK, USA

Nicole E. Walczyk School of Biological Sciences, University of Sydney, Sydney, Australia

Department of Chemistry and Biomolecular Sciences, Macquarie University, Macquarie, Australia

Fei Wang Shanghai Key Laboratory of Bio-Energy Crops, School of Life Sciences, Shanghai University, No. 333 Nanchen Road, Shanghai, P. R. China, 200444, China

Gang Wang Shanghai Key Laboratory of Bio-Energy Crops, School of Life Sciences, Shanghai University, No. 333 Nanchen Road, Shanghai, P. R. China, 200444, China

Guifeng Wang Shanghai Key Laboratory of Bio-Energy Crops, School of Life Sciences, Shanghai University, No. 333 Nanchen Road, Shanghai, P. R. China, 200444, China

Jean-Luc Wolfender School of Pharmaceutical Sciences, EPGL, University of Geneva, University of Lausanne, Quai Ernest-Ansermet 30, 1211 Geneva, Switzerland
e-mail: Jean-Luc.Wolfender@unige.ch

Eleanore T. Wurtzel Department of Biological Sciences, Lehman College and The Graduate School, The City University of New York, 250 Bedford Park Blvd. West, Bronx, New York, NY 10468, USA

Joshua S. Yuan Department of Plant Pathology and Microbiology, Texas A & M University, 2123 TAMU, College Station, TX 77843, USA
e-mail: syuan@tamu.edu

Hélène Zuber INRA (National Institute for Agronomic Research), UMR 1347 Agroécologie, BP 86510, 21065 Dijon, France

Abbreviations

ABA	abscisic acid
ABI/abi	abscisic acid insensitive
AB-QTL	advanced backcross QTL
ABRE	ABA response element
ACCase	acetyl-CoA carboxylase
ACP	acyl carrier protein
ADH	alcohol dehydrogenase
ADP	adenosine diphosphate
ADPG	adenosine diphosphate glucose
AdoMet	*S*-adenosylmethionine
AEP	Association Européenne des Protéagineux
AFP	aspartic acid family pathway
AFLP	amplified fragment length polymorphism
AGL	AGAMOUS
AGPase	ADP-glucose pyrophosphorylase
ANOVA	analysis of variance
ANR	anthocyanidins reductase
AP	APETALA
APCI	atmospheric pressure chemical ionization
AP-MS	affinity purification mass spectrometry
APPI	atmospheric pressure photoionization
APS	5'-adenylyl sulphate
APX	ascorbate peroxidase
ARA	arachidonic acid
ARF	auxin response factor
Arg	arginine
ASE	accelerated solvent extraction
Asp	asparagine
ATP-PFK	ATP-dependent phosphofructokinase
ATP	adenosine triphosphate
BAC	bacterial artificial chromosome
BAC1	basic amino acid carrier1

BAR	bio-array resource
BCCP	biotin carboxyl carrier protein
bHLH	basic helix-loop-helix
BIFC	bimolecular fluorescence complementation
BiP	chaperonin of the binding protein
BL-SOM	batch-learning self-organizing method
BN	blue native
bp	base pair
BR	brassinosteroid
Bt	*Bacillus thuringiensis*
BTPC	bacterial-type PEPC
BY	bright yellow
bZIP	basic leucine-zipper
C	carbon
Ca	calcium
CA	correlation-based analysis
CAM	crassulacean acid metabolism
CaMV	cauliflower mosaic virus
CAPS	cleaved amplified polymorphic sequence
CAT	catalase
CBA	constraint-based analysis
CCA	canonical correlation analysis
CE	capillary electrophoresis
Cd	cadmium
CDK	cyclin-dependent kinase
cDNA	complementary DNA
CDPK	calcium-dependent protein kinase
CGA	chlorogenic acids
C3'H	*p*-coumaroyl ester 3'-hydroxylase
C4H	cinnamate 4-hydroxylase
CHI	chalcone isomerase
CHS	chalcone synthase
CK	cytokinin
CL	4-Coumarate-CoA ligase
CLE	carbon labeling experiment
CLSM	confocal laser scanner microscopy
C-N	carbon-nitrogen
CNA	cell net analyzer
CNBr	cyanogen bromide
CO_2	carbon dioxide
C-O	carbon-oxygen
Co	cobalt
CoA	coenzyme A
COBRA	constraint-based reconstruction and analysis

COS	castor oil seeds
COSY	correlation spectroscopy
CQA	caffeoyl quinic acid or caffeoyl quinate
CRTISO	carotenoid isomerase
Cu	copper
CW	cell wall
Cy	cyanine
Cys	cysteine
Da	dalton
DAF	days after flowering
DAGAT	diacylglycerol acyltransferase
DAHP	3-deoxy-D-arabino-heptulosonate-7-P
DAP	days after pollination
DArT	diversity arrays technology
DBE	debranching enzyme
DBPCFC	double-blind placebo-controlled food challenge
ddNTPs	dideoxy nucleoside triphosphates
DFR	dihydroflavonol 4-reductase
DG	diacylglycerol
1-D	one-dimensional
1-DGE	one-dimensional gel electrophoresis
2-D	two-dimensional
3-D	three-dimensional
2-DGE	two-dimensional gel electrophoresis
DHA	docosahexaenoic acid
DIGE	difference gel electrophoresis
DIMS	direct infusion mass spectrometry
DM	dry mass
DMSO	dimethylsulfoxide
DNA	deoxyribose nucleic acid
DOF	DNA-binding with one finger
DP	dirigent protein
DPA	days post anthesis
dsRNA	double-stranded RNA
DTT	dithiothreitol
DW	dry weight
DXR	1-deoxy-D-xylulose-5-phosphate reductoisomerase
DXS	1-deoxy-D-xylulose-5-phosphate synthase
EDTA	ethylenediaminetetraacetic acid
EEL	enhanced EM level
eFP	electronic fluorescent pictographic
ELISA	enzyme-linked immunosorbent assay
EM	electron microscopy

EMA	elementary modes analysis
EMM	elementary mode
EMS	ethyl methane sulphonate
EMU	elementary metabolite unit
ENR	enoyl-ACP reductase
EPA	eicosapentaenoic acid
EPA	extreme pathway analysis
EPSP	5-enolpyruvyl shikimate-3-P
ER	endoplasmic reticulum
EREBP	ethylene-responsive element binding protein
ERF	ethylene response factor
ESI	electrospray ionization
EST	express sequence tag
ETC	electron transport chain
eV	electron volt
EXP	expansin
FA	fatty acid
FAE	fatty acid elongase
FAH	ferulic acid 5-hydroxylase
FAIMS	field asymmetric waveform ion mobility spectrometry
FAO	Food and Agriculture Organization
FAS	fatty acid synthase
FAT	acyl-ACP thioesterases
FBA	flux balance analysis
FDR	false discovery rate
Fe	iron
F3H	flavonoid 3-hydoxylase
FFZE	free flow zonal electrophoresis
FHA	forkhead-associated
FID	flame ionisation detector
FQA	feruloyl quinic acid
FT-ICR	Fourier transform-ion cyclotoron resonance
FTIR	Fourier transform infrared spectroscopy
FUS	FUSCA
FVA	flux variability analysis
γ	gamma
GA	gibberellin
GA3	gibberellic acid
GABA	gamma-aminobutyric acid
GARE	GA-responsive element
Gbp	giga base pair
GC	gas chromatography

GES	genomic selection
GFP	green fluorescent protein
GGPPS	geranylgeranyl pyrophosphate synthase
GL	GLABRA
Glc	glucose
Glc-1-P	glucose-1-phosphate
Glc-6-P	glucose-6-phosphate
Gln	glutamine
Glu	glutamic acid
Gly	glycine
GM	genetically modified
GO	gene ontology
GOGAT	glutamate-oxoglutarate amino transferase
G-3-P	glycerol-3-phosphate
GPAT	glycerol-3-P acyltransferase
GPC	gel permeation chromatography
GRP	glycine-rich protein
GS	glutamine synthetase
GST	glutathione-*S*-transferase
GUI	graphical user interface
^{1}H	proton
HAP	heme-associated protein
HCA	hierarchical cluster analysis
Hcy	homocysteine
HD	3-hydroxyacyl-ACP dehydratase
HD-ZIP	homeodomain leucine zipper
HEAR	high erucic acid rapeseed
HEPES	4-(2-hydroxyethyl)-1-piperazineethanesulfonic acid
HI	harvest index
HILIC	hydrophilic interaction liquid chromatography
HMBC	heteronuclear multiple bond correlation
H_2O_2	hydrogen peroxide
HPLC	high performance liquid chromatography
HPR	hydroxypyruvate reductase
HQT	hydroxycinnamoyl-CoA shikimate/quinate hydroxycinnamoyl transferase
HR-MS	high-resolution mass spectrometry
HR-MAS	high-resolution magic angle spinning
H/S	hexose and sucrose
HSF	heat shock factor
HSP	heat shock protein
HSQC	heteronuclear single quantum coherence spectroscopy
HYD	hydroxylase enzyme

IAA	indole-3-acetic acid
ICAT	isotopic-coded affinity tags
ICP-AES	inductively coupled plasma-optical emission spectrometry
i.d.	internal diameter
ID	identification
IEC	ion exchange chromatography
IEF	isoelectric focusing
IgE	immunoglobulin E
IgG	immunoglobulin G
IL	introgression line
IMAC	immobilized metal ion affinity chromatography
INPPO	International Plant Proteomics Organization
IPA	isopropanol
iPP	inorganic pyrophosphatase
IPPI	isopentenyl pyrophosphate isomerase
IRGSP	International Rice Genome Sequencing Project
IRRI	International Rice Research Institute
ISAC	immuno solid-phase allergen chip
ISSR	inter-simple sequence repeat
IT	ion trap
iTRAQ	isobaric tag for relative and absolute quantitation
JA	jasmonic acid
K	potassium
KAR	3-ketoacyl-ACP reductase
KAS	3-ketoacyl-ACP synthase
KCS	keto-acyl-CoA synthase
kDa	kilo dalton
KEGG	Kyoto encyclopedia of genes and genomes
kV	kilo volt
LC	liquid chromatography
LCM	laser-capture microdissection
LC-MS	liquid chromatography-mass spectrometry
LC-MS/MS	liquid chromatography-tandem mass spectrometry
LCYB	lycopene β-cyclase
LCYE	lycopene ε-cyclase
lcyE	lycopene epsilon cyclase
LD	linkage disequilibrium
LDOX	leucoanthocyanidin dioxygenase
LEA	late embryogenesis abundant
LEC	LEAFY COTYLEDON
Lys	lysine
LP	linear programming

lpa	low phytic acid
LPAAT	lysophosphatidic acid acyltransferase
LTP	lipid transfer protein
LTQ	linear trap quadrupole
mAB	monoclonal antibody
MABC	marker-assisted backcrossing
MAGIC	mutant-assisted gene identification and characterization
MALDI	matrix-assisted laser desorption ionization
MARS	marker-assisted recurrent selection
MAS	marker-assisted selection
MAT	malonyl-CoA:ACP malonyltransferase
MATE	multidrug and toxic compound extrusion-type transporter
MAYG	mapping as you go
MCF	medium chain fatty acid
MDH	malate dehydrogenase
ME	malic enzyme
MeCN	acetonitrile
MeOH	methanol
Met	methionine
MetNetDB	metabolic networking database
MeV	multiexperiment viewer
MFA	metabolic flux analysis
miRNA	micro ribonucleic acid
Mn	manganese
MPSS	massively parallel signature sequencing
mRNA	messenger ribonucleic acid
MS	mass spectrometry
MS/MS	tandem mass spectrometry
MtGEA	*Medicago truncatula* gene expression atlas
MudPIT	multidimensional protein identification technology
nESI	nano electro spray ionization
n	nano
N_2	atmospheric nitrogen
NAD	nicotinamide adenine dinucleotide
NADH	nicotinamide adenine dinucleotide hydride
NADPH	nicotinamide adenine dinucleotide phosphate (reduced form)
NF-Y	NUCLEAR FACTOR-Y
NGS	next generation sequencing
NH_3	ammonia
NI	negative ion
NIL	near-isogenic line
NIR	near-infrared reflectance
NIRS	near-infrared spectrometry

NMR	nuclear magnetic resonance
NO	nitric oxide
NR	nonredundant
nsLTP	nonspecific lipid transfer protein
O_2	oxygen
O2	OPAQUE2
OAA	oxaloacetate
OB	oil body
2-OG	2-oxoglutarate
O-P	oxygen-phosphorus
OPPP	oxidative pentose phosphate pathway
OPT	oligopeptide transporter
ORF	open reading frame
Orn	ornithine
Os	*Oryza sativa*
OSC	orthogonal signal correction
P	phosphate
P_2	diphosphate
PA	proanthocyanidin
PAGE	polyacrylamide gel electrophoresis
PAL	phenylalanine ammonia lyase
PAPS	3'-phosphoadenylyl sulfate
PAT	phosphinothricin acetyl transferase
PBF	prolamin box-binding factor
PCA	principal component analysis
PCD	programmed cell death
PcG	polycomb complex
PCR	polymerase chain reaction
PDA	photodiode array
PDC	pyruvate dehydrogenase complex
PDAT	phospholipid:diacylglycerol acyltransferase
PDI	protein disulfide isomerase
PDH	pyruvate dehydrogenase
PDS	phytoene desaturase
PEP	phospho*enol*pyruvate
PEPC	phospho*enol*pyruvate carboxylase
PET	potential evapotranspiration
PFK	phosphofructokinase
PGA	phosphoglyceric acid
PGDB	pathway/genome database
PGM	phosphoglucomutase
*p*I	isoelectric point
PI	positive ion

PIF	phytochrome interacting factor
PIMT	protein L-isoaspartyl methyltransferase
PIP5K	phosphatidylinositol-4-monophosphate 5-kinase
PKL	PICKLE
PKS	polyketide synthases
PLS	partial least squares
PMF	peptide mass fingerprint
PMN	plant metabolic network
PO	plant ontology
POP	polyphenoloxydase
PP2C	protein phosphatase 2C
PPCK	PEPC protein kinase
PPDK	pyruvate-Pi-dikinase
PPi	inorganic pyrophosphate
PPi-PFK	inorganic pyrophosphate phosphofructokinase
ppm	parts per million
PPT	phosphoenolpyruvate translocator
PR	pathogenesis-related
PS	photosystem
PSY	phytoene synthase
PTM	post-translational modification
PTPC	plant-type PEPC
PTS	peroxisome targeting signal
PUFA	polyunsaturated fatty acid
PvAlf	*Phaseolus vulgaris* abscisic acid-insensitive 3-like factor
q	quantitative
Q	quadrupole
QPM	quality protein maize
QT	quantitative trait
QLIT	quadrupole-linear ion trap
QTL	quantitative trait loci
Q-TOF	quadrupole time of flight
QTT	quantitative trait transcript
RAP	rice annotation project
RIL	recombinant-inbred line
RISC	RNA-induced silencing complex
RNA	ribonucleic acid
RNAi	RNA interference
RNA-seq	RNA sequencing
ROM	repressor of maturation
ROS	reactive oxygen species
RP	reverse phase
RSR	rice starch regulator

RuBisCO	ribulose 1,5-bisphosphate carboxylase/oxygenase
RT	retention time
RT-PCR	reverse transcriptase-polymerase chain reaction
SAD	stearoyl-ACP desaturase
SAGE	serial analysis of gene expression
SAM	shoot apical meristem
SAUR	small auxin-up RNA
SBE	starch branching enzyme
SBRT	systems biology research tool
SCAR	sequence-characterized region
SCT	sinapoylglucose:choline sinapoyltransferase
SCX	strong cation exchange
SDS	sodium dodecyl sulfate
SDS-PAGE	sodium dodecyl sulfate-polyacrylamide gel electrophoresis
SDH	succinate dehydrogenase
SEC	size exclusion chromatography
SGT	UDP-glucose:sinapate glucosyltransferase
SHB	short hypocotyl under blue
siRNA	small interfering RNA
SMM	*S*-methylmethionine
SNA	stoichiometric network analysis
SNP	single nucleotide polymorphism
SnRK1	sucrose nonfermenting-1-related protein kinase
SOM	self-organizing maps
SP	storage protein
SPME	solid-phase micro-extraction
SPT	spatula
SS	starch synthase
SSP	seed storage protein
SSR	microsatellites or simple sequence repeat
STR	short tandem repeat
STS	sequence tag site
SURE	sucrose response element
SUS	sucrose synthase
S.W.	south west
SXRF	synchrotron X-ray fluorescence
TAG	triacylglycerol
TAIR	the *Arabidopsis* information resource
TALEN	transcription activator-like effector nucleases
TAP	tandem affinity purification
TC	tentative contig
TCA	tricarboxylic acid
T-DNA	transfer DNA

TE	transposable element
TEM	transmission electron microscopy
TF	transcription factor
TIC	total ion chromatogram
TILLING	targeting-induced local lesions in genomes
TIGR	the institute of genome research
TOCSY	total correlation spectroscopy
TOF	time-of-flight
Trp	tryptophan
TST	thiosulfate sulfurtransferase
TT	transparent testa
TTG	transparant testa glabrous
UB	ubiquitin
UDP	uridine diphosphate
UGT	UDP-dependent glycosyltransferase
UHPLC	ultra-high performance liquid chromatography
UPLC	ultra performance liquid chromatography
UTR	untranslated region
UV	ultra violet
VANTED	visualization and analysis of networks containing experimental data
VIT	vacuolar iron and transparent testa transporter
VLCFA	very long chain fatty acid
VLCPUFA	very long-chain polyunsaturated fatty acid
VP	viviparous
WAA	weeks after anthesis
WAF	weeks after flowering
WRI/wri	wrinkled
WS	wax synthase
XTH	xyloglucan transferase/hydrolase
Y2H	yeast-two-hybrid
ZDS	ζ-carotene desaturase
ZFN	zinc-finger nucleases
Z-ISO	ζ-carotene isomerase
Zn	zinc
ZR	zeatin riboside

Part I
Introduction

Chapter 1
Seed Development: A Comparative Overview on Biology of Morphology, Physiology, and Biochemistry Between Monocot and Dicot Plants

Paolo A. Sabelli

Abstract By and large, the most valuable crops and thoroughly investigated model angiosperm species belong to dicotyledonous and monocotyledonous plants. In both groups, development of the seed plays a fundamental role in the reproduction and in determining the economic value of different species. While investigation of seed development is far from being applied to every monocot and dicot species, the available information indicates that a large degree of variability in seed developmental patterns exists within both monocots and dicots. At the same time, however, specific features characterizing the morphology, physiology, and biochemistry of the seed can be identified in these two large taxa. In this chapter, a comparative assessment of the salient features characterizing seed development in monocots and dicots is presented.

Keywords Dicot · Embryogenesis · Endosperm · Monocot · Seed development · Seed physiology

1.1 Introduction

Monocotyledonous (Liliopsida or monocots) and dicotyledonous plants (Magnoliopsida or dicots) are two major taxonomic groups within the Angiosperms (flowering plants). They are found in most environments and comprise the vast majority of cultivated plants. In general terms, the process of sexual reproduction in these plants results in the formation of seeds, which are mature, fertilized ovules (although seed development can also occur in the absence of fertilization as in the notable case of agamospermy—a type of apomixis; Linkies et al. 2010). In a simplified view, the seed contains three major structures: the newly formed individual sporophyte (i.e., the embryo); a nutrients compartment (i.e., the endosperm and/or one or two cotyledons) devoted to supporting embryo development and, sometimes, early post-embryonic development of the sporophyte; and a protective structure (i.e., the seed coat). Species belonging to the monocots and dicots share with other angiosperms

P. A. Sabelli (✉)
School of Plant Sciences, University of Arizona, 303 Forbes Hall, Tucson, AZ 85721, USA
e-mail: psabelli@ag.arizona.edu

G. K. Agrawal, R. Rakwal (eds.), *Seed Development: OMICS Technologies toward Improvement of Seed Quality and Crop Yield*,
DOI 10.1007/978-94-007-4749-4_1, © Springer Science+Business Media Dordrecht 2012

two key features concerning their reproduction and the development of the seed: first, the ovule is contained within an ovary; and second, in a double fertilization process, fertilization of the egg within the ovule (generating the zygote, which develops into the embryo) is accompanied by another fertilization event that generates a triploid tissue, termed endosperm, which nourishes the embryo and supports its development. The most obvious difference between the seeds of monocot and dicot species is the presence of one or two embryonic leaves (or cotyledons), respectively. Owing to the tremendous variation existing in terms of anatomical, morphological, developmental, and metabolic features within the extremely large groups of monocot and dicot plants (Baskin and Baskin 2007; Bewley and Black 1994; Finch-Savage and Leubner-Metzger 2006; Kigel and Galili 1995; Martin 1946; see also the website of the Leubner Lab, http://www.seedbiology.de, accessed on 02/17/2011, for an excellent online resource on seed biology, morphology, and classification), it may be arduous to identify aspects of seed biology that are truly specific to either one of these two taxa. In addition, the most important crop species belong to legumes and grasses, whose seed developmental patterns may be considered, for certain aspects, more exceptions than representative examples of the dicots and monocots to which they respectively belong. Yet, the most intensely investigated plant model species, as well as the most widespread and valuable crops in the world today belong to only a few subgroups within the monocots and dicots. In this introductory chapter, specific examples drawn from the most important species will be described to illustrate salient aspects of seed development in monocots and dicots and the main differences between them (summarized in Table 1.1). The goal is to provide an overview on seed development for the novice and the reader unfamiliar with this subject while helping provide a conceptual framework for in-depth discussion in the following chapters.

1.2 Seed Development: Key Events

The developed, mature seed is a compartmentalized structure that is essential for the sexual reproduction of Angiosperms and other Spermatophytes. It contains a fully formed embryo, as well as other filial structures, such as the cotyledon(s) and/or the endosperm, which are stocked with storage compounds and support the development of the embryo and, in some species, of the seedling upon seed germination as well. Both the embryo and storage compartments develop and are contained within a protective structure of maternal origin, the seed coat. Seed development is a complex and highly coordinated process that involves the integration of many genetic, physiological, metabolic, and signaling pathways, which also are affected by endogenous and environmental signals and stimuli. The development of the seed can be viewed as a discontinuous, stepwise process where several different phases occur in succession to ensure the formation of a functional reproductive unit. Generally, angiosperm seed development starts with a double fertilization event, in which one of the two haploid pollen nuclei fertilizes the haploid egg cell while the other

Table 1.1 Comparison of some features of seed development between monocots and dicots

Process or seed structure	Dicots	Monocots	Notes
Embryo origin	Primarily axial cell	Primarily axial cell	Some embryo tissues can originate from the basal cell in certain species
Suspensor origin	Basal cell	Basal cell	The suspensor is an ephemeral structure that connects the developing proembryo to maternal nutrient supplies
Number of cotyledons	Two, laterally positioned relative to the shoot apex. They derive from opposite cells in the terminal quadrant of the proembryo	One. It has a variable origin from terminal quadrant cells of the proembryo. It grows more aggressively than the shoot apex, which, at maturity, becomes displaced laterally. The cotyledon is termed scutellum in the Poaceae	The scutellum is an absorptive, persistent sheath-like structure enveloping the shoot apex. It mediates nutrient transfer from the endosperm to the seedling upon germination
Coleoptile	Absent	Present in the Poaceae	The coleoptile is a hollow structure covering the plumule and the first foliage leaf
Coleorhiza	Absent	Present in the Poaceae	The coleorhiza is a cap-like structure enveloping the radicle
Epiblast	Absent	Present in some Poaceae	The epiblast is a small outgrowth of the coleorhiza
Endosperm (during early seed development)	Present	Present	Triploid tissue generated at fertilization, which supports embryo development
Endosperm (at seed maturity)	Depending on the species, it can be present (e.g., castor bean, fenugreek, carob, honey locust, Arabidopsis, lettuce) or absent (e.g., soybean, pea, Phaseolus, rapeseed)	Present (e.g., Poaceae)	In coconut, a monocot, part of the endosperm is acellular and liquid
Type of endosperm development	Nuclear (most common) or cellular	Nuclear (most common), cellular or helobial	Nuclear endosperm development typically involves acytokinetic mitosis, mitosis coupled to cell division and, in the Poaceae, endoreduplication

Table 1.1 (continued)

Process or seed structure	Dicots	Monocots	Notes
Aleurone layer	Generally absent (but present in endospermic legumes)	Present. The aleurone is made of up to four cell layers in the Poaceae	Aleurone cells generally surround endosperm storage cells. In the Poaceae, aleurone cells are the only live endosperm cell type at maturity
Endosperm storage cells	Present or absent	Present	Depending on the species, storage cells may be living or nonliving cells
Endosperm endoreduplication	Absent	Present in the Poaceae	During endoreduplication, DNA is replicated without mitosis or cytokinesis, resulting in polyploid cells
Perisperm	Present in a few nonendospermic species such as black pepper, sugar beet, yucca and pseudocereals like quinoa and amaranth	Absent	The perisperm is a maternal, diploid tissue derived from the nucellus
Transfer cells	Present in most species	Present in most species	Transfer cells mediate nutrients uptake at the seed symplast/apoplast interface
Primary site of storage metabolites accumulation at maturity	Cotyledons, and endosperm if present	Endosperm	Endosperm cells can store carbohydrates as starch granules (e.g., Poaceae) or cell wall carbohydrates (e.g., palms and some endospermic legumes)
Storage metabolites	Primarily starch, proteins and lipids	Primarily carbohydrates and proteins	Storage metabolites usually accumulate as granular structures within cells, structural components of the cell wall, or mineral inclusion in protein bodies. The endosperm of certain palms (monocots) is rich in lipids
Storage carbohydrates	Primarily starch in legumes (generally around 40–50 % of dry weight) although some species (e.g., lupin and soybean) are relatively starch-poor	Primarily starch in the Poaceae (around 65–75 % of dry weight)	Seeds of legumes generally contain higher amounts of disaccharides and oligosaccharides than those of cereals

Table 1.1 (continued)

Process or seed structure	Dicots	Monocots	Notes
Prevalent storage protein classes	Globulins (7S and 11S) and albumins (2S). Generally poor in cysteine and methionine	Prolamins (in the Poaceae) but also globulins. Generally poor in lysine	Storage proteins can be distinguished by their solubility, mass and amino acid composition
Pathway for storage protein trafficking	Best characterized in legumes, in which proteins are deposited into vacuoles via passage through the ER and Golgi apparatus	Best characterized in the Poaceae, which display several distinct pathways	Storage proteins are eventually deposited into protein bodies
Seed coat	In some species like legumes, it plays a transient storage tissue role	In the Poaceae, it comprises the pericarp (fruit coat deriving from ovary walls and other flower tissues) fused to the testa	Maternal structure derived from the ovule integument (s), which confers protection and dormancy to the seed

fertilizes the diploid central cell within the female gametophyte inside the ovule. This syngamic process generates a diploid zygote and triploid primary endosperm nucleus, respectively. These filial entities proliferate to form most of the cells comprising the embryo and the endosperm within the maternally-derived seed coat. As filial seed structures at early stages of development are invariably photosynthetically inactive, they must derive the nutrients and energy required to sustain cell proliferation and tissue growth from maternal resources both present within the ovule and other tissues. Thus, early seed development is also characterized by the absorption of storage maternal gametophytic structures, such as the nucellus [through a cell degradation process that involves programmed cell death (PCD)], while developing filial structures for the appropriation of nutrients from maternal cells, such as the suspensor and the endosperm (see below). While symplastic (plasmodesmata) connections are present among cells within both the maternal and filial compartments, they are lacking between filial and maternal seed tissues, and thus alternative strategies (i.e., apoplastic transport) must be developed to ensure the flux of nutrients from maternal tissues to filial structures. This initial developmental phase generally culminates with the differentiation of dedicated transfer cells in particular regions of filial tissues and/or the seed coat, to support intense metabolic fluxes from the mother plant toward the embryo and/or endosperm that characterize the next phase of seed development, the seed maturation phase. During the maturation phase, the differentiation of storage cells takes place, which become biosynthetically highly active, accumulate vast amounts of storage compounds, and expand in volume considerably. This phase of seed development begins with a transition from maternally-controlled to filially-controlled growth. In nonendospermic species (that is, species that do not retain a sizable endosperm at maturity), the endosperm becomes absorbed by the growing embryo. As filial structures grow inside the developing seed, the seed coat also grows and differentiates in a coordinated manner. Given the high degree of coordination between development of different tissues, compartments and between various pathways, extensive cross-talk between filial and maternal tissues plays a paramount role in seed development (Nowack et al. 2010). The later stages of the maturation phase generally involve physiological changes that allow the embryo to tolerate desiccation, and often induce a quiescent state (i.e., dormancy) that prevents germination for a period of time.

1.3 Comparative Morphology of Seed Development

1.3.1 Embryogenesis

Embryogenesis is the process by which the plant embryo is formed and developed starting with egg cell fertilization. This involves zygotic cell proliferation, cell differentiation and patterning, specification of the apical-basal axis with the formation of the shoot and root meristems, and the formation of one (as in monocots) or two

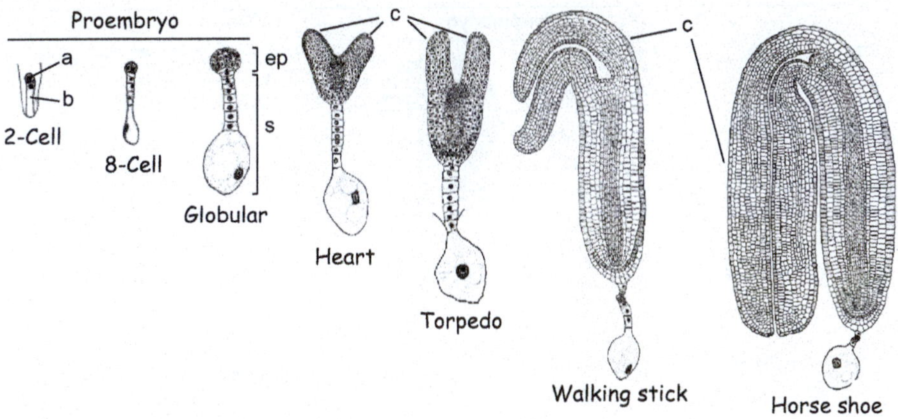

Fig. 1.1 Major phases in the embryogenesis of a typical dicot, *Capsella bursa-pastoris*. Abbreviations: *a*, axial cell; *b*, basal cell; *c*, cotyledons; *ep* embryo proper; *s*, suspensor. See main text for details. Modified from Schaffner (1906). (Credits: The Ohio State University and the Ohio Academy of Science)

(as in dicots) cotyledons containing nutrients for the developing embryo (and, in some species, the germinating seedling). The later stages of embryo development generally involve the establishment of a dormant state that allows its prolonged survival until environmental conditions are suitable for the germination of the seed.

In both monocots and dicots, the early stages of embryo development (up to the eight- or 16-celled proembryo) are relatively well conserved. However, at later developmental stages and at maturity monocots embryos are markedly different from those of dicots (Lakshamanan 1972; Table 1.1, Figs. 1.1 and 1.2). Thus, after an initial phase of proembryonic development up to the globular stage, which has common characteristics in both monocots and dicots, with the appearance of the cotyledon(s) specific patterns and structures begin to differentiate the seeds in these two groups. *Capsella bursa-pastoris* provides a classic example of embryogenesis in dicots and, relatively to its early stages, in monocots as well (Natesh and Rau 1984; Raghavan 1986, 1997; Schaffner 1906). Generally, the zygote undergoes a 'resting' period before dividing, the duration of which varies widely with the species. During such a period, the zygote becomes highly polarized in its cytoplasmic content, with its chalazal-end being rich in organelles and with a prominent nucleus (indicating high metabolic activity), and a vacuolated micropylar end.

In plants such as *Capsella*, the resting zygote also elongates along the micropyle-chalaza direction (Natesh and Rau 1984). But in other species it can decrease in size considerably relative to that of the unfertilized egg cell (Natesh and Rau 1984). In the vast majority of cases, the first zygotic division occurs asymmetrically and transversally into a small and cytoplasmic-dense axial (also known as terminal or apical) cell projected towards the chalaza and a large, vacuolated basal cell anchored to the ovule integuments at the micropylar end of the embryo sac. This process involves extensive cytoskeletal rearrangements (Kropf 1994; Webb and

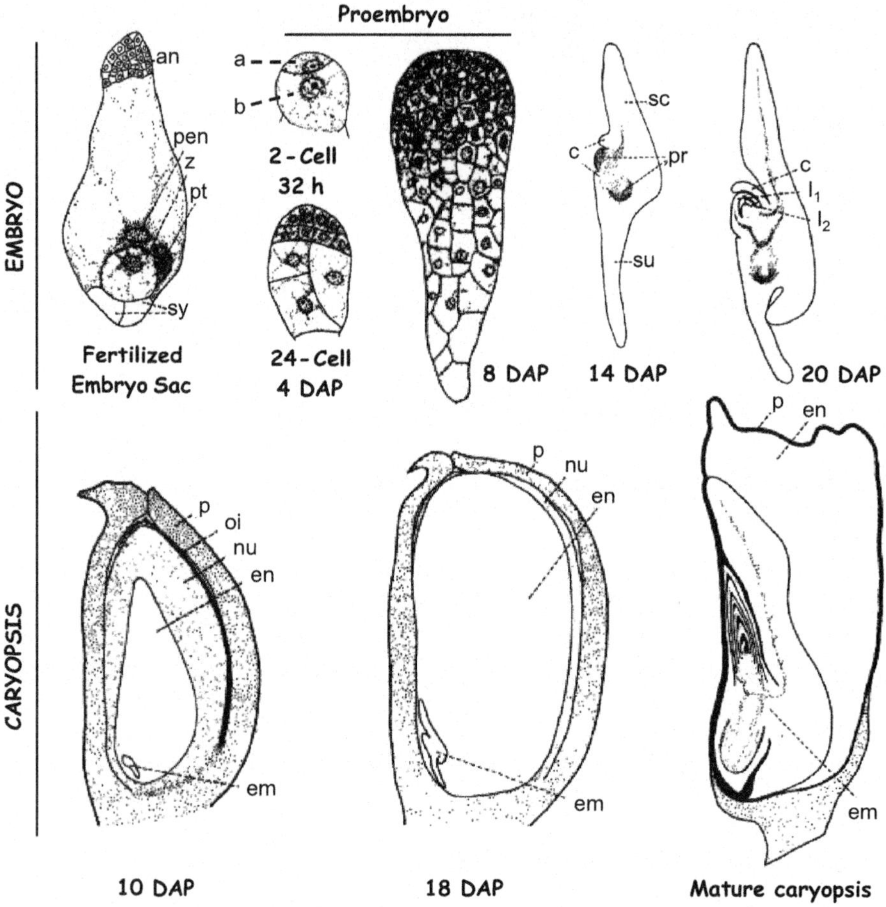

Fig. 1.2 Major stages of embryo (*top row*) and caryopsis (*bottom row*) development in maize. Abbreviations: *a*, axial cell; *b*, basal cell; *an*, antipodals; *c* coleoptile; *em* embryo; *en* endosperm; l1 and l2, primordia of first and second seedling leaves; *h* hour; *nu* nucellus; *oi* outer integument; pen, primary endosperm nucleus; *p* pericarp; *pr* plumule-radicle axis; *pt* pollen tube remnants; *sc* scutellum; *su* suspensor; *sy* synergids; *z* zygote; *DAP* days after pollination. See main text for details. [Modified from Randolph (1936) with permission from the publisher]

Gunning 1991). The polarization of the zygote together with the initial asymmetrical division is the first sporophytic event that leads to the establishment of the apical-basal bipolarity and the patterning typical of the embryo. Several different types of cell division patterns during early embryo development have been described in angiosperms (Natesh and Rau 1984; Raghavan and Sharma 1995). In the majority of cases, however, the axial cell divides longitudinally (as in *Capsella bursa-pastoris* or in *Arabidopsis* among dicots and in the Poaceae among monocots) or transversally (such as in *Nicotiana*). In many plants, the cells constituting the proembryo are generated to a large extent or even entirely by the proliferation of the axial cell.

Proliferation of the basal cell primarily generates the suspensor. However, in other species including members of the Asteraceae (such as *Lactuca sativa*), Rosaceae and others among dicots, and in the Liliaceae and Poaceae among monocots, both the axial and basal cells contribute more evenly to generating embryo structures (Natesh and Rau 1984). The functional determination of otherwise seemingly homogeneous proembryonic cells occurs early in development as shown in *Capsella bursa-pastoris* in which the upper cell tier of the eight-cell proembryo will generate the cotyledons and the upper part of the stem, whereas the lower tier will generate the hypocotyl (Fig. 1.1). The octant cells of *Capsella* divide periclinally, which results in the globular embryo. This stage is phased into, through a lateral enlargement of the distal pole, the heart embryo, which is characterized by the emergence of two cotyledon primordia.

The following stage of embryo development, the torpedo embryo, is brought about by cell proliferation and differentiation of the basal cell tier. As the embryo continues to grow, it bends over due to physical constraints within the ovule, thereby assuming the typical walking-stick and, later at maturity, horse-shoe shapes. Meanwhile, repeated transversal divisions of the basal cell generate the suspensor and the hypophysis—the uppermost suspensor cell that becomes gradually incorporated into the embryo proper. The suspensor is typically characterized by a filamentous shape and, in species such as *Capsella bursa-pastoris* and *Arabidopsis thaliana*, terminates at the micropylar end with an enlarged, basal cell. The suspensor differentiates and develops relatively more quickly than the embryo, reaching full size by the globular or heart stage. Subsequently, suspensor cells degenerate through a PCD mechanism and mature seeds in most species are devoid of suspensor. The hypophysis is the only persistent cell of suspensor origin, and has important morphogenetic roles as it generates the cortex, cap and epidermis of the root. Although ephemeral in its nature, the suspensor plays several important roles in the development of the embryo: It physically supports the embryo proper and connects it to maternal tissue; it absorbs nutrients from the endosperm and maternal tissues and transfers them to the embryo proper; and it is a highly active biosynthetic structure in its own right. Several structural features make suspensor cells well apt for these tasks: they have cell wall (CW) projections and sometimes haustorias; they are rich in plasmodesmata and different types of organelles and have an extensive endoplasmic reticulum (ER); and, contrary to the cells of the embryo proper, they lack a waxy cuticle layer, which facilitates absorption of nutrients from surrounding seed tissues (Kawashima and Goldberg 2010; Sharma 2009).

Embryogenesis in monocots has been investigated in less detail than in dicots, and studies have largely focused on the Poaceae, whose embryos display specific features (Fig. 1.2). Generally in monocots, as shown in *Sagittaria sagittaefolia*, morphogenetic specification occurs soon after the first zygotic division: the cotyledon and the shoot apex originate from the axial cell, while the basal cell generates the hypocotyl-shoot axis and the suspensor (Sharma 2009). A major difference between embryogenesis in monocots and dicots concerns the origin of the cotyledon(s) from the terminal quadrant cells of the eight-celled proembryo. Whereas, in dicots, the cotyledons typically arise from two opposite cells, in monocots the origin of the single cotyledon is more variable and may involve contributions from all four cells of the quadrant (Lakshamanan 1972). Due to its slow growth rate, the shoot apex is displaced laterally from the axis by the rapidly growing cotyledon. In addition, the embryo of nongrass monocots has neither the coleoptile nor the coleorhiza. The distinct developmental morphologies of the embryos of maize among Poaceae monocots and that of the model dicot *Arabidopsis* reflect fundamental differences in the orientation of the embryo axis (which is oblique in maize), the presence of one (maize) or two (*Arabidopsis*) planes of bilateral symmetry, and the lack in maize of central-peripheral axis. The action of gene networks and auxin involved in the specification of the embryo axis, cell fate, and planes of symmetry has been reviewed elsewhere (Chandler et al. 2008).

Early embryo development (including the proembryo stage) in the Poaceae is similar to that in dicots and other monocots. Afterwards, however, it follows rather atypical patterns, even among monocot plants. In some species, such as maize, cell division patterns become unpredictable early in the proembryo (Randolph 1936). As already mentioned, the embryo develops less rapidly than the single cotyledon, which is termed scutellum, and as a result it becomes displaced laterally off the embryo axis. The scutellum attaches to the axis at the scutellar node, which defines the separation between an upper axis region, the epicotyl, which comprises the shoot apex and leaf primordia (plumule), and a lower region that comprises the radicle. In some species, such as wheat, there is a small flag-like structure opposite the scutellum, called epiblast. Differently from other monocots, the origin of the radicle in the grasses is endogenous, from a central domain of the embryo (Natesh and Rau 1984; Raghavan 1986). Both the epicotyl and the radicle are protected by sheath-like structures, termed coleoptile and coleorhiza, respectively. During germination, the plumule and the radicle break through the coleoptile and coleorhiza.

1.3.2 Endosperm Development

The endosperm originates from the second fertilization event, which involves the fusion of one sperm nucleus with two polar nuclei within the central cell of the female gametophyte. As a result, the primary endosperm nucleus and cell, as well as its derivatives, are triploid, containing two complements of the maternal genome and one complement of the paternal genome. The endosperm has long been known to provide nutrients to the developing embryo and thus supporting its development. In some species it also supports seedling development during and after germination. However, more recent research has highlighted a key role for the endosperm in integrating different signals and cross-talk among the major seed compartments, as well as representing a conduit for epigenetic mechanisms controlling seed development (Berger et al. 2006; Raghavan 1986).

Three major patterns of endosperm development have been identified and termed nuclear, cellular, and helobial (Lopes and Larkins 1993). Both the nuclear and cellular type of endosperm development occurs among monocots and dicots, whereas the helobial type is found only in monocots. The nuclear type of endosperm development is the most frequently encountered among angiosperms, and the best characterized. It consists of an early proliferative phase in which the mitotic cell cycle is uncoupled from cell division, resulting in a single cell containing many nuclei (i.e., a syncytium). Usually, but not in all cases, further development of the nuclear type endosperm occurs through a resumption of the cell division program. In the cellular type, the syncytial phase does not take place and cell proliferation occurs through a standard mitotic cell cycle coupled to cytokinesis. Helobial endosperm development is a combination of the nuclear and cell type in which, beginning with an unequal division of the primary endosperm cell, each of the two progeny cells develops following a distinct program by either conventional cell division or acytokinetic mitosis (Vijayaraghavan and Prabhakar 1984).

Generally speaking, development of the nuclear endosperm type follows rather conserved steps in endospermic species, which comprise plants in which the endosperm is retained at seed maturity. These steps include early formation of a syncytium, cell proliferation, cell expansion coupled to accumulation of storage compounds, and maturation, which involves dehydration and dormancy. In so-called nonendospermic species instead, the endosperm is absorbed by the rapidly developing embryo and is virtually missing at maturity. Interestingly, the pattern of takeover of the endosperm by the growing embryo is developmentally fixed in each nonendospermic species, but varies considerably between species. Thus, the embryo can absorb the endosperm before, during, or after endosperm cellularization, as shown in pea, bean, or sunflower, respectively (DeMason 1997). That is, the process of cellularization of the endosperm does not appear to impinge on its absorption by the growing embryo in a universal way. Regardless of these distinctions, however, the early phases of nuclear endosperm development, up to its cellularization, are very similar in monocot and dicot species (Olsen 2004; Sabelli and Larkins 2009a). After fertilization, the primary endosperm nucleus generates a syncytium by acytokinetic mitosis. Up to several thousand nuclei can thus populate the primary endosperm cell. These nuclei are initially distributed in a thin layer of cytoplasm at the periphery of the cell because of the expanding central vacuole (Olsen 2004). Cellularization of the syncytium ensues through a reorganization of the microtubular cytoskeleton that results in the formation of open-ended alveoli around each nucleus (Brown et al. 1994). Deposition of periclinal CWs completes the formation of the first layer of endosperm cells. Reiteration of alveolation coupled to periclinal CW formation allows the inward formation of additional cell layers until virtually the entire central cavity is filled with cells. In cereals, alveolation occurs through a synchronous mitotic process, whereas in *Arabidopsis* it takes place in a sequence of nuclear domains spatially arranged along the micropylar-chalazal poles in a wave-like fashion (Olsen 2004). However, it is not clear whether this distinction can be generalized to all monocots and dicots. Cellularization is generally completed by the torpedo stage in *Arabidopsis*, and by 3–6 days after pollination (DAP) in cereals.

The next phase of endosperm development involves cell proliferation through mitotic division coupled to cytokinesis, which generates the vast majority of the cells that make up the fully developed endosperm. In the Poaceae family among the monocots, a phase of reiterated genome duplication in the absence of mitosis or cytokinesis, termed endoreduplication or endocycle, follows the cell division phase. This endoreduplication phase results in polyploid cells and is associated with increases in nuclear, cell, endosperm and caryopsis sizes, and the accumulation of storage metabolites (Kowles 2009; Sabelli 2012; Sabelli and Larkins 2008, 2009a, b). It is not clear what the function of endoreduplication in the cereal endosperm might be, and this remains a controversial subject. Several different functions have been proposed, including a role in supporting increased gene expression, the accumulation of storage metabolites in DNA molecules, and increased rate of tissue growth without cell proliferation among others. But, the evidence in their support is still inconclusive. Following endoreduplication, most endosperm cells in cereals undergo PCD (Nguyen et al. 2007; Sabelli 2012; Young and Gallie 2000), while aleurone cells remain alive.

The diversification of strategies between cereal monocots and nonendospermic dicots, particularly with regard to the development of primary storage tissue (endosperm in the former and cotyledons in the latter) and the transition into the phase of accumulation of storage compounds involves a complex array of differences in genetic, metabolic, signaling and gene expression programs, which genomic approaches are beginning to dissect (Agarwal et al. 2011; Gallardo et al. 2007; Melkus et al. 2009).

1.3.3 Seed Coat Development

In most species, including many dicots and monocots, the seed coat (also known as testa) generally originates from the ovule integuments, two structures surrounding the nucellus (Beeckman et al. 2000; Schneitz et al. 1995), though single-integument ovules can be found in members of certain families. The seed coat provides a mechanical shield protecting the embryo and the endosperm from the environment, but also plays additional important roles in regulating phloem unloading of assimilates in developing seeds (Thorne 1985), fluids and gases exchanges with the environment, and in controlling seed dormancy and germination (Bewley and Black 1994; Debeaujon et al. 2007; Moïse et al. 2005). In *Arabidopsis* and Fabaceae species, development of the testa is coordinated with both embryogenesis and endosperm development. Generally, seed coat development and maturation precede that of filial tissues. Abscisic acid (ABA) plays an important role in seed coat development in these species. The five cell layers that define the ovule integuments in *Arabidopsis* for the most part undergo distinct developmental programs following fertilization, which involve the accumulation of different sets of metabolites. Several cell layers become crushed together at maturity.

In legumes, the seed coat plays an important role in supplying filial tissues with nutrients from the mother plant. The CW invertase, localized to seed coat regions, is crucial to establishing and sustaining a sucrose gradient that facilitates carbon uptake, as hexoses, by the endosperm and embryo (see below). In certain species, seed coat maturation and integrity is also important for seed coat-dependent dormancy and for germination.

In the caryopsis of the Poaceae, the fruit coat (termed pericarp) and the testa are fused. In barley, after an initial phase of cell division during the first two days after flowering (DAF), pericarp differentiation involves cell elongation along the longitudinal axis between 3 and 10 DAF coupled to PCD, and it coincides with the cellularization of the endosperm (Radchuk et al. 2011). A similar process occurs in the placenta-chalazal domain of the maternal tissue of the maize caryopsis, which also depends on fertilization, further illustrating a dependence of maternal tissue differentiation and growth on development of the filial tissue (Kladnik et al. 2004). PCD in the pericarp may contribute to redistribution of nutrients, relaxation of physical constraints of the maternal tissue to allow inner growth of the filial tissue, and the re-activation, together with PCD in the nucellus and the nucellar projections, of post-phloem transport functions to allow passage of solutes (Kladnik et al. 2004; Radchuk et al. 2011). Crosstalk among embryo, endosperm, and seed coat appears to be complex, but gene networks that coordinate development of these three seed compartments are being elucidated (Debeaujon et al. 2007; Haughn and Chaudhury 2005; Ingram 2010; Moïse et al. 2005; Ohto et al. 2009).

1.4 Comparative Physiology: Phytohormones, Signaling, and Metabolism

Major phytohormones, such as auxins [primarily indole-3-acetic acid (IAA)], cytokinins (CKs), ABA, gibberellins (GAs), and ethylene play important roles in regulating many aspects of seed development. These include growth and development of seed tissues, the arrest of growth preceding seed maturation, the biosynthesis and accumulation of storage compounds, desiccation tolerance and dormancy, and the utilization of storage reserves to support germination. In general terms, higher levels of CKs are associated with cell proliferation in the embryo and/or endosperm, auxins are associated with cell differentiation and expansion, ABA has long been established in regulating maturation, desiccation tolerance and dormancy, but it also has a role in arresting seed growth and stimulating storage metabolite biosynthesis (Gutierrez et al. 2007; Nambara and Marion-Poll 2003). ABA is counteracted by ethylene, whereas GAs are generally associated with germination of the seed. These features are common to members of both monocots and dicots, but significant variation among species in both groups has made it difficult to provide a unifying picture (Bewley and Black 1994). This problem has been compounded by the existence of substantial cross-talk and feedback between, for example, osmotic state of the cells, various environmental stresses, nutrients (such as sucrose), and phytohormones. However, as discussed in the previous sections, seed developmental patterns differ between these two taxa. It is, therefore, predictable that they are the outcome of distinct regulatory networks that involve specific interactions among genes, hormones, and metabolites. While comparative studies between dicot and monocot species are far from being comprehensive, several different features have been highlighted that distinguish the physiology of seed development in monocots and dicots.

During the development of cereal seeds, both auxins and CKs display transient increases. The major CKs, zeatin and zeatin riboside (ZR), increase early during seed development at a stage corresponding to the peak in mitotic activity of the endosperm (for example, between 3 and 8 DAP in maize; Morris 1997). The majority of CKs are synthesized in the cereal grain. The decrease in CKs levels observed at later stages of seed development is primarily due to cytokinin oxidase activity. The peak in IAA usually follows that of CKs and correlates with the expansion/endoreduplication phase of endosperm development (Lur and Setter 1993a, b). In addition, GA and ABA are also involved in cell differentiation and grain filling (Nguyen et al. 2007; Yang et al. 2006). ABA also appears to inhibit the cell cycle while inducing seed maturation, possibly by upregulating inhibitors of cyclin dependent kinases, the master cell cycle regulators (Barrôco et al. 2006; Coelho et al. 2005; Sreenivasulu et al. 2010). However, there appears to be a division of labor for ABA between the endosperm, where it influences the synthesis and accumulation of storage compounds, and the embryo, where it is involved in conferring desiccation tolerance (Sreenivasulu et al. 2006).

Phytohormones appear to be also involved in PCD, which is prevalent during development of the pericarp and the endosperm in cereal monocots. Both anatomical and transcriptome analyses associated with PCD in barley indicate that there are marked differences between maternal (i.e., nucellus, nucellar projections, and pericarp) and filial (i.e., endosperm) tissues. These primarily correlative studies suggest that distinct sets of genes/enzymes and hormones play different roles in regulating PCD in maternal and filial compartments of the seed. For example, maternal PCD involves a thorough disruption of the cell culminating with the degradation of the CW and the remobilization of the cellular content. This process seems to require various CW degrading enzymes and also hydrolases, such as α-amylase. Jasmonic acid (JA) and ethylene also appear to be involved. On the contrary, during PCD in the filial endosperm, CW degradation and cell disruption are not prominent, α-amylase activity appears suppressed, and while ABA and ethylene are required, JA biosynthesis is inhibited (Nguyen et al. 2007; Sreenivasulu et al. 2006, 2010; Young and Gallie 2000).

In legumes, the developing seed coat functions early as a transient storage organ, accumulating starch and proteins that later are redistributed to the growing embryo. There is evidence suggesting that ABA, ADP-glucose pyrophosphorylase (AGPase) and other enzymes related to sink strength play important roles in seed coat development (Weber et al. 2005). The CW-bound invertase activity, localized in the sucrose-unloading area of the seed coat in *Vicia faba* or in the basal transfer cell layer of the maize endosperm, is responsible for converting sucrose into hexose sugars that can be utilized for cell division by the growing embryo and/or endosperm. A critical role for invertase has been proposed in regulating the supply of hexose sugars to fuel cell division in the developing embryo, thereby sustaining sink strength early in seed development (Weber et al. 1997). However, as the seed transitions into the maturation phase and filial control, sucrose increases at the expense of hexoses, possibly as a result of the inhibition of CW invertase by ABA-induced inhibitors (Weber et al. 2005).

Complex metabolic changes occur during seed development. Convincing evidence implicates glucose and sucrose, ABA, and sucrose nonfermenting-1-related protein kinase (SnRK1 kinase) as key players in a signaling network that coordinates the utilization of difference sources of carbon, nutrients allocation and processing, the balance between anabolic *versus* catabolic processes, the regulation of energy states, and metabolic homeostasis in developing seeds (Melkus et al. 2009; Radchuk et al. 2010; Weber et al. 2005). As already mentioned, in addition to its well-known function as nutrient sugar, sucrose also plays an important signaling role in stimulating the accumulation of storage compounds and grain filling (Koch 2004; Weber et al. 2005). During the transition from maternal to filial control, increased concentrations of sucrose stimulate expression of key sugar metabolism genes, such as sucrose synthase (SUS) and AGPase, as well as developmental events associated with grain filling such as reprogramming gene expression, the cell proliferation/cell expansion switch and endoreduplication (Koch 1996; Sreenivasulu et al. 2010; Weber et al. 2005). Because the cleavage of sucrose into fructose and UDP-glucose is a reaction that can be inhibited, and reversed, by high concentrations of hexose, sustained sink strength is facilitated by the prompt utilization of hexose products for starch biosynthesis.

1.5 Comparative Biochemistry

1.5.1 Storage Compounds

The seeds of both dicots and monocots take up sucrose and nitrogen (primarily in the form of amino acids) from the mother plant and utilize them to synthesize different storage compounds such as starch, protein, lipids, and CW storage carbohydrates like cellulose (Table 1.2), and other minor components.

In general, the main seed storage tissues in both groups of plants are of filial origin but can be further distinguished between embryonic (such as in dicots) and endospermic (such as in the Poaceae among the monocots; Ruan and Chourey 2006). Nutrients generally move symplastically through the plasmodesmata from phloem unloading sites in maternal tissues as well as within filial (storage) tissues. However, an important aspect of nutrients uptake in the seed is the fact that filial and maternal tissues are not symplastically connected, as plasmodesmata are absent on the cells at the maternal/filial tissue interface. Thus, movement of nutrients from maternal to filial cells generally requires transport *via* the noncellular space, the apoplast, through tightly regulated nutrients membrane efflux from maternal cells to the apoplast and membrane influx from the apoplast to filial cells. In addition, in some species (such as wheat and rice), specific cell layers become lignified during the seed-filling stage thereby spatially limiting apoplastic transport to localized

Table 1.2 Chemical composition of seeds of some important grain crops. [Modified from Chibbar et al. (2004) with permission from Elsevier]

CROPS	Composition (Percentage dry weight)		
	Carbo-hydrate	Protein	Lipid
Cereals (Monocots)			
Wheat (*Triticum aestivum*)	82	14	2
Barley (*Hordeum vulgare*)	80	9	1
Corn (*Zea mays*)	84	10	5
Rice (*Oryza sativa*)	88	8	2
Oat (*Avena sativa*)	67	28	1
Sorghum (*Sorghum vulgare*)	82	12	4
Rye (*Secale cereale*)	82	14	2
Pulses (Dicots)			
Lentil (*Lens culinaris*)	65	27	1
Pea (*Pisum sativum*)	68	27	2
Chickpea (*Cicer arietinum*)	65	23	5
Common bean (*Phaseolus vulgaris*)	70	24	2
Mung bean (*Vigna radiata*)	69	26	1
Pigeon pea (*Cajanus cajan*)	65	21	1
Soybean (*Glycine max*)	32	38	20
Cowpea (*Vigna unguiculata*)	60	24	1
Lupin (*Lupinus albus*)	37	38	20
Oilseeds (Dicots)			
Rapeseed (*Brassica napus*)	25	23	48
Sunflower (*Helianthus annuus*)	48	20	29
Flax (*Linum usitatissum*)	32	26	38
Peanut (*Arachis hypogea*)	25	27	45
Sesame (*Sesamum indicum*)	19	20	54
Hemp (*Cannabis sativa*)	27	29	41

regions. The sites, where apoplastic transport takes place, are commonly equipped with membrane transporter proteins and enzymes dedicated to sucrose breakdown and its re-synthesis or the transport and synthesis of other storage compound precursors. In many cases, special cell types known as transfer cells differentiate at the maternal/filial tissue boundaries to facilitate transport of nutrients (Offler et al. 2003; Patrick and Offler 2001). From an ultrastructural point of view, these cells are characterized by increased cell surface due to extensive secondary CW ingrowths. Depending on the species, transfer cells can be localized in filial or maternal tissues, or in both (Davis et al. 1990; Offler et al. 2003).

1.5.2 Carbohydrates

During the cell-division phase of filial tissue development, the prominent source of carbon for both monocot and dicot seeds are hexoses. Hexoses are derived from the degradation of phloem-unloaded sucrose by CW-bound invertase (sucrose + $H_2O \rightarrow$

glucose + fructose). In addition to being a source of carbon, hexoses may play an important role in signaling filial tissues to sustain their engagement in cell division (Cheng and Chourey 1999; Weber et al. 1997). In cereals, hexoses, once taken up by the filial tissues, are generally reconverted into sucrose by SUS phosphate (UDP-glucose + Fructose-6-P ↔ sucrose-6'-P + UDP). In both dicot and monocot seeds, the switch to the cell-expansion/storage phase is accompanied by increased levels of sucrose, which is utilized to synthesize starch, storage proteins (SPs), and lipids (Ruan and Chourey 2006; Weber et al. 1997, 2005). Coincident with this, invertase activity drops. Starch biosynthesis in cereals involves several enzymatic steps, the most important being catalyzed by SUS (sucrose + UDP ↔ fructose + UDP-glucose), AGPase (glucose-1P+ ATP ↔ ADP-glucose + PPi), starch synthase, and starch branching and de-branching enzymes (Hannah 2007; James et al. 2003). Although there are exceptions, in general legume seeds accumulate more disaccharides and oligosaccharides (i.e., raffinose, stachyose, verbascose, and α-galactosides) than cereal grains, in which starch represents up to about 75 % of the dry weight (DW; Chibbar et al. 2004; Table 1.2). In addition, there is evidence to suggest that the SUS pathway in legumes plays an important role in coordinating amino acid and lipids biosynthesis with that of storage carbohydrates (Ruan and Chourey 2006).

1.5.3 Storage Proteins

Monocots and dicots synthesize and accumulate large amounts of SPs during the maturation phase of seed development (Coleman and Larkins 1999: Shewry et al. 1995). In developmental terms, as already mentioned, seed maturation requires halting the cell-proliferation phase, and appears to be brought about by a signaling network that involves sugars and ABA. Monocots seeds accumulate SPs primarily in the endosperm, whereas in the case of dicots SPs accumulate mostly in the cotyledons (Table 1.1). Seed SPs have best been characterized in the Poaceae among monocots and in legumes among dicots. Based on their solubility properties, seed SPs have been historically classified as albumins (soluble in water), globulins (soluble in dilute salt solutions), prolamins (soluble in alcohol), and glutelins (soluble in diluted acids or bases; Osborne 1924). The SPs represent 10–45 % of the seed DW in dicots and comprise mostly globulins and albumins. They have also been classified with respect to their molecular masses expressed as sedimentation coefficients (S) determined by sedimentation equilibrium ultracentrifugation (Shewry 2004a). Thus, dicot seeds primarily contain 7S and 11S globulins and 2S albumin, which often have also been termed according to the taxonomic group/species of origin. Globulins are typically polymeric proteins composed of nonidentical subunits: 7S globulins are trimeric and with a molecular weight (MW) of ~150–200 kDa, whereas 11S globulins are hexameric with a MW of ~300–400 kDa. The 2S albumins generally comprise two subunits of 8–10 kDa and 4–5 kDa, linked by disulfide bonds. Both polypeptides derive from proteolytic digestion of a single precursor. The relative proportions of 7S and 11S globulins and 2S albumins can vary widely among different species and even within a species. The nutritional quality of the seed is influenced considerably by the relative proportions of the different classes of SPs as they have distinct amino acid compositions. The SPs of dicot seeds are poor in the essential amino acids cysteine and methionine. In addition, members of these protein classes are known allergens in several species.

The seed SPs in monocots have best been characterized in cereal grains, where they represent about 10–12 % of the seed DW. In most Poaceae, SPs are primarily made of prolamins, though globulins are prevalent in some species, such as oats and rice (Coleman and Larkins 1999; Shewry and Halford 2002; Shewry et al. 1995). Prolamins include monomeric and polymeric forms, the latter stabilized by interchain disulfide bonds. Like in the case of legumes, SPs in the Poaceae are often named according to the species of origin. Prolamins are generally rich in proline and glutamine, and poor in the essential amino acids lysine and tryptophan. Their unusual amino acid composition is due to the presence of repeated motifs rich in proline and glutamine. The prolamin fraction is usually made of many proteins, and its composition and the features of individual components influence profoundly not only the nutritional value of the seed, but also its technological properties (e.g., bread- and pasta-making in the case of wheat).

Mature seed SPs are normally found in dense organelle known as protein bodies. While this feature is common to monocots and dicots, different pathways determine their intracellular trafficking and targeting (Herman and Larkins 1999; Shewry 2004b; Vitale and Denecke 1999). SPs are synthesized on the rough ER and cotranslationally targeted to the ER lumen by a signal peptide located at their *N*-terminus. Inside the ER lumen, they undergo a number of modifications including cleavage of the signal peptide, chaperone-assisted folding, the formation of intrachain and chain-to-chain disulfide bonds, and polymerization through noncovalent bonds.

The 2S albumins as well as 7S and 11S globulin are targeted through the ER and the Golgi apparatus to specialized storage vacuoles, often undergoing further processing, and form intravacuolar dense bodies. Protein bodies may eventually result from division of the vacuoles.

The prolamins of cereals, instead, undergo relatively little processing in the endomembrane system, although they may form large polymers through disulfide bonds. Depending on the species, protein body formation in the Poaceae occurs through distinct pathways, which may include the direct budding off of ER vesicles containing protein accretions, and the ER/Golgi apparatus/vacuole pathway, or a combination of both.

1.5.4 Other Compounds

Although carbohydrates and SPs are the two major classes of storage compounds containing carbon and nitrogen, seeds also accumulate other important types of elements and molecules, such as lipids, phosphate, minerals, and protease inhibitors. Space limitations prevent their discussion here, but they have been comprehensively reviewed in several publications (Bewley and Black 1994; Browse 1997; Raboy 1997; Wilson and Larkins 1984; Weselake 2005).

1.6 Dormancy and Desiccation Tolerance

The term dormancy refers to a state of embryonic quiescence that is established during seed maturation during/after the accumulation of storage compounds that effectively prevents germination (even under favorable conditions; Baskin and Baskin 2004; Finch-Savage and Leubner-Metzger 2006). Dormancy can be determined by conditions internal to the embryo (endogenous dormancy) or it can depend on the tissues that surround it (exogenous dormancy), or a combination of both. Dormancy and the release from it are complex processes that may take various different forms resulting from a vast array of interactions between genetic and environmental factors (Finch-Savage and Leubner-Metzger 2006). To further complicate matters, the behavior of developing seeds with regard to their ability to stay dormant or undergo germination covers a wide-ranging spectrum, from fully-dormant to nondormant seeds, sometimes even within one species. It is well established that ABA functions as a critical physiological inducer and maintenance factor of dormancy, whereas GAs, brassinosteroids, ethylene, reactive oxygen species (ROS), and nitrogen-containing compounds, such as nitrate and nitric oxide (NO) play important roles in the release from it and promoting seed germination (Finkelstein et al. 2008). In particular, a delicate balance between biosynthesis and degradation of ABA and GAs appears to be critical for tipping the ABA/GAs concentration level ratio towards either dormancy or release from it. Equally important in determining the dormant state of a seed is its sensitivity to the levels of ABA and GAs, and this characteristic changes dynamically during seed maturation. In addition to physiological factors, and often in combination with them, physical factors (such as a water-impermeable seed coat) can also contribute significantly to dormancy (Debeaujon et al. 2007; Finch-Savage and Leubner-Metzger 2006).

During the late-maturation stages, seeds undergo dehydration, sometimes rapidly, but remain viable. This seed property is termed desiccation tolerance (Berjak 2006). ABA is known to induce a desiccation-tolerant state through a mechanism that in part appears to impinge on the ability of the seed to counteract oxidative stress and ROS. Also, during desiccation tolerance, ABA induces the accumulation of a class of proteins, known as late embryogenesis abundant (LEA) proteins, which, in combination with the accumulation of certain sugars, determines the so-called "glassy state" (Finkelstein et al. 2008). LEA proteins appear to be ubiquitous in both monocots and dicots (Shih et al. 2008).

1.7 Concluding Remarks

Monocot and dicot seeds develop through a series of broad processes that are essentially common to both taxa. These include a morphogenetic phase characterized by intense cell proliferation, a transition to an accumulation phase of storage metabolites that involves a switch from maternal to filial control, and a maturation phase, during which the accumulation of storage reserves is completed, a quiescent state of the embryo is frequently established, and the seed undergoes dehydration. However, the details of seed development vary considerably among different species even within a taxonomic group. Yet, several features distinguish the development of the seed in monocots and dicots, the most obvious one being the presence of one or two cotyledons, respectively. This implies the establishment of specific cell fate patterns early in embryogenesis, which, in turn, affect the morphology of the developing embryo in different ways in these two groups. As discussed here, the presence of one or two cotyledons in the embryo is associated with different morphological, physiological, and biochemical features characterizing developing seeds, at least as far as the best characterized monocot and dicot species are concerned. However, it remains to be established whether these differences hold true across the large number of members constituting these two diverse taxa.

Acknowledgment This work was supported by the U.S. Department of Energy Grant DE-FG02-96ER20242.

References

Agarwal P, Kapoor S, Tyagi AK (2011) Transcription factors regulating the progression of monocot and dicot seed development. Bioessays 33:189–202

Barrôco RM, Peres A, Droual AM, De Veylder L, Nguyen LSL, De Wolf J, Mironov V, Peerbolte R, Beemster GTS, Inzé D, Broekaert WF, Frankard V (2006) The cyclin-dependent kinase inhibitor Orysa; KRP1 plays an important role in seed development of rice. Plant Physiol 142:1053–1064

Baskin CC, Baskin JM (2007) A revision of Martin's seed classification system, with particular reference to his dwarf-seed type. Seed Sci Res 17:11–20

Baskin JM, Baskin CC (2004) A classification system for seed dormancy. Seed Sci Res 14:1–16

Beeckman T, De Rycke R, Viane R, Inzé D (2000) Histological study of seed coat development in *Arabidopsis thaliana*. J Plant Res 113:139–148

Berger F, Grini PE, Schnittger A (2006) Endosperm: an integrator of seed growth and development. Curr Opin Plant Biol 9:664–670

Berjak P (2006) Unifying perspectives of some mechanisms basic to desiccation tolerance across life forms. Seed Sci Res 16:1–15

Bewley J, Black M (1994) Seeds: physiology of development and germination, 2nd ed. Plenum, New York. ISBN 0-306-44747-9

Brown RC, Lemmon BE, Olsen OA (1994) Endosperm development in barley: microtubule involvement in the morphogenetic pathway. Plant Cell 6:1241–1252

Browse J (1997) Synthesis and storage of fatty acids. In: Larkins BA, Vasil IK (eds) Cellular and molecular biology of plant seed development. Kluwer Academic, Dordrecht, pp 407–440

Chandler J, Nardmann J, Werr W (2008) Plant development revolves around axes. Trends Plant Sci 13:78–84

Cheng WH, Chourey PS (1999) Genetic evidence that invertase-mediated release of hexoses is critical for appropriate carbon partitioning and normal seed development in maize. Theor Appl Genet 98:485–495

Chibbar RN, Ganeshan S, Baga M, Khandelwal RL (2004) Carbohydrate metabolism. In: Wringley C (ed) Encyclopedia of grain. Science. Elsevier, Oxford, pp 168–179

Coelho CM, Dante RA, Sabelli PA, Sun Y, Dilkes BP, Gordon-Kamm WJ, Larkins BA (2005) Cyclin-dependent kinase inhibitors in maize endosperm and their potential role in endoreduplication. Plant Physiol 138:2323–2336

Coleman CE, Larkins BA (1999) The prolamins of maize. In: Shewry PR, Case R (eds) Seed proteins. Kluwer Academic, Dordrecht, pp 109–139

Davis RW, Smith JD, Cobb BG (1990) A light and electron microscope investigation of the transfer cell region of maize caryopses. Can J Botany 68:471–479

Debeaujon I, Lepiniec L, Pourcel L, Routaboul JM (2007) Seed coat development and dormancy. In: Bradford KJ, Nonogaki H (eds) Seed development, dormancy and germination. Blackwell, Oxford, pp 25–49

DeMason DA (1997) Endosperm structure and development. In: Larkins BA, Vasil IK (eds) Cellular and molecular biology of plant seed development. Kluwer Academic, Dordrecht, pp 73–115

Finch-Savage WE, Leubner-Metzger G (2006) Seed dormancy and the control of germination. New Phytol 171:501–523

Finkelstein R, Reeves W, Ariizumi T, Steber C (2008) Molecular aspects of seed dormancy. Annu Rev Plant Biol 59:387–415

Gallardo K, Firnhaber C, Zuber H, Hericher D, Belghazi M, Henry C, Kuster H, Thompson RD (2007) A combined proteome and transcriptome analysis of developing *Medicago truncatula* seeds. Mol Cell Proteomics 6:2165–2179

Gutierrez L, Van Wuytswinkel O, Castelain M, Bellini C (2007) Combined networks regulating seed maturation. Trends Plant Sci 12:294–300

Hannah LC (2007) Starch formation in the cereal endosperm. In: Olsen OA (ed) Plant cell monographs. Endosperm: development and molecular biology, vol 8. Springer, Berlin, pp 179–193

Haughn G, Chaudhury A (2005) Genetic analysis of seed coat development in *Arabidopsis*. Trends Plant Sci 10:472–477

Herman EM, Larkins BA (1999) Protein storage bodies and vacuoles. Plant Cell 11:601–614

Ingram GC (2010) Family life at close quarters: communication and constraint in angiosperm seed development. Protoplasma 247:195–214

James MG, Denyer K, Myers AM (2003) Starch synthesis in the cereal endosperm. Curr Opin Plant Biol 6:215–222

Kawashima T, Goldberg RB (2010) The suspensor: not just suspending the embryo. Trends Plant Sci 15:23–30

Kigel J, Galili G (1995) (eds) Seed development and germination. CRC Press, New York. ISBN 0824792297

Kladnik A, Chamusco K, Dermastia M, Chourey P (2004) Evidence of programmed cell death in post-phloem transport cells of the maternal pedicel tissue in developing caryopsis of maize. Plant Physiol 136:3572–3581

Koch KE (1996) Carbohydrate-modulated gene expression in plants. Annu Rev Plant Physiol Plant Mol Biol 47:509–540

Koch K (2004) Sucrose metabolism: regulatory mechanisms and pivotal roles in sugar sensing and plant development. Curr Opin Plant Biol 7:235–246

Kowles RV (2009) The importance of DNA endoreduplication in the developing endosperm of maize. Maydica 54:387–399

Kropf DL (1994) Cytoskeletal control of cell polarity in a plant zygote. Dev Biol 165:361–371

Lakshamanan KK (1972) The monocot embryo. In: Varghese TM, Grover RK (eds) Vistas in plant sciences, vol 2. International Biological Sciences, Hissar

Linkies A, Graeber K, Knight C, Leubner-Metzger G (2010) The evolution of seeds. New Phytol 186:817–831

Lopes MA, Larkins BA (1993) Endosperm origin, development, and function. Plant Cell 5:1383–1399

Lur HS, Setter TL (1993a) Endosperm development of maize *defective-kernel* (*Dek*) mutants. Auxin and cytokinin levels. Ann Bot 72:1–6

Lur HS, Setter TL (1993b) Role of auxin in maize endosperm development. Timing of nuclear DNA endoreduplication, zein expression, and cytokinin. Plant Physiol 103:273–280

Martin AC (1946) The comparative internal morphology of seeds. Am Mid Nat 36:513–660
Melkus G, Rolletschek H, Radchuk R, Fuchs J, Rutten T, Wobus U, Altmann T, Jakob P, Borisjuk L (2009) The metabolic role of the legume endosperm: a noninvasive imaging study. Plant Physiol 151:1139–1154
Moïse JA, Han S, Gudynaite-Savitch L, Johnson DA, Miki BLA (2005) Seed coats: structure, development, composition, and biotechnology. In Vitro Cell Dev Biol-Plant 41:620–644
Morris RO (1997) Hormonal regulation of seed development. In: Larkins BA, Vasil IK (eds) Cellular and molecular biology of plant seed development. Kluwer Academic Publishers, Dordrecht, pp 117–149
Nambara E, Marion-Poll A (2003) ABA action and interaction in seeds. Trends Plant Sci 8:213–217
Natesh S, Rau MA (1984) The embryo. In: Johri LBM (ed) Embryology of angiosperms. Springer-Verlag, Berlin, pp 377–443
Nguyen HN, Sabelli PA, Larkins BA (2007) Endoreduplication and programmed cell death in the cereal endosperm. In: Olsen OA (ed) Plant cell monographs. Endosperm: development and molecular biology, vol 8. Springer-Verlag, Berlin/Heidelberg, pp 21–43
Nowack MK, Ungru A, Bjerkan KN, Grini PE, Schnittger A (2010) Reproductive cross-talk: seed development in flowering plants. Biochem Soc Trans 38:604–612
Offler CE, McCurdy DW, Patrick JW, Talbot MJ (2003) Transfer cells: cells specialized for a special purpose. Annu Rev Plant Biol 54:431–454
Ohto MA, Floyd SK, Fischer RL, Goldberg RB, Harada JJ (2009) Effects of APETALA2 on embryo, endosperm, and seed coat development determine seed size in *Arabidopsis*. Sex Plant Reprod 22:277–289
Olsen OA (2004) Nuclear endosperm development in cereals and *Arabidopsis thaliana*. Plant Cell 16:S214–S227
Osborne TB (1924) The vegetable proteins, 2nd edn. Longmans, Green & Co, London, UK
Patrick JW, Offler CE (2001) Compartmentation of transport and transfer events in developing seeds. J Exp Bot 52:551–564
Raboy V (1997) Accumulation and storage of phopsphate and minerals. In: Larkins BA, Vasil IK (eds) Cellular and molecular biology of plant seed development. Kluwer Academic Publishers, Dordrecht, pp 441–477
Radchuk R, Emery RJN, Weier D, Vigeolas H, Geingenberger P, Lunn JE, Feil R, Weschke W, Weber H (2010) Sucrose non-fermenting kinase 1 (SnRK1) coordinates metabolic and hormonal signals during pea cotyledon growth and differentiation. Plant J 61:324–338
Radchuk V, Weier D, Radchuk R, Weschke W, Weber H (2011) Development of maternal seed tissue in barley is mediated by regulated cell expansion and cell disintegration and coordinated with endosperm growth. J Exp Bot 62:1217–1227
Raghavan V (1986) Embryogenesis in angiosperms: a developmental and experimental study. Cambridge University Press, Cambridge
Raghavan V (1997) Molecular embryology of flowering plants. Cambridge University Press, Cambridge, UK
Raghavan V, Sharma KK (1995) Zygotic embryogenesis in gymnosperms and angiosperms. In: Torpe TA (ed) In vitro embryogenesis in plants. Kluwer Academic Publishers, Dordrecht, pp 73–115
Randolph LF (1936) Developmental morphology of the caryopsis in maize. J Agric Res 53:881–916
Ruan YL, Chourey PS (2006) Carbon partitioning in developing seeds. In: Basra AS (ed) Handbook of seed science and technology. Food Products Press, Binghamton, pp 125–152
Sabelli PA (2012) Replicate and die for your own good: endoreduplication and cell death in the cereal endosperm. J Cereal Sci 56:9–20
Sabelli PA, Larkins BA (2008) The endoreduplication cell cycle: regulation and function. In: Verma DPS, Hong Z (eds) Cell division control in plants. Springer-Verlag, Berlin-Heidelberg, pp 75–100

Sabelli PA, Larkins BA (2009a) The contribution of cell cycle regulation to endosperm development. Sex Plant Reprod 22:207–219
Sabelli PA, Larkins BA (2009b) The development of endosperm in grasses. Plant Physiol 149:14–26
Schaffner M (1906) The embryology of the shepherd's purse. Ohio Nat 7:1–8
Schneitz K, Hülskamp M, Pruitt RE (1995) Wild-type ovule development in *Arabidopsis thaliana*: a light microscope study of cleared whole-mount tissue. Plant J 7:731–749
Sharma HP (2009) Plant embryology: classical and experimental. Alpha Science Intl Ltd., USA
Shewry PR (2004a) Protein chemistry of dicotyledonous grains. In: Wrigley C (ed) Encyclopedia of Grain Science. Elsevier, Oxford, pp 466–472
Shewry PR (2004b) Protein synthesis and deposition. In: Wrigley C (ed) Encyclopedia of grain science. Elsevier, Oxford, pp 472–480
Shewry PR, Halford NG (2002) Cereal seed storage proteins: structures, properties and role in grain utilization. J Exp Bot 53:947–958
Shewry PR, Napier JA, Tatham AS (1995) Seed storage proteins: structures and biosynthesis. Plant Cell 7:945–956
Shih MD, Hoekstra FA, Hsing YI (2008) Late embryogenesis abundant proteins. Adv Bot Res 48:211–255
Sreenivasulu N, Borisjuk L, Junker BH, Mock HP, Rolletschek H, Seiffert U, Weschke W, Wobus U (2010) Barley grain development:Toward an integrative view. Int Rev Cell Mol Biol 281: 49–89
Sreenivasulu N, Radchuk V, Strickert M, Miersch O, Weschke W, Wobus U (2006) Gene expression patterns reveal tissue-specific signaling networks controlling programmed cell death and ABA-regulated maturation in developing barley seeds. Plant J 47:310–327 (Erratum in: Plant J 47 (2006) 987)
Thorne JH (1985) Phloem unloading of C and N assimilates in developing seeds. Annu Rev Plant Physiol 36:317–343
Vijayaraghavan MR, Prabhakar K (1984) The endosperm. In: Johri LBM (ed) Embryology of angiosperms. Springer-Verlag, Berlin, pp 319–376
Vitale A, Denecke J (1999) The endoplasmic reticulum-gateway of the secretory pathway. Plant Cell 11:615–628
Webb MC, Gunning BES (1991) The microtubular cytoskeleton during development of the zygote, proembryo and free-nuclear endosperm in *Arabidopsis thaliana* (L.) Heynh. Planta 184:187–195
Weber H, Borisjuk L, Wobus U (1997) Sugar import and metabolism during seed development. Trends Plant Sci 2:169–174
Weber H, Borisjuk L, Wobus U (2005) Molecular physiology of legume seed development. Annu Rev Plant Biol 56:253–279
Weselake RJ (2005) Storage lipids. In: Murphy DJ (ed) Plant lipids: biology, utilization and manipulation. Blackwell Publishing, Oxford, pp 162–221
Wilson DR, Larkins BA (1984) Zein gene organization in maize and related grasses. J Mol Evol 20:330–340
Yang J, Zhang J, Liu K, Wang Z, Liu L (2006) Abscisic acid and ethylene interact in wheat grains in response to soil drying during grain filling. New Phytol 171:293–303
Young TE, Gallie DR (2000) Programmed cell death during endosperm development. Plant Mol Biol 44:283–301

Chapter 2
Proteomics Reveals A Potential Role of the Perisperm in Starch Remobilization During Sugarbeet Seed Germination

Julie Catusse, Claudette Job and Dominique Job

Abstract Sugarbeet is a crop of high economic importance because it is one of the two main sources of plant sugar, the other being sugarcane. The sugarbeet seeds have the peculiarity of containing at maturity a large starchy storage tissue, called the perisperm. In contrast to the well-documented cereal endosperm, the physiology of this perisperm is completely unknown. Here, we used proteomics of perisperm isolated either from dry mature or imbibed sugarbeet seeds to unravel the mechanisms of starch remobilization during germination. We also carried out a comparative proteomics analysis with the perisperm isolated from the dry mature sugarbeet seeds. We observed an accumulation of α-amylase in the perisperm isolated from imbibed whole seeds but not from the isolated imbibed perisperm alone, suggesting a role of the embryo in triggering the accumulation of this starch-mobilizing enzyme in the perisperm during germination. In this way, the mechanisms occurring in the sugarbeet seed perisperm during germination would appear to be similar to those documented for the endosperm of cereals. In contrast, an accumulation of β-amylase and α-glucosidase was observed in the isolated imbibed perisperm, suggesting that the embryo was not mandatory for induction of these enzymes in the perisperm during imbibition.

Keywords Amylases · Perisperm · Proteomics · Seed germination · Starch · Sugarbeet

D. Job (✉) · J. Catusse · C. Job
CNRS/Bayer CropScience Joint Laboratory (UMR5240), Bayer CropScience,
14-20 rue Pierre Baizet, 69263, Lyon cedex, France
e-mail: dominique.job@bayer.com

J. Catusse
Quintiles – Navigating the new health,
Parc d'innovation, rue Jean Dominique Cassini, 67404, illkirch cedex, France

G. K. Agrawal, R. Rakwal (eds.), *Seed Development: OMICS Technologies toward Improvement of Seed Quality and Crop Yield*,
DOI 10.1007/978-94-007-4749-4_2, © Springer Science+Business Media Dordrecht 2012

2.1 Introduction

2.1.1 Seed Physiology

In the botanical sense, the seed is the organ issued from the double fertilization of the ovule by the pollen grain. It contains a zygotic embryo that will form upon germination the new plant and a storage tissue, usually a triploid endosperm, but that may derive from the maternal nucellus in some species such as sugarbeet (*Beta vulgaris* L.), and that will contribute to successful seedling establishment following germination. In their mature form, most seeds ensure the dispersion of plant species in the environment (Bewley and Black 1994; Linkies et al. 2010; http://www.seedbiology.de/; accessed April 28th 2011). Embryos from seeds of most plants belonging to the orthodox family have the remarkable property of being able, through intense desiccation during maturation, to interrupt their metabolic activity on the mother plant and to restart this activity after a simple imbibition during germination (Angelovici et al. 2010; Boudet et al. 2006; Buitink et al. 2000).

The seed is thus one of the rare and remarkable examples in eukaryotes where there is disruption of development and preservation of the competence of an embryo in the dry state over long periods of time. Remarkably, radiocarbon dating allowed the determination of the age of date (*Phoenix dactylifera* L.) seeds at about 2000 years (Sallon et al. 2008), sacred lotus (*Nelumbo nucifera*) seeds at 1300 years (Shen-Miller 2002; Shen-Miller et al. 1995), or canna (*Canna compacta*) seeds at 600 years (Lerman and Cigliano 1971). This implies the existence of specific molecular and biochemical mechanisms to maintain the state of metabolic quiescence of mature dry seeds while preserving their viability (longevity), but also to ensure the restart of cellular metabolism during germination (Bewley and Black 1994; Catusse et al. 2011; Holdsworth et al. 2008a; Rajjou and Debeaujon 2008; Rajjou et al. 2008). These mechanisms determine the potential of seed germination, that is to say the success with which the new plant will be established. A better understanding of these mechanisms would allow the development of molecular and biochemical markers reflecting this potential, which could be used as quality markers in the seed sector for the marketing of high-vigor seed lots (Catusse et al. 2011). Knowing that our food depends on agricultural production, the concept of seed quality is central in a socio-economic context where agriculture is heavily industrialized, refocused on a few crops, and where it is expected that there will be 9 billion people to feed within the next 40 years (The U.S. Census Bureau: http://www.census.gov/ipc/www/idb/ (accessed April 28th 2011)).

In angiosperms, seed formation results from a double fertilization (Dumas and Rogoswki 2008). Initially, the male gametophyte (pollen grain) germinates on the stigma of the flower, thus forming a pollen tube carrying two sperm nuclei to the egg. There is then a double fertilization, one giving the zygotic embryo, the other a storage tissue (triploid endosperm) by fusion of a sperm nucleus and two polar nuclei from the female gametophyte. The embryo can be considered as a miniature plant with a root (radicle), a draft of stem (hypocotyl), and cotyledons. By definition,

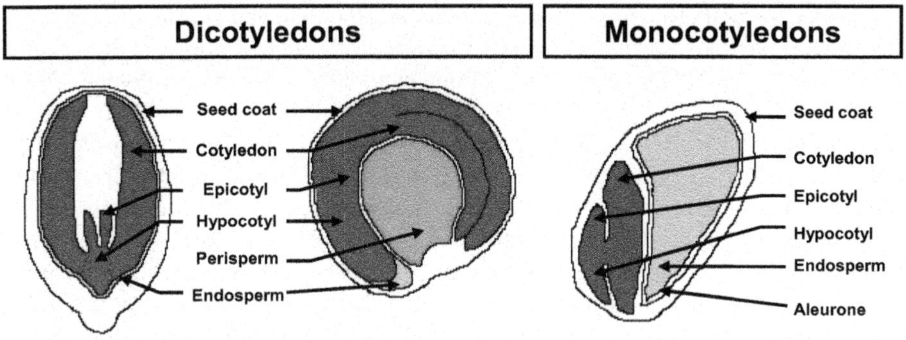

Fig. 2.1 Schematic structure of dicot and monocot seeds

embryos of seeds of dicotyledonous plants have two cotyledons (e.g., legumes and oilseeds), while there is only one cotyledon in seeds of monocotyledons (e.g., rice, maize, and wheat). Seeds can be classified by the presence or absence of a mature well-formed endosperm. In the latter case, the initial endosperm is readily consumed during development and other structures, as the cotyledons (e.g., *Arabidopsis*) but also perisperm (e.g., sugarbeet), becomes the main storage organs. In contrast, some seeds such as cereals contain at maturity a large dead starchy endosperm surrounded by a few living cell layers of endosperm, called the aleurone layer (Fig. 2.1). The death of the starchy endosperm cells is a developmental phenomenon (programmed cell death) rather than one induced by water loss (Bethke et al. 1991; Golovina et al. 2000). However, cells of the aleurone layer can express a specific developmental program allowing them to survive desiccation. The barley (*Hordeum vulgare*) aleurone layer can be separated from the other seed tissues and maintained in culture, allowing studying the effect of added signaling molecules in an isolated system (Bethke et al. 1997; Finnie et al. 2011).

Two main phytohormones, ABA and GAs control seed formation, dormancy, and germination; ABA being an inhibitor of germination involved in the development of the embryo, accumulation of seed reserves, and maintenance of dormancy, while GAs stimulate seed germination and mobilization of the stored reserves (Hilhorst and Karssen 1992; Ho et al. 2003; Koornneef and van der Veen 1980; Koornneef et al. 2002; Kucera et al. 2005; Lovegrove and Hooley 2000).

2.1.2 Proteomics in Addressing Biological Questions

With the completion of the genome sequence for several organisms and the accelerated development of analytical methods for protein characterization, proteomics has become an indispensable approach to functional genomics (Bradshaw and Burlingame 2005; Rabilloud et al. 2010; Walther and Mann 2010), notably in plants (Agrawal and Rakwal 2006, 2008, 2011; Agrawal et al. 2006, 2011a, b; Job et al. 2011; Wienkoop et al. 2010; http://www.inppo.com/; accessed April 28th 2011). In particular, several proteomics studies on the development and germination of seeds have been published in recent years (Agrawal et al. 2008; Bourgeois et al. 2009; Bykova et al. 2011; Dam et al. 2009; Finnie et al. 2002, 2004; Gallardo et al. 2001, 2003, 2007; Hajduch et al. 2005, 2006, 2010; Irar et al. 2010; Miernyk and Hajduch 2011; Méchin et al. 2004; Müller et al. 2010; Nadaud et al. 2010; Nautrup-Pedersen et al. 2010; Rajjou et al. 2006; Sheoran et al. 2005; Wang et al. 2010; Yang et al. 2007). Moreover, the study of biosynthetic pathways responsible for the accumulation of the various seed storage compounds is of major importance for at least two reasons: (i) these reserves support the early growth of seedlings after germination, hence contributing to seedling vigor; and (ii) they are widely used as food and feed. Many biotechnological applications focus on improving the nutritional value of seeds and also to promote other non-nutritional uses (e.g., biofuel production) or the use of seeds as plants for the production/storage of recombinant proteins (Boothe et al. 2010; Gressel 2008; Job 2002).

2.1.3 Seed Proteins: A brief Historical Event

The systematic study of seed proteins dates from the nineteenth century. Particularly the work of Osborne (Osborne 1924) on proposed classification of proteins is still in use, according to their solubility properties in water (albumins), dilute salt solutions (globulins), alcohol-water mixtures (prolamins), or dilute acid or alkaline solutions (glutelins). Besides all these proteins, mature dry seeds also contain stored mRNA during maturation. Rajjou et al. (2004) showed that germination of *Arabidopsis* seeds could be observed even in the presence of an excess of a fungal toxin α-amanitin, which is a potent specific inhibitor of transcription elongation by eukaryotic DNA-dependent RNA polymerase II (Jendrisak 1980; de Mercoyrol et al. 1989). On the contrary, *Arabidopsis* seed germination (radicle protrusion) was completely inhibited in the presence of cycloheximide, an inhibitor of protein synthesis in eukaryotic cells that blocks translation elongation (Rajjou et al. 2004). These findings have been confirmed both in *Arabidopsis* (Kimura and Nambara 2010) and rice (He et al. 2011). These results disclosed that the germination potential is largely programmed during the process of seed maturation on the mother plant. They also highlighted the usefulness of a proteomics approach for studying the germination process since germination can be evidenced in the absence of *de novo* transcription, thus illustrating the role of stored proteins and of translation of mRNAs also stored in the seeds (Catusse et al. 2008a; He et al. 2011; Holdsworth et al. 2008b; Rajjou et al. 2004).

2.1.4 Sugarbeet and its Importance

Sugarbeet (*Beta vulgaris*), a dicotyledonous biennal plant belonging to the *Amaranthaceae*, can be grown commercially in a wide variety of temperate climates.

It has a high economic importance because it is one of the two main sources of plant sugar (the total world production in 2005 was about 250 million metric tons, http://www.sugarindustrybiotechcouncil.org/sugar-beet-faq/; accessed April 28th 2011), the other being sugarcane (*Saccharum officinarum* L.). The agronomic productivity of this direct seeded sugarbeet crop is determined significantly by the uniformity of seedling emergence in the field (de Los Reyes and McGrath 2003; Elamrani et al. 1992; Job et al. 1997; Mukasa et al. 2003). Besides food applications, its culture is attracting renewed interest, particularly in Europe, for the production of bioethanol (http://www.biofuelstp.eu/bioethanol.html; accessed April 28th 2011). The plants from which descend the current varieties and that were described as 'versatile garden plants', originated in the Middle East and their discovery dates back about 800 years BC. The high sugar content of their roots was noticed at the end of the sixteenth century by the French agronomist Olivier de Serres (years 1539–1619; http://www.bookrags.com/research/olivier-de-serres-scit-0312/; accessed April 28th 2011). Then, the German chemist Andreas Marggraf (1709–1782) used alcohol to extract the juices from several plants, including, in particular, sugarbeet. Andreas Marggraf is internationally recognized as the founding father of this technique (http://en.wikipedia.org/wiki/Andreas_Sigismund_Marggraf; accessed April 28th 2011). Fifty-four years later, in the year 1801, the world's first beet sugar factory opened at Cunern in Silesia (now Konary, Poland) and produced about 70 kg of sugar per day. Original forms contained only about 4 % sugar but careful selection and breeding have raised this to a content of 20 % (http://en.wikipedia.org/wiki/Sugar_beet; accessed July 11th 2011).

Sugarbeet seeds contain a large starchy storage tissue, called the perisperm. The perisperm is a dead tissue at maturity and consists of a network of thin-walled cells filled with starch grains in the form of globoids (Artschwager 1927). The perisperm exists in all seeds, but for the majority of them, it is consumed during embryogenesis or seed development. Only in certain cases (like sugarbeet), this tissue becomes the major storage tissue of the seed. In watermelon (*Cucumis melo*), the perisperm surrounds the embryo and forms an envelope that intervenes in the control of germination. This envelope is composed of two- to four-cell layers of perisperm and a single layer of endosperm. In the dry seed, the role of this envelope is to prevent the diffusion of solutes, but not water (Welbaum and Bradford 1990). Although in the case of the grain of coffee species the storage tissue was originally suggested to correspond to perisperm (Houk 1938), it is now identified as endosperm (Mendes 1941). Here, during development the grain is dominated by a well-developed maternal perisperm tissue up to approximately the halfway stage of maturation, following which the locular space is progressively filled with endosperm up to full grain maturity (Rogers et al. 1999). In contrast, among *Chenopodiaceae* and *Amaranthaceae*, the perisperm is located in a central position and is mainly dedicated to the accumulation of starch reserves (Fig. 2.1). This has been demonstrated in seeds of sugarbeet (Artschwager 1927), *Amaranthus hypochondriacus* (Coimbra and Salema 1994), *Chenopodium quinoa* (Prego et al. 1998), and some *Salicornioideae* (Shepherd et al. 2005), a subfamily of *Amaranthaceae*. In contrast to the well-documented cereal endosperm (Ritchie et al. 2000), the physiology of this perisperm is completely unknown. In particular, it is presently unknown if the mechanisms of storage compound (starch, proteins) remobilization established for the cereal endosperm during germination can apply to the perisperm.

Starch can be degraded by hydrolysis in reactions catalyzed by the amylase family (Buchanan et al. 2000; Hills 2004; Machovic and Janecek 2006; Sun and Henson 1991; Yamasaki 2003). The α-amylases are endoenzymes that break α-(1,4) bonds at random within polysaccharide chains to produce shorter chains called limit dextrins, glucose, and maltose. The β-amylases are exohydrolases that hydrolyze α-(1,4) bonds in polysaccharides to remove successive units maltose (disaccharides of α-glucose) from nonreducing ends of α-1,4-linked poly- and oligoglucans until the first α-1,6-branching point along the substrate molecule is encountered. In germinating barley seeds, α-glucosidases can initiate the attack of starch granules and exert a marked synergistic effect on the initial stages of native starch granule degradation by α-amylases (Sun and Hanson 1990). Their roles are to degrade maltose and limit dextrin. All these three enzymes are particularly active during germination, and their expressions are induced by GAs secreted by the embryo (Bak-Jensen et al. 2007; Bethke et al. 1997; Buchanan et al. 2000; Fincher 1989; Gubler et al. 1987; Hills 2004; Jones and Jacobsen 1991; Laidman 1983; Sun and Gubler 2004; Zentella et al. 2002). In the presence of endogenous GAs, α-amylase is synthesized *de novo* in the aleurone layer surrounding the endosperm. In contrast, a precursor form of β-amylase is present in seeds in the dry state as a proenzyme devoid of enzyme activity, which is activated during germination after cleavage of a peptide sequence at the C-terminal part of the enzyme. Moreover, if the α-glucosidases are already partly accumulated in the dry seeds, they are for the most part newly synthesized in response to GAs in the aleurone layer and transported to the endosperm (Buchanan et al. 2000). Unlike for cereals, the mechanisms of regulation of starch degradation during germination are not known in species such as sugarbeet, which store starch reserves in a perisperm tissue. Also, the existence of an aleurone layer similar to that of cereals and *Arabidopsis* (Bethke et al. 1997; Finnie et al. 2011; Sun and Gubler 2004; Tasleem-Tahir et al. 2011) has not been described in these seed species, containing a large perisperm at maturity.

The α-glucosidases have previously been characterized in sugarbeet seeds (Chiba et al. 1978; Matsui et al. 1978; Yamasaki and Suzuki 1980). Also, the initiation of starch metabolism has already been documented during sugarbeet seed germination (Lawrence et al. 1990). This metabolism results in the transfer of sugars from the perisperm to the various tissues of the embryo upon imbibition and seedling growth (Lawrence et al. 1990). In particular, these authors reported a rise of α-glucosidase activity just before root extrusion. Although the work of Lawrence et al. (1990) firmly established an increase in certain enzyme activities (α-amylase, β-amylase, and α-glucosidase) in the sugarbeet seed perisperm during germination, the mechanisms responsible for such an induction were not addressed. At least two possibilities can be envisaged to account for the observed behavior. In the first one, the starch metabolizing enzymes would have been stored in an inactive precursor form in the perisperm during maturation and they became activated during the course of germination, for example owing to specific protein modifications occurring in the precursor protein. In the second, as the mature perisperm is a dead tissue and hence presumably incapable of supporting protein synthesis, *de novo* synthesis of these enzymes could occur in the embryo upon seed imbibition followed by their transport in the perisperm. To address these questions, we presently studied the modifications occurring in the sugarbeet seed perisperm proteome during germination.

2.2 Experimental Strategies

2.2.1 Materials and Experimental Design

Fruits of a triploid monogerm sugarbeet seed lot (ref. 302–688C) were cleaned, polished, and calibrated after harvest according to commercial standards (KWS SAAT AG, Einbeck, Germany). They were previously used in proteomics (Catusse et al. 2008b, 2011) and transcriptomics (Pestsova et al. 2008) studies. Germination assays were carried out with triplicates of 400 seeds (true seeds plus pericarp) placed in covered plastic boxes (13 × 18 × 6 cm) on pleated filter paper wetted with 30 mL of distilled water and incubated at 10 °C in the dark. Germination started at about 60 h of imbibition and about 90 % of the imbibed seeds had germinated after 3 days imbibition.

Two protocols were used to characterize the role of perisperm during germination. In the first one, the whole seeds were incubated for 3 days as above, and the perisperm tissue was collected. In the second, the perisperm was first isolated from the dry mature seeds, and then imbibed for 3 days as above.

To isolate the perisperm from the dry mature seeds, the operculum, which is the ovary cap of the fruit at the upper part of the pericarp was removed from the fruit by prying it off with high precision reverse action tweezers and a scalpel, starting just in front of the position where is located the radicle (Hermann et al. 2007). Then, the botanical true seed appears as being layered in the remaining pericarp and it can be carefully removed from the pericarp. Tissues were then dissected using a scalpel under a binocular microscope. In particular, by the gentle action of the scalpel, the perisperm can be detached from the embryo, without contamination from adjacent tissues. The dissection of 100 seeds gave about 100 mg perisperm tissue. The isolation of the perisperm tissue from 3 days imbibed seeds was much easier, since at this imbibition time the tissue was fully hydrated and presented a viscous aspect.

2.2.2 Preparation of Protein Extracts, 2-DGE, Protein Staining, and Gel Analyses

Protein extracts were prepared from 100 seeds or perisperm tissue using a low ionic-strength buffer containing 50 mM HEPES (pH 8.0) and 1 mM EDTA (pH 9.0), yielding the albumin fraction as described (Catusse et al. 2008b). The 2-DGE was

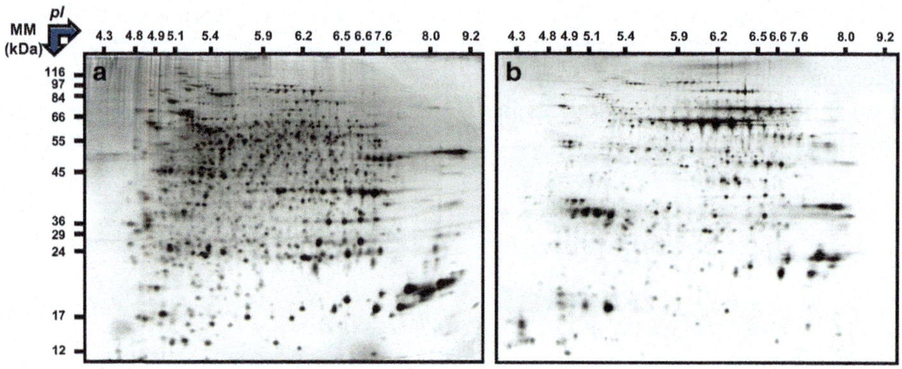

Fig. 2.2 Proteome of the dry mature sugar beet seed. **a** 2-DGE analysis of total soluble proteins (100 µg) from whole dry mature seeds (759 proteins identified; Catusse et al. 2008b). **b** 2-DGE analysis of total soluble perisperm proteins (100 µg) isolated from dry mature sugarbeet seeds (172 proteins identified; Catusse et al. 2008b)

carried out using an amount of 100 µg protein as described (Catusse et al. 2008b). Three gels loaded with biological replicates were run for each treatment. Following protein staining with silver nitrate, image analysis of the scanned 2-D gels was carried out using the ImageMaster 2-D Elite version 4.01 software (Amersham Biosciences; Catusse et al. 2008b).

2.3 Results and Discussion

Protein extracts from perisperm tissues of mature dry and imbibed seeds were prepared as described above and separated by 2-DGE (Fig. 2.2). Among the 182 proteins identified in the perisperm from dry mature seeds (Fig. 2.2b), 120 proteins are specifically expressed in this tissue, testifying of the purity of this tissue preparation (Catusse et al. 2008b).

We then characterized changes in protein accumulation levels occurring in the perisperm during imbibition using the two protocols described under EXPERIMENTAL STRATEGIES. A number of spots displayed varying accumulation patterns, as depicted in Fig. 2.3, in which either the accumulation pattern was similar or different when comparing the isolated imbibed perisperm or the perisperm isolated from whole imbibed seeds. Here, we concentrate on the spots related to starch metabolism as this metabolism has been very well detailed in the endosperm of cereal seeds during germination (Buchanan et al. 2000).

Highly abundant proteins (window b of Fig. 2.3) corresponding to soluble starch synthase dominated the proteome of the perisperm isolated from the dry mature sugarbeet seeds. As this enzyme catalyzes starch synthesis during seed maturation, it is therefore not surprising to observe its disappearance in the perisperm isolated

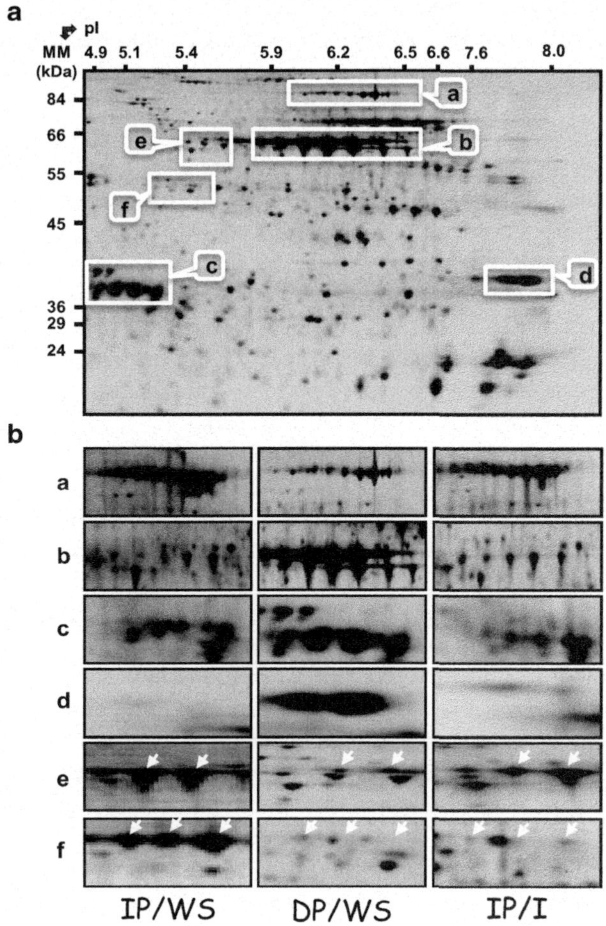

Fig. 2.3 Accumulation patterns of specific proteins in the sugarbeet perisperm during germination corresponding to 3 days imbibition on water. The perisperm was isolated from whole mature dry or imbibed seeds. For comparison the perisperm was also isolated from whole dry mature seeds then imbibed for 3 days on water. 2-DGE analysis of total soluble perisperm proteins (100 μg) was carried out as described under EXPERIMENTAL STRATEGIES. **a** Total soluble perisperm proteins isolated from dry mature sugarbeet seeds. The indicated portions of the gel (a, b, c, d, e and f) are reproduced in **b**. **b** Enlarged windows (a–f) of 2-D gels as shown in **a** for the perisperm isolated from imbibed whole seeds (left; IP/WS), the perisperm isolated from whole dry mature seeds (middle; DP/WS), or from the perisperm isolated from whole dry mature seeds then incubated for 3 days on water (right; IP/I). Window a, the train of spots corresponds to α-glucosidase; Window b, the train of spots corresponds to soluble starch synthase; Windows c and d; the major spots correspond to granule-bound starch synthase; Window e, the white arrows correspond to β-amylase; Window f, the white arrows correspond to α-amylase. Proteins were previously identified by mass spectrometry (Catusse et al. 2008b)

from the imbibed whole seeds (window b of Fig. 2.3). The same behavior was observed with the imbibed isolated perisperm only (window b of Fig. 2.3). In addition to the soluble starch synthases, we also found granule-bound starch synthases in several spots (window c of Fig. 2.3). All of them showed the same accumulation behavior than seen for soluble starch synthase in imbibed perisperm isolated from whole seeds and imbibed isolated perisperm. From these data, we conclude that the perisperm is autonomous for the specific degradation of such proteins, meaning that this degradation does not require the synergic action of the embryo. This finding is in agreement with our previous work documenting the existence of several proteinases and proteasome subunits in the mature dry perisperm (Catusse et al. 2008b).

In the perisperm isolated from imbibed whole seeds, we observed a strong accumulation of α-amylase containing spots (white arrows within window f of Fig. 2.3b). However, such an increased protein accumulation was not seen with the imbibed isolated perisperm (window f of Fig. 2.3).

Two spots containing β-amylases were detected in very low amounts in the perisperm isolated from the dry mature sugarbeet seeds (white arrows within window e of Fig. 2.3b). Their accumulation increased substantially both in the perisperm isolated from imbibed whole seeds and in the imbibed isolated perisperm (window e of Fig. 2.3).

The α-glucosidases were present in the perisperm of the mature dry seeds and their protein abundances increased both in the perisperm isolated from the imbibed whole seeds and the imbibed isolated perisperm (window a of Fig. 2.3).

As mentioned above, the enzymes involved in starch degradation could either be stored in the perisperm in the form of zymogens, or be newly synthesized in the embryo and then transported. In our study, it appears that the first assumption might not apply. Indeed, for α-amylases, β-amylases, and α-glucosidases, we did not observe on the 2-D gels the transition described for the β-amylase of barley endosperm, that is the transition from one form of the enzyme of high molecular weight in the proteome of the perisperm isolated from the dry seeds to a form of lower molecular weight in the proteome of imbibed perisperm. Moreover, *de novo* synthesis of the enzymes in the perisperm tissue is quite unlikely since it is a dead tissue (Artschwager 1927). The remaining possibility to explain the increase in volume of the spots of amylases and glucosidases in this tissue is *de novo* synthesis of these proteins in the embryo. This mechanism would be analogous to that established for starch mobilization in the endosperm of monocots.

2.4 Concluding Remarks

This study presents for the first time data on the accumulation of enzymes of starch mobilization (e.g., α-amylase, β-amylase, and α-glucosidase) in the perisperm of imbibed sugarbeet seeds. We noted important differences between the accumulation patterns of α-amylase in the perisperm isolated from imbibed whole seeds and the isolated imbibed perisperm alone. This observation suggests a positive role of the embryo in triggering the accumulation of this starch-mobilizing enzyme (s) in the perisperm during germination. In this way, the mechanisms occurring in the sugar-beet seed perisperm during germination would appear to be similar to those documented for the endosperm of cereals (Bethke et al. 1997; Finnie et al. 2011; Sun and Gubler 2004). However, an accumulation of β-amylase and α-glucosidase was also observed in the isolated imbibed perisperm alone, suggesting that the existence of embryo was not mandatory for induction of these enzymes in the perisperm during imbibition. It is also noted that the catabolic processes leading to the disappearance of the highly-abundant soluble and granule-bound starch synthase can occur to the same extent in the perisperm isolated from imbibed whole seeds and the isolated imbibed perisperm alone, indicative of some autonomy of the perisperm concerning proteome changes during imbibition. Pending questions concern the role of GAs in the observed modifications in accumulation levels of the starch degradation enzymes and the existence of an equivalent of the cereal aleurone layer in sugarbeet seeds. Further work will allow addressing these questions.

Acknowledgments This work was supported by the French Ministry of Research. We thank Juliane Meinhard, Andreas Menze, and Uwe Fischer for helpful discussions, and Emeline Leydier, Camille Bechetoille, and Bernadette Claus for skillful technical assistance.

References

Agrawal GK, Jwa NS, Iwahashi Y, Yonekura M, Iwahashi H, Rakwal R (2006) Rejuvenating rice proteomics: facts, challenges, and visions. Proteomics 6:5549–5576

Agrawal GK, Hajduch M, Graham K, Thelen JJ (2008) In-depth investigation of the soybean seed-filling proteome and comparison with a parallel study of rapeseed. Plant Physiol 148:504–518

Agrawal GK, Bourguignon J, Rolland N, Ephritikhine G, Ferro M, Jaquinod M, Alexiou KG, Chardot T, Chakraborty N, Jolivet P, Doonan JH, Rakwal R (2011a) Plant organelle proteomics: collaborating for optimal cell function. Mass Spectrom Rev 30:772–853

Agrawal GK, Job D, Zivy M, Agrawal VP, Bradshaw R, Dunn MJ, Haynes PA, van Wijk KJ, Kikuchi S, Renaut J, Weckwerth W, Rakwal R (2011b) Time to articulate a vision for the future of plant proteomics—a global perspective. An initiative for establishing the international plant proteomics organization (INPPO). Proteomics 11:1559–1568

Angelovici R, Galili G, Fernie AR, Fait A (2010) Seed desiccation: a bridge between maturation and germination. Trends Plant Sci 15:211–218

Artschwager E (1927) Development of flowers and seed in the sugar beet. J Agric Res 34:1–25

Bak-Jensen KS, Laugesen S, Østergaard O, Finnie C, Roepstorff P, Svensson B (2007) Spatio-temporal profiling and degradation of alpha-amylase isozymes during barley seed germination. FEBS J 274:2552–2265

Bethke PC, Lonsdale JE, Fath A, Jones RL (1991) Hormonally regulated programmed cell death in barley aleurone cells. Plant Cell 11:1033–1045

Bethke PC, Schuurink R, Jones RL (1997) Hormonal signalling in cereal aleurone. J Exp Bot 48:1337–1356

Bewley J, Black M (1994) Seeds: physiology of development and germination. In: Bewley J, Black M (eds). Plenum, New York, pp 460

Boothe J, Nykiforuk C, Shen Y, Zaplachinski S, Szarka S, Kuhlman P, Murray E, Morck D, Moloney MM (2010) Seed-based expression systems for plant molecular farming. Plant Biotechnol J 8:588–606

Boudet J, Buitink J, Hoekstra FA, Rogniaux H, Larré C, Satour P, Leprince O (2006) Comparative analysis of the heat stable proteome of radicles of *Medicago truncatula* seeds during germination identifies late embryogenesis abundant proteins associated with desiccation tolerance. Plant Physiol 140:1418–1436

Bourgeois M, Jacquin F, Savois V, Sommerer N, Labas V, Henry C, Burstin J (2009) Dissecting the proteome of pea mature seeds reveals the phenotypic plasticity of seed protein composition. Proteomics 9:254–271

Bradshaw R, Burlingame A (2005) From proteins to proteomics. IUBMB Life 57:267–272

Buchanan BB, Gruissem W, Jones RL (2000) Biochemistry and molecular biology of plants. In: Buchanan BB, Gruissem W, Jones RL (eds). American Society of Plant Physiologists, Rockville

Buitink J, Leprince O, Hoekstra FA (2000) Dehydration induced redistribution of amphiphilic molecules between cytoplasm and lipids is associated with desiccation tolerance in seeds. Plant Physiol 124:1413–1426

Bykova NV, Hoehn B, Rampitsch C, Banks T, Stebbing JA, Fan T, Knox R (2011) Redox-sensitive proteome and antioxidant strategies in wheat seed dormancy control. Proteomics 11:865–882

Catusse J, Job C, Job D (2008a) Transcriptome- and proteome-wide analyses of seed germination. C R Biologies:815–822

Catusse J, Strub JM, Job C, Van Dorsselaer JM, Job D (2008b) Proteome-wide characterization of sugarbeet seed vigor and its tissue specific expression. Proc Natl Acad Sci U S A 105:10262–10267

Catusse J, Meinhard J, Job C, Strub JM, Fischer U, Pestsova E, Westhoff P, Van Dorsselaer A, Job D (2011) Proteomics reveals potential biomarkers of seed vigor in sugarbeet. Proteomics 11:1569–1580

Chiba S, Inomata S, Matsui H, Shimomura T (1978) Purification and properties of an α-glucosidase (glucoamylase) in sugar beet seed. Agric Biol Chem 42:241–245

Coimbra S, Salema R (1994) *Amaranthus hypochondriacus*: seed structure and localization of seed reserves. Ann Bot 74:373–379

Dam S, Laursen BS, Ornfelt JH, Jochimsen B, Staerfeldt HH, Friis C, Nielsen K, Goffard N, Besenbacher S, Krusell L, Sato S, Tabata S, Thøgersen IB, Enghild JJ, Stougaard J (2009) The proteome of seed development in the model legume *Lotus japonicus*. Plant Physiol 149:1325–1340

de Los Reyes BG, McGrath JM (2003) Cultivar-specific seedling vigor and expression of a putative oxalate oxidase germin-like protein in sugar beet (*Beta vulgaris* L.). Theor Appl Genet 107:54–61

de Mercoyrol L, Job C, Job D (1989) Studies on the inhibition by alpha-amanitin of single-step addition reactions and productive RNA synthesis catalysed by wheat germ RNA polymerase II. Biochem J 258:165–169

Dumas C, Rogoswki P (2008) Fertilization and early seed formation. C R Biologies 331:715–725

Elamrani AJ, Raymond P, Saglio P (1992) Nature and utilization of seed reserves during germination and heterotrophic growth of young sugar beet seedlings. Seed Sci Res 2:1–8

Fincher GB (1989) Molecular and cellular biology associated with endosperm mobilization in geminating cereal grain. Annu Rev Plant Physiol Plant Mol Biol 40:305–346

Finnie C, Melchior S, Roepstorff P, Svensson B (2002) Proteome analysis of grain filling and seed maturation in barley. Plant Physiol 129:1308–1319

Finnie C, Maeda K, Østergaard O, Bak-Jensen KS, Larsen J, Svensson B (2004) Aspects of the barley seed proteome during development and germination. Biochem Soc Trans 32:517–519

Finnie C, Andersen B, Shahpiri A, Svensson B (2011) Proteomes of the barley aleurone layer: a model system for plant signalling and protein secretion. Proteomics 11:1595–605

Gallardo K, Job C, Groot SPC, Puype M, Demol H, Vandekerckhove J, Job D (2001) Proteomic analysis of *Arabidopsis* seed germination and priming. Plant Physiol 126:835–848

Gallardo K, Le Signor C, Vandekerckhove J, Thompson RD, Burstin J (2003) Proteomics of *Medicago truncatula* seed development establishes the time frame of diverse metabolic processes related to reserve accumulation. Plant Physiol 133:664–682

Gallardo K, Firnhaber C, Zuber H, Hericher D, Belghazi M, Henry C, Kuster H, Thompson RD (2007) A combined proteome and transcriptome analysis of developing *Medicago truncatula* seeds. Mol Cell Proteomics 6:2165–2179

Golovina EA, Hoekstra FA, van Aelst AC (2000) Programmed cell death or desiccation tolerance: two possible routes for wheat endosperm cells. Seed Sci Res 10:365–379

Gressel J (2008) Transgenics are imperative for biofuel crops. Plant Sci 174:246–263

Gubler F, Ashford AE, Jacobsen JV (1987) The release of α-amylase through gibberellin-treated barley aleurone cell walls. Planta 172:155–161

Hajduch M, Ganapathy A, Stein JW, Thelen JJ (2005) A systematic proteomic study of seed filling in soybean. Establishment of high-resolution two-dimensional reference maps, expression profiles, and an interactive proteome database. Plant Physiol 137:1397–1419

Hajduch M, Casteel JE, Hurrelmeyer KE, Song Z, Agrawal GK, Thelen JJ (2006) Proteomic analysis of seed filling in *Brassica napus*. Developmental characterization of metabolic isozymes using high-resolution two-dimensional gel electrophoresis. Plant Physiol 141:32–46

Hajduch M, Hearne LB, Miernyk JA, Casteel JE, Joshi T, Agrawal GK, Song Z, Zhou M, Xu D, Thelen JJ (2010) Systems analysis of seed filling in *Arabidopsis thaliana*: using general linear modeling to assess concordance of transcript and protein expression. Plant Physiol 152:2078–2087

He D, Han C, Yao J, Shen S, Yang PF (2011) Constructing the metabolic and regulatory pathways in germinating rice seeds through proteomic approach. Proteomics 11:2693–2713

Hermann K, Meinhard J, Dobrev P, Linkies A, Pesek B, Hess B, Machacova I, Fischer U, Leubner-Metzger G (2007) 1-Aminocyclopropane-1-carboxylic acid and abscisic acid during the germination of sugar beet (*Beta vulgaris* L.): a comparative study of fruits and seeds. J Exp Bot 58:3047–3060

Hilhorst HWM, Karssen CM (1992) Seed dormancy and germination: the role of abscisic acid and gibberellins and the importance of hormone mutants. Plant Growth Regul 11:225–238

Hills MJ (2004) Control of storage-product synthesis in seeds. Curr Opin Plant Biol 7:302–308

Ho THD, Gomez-Cadenas A, Zentella R, Casaretto J (2003) Crosstalk between gibberellin and abscisic acid in cereal aleurone layer. Plant Growth Regul 22:185–194

Holdsworth MJ, Bentsink L, Soppe WJJ (2008a) Molecular networks regulating *Arabidopsis* seed maturation, after-ripening, dormancy and germination. New Phytol 179:33–54

Holdsworth MJ, Finch-Savage WE, Grappin P, Job D (2008b) Postgenomics dissection of seed dormancy and germination. Trends Plant Sci 13:7–13

Houk WG (1938) Endosperm and perisperm of coffee with notes on the morphology of the ovule and seed development. Am J Bot 25:56–60

Irar S, Brini F, Goday A, Masmoudi K, Pages M (2010) Proteomic analysis of wheat embryos with 2-DE and liquid-phase chromatography (ProteomeLab PF-2D)—a wider perspective of the proteome. J Proteomics 73:1707–1721

Jendrisak JJ (1980) The use of α-amanitin to inhibit *in vivo* RNA synthesis and germination in wheat embryos. J Biol Chem 255:8529–8533

Job C, Kersulec A, Ravasio L, Chareyre S, Pépin R, Job D (1997) The solubilization of the basic subunit of sugarbeet seed 11-S globulin during priming and early germination. Seed Sci Res 7:225–243

Job D (2002) Plant biotechnology in agriculture. Biochimie 84:1105–1110

Job D, Haynes PA, Zivy M (2011) Special issue plant proteomics. In: Job D, Haynes PA, Zivy M (eds). Proteomics 11:1557–1850

Jones RL, Jacobsen JV (1991) Regulation of synthesis and transport of secreted proteins in cereal aleurone. Int Rev Cytol 126:49–88

Kimura M, Nambara E (2010) Stored and noesynthesized mRNA in *Arabidopsis* seeds: effect of cycloheximide and controlled deterioration treatment on the resumption of transcription during imbibition. Plant Mol Biol 73:119–129

Koornneef M, van der Veen JH (1980) Induction and analysis of gibberellin sensitive mutants in *Arabidopsis thaliana* (L.) Heynh. Theor Appl Genet 58:257–263

Koornneef M, Bentsink L, Hilhorst H (2002) Seed dormancy and germination. Curr Opin Plant Biol 5:33–36

Kucera B, Cohn MA, Leubner-Metzger G (2005) Plant hormone interactions during seed dormancy release and germination. Seed Sci Res 15:281–307

Laidman DL (1983) The role of gibberellins in controlling cellular processes in germinating cereals. Biochem Soc Trans 11:534–547

Lawrence DM, Halmer P, Bowles DJ (1990) Mobilisation of storage reserve during germination and early seedling growth of sugar beet. Physiol Plant 78:421–429

Lerman JC, Cigliano EM (1971) New carbon-14 evidence for six hundred years old *Canna compacta* seed. Nature 232:568–570

Linkies A, Graeber K, Knight C, Leubner-Metzger G (2010) The evolution of seeds. New Phytol 186:817–831

Lovegrove A, Hooley R (2000) Gibberellin and abscisic acid signaling in aleurone. Trends Plant Sci 5:102–110

Machovic M, Janecek S (2006) Starch-binding domains in the post-genome era. Cell Mol Life Sci 63:2710–2724

Matsui H, Chiba S, Shimomura T (1978) Substrate-specificity of an alpha-glucosidase in sugar-beet seed. Agric Biol Chem 42:1855–1860

Méchin V, Balliau T, Château-Joubert S, Davanture M, Langella O, Négroni L, Prioul JL, Thévenot C, Zivy M, Damerval C (2004) A two-dimensional proteome map of maize endosperm. Phytochemistry 65:1609–1618

Mendes AJT (1941) Cytological observations in Coffea. VI. Embryo and endosperm development in *Coffea arabica* L. Am J Bot 28:784–789

Miernyk JA, Hajduch M (2011) Seed proteomics. J Proteomics 74:389–400

Mukasa Y, Takahashi H, Taguchi K, Ogata N, Okazaki K, Tanaka M (2003) Accumulation of soluble sugar in true seeds by priming of sugar beet seeds and the effects of priming on growth and yield of drilled plants. Plant Prod Sci 6:74–82

Müller K, Job C, Belghazi M, Job D, Leubner-Metzger G (2010) Proteomics reveal tissue-specific features of the cress (*Lepidium sativum* L.) endosperm cap proteome and its hormone-induced changes during seed germination. Proteomics 10:406–416

Nadaud I, Girousse C, Debiton C, Chambon C, Bouzidi MF, Martre P, Branlard G (2010) Proteomic and morphological analysis of early stages of wheat grain development. Proteomics 10:2901–2910

Nautrup-Pedersen, G, Dam S, Laursen BS, Siegumfeldt AL, Nielsen K, Goffard N, Staerfeldt HH, Friis C, Sato S, Tabata, S, Lorentzen A, Roepstorff P, Stougaard J (2010) Proteome analysis of pod and seed development in the model legume *Lotus japonicus*. J Proteome Res 9:5715–5726

Osborne TB (1924) The vegetable proteins, 2nd edn. Longmans, Green & Co, London

Pestsova E, Meinhard J, Menze A, Fischer U, Windhovel A, Westhoff P (2008) Transcript profiles uncover temporal and stress-induced changes of metabolic pathways in germinating sugar beet seeds. BMC Plant Biol 8:122

Prego I, Maldonado S, Otegui M (1998) Seed structure and localization of reserves in *Chenopodium quinoa*. Ann Bot 82:481–488

Rabilloud T, Chevallet M, Luche S, Lelong C (2010) Two-dimensional gel electrophoresis in proteomics: past, present and future. J Proteomics 73:2064–2077

Rajjou L, Gallardo K, Debeaujon I, Vandekerckhove J, Job C, Job D (2004) The effect of α-amanitin on the *Arabidopsis* seed proteome highlights the distinct roles of stored and neosynthesized mRNAs during germination. Plant Physiol 134:1598–1613

Rajjou L, Belghazi M, Huguet R, Robin C, Moreau A, Job C, Job D (2006) Proteomic investigation of the effect of salicylic acid on *Arabidopsis* seed germination and establishment of early defense mechanisms. Plant Physiol 141:910–923

Rajjou L, Debeaujon I (2008) Seed longevity: survival and maintenance of high germination ability of dry seeds. C R Biologies 331:796–805

Rajjou L, Lovigny Y, Groot SPC, Belghazi M, Job C, Job D (2008) Proteome-wide characterization of seed aging in *Arabidopsis*: a comparison between artificial and natural aging protocols. Plant Physiol 148:620–641

Ritchie S, Swanson SJ, Gilory S (2000) Physiology of the aleurone layer and starchy endosperm during grain development and early seedling growth: new insights from cell and molecular biology. Seed Sci Res 10:193–212

Rogers WJ, Michaux S, Bastin M, Bucheli P (1999) Changes to the content of sugars, sugar alcohols, myo-inositol, carboxylic acids and inorganic anions in developing grains from different varieties of Robusta (*Coffea canephora*) and Arabica (*C. arabica*) coffees. Plant Sci 149:115–123

Sallon S, Solowey E, Cohen Y, Korchinsky R, Egli M, Woodhatch I, Simchoni O, Kislev M (2008) Germination, genetics, and growth of an ancient date seed. Science 320:1464

Shen-Miller J, Mudgett MB, Schopf JW, Clarke S, Berger R (1995) Exceptional seed longevity and robust growth: ancient sacred *Lotus* from China. Am J Bot 82:1367–1380

Shen-Miller J (2002) Sacred *Lotus*, the long-living fruits of China antique. Seed Sci Res 12:131–143

Sheoran IS, Olson DJ, Ross AR, Sawhney VK (2005) Proteome analysis of embryo and endosperm from germinating tomato seeds. Proteomics 5:3752–3764

Shepherd KA, Macfarlane TD, Colmer TD (2005) Morphology, anatomy and histochemistry of *Salicornioideae* (*Chenopodiaceae*) fruits and seeds. Ann Bot 95:917–933

Sun TP, Gubler F (2004) Molecular mechanism of gibberellin signaling in plants. Annu Rev Plant Biol 55:197–223

Sun Z, Henson CA (1990) Degradation of native starch granules by barley α-glucosidases. Plant Physiol 94:320–327

Sun ZT, Henson CA (1991) A quantitative assessment of the importance of barley seed α-amylase, β-amylase, debranching enzyme, and α-glucosidase in starch degradation. Arch Biochem Biophys 284:298–305

Tasleem-Tahir A, Nadaud I, Girousse C, Martre P, Marion D, Branlard G (2011) Proteomic analysis of peripheral layers during wheat (*Triticum aestivum* L.) grain development. Proteomics 11:371–379

Walther TC, Mann M (2010) Mass spectrometry-based proteomics in cell biology. J Cell Biol 190:491–500

Wang K, Han XF, Dong K, Gao LY, Li HY, Ma WJ, Yan YM, Ye XG (2010) Characterization of seed proteome in *Brachypodium distachyon*. J Cereal Sci 52:177–186

Welbaum GE, Bradford KJ (1990) Water relations of seed development and germination in muskmelon (*Cucumis melo* L.): V. Water relations of imbibition and germination. Plant Physiol 92:1046–1052

Wienkoop S, Baginsky S, Weckwerth W (2010) *Arabidopsis thaliana* as a model organism for plant proteome research. J Proteomics 73:2239–2248

Yamasaki Y (2003) Beta-amylase in germinating millet seeds. Phytochemistry 64:935–939

Yamasaki Y, Suzuki Y (1980) Two forms of α-glucosidase from sugar-beet seeds. Planta 148:354–361

Yang P, Li X, Wang X, Chen H, Chen F, Shen S (2007) Proteomic analysis of rice (*Oryza sativa*) seeds during germination. Proteomics 7:3358–3368

Zentella R, Yamauchi D, Ho TH (2002) Molecular dissection of the gibberellin/abscisic acid signaling pathways by transiently expressed RNA interference in barley aleurone cells. Plant Cell 14:2289–2301

Chapter 3
Omics Platforms: Importance of Twenty-First Century Genome-Enabled Technologies in Seed Developmental Research for Improved Seed Quality and Crop Yield

Ryan D. Syrenne, Weibing Shi, Charles N. Stewart and Joshua S. Yuan

Abstract As the global population increases, so does the demand for caloric intake provided from agricultural crops. Agricultural food production depends predominantly on those species which propagate through seed formation, such as cereals and legumes. The available seed supply of agriculturally important crop varieties with enhanced disease resistance and improved yield is critical for improving food crop production. Traditional techniques used to identify favorable crop characteristics for use in plant breeding are often inadequate in determining specific gene-trait associations. This has resulted in a shift towards integrating plant breeding with new omics technologies. Twenty-first century omics technologies take advantage of many recently released crop genome sequences to investigate gene-function through four disciplines: First, genomics characterizes genome wide expression of DNA; secondly, proteomics studies global protein function and expression; thirdly, transcriptomics is the study of RNA regulation; and fourthly, interactomics is the analysis of complex protein-protein interactions. Examining gene-function of important seed characteristics, through use of omics technology, could reveal critical components that could be exploited for improving seed quality and yield.

Keywords Genomics · Interactomics · Improvement · Proteomics · Seed · Transcriptomics

---

J. S. Yuan (✉) · R. D. Syrenne · Weibing Shi
Department of Plant Pathology and Microbiology, Texas A & M University,
2123 TAMU, College Station, TX 77843, USA
e-mail: syuan@tamu.edu

R. D. Syrenne · Weibing Shi
Institute for Plant Genomics and Biotechnology, Texas A & M University, 2123 TAMU,
College Station, TX 77843, USA

C. N. Stewart
Department of Plant Science, University of Tennessee, Knoxville, TN 37996, USA

G. K. Agrawal, R. Rakwal (eds.), *Seed Development: OMICS Technologies toward Improvement of Seed Quality and Crop Yield*,
DOI 10.1007/978-94-007-4749-4_3, © Springer Science+Business Media Dordrecht 2012

3.1 Introduction

As the global population increases—with an estimated 2 billion more people by 2050 to total 9 billion—so does the demand for caloric intake provided from agricultural crops (Expert Meeting on How to Feed the World et al. 2009). World food production is largely dependent on plant species that propagate through seed formation, such as cereals and legumes, and consequently must rise to meet higher global demand (Borlaug 1983). Advancement in scientific methods and tools has necessitated the need for new technological systems to tackle computational based annotation and comparative genome annotation for the rapid evolution in high-throughput DNA sequencing technologies (Zhang et al. 2010b). As more genome sequences are released, it becomes increasingly important to develop technologies that can rapidly investigate gene function in crop species that correspond to important plant physiological characteristics (Springer and Jackson 2010). Increasingly, application of state-of-the-art omics technologies (i.e., genomics, proteomics, transcriptomics, and metabolomics) are becoming available that hold the promise to detect tissue-specific changes with increasing sensitivity. These methods permit the simultaneous analysis of thousands of genes, proteins or metabolites (Kroeger 2006). This is more relevant than ever, potentially improving overall human wellness through several avenues.

Development of a modern seed industry system is important for providing farmers with consistent, high quality seeds in order to provide optimum crop yield. Coupled with new genome insight, integration of new omics technologies with traditional breeding strategies are vital for providing plant breeders with tools required for commercial criteria of seed production (Flavell 2010). A critical component for improved food crop development and production, especially in developing nations, is the available supply of seed for agriculturally important crop varieties with enhanced disease resistance and improved yield and quality. Traditional trait mapping of favorable crop characteristics used to efficiently breed plants with alleles of interest is often inadequate for precisely identifying favorable gene-trait associations (Flavell 2010). This has resulted in a paradigm shift towards incorporating twenty-first century omics technologies with traditional plant breeding to locate specific gene functions that are important for seed improvement. Providing developing nations with the capacity to produce rapidly improved seed through integrated omics technologies and plant breeding is vital for establishing a mature seed industry that is accessible by the public sector.

Concomitantly to developing an established modern seed industry, many characteristics of economically important crops could potentially improve through state-of-the-art omics technologies. Examining gene function of certain seedling developmental stages, such as grain filling or embryogenesis, could reveal critical components regulating important metabolic processes that could be exploited for improving seed quality (Thompson et al. 2009). Characteristics important for seed improvement include seed dormancy, pathogen and insect resistance, increased abiotic stress tolerance, and carbohydrate accumulation and biosynthetic pathways. The aim of this chapter is to provide an introduction to the so-called omics technologies developing and applying in twenty-first century, their synergistic relationship with plant breeding, and to elucidate how these technologies have provided a fundamental understanding of seed physiology for improved seed quality and crop yield.

3.2 Biotechnologies Role in Improving Seed Quality and Development

3.2.1 Plant Breeding

The core supply of improved crop varieties has traditionally arisen through farmers selecting and breeding plants with favorable characteristics and then trading those seeds amongst peers. As a result, farmers were limited to select traits observed in the field and improve upon only the natural genetic variation found in the crop species. Because traditional plant breeding at its heart involves the transfer of large genetic portions of parental plants to offspring, specific gene-function association with observed phenotypes is difficult (Flavell 2010; Langridge and Fleury 2011; Varshney et al. 2005). Cross pollination between favorable and unfavorable traits among agricultural crops is difficult for rural farmers to control and often diminishes hybrid vigor, also known as heterosis, resulting in inconsistent seed quality (Guei et al. 2010). The omics platforms can provide support for seed improvement through two aspects. First, crop improvement is essentially dependent upon the natural genetic variation among the species; and breeding programs aim to extend or increase that amount of genetic variation. New omics technologies allow breeders to characterize allelic content of crop germplasm in detail, increasing the ability to discover novel genes and pathways responsible for certain traits (Flavell 2010; Langridge and Fleury 2011). Second, plant breeding requires selection of traits conferred through increased genetic variation. Intensively bred species such as cereal crops often evaluate a million lines every year to maintain high natural genetic variation (Langridge and Fleury 2011) analysis for selection strategies, in part, because of the large computational-based annotation capability and revolutions in DNA sequencing technologies.

3.2.2 Genetic Engineering

For the past millennia, farmers have selected and crossed crops possessing desired traits to improve yield or provide resilient plants through hybrid vigor. Selective pressure towards desirable traits to generate improved crops fundamentally lies in the transfer of allelic material between parent and offspring. This is indirect genetic engineering at its infancy, and has remained with us since the domestication food crops. As human population grew, demand for food and livestock feed increased and crops high in starch, such as cereals and other grain crops, became immensely important (Wan et al. 2008). Contemporary omics platforms and advances in computational biology and other bioinformatics skills enabled the ability to integrate complex gene expression, crosstalk, and protein function of seeds in different developmental stages (Lonneke and van der Geest 2002). Mutant study analysis of seed gene-function through introduction of recombinant DNA, such as T-DNA, to create knockout or over expressed genes has played a significant role in elucidating complex biochemical and metabolic pathways responsible for the dynamic changes that occur during seed development (Lonneke and van der Geest 2002). A more direct approach to colloquially defined genetic engineering involves the insertion of genes into germplasm outside the scope of sexual crossing. The annotated genome of the model reference *Arabidopsis thaliana* has provided a wealth of information for seed biology (North et al. 2010). Inserting genes with known functions into economically important crops and subsequent screening for favorable phenotypes is a classic reverse genetics approach, using genetic engineering, in developing improved seeds. When complimented with new omics technology, use of recombinant DNA can offer exquisite detail of gene-functions, protein interactions, and metabolic biosynthetic pathways important for developing seeds.

3.2.3 Understanding Fundamental Plant Biology for Efficient Concomitant Development of Biotechnology

Twenty-first century omics technologies have enabled in-depth understanding of seed specific cellular processes by observing the abundance of various biological molecules as a function of environmental conditions and genomic expression (Zhang et al. 2010b). High-throughput analysis of various seed molecular components in model plant species (such as *Arabidopsis*) has provided an invaluable tool for development of improved crop cultivars. Common to all seeds are the storage reserves required for seed germination and survival. Ninety percent of seed DW consists of starch (carbohydrates), oils (triacylglycerols), and SPs, and represent the major components of economic value (Ruuska et al. 2002). Advances in fundamental biology have enhanced our understanding of gene function, metabolic regulatory networks, and protein function for economically important seed characteristics. Traditionally, biotechnology has sought to exploit this basic understanding to provide improved seed development for enhanced crop yield through integration of plant breeding and genetic engineering. Recently, these efforts have led to the generation of several new omics technology platforms, system biology, and synthetic biology strategies to facilitate real world application of biotechnology for improved seed development.

3.2.4 Twenty-First Century Omics Platforms

Seed quality improvement and development of high yielding crop varieties revolves around the integration of high-throughput computational data and functional understanding of cell-wide biological processes. The term omics is collectively used to describe the comprehensive analysis of all biological components of a given system, and often consists of systems biology approaches to assimilate tremendous amounts of data generated by omics technology (Yuan et al. 2008). Generally, shared features of omics technologies and approaches include; holistic and data driven top-down methodologies, the attempt to understand biological gene expression, metabolism, and protein function as one integrated system, and the generation of large datasets (Zhang et al. 2010b). Advances in molecular biology techniques and technologies have resulted in subsequent generation of copious amounts of data, making possible the study of genome-wide expression of DNA (genomics), global protein function and expression (proteomics), RNA regulation and analysis (transcriptomics), and complex protein-protein interactions (interactomics; Singh and Nagaraj 2006). Related to seed improvement and development, omics technologies have allowed detailed study of many facets governing seed biology. In legumes, omics technologies have been used to dissect gene expression at different levels of resolution (organelle, whole seeds), study seed developmental phenomena such as embryogenesis and germination, and understand how biotic and abiotic stresses influence phenotypic traits (Thompson et al. 2009). The remainder of the chapter will describe the evolution and current application towards seed improvement and development of several omics technologies: genomics, transcriptomics, proteomics, and interactomics.

Genomics Seed development and physiology involves many events occurring in different cellular compartments, such as the embryo, endosperm, and seed coat. Therefore, a mosaic of gene expression programs occurs in parallel among seed tissues during different developmental stages (Le et al. 2007). An organism's genome is defined by all the hereditary information encoded in its DNA. Genomics is the study of how the entire set of genes in a given genome are expressed and regulated. Because genomics technologies rely on the sequence of bases in DNA, sequencing technologies have played a crucial role in unraveling the genome.

The original sequencing technology, termed 'Sanger sequencing' for its inventor, was used to determine nucleotide sequences in DNA. Sanger chemistry uses specifically fluorescence labeled nucleotides, Dideoxy nucleoside triphosphates (ddNTPs), which are incorporated into newly synthesized DNA fragments using an original DNA template and DNA polymerase. This allows automated sequencing machines the capability to read through a DNA template during DNA synthesis to determine the order of nucleotides (Zhang et al. 2011). However, even through a series of technological improvements, Sanger sequencing is unable to read fragments of DNA longer than 1 kb which severely limits this technology for sequencing genomes. This is mainly due to the fact that beyond a certain distance from the sequencing primer, very few products are produced because of chain termination.

In order to sequence longer sections of DNA in attempt to obtain the entire genome, a new technology had to be developed. Shotgun sequencing, a DNA sequencing approach developed during the Human Genome Project, allowed the capacity to sequence larger portions of DNA. Shotgun sequencing strategies involve shearing genomic DNA into smaller fragments, either enzymatically or mechanically, and cloning into sequencing vectors such as Bacterial Artificial Chromosomes (BAC) where they can be sequenced individually. Cloning large fragments of DNA into BACs is then used to create a genomic library for a set of recombinant clones that contain the entire DNA for a given organism. These libraries with large genomic inserts are important for genome sequencing and have become invaluable genomic tools because of BACs inherent ability to maintain large DNA fragments for easy manipulation (Shizuya et al. 1992).

Next-generation genome sequencing technology, or massively parallel sequencing, uses the core philosophy of shotgun sequencing approaches for generating cost effective and efficient methods of genome sequencing (Zhang et al. 2011). Next-generation sequencing (NGS) technology reads the DNA template along the entire genome by first fragmenting the DNA, and then ligating the DNA fragments to adaptors that are randomly read during DNA synthesis. However, because the read lengths are typically shorter than those produced by Sanger sequencing (50–500 base pairs), increasing coverage is important. Coverage is the total amount of fragmented DNA sequences that overlap within a specific region and is important to accurately assemble the fragments (Zhang et al. 2011). Because of the prohibitive costs associated with sequencing and assembling large eukaryotic genomes, next-generation massively parallel sequencing using only the Illumina Genome Analyzer was first carried out for the giant panda genome (Li et al. 2010a).

Comparative genomics is an approach used in conjunction with the aforementioned technologies to exploit similarities between sequenced model species, such as *Arabidopsis*, and species of interest in order to identify novel genes and infer gene function (Buckley 2007). So far, there are more than 25 plant model organisms were applied for whole genome sequencing, such as beans, rice, grasses, corn, maize, grape, sorghum, banana, wheat, etc. (http://www.arabidopsis.org/). The release of these whole genomes sequencing data makes the life much easier for molecular breeding. Because closely related cultivars generally used for crossing material lack sufficient known DNA polymorphisms due to their genetic relatedness, and next-generation sequencing allows the identification of a massive number of DNA polymorphisms, such as nucleotide polymorphisms (SNPs) and insertions-deletions between highly homologous genomes. Most recently, Arai-Kichise et al. (2011) performed whole-genome sequencing of a landrace of japonica rice, Omachi, which is used for sake brewing and is an important source for modern cultivars. In this study, they identified 132,462 SNPs, 16,448 Insertions, and 19,318 deletions between the Omachi and Nipponbare genomes (Arai-Kichise et al. 2011). This is incredible for conventional molecular breeding scientists. There are also numerous whole-genome sequencing works that have been done, and all of these efforts will dramatically reform the situation of molecular breeding and seed development.

Genomics technologies have elucidated many complex physiological and developmental characteristics exhibited during different stages of seed growth. Information gained through genomics studies of seeds have led to various discoveries that could prove useful in applications for improved seed development. Gene expression studies have led to the discovery of genes that contribute to cell wall weakening during germination, such as expansin, and other genes such as those involved in energy metabolism (Bradford et al. 2000; Lonneke and van der Geest 2002).

Transcriptomics The measure of all mRNA molecules or transcripts produced in a given cell, often called genome-wide expression profiling, is known as transcriptomics (Zhang et al. 2010b). Transcriptomics have allowed the development of analyses that relate the abundance of mRNA molecules, or transcripts, to gene expression and regulation under varying environmental conditions. As a result, transcriptomics is the dynamic link between genomics and proteomics and thus can elaborate on the complex cellular processes responsible for adapting to environmental conditions (Singh and Nagaraj 2006). Seed development is associated with enormous differential gene expression depending on growth stage, setting the importance of transcriptomics in context for seed development and improvement.

Like genomics, transcriptomics has seen a revolution in technological improvements. Microarray technology has been at the forefront for studying simultaneous changes in gene expression and quantification across the entire genome. Microarray technology essentially works through the principle of DNA hybridization. That is, when mRNA molecules are fluorescently labeled and mixed with individual genes of an organism on a microarray plate, single mRNA molecules will hybridize with only one complimentary DNA strand. The amount of mRNA that attaches to its complimentary DNA strand is proportional to the abundance of mRNA transcripts in the sample. In its infancy, microarray technology used only one labeled probe to attach to mRNA molecules, which allowed only the expression of genes under one condition of the sample to be analyzed. Two-color microarray technology, developed by Patrick Brown, allowed the differential quantification and gene expression of two samples (test and control) based on two different labeled probes, green cyanine (Cy)3 and red Cy 5 (Schena et al. 1995).

Affymetrix microarrays, as compared to two-color systems, use one color-labeled fluorescent probe to detect mRNA hybridizations with cRNA. Several advantages lie with the Affymetrix system over traditional two-color microarrays: first, the fabrication process for Affymetrix GeneChips allows for very high density arrays to be produced, and second, because a single sample is hybridized to each array more precision in measurement is achieved by minimizing array-array variability often seen in two-color microarrays (de Reynies et al. 2006; Woo et al. 2004). Also, it was found that single color Affymetrix GeneChips were capable of higher levels of reproducibility compared to two-color microarrays (de Reynies et al. 2006).

Concomitant to the two-color microarray systems for quantifying transcripts in a cell, in fact published in the same article of Science, another technique for measuring gene expression based on mRNA quantification was developed, called Serial Analysis of Gene Expression (SAGE). Two principles define SAGE: first, short nucleotide sequence tags contain enough information to identify individual transcripts, even among thousands; and second, joining sequence tags into a string followed by sequencing in a single clone allows for efficient analysis of transcripts (Velculescu et al. 1995). Quantification and identification of novel expressed genes can also be accomplished through using SAGE, as well as the analysis of eukaryotic genomes other than the human genome.

Specific hybridizations of probes to microarrays or GeneChips, and counting of tags on DNA fragments used in SAGE are subject to many sources of variation. In tackling the shortcomings of these technologies, Brenner et al. (2000) developed a technology that does not require physical separation of DNA fragments, termed Massively Parallel Signature Sequencing (MPSS). MPSS benefits reside in the ability to handle highly variable mixtures of nucleic acid fragments and subsequent cloning onto a million microbeads (Brenner et al. 2000). Millions of template microbeads are then analyzed and sequenced through flow cytometry using a fluorescent imaging system. Since MPSS does not require separate isolation and processing of templates, it is suited well for gene expression analysis.

Technical advancements in high-throughput mRNA transcript analysis through technologies such as microarray, Affymetrix, and multiple parallel signature sequencing have greatly contributed to our understanding of biological gene function. Recently, the tools developed for massively parallel sequencing have brought about another major breakthrough in the field of transcriptomics, known as RNA sequencing (RNA-seq; Malone and Oliver 2011). While microarray technology was used for genome-wide expression studies during the 1990s, new RNA-seq technology offers several select advantages, making it an attractive alternative. RNA-seq technologies, such as those offered by Illumina, sample the amount of starting transcripts present in the tissue by direct sequencing of cDNA in order to derive information about RNA (Malone and Oliver 2011). Molecular hybridization to a probe for capture and determination of transcript abundance is often irrespective of the actual transcript expression level, resulting in reproducibility errors and limiting the accuracy of gene expression experiments (Marioni et al. 2008). RNA-seq experiments using Illumina are, on the other hand, highly reproducible with little technical variation when compared to traditional gene expression assays (Marioni et al. 2008). Also, since RNA-seq provides direct information about the sequence, it is useful for measuring gene expression for organisms which there is no full genome sequence (Malone and Oliver 2011).

Recently, transcriptomics has become more and more of a hot topic. For instance, Yang and coworkers presented the first comprehensive characterization of the *O. longistaminata* root transcriptome using 454 pyrosequencing, and revealed expressed genes across different plant species, and also identified 15.7 % of unknown ESTs (Yang et al. 2010a). These sequence data are publicly available and will facilitate gene discovery and seed development studies. RNA-seq technology also used to explore the transcriptome of a single plant cell type, the *Arabidopsis* mal meiocyte, detecting the expression of approximately 20,000 genes (Yang et al. 2011b). In this study, Yang et al. (2011b) identified more than 1000 orthologous gene clusters that are also expressed in meiotic cells of mouse and fission yeast. These RNA-seq transcriptome data provide an overview of gene expression in male mieocytes and invaluable information for future functional studies.

Transcriptomics data has provided a valuable resource for molecular studies of many important agricultural crops, such as cereals and legumes. Affymetrix arrays have proved invaluable for developing the transcriptome and elucidating seed developmental changes for grain development of wheat (Wan et al. 2008). RNA-seq using Illumina has provided a high quality in-depth view of the Soybean transcriptome during seed developmental stages and has potentially identified novel genes previously undetected (Severin et al. 2010). Combined microarrays, Affymetrix fluorescent fluorophoresGeneChip's, MPSS, and next-generation RNA-Seq have contributed greatly to our current understanding of the seed transcriptome.

Proteomics Proteins are essential components for organisms since they comprise many of the biological building components, catalytic enzymes for metabolic pathways, and signal transduction elements in the cell (Zhang et al. 2010b). When integrated with previously mentioned omics technologies, proteomics provides powerful insight into referencing protein properties and functions for improved seed development and crop yield (Tyers and Mann 2003). Proteomics analysis is critical in improved seed development for incorporation into modern plant breeding strategies because proteins directly affect phenotypic traits as opposed to gene sequence (Singh and Nagaraj 2006).

The ability to detect and measure protein molecules has traditionally relied on several methods to separate and observe protein structure: two-dimensional PAGE (2-DGE) and mass spectrometry (Zhang et al. 2010b). 2-DGE separates proteins based on two distinct properties: mass and isoelectric point (*pI*). Basically, the protein is first separated based on its isoelectric point; a property wherein proteins will remain charged at all pH values other than their *pI*. This is accomplished through use of a pH gradient. After proteins are separated based on isoelectric point, they are washed and treated with sodium dodecyl sulfate (SDS) to denature the protein and yield a linear, negatively charged polypeptide. Separation of the protein is then done by applying an electric current to the gel, where proteins are separated by size similar to traditional gel electrophoresis (for review see, Righetti et al. 2008).

Difference gel electrophoresis (DIGE) is similar to 2-DGE, but involves labeling the amino group of lysines in the polypeptides with fluorescent fluorophores (Cy2, Cy3, and Cy5; Wu et al. 2006). DIGE allows the comparison of proteins for different treated samples on the same gel, thus removing inter-gel variation commonly associated with traditional 2-DGE. Even though 2-DGE has undergone several technical innovations, protein coverage still lacks compared to gel-free proteomics techniques, such as mass spectrometry (MS)-based shotgun proteomics.

MS-based shotgun proteomics approach for protein identification and quantification are quickly becoming the preferred method, namely for their high-throughput platforms and ability to analyze complex samples at high resolution (Webb-Robertson and Cannon 2007). Traditionally, proteomics studies relied on protein separation via 2-DGE followed by identification of excised gel bands using MS. However, this method is limited by the fact that only the most abundant proteins in the sample are analyzed, which leads to an issue of coverage (Tyers and Mann 2003). Gel-free analysis of complex mixtures now dominates the repertoire of analytical techniques used for protein identification of low-abundance peptides. Liquid chromatography (LC) tandem MS (LC-MS/MS) is one such preferred analytical technique for protein identification.

Multidimensional protein identification technology (MudPIT) is quickly becoming a common approach for protein identification using gel-free based LC-MS/MS methods. Relative protein quantitation using MudPIT technology takes essentially two different approaches: label and label free (Cooper et al. 2010). The most popular stable-isotope labeling uses isotope-coded affinity tags (ICAT) technology for protein quantitation (Wu et al. 2006). Essentially, isobaric tags and prototypical isotope labeling results in a shift of labeled protein mass and allows for comparison between differentially labeled ions, ion peak height, and the resulting mass spectra produced (Cooper et al. 2010). Label-free approaches are also used to determine amounts of peptide in sample mixtures, and are based on the relationship between protein abundance, determined by the mass spectral peak analysis, and sampling statistical analysis (Bridges et al. 2007; Cooper et al. 2010). MudPIT technology first requires site specific enzymatic digestion of protein followed by separation of digested fragments using liquid chromatography, typically with a C18 reverse-phase (RP) column, based on cation exchange (Wu et al. 2006; Bridges et al. 2007). Protein fragments are then subjected to MS/MS for protein identification and quantification.

Right now, proteomics is a very popular technology and platform, and has been widely applied in plant tissue specific studies alone or together with other omics technologies. In previous studies, proteomics approach has been used to study the mechanism of action of grape seed proanthocyanidin extracts on arterial remodeling in diabetic rats (Li et al. 2010b). It also used to identify pathogenesis-related (PR) proteins in rice (Yang et al. 2011c), soybean (Yang et al. 2011a), etc. Some scientists used proteomics technology to identify the stress-related proteins (Lee et al. 2011; Wen et al. 2010; Zhou et al. 2011). Integration of proteomics and transcriptomics, even metabolomics technologies provides deep and new insights into the plant systems biology (Fig. 3.1). For instance, Barros and colleagues applied three profiling technologies to compare the transcriptome, proteome, and metabolome of two transgenic maize lines with the respective control line, and revealed that the environment was shown to play an important effect in the protein, gene expression and metabolite levels of the maize samples (Barros et al. 2010).

In addition, shotgun proteomics technology has also been used to help develop the proteome and understand the metabolic processes of seed filling in soybean. It was found by Agrawal and coworkers, that when using 2-DGE, membrane proteins were under represented in soybean samples when compared to MudPIT technology (Agrawal et al. 2008). However, proteins involved in metabolism, protein transport, and energy were equally represented using both 2-D gel technologies and MudPIT, highlighting the respective strengths of each technology for application in improved seed development (Agrawal et al. 2008).

Interactomics Understanding genome-wide mRNA expression profiling has provided exceptional information regarding gene-function analyses studies, but lacks

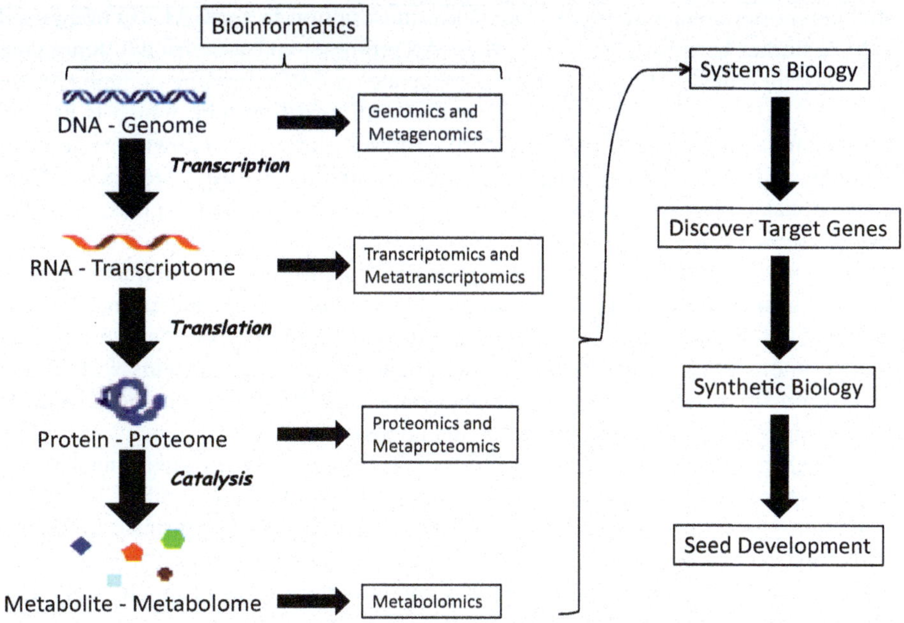

Fig. 3.1 Overview of 21st 'omics' technologies for seed development

the ability to provide information regarding gene product destination, and post-transcriptional modification (PTM) behaviors (Singh and Nagaraj 2006). The large scale detection of aggregated protein complexes in similar functional pathways, coupled with protein-protein interactions is collectively defined as interactomics (Zhang et al. 2010a). Protein-protein interactions are important considerations for understanding terminal gene function of different seed developmental stages and metabolic processes, such as embryogenesis and grain filling. Understanding how proteins interact could provide the fundamental knowledge necessary for improved seed development and crop yield.

The recent advent of the systems biology age and the accumulation of tremendous amounts of data concerning protein-protein interaction networks have resulted in several major technologies. Yeast two hybrid (Y2H) systems and affinity purification MS (AP-MS) have underpinned major technological breakthroughs for understanding protein interaction networks (Zhang et al. 2010a).

Y2H systems, first introduced in 1989, have seen a major evolution in technical improvement, resulting in adaptation for the various needs in studying protein-protein interaction networks (Zhang et al. 2010a). Y2H is a molecular biology technique used to identify association between proteins based on interaction of transcription factor components to express a reporter gene, such as green fluorescent protein (GFP). Fundamentally, Y2H relies on the ability of fragmented transcription factors to function when concatenated under close proximity to activate downstream reporter gene expression. The yeast transcription factor GAL4 is often used in Y2H studies because it binds with the UAS promoter and contains two functional domains; the transcription activation domain and the DNA-binding domain (Zhang et al. 2010a). Y2H fuses these two domains with a bait protein, which is usually known, and 'prey' protein typically derived from a library of known or unknown proteins, and relies on the interaction with bait and target 'prey' proteins to form a functional GAL4 transcription factor to activate UAS-driven expression of the reporter genes.

However, limitations of the Y2H system for identify protein interactions have resulted in incomplete data for developing a comprehensive interactome. For example, Y2H systems have a tendency for increasing high false-positive rates for protein interactions, in particular for membrane proteins and proteins with strong expression signals in the cytoplasm (Zhang et al. 2010a). This is due because Y2H interaction systems are often limited to the nucleus.

AP-MS has proven more successful for identifying protein-protein interactions and components of protein complexes through highlighting several of the major limitation present in Y2H techniques (Zhang et al. 2010a). An intrinsic limitation with Y2H technology is that only two proteins can be examined for interaction, whereas AP-MS is capable of elucidating protein interactions of much larger protein complexes.

AP-MS combine's affinity purification of bait protein's labeled with affinity tags, such as his-tag, with MS to identify novel protein complexes (Van Leene et al. 2008; Zhang et al. 2010a). In AP-MS, multi-component protein complexes are isolated from cell's *via* affinity purification of the labeled tags and then subject to downstream analysis by MS or MS/MS. Several advantages of AP-MS techniques for identifying protein-protein interaction complexes are apparent: first, AP-MS is conducted under native physiological conditions, which represents a more accurate picture of *in vivo* protein binding compared to Y2H; and second, the dynamics of protein interactions can be observed under various conditions (Zhang et al. 2010a).

Bimolecular fluorescent complementation (BiFC) is another molecular biological approach often used to examine peptide interactions. BiFC approaches are based on the idea that when two proteins are close together, the probability of interaction is high. Basically, BiFC uses a split, nonfunctional, fluorescent reporter protein combined to a bait and target protein. When these two nonfunctional fluorescent reporter components are reconstituted, through close proximal interaction of bait and labeled proteins, a functional GFP complex is formed to emit an observable signal. Self-fluorescence of plants generally precludes the use of GFP, and instead, red fluorescent or yellow fluorescent proteins are used as reporters (Zhang et al. 2010a). Although not considered a high-throughput system, and thus a major drawback as a next-generation protein interaction analytical tool, BiFC advantages lie in its high sensitivity and its ability to be used to explore protein interactions in various cellular compartments.

3.3 Concluding Remarks

The twenty-first century omics technologies are an essential component for improved seed development and crop yield. Genomics, transcriptomics, proteomics, and interactomics have provided biologists with comprehensive data concerning global gene function and identification, dynamic changes between gene expression, protein identification and function, and interaction of entire protein networks. When combined, these information technologies present a snapshot of entire cellular functional pathways, such as genome structure and metabolic regulatory networks. Combined, the twenty-first century omics are capable of characterizing a comprehensive picture of seed developmental physiology and morphology, vital for transcending the gap between fundamental biology knowledge and application towards improved seeds important for establishing modern seed industry and plant breeding programs for increased food production.

References

Agrawal GK, Hajduch M, Graham K, Thelen JJ (2008) In-depth investigation of the soybean seed-filling proteome and comparison with a parallel study of rapeseed. Plant Physiol 148:504–518

Arai-Kichise Y, Shiwa Y, Nagasaki H, Ebana K, Yoshikawa H, Yano M, Wakasa K (2011) Discovery of genome-wide DNA polymorphisms in a landrace cultivar of *Japonica* rice by whole-genome sequencing. Plant Cell Physiol 52:274–282

Barros E, Lezar S, Anttonen MJ, Van Dijk JP, Röhlig RM, Kok EJ, Engel KH (2010) Comparison of two GM maize varieties with a near-isogenic non-GM variety using transcriptomics, proteomics and metabolomics. Plant Biotechnol J 8:436–451

Borlaug NE (1983) Contributions of conventional plant breeding to food production. Science 219:689–693

Bradford KJ, Chen F, Cooley MB, Dahal P, Downie B, Fukunaga KK, Gee OH, Gurusinghe S, Mella RA, Nonogaki H, Wu CT, Yang H, Yim KO (2000) Gene expression prior to radicle emergence in imbibed tomato seeds. In: seed biology: advances and applications. Proceedings of the sixth international workshop on seeds, Merida, Mexico, 1999, Black M, Bradford KJ, Vazquez-Ramos J (eds), CABI, CABI International, Wallingford, pp 231–251

Brenner S, Johnson M, Bridgham J, Golda G, Lloyd DH, Johnson D, Luo S, McCurdy S, Foy M, Ewan M, Roth R, George D, Eletr S, Albrecht G, Vermaas E, Williams SR, Moon K, Burcham T, Pallas M, DuBridge RB, Kirchner J, Fearon K, Mao J, Corcoran K (2000) Gene expression analysis by massively parallel signature sequencing (MPSS) on microbead arrays. Nat Biotechnol 18:630–634

Bridges SM, Magee GB, Wang N, Williams WP, Burgess SC, Nanduri B (2007) ProtQuant: a tool for the label-free quantification of MudPIT proteomics data. BMC Bioinformatics 8(Suppl 7):S24

Buckley BA (2007) Comparative environmental genomics in non-model species: using heterologous hybridization to DNA-based microarrays. J Exp Biol 210:1602–1606

Cooper B, Feng J, Garrett WM (2010) Relative, label-free protein quantitation: spectral counting error statistics from nine replicate MudPIT samples. J Am Soc Mass Spectrom 21:1534–1546

de Reynies A, Geromin D, Cayuela JM, Petel F, Dessen P, Sigaux F, Rickman DS (2006) Comparison of the latest commercial short and long oligonucleotide microarray technologies. BMC Genomics 7:51

Expert Meeting on How to Feed the World, i., Food and Agriculture Organization of the United Nations (2009) Proceedings of the expert meeting on how to feed the world in 2050 (electronic resource), FAO Headquarters, Rome, 24–26 June 2009

Flavell R (2010) From genomics to crop breeding. Nat Biotechnol 28:144–145

Guei RG, Food and Agriculture Organization of the United Nations (2010) Promoting the growth and development of smallholder seed enterprises for food security crops: case studies from Brazil, Coȉ,te d'Ivoire and India/coordinated by Robert G. Guei. Rome, Food and Agriculture Organization of the United Nations

Kroeger M (2006) How omics technologies can contribute to the '3R' principles by introducing new strategies in animal testing. Trends Biotechnol 24:343–346

Langridge P, Fleury D (2011) Making the most of 'omics' for crop breeding. Trends Biotechnol 29:33–40

Le BH, Wagmaister JA, Kawashima T, Bui AQ, Harada JJ, Goldberg RB (2007) Using genomics to study legume seed development. Plant Physiol 144:562–574

Lee DG, Park KW, An JY, Sohn YG, Ha JK, Kim HY, Bae DW, Lee KH, Kang NJ, Lee BH, Kang KY, Lee JJ (2011) Proteomics analysis of salt-induced leaf proteins in two rice germplasms with different salt sensitivity. Can J Plant Sci 91:337–349

Li R, Fan W, Tian G, Zhu H, He L, Cai J, Huang Q, Cai Q, Li B, Bai Y, Zhang Z, Zhang Y, Wang W, Li J, Wei F, Li H, Jian M, Li J, Zhang Z, Nielsen R, Li D, Gu W, Yang Z, Xuan Z, Ryder OA, Leung FCC, Zhou Y, Cao J, Sun X, Fu Y, Fang X, Guo X, Wang B, Hou R, Shen F, Mu B, Ni P, Lin R, Qian W, Wang G, Yu C, Nie W, Wang J, Wu Z, Liang H, Min J, Wu Q, Cheng S, Ruan J, Wang M, Shi Z, Wen M, Liu B, Ren X, Zheng H, Dong D, Cook K, Shan G, Zhang H, Kosiol C, Xie X, Lu Z, Zheng H, Li Y, Steiner CC, Lam TTY, Lin S, Zhang Q, Li G, Tian J, Gong T, Liu H, Zhang D, Fang L, Ye C, Zhang J, Hu W, Xu A, Ren Y, Zhang G, Bruford MW, Li Q, Ma L, Guo Y, An N, Hu Y, Zheng Y, Shi Y, Li Z, Liu Q, Chen Y, Zhao J, Qu N, Zhao S, Tian F, Wang X, Wang H, Xu L, Liu X, Vinar T, Wang Y, Lam TW, Yiu SM, Liu S, Zhang H, Li D, Huang Y, Wang X, Yang G, Jiang Z, Wang J, Qin N, Li L, Li J, Bolund L, Kristiansen K, Wong GKS, Olson M, Zhang X, Li S, Yang H, Wang J, Wang J (2010a) The sequence and *de novo* assembly of the giant panda genome. Nature 463:311–317

Li XL, Li BY, Gao HQ, Cheng M, Xu L, Li XH, Zhang WD, Hu JW (2010b) Proteomics approach to study the mechanism of action of grape seed proanthocyanidin extracts on arterial remodeling in diabetic rats. Int J Mol Med 25:237–248

Lonneke A, van der Geest HM (2002) Seed genomics: germinating opportunities. Seed Sci Res 12:145–153

Malone J, Oliver B (2011) Microarrays, deep sequencing and the true measure of the transcriptome. BMC Biol 9:34

Marioni JC, Mason CE, Mane SM, AStephens M, Gilad Y (2008) RNA-seq: an assessment of technical reproducibility and comparison with gene expression arrays. Genome Res 18:1509–1517

North H, Baud S, Debeaujon I, Dubos C, Dubreucq B, Grappin P, Jullien M, Lepiniec L, Marion-Poll A, Miquel M, Rajjou L, Routaboul JM, Caboche M (2010) *Arabidopsis* seed secrets unravelled after a decade of genetic and omics-driven research. Plant J 61:971–981

Righetti PG, Antonioli P, Simò C, Citterio A (2008) Gel-based proteomics. In: Agrawal GK, Rakwal R (eds) Plant proteomics: technologies, strategies, and applications. Wiley, Hoboken, pp 11–30

Ruuska SA, Girke T, Benning C, Ohlrogge JB (2002) Contrapuntal networks of gene expression during *Arabidopsis* seed filling. Plant Cell 14:1191–1206

Schena M, Shalon D, Davis RW, Brown PO (1995) Quantitative monitoring of gene-expression patterns with a complementary-DNA microarray. Science 270:467–470

Severin A, Woody J, Bolon YT, Joseph B, Diers B, Farmer A, Muehlbauer G, Nelson R, Grant D, Specht J, Graham M, Cannon S, May G, Vance C, Shoemaker R (2010) RNA-Seq Atlas of *Glycine max*: a guide to the soybean transcriptome. BMC Plant Biol 10:160

Shizuya H, Birren B, Kim UJ, Mancino V, Slepak T, Tachiiri Y, Simon M (1992) Cloning and stable maintenance of 300-kilobase-pair fragments of human DNA in *Escherichia coli*using an F-factor-based vector. Proc Natl Acad Sci U S A 89:8794–8797

Singh OV, Nagaraj NS (2006) Transcriptomics, proteomics and interactomics: unique approaches to track the insights of bioremediation. Brief Funct Genomics Proteomics 4:355–362
Springer NM, Jackson SA (2010) Realizing the potential of genomics for crop improvement. Brief Funct Genomics 9:93–94
Thompson R, Burstin J, Gallardo K (2009) Post-genomics studies of developmental processes in legume seeds. Plant Physiol 151:1023–1029
Tyers M, Mann M (2003) From genomics to proteomics. Nature 422:193–197
Van Leene J, Witters E, Inzé D, De Jaeger G (2008) Boosting tandem affinity purification of plant protein complexes. Trends Plant Sci 13:517–520
Varshney RK, Graner A, Sorrells ME (2005) Genomics-assisted breeding for crop improvement. Trends Plant Sci 10:621–630
Velculescu VE, Zhang L, Vogelstein B, Kinzler KW (1995) Serial analysis of gene expression. Science 270:484–487
Wan YF, Poole RL, Huttly AK, Toscano-Underwood C, Feeney K, Welham S, Gooding MJ, Mills C, Edwards KJ, Shewry PR, Mitchell RAC (2008) Transcriptome analysis of grain development in hexaploid wheat. BMC Genomics 9:121
Webb-Robertson BJM, Cannon WR (2007) Current trends in computational inference from mass spectrometry-based proteomics. Brief Bioinformatics 8:304–317
Wen FP, Zhang ZH, Bai T, Xu Q, Pan YH (2010) Proteomics reveals the effects of gibberellic acid GA3 on salt-stressed rice (*Oryza sativa* L.) shoots. Plant Sci 178:170–175
Woo Y, Affourtit J, Daigle S, Viale A, Johnson K, Naggart J, Churchill G (2004) A Comparison of cDNA, oligonucleotide, and Affymetrix GeneChip gene expression microarray platforms. J Biomol Tech 15:276–284
Wu WW, Wang G, Baek SJ, Shen RF (2006) Comparative study of three proteomic quantitative methods, DIGE, cICAT, and iTRAQ, using 2D gel- or LC-MALDI TOF/TOF. J Proteome Res 5:651–658
Yang H, Huang YP, Zhi HJ, Yu DY (2011a) Proteomics-based analysis of novel genes involved in response toward soybean mosaic virus infection. Mol Biol Rep 38:511–521
Yang H, Lu P, Wang Y, Ma H (2011b) The transcriptome landscape of *Arabidopsis* male meiocytes from high-throughput sequencing: the complexity and evolution of the meiotic process. Plant J 65:503–516
Yang L, Su N, Wu M, Wang C, Hu H (2011c) Proteomics to identify pathogenesis-related proteins in rice roots under water deficit. Biologia 66:477–483
Yuan JS, Galbraith DW, Dai SY, Griffin P, Stewart CN Jr (2008) Plant systems biology comes of age. Trends Plant Sci 13:165–171
Zhang Y, Gao P, Yuan JS (2010a) Plant protein-protein interaction network and interactome. Curr Genomics 11:40–46
Zhang W, Li F, Nie L. (2010b) Integrating multiple 'omics' analysis for microbial biology: application and methodologies. Microbiol 156:287–301
Zhang J, Chiodini R, Badr A, Zhang G (2011) The impact of next-generation sequencing on genomics. J Genet Genomics 38:95–109
Zhou L, Bokhari SA, Dong CJ, Liu JY (2011) Comparative proteomics analysis of the root apoplasts of rice seedlings in response to hydrogen peroxide. PLoS One 6:e16723

Part II
Transcriptomics

Chapter 4
Rice Seed Development: Highly Resolved Transcriptomic Views

Tie Liu, S. Lori Tausta, Neeru Gandotra and Timothy Nelson

Abstract Rice is an experimental model for monocot grain crop biology. Its seeds provide a major portion of the diets of many human cultures. Recently, microarray and RNA-sequencing methods have produced highly resolved genome-wide inventories of RNA transcripts present in cell types and regions of rice seeds, including the development of embryo and endosperm tissues and the changes that accompany germination. This chapter describes these rice seed transcriptome inventories and highlights the novel biological insights that they provide regarding the regional specialization of biological processes in endosperm and embryo. The spatial and temporal patterns of expressed genes support some of the anatomical homologies that have long been proposed for regions of the monocot embryo and endosperm.

Keywords Embryo · Endosperm · Germination · Rice · Transcriptome · Seed

4.1 Introduction

Rice (*Oryza sativa* L.) seeds are the essential food staple of many cultures. In Asia, rice accounts for half the calories consumed (Khush 1997, 2005; Kovach et al. 2007). Rice has become an experimental model for monocots, thanks to its small sequenced genome, diverse germplasm collections, high seed number per plant, and amenability to transgenic manipulation (Agrawal and Rakwal 2006; Goff et al. 2002; International Rice Genome Sequencing Project 2005; Krishnan et al. 2009; Yu et al. 2005). Rice is a useful representative of the cereal grain crops, since the genomes of rice, corn, wheat, oats, rye, sorghum, and barley exhibit considerable synteny (Gaut 2002).

T. Nelson (✉) · T. Liu · S. L. Tausta · N. Gandotra
Department of Molecular, Cellular & Developmental Biology, Yale University,
P. O. Box 203104, New Haven, CT 06520-8104, USA
e-mail: timothy.nelson@yale.edu

T. Liu
Department of Plant Biology, Carnegie Institute of Washington, 260 Panama Street,
Stanford, CA 94305, USA

G. K. Agrawal, R. Rakwal (eds.), *Seed Development: OMICS Technologies toward Improvement of Seed Quality and Crop Yield,*
DOI 10.1007/978-94-007-4749-4_4, © Springer Science+Business Media Dordrecht 2012

The seeds of rice and other monocot grain crops include an embryo with a flattened morphology that is optimized for its association with a massive endosperm. The developmental and metabolic processes that produce the distinct features and resources of the rice seed and that utilize them during rice seed germination are reflected in the dynamic changes in inventories of mRNAs, proteins, and other molecules from fertilization to germination. The cellular resolution of transcriptome dynamics in various regions of developing and germinating grain seeds has improved, thanks to recent technical advances. The first mRNA inventories represented whole seed or endosperm samples, and generally were limited to specific gene sets or partial transcriptomes. Recent sampling methods, such as laser microdissection, provide finer resolution, selecting individual cell types or highly defined sub regions of organs and tissues (Asano et al. 2002). Similarly, recent improvements in the rice genome sequence and annotation have enabled whole genome transcriptome inventories of these spatially restricted samples. Together, these advances provide highly resolved views of the transcriptome dynamics that underlie the changing functions and activities of developing and germinating seed. With cellular resolution, they provide inventories of genes that are coexpressed at key times and places in seed processes and therefore provide candidates for roles in specific processes and molecular interactions.

4.2 Rice Seed Morphology, Development, and Germination

4.2.1 Morphology

Rice seeds, like those of most grains, include an embryo flattened around a substantial endosperm. The mature embryo extends from the plumule (a region including the embryonic leaves and shoot meristem) to the radicle, the embryonic primary root (Fig. 4.1a and b). The rice plumule contains three leaves and is enclosed by the coleoptile. The radicle is ensheathed by the coleorhiza. Together they mark the extremes of the embryonic axis. The coleoptile is the first organ to emerge from the globular embryo and is soon surrounded by the scutellum and epiblast. The coleoptile appears to have evolved to protect the emerging leaves during germination and early seedling growth, although the derivation of the coleoptile is controversial. From histological examinations, the coleoptile has variously been suggested to be: (i) a cotyledon; (ii) sheath or ligule tissue from a cotyledonary leaf; (iii) a modified foliar leaf; (iv) an axillary bud; or (v) a palea, a leaf-like bract that protects the grass floret (Brown 1960). The scutellum is a circular or oval-shaped tissue in transverse sections near the base (Fig. 4.1a) and appears as a fleshy, shield-like structure in longitudinal sections (Fig. 4.1b). The embryonic axis is bounded on the inner side by the scutellum, which is generally agreed to correspond to a cotyledon, has rudimentary vasculature and is differentiated independently of the shoot apical meristem (SAM; Hinton 1948; Satoh et al. 1999). The epidermal layer of the scutellum is composed of palisade-shaped cells and is the site of lipid and protein storage

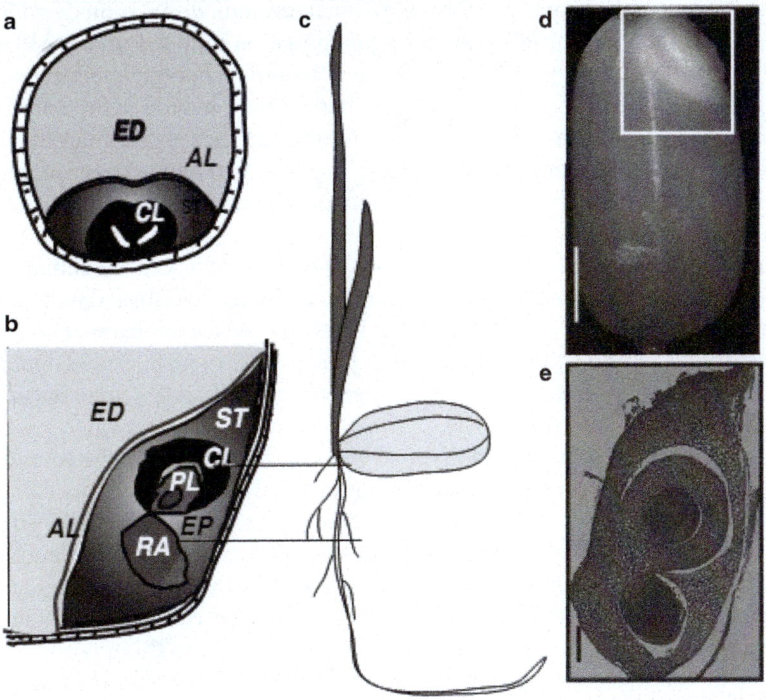

Fig. 4.1 Anatomy of a germinating rice grain. **a–c** Summary diagram of cell types and regions of rice seeds. **a** Cross and **b** Longitudinal section through a germinating rice grain after 12 h of imbibition. Abbreviations: *AL* aleurone; *CL* coleoptile; *ED* endosperm; *EP* epiblast; *PL* plumule; *RA* radicle; and *ST* scutellum. Not drawn to scale. **d–e** Light micrograph and longitudinal section of a germinating rice seed. White frame in **d** shows section region. **d** Bar = 1 mm; **e** Bar = 200 μM

and of considerable seed-specific gene expression (Miyoshi et al. 2002; Sugimoto et al. 1998). The epiblast is a flap of tissue opposite the scutellum to which no specific function has yet been attributed. The epiblast has no vasculature, is not present in all grasses, and seems to be dependent on the SAM for differentiation (Hinton 1948; Itoh et al. 2005). The epiblast has been suggested to be a second cotyledon, a cotyledonary outgrowth, the ligular region of a cotyledon, a scutellar appendage, or an outgrowth from the scutellum or the coleorhizae (Brown 1960). The comparison of spatially resolved transcriptomes from seed regions should provide a new basis for sorting out the homologies and roles of monocot embryonic structures that have been proposed based on histology and developmental anatomy.

4.2.2 Embryo and Endosperm Development

Embryo and endosperm development proceed adjacently but relatively independently. Embryo formation begins with an unequal cell division of the fertilized e.g., followed by rapid divisions that form a globular embryo with a protoderm layer by 2 days (Jones and Rost 1989). Organ development is initiated at 4 DAP, starting with the scutellum, followed closely by the coleoptile, shoot apical meristem, and radicle. Finally, the leaf primordia are formed in the plumule. The embryo itself accumulates starch grains, first in the coleorhiza and then throughout the embryo except for meristematic or procambial cells. Lipid and protein bodies accumulate in the scutellum from about 6 DAP. The coleoptile also acquires protein bodies (Jones and Rost 1989).

The endosperm develops initially as a syncytium, followed by cellularization, cellular differentiation, and maturation (Sabelli and Larkins 2009). Grain filling occurs largely when embryo development is finished. The switch from cellularization to grain filling in rice is associated with a global shift in gene expression pattern, as observed also in seeds of barley and wheat (Drea et al. 2005; Sreenivasulu et al. 2004).

The aleurone is the outermost layer of endosperm cells and is the source of hydrolytic enzymes required during germination for release of the stored sugar in the endosperm (Becraft and Yi 2011). In maize, the *calpain-like dek1* (*defective kernel 1*) is one of the few identified regulators of the development of aleurone cells (Gruis et al. 2006; Lid et al. 2002). The expression of its rice homolog (*Osdek1*, Os02g47970) has not yet been characterized. Aleurone cells store protein, lipid, sucrose, and minerals, but little starch. The epidermis of scutellum is similar to the aleurone in that it also provides hydrolytic enzymes for the breakdown of starch in the endosperm and appears to be specialized for synthesis and exportation of protein (Negbi 1984).

Maturation of the seed includes the programmed death of endosperm cells (except aleurone), followed by embryo dormancy, and desiccation of the embryo and endosperm. Desiccation achieves the removal of almost all water, thus allowing the seed to survive adverse environmental conditions (Bewley and Black 1994). The transcriptome dynamics during the processes of grain filling, desiccation, and dormancy have been described in several seed plants, although not yet fully in rice. The overlap in developmental timing of these processes makes it challenging to associate particular transcripts with a single process based solely on their spatial and temporal dynamics. Many seed desiccation-related genes are also responsive to environmental dehydration stress (Cooper et al. 2003; Sivamani et al. 2000).

4.2.3 Dynamic Processes during Seed Germination

The processes of rice seed germination occur in the sequence imbibition > activation > post germination growth (Bewley and Black 1994; Yoshida 1981). During imbibition, the seed rapidly absorbs water to a content of 25–30 % (Yoshida 1981). The aleurone and scutellum become activated by imbibition and produce proteins for starch breakdown. As early as one hour after imbibition, rice embryos are active in RNA synthesis (Howell et al. 2009). The metabolically inactive embryo resumes growth and germination, which is considered complete when the embryo emerges from the seed coat.

Numerous mRNAs are stored in the desiccated seed ("stored" or "long-lived" mRNAs), and only translation is necessary for germination to proceed (Bewley and Black 1994; Dure and Waters 1965; Holdsworth et al. 2008; Marcus and Feeley 1964; Rajjou et al. 2004). The transcriptome has been shown to change during the dry seed stage (Bove et al. 2005; Comai and Harada 1990; Leubner-Metzger 2005), even before the dramatic changes triggered by imbibition.

4.3 Seed Transcriptomics

Several recent projects have produced comprehensive transcriptome datasets from whole rice seeds, embryos, or regions of embryos at successive developmental stages (Gibbings et al. 2003; Howell et al. 2009; Jain et al. 2007; Li et al. 2007; Ma et al. 2005; Meyers et al. 2004; Nakano et al. 2006; Wang et al. 2010b; Xue et al. 2009; Zhu et al. 2003). A recent Rice Atlas project produced quantitative transcriptome datasets from a variety of selected rice embryo tissues isolated by laser microdissection from dry seed or after 12 h or 24 h of imbibition (Jiao et al. 2009). Detailed transcriptome information from this project is cited below as "Rice Atlas" and is available at http://bioinformatics.med.yale.edu/riceatlas/. The transcriptomes of embryo organs included 10,733 of the 30,731 rice genes represented on the whole genome Rice Atlas microarrays (Jiao et al. 2009). In the following sections, we focus on the dynamic changes in the transcriptomes of regions of the developing embryo that are associated with key features of seed development and germination.

4.3.1 Dynamics of GA3- and ABA-Related Transcripts

ABA and gibberellic acid (GA3) provide opposing signals that coordinate seed development and germination. ABA induces dormancy and inhibits germination (Kucera et al. 2005; Leubner-Metzger 2005). Maternal tissues provide ABA to the seed early in development. GA3 provides a later signal associated with imbibition that promotes growth activities. The balance between ABA remaining in the dry seed and GA3 induced by imbibition appears to regulate the developmental switch between dormancy and germination (Holdsworth et al. 1999; Ogawa et al. 2003). Several ABA-related genes are expressed in highly region-specific patterns within the germinating embryo. Transcripts from the rice orthologs of the *Arabidopsis* ABA-related genes *ABI1-ABI5* (Finkelstein and Somerville 1990; Finkelstein et al. 2008) are particularly region-specific. The *ABI5* ortholog *Os05g28350* accumulates at 24 h post imbibition only in scutellum and coleoptile, while the *ABI5* ortholog *Os08g43600* accumulates in scutellum in dry seed, in plumule at 12 h post imbibition, and in radicle at 24 h post imbibition (Rice Atlas), similar to the pattern for the

Arabidopsis ABI5 (Brocard et al. 2002). In the plumule, transcripts for two ABA receptors (*Os03g18600* and *Os01g61210*, homolog to *RCAR10/PYL4* of *Arabidopsis*) increase specifically from dry seed to 24 h post imbibition (Klinger et al. 2010; Park et al. 2009), while transcripts for their likely ABA signaling partners, *PP2C* and *SnRK* family members, accumulate less specifically, suggesting that organ-specific responses may be regulated by specific expression of the receptors alone.

Many GA-responsive genes contain GAREs (GA-response elements) as *cis*-promoter elements. Included among these are genes for degradative enzymes, which are important during the mobilization of stored carbohydrate and protein. The scutellum is a source of GA3 and of at least one GA-induced transcription factor (TF), namely, *OsGAMYB*, during the early stages of imbibition (Kaneko et al. 2004). The transcripts for 15 out of 28 (54 %) GARE-containing protease genes increase to a maximum at 12 or 24 h after imbibition in the scutellum (Rice Atlas). A subset of four hydrolase genes exhibits a contrasting pattern of very high initial expression that declines during imbibition.

4.3.2 Dynamics of Transcripts Related to other Hormones

Auxin The transcriptional changes among families of auxin-responsive genes have been profiled during five stages of seed development in field-grown Indica rice. Those gene families include the *GH3* (Jain et al. 2006b), *Aux/IAA* (Jain et al. 2006a), *ARF* (Auxin response factor; Jain and Khurana 2009), and *SAUR* (Small Auxin-Up RNA; Jain et al. 2006c). At least nine *GH3*, 13 *Aux/IAA*, 18 *ARF*, and 17 *SAUR* genes are differentially expressed through developmental stages (Jain and Khurana 2009), many in a tissue-specific and stage-specific manner. Most ARF family members (17/21) have significant expression in one or a few embryo organs during germination (see Rice Atlas; Jiao et al. 2009). Since the targets of ARF TFs generally have an AuxRE *cis*-element, it should be possible to build candidate lists of ARFs and their cognate seed targets based on cell- or region-specific coexpression. Transcripts for Aux/IAA factors increase with germination, reflecting the diverse roles of auxin signals in seedling growth and development (Howell et al. 2009).

The *ARFs* and *TIR1-like F-box* genes are regulated at the posttranscriptional level by microRNAs (miRNAs; Bishopp et al. 2006). Two of the 19 rice Argonaute genes, which encode components of miRNA-dependent gene silencing, exhibit increases in expression specifically in developing rice seeds (Kapoor et al. 2008), suggesting that this mode of regulation plays a role in seed development. Of the several miRNAs that accumulate specifically in the rice embryo or endosperm (Xue et al. 2009), one (*Osa-miR167*) targets *ARF6* and *ARF8* and increases with auxin treatment.

Cytokinin The CKs affect endosperm development through the stimulation of cell division (Yang et al. 2002). The loss of cytokinin oxidase/dehydrolase (*OsCKX2*), which degrades CK, results in an increased number of grains and an enhancement of grain production (Ashikari et al. 2005). In *Arabidopsis*, CK-responsive genes are significantly upregulated in developing endosperm (Brugière et al. 2008; Lur and Setter 1993; Yang et al. 2002). In rice, CK-related transcripts increase both in developing endosperm and germinating embryo. The CK response factor 2 (*Os01g12440*) and the type A response regulator (*Os02g58350*) increase strikingly in the embryo with imbibition (Rice Atlas).

Ethylene Ethylene affects the release of dormancy and germination. The binding of ethylene to membrane receptors activates *AP2/EREBP* TFs (APETALA2/ethylene-responsive element binding proteins), a diverse family with roles in organ identity determination, cell fate specificity, stress response, and other processes (Dietz et al. 2010; Kizis et al. 2001; Riechmann et al. 1996). Some members are expressed in all embryo cell types with maxima at 12 h post imbibition in scutellum, followed by the other organs with tendency to peak at 24 h, suggesting that these genes are involved in later phases of germination (Rice Atlas).

Brassinosteroids At least 13 genes related to brassinosteroid (BR) signaling have significant expression in the dry seed and imbibition stages, and all are expressed nonspecifically throughout embryo tissue types and regions (Rice Atlas). The non-specific pattern is consistent with the diverse roles of BRs in the development of seeds and other plant parts (Tanabe et al. 2005; Wu et al. 2008). BR-insensitive mutants bear shortened and small seeds (Tanabe et al. 2005), and transgenic rice with elevated BR levels show increased number of tillers and seeds (Wu et al. 2008). BRs also regulate rice seed germination and can stimulate aged rice seeds to germinate (Takeuchi et al. 1995).

4.3.3 Dynamics of Transcripts for Carbohydrate Metabolism

The transcriptional patterns of genes for carbohydrate synthesis and degradation are particularly dynamic during seed development and germination. During grain filling, sucrose is imported and converted into the endosperm starch that fuels seedling growth (Fincher 1989). The genes for starch synthesis enzymes, including *AGPS2/shrunken2* and others, are upregulated in developing endosperm during its period of rapid DW increase (Ohdan et al. 2005). Members of the same gene families exhibit increases in expression after fertilization, each with slightly different kinetics. The developing endosperm uses sucrose and starch synthesis genes and isozymes distinct from those active in leaf metabolism (Bak-Jensen et al. 2007; Ohdan et al. 2005). Two rice sucrose synthase (*SUS*) genes are expressed early in grain filling, another one later in filling, and additional members only in stem or root, but not in endosperm (Zhu et al. 2003). A comparison of the transcriptomes and metabolomes of normal and heat-treated rice grains linked decreases in metabolites (e.g., glucose-1-P) needed for starch accumulation with the down regulation of the genes encoding particular anabolic enzymes (e.g., starch-branching enzyme) or upregulation of genes encoding catabolic enzymes (e.g., α-amylase; Yamakawa and Hakata 2010).

Imbibition triggers the expression of metabolic genes in the embryo organs that enlarge first, the coleoptile and radicle (Rice Atlas), followed by the plumule. Some gene family members are specific to embryo organ. Genes for sucrose metabolism are differentially expressed among rice embryonic organs undergoing imbibition (Rice Atlas). Of the six rice *SUS* genes, *SUS2* (*Os06g09450*) is particularly region specific, accumulating in coleoptile at 24 h post imbibition (Rice Atlas), consistent with its absence in dry whole seed and appearance with germination (Hirose et al. 2008). In coleoptile and radicle, transcripts for ~70 % of annotated sucrose metabolic genes increase, while in plumule only ~ 30 % increase (Rice Atlas).

Many rice genes currently annotated as 'response to water' and 'embryonic development' are expressed in developing rice endosperm with dynamics similar to genes for carbohydrate metabolism (Wang et al. 2010b). Water-response genes might be associated with rice desiccation- and germination-specific processes, while embryonic development transcripts may just reflect the derivation of annotation from *Arabidopsis* orthologs. For example, genes, that in *Arabidopsis* may have embryo-specific functions, might be active in rice in both the aleurone and scutellum.

4.3.4 Dynamics of Transcripts for Storage and Desiccation-Tolerance Proteins

The transcripts for storage and dessication-tolerance proteins have the same dynamics as their well-documented patterns of protein accumulation during endosperm and embryo development. In rice endosperm, SPs include the glutelins (12 genes and ~75 % of the total protein), cupin (one gene), globulin (four genes), prolamin (two genes), and albumin (five genes; Takaiwa et al. 1991). Genes for cupin, seed storage lipid transfer protein, seed maturation protein, and glycine-rich proteins (GRPs) are expressed endosperm-specifically and with distinct developmental dynamics (Wang et al. 2010b). For example, seed maturation proteins are highly expressed at 14 and 21 DAP, but not at 7 DAP, whereas the opposite is true for cupin (Wang et al. 2010a). Genes encoding glycosylation enzymes that act on SPs and polysaccharides are upregulated during endosperm development.

In embryos, transcripts for desiccation-tolerance proteins appear late in the maturation period. Transcripts for the LEA, germins, and germin-like proteins are abundant in all regions of dry seeds and disappear with imbibition (Rice Atlas), as do the proteins (Rorat 2006). Transcripts for LEA dehydrin (*Os01g50700*) are most abundant in coleoptile, plumule, and scutellum, and decline abruptly at 12 h post imbibition for all embryo organs, a general pattern shared by other *LEA* genes, including *OsLEA8* (*Os05g50710*) and *OsLEA30* (*Os03g06360*; Rice Atlas; Wang et al. 2007).

4.3.5 Dynamics of Transcripts for Cell Wall Extension and Modification Genes

Imbibition triggers a global increase in transcripts for genes associated with the CW synthesis and expansion as the seedling elongates, including those for GRPs, expansins (EXPs) and xyloglucan transferase/hydrolases (XTHs). GRP transcripts are most abundant in plumule of dry seed, in scutellum and epiblast 12 h post imbibition, and in coleoptile and radicle 24 h post imbibition. Similar expression patterns appear for the expansins and xyloglucan genes (Rice Atlas). Of the over 22 EXP rice genes, there are four EXP-like and one EXP-related gene (Lee and Kende 2001, 2002), and seven predominate (*EXPB2*, *EXPA5*, *EXPA7*, *EXPA10*, *EXPA12*, *EXP16*, and *EXPA29*; Rice Atlas). Transcripts from the 29-member *OsXTH* gene family are abundant in all cell types of dry and germinating seeds (Rice Atlas), consistent with previous studies of these gene families (Huang et al. 2000; Yokoyama et al. 2004).

4.3.6 Dynamics of Transcripts for TFs

The global changes in function and composition of the seed through its development and germination are reflected in the dynamics and locations of specific TFs that regulate particular developmental and metabolic processes. A recent review compares the *Arabidopsis* TFs known to be essential for seed formation with those of rice (Agarwal et al. 2011). While there are many similarities in TF-expression patterns in rice and *Arabidopsis* seeds (e.g., post germination induction of HAP TFs), the patterns for many corresponding TFs differ. Global changes in TFs correspond to biological shifts. The 'transcription' GO category is well represented in early endosperm transcriptomes, but declines during the processes of grain filling and cell death (Wang et al. 2010b). The opposite occurs as the seeds imbibe water and begin germination-associated metabolism. The Rice Atlas datasets (Jiao et al. 2009) document the embryo germination expression dynamics of 1,174 genes from 19 predicted TF families. Approximately 55 % (644) changed significantly in at least one embryo organ during imbibition, including the plumule (66 %, 441), radicle (62 %, 404), scutellum (57 %, 370), coleoptile (55 %, 352), and epiblast (54 %, 346). TFs are thus generally enriched in the plumule and radicle, the two organs destined to become the vegetative seedling. The dynamics are similarly striking in studies of whole seed TF transcripts through times points of germination (Howell et al. 2009).

MADS box TFs The 54 of 75 rice MADS-box genes are expressed in more than one embryo organ of dry seeds and post imbibition seeds (Rice Atlas), and 12 appear to be expressed only in seeds (Arora et al. 2007). The rice orthologs of *Arabidopsis* ABC class genes displayed an increase in coleoptile, plumule, and/or radicle (Rice Atlas), including *Os01g66030* and *Os05g34940* (*PISTILLATA* orthologs), *Os06g49840* (*APETALA3* ortholog), and *Os09g32948* (*SEPALLATA1* ortholog). This suggests that they have roles in the establishment and maintenance of the vegetative organs in rice, consistent with previous observations (Arora et al. 2007).

bZip (basic leucine zipper) TFs The bZip TFs have a key role in regulating the accumulation of endosperm SPs (Oñate et al. 1999; Schmidt et al. 1992; Vicente-

Carbajosa et al. 1998). Nine rice bZip family members are preferentially expressed during seed development (Nijhawan et al. 2008). Two rice bZip genes with major roles in seed development (*RITA-1* and *REB*) are also expressed in non seed tissues (Izawa et al. 1994; Nakase et al. 1997).

C_2H_2 Zinc-finger TFs A significant subset of the rice C_2H_2 zinc-finger TF family (12/189) is seed specific in expression (Agarwal et al. 2007).

ABI3/VP1 TFs The rice genome has 63 predicted ABI3/VP1 TFs, some of which have been associated with morphogenesis, maturation, dormancy, and germination in seed development. Transcripts for 33 genes increase during germination (Rice Atlas); the majority, 24 and 22, of them are highly expressed respectively in radicle and plumule. As an exception, the transcript level of *Os11g05740* peaks in dry seed and declines during imbibition, and this is consistent with previous observations of its dynamics (Bassel et al. 2006; Nambara et al. 1995).

MYB TFs The transcript levels of many MYB family genes are modulated during germination. Transcripts for 85 MYB genes are elevated at the very early stage of rice seed germination (Rice Atlas), including many identified in previous studies, such as *OsMybS1* (*Os01g34060*), and *OsMybS3* (*Os10g41200*; Lu et al. 2002). Several of the germination-induced *MYB* genes respond to GA and regulate *α-amylase* gene expression (Assmann 2005).

bHLH TFs Fifteen basic helix-loop-helix (bHLH) genes are induced during rice germination (Rice Atlas), all belonging to the relatively conserved *OsbHLH* subfamily. All are highly induced during the imbibition stage from their dry seed levels; ten increase in all seed tissues. Among rice *bHLH* genes, transcripts for *OsSPCH1* accumulate primarily in coleoptile, while transcripts for *OsMUTE* accumulate in all organ tissues of dry and developing seeds, despite the documented specific roles of their *Arabidopsis* orthologs in stomatal development (Barton 2007; Liu et al. 2009).

HDZIP TFs At least 37 homeodomain leucine zipper (HD-ZIP) genes accumulate predominantly during one or more stages of seed development (Jain et al. 2008). In *Arabidopsis*, the HD-ZIPIII genes *REVOLUTA (REV), PHABULOSA (PHB), PHAVOLUTA (PHV)*, and *ATHB8* have roles in the establishment of the embryonic SAM and in the development of organ polarity and leaf venation (McConnell and Barton 1998; McConnell et al. 2001). In rice, transcripts for the four HD-ZIPIII genes *OSHB1* to *OSHB4* (*Os03g01890*, *Os03g43930*, *Os10g41230*, and *Os01g10320*, respectively) accumulate in the SAM, leaf primordial, and vascular tissue (Itoh et al. 2008). These transcripts are most abundant in plumule and radicle (Rice Atlas), consistent with in situ hybridization data (Itoh et al. 2008).

CCAAT-HAP TFs In plants, a substantial gene family encodes the three interacting HAP2, HAP3, and HAP5 subunits (Agarwal et al. 2011), which resemble the yeast HAP or mammalian NUCLEAR FACTOR-Y (NF-Y) factors (Gusmaroli et al. 2001; Lotan et al. 1998; Mantovani 1999). In *Arabidopsis* the HAP3 genes *LEAFY COTYLEDON1 (LEC1)* and *LEC1-LIKE* influence embryogenesis (Kwong et al. 2003; Lotan et al. 1998). Their rice homologs *OsHAP3D* and *OsHAT3E* are actively transcribed during embryogenesis (Thirumurugan et al. 2008), and 17 out of 26 rice HAP genes exhibit distinct developmental dynamics and cellular patterns in the embryo (Rice Atlas), suggesting they may regulate distinct aspects of embryogenesis.

TCP TFs The transcripts of 16 rice TCP-domain TFs accumulate significantly in dry seeds and germinating tissues (Rice Atlas). Most exhibit maximum expression in the early seed stages (Sharma et al. 2010).

Dof TFs Plant DOF (DNA-binding-with-one-finger) domain TFs have roles in the regulation of diverse processes, such as endosperm SP synthesis, seed germination, light regulation, phytohorm one and defense responses, and guard cell-specific gene regulation (Yanagisawa 2002). In rice, 36 DOF proteins have been identified, of which 22 are induced specifically in seeds (Rice Atlas).

The tissue- and stage-specific inventories of TFs described above should facilitate the association of seed-relevant cognate pairs of TFs and their potential targets from candidate pools. For example, the seed-expressed member of a duplicated pair of genes with a target *cis*-regulatory region was identified by its cell-specific expression pattern (Ray et al. 2007). Cognate pairs of TFs and potential gene targets were also identified among a group of genes sharing developmental dynamics during grain filling (Zhu et al. 2003). The highly resolved coexpression analysis for classes of TFs should also facilitate the analysis and modeling of regulatory networks for seed development and germination.

Many of the cited studies revealed that some seed development/germination TFs are also induced by abiotic stresses, such as heat and anoxia. For example, at least 11 rice bZIP genes are expressed both during panicle/seed development and more generally under abiotic stress conditions (Nijhawan et al. 2008).

4.3.7 Dynamics of Transcripts Associated with Light Regulation

Light affects the germination and postgermination growth of rice through phytochrome. Transcripts for phytochromes A, B, and C (*OsPHYA*, *OsPHYB*, and *OsPHYC*) accumulate with distinct dynamics in dry and imbibed seeds (Takano et al. 2001; Rice Atlas). Transcripts for *OsPHYB* (*Os03g19590*) and *OsPHYC* (*Os03g54084*) are present in dry seed, but increase in all cell types during imbibition. Transcripts for the blue-light photoreceptor *PHR2* (*Os03g22330*) are present in dry seeds, suggesting that it may have a role in the control of seed germination as well.

Comprehensive surveys have identified many genes associated with light-regulated growth and development (Jiao et al. 2005; Ma et al. 2003). Over 300 genes have been identified *via* light-induced transcriptome inventories (Jiao et al. 2005, 2007). Twenty with growth defects in corresponding T-DNA mutant lines (Jung et al. 2008) are expressed specifically in one or more embryo regions (Rice Atlas). For example:transcripts are evident in coleoptile for *Os01g08460* (TMS membrane protein, dwarf phenotype); scutellum is rich in transcripts for *Os01g45274* (carbonic anhydrase, oxidative stress phenotype) and *Os05g47540* (phosphoethanolamine *N*-methyltransferase, slender phenotype); plumule is rich in those for *Os04g37619* (zeaxanthin epoxidase, Aba1; dwarf phenotype) and *Os07g32880* (ATP synthase gamma chain, pale green phenotype); radicle in that for *Os08g40160* (thylakoid lumen protein, dwarf phenotype); and both coleoptile and plumule are rich in transcripts for *Os01g40710* (expressed protein, pale green phenotype).

4.3.8 Dynamics of Transcripts Associated with Submergence and Anaerobiosis

Rice is the one of the few plant species that can tolerate long-term flooding, a significant abiotic stress. Submergence, hypoxia, and anoxia inhibit seed germination of the majority of higher plants. However, most rice seeds can germinate even when deprived of oxygen completely (anaerobiosis). *Submergence-1* (*SUB1 A*), a quantitative trait locus associated with several ERF genes, has been associated with high tolerance to anaerobiosis (Fukao et al. 2006; Siangliw et al. 2003; Xu et al. 2006). A survey of Sub1 A-regulated genes revealed the AP2/ERF TF gene family members may form a complex that is expressed during both submergence tolerance and anaerobic germination (Jung et al. 2010). Transcriptome analysis of anoxia-treated rice seed associated the EXP gene family members EXPA7 and EXPB12 with anoxia-induced coleoptile elongation (Lasanthi-Kudahettige et al. 2007). Anoxia also upregulates genes encoding ethylene signaling response and HSPs, but reduces the expression of cytochrome P450 enzymes which use oxygen as a substrate (Lasanthi-Kudahettige et al. 2007). A comprehensive study of oxygen-response genes in rice associated 10 metabolites and over 1,100 genes with tolerance to anaerobiosis; 13 metabolites and 730 genes were defined as anaerobic responders (Narsai et al. 2009). Other types of stress-response genes, such as heat shock responsive genes (Sarkar et al. 2009) and many of those induced by flooding and anoxia, are utilized during normal seed development.

4.4 Comparative Transcriptome Evidence for Monocot Embryo Organ Homologies

As described in the INTRODUCTION above, various homologies have been proposed for the specialized structures of the monocot embryo, including epiblast, scutellum, plumule, and coleoptile, based on comparative anatomy and development. Can comparative transcriptomics clarify these relationships? Although endosperm and embryo transcriptomes differ significantly at all stages (Anderson et al. 2003; Zhu et al. 2003), there are relatively few regional differences within the embryo until root/shoot functions are activated by germination (Rice Atlas). Several of these embryo regional differences warrant comment. Transcripts of genes from several metabolic pathways exhibit a consistent pattern of differential expression in embryo regions. The transcriptome patterns for pathways such as glycolysis are most similar for the shoot-related regions coleoptile and plumule, when compared to the likely cotyledon homologs, scutellum, and epiblast. At all stages measured, the glycolysis pathway transcriptomes of scutellum and epiblast are consistently different from coleoptile and plumule (~45 % of pathway members), but much less different from radicle (~10 %). Scutellum, epiblast, and radicle exhibit similar expression patterns for members of the HSF, MADS, MYB, and TCP TF gene families. Radicle differs from all other embryo regions in its use of members of bHLH and bZIP TF families.

The lack of extensive differentiation among regional transcriptomes is consistent with observations that a variety of embryo mutants with defective radicle regions can organize roots upon germination, despite loss of the presumptive sites for these (Hong et al. 1995), although mutants lacking plumule are shootless. As plumule and radicle develop, their transcriptomes begin to overlap with leaf and root, but at earlier stages there is little to distinguish them from other embryo regions. For example, at 48 h post germination, the embryo radicle expresses many genes associated with mature root (Wang et al. 2010a), but at earlier stages resembles all other embryo regions. At this same 48 h stage, the plumule and radicle transcriptomes overlap considerably, reflecting the rapid growth of both organs (Wang et al. 2010a); the cellulose synthase family is particularly active in both (Wang et al. 2010b). Transcriptome datasets have revealed some similarities in patterns for aleurone and scutellum, as might be expected if their functions overlap. Rice scutellum accumulates transcripts for the aleurone-related genes *Osdek1*, *Ossal1* (*Os06g43590*, homolog of maize supernumerary aleurone1, *sal1*; Shen et al. 2003), and *Os02g53510* (Kuwano et al. 2011).

4.5 Concluding Remarks

The transcriptomes obtained recently for stages and regions of rice seed development provides a depth and breadth of information that is only beginning to be exploited. Initially, it provides an independent confirmation for expression patterns reported previously in studies of single genes or gene families. Using reliable transcriptomic datasets, genes with known functions in processes, such as starch metabolism and desiccation tolerance can serve as discovery tools to identify functionally related genes *via* coexpression patterns highly resolved by stage, region, and even cell type. This has the potential to associate cognate members of functional complexes, targets of TFs and other binding proteins, members of pathways and networks, and other partner relationships that imply expression in the same cell at the same time. The global patterns of transcription revealed in these datasets provide support for the relatedness of coleoptile to plumule and of scutellum to epiblast (and perhaps the relation of both to cotyledons). However, it is striking that there is so little transcriptome differentiation among rice embryo regions until postgermination growth. Further exploration may reveal where the key differences reside that confer anatomical identity.

Acknowledgments Work on the Rice Atlas transcriptome project was supported by US National Science Foundation Plant Genome Program grant DBI-0325821 to TN. The Rice Atlas data cited in this chapter is tabulated for browsing at http://plantgenomics.biology.yale.edu/riceatlas/.

References

Agrawal GK, Rakwal R (2006) Rice proteomics: a cornerstone for cereal food crop proteomes. Mass Spectrom Rev 25:1–53

Agarwal P, Arora R, Ray S, Singh AK, Singh VP, Takatsuji H, Kapoor S, Tyagi AK (2007) Genome-wide identification of C2H2 zinc-finger gene family in rice and their phylogeny and expression analysis. Plant Mol Biol 65:467–485

Agarwal P, Kapoor S, Tyagi AK (2011) Transcription factors regulating the progression of monocot and dicot seed development. Bioessays 33:189–202

Anderson A, Hudson M, Chen W, Zhu T (2003) Identification of nutrient partitioning genes participating in rice grain filling by singular value decomposition (SVD) of genome expression data. BMC Genomics 4:26

Arora R, Agarwal P, Ray S, Singh AK, Singh VP, Tyagi AK, Kapoor S (2007) MADS-box gene family in rice: genome-wide identification, organization and expression profiling during reproductive development and stress. BMC Genomics 8:242

Asano T, Masumura T, Kusano H, Kikuchi S, Kurita A, Shimada H, Kadowaki K (2002) Construction of a specialized cDNA library from plant cells isolated by laser capture microdissection: toward comprehensive analysis of the genes expressed in the rice phloem. Plant J 32:401–408

Ashikari M, Sakakibara H, Lin S, Yamamoto T, Takashi T, Nishimura A, Angeles ER, Qian Q, Kitano H, Matsuoka M (2005) Cytokinin oxidase regulates rice grain production. Science 309:741–745

Assmann SM (2005) G protein signaling in the regulation of rice seed germination. Sci STKE 2005:cm12

Bak-Jensen KS, Laugesen S, Østergaard O, Finnie C, Roepstorff P, Svensson B (2007) Spatio-temporal profiling and degradation of alpha-amylase isozymes during barley seed germination. FEBS J 274:2552–2265

Barton MK (2007) Making holes in leaves: promoting cell state transitions in stomatal development. Plant Cell 19:1140–1143

Bassel GW, Mullen RT, Bewley JD (2006) ABI3 expression ceases following, but not during, germination of tomato and *Arabidopsis* seeds. J Exp Bot 57:1291–1297

Becraft P, Yi G (2011) Regulation of aleurone development in cereal grains. J Exp Bot 62:1669–1675

Bewley J, Black M (1994) Seeds: physiology of development and germination. Plenum, New York, pp 460

Bishopp A, Mähönen AP, Helariutta Y (2006) Signs of change: hormone receptors that regulate plant development. Development 133:1857–1869

Bove J, Lucas P, Godin B, Ogé L, Jullien M, Grappin P (2005) Gene expression analysis by cDNA-AFLP highlights a set of new signaling networks and translational control during seed dormancy breaking in *Nicotiana plumbaginifolia*. Plant Mol Biol 57:593–612

Brocard IM, Lynch TJ, Finkelstein RR (2002) Regulation and role of the *Arabidopsis* abscisic acid-insensitive 5 gene in abscisic acid, sugar, and stress response. Plant Physiol 129:1533–1543

Brown WV (1960) The morphology of the grass embryo. Phytomorphology 10:215–223

Brugière N, Humbert S, Rizzo N, Bohn J, Habben JE (2008) A member of the maize isopentenyl transferase gene family, *Zea mays isopentenyl transferase 2* (ZmIPT2), encodes a cytokinin biosynthetic enzyme expressed during kernel development. Cytokinin biosynthesis in maize. Plant Mol Biol 67: 15–229

Comai L, Harada JJ (1990) Transcriptional activities in dry seed nuclei indicate the timing of the transition from embryogeny to germination. Proc Natl Acad Sci U S A 87:2671–2674

Cooper B, Clarke JD, Budworth P, Kreps J, Hutchison D, Park S, Guimil S, Dunn M, Luginbühl P, Ellero C, Goff SA, Glazebrook J (2003) A network of rice genes associated with stress response and seed development. Proc Natl Acad Sci U S A 100:4945–4950

Dietz K, Vogel MO, Viehhauser A (2010) AP2/EREBP transcription factors are part of gene regulatory networks and integrate metabolic, hormonal and environmental signals in stress acclimation and retrograde signalling. Protoplasma 245:3–14

Drea S, Leader DJ, Arnold BC, Shaw P, Dolan L, Doonan JH (2005) Systematic spatial analysis of gene expression during wheat caryopsis development. Plant Cell 17:2172–2185

Dure L, Waters L (1965) Long-lived messenger RNA: evidence from cotton seed germination. Science 147: 410–412

Fincher GB (1989) Molecular and cellular biology associated with endosperm mobilization in geminating cereal grain. Annu Rev Plant Physiol Plant Mol Biol 40:305–346

Finkelstein RR, Somerville CR (1990) Three classes of abscisic acid (ABA)-insensitive mutations of *Arabidopsis* define genes that control overlapping subsets of ABA responses. Plant Physiol 94:1172–1179

Finkelstein R, Reeves W, Ariizumi T, Steber C (2008) Molecular aspects of seed dormancy. Annu Rev Plant Biol 59:387–415

Fukao T, Xu K, Ronald PC, Bailey-Serres J (2006) A variable cluster of ethylene response factor-like genes regulates metabolic and developmental acclimation responses to submergence in rice. Plant Cell 18:2021–2034

Gaut B (2002) Evolutionary dynamics of grass genomes. New Phytol 154:15–28

Gibbings JG, Cook BP, Dufault MR, Madden SL, Khuri S, Turnbull CJ, Dunwell JM (2003) Global transcript analysis of rice leaf and seed using SAGE technology. Plant Biotechnol J 1:271–285

Goff SA, Ricke D, Lan TH, Presting G, Wang R, Dunn M, Glazebrook J, Sessions A, Oeller P, Varma H, Hadley D, Hutchison D, Martin C, Katagiri F, Lange BM, Moughamer T, Xia Y, Budworth P, Zhong J, Miguel T, Paszkowski U, Zhang S, Colbert M, Sun WL, Chen L, Cooper B, Park S, Wood TC, Mao L, Quail P, Wing R, Dean R, Yu Y, Zharkikh A, Shen R, Sahasrabudhe S, Thomas A, Cannings R, Gutin A, Pruss D, Reid J, Tavtigian S, Mitchell J, Eldredge G, Scholl T, Miller RM, Bhatnagar S, Adey N, Rubano T, Tusneem N, Robinson R, Feldhaus J, Macalma T, Oliphant A, Briggs S (2002) A draft sequence of the rice genome (*Oryza sativa* L. ssp. japonica). Science 296:92–100

Gruis DF, Guo H, Selinger D, Tian Q, Olsen O (2006) Surface position, not signaling from surrounding maternal tissues, specifies aleurone epidermal cell fate in maize. Plant Physiol 141:898–909

Gusmaroli G, Tonelli C, Mantovani R (2001) Regulation of the CCAAT-Binding NF-Y subunits in *Arabidopsis thaliana*. Gene 264:173–185

Hinton JJC (1948) Parboiling treatment of rice. Nature 162:913–915

Hirose T, Scofield G, Terao T (2008) An expression analysis profile for the entire sucrose synthase gene family in rice. Plant Sci 174:534–543

Holdsworth MJ, Kurup S, McKibbin R (1999) Molecular and genetic mechanisms regulating the transition from embryo development to germination. Trends Plant Sci 4:275–280

Holdsworth MJ, Finch-Savage WE, Grappin P, Job D (2008) Postgenomics dissection of seed dormancy and germination. Trends Plant Sci 13:7–13

Hong SK, Aoki T, Kitano H, Satoh H, Nagato Y (1995) Phenotypic diversity of 188 rice embryo mutants. Dev Genet 16:298–310

Howell KA, Narsai R, Carroll A, Ivanova A, Lohse M, Usadel B, Millar AH, Whelan J (2009) Mapping metabolic and transcript temporal switches during germination in rice highlights

specific transcription factors and the role of RNA instability in the germination process. Plant Physiol 149:961–980
Huang J, Takano T, Akita S (2000) Expression of alpha-expansin genes in young seedlings of rice (*Oryza sativa* L.). Planta 211:467–473
International Rice Genome Sequencing Project (2005) The map-based sequence of the rice genome. Nature 436:793–800
Itoh J, Nonomura K, Ikeda K, Yamaki S, Inukai Y, Yamagishi H, Kitano H, Nagato Y (2005) Rice plant development: from zygote to spikelet. Plant Cell Physiol 46:23–47
Itoh J, Hibara K, Sato Y, Nagato Y (2008) Developmental role and auxin responsiveness of class III homeodomain leucine zipper gene family members in rice. Plant Physiol 147:1960–1975
Izawa T, Foster R, Nakajima M, Shimamoto K, Chua NH (1994) The rice bZIP transcriptional activator RITA-1 is highly expressed during seed development. Plant Cell 6:1277–1287
Jain M, Khurana JP (2009) Transcript profiling reveals diverse roles of auxin-responsive genes during reproductive development and abiotic stress in rice. FEBS J 276:3148–3162
Jain M, Kaur N, Tyagi AK, Khurana JP (2006a) The auxin-responsive GH3 gene family in rice (*Oryza sativa*). Funct Integr Genomics 6:36–46
Jain M, Kaur N, Garg R, Thakur JK, Tyagi AK, Khurana JP (2006b) Structure and expression analysis of early auxin-responsive Aux/IAA gene family in rice (*Oryza sativa*). Funct Integr Genomics 6:47–59
Jain M, Tyagi AK, Khurana JP (2006c) Genome-wide analysis, evolutionary expansion, and expression of early auxin-responsive SAUR gene family in rice (*Oryza sativa*). Genomics 88:360–371
Jain M, Nijhawan A, Arora R, Agarwal P, Ray S, Sharma P, Kapoor S, Tyagi AK, Khurana JP (2007) F-box proteins in rice. Genome-wide analysis, classification, temporal and spatial gene expression during panicle and seed development, and regulation by light and abiotic stress. Plant Physiol 143:1467–1483
Jain R, Katavic V, Agrawal GK, Guzov VM, Thelen JJ (2008) Purification and proteomic characterization of plastids from *Brassica napus* developing embryos. Proteomics 8:3397–3405
Jiao Y, Ma L, Strickland E, Deng XW (2005) Conservation and divergence of light-regulated genome expression patterns during seedling development in rice and *Arabidopsis*. Plant Cell 17:3239–3256
Jiao Y, Lau OS, Deng XW (2007) Light-regulated transcriptional networks in higher plants. Nat Rev Genet 8:217–230
Jiao Y, Tausta SL, Gandotra N, Sun N, Liu T, Clay NK, Ceserani T, Chen M, Ma L, Holford M, Zhang HY, Zhao H, Deng XW, Nelson T (2009) A transcriptome atlas of rice cell types uncovers cellular, functional and developmental hierarchies. Nat Genet 41:258–263
Jones TJ, Rost TL (1989) Histochemistry and ultrastructure of rice (*Oryza sativa*) zygotic embryogenesis. Am J Bot 76:504–520
Jung KH, An G, Ronald PC (2008) Towards a better bowl of rice: assigning function to tens of thousands of rice genes. Nat Rev Genet 9:91–101
Jung KH, Seo YS, Walia H, Cao P, Fukao T, Canlas PE, Amonpant F, Bailey-Serres J, Ronald PC (2010) The submergence tolerance regulator Sub1A mediates stress-responsive expression of AP2/ERF transcription factors. Plant Physiol 152:1674–1692
Kaneko M, Inukai Y, Ueguchi-Tanaka M, Itoh H, Izawa T, Kobayashi Y, Hattori T, Miyao A, Hirochika H, Ashikari M, Matsuoka M (2004) Loss-of-function mutations of the rice GAMYB gene impair alpha-amylase expression in aleurone and flower development. Plant Cell 16:33–44
Kapoor M, Arora R, Lama T, Nijhawan A, Khurana JP, Tyagi AK, Kapoor S (2008) Genome-wide identification, organization and phylogenetic analysis of Dicer-like, Argonaute and RNA-dependent RNA polymerase gene families and their expression analysis during reproductive development and stress in rice. BMC Genomics 9:451
Khush GS (1997) Origin, dispersal, cultivation and variation of rice. Plant Mol Biol 35:25–34
Khush GS (2005) What it will take to feed 5.0 billion rice consumers in 2030. Plant Mol Biol 59:1–6

Kizis D, Lumbreras V, Pagès M (2001) Role of AP2/EREBP transcription factors in gene regulation during abiotic stress. FEBS Lett 498:187–189

Klinger JP, Batelli G, Zhu JK (2010) ABA receptors: the START of a new paradigm in phytohormone signalling. J Exp Biol 61:3199–3210

Kovach MJ, Sweeney MT, McCouch SR (2007) New insights into the history of rice domestication. Trends Genet 23:578–587

Krishnan HB, Oehrle NW, Natarajan SS (2009) A rapid and simple procedure for the depletion of abundant storage proteins from legume seeds to advance proteome analysis: a case study using *Glycine max*. Proteomics 9:3174–3188

Kucera B, Cohn MA, Leubner-Metzger G (2005) Plant hormone interactions during seed dormancy release and germination. Seed Sci Res 15:281–307

Kuwano M, Masumura T, Yoshida K (2011) A novel endosperm transfer cell-containing region-specific gene and its promoter in rice. Plant Mol Biol 76:47–56

Kwong RW, Bui AQ, Lee H, Kwong LW, Fischer RL, Goldberg RB, Harada JJ (2003) LEAFY COTYLEDON1-LIKE defines a class of regulators essential for embryo development. Plant Cell 15:5–18

Lasanthi-Kudahettige R, Magneschi L, Loreti E, Gonzali S, Licausi F, Novi G, Beretta O, Vitulli F, Alpi A, Perata P (2007) Transcript profiling of the anoxic rice coleoptile. Plant Physiol 144:218–231

Lee Y, Kende H (2001) Expression of β-expansins is correlated with internodal elongation in deepwater rice. Plant Physiol 127:645–654

Lee Yi, Kende H (2002) Expression of alpha-expansin and expansin-like genes in deepwater rice. Plant Physiol 130:1396–1405

Leubner-Metzger G (2005) Beta-1,3-Glucanase gene expression in low-hydrated seeds as a mechanism for dormancy release during tobacco after-ripening. Plant J 41:133–145

Li M, Xu W, Yang W, Kong Z, Xue Y (2007) Genome-wide gene expression profiling reveals conserved and novel molecular functions of the stigma in rice. Plant Physiol 144:1797–1812

Lid SE, Gruis D, Jung R, Lorentzen JA, Ananiev E, Chamberlin M, Niu X, Meeley R, Nichols S, Olsen OA (2002) The defective *kernel 1* (*dek1*) gene required for aleurone cell development in the endosperm of maize grains encodes a membrane protein of the calpain gene superfamily. Proc Natl Acad Sci USA 99:5460–5465

Liu T, Ohashi-Ito K, Bergmann DC (2009) Orthologs of *Arabidopsis thaliana* stomatal bHLH genes and regulation of stomatal development in grasses. Development 136:2265–2276

Lotan T, Ohto M, Yee KM, West MAL, Lo R, Kwong RW, Yamagishi K, Fischer RL, Goldberg RB, Harada JJ (1998) *Arabidopsis* LEAFY COTYLEDON1 is sufficient to induce embryo development in vegetative cells. Cell 93:1195–1205

Lu CA, Ho TD, Ho SL, Yu SM (2002) Three novel MYB proteins with one DNA binding repeat mediate sugar and hormone regulation of alpha-amylase gene expression. Plant Cell 14:1963–1980

Lur HS, Setter TL (1993) Role of auxin in maize endosperm development. Timing of nuclear DNA endoreduplication, zein expression, and cytokinin. Plant Physiol 103:273–280.

Ma L, Zhao H, Deng XW (2003) Analysis of the mutational effects of the COP/DET/FUS loci on genome expression profiles reveals their overlapping yet not identical roles in regulating *Arabidopsis* seedling development. Development 130:969–981

Ma L, Chen C, Liu X, Jiao Y, Su N, Li L, Wang X, Cao M, Sun N, Zhang X, Bao J, Li J, Pedersen S, Bolund L, Zhao H, Yuan L, Wong GKS, Wang J, Deng XW, Wang J (2005) A microarray analysis of the rice transcriptome and its comparison to *Arabidopsis*. Genome Res 15:1274–1283

Mantovani R (1999) The molecular biology of the CCAAT-binding factor NF-Y. Gene 239:15–27

Marcus A, Feeley J (1964) Activation of protein synthesis in the imbibition phase of seed germination. Proc Natl Acad Sci USA 51:1075–1079

McConnell JR, Barton MK (1998) Leaf polarity and meristem formation in *Arabidopsis*. Development 125:2935–2942

McConnell JR, Emery J, Eshed Y, Bao N, Bowman J, Barton MK (2001) Role of PHABULOSA and PHAVOLUTA in determining radial patterning in shoots. Nature 411:709–713

Meyers BC, Galbraith DW, Nelson T, Agrawal V (2004) Methods for transcriptional profiling in plants. Be fruitful and replicate. Plant Physiol 135:637–652

Miyoshi K, Kagaya Y, Ogawa Y, Nagato Y, Hattori T (2002) Temporal and spatial expression pattern of the *OSVP1* and *OSEM* genes during seed development in rice. Plant Cell Physiol 43:307–313

Nakano M, Nobuta K, Vemaraju K, Tej SS, Skogen JW, Meyers BC (2006) Plant MPSS databases: signature-based transcriptional resources for analyses of mRNA and small RNA. Nucl Acids Res 34:D731–D735

Nakase M, Aoki N, Matsuda T, Adachi T (1997) Characterization of a novel rice b ZIP protein which binds to the alpha-globulin promoter. Plant Mol Biol 33:513–522

Nambara E, Keith K, McCourt P, Naito S (1995) A regulatory role for the *ABI3* gene in the establishment of embryo maturation in *Arabidopsis thaliana*. Development 121:629–636

Narsai R, Howell KA, Carroll A, Ivanova A, Millar AH, Whelan J (2009) Defining core metabolic and transcriptomic responses to oxygen availability in rice embryos and young seedlings. Plant Physiol 151:306–322

Negbi M (1984) The structure and function of the scutellum of the Gramineae. Bot J Linn Soc 88:205–222

Nijhawan A, Jain M, Tyagi AK, Khurana JP (2008) Genomic survey and gene expression analysis of the basic leucine zipper transcription factor family in rice. Plant Physiol 146:333–350

Ogawa M, Hanada A, Yamauchi Y, Kuwahara A, Kamiya Y, Yamaguchi S (2003) Gibberellin biosynthesis and response during *Arabidopsis* seed germination. Plant Cell 15:1591–1604

Ohdan T, Francisco PB, Sawada T, Hirose T, Terao T, Satoh H, Nakamura Y (2005) Expression profiling of genes involved in starch synthesis in sink and source organs of rice. J Exp Bot 56:3229–3244

Oñate L, Vicente-Carbajosa J, Lara P, Díaz I, Carbonero P (1999) Barley BLZ2, a seed-specific bZIP protein that interacts with BLZ1 *in vivo* and activates transcription from the GCN4-like motif of B-hordein promoters in barley endosperm. J Biol Chem 274:9175–9182

Park J, Knoblauch M, Okita TW, Edwards GE (2009) Structural changes in the vacuole and cytoskeleton are key to development of the two cytoplasmic domains supporting single-cell C(4) photosynthesis in *Bienertia sinuspersici*. Planta 229:369–382

Rajjou L, Gallardo K, Debeaujon I, Vandekerckhove J, Job C, Job D (2004) The effect of α-amanitin on the *Arabidopsis* seed proteome highlights the distinct roles of stored and neosynthesized mRNAs during germination. Plant Physiol 134:1598–1613

Ray S, Agarwal P, Arora R, Kapoor S, Tyagi AK (2007) Expression analysis of calcium-dependent protein kinase gene family during reproductive development and abiotic stress conditions in rice (*Oryza sativa* L. ssp. indica). Mol Genet Genomics 278:493–505

Riechmann JL, Krizek BA, Meyerowitz EM (1996) Dimerization specificity of *Arabidopsis* MADS domain homeotic proteins APETALA1, APETALA3, PISTILLATA, and AGAMOUS. Proc Natl Acad Sci USA 93:4793–4798

Rorat T 2006 Plant dehydrins-tissue location, structure and function. Cell Mol Biol Lett 11:536–556

Sabelli PA, Larkins BA (2009) The contribution of cell cycle regulation to endosperm development. Sex Plant Reprod 22:207–219

Sarkar NK, Kim YK, Grover A (2009) Rice sHsp genes: genomic organization and expression profiling under stress and development. BMC Genomics 10:393

Satoh N, Hong SK, Nishimura A, Matsuoka M, Kitano H, Nagato Y (1999) Initiation of shoot apical meristem in rice: characterization of four *SHOOTLESS* genes. Development 126:3629–3636

Schmidt RJ, Ketudat M, Aukerman MJ, Hoschek G (1992) Opaque-2 is a transcriptional activator that recognizes a specific target site in 22-kD zein genes. Plant Cell 4:689–700

Sharma R, Kapoor M, Tyagi AK, Kapoor S (2010) Comparative transcript profiling of TCP family genes provide insight into gene functions and diversification in rice and *Arabidopsis*. J Plant Mol Biol Biotechnol 1:24–38

Shen B, Li C, Min Z, Meeley RB, Tarczynski MC, Olsen OA (2003) sal1 determines the number of aleurone cell layers in maize endosperm and encodes a class E vacuolar sorting protein. Proc Natl Acad Sci USA 100:6552–6557

Siangliw M, Toojinda T, Tragoonrung S, Vanavichit A (2003) Thai jasmine rice carrying QTLch9 (*Sub*QTL) is submergence tolerant. Ann Botany 91:255–261

Sivamani E, Bahieldin A, Wraith J, Al-Niemi T, Dyer WE, Ho TD, Qu R (2000) Improved biomass productivity and water use efficiency under water deficit conditions in transgenic wheat constitutively expressing the barley *HVA1* gene. Plant Sci 155:1–9

Sreenivasulu N, Altschmied L, Radchuk V, Gubatz S, Wobus U, Weschke W (2004) Transcript profiles and deduced changes of metabolic pathways in maternal and filial tissues of developing barley grains. Plant J 37:539–553

Sugimoto N, Takeda G, Nagato Y, Yamaguchi J (1998) Temporal and spatial expression of the α-amylase gene during seed germination in rice and barley. Plant Cell Physiol 39:323–333

Takaiwa F, Oono K, Wing D, Kato A (1991) Sequence of three members and expression of a new major subfamily of glutelin genes from rice. Plant Mol Biol 17:875–885

Takano M, Kanegae H, Shinomura T, Miyao A, Hirochika H, Furuya M (2001) Isolation and characterization of rice phytochrome A mutants. Plant Cell 13:521–534

Takeuchi Y, Omigawa Y, Ogasawara M, Yoneyama K, Konnai M, Worsham AD (1995) Effects of brassinosteroids on conditioning and germination of clover broomrape (*Orobanche minor*) seeds. Plant Growth Reg 16:153–160

Tanabe S, Ashikari M, Fujioka S, Takatsuto S, Yoshida S, Yano M, Yoshimura A, Kitano H, Matsuoka M, Fujisawa Y, Kato H, Iwasaki Y (2005) A novel cytochrome P450 is implicated in brassinosteroid biosynthesis via the characterization of a rice dwarf mutant, dwarf11, with reduced seed length. Plant Cell 17:776–790

Thirumurugan T, Ito Y, Kubo T, Serizawa A, Kurata N (2008) Identification, characterization and interaction of HAP family genes in rice. Mol Genet Genomics 279:279–289

Vicente-Carbajosa J, Oñate L, Lara P, Díaz I, Carbonero P (1998) Barley BLZ1: a bZIP transcriptional activator that interacts with endosperm-specific gene promoters. Plant J 13:629–640

Wang XS, Zhu HB, Jin GL, Liu HL, Wu WR, Zhu J (2007) Genome-scale identification and analysis of *LEA* genes in rice (*Oryza sativa* L.). Plant Sci 172:414–420

Wang L, Guo K, Li Y, Tu Y, Hu H, Wang B, Cui X, Peng L (2010a) Expression profiling and integrative analysis of the *CESA/CSL* superfamily in rice. BMC Plant Biol 10:282

Wang L, Xie W, Chen Y, Tang W, Yang J, Ye R, Liu L, Lin Y, Xu C, Xiao J, Zhang Q (2010b) A dynamic gene expression atlas covering the entire life cycle of rice. Plant J 61:752–766

Wu C, Trieu A, Radhakrishnan P, Kwok SF, Harris S, Zhang K, Wang J, Wan J, Zhai H, Takatsuto S, Matsumoto S, Fujioka S, Feldmann KA, Pennell RI (2008) Brassinosteroids regulate grain filling in rice. Plant Cell 20:2130–2145

Xu K, Xu X, Fukao T, Canlas R, Maghirang-Rodriguez R, Heuer S, Ismail AM, Bailey-Serres J, Ronald PC, Mackill DJ (2006) *Sub1 A* is an ethylene-response-factor-like gene that confers submergence tolerance to rice. Nature 442:705–708

Xue LJ, Zhang JJ, Xue HW (2009) Characterization and expression profiles of miRNAs in rice seeds. Nucl Acids Res 37:916–930

Yamakawa H, Hakata M (2010) Atlas of rice grain filling-related metabolism under high temperature: Joint analysis of metabolome and transcriptome demonstrated inhibition of starch accumulation and induction of amino acid accumulation. Plant Cell Physiol 51:795–809

Yanagisawa S (2002) The Dof family of plant transcription factors. Trends Plant Sci 7:555–560

Yang J, Zhang J, Huang Z, Wang Z, Zhu Q, Liu L (2002) Correlation of cytokinin levels in the endosperms and roots with cell number and cell division activity during endosperm development in rice. Ann Bot 90:369–377

Yokoyama R, Rose JKC, Nishitani K (2004) A surprising diversity and abundance of xyloglucan endotransglucosylase/hydrolases in rice. Classification and expression analysis. Plant Physiol 134:1088–1099

Yoshida S (1981) Fundamentals of rice crop science. International Rice Research Institute, Manila

Yu J, Wang J, Lin W, Li S, Li H, Zhou J, Ni P, Dong W, Hu S, Zeng C, Zhang J, Zhang Y, Li R, Xu Z, Li S, Li X, Zheng H, Cong L, Lin L, Yin J, Geng J, Li G, Shi J, Liu J, Lv H, Li J, Wang J, Deng Y, Ran L, Shi X, Wang X, Wu Q, Li C, Ren X, Wang J, Wang X, Li D, Liu D, Zhang X, Ji Z, Zhao W, Sun Y, Zhang Z, Bao J, Han Y, Dong L, Ji J, Chen P, Wu S, Liu J, Xiao Y, Bu D, Tan J, Yang L, Ye C, Zhang J, Xu J, Zhou Y, Yu Y, Zhang B, Zhuang S, Wei H, Liu B, Lei M, Yu H, Li Y, Xu H, Wei S, He X, Fang L, Zhang Z, Zhang Y, Huang X, Su Z, Tong W, Li J, Tong Z, Li S, Ye J, Wang L, Fang L, Lei T, Chen C, Chen H, Xu Z, Li H, Huang H, Zhang F, Xu H, Li N, Zhao C, Li S, Dong L, Huang Y, Li L, Xi Y, Qi Q, Li W, Zhang B, Hu W, Zhang Y, Tian X, Jiao Y, Liang X, Jin J, Gao L, Zheng W, Hao B, Liu S, Wang W, Yuan L, Cao M, McDermott J, Samudrala R, Wang J, Wong GKS, Yang H (2005) The Genomes of *Oryza sativa*: a history of duplications. PLoS Biol 3:e38

Zhu T, Budworth P, Chen W, Provart N, Chang HS, Guimil S, Su W, Estes B, Zou G, Wang X (2003) Transcriptional control of nutrient partitioning during rice grain filling. Plant Biotechnol J 1:59–70

Chapter 5
A Transcriptional Roadmap for Seed Development in Maize

Guifeng Wang, Gang Wang, Fei Wang and Rentao Song

Abstract Maize (*Zea mays* L.) is an important source for food, feed, and fuel with rapidly increasing global demand. According to the Food and Agriculture Organization, it is amongst the top three crops in terms of its production and consumption throughout the world. To date, several studies have unraveled many aspects of the physiological, biochemical, transcriptomics, and proteomics properties of maize seed development. Those studies have helped in better understanding the underlying genetic control of this important event in maize, and cereal seeds in general. In this chapter, we discuss the transcriptomic behavior of maize plants during seed developmental phase and highlight the regulatory roadmap for the synthesis of starch, fatty acids, and storage proteins in seeds. Genes associated with seed development will not only provide information for understanding the transcriptional network during seed development, but also will provide a blueprint for future modification of seed quality and seed vigor through genetic engineering in maize, and other important cereal crops.

Keywords Embryogenesis · Endosperm · Maize · Starch synthesis · Storage protein · Transcriptomics · Transcriptional regulation

5.1 Introduction

Seed development is one of the key biological processes in the life cycle of higher plants, not only for facilitating the species survival and distribution but also for accumulating storage compounds. Seed formation is a complicated and delicate process that requires the coordinated development of three distinctly-originated tissues, embryo, endosperm, and integument (Baud et al. 2008). The zygotic tissues (embryo (diploid, *2n*) and endosperm (triploid, *3n*)) are derived from the double fertilization event, which is a unique characteristic of higher plants (Berger 1999; Chaudhury et al. 1998). However, the integuments of maternal origin mainly constitute the seed coat (diploid, *2n*) and protect the embryo and endosperm. Seed is one

R. Song (✉) · G. Wang · G. Wang · F. Wang
Shanghai Key Laboratory of Bio-Energy Crops, School of Life Sciences, Shanghai University, No. 333 Nanchen Road, Shanghai, P. R. China, 200444, China
e-mail: rentaosong@staff.shu.edu.cn

G. K. Agrawal, R. Rakwal (eds.), *Seed Development: OMICS Technologies toward Improvement of Seed Quality and Crop Yield*,
DOI 10.1007/978-94-007-4749-4_5, © Springer Science+Business Media Dordrecht 2012

of the major food sources for human society, and serves as industrial raw materials. Due to its economical and nutritional values, the seed developmental aspects have been thoroughly analyzed cytologically and genetically (Chaudhury et al. 2001; Olsen 2001).

In the present time, global agriculture is suffering from multiple challenges like—shortening of agricultural lands, energy hikes, climate change, etc., which ultimately results in severe 'food crisis'. According to Fedoroff (2010), the 'food crisis' alone has sparked riots in many countries in every continent (i.e., Africa). So, to fight these challenges, and have a better and sustainable food supply; we should find potent crops with higher vigor and superior adaptability for future environment. Maize (*Zea mays* L.; also called corn), one of the most important crops, belongs to the cereal family and has long been a model species for genetic studies. Maize is a highly photosynthetic-efficient C4 grass and has recently been proposed for use as a key model system for gene discovery relating to biomass yield and quality in the bioenergy grasses (Carpita and McCann 2008). To meet the dramatically growing, global demand for grain, it is estimated that an additional 200 million tonnes/year of corn and wheat (*Triticum aestivum* L.) would have to be produced by 2017 (Edgerton 2009). It is widely considered that gene engineering strategies are powerful approaches to increase crop productivity. Therefore, a complete list of genes expressed in developing seeds will help in constructing the gene transcriptional regulation network and provide a blueprint for future modification of grain quantity and quality in cereal crops.

Thanks to the application of high-throughput omics technologies to crops and to the knowledge gained from the model plants (like *Arabidopsis* and rice), our current understanding of the genetic control underlying cereal seed development has been greatly enhanced. During the past decade, several studies have unraveled many aspects of the biochemical, transcriptomics, and proteomics properties of maize seed development (Alexandrov et al. 2009; Guo et al. 2003; Lai et al. 2004; Liu et al. 2008; Méchin et al. 2004, 2007; Prioul et al. 2008; Verza et al. 2005). In addition, a collection of at least 300 endosperm-defective mutants of maize identified in nature or created in the laboratory, is also a significant resource for the molecular characterization of genes involved in this biological process (Neuffer and Sheridan 1980). In this chapter, we mainly discuss the transcriptomics-driven findings, and our current understanding of maize seed development.

5.2 A Portrait of Maize Seed Development

For most grass crop species, the value of grain is largely determined by the quantity and quality of the endosperm. The endosperm originates from the fertilization of two polar nuclei in the central cell of the embryo sac by one sperm cell nucleus (triploid, *3n*, 3C), whereas the embryo originates from fertilization of the egg cell by the second sperm cell nucleus (diploid, *2n*, 2C). The seed development of maize

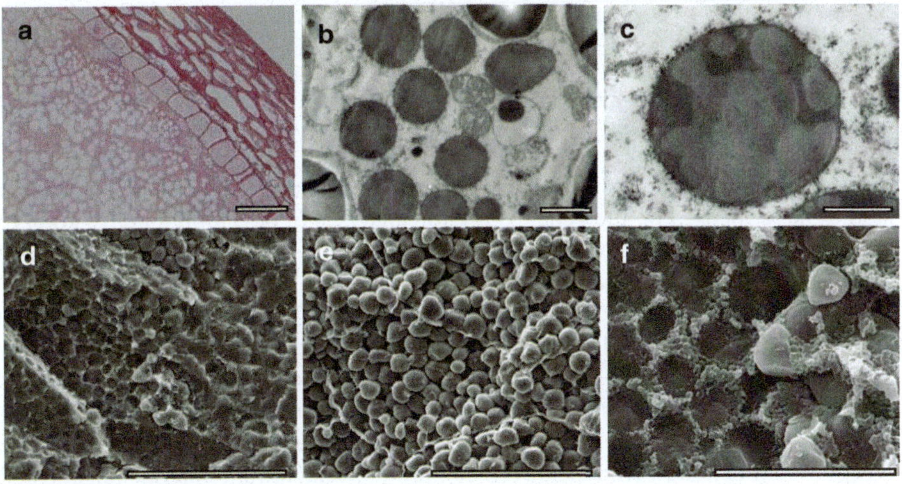

Fig. 5.1 Cytological images of maize kernels. **a** Transverse section of resin-embedded seed at 25 DAP; bar, 0.05 mm. **b** Transmission electron microscopy image of developing seed at 25 DAP; bar, 1.0 μm. **c** Magnified section of **b**; bar, 500 nm. **d** and **e** The peripheral and central parts of the mature seed, respectively; bar, 100 μm. **f** Close to **d**; bar, 30 μm

contains several distinct but considerably overlapping phases (Fig. 5.1), including early development (double fertilization, syncytium formation, and cellularization), differentiation (formation of the main cell types), periods of mitosis and endoreduplication, accumulation of storage compounds, and maturation (programmed cell death, dormancy, and desiccation) (Sabelli and Larkins 2009).

During genesis, the embryo follows a sequence of developmental phases: cell division, cell enlargement, deposition of reserves, maturation, and desiccation of the seed. The embryo proper is formed consisting of histodifferentiation, followed by acquisition of polarity and bilateral symmetry and, morphogenesis. Finally, storage reserves accumulation, dehydration, and quiescence take place during the maturation phase (Fontanet and Vicient 2008; Kiesselbach 1998). Noticeably, the expression of metabolic enzymes is most active around 20 DAP, when morphogenesis and cell division processes basically stop and, storage and desiccation processes begin.

In contrast to the dicot plants, the cereal endosperm is not a transient tissue absorbed at later stages of seed development, but instead provides a continuous favorable environment for the developing embryo. The endosperm development includes several partially overlapping stages. The first stage of development extends up to 12 DAP and characteristically involves cell division resulting in rapid expansion of the endosperm. Subsequently, 12–40 DAP is called the 'kernel filling' stage, referring to the activity of several metabolic pathways that convert imported sucrose and amino acids into starch and storage proteins. Finally, 40–70 DAP is the desiccation stage and leads to the formation of mature seeds (Olsen 2001; Prioul et al. 2008; Wang et al. 2010).

5.2.1 Early Seed Development

After double fertilization, the fertilized egg cell enters into the embryogenesis process, differentiating into a mature embryo containing a number of cell types and tissues with different biological functions. It is important to understand the transcriptional regulation networks from early embryogenesis to maturity of seeds. Several studies have focused on the gene expression analyses of developing maize embryo using a transcriptomics approach, providing invaluable insights into the mechanism underlying early embryo development (Lee et al. 2002; Liu et al. 2008).

Genome-wide analysis of embryo development has been performed on two elite maize inbred lines B73 and X178 (Lee et al. 2002; Liu et al. 2008). During the early kernel embryogenesis, cell division and enlargement were the dominating activities for the formation of tissues and organs. The highly-expressed genes in the growing embryo during 1 through 5 DAP fell into several abundant functional categories: the biogenesis of cellular components (30 %), protein-related function (29 %), and cell cycle and DNA processing (20 %). These genes could also be grouped into three pathways: CDC2, SCF, and CDC5 (Liu et al. 2008). In the CDC2 pathway, one gene encoding the CDC2 protein, a member of the Ser/Thr protein kinase, might control the G1-S-G2-M checkpoint transition of the cell cycle during the cell-division process. A gene encoding a β-glucosidase was differentially expressed in this stage. It supports the hypothesis that β-glucosidase might stimulate tyrosine dephosphorylation, activating CDC2 kinase and indirectly promoting cell division. The SCF complex (SKP1, Cullin, and F-box) could associate with E3 and help in the recruitment and degradation of target proteins during the G1-S transition of the cell cycle (Vodermaier 2004). Genes encoding SKP1, Cullin, and F-box were differentially expressed, suggesting that the SCF complex pathway is also involved in the regulation of cell division during the maize kernel formation. Another gene encoding a putative CDC5 was upregulated from 5 to 15 DAP and then downregulated. The CDC5 was considered as an activator of the Anaphase-Promoting Complex, which regulates the G2-M transition of cell cycle through targeting cyclins for degradation (Lei et al. 2000), indicating that CDC5 could enhance the activity of cell division during the early kernel development.

The maternal-to-zygotic transition is one of the interesting developmental events of early seed development. It happens after fertilization and causes extensively gene expression patterns, leading to the establishment of an embryo-specific developmental program (Grimanelli et al. 2005). In animals and plants, early embryonic development is largely dependent on maternal control during gametogenesis. However, the actual timing of zygotic genome activation is unknown. Grimanelli et al. (2005) used maize microarrays to compare the expression profile between unfertilized ovules from apomictic maize-Tripsacum hybrid plants with a developing proembryo (but no endosperm) and ovules from sexual reproduction maize plants with neither endosperm nor embryo at 3 and 7 DAP. Of the 7,300 unique sequences, 111 differentially-expressed genes were identified. Of these, 77 genes were differentially expressed between both the samples, and the expressions of another 34 genes were significantly different from null expression. The results indicate that the maternal-to-zygotic transition in maize does not occur until at least 3 days after the initiation of embryo development. Unlike the early embryogenesis, early endosperm development in maize correlates with significant novel transcriptional regulation in the ovule, which suggests that different mechanisms are involved in the initiation of endosperm and embryo development.

5.2.2 Seed Metabolism

Accumulation of Starch and Other Carbohydrates Starch, a renewable carbohydrate polymer, comprises approximately 70 % of the DW of maize kernel. Toward the prospect of designing starch for increased quantity and quality by genetic manipulation of genes involved in starch biosynthesis, the pathways of starch metabolism in cereal endosperms have been studied intensively and thoroughly. A full list of genes involved in starch biosynthesis in maize genome has been summarized in Table 5.1.

Sucrose, originating from the source leaves, is imported into the developing cob tissue of maize and converted into a mixture of fructose and UDP-glucose in the cytosol of endosperm cells. This conversion was accomplished by SUS, which can cleave sucrose to generate UDP-glucose. Four genes encoding SUS were significantly upregulated from 10 DAP, which may imply their essential functions in supplying raw materials for starch synthesis (Liu et al. 2008). Then UDP-glucose is converted into activated hexoses and glucose-1-P, followed by synthesized the starch precursor ADP-glucose, which is then transported into the plastid (Glawischnig et al. 2002). The transcripts of identified genes, involved in starch and soluble carbohydrate metabolism, represent 15–23 % of the expressed sequence tags (ESTs) expressed in the kernel cDNA libraries at 10, 14, and 21 DAP (Prioul et al. 2008). Starch biosynthesis was initiated at a very early stage of the kernel development, as illustrated by upregulation of starch synthase III b1 and b2 at 3 DAP. The AGPase catalyzes the first reaction in starch biosynthesis, producing the activated glucosyl donor ADP-glucose (James et al. 2003). Of the six AGPase listed in Table 5.1, their expression patterns in seed development could be divided into two groups. The first group consists of those genes, which were stably expressed during the developmental process, such as *AGPL1* and *AGPS1*. The second group generally tended to increase and maintained high expression intensity, such as *AGPL2* (*Shrunken2*), *AGPL3*, and *AGPS2* (*Brittle2*). Soluble and granule-bound starch synthases utilize ADP-glucose to specifically elongate amylopectin and amylose, respectively. All starch synthase isoforms (Table 5.1) were differentially expressed during maize seed development. Except upregulation of *SSIIIb1* and *SSIIIb2* genes at the early stage, expression levels of the remaining genes increased gradually and peaked between 15 and 18 DAP. Starch branching enzymes (SBEs) generate α-1,6 linkages by cleaving internal α-1,4 bonds and transferring the released reducing ends to

Table 5.1 Genes involved in maize starch synthesis pathway

Enzyme		Gene name	Chromosome	cDNA access number	Protein ID	Amino acid	Mutant
ADP-glucose pyrophosphorylase	Large subunit	ZmAGPL1	6	Z38111	CAA86227	521	
		ZmAGPL2	3	S48563	AAB52952	522	sh2 (shrunken)
		ZmAGPL3	1	DQ406819	ABD66656	505	
		ZmAGPL4	7	EF694838	ABS19874	514	ZmAGPLS3
	Small subunit	ZmAGPS1	2	AY032604	AAK39640	510	Agpsemzm
		ZmAGPS2	4	AF330035	AAK69627	475	bt2 (brittle-2)
Starch synthase	Granule-bound	ZmGBSSI	9	AF267643	CAA27574	605	wx1 (waxy)
		ZmGBSSIIa	7	EF471312	ABO71782	609	
	Soluble	ZmSSI	9	AF036891	AAB99957	640	ss1
		ZmSSIIa	6	AF019296	AAD13341	732	su2 (sugary2)
		ZmSSIIb	6	AF019297	AAD13342	698	
		ZmSSIIIa	10	AF023159	AAC14014	1,674	du1
		ZmSSIIIb1	10	EF472250	ABP35815	1,191	
		ZmSSIIIb2	10	EF472251	ABP35816	1,188	
Starch branching enzyme		ZmBEI	5	U17897	AAC36471	823	sbe1
		ZmBEIIa	2	U65948	AAB67316	814	sbe2a
		ZmBEIIb	5	L08065	AAC33764	799	ae1 (amylose-extender)
Starch debranching enzyme	Isoamylase	ZmISA1	4	U18908	AAB97167	789	su1 (sugary1)
		ZmISA2	6	AY172633	AAO17048	799	isa2
		ZmISA3	2	AY172634	AAO17049	694	isa3
	Pullulanas	ZmPUL	2	AF080567	AAD11599	962	zpu1

C6 hydroxyls. The *SBEIIa* gene was expressed at the early stage, while *SBEI* and *SBEIIb* were highly expressed at the middle stage of seed development. Two types of debranching enzymes (DBEs) families exist in plants, isoamylase and pullulanase types. Both types hydrolyze α-1,6 linkages, but they differ in substrate specificity (James et al. 2003). Three isoamylase (*ISO*) genes share a similar expression pattern, whose expressions were increased gradually and reached to maxima around 18 DAP. Together, the expression pattern of these genes and overrepresentation of starch and sucrose metabolism imply that starch accumulation occurs in a large quantities from 10 DAP, peaks at 15 DAP, and remains steady thereafter (Liu et al. 2008; Yan et al. 2009).

Synthesis of Fatty Acids and Lipids Maize oil is one of the most valuable coproducts from industrial processing of maize grain through milling and the oil is of high-quality for human consumption. It is stable because the oil contains quite little (less than 1.0 %) linolenic acid, which is susceptible to oxidation (Shen et al. 2010). In plant, oil is synthesized from glycerol-3-P and fatty acyl-CoA in the ER as triacylglycerols (TAGs). In maize seeds, lipid reserves are synthesized and stored primarily in a specialized tissue of the embryo, known as scutellum.

A bunch of key genes are involved in fatty acid (FA) biosynthesis and oil storage. In *Arabidopsis*, the plastidial FA biosynthetic pathway includes at least 24 enzymes or subunits, including acetyl-CoA carboxylase (ACCase), enoyl-acyl carrier protein (ACP) reductase, malonyl-CoA:ACP transacylase, beta-ketoacyl-ACP synthase I, and FA desaturase. The majority of these genes were significantly upregulated from 10 DAP in the maize kernel development (Lee et al. 2002; Liu et al. 2008). ACCase is a major control point of the pathway. The transcripts of ACCase accumulated to a high level in whole kernels as early as 5 DAP and remained high in embryos until 20 DAP. The expression of malonyl-CoA: ACP transacylase, -ketoacyl-ACP synthase III, enoyl-ACP reductase, and ACP reached to their highest levels between 15 to 20 DAP. A seed-specific desaturase (stearoyl-ACP desaturase) was also found to express to its maximum level between 15 DAP and 20 DAP, but decreased significantly at later developmental stages (Lee et al. 2002). Storage lipids in maize kernels were transported into lipid bodies for storage. The genes encoding the lipid body proteins, including steroleosin, oleosin, and caleosin, were significantly upregulated starting from 25 DAP and continued to accumulate till 35 DAP. Transcripts of oleosin 16, 17, and 18 started to accumulate till 15 DAP and maintained until 25 DAP, followed by slight reduction thereafter (Liu et al. 2008). These divergent expression patterns indicate that the expressions of genes related to FA synthesis occurred prior to lipid body proteins for oil storage. In maize embryo, the rate of oil synthesis peaks between 15 and 25 DAP and the concentration of oil reaches its optimal level at 30 DAP (Curtis et al. 1968). It is interesting to note that many transcripts encoding the appropriate enzymes and proteins involved in oil biosynthesis accumulate several days before the maximal storage oil accumulation. Therefore, it is valuable to understand the transcriptional control underlying oil and other storage reserves biosynthetic pathways during maize seed development by large-scale gene expression analysis.

Table 5.2 Summary of zein genes in the maize genome

Class		Protein molecular weight (kDa)	Map location	Copy number	cDNA access number (expressed)
α-zein	z1A	19	4.02/4.04	9 and 3	NM_001152088
					NM_001137771
					NM_001112275
					NM_001165463
					NM_001174232
					NM_001204324
					NP_001105745
					EU116452
	z1B	19	7.01	8	NM_001149065
					BT016489
	z1C	22	4.02/4.03	15 and 1	NM_001111587
					NM_001111588
					NM_001112277
					NM_001112530
					NM_001152106
					BT061059
	z1D	19	1.06	5	NM_001112176
					BT016174
β-zein		15	6.01	1	NM_001112534
γ-zein		16	2.07	1	AF371262
		27	7.05	1	AF371261
		50	7.05	1	BT062750
δ-zein		10	9.03	1	AF371266
		18	6.04	1	NM_001111584

Amino Acid Metabolism and Storage Proteins Biosynthesis The total content of amino acids and their composition of seed storage proteins (SSPs) significantly impact the nutritional value of maize seeds. SSPs are specifically synthesized and highly accumulated in the endosperm tissue. According to their solubility in water or alcohol, SSPs are classified into albumin, globulin, prolamin, and glutelin (Kawakatsu and Takaiwa 2010).

The metabolism of amino acids is tightly correlated with the nitrogen use efficiency from the environment, which is a key agronomic trait for crops, such as maize. It has been demonstrated that the translocation of glutamic acid (Glu) and glutamine (Gln) to the ear, and the subsequent synthesis of asparagine (Asp) are important metabolic processes that highly influence kernel productivity (Seebauer et al. 2004). Transcriptomics study has revealed that the genes involved in amino acid biosynthesis accumulate to their maximal levels in kernels at 10 DAP and in embryos at 15 DAP, including chorismate synthase, putative diaminopimelate decarboxylase, cystathionine gammalyase (Lee et al. 2002). Recently, Canas et al. (2009) systematically investigated changes in metabolite content, enzyme activities, and expression levels for marker genes of amino acid biosynthesis and interconversion in the cob and kernels during seed filling (Table 5.2). *ZmAS1* was expressed in both the cob and kernels, while the expression of *ZmAS2* and *ZmAS3* was almost restricted to developing kernels. *ZmAspAT1.1*, *ZmAspAT1.2*, *ZmAspAT1.3*, and *ZmAspAT2.1* were all constitutively expressed in the cob and kernels. The expression of *ZmAspAT2.2* was not detected in either the cob or kernels. The expression profile of *ZmAlaAT1* was similar to that of *ZmAspAT1.3*. *ZmAlaAT2* was specifically expressed in both the cob and kernels. The expression levels of *ZmP5CS1* and *ZmP5CS3* genes were more abundant in cob than that in kernel under both low- and high-nitrogen conditions, while the *ZmP5CS2* expression was even in both cob and kernel. The transcript levels of *ZmGDH1* and *ZmGDH2* were more abundant in cob than kernel. These expression data provide an insight into the genetic and environmental control of nitrogen metabolism and potential candidate genes for improving efficient utilization of maize nitrogen.

Prolamins, refer to as zeins, are predominantly-expressed SPs in maize seeds, which comprise approximately 50 % of the total proteins in mature kernels. Based on the solubility and structural relationships, zein proteins could be classified as α, β, γ, and δ zeins (Table 5.3). The majority of zeins was differentially expressed, significantly increased at 15 DAP and decreased dramatically after 40 DAP, during maize seed development (Liu et al. 2008; Luo et al. 2008). The α-zeins, which are encoded by a large super gene family, comprise the major fraction of maize zeins. Based on the sequence similarity and copy number, the α-zein gene family could be further divided into four gene families: z1A, z1B, z1C, and z1D (Table 5.3; Heidecker and Messing 1986). Our laboratory combined quantitative reverse transcriptase-polymerase chain reaction (RT-PCR) and random cDNA clones sequencing to successfully profile the expression of α-zein super gene family during endosperm development (Feng et al. 2009). There were 8 out of 12 members of the z1A family, 2 out of 8 members of the z1B family, 6 out of 16 members of the *z1C* family, and 2 out of 5 members of the *z1D* gene family expressed in endosperm (Table 5.3). Noticeably, an apparent 'up and down' expression rhythm was observed with all α-zein sub-families. The fluctuation interval varied among the different families. The z1A and z1B had an interval of 4 days, while the fluctuation interval of z1C and z1D was 6 days. This result suggests that there may be a feedback loop regulating α-zein synthesis during endosperm development. Moreover, different subfamilies showed different expression patterns. For the z1A family, expression levels gradually increased at 12 DAP, reached their peak at 18 DAP, and then decreased with fluctuation until 34 DAP. For the z1B genes, expression peaked at 22 DAP, and then decreased with a fluctuation until 34 DAP. The expression patterns of z1C and z1D gene families were similar with a peak at 24 DAP. The expression divergence among the different members indicates that a fast evolution rate of the maize α-zeins gene family (Feng et al. 2009).

Compared with zeins, the expression profiles of globulin members were strikingly different. The transcript of α-Globulin was decreased gradually and reached the lowest level at 40 DAP. The expression of *globulin-1* increased gradually with increase in kernels growth and maturation, while *globulin-2* maintained its expression at a relatively high level (Luo et al. 2008). In addition, *LEA* (*MZ00026333*, *MZ00028039*, and *MZ00040778*) and other seed maturation genes (*MZ00007111*

Table 5.3 Maize (*Zea mays*) genes encoding enzymes involved in amino acid biosynthesis and interconversion

Enzyme	Gene name	Chromosome	cDNA access number	Protein ID	Predicted subcellular localization
Asparagine synthetase	ZmAS1	Unknown	FM212633	CAR82078	Cytoplasm
	ZmAS2	3	FM212634	CAR82079	Cytoplasm
	ZmAS3	1	FM212635	CAR82080	Cytoplasm
	ZmAS4	9	FM212636	CAR82081	Cytoplasm
Glutamine synthetase	ZmGln1	10	NM_001112255	NP_001105725	Chloroplast
	ZmGln2	1	NM_001111973	NP_001105443	Cytoplasm
	ZmGln3	9	NM_001111827	NP_001105297	Cytoplasm
	ZmGln4	5	NM_001111826	NP_001105296	Cytoplasm
	ZmGln5	4	NM_001111974	NP_001105444	Cytoplasm
	ZmGln6	1	NM_001112068	NP_001105538	Cytoplasm
Aspartate aminotransferase	ZmAspAT1.1	3	NM_001147752	NP_001141224	Chloroplast/mitochondria
	ZmAspAT1.2	3	NM_001155533	NP_001149005	Chloroplast
	ZmAspAT1.3	5	NM_001148497	NP_001141969	Mitochondria
	ZmAspAT2.1	8	NM_001148831	NP_001142303	Chloroplast
	ZmAspAT2.2	3	NM_001150297	NP_001143769	Cytoplasm
Alanine aminotransferase	ZmAlaAT1	5	NM_001155755	NP_001149227	Cytoplasm
	ZmAlaAT2	1	BT041292	ACF86297	Cytoplasm
	ZmAlaAT3	7	BT017492	–	Mitochondria
	ZmAlaAT4	2	BT062415	ACN27112	Mitochondria
Pyrroline-5-carboxylate synthetase	ZmP5CS1	6	AY108698	–	Cytoplasm
	ZmP5CS2	8	BT056094	ACL54701	Cytoplasm/mitochondria
	ZmP5CS3	8	AY105865	ACG25484	Cytoplasm
Glutamate dehydrogenase	ZmGDH1	1	NM_001111831	NP_001105301	Mitochondria/cytoplasm
	ZmGDH2	10	NM_001138715	NP_001132187	Mitochondria/cytoplasm

All data derived from the Genbank. The protein subcellular localization was peridicted using WoLF PSORT and *MZ00029271*) were upregulated from 25 or 35 DAP, suggesting that such genes might play a role in the process of kernel maturation (Liu et al. 2008).

5.3 Transcriptional Regulation of Maize Seed Development

Genes involved in maize seed development, especially in storage compounds metabolism, have been well documented (Kawakatsu and Takaiwa 2010). However, the potential transcriptional regulators underlying this process are largely unknown. The TFs play key roles in regulating gene expression at the transcriptional level, and thereby control many biological processes. Compared to 2,383 TFs in rice and 1,922 TFs in *Arabidopsis*, 1,965 putative TFs have been reported in the maize genome (Soderlund et al. 2009). Transcriptomics approach has also proved very useful in cataloging TFs expressed during seed development in maize. Verza and coworkers reported expression of 365 TFs out of approximately 11,000 estimated genes expressed in the developing maize endosperm (Verza et al. 2005). Liu et al. (2008) profiled genome-wide differentially-expressed gene expressions during maize kernel development and found 106 TFs. In our laboratory, we identified 57 putative TFs from 6,630 unique ESTs isolated from a normalized maize seed cDNA library (Wang et al. 2010). The TFs belong to seven different super families, specifically 17 Zinc-finger, 13 basic leucine-zipper (bZIP), 8 bHLH, 6 MADS, 7 MYB, 3 Homeodomain, and 3 AP2/EREBP. One third (19) of the maize TFs have their putative orthologs in *Arabidopsis*. The majority of such orthologs (15 out of 19) has a similar expression patterns in maize and *Arabidopsis*, suggesting that they might have a conserved function in seed development. A transcriptional regulation roadmap for storage protein, starch, and FA biosynthesis in cereal seed is summarized in Fig. 5.2. Systematic functional characterization of these identified TFs will provide information for understanding the transcriptional networks of seed development.

Genes encoding SSPs are specifically expressed at high levels during seeds maturation, resulting in high accumulation of these proteins in the seeds. Their expression patterns and a tight correlation between gene expression and protein synthesis have suggested that SSPs synthesis is regulated primarily at transcriptional level (Goldberg et al. 1989). Isolation and comparison of numerous SSP gene promoters have resulted in identification of several *cis*-elements potentially involved in transcriptional regulation of the SSP genes. The endosperm box (5'-TGACA TGT AAAG TGAATAAGA TGAGTCA TG-3') also called the prolamin element, is well conserved around 300 bp upstream of the transcription start site in cereal prolamin gene promoters (Forde et al. 1985). It consists of two independent *cis*-elements, the prolamin box [P box, TG(T/C/A)AAAG] and the GCN4 motif [TGA(G/C)TCA]. Another two elements, the AACA (AACAAAC) and ACGT (ACGTG) motifs were identified within glutelin promoters (Takaiwa et al. 1996). The GCN4 is a critical *cis*-element both for determining qualitative expression of the prolamin promoters and for defining tissue specific expression. The

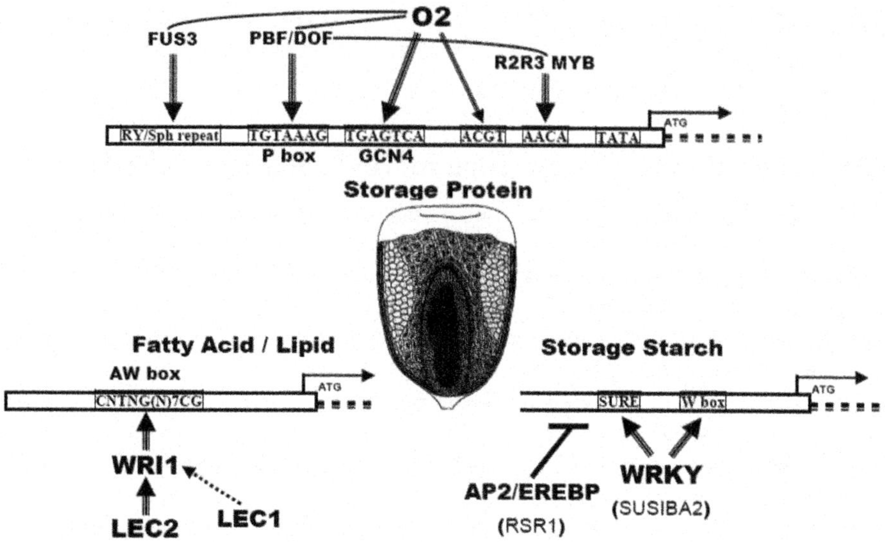

Fig. 5.2 A transcriptional regulation roadmap for SP, starch, and FA biosynthesis in the cereal seed

remaining three motifs (AACA motif, P box, and ACGT motif) are also quantitative *cis*-elements (reviewed by Kawakatsu and Takaiwa 2010).

A maize endosperm defective mutant opaque 2 (*O2*) is one of the seed mutants with a severe reduction in the content of 22 kDa α-zein. The *O2* encodes a bZIP TF, which could specifically bind to the GCN4 motif in the 22 kDa α-zein promoter and regulate its expression (Schmidt et al. 1992). Its orthologs have been identified from wheat (SPA; Albani et al. 1997), barley (BLZ2; Oñate et al. 1999), and rice (RISBZ1; Onodera et al. 2001). Similar to yeast GCN4, these bZIP genes regulate the expression of a set of genes involved in amino acid biosynthesis. They can also bind to ACGT motif in the 22 kDa α-zein promoter and in the b32 promoter (Schmidt et al. 1992). In maize, the P box is specifically recognized by prolamin box-binding factor (PBF) (Vicente-Carbajosa et al. 1997). The PBF encodes plant-specific DOF TFs, which contain a Cys2–Cys2 zinc finger motif. Its orthologs have also been found in other cereals, such as wheat WPBF/putative ESBF-I (Conlan et al. 1999), barley BPBF (Mena et al. 1998), and rice RPBF/OsDOF3 (Yamamoto et al. 2006). Evidences have demonstrated that *O2*-like bZIPs can interact with PBF-like DOFs *in vitro* or *in vivo*, such as *O2*-PBF (maize), BLZ2-BPBF (barley), and RISBZ1-RPBF (rice). Therefore, they are likely to act additively and coordinately regulate the SSPs gene expression (Kawakatsu and Takaiwa 2010). In barley, a R2R3-type MYB TF (HvGAMYB) can bind to the AACA motif in the *Hor2* B-hordein gene promoter (Diaz et al. 2002). Moreover, GAMYB and DOF TFs physically interact with each other. In dicots, FUSCA3 (FUS3) plays a critical role in the regulation of SSP gene expression by binding to RY/Sph repeats. The barley ortholog HvFUS3 can also regulate its SSP gene expression by recognizing the RY repeats within the promoter. Similarly, HvFUS3 and BLZ2 physically interact and additively regulate the SSP gene expression, too (Moreno-Risueno et al. 2008).

In *Arabidopsis*, LEAFY COTYLEDON1 (LEC1) and WRINKLED1 (WRI1) have been identified as two key TFs involved in the regulation of FA biosynthesis and oil accumulation (Cernac and Benning 2004; Mu et al. 2008). The LEC1 encodes a HAP3 subunit of the CCAAT-binding factor and also plays an important role in embryo development. The WRI1 encodes an AP2/EREBP TF and its overexpression resulted in an increased TAG level in both seeds and leaves (Cernac and Benning 2004). In maize, a LEC1 functionally equivalent *ZmLEC1* (functionally complement the *Arabidopsis lec1* mutant phenotype) has been cloned and its overexpression increases seed oil by up to 48 % compared to control, but retarded seed germination and leaf growth (Shen et al. 2010). Furthermore, a maize *WRI1* (*ZmWRI1*) gene was cloned having 43 % identity at the amino acids level with *Arabidopsis* AtWRI1 (Shen et al. 2010). Overexpression of *ZmWRI1* resulted in increased oil content in seed similar to that reported for overexpression of *ZmLEC1* without affecting the germination, seedling growth, or grain yield (Shen et al. 2010). In *Arabidopsis*, expression analyses and genetic experiments indicate that WRI1 functions downstream of LEC1 (Maeo et al. 2009; Mu et al. 2008). However, Baud et al. (2007) showed that WRI1 is a direct target of LEC2, which specifies the regulatory action of WRI1 for seed maturation towards FA metabolism. ZmWRI1 showed an expression pattern similar to *ZmLEC1*, and was upregulated by approximately two-fold in *ZmLEC1*-expressing embryos. Moreover, the upregulation of *ZmWRI1* by *ZmLEC1* was confirmed by coexpression of the ZmLEC1 protein with a ZmWRI1 promoter: GUS reporter in *the* maize culture cell (Shen et al. 2010). Thus, ZmWRI1 also functions downstream of ZmLEC1, and *ZmLEC1* regulates the expression of *ZmWRI1* directly or indirectly. Recently, *in vitro* evidence showed that WRI1 can directly regulate a set of genes involved in FA biosynthesis, such as a subunit of pyruvate kinase (Pl-PKb1), ACCase (BCCP2), ACP (ACP1), and ketoacyl-ACP synthase (KAS1), by binding to two separate AW-box [CNTNG(N)7CG] motifs within the 200 bp fragment upstream of ATG (Maeo et al. 2009). Taken together, WRI1 is a master gene promoting carbon to oil during seed development, by directly *trans*-activating genes involved in FA biosynthesis and genes for assembly and storage of TAG.

Compared to the regulatory networks of SSP and oil biosynthesis in developing seed, the transcriptional regulation of storage starch synthesis remains poorly understood. The starch biosynthesis is a flexible but complex network in the cytosol and plastid. Sugar signaling regulates the transcription of many of the starch biosynthesis genes, including three critical steps: (i) sugar/signal sensing; (ii) signal transduction; and (iii) action at target genes. Sucrose is well known to be a general inducer of storage starch biosynthesis (Wobus and Weber 1999). Moreover, sucrose response elements (SURE), like *cis*-element in plant sugar signaling, are present in the promoters of maize *sbeI*, sorghum *sbeIIb*, and barley *sbeIIb* and *isa1* genes (Mutisya et al. 2006). The SUSIBA2 belongs to the WRKY proteins and was shown to bind to SURE and W-box elements in the barley *isa1* gene promoter, inducing *isa1* transcription (Sun et al. 2003). More recently, rice starch regulator1 (RSR1), an AP2/EREBP family TF, was found to negatively regulate the expression of type I (in seeds; sink tissues) starch synthesis genes, and is important for starch content and structure, starch granule morphology, and gelatinization properties, too (Fu and Xue 2010).

5.4 Concluding Remarks

The post-genomics era has provided us powerful tools and capabilities to study seed biology in-depth in a high-throughput manner. *Via* transcriptomics, we are able to understand when and where all genes in an organism are expressed. We have summarized the transcriptomics-driven progress of seed development in maize, focusing on the three major components of the kernels, including starch, oil, and protein. Genes involved in the three pathways are largely known with their expression patterns during seed development. Moreover, schematic transcriptional regulation roadmap for SSP, FA, and storage starch biosynthesis were also provided. Given that little is known about transcriptional regulation of the starch pathway, this field needs more efforts to put into and is predicted to be an exciting avenue in the future. Identification of master regulators, especially TFs, has a potential impact for a wide range of applications in cereal seeds. A dramatic increase in use of plant oils for the production of chemicals and bioenergy has drawn interest in improved seed FA accumulation and yield in cereals. One successful example is the overexpression of a key gene *ZmWRI1* regulating FA biosynthesis that increases seed oil by up to 46 % without affecting germination, seedling growth, or grain yield (Shen et al. 2010).

On the technology, RNA-Seq is a recently developed global technique for genome-wide transcript profiling (Marioni et al. 2008; Wang et al. 2009). In contrast to hybridization- or sequence-based approaches, RNA-Seq offers several key advantages and minor challenges. It enabled us to generate an unprecedented global view of transcripts expression and their organization for a number of species and cell types. With the aid of novel technologies and use of systems biology approach, growing information will eventually provide us details and inspirations to draw a coherent and delicate picture of the mechanisms underlying seed development.

Acknowledgments Authors acknowledge and thank the National Natural Sciences Foundation of China (31000747, 30671303, 30700472, and 30900900) for supporting the research.

References

Albani D, Hammond-Kosack MC, Smith C, Conlan S, Colot V, Holdsworth M, Bevan MW (1997) The wheat transcriptional activator SPA: a seed-specific bZIP protein that recognizes the GCN4-like motif in the bifactorial endosperm box of prolamin genes. Plant Cell 9:171–184

Alexandrov NN, Brover VV, Freidin S, Troukhan ME, Tatarinova TV, Zhang H, Swaller TJ, Lu YP, Bouck J, Flavell RB, Feldmann KA (2009) Insights into corn genes derived from large-scale cDNA sequencing. Plant Mol Biol 69:179–194

Baud S, Mendoza MS, To A, Harscoet E, Lepiniec L, Dubreucq B (2007) WRINKLED1 specifies the regulatory action of LEAFY COTYLEDON2 towards fatty acid metabolism during seed maturation in *Arabidopsis*. Plant J 50:825–838

Baud S, Dubreucq B, Miquel M, Rochat C, Lepiniec L (2008) Storage reserve accumulation in *Arabidopsis*: metabolic and developmental control of seed filling. The *Arabidopsis* Book, 1–24

Berger F (1999) Endosperm development. Curr Opin Plant Biol 2:28–32

Canas RA, Quillere I, Christ A, Hirel B (2009) Nitrogen metabolism in the developing ear of maize (*Zea mays*): analysis of two lines contrasting in their mode of nitrogen management. New Phytol 184:340–352

Carpita NC, McCann MC (2008) Maize and sorghum: genetic resources for bioenergy grasses. Trends Plant Sci 13:415–420

Cernac A, Benning C (2004) WRINKLED1 encodes an AP2/EREB domain protein involved in the control of storage compound biosynthesis in *Arabidopsis*. Plant J 40:575–585

Chaudhury AM, Craig S, Dennis E, Peacock W (1998) Ovule and embryo development, apomixis and fertilization. Curr Opin Plant Biol 1:26–31

Chaudhury AM, Koltunow A, Payne T, Luo M, Tucker MR, Dennis ES, Peacock WJ (2001) Control of early seed development. Annu Rev Cell Dev Biol 17:677–699

Conlan RS, Hammond-Kosack M, Bevan M (1999) Transcription activation mediated by the bZIP factor SPA on the endosperm box is modulated by ESBF-1 *in vitro*. Plant J 19:173–181

Curtis PE, Leng ER, Hageman RH (1968) Developmental changes in oil and fatty acid content of maize strains varying in oil content. Crop Sci 8:689–693

Diaz I, Vicente-Carbajosa J, Abraham Z, Martínez M, Isabel-La MI, Carbonero P (2002) The GAMYB protein from barley interacts with the DOF transcription factor BPBF and activates endosperm-specific genes during seed development. Plant J 29:453–464

Edgerton MD (2009) Increasing crop productivity to meet global needs for feed, food, and fuel. Plant Physiol 149:7–13

Fedoroff NV (2010) The past, present and future of crop genetic modification. Nat Biotechnol 27:461–465

Feng L, Zhu J, Wang G, Tang Y, Chen H, Jin W, Wang F, Mei B, Xu Z, Song R (2009) Expressional profiling study revealed unique expressional patterns and dramatic expressional divergence of maize alpha-zein super gene family. Plant Mol Biol 69:649–659

Fontanet P, Vicient CM (2008) Maize embryogenesis. Methods Mol Biol 427:17–29

Forde BG, Heyworth A, Pywell J, Kreis M (1985) Nucleotide sequence of a B1 hordein gene and the identification of possible upstream regulatory elements in endosperm storage protein genes from barley, wheat, and maize. Nucl Acids Res 13:7327–7339

Fu FF, Xue HW (2010) Coexpression analysis identifies rice starch regulator1, a rice AP2/EREBP family transcription factor, as a novel rice starch biosynthesis regulator. Plant Physiol 154:927–938

Glawischnig E, Gierl A, Tomas A, Bacher A, Eisenreich W (2002) Starch biosynthesis and intermediary metabolism in maize kernels. Quantitative analysis of metabolite flux by nuclear magnetic resonance. Plant Physiol 130:1717–1727

Goldberg RB, Barker SJ, Perez-Grau L (1989) Regulation of gene expression during plant embryogenesis. Cell 56:149–160

Grimanelli D, Perotti E, Ramirez J, Leblanc O (2005) Timing of the maternal-to-zygotic transition during early seed development in maize. Plant Cell 17:1061–1072

Guo M, Rupe MA, Danilevskaya ON, Yang X, Hu Z (2003) Genome-wide mRNA profiling reveals heterochronic allelic variation and a new imprinted gene in hybrid maize endosperm. Plant J 36:30–44

Heidecker G, Messing J (1986) Structural analysis of plant genes. Annu Rev Plant Physiol 37:439–446

James MG, Denyer K, Myers AM (2003) Starch synthesis in the cereal endosperm. Curr Opin Plant Biol 6:215–222

Kawakatsu T, Takaiwa F (2010) Cereal seed storage protein synthesis: fundamental processes for recombinant protein production in cereal grains. Plant Biotechnol J 8:939–953

Kiesselbach TA (1998) The structure and reproduction of corn. 50th anniversary edition. Cold Spring Harbor Laboratory, Cold Spring Harbor

Lai J, Dey N, Kim CS, Bharti AK, Rudd S, Mayer KF, Larkins BA, Becraft P, Messing J (2004) Characterization of the maize endosperm transcriptome and its comparison to the rice genome. Genome Res 14:1932–1937

Lee JM, Williams ME, Tingey SV, Rafalski JA (2002) DNA array profiling of gene expression changes during maize embryo development. Funct Integr Genomics 2:13–27

Lei XH, Shen X, Xu XQ, Bernstein HS (2000) Human Cdc5, a regulator of mitotic entry, can act as a site-specific DNA binding protein. J Cell Sci 113:4523–4531

Liu X, Fu J, Gu D, Liu W, Liu T, Peng Y, Wang J, Wang G (2008) Genome-wide analysis of gene expression profiles during the kernel development of maize (*Zea mays* L.). Genomics 91:378–387

Luo M, Liu J, Lee RD, Guo BZ (2008) Characterization of gene expression profiles in developing kernels of maize (*Zea mays*) inbred Tex6. Plant Breeding 127:569–578

Maeo K, Tokuda T, Ayame A, Mitsui N, Kawai T, Tsukagoshi H, Ishiguro S, Nakamura K (2009) An AP2-type transcription factor, WRINKLED1, of *Arabidopsis thaliana* binds to the AW-box sequence conserved among proximal upstream regions of genes involved in fatty acid synthesis. Plant J 60:476–487

Marioni JC, Mason CE, Mane SM, AStephens M, Gilad Y (2008) RNA-seq: an assessment of technical reproducibility and comparison with gene expression arrays. Genome Res 18:1509–1517

Méchin V, Balliau T, Château-Joubert S, Davanture M, Langella O, Négroni L, Prioul JL, Thévenot C, Zivy M, Damerval C (2004) A two-dimensional proteome map of maize endosperm. Phytochemistry 65:1609–1618

Méchin V, Thévenot C, Le Guilloux M, Prioul JL, Damerval C (2007) Developmental analysis of maize endosperm proteome suggests a pivotal role for pyruvate orthophosphate dikinase. Plant Physiol 143:1203–1219

Mena M, Vicente-Carbajosa J, Schmidt RJ, Carbonero P (1998) An endosperm-specific DOF protein from barley, highly conserved in wheat, binds to and activates transcription from prolamin-box of a native B-hordein promoter in barley endosperm. Plant J 16:53–62

Moreno-Risueno MA, Gonzalez N, Diaz I, Parcy F, Carbonero P, Vicente-Carbajosa J (2008) FUSCA3 from barley unveils a common transcriptional regulation of seed-specific genes between cereals and *Arabidopsis*. Plant J 53:882–894

Mu J, Tan H, Zheng Q, Fu F, Liang Y, Zhang J, Yang X, Wang T, Chong K, Wang XJ, Zuo J (2008) *LEAFY COTYLEDON1* is a key regulator of fatty acid biosynthesis in *Arabidopsis*. Plant Physiol 148:1042–1054

Mutisya J, Sun C, Palmqvist S, Baguma Y, Odhiambo B, Jansson C (2006) Transcriptional regulation of the *sbeIIb* genes in sorghum (*Sorghum bicolor*) and barley (*Hordeum vulgare*): Importance of the barley *sbeIIb* second intron. J Plant Physiol 163:770–780

Neuffer MG, Sheridan WF (1980) Defective kernel mutants of maize. I. Genetic and lethality studies. Genetics 95:929–944

Olsen OA (2001) Endosperm development: cellularization and cell fate specification. Annu Rev Plant Physiol Plant Mol Biol 52:233–267

Onodera Y, Suzuki A, Wu CY, Washida H, Takaiwa F (2001) A rice functional transcriptional activator, RISBZ1, responsible for endosperm-specific expression of storage protein genes through GCN4 motif. J Biol Chem 276:14139–14152

Oñate L, Vicente-Carbajosa J, Lara P, Díaz I, Carbonero P (1999) Barley BLZ2, a seed-specific bZIP protein that interacts with BLZ1 *in vivo* and activates transcription from the GCN4-like motif of B-hordein promoters in barley endosperm. J Biol Chem 274:9175–9182

Prioul JL, Méchin V, Lessard P, Thévenot C, Grimmer M, Chateau-Joubert S, Coates S, Hartings H, Kloiber-Maitz M, Murigneux A, Sarda X, Damerval C, Edwards KJ (2008) A joint transcriptomic, proteomic and metabolic analysis of maize endosperm development and starch filling. Plant Biotechnol J 6:855–869

Sabelli PA, Larkins BA (2009) The contribution of cell cycle regulation to endosperm development. Sex Plant Reprod 22:207–219

Schmidt RJ, Ketudat M, Aukerman MJ, Hoschek G (1992) Opaque-2 is a transcriptional activator that recognizes a specific target site in 22-kD zein genes. Plant Cell 4:689–700

Seebauer JR, Moose SP, Fabbri BJ, Crossland LD, Below FE (2004) Amino acid metabolism in maize earshoots. Implications for assimilate preconditioning and nitrogen signaling. Plant Physiol 136:4326–4334

Shen B, Allen WB, Zheng P, Li C, Glassman K, Ranch J, Nubel D, Tarczynski MC (2010) Expression of *ZmLEC1* and *ZmWRI1* increases seed oil production in maize. Plant Physiol 153:980–987

Soderlund C, Descour A, Kudrna D, Bomhoff M, Boyd L, Currie J, Angelova A, Collura K, Wissotski M, Ashley E, Morrow D, Fernandes J, Walbot V, Yu Y (2009) Sequencing, mapping, and analysis of 27,455 maize full-length cDNAs. PLoS Genet 5:e1000740

Sun C, Palmqvist S, Olsson H, Boren M, Ahlandsberg S, Jansson C (2003) A novel WRKY transcription factor, SUSIBA2, participates in sugar signaling in barley by binding to the sugar-responsive elements of the iso1 promoter. Plant Cell 15:2076–2092

Takaiwa F, Yamanouchi U, Yoshihara T, Washida H, Tanabe F, Kato A, Yamada K (1996) Characterization of common *cis*-regulatory elements responsible for the endosperm-specific expression of members of the rice glutelin multigene family. Plant Mol Biol 30:1207–1221

Verza NC, E Silva TR, Neto GC, Nogueira FT, Fisch PH, de Rosa VE Jr, Rebello MM, Vettore AL, da Silva FR, Arruda P (2005) Endosperm-preferred expression of maize genes as revealed by transcriptome-wide analysis of expressed sequence tags. Plant Mol Biol 59:363–374

Vicente-Carbajosa J, Moose SP, Parsons RL, Schmidt RJ (1997) A maize zinc-finger protein binds the prolamin box in zein gene promoters and interacts with the basic leucine zipper transcriptional activator Opaque2. Proc Natl Acad Sci U S A 94:7685–7690

Vodermaier HC (2004) APC/C and SCF: controlling each other and the cell cycle. Curr Biol 14:R787–R796

Wang GF, Wang H, Zhu J, Zhang J, Zhang XW, Wang F, Tang YP, Mei B, Xu ZK, Song RT (2010) An expression analysis of 57 transcription factors derived from ESTs of developing seeds in maize (*Zea mays*). Plant Cell Rep 29:545–559

Wang Z, Gerstein M, Snyder M (2009) RNA-Seq: a revolutionary tool for transcriptomics. Nat Rev Genet 10:57–63

Wobus U, Weber H (1999) Sugars as signal molecules in plant seed development. Biol Chem 380:937–944

Yamamoto MP, Onodera Y, Touno SM, Takaiwa F (2006) Synergism between RPBF Dof and RISBZ1 bZIP activators in the regulation of rice seed expression genes. Plant Physiol 141:1694–1707

Yan HB, Pan XX, Jiang HW, Wu GJ (2009) Comparison of the starch synthesis genes between maize and rice: copies, chromosome location and expression divergence. Theor Appl Genet 119:815–825

Chapter 6
Using Transcriptomics to Reveal Gene Networks of Seed Development in *Arabidopsis*

Richard C. Macknight, Rowan P. Herridge and Robert C. Day

Abstract To obtain a complete understanding of how a seed is made, it is important to identify and characterize the biological processes that occur in the different seed tissues at different stages of development. *Arabidopsis* is an important model plant amenable to the study of most aspects of plant biology; however, it offers considerable challenges when studying seed development due to the small size of the seeds. The development of new technologies, such as laser microdissection for isolating individual tissue types and RNA amplification methods that allow microarray analysis to be performed starting with nanogram quantities of RNA, has provided detailed transcriptome data. This chapter summarizes the transcriptome data that is available for *Arabidopsis* seeds and provides an example of how this data can be analyzed to identify gene networks involved in the different biological processes within a developing seed.

Keywords *Arabidopsis thaliana* · Gene networks · Interactome · Laser microdissection · Seed development · Transcriptome

6.1 Introduction

Seed development begins with double fertilization where the haploid egg cell and the double haploid central cell are both fertilized by identical haploid sperm cells contributed from a single pollen grain. This generates the diploid embryo and the triploid endosperm, respectively. The embryo and the endosperm grow rapidly in a coordinated manner and that is heavily influenced by the surrounding maternal integument tissues, which later form the seed coat (Garcia et al. 2005; Olsen 2004). Developing seed is therefore comprised of three genetically distinct compartments; the maternally diploid seed coat, the diploid embryo, and the endosperm. During early stages of seed development, tissues form and become established, using maternal resources mainly for rapid cell division and growth. This proliferative phase of endosperm development is fundamental for setting the final size of the mature seed (Scott et al. 1998). Once cell division slows, the seed enters a maturation phase

R. C. Macknight (✉) · R. P. Herridge · R. C. Day
Department of Biochemistry, University of Otago, Dunedin 9054, New Zealand
e-mail: richard.macknight@otago.ac.nz

G. K. Agrawal, R. Rakwal (eds.), *Seed Development: OMICS Technologies toward Improvement of Seed Quality and Crop Yield,*
DOI 10.1007/978-94-007-4749-4_6,

during which resources are reallocated for the synthesis of storage compounds, such as starch, followed by accumulation of oils and proteins. Biosynthetic activity then slows as the seed moves through a late-maturation phase and prepares to desiccate prior to dormancy (Baud et al. 2002; Fait et al. 2006; Gutierrez et al. 2007).

To obtain a complete understanding of the different developmental and biochemical processes that occur during seed development, it is important to know which genes are being expressed in the different seed tissues throughout development. This chapter provides an introduction to the transcriptomic data available for *Arabidopsis* seed development, and examples of how this data can be analyzed to identify groups of key genes and networks involved in particular processes.

6.2 Analyzing the *Arabidopsis* Seed Transcriptome Data

The Bio-Array Resource (BAR) for Plant Biology from the University of Toronto, Canada, is an excellent web-based resource for examining and analyzing the available *Arabidopsis* Affymetrix Genechip microarray data (http://bar.utoronto.ca/; Winter et al. 2007). *Arabidopsis* eFP (electronic Fluorescent Pictographic) browser provides a pictorial representation of the expression pattern of your gene-of-interest based on microarray data provided by the *Arabidopsis* research community.

Microarray data covering a time course of seed development, which consists of seeds at eight stages of development, has been provided by Schmid et al. (2005). For the early stages (globular to heart stages of embryo development), RNA was isolated from whole siliques, whereas at later stages of development the developing seeds were dissected out (Schmid et al. 2005). Figure 6.1 provides an example of the expression of a gene involved in the regulation of the cell cycle, *CDKB1;2* (*At2g38620*), that encodes a cyclin-dependent kinase (CDK). CDKs form heterodimers with cyclins and regulate the progression through the cell cycle (De Veylder et al. 2007; Inze and De Veylder 2006). There are 12 CDKs in *Arabidopsis* and the comprehensive data provided on the eFP browser provides a quick way of identifying which of these genes are expressed during seed development. As can be seen in Fig. 6.1a, CDKB1;2 is expressed at higher levels during the early stages of seed development. However, since the data is from whole siliques or seeds, it is not possible to tell which tissue compartments it is expressed in from this data.

Addressing where in the seed particular genes are expressed has required the adoption of the new technology of laser microdissection. Laser microdissection involves first fixing and embedding tissue of interest, which is then sectioned and placed on a microscope slide so that the desired cells can be identified. Laser microdissection is then used to either cut out these cells or specifically attach them to a membrane for subsequent analysis (Casson et al. 2008; Day 2010; Day et al. 2005, 2007a; Nelson et al. 2006). To discover the transcriptome of the dissected cells, RNA is isolated and, because of the small quantities of RNA recovered, an amplification step is usually used (Day et al. 2007b). This RNA can either be used to probe a microarray or increasingly to generate libraries for next generation sequencing (e.g., Hsieh et al. 2011; Jiao et al. 2009). The first studies to use laser microdissec-

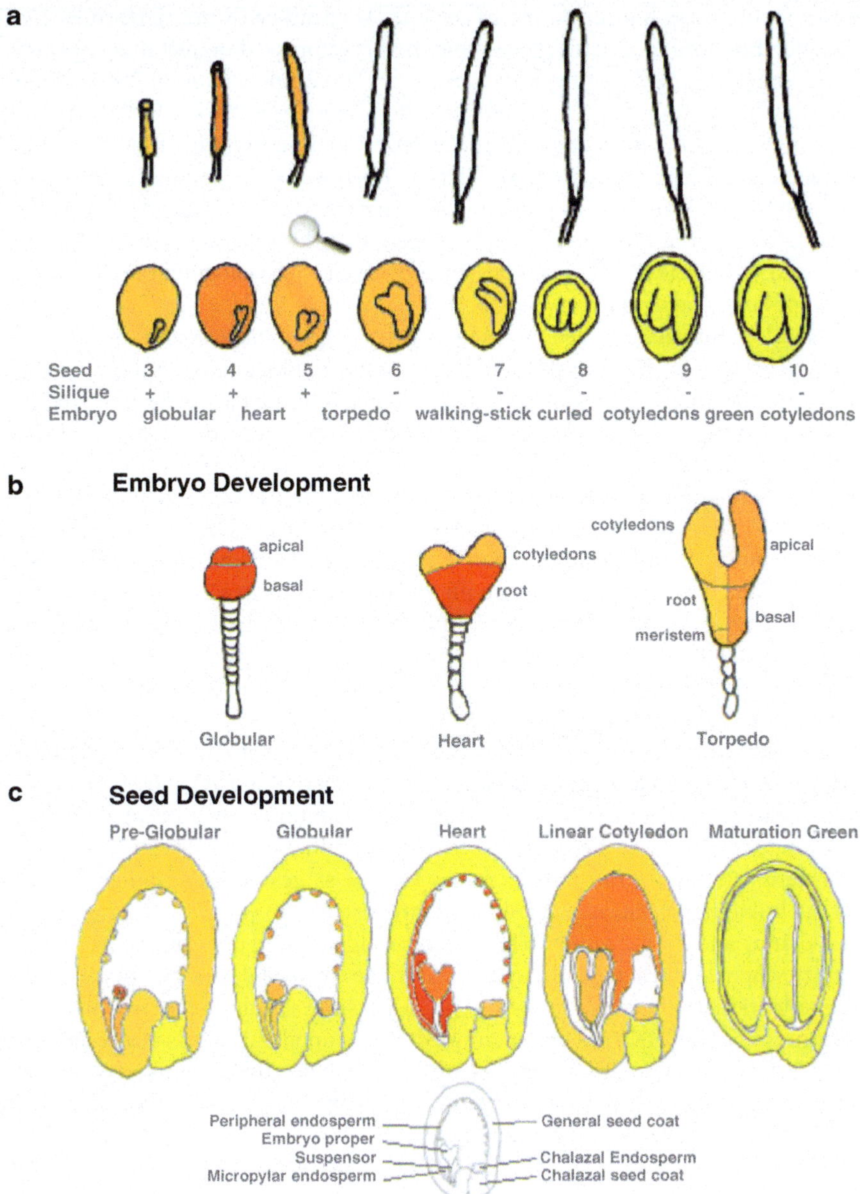

Fig. 6.1 Visualization of the seed microarray expression data. Pictorial representation of *CDKB1;2* (At2g38620) expression-based on the *Arabidopsis* affymetrix microarray data visualized using the eFP brower (http://bar.utoronto.ca/). **a** Expression data from a time course of seed development, with the early stages (globular to heart stages of embryo development) from whole siliques and later stages from dissected developing seeds (data from Schmid et al. 2005). **b** Expression data from different embryo regions laser-dissected from globular, heart, and torpedo stage embryos (data from Spencer et al. 2007). **c** Expression data from laser-microdissected seed tissue (general and chalazal seed coat; peripheral, micropylar, and charlazal endosperm; suspensor and embryo proper) at five stages of development (preglobular, globular, heart, linear cotyledon, and mature green). (Data from Goldberg/Harada laboratories; Le et al. 2010)

tion to identify the transcriptome of specific seed tissues were performed by the Lindsey laboratory to identify genes specifically expressed in different regions of a developing *Arabidopsis* embryo (Casson et al. 2005; Spencer et al. 2007). Embryo sacs containing globular, heart, and torpedo stage embryos were dissected from siliques, processed, and sectioned. Tissue from specific regions of the embryo was isolated using laser-capture microdissection. RNA was then isolated, amplified, and analyzed by DNA microarray using the Affymetrix ATH1 GeneChips (Casson et al. 2005; Spencer et al. 2007). This data reveals that our example gene *CDKB1;2* is expressed at high levels in globular stage of embryo development and in the root region of heart stage embryos (Fig. 6.1b).

A similar approach has been used to identify genes expressed during early endosperm development. Day et al. (2007b, 2008) used laser microdissection to isolate endosperm from developing seeds 4 DAP. Transcriptome data were obtained using two-color microarrays rather than Affymetrix Genechips and this data can be downloaded from the NCBI Gene Expression Omnibus (GSE:6703). Similar datasets are easily viewed and analyzed using freely available programs, such as the MultiExperiment Viewer (MeV) software package. In this data, *CDKB1;2* was identified as being preferentially expressed in the endosperm compared with whole siliques, and suggested that this gene, along with a number of other cell-cycle-related genes, might play an important role in controlling syncytial cell division of the endosperm (Day et al. 2008, 2009). This stage of endosperm development, where the endosperm proliferates rapidly through repeated rounds of mitosis without cytokinesis, is important in determining final seed size.

The most comprehensive analysis of the seed transcriptome, however, has been performed during collaboration between the Goldberg (UCLA) and Harada (UCD) laboratories (Le et al. 2010). Using laser microdissection, they isolated a range of seed tissues (general and chalazal seed coat; peripheral, micropylar, and chalazal endosperm; suspensor and embryo proper) at five stages of development (pre-globular, globular, heart, linear cotyledon, and mature green) (Fig. 6.1c). They also isolated ovules and 24 h zygotes. This outstanding resource provides a way to easily identify where and when genes are expressed during seed development (http://seed-genenetwork.net). The data is also easily accessible *via* the BAR web site. When the expression of *CDKB1;2* was examined, it was apparent that as suggested by the data from Day et al. (2008) and Casson et al. (2005), this gene is predominately expressed in the endosperm at heart stage with some additional expression in the embryo and possibly the seed coat (Fig. 6.1c).

6.3 Identifying Gene Networks

The large amount of available microarray data can also facilitate the identification of genes that are functionally related to your gene-of-interest. This can be particularly useful if your gene is one of the many whose function is yet to be determined.

Since genes encoding components of a particular process are often expressed in a coordinated manner, one strategy is to identify the genes that are coordinately expressed with a gene-of-interest. These genes can be identified using the Expression Angler function of the BAR website (Toufighi et al. 2005). To illustrate the power of this program, we looked for the genes that are coordinately regulated with the cell-cycle gene *CDKB1;2*, which was used as our example gene in Fig. 6.1. Consistent with its function in regulating the cell cycle, other cell-cycle genes were identified as being coordinately expressed with *CDKB1;2*, such as the related cyclin-dependent kinase, *CDKB2;2* and the two cyclins (*CYC3* and *CYCB1;4*). In addition, this analysis identified a number of cytoskeleton-associated genes, such as kinesins. Kinesins are motor proteins that are involved in transporting cargo along the microtubules and are required for the reorganization of the cell during cell division (Ambrose and Cyr 2007; Asada and Collings 1997; Chen et al. 2002). The cytoskeleton performs essential functions during the cell cycle, thus coordinated expression of cell cycle and cytoskeleton-associated genes makes biological sense (Day et al. 2009). Interestingly, the second gene on the list of coordinately expressed genes is *At5g33300*, whose function has not been established. This gene encodes a kinesin-related protein and our analysis supports a role in cell division/cytokinesis thus providing a novel candidate for further study.

6.4 Interactome

Proteins rarely work in isolation, rather they function in protein complexes, act in signaling pathways by modifying the properties of other proteins, or are involved in transport of other proteins within the cell. Therefore, to fully understand how a protein functions requires an understanding of what other proteins it interacts with, and the nature of these interactions. In addition, since 30 % of the *Arabidopsis* proteome has no known function, discovering the proteins they interact with can provide vital clues as to their function (Wigge and Weigel 2001). An interactome refers to all the protein interactions in an organism and a number of different methods are being employed to define the *Arabidopsis* interactome. Currently, there are about 6000 experimentally reported protein interactions in *Arabidopsis* (Lin et al. 2011). These have been shown using various techniques, such as Y2H, BiFC (Boruc et al. 2010), and tandem affinity purification (TAP) of protein complexes followed by identification using matrix-assisted laser desorption ionization (MALDI) MS (Van Leene et al. 2010). These methods all suffer from false positive and negatives to varying degrees. In addition to the experimentally determined interactions, ~1,50,000 interactions have been predicted based on the interactions of homologs from other species, in particular yeast, *Caenorhabditis elegans*, drosophila, and humans, where large-scale Y2H and affinity purification/MS studies have been undertaken (Formstecher et al. 2005; Ho et al. 2002; Li et al. 2004; Krogan et al. 2006; Rual et al. 2005; Stelzl et al. 2005). In addition to identifying the coexpressed genes,

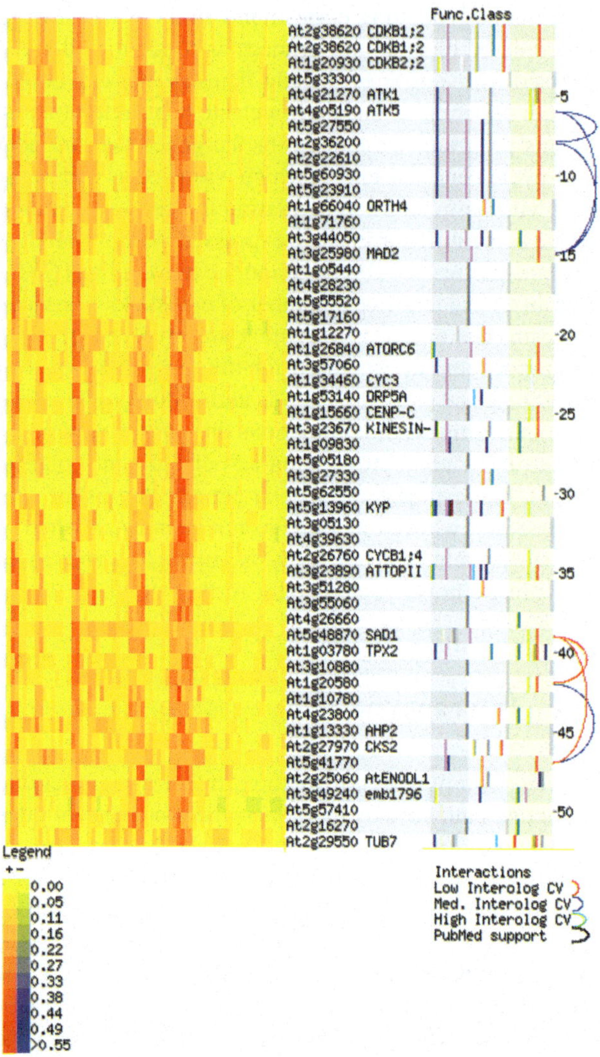

Fig. 6.2 Identification of coexpressed genes. Genes that have a similar expression pattern to CDKB1;2 in the *Arabidopsis* seed microarray data were identified using the Expression Angler function (http://bar.utoronto.ca/). Only expression in the Harada/Goldberg seed tissues is shown although all seed data were used in the analysis. *Yellow* indicates low expression and *red*, high expression. See the Expression Angler website for definition of the functional classes of the genes

the output from the BAR Expression Angler can be used to visualize known and/or potential interactions between the encoded proteins. For example, in Fig. 6.2 the lines connecting At4g05190/ATK5 with At2g36200 (a kinesin-5 protein) and At3g25980/MAD2 (a DNA binding protein) indicates that these proteins might interact with each other. Another group of interacting and coexpressed genes is also shown in Fig. 6.2, At1g20580 (a snRNP family protein), At5g48870/SAD1 (similar to Sm-like snRNP protein), and At5g41770 (also involved in RNA processing). The evidence that these proteins interact and that their genes are coexpressed provides support for them being involved in the same or related processes.

Another way to use the transcriptomic data to identify the network of genes involved in a particular process within a tissue is to combine this data with the interactome data. Since in any particular tissue or cell-type, predicted interactions cannot occur if the proteins are not present, combining the interactome data with the transcriptomic data provides an indication as to which of the possible interactions might actually occur (this assumes a correlation between transcript and protein levels, which is not always true). This is illustrated in Fig. 6.3, where the putative interactions between the core cell cycle proteins are shown along with the level of expression of the genes encoding these proteins in the chalazal endosperm (heart stage). From Fig. 6.3, it can be seen that while there are a large number of components that could be involved in regulating cell division of the chalazal endosperm, the majority of the core cell-cycle genes are not expressed. Another useful tool available is the "analyze" function, available on the Harada-Goldberg website (http://seedgenenetwork.net; accessed March 2011). This function enables users to identify genes expressed in the various seed tissues; certain criteria can be added, such as the level of expression and the absence of expression in particular stages/tissues. This type of analysis is a powerful way of identifying key molecular networks that are involved during important developmental stages in specific seed compartments.

6.5 Further Analysis of Candidate Genes

Despite the obvious power of high-throughput technologies, some caveats should be considered when examining these data sets; especially when the data has been generated from small amounts of starting tissue. RNA-amplification methods can amplify particular transcripts more or less efficiently than others, so where possible, comparisons should be limited to similarly amplified samples (Day et al. 2007b). Alternatively, technical artifacts and false positives can be identified using complimentary approaches with independent biological samples. Many microarray datasets have expression results validated by quantification of specific transcripts of interest by PCR. Another widely-used approach to confirm expression on a gene-by-gene basis is the use of reporter constructs (Fig. 6.4). These have the advantage of enabling expression analysis across the whole of the developmental program but require the generation transgenic plants. In some cases reporter constructs may also give erroneous results due to non-inclusion regulatory elements, such as those present within introns or 3' regions.

The identified candidate gene networks associated with particular tissues and stages of seed development may provide an excellent starting point for a rational approach for seed improvement. The excellent genetic resource available for *Arabidopsis*, such T-DNA mutant lines and simple methods for generating transgenic lines, means that *Arabidopsis* is the plant of choice for the characterization of gene networks involved in seed biology.

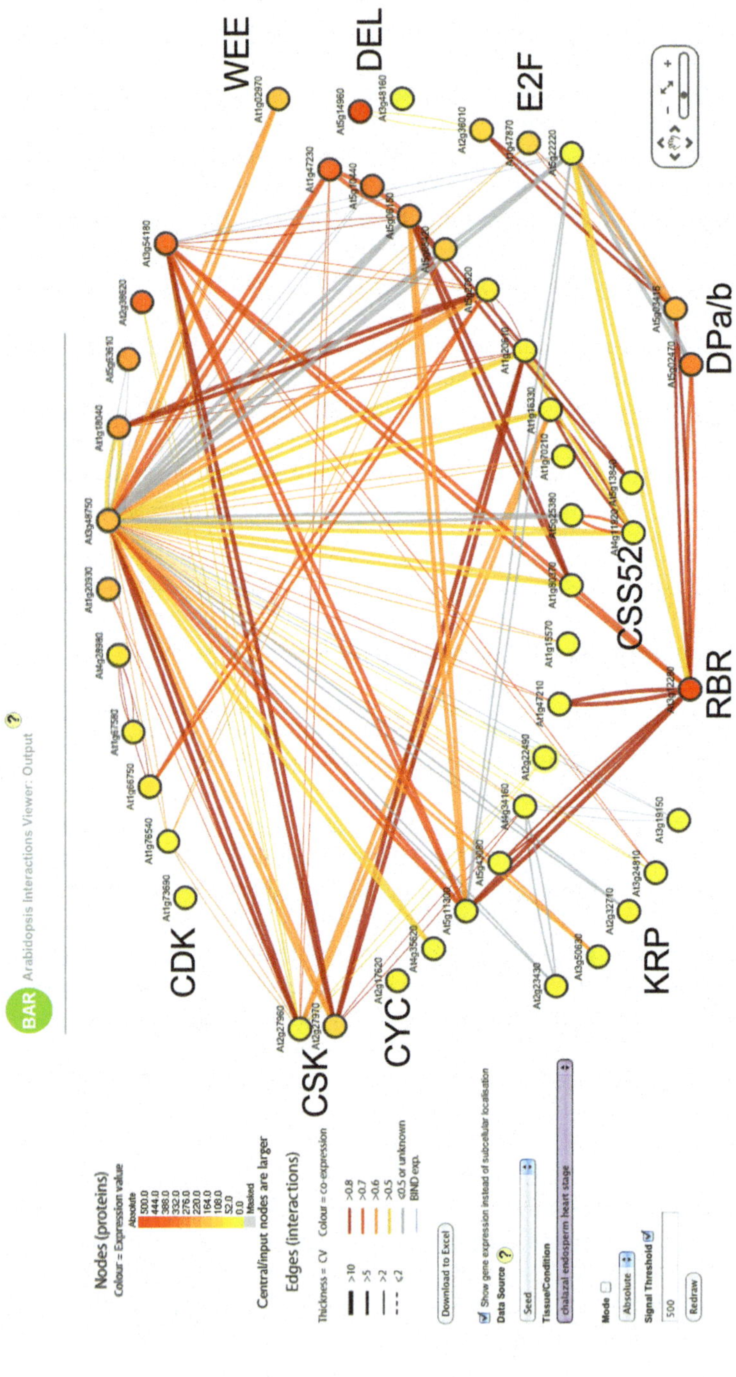

Fig. 6.3 Identification of the network of interacting proteins together and expression of the corresponding genes. Output of the *Arabidopsis* interactions viewer (http://bar.utoronto.ca/) showing the expression of the core-cell cycle genes in the chalazal endosperm at heart stage of seed development and the network of protein-protein interaction of the encoded proteins. The list of core cell cycle genes and interaction data can be found in Boruc et al. (2010)

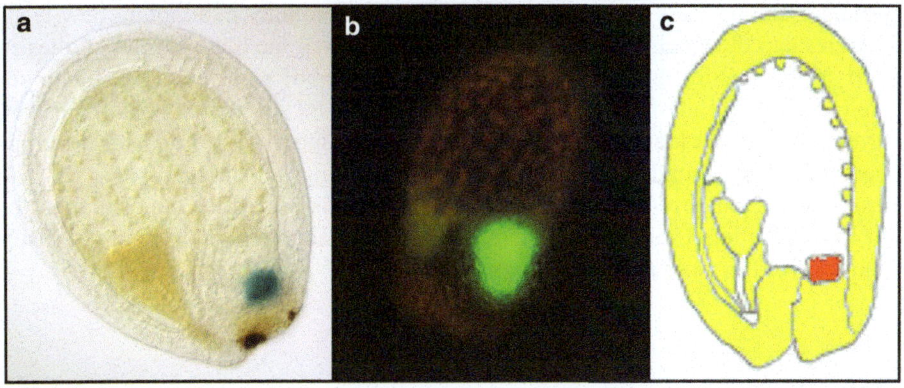

Fig. 6.4 Confirming tissue-specific expression patterns using reporter genes and tissue-specific microarray data. The gene At4g21080 was identified as being seed specific and preferentially expressed in the endosperm (Day et al. 2008). The result was confirmed using a promoter-GUS reporter construct **(a)**, and a promoter-GFP construct **(b)** in transgenic *Arabidopsis* plants, revealing expression specifically in the chalazal endosperm. **(c)** The chalazal-specific expression pattern of At4g21080 is supported by the recent tissue-specific transcriptome data of Goldberg/Harada visualized using the *Arabidopsis* eFP (http://bar.utoronto.ca/)

6.6 Advantages of Using Next-Generation Sequencing for Transcriptome Analysis

NGS is now a good alternative to microarrays for analyzing a transcriptome (Lister et al. 2009). Whilst still expensive, a sequencing approach has the advantage that analysis is not limited to those genes represented on an array and it is a more sensitive method for detecting genes expressed at low levels. Library construction for sequencing is also conducive to RNA amplification thus maintaining the ability to carry out tissue specific analysis. Furthermore, because the output is a sequence, it is also possible to track parent-of-origin based on sequence polymorphisms between parents (Hsieh et al. 2011). This ability is enabling new insight into the mechanism of parent-of-origin dependent regulation and allelic imbalance operating in endosperm, thought to be responsible for its role as a hybridization barrier and as a regulator of growth. Similar approaches should also facilitate research into the molecular determinants of hybrid vigor in the developing embryo/early seedlings.

6.7 Concluding Remarks

The extensive resources available for *Arabidopsis* make it a powerful tool to understand the basic molecular mechanisms underlying plant development. These process can then be targeted to improve seed quality and yield in crop plants. The ground-breaking work done by the *Arabidopsis* community in adopting new technologies and developing data processing pipelines will facilitate similar studies in crop systems. With the rapid growth in the number of plant genomes being sequenced, translation of research between different plant species is becoming easier. Orthologs of genes identified as being essential for aspects of *Arabidopsis* development can easily found in other systems thus enabling translational research. The ease with which *Arabidopsis* can be transformed also provides opportunities for the rapid functional study of candidate genes identified in crop plants but not easily characterized in more recalcitrant systems.

References

Ambrose JC, Cyr R (2007) The kinesin atk5 functions in early spindle assembly in *Arabidopsis*. Plant Cell 19:226–236

Asada T, Collings D (1997) Molecular motors in higher plants. Trends Plant Sci 2:29–37

Baud S, Boutin JP, Miquel M, Lepiniec L, Rochat C (2002) An integrated overview of seed development in *Arabidopsis thaliana* ecotype WS. Plant Physiol Biochem 40:151–160

Boruc J, Van den Daele H, Hollunder J, Rombauts S, Mylle E, Hilson P, Inze D, De Veylder L, Russinova E (2010) Functional modules in the *Arabidopsis* core cell cycle binary protein-protein interaction network. Plant Cell 22:1264–1280

Casson S, Spencer M, Walker K, Lindsey K (2005) Laser-capture microdissection for the analysis of gene expression during embryogenesis of *Arabidopsis*. Plant J 42:111–123

Casson SA, Spencer MW, Lindsey K (2008) Laser-capture microdissection to study global transcriptional changes during plant embryogenesis. Methods Mol Biol 427:111–120

Chen CB, Marcus A, Li WX, Hu Y, Calzada JPV, Grossniklaus U, Cyr RJ, Ma H (2002) The *Arabidopsiatk1* gene is required for spindle morphogenesis in male meiosis. Development 129:2401–2409

Day RC (2010) Laser microdissection of paraffin-embedded plant tissues for transcript profiling. In: Hennig L, Kohler C (eds) Laser microdissection of paraffin-embedded plant tissues for transcript profiling. Humana, Totowa, pp 321–346

Day R, Grossniklaus U, Macknight R (2005) Be more specific! Laser-assisted microdissection of plant cells. Trends Plant Sci 10:397–406

Day RC, McNoe LA, Macknight RC (2007a) Transcript analysis of laser microdissected plant cells. Physiol Plant 129:267–282

Day RC, McNoe L, Macknight RC (2007b) Evaluation of global RNA amplification and its use for high-throughput transcript analysis of laser-microdissected endosperm. Int J Plant Genomics 2007:1–18

Day RC, Herridge RP, Ambrose BA, Macknight RC (2008) Transcriptome analysis of proliferating *Arabidopsis* endosperm reveals biological implications for the control of syncytial division, cytokinin signaling, and gene expression regulation. Plant Physiol 148:1964–1984

Day RC, Müller S, Macknight RC (2009) Identification of cytoskeleton-associated genes expressed during *Arabidopsis* syncytial endosperm development. Plant Signal Behav 4:883–886

De Veylder L, Beeckman T, Inze D (2007) The ins and outs of the plant cell cycle. Nat Rev Mol Cell Biol 8:655–665

Fait A, Angelovici R, Less H, Ohad I, Urbanczyk-Wochniak E, Fernie AR, Galili G (2006) *Arabidopsis* seed development and germination is associated with temporally distinct metabolic switches. Plant Physiol 142:839–854

Formstecher E, Aresta S, Collura V, Hamburger A, Meil A, Trehin A, Reverdy C, Betin V, Maire S, Brun C, Jacq B, Arpin M, Bellaiche Y, Bellusci S, Benaroch P, Bornens M, Chanet R, Chavrier

P, Delattre O, Doye V, Fehon R, Faye G, Galli T, Girault JA, Goud B, de Gunzburg J, Johannes L, Junier MP, Mirouse V, Mukherjee A, Papadopoulo D, Perez F, Plessis A, Rosse C, Saule S, Stoppa-Lyonnet D, Vincent A, White M, Legrain P, Wojcik J, Camonis J, Daviet L (2005) Protein interaction mapping: a *Drosophila* case study. Genome Res 15:376–384

Garcia D, Fitz Gerald JN, Berger F (2005) Maternal control of integument cell elongation and zygotic control of endosperm growth are coordinated to determine seed size in *Arabidopsis*. Plant Cell 17:52–60

Gutierrez L, Van Wuytswinkel O, Castelain M, Bellini C (2007) Combined networks regulating seed maturation. Trends Plant Sci 12:294–300

Ho Y, Gruhler A, Heilbut A, Bader GD, Moore L, Adams SL, Millar A, Taylor P, Bennett K, Boutilier K, Yang LY, Wolting C, Donaldson I, Schandorff S, Shewnarane J, Vo M, Taggart J, Goudreault M, Muskat B, Alfarano C, Dewar D, Lin Z, Michalickova K, Willems AR, Sassi H, Nielsen PA, Rasmussen KJ, Andersen JR, Johansen LE, Hansen LH, Jespersen H, Podtelejnikov A, Nielsen E, Crawford J, Poulsen V, Sorensen BD, Matthiesen J, Hendrickson RC, Gleeson F, Pawson T, Moran MF, Durocher D, Mann M, Hogue CWV, Figeys D, Tyers M (2002) Systematic identification of protein complexes in *Saccharomyces cerevisiae* by mass spectrometry. Nature 415:180–183

Hsieh TF, Shin J, Uzawa R, Silva P, Cohen S, Bauer MJ, Hashimoto M, Kirkbride RC, Harada JJ, Zilberman D, Fischer RL (2011) Regulation of imprinted gene expression in *Arabidopsis* endosperm. Proc Natl Acad Sci U S A 108:1755–1762

Inze D, De Veylder L (2006) Cell cycle regulation in plant development. Annu Rev Genet 40:77–105

Jiao Y, Tausta SL, Gandotra N, Sun N, Liu T, Clay NK, Ceserani T, Chen M, Ma L, Holford M, Zhang HY, Zhao H, Deng XW, Nelson T (2009) A transcriptome atlas of rice cell types uncovers cellular, functional and developmental hierarchies. Nat Genet 41:258–263

Krogan NJ, Cagney G, Yu HY, Zhong GQ, Guo XH, Ignatchenko A, Li J, Pu SY, Datta N, Tikuisis AP, Punna T, Peregrin-Alvarez JM, Shales M, Zhang X, Davey M, Robinson MD, Paccanaro A, Bray JE, Sheung A, Beattie B, Richards DP, Canadien V, Lalev A, Mena F, Wong P, Starostine A, Canete MM, Vlasblom J, Wu S, Orsi C, Collins SR, Chandran S, Haw R, Rilstone JJ, Gandi K, Thompson NJ, Musso G, St Onge P, Ghanny S, Lam MHY, Butland G, Altaf-Ui AM, Kanaya S, Shilatifard A, O'Shea E, Weissman JS, Ingles CJ, Hughes TR, Parkinson J, Gerstein M, Wodak SJ, Emili A, Greenblatt JF (2006) Global landscape of protein complexes in the yeast *Saccharomyces cerevisiae*. Nature 440:637–643

Le BH, Cheng C, Bui AQ, Wagmaister JA, Henry KF, Pelletier J, Kwong L, Belmonte M, Kirkbride R, Horvath S, Drews GN, Fischer RL, Okamuro JK, Harada JJ, Goldberg RB (2010) Global analysis of gene activity during *Arabidopsis* seed development and identification of seed-specific transcription factors. Proc Natl Acad Sci U S A 107:8063–8070

Li SM, Armstrong CM, Bertin N, Ge H, Milstein S, Boxem M, Vidalain PO, Han JD, Chesneau A, Hao T, Goldberg DS, Li N, Martinez M, Rual JF, Lamesch P, Xu L, Tewari M, Wong SL, Zhang LV, Berriz GF, Jacotot L, Vaglio P, Reboul J, Hirozane-Kishikawa T, Li Q, Gabel HW, Elewa A, Baumgartner B, Rose DJ, Yu H, Bosak S, Sequerra R, Fraser A, Mango SE, Saxton WM, Strome S, van den Heuvel S, Piano F, Vandenhaute J, Sardet C, Gerstein M, Doucette-Stamm L, Gunsalus KC, Harper JW, Cusick ME, Roth FP, Hill DE, Vidal M (2004) A map of the interactome network of the metazoan C. elegans. Science 303:540–543

Lin MZ, Shen XL, Chen X (2011) Pair: the predicted *Arabidopsis* interactome resource. Nucl Acids Res 39:D1134–D1140

Lister R, Gregory BD, Ecker JR (2009) Next is now: new technologies for sequencing of genomes, transcriptomes, and beyond. Curr Opin Plant Biol 12:107–118

Nelson T, Tausta SL, Gandotra N, Liu T (2006) Laser microdissection of plant tissue: what you see is what you get. Annu Rev Plant Biol 57:181–201

Olsen OA (2004) Nuclear endosperm development in cereals and *Arabidopsis thaliana*. Plant Cell 16:S214–S227

Rual JF, Venkatesan K, Hao T, Hirozane-Kishikawa T, Dricot A, Li N, Berriz GF, Gibbons FD, Dreze M, Ayivi-Guedehoussou N, Klitgord N, Simon C, Boxem M, Milstein S, Rosenberg

J, Goldberg DS, Zhang LV, Wong SL, Franklin G, Li SM, Albala JS, Lim JH, Fraughton C, Llamosas E, Cevik S, Bex C, Lamesch P, Sikorski RS, Vandenhaute J, Zoghbi HY, Smolyar A, Bosak S, Sequerra R, Doucette-Stamm L, Cusick ME, Hill DE, Roth FP, Vidal M (2005) Towards a proteome-scale map of the human protein-protein interaction network. Nature 437:1173–1178

Schmid M, Davison TS, Henz SR, Pape UJ, Demar M, Vingron M, Scholkopf B, Weigel D, Lohmann JU (2005) A gene expression map of *Arabidopsis thaliana* development. Nat Genet 37:501–506

Scott RJ, Spielman M, Bailey J, Dickinson HG (1998) Parent-of-origin effects on seed development in *Arabidopsis thaliana.* Development 125:3329–3341

Spencer MWB, Casson SA, Lindsey K (2007) Transcriptional profiling of the *Arabidopsis* embryo. Plant Physiol 143:924–940 (Erratum in: Plant Physiol 143:1982)

Stelzl U, Worm U, Lalowski M, Haenig C, Brembeck FH, Goehler H, Stroedicke M, Zenkner M, Schoenherr A, Koeppen S, Timm J, Mintzlaff S, Abraham C, Bock N, Kietzmann S, Goedde A, Toksoz E, Droege A, Krobitsch S, Korn B, Birchmeier W, Lehrach H, Wanker EE (2005) A human protein-protein interaction network: a resource for annotating the proteome. Cell 122:957–968

Toufighi K, Brady SM, Austin R, Ly E, Provart NJ (2005) The botany array resource: e-northerns, expression angling, and promoter analyses. Plant J 43:153–163

Van Leene J, Hollunder J, Eeckhout D, Persiau G, Van de Slijke E, Stals H, Van Isterdael G, Verkest A, Neirynck S, Buffel Y, De Bodt S, Maere S, Laukens K, Pharazyn A, Ferreira PC, Eloy N, Renne C, Meyer C, Faure JD, Steinbrenner J, Beynon J, Larkin J, Van de Peer Y, Hilson P, Kuiper M, De Veylder L, Van Onckelen H, Inze D, Witters E, De Jaeger G (2010) Targeted interactomics reveals a complex core cell cycle machinery in *Arabidopsis thaliana.* Mol Syst Biol 6:397

Wigge PA, Weigel D (2001) *Arabidopsis* genome: life without notch. Current Biol 11:R112–R114

Winter D, Vinegar B, Nahal H, Ammar R, Wilson GV, Provart NJ (2007) An "electronic fluorescent pictograph" browser for exploring and analyzing large-scale biological data sets. PLoS One 2:e718

Chapter 7
The *Medicago truncatula* Gene Expression Atlas (MtGEA): A Tool for Legume Seed Biology and Biotechnology

Jerome Verdier, Vagner A. Benedito and Michael K. Udvardi

Abstract Legumes account for approximately one third of world primary crop production and are vital sources of protein, oil, carbohydrates, fiber, and minerals for humans, livestock, and industrial processing. Grain nutrient composition varies between legume species as well as between genotypes of the same species. The major metabolic pathways responsible for storage protein, lipid, and starch biosynthesis are well characterized in several plant species, although it remains unclear how partitioning between these pathways and their end products is regulated. Seed development is a complex process that involves coordinated expression and regulation of thousands of genes in different cell and tissue types. The *Medicago truncatula* Gene Expression Atlas (MtGEA) provides genome-wide expression data for all major organs of *M. truncatula*, including a rich time series for seed development. This chapter describes how the MtGEA provides a comprehensive view of the genetic and molecular processes of seed development. MtGEA is playing an instrumental role in identifying regulatory and metabolic genes potentially determining seed composition and seed quality. Therefore, MtGEA is a valuable resource for seed biotechnology.

Keywords Biotechnology · Legume · *Medicago* · Seed development · Gene expression · Transcriptome

7.1 Introduction

7.1.1 Legume Grains

Legumes (or pulses, the *Fabaceae* family) are the second most important crop family, after cereals. Legumes account for approximately one third of world primary crop production and are vital sources of protein, oil, carbohydrates, fiber,

M. K. Udvardi (✉) · J. Verdier
Plant Biology Division, The Samuel Roberts Noble Foundation, Ardmore, OK, USA
e-mail: mudvardi@noble.org

V. A. Benedito
West Virginia University, Morgantown, WV, USA

G. K. Agrawal, R. Rakwal (eds.), *Seed Development: OMICS Technologies toward Improvement of Seed Quality and Crop Yield*,
DOI 10.1007/978-94-007-4749-4_7,

and minerals for human nutrition, as well as livestock feed and industrial feedstock, including for biofuels. Food legumes of global importance include common bean (*Phaseolus vulgaris*), soybean (*Glycine max*), pea (*Pisum sativum*), chickpea (*Cicer arietinum*), broad (or fava) bean (*Vicia faba*), pigeon pea (*Cajanuscajan*), cowpea (*Vigna unguiculata*), lentils (*Lens esculenta*), lupins (*Lupinus* spp.), peanut (*Arachis hypogaea*), mung bean (*V. radiata*), and bambara groundnut (*V. subterranea*). Alfalfa (*Medicago sativa*), clovers (*Trifolium* spp.), vetches (*Vicia* spp.), *Leucaena* spp., and other legumes are important forage plants for grazing livestock. Apart from their use as food and feed, legumes are used in soil conservation, phytoremediation, lumber production, ornamental gardening, and for extraction of gums, resins, and food additives.

Worldwide, human nutrition is heavily reliant on seeds, especially as major sources of carbohydrates, lipids, and proteins. Thus, seed production is the major activity in agriculture that guarantees food supply. Nitrogen is often the most critical nutrient limiting crop yields and the most energetically and financially expensive fertilizer. Legumes play a key role in sustainable agriculture because they are capable of reducing atmospheric nitrogen (N_2) to ammonia (NH_3) *via* symbiotic associations with bacteria called rhizobia, which make them largely independent of fertilizer nitrogen and its contingent ecological and economic costs. Considering the projected worldwide population growth in the next decades, the demand for food production will increase by 70 % between now and 2050 (FAO 2009; van der Mensbrugghe et al. 2009). Therefore, the most critical challenge for agriculture is to maintain and improve global food security as population grows, despite climate change, near saturation of use of the available arable lands, land degradation, and the projected shortages of water, rock phosphate, and energy for nitrogen-fertilizer production, mechanized farming, and transportation (Beddington 2010).

Legume grains are of high value for human nutrition, given their high protein content, polyunsaturated oils, high fiber, high contents of amylose starch (Guillon and Champ 2002), minerals (especially iron), and a wide range of secondary compounds with health-promoting properties (i.e., tocopherol and isoflavones) and other flavonoids (Dixon and Sumner 2003; Grela and Günter 1995; Veitch 2007). Moreover, although the legume grains are low in the essential sulfur amino acid methionine, they are usually high in lysine compared to cereals. In developing countries, legume grains supply up to 80 % of the human dietary protein because of the high costs and low availability of animal protein.

Grain nutrient composition varies between legume species as well as between genotypes of the same species (le Signor et al. 2005). Seeds of soybean and peanut contain more oil than protein, i.e., peanut typically contains 50 % oil *versus* 25 % protein. Other legumes produce seeds with high protein and starch, but little oil. For example: (i) broad beans contain 29 % protein, 44 % starch, 21 % fiber, and only 2 % oil; (ii) seeds of yellow lupin consist of up to 48 % protein with little oil; and (iii) pea seeds are made up of 50 % starch, 25 % protein, 15 % fiber, and low lipids. Major goals of legume breeding and improvement programs include improvement of plant yield, increased resistance to pathogens, tolerance of abiotic stress, and improvement of the nutritional quality of seeds, as described below.

7.1.2 Desirable Traits in Legume Grains

In order to respond effectively to the growing world population, legume grain breeding programs aim to improve yield and nutritional quality of seeds. Legume grains are known for their high protein content but low methionine, which is an essential amino acid for humans. Transgenic approaches are available to introduce metabolic genes into legume genomes to enhance methionine production in seeds (Molvig et al. 1997). The transgenic approach is problematic, however, given current concerns over human food and commercial embargoes in many countries. Alternatively, nontransgenic approaches can also be pursued to improve seed quality, such as searching for natural genotypes with increased methionine content to be used in traditional breeding programs. In this case, a better understanding of the molecular and genetic basis of seed-storage metabolism will help to design breeding strategies aimed at seed-quality improvement.

Another problem generally associated with legume seeds is the presence of antinutritional compounds, such as phytate (phytic acid), a hexaphosphorylated myo-inositol compound. Phytate is thought to act as a seed reservoir of carbon and phosphate to be used during germination, and contains up to 80 % of the phosphate pool in seed (Raboy 1990). Phytate is formed during seed development and accumulates in single-membrane storage organelles (protein bodies) as insoluble crystals formed by complexation between phytate phosphate groups and minerals. Monogastric mammals, including humans, lack enzymes in their digestive tract to hydrolyze phytate, which consequently acts as an antinutrient by chelating minerals during digestion, making them unavailable to the ingesting organism. Phytate biosynthesis is not completely understood, although low-phytic acid (*lpa*) mutants have been isolated (reviewed by Raboy 2007). Interestingly, these are not especially compromised in germination (Campion et al. 2009), which makes low-phytate lines an attractive target of breeding for improved mineral availability in seed.

Although polyphenols (tannins) can have health benefits, by virtue of their antioxidant activity, they also have antinutrient effects in some legume seeds, as they chelate iron and proteins and make them less available (Ariza-Nieto et al. 2007; Hagerman and Butler 1981). On the other hand, polyphenols are positively correlated with resistance to biotic stress (for reviews see, Dixon et al. 2005; Dixon and Pasinetti 2010). Hence, lowering tannin levels in grains may lead to lower agronomical performance.

Some legume seeds have antisocial effects, i.e. they cause flatulence, triggered by galacto oligosaccharides present in the seed coat, such as raffinose, stachyose, and verbascose (McPhee et al. 2002). These sugars cannot be digested by monogastric animals, and ferment in the digestive tract, where they generate gas. Such sugars can account for up to 6 % of the grain fresh weight and, therefore, represent a significant nutritional loss to herbivores. From the point of view of the plant, however, galacto oligosaccharides contribute to drought tolerance and germination vigor in legume seeds that store carbohydrates as reserves. Nonetheless, removal of such sugars from seed seems a reasonable target for breeding to improve seed quality for human and animal consumption.

7.1.3 *Medicago truncatula* as a Model Legume

Crop legumes are generally poor model systems for basic genetic and genomic research: (i) some cultivated legumes are tetraploid (e.g. peanut and soybean); (ii) many have large genomes (e.g., pea and broad beans); and (iii) most are recalcitrant to genetic transformation or difficult to regenerate *in vitro* (e.g., common bean, pea, and soybean).

Medicago truncatula (henceforth, *Medicago*) is a model legume species for genetic and functional genomic research (Cannon et al. 2009; Young and Udvardi 2009). *Medicago* has a number of features that make it an excellent model for other legume species. First, as a member of the Papilionoideae (synonymous with Faboideae), the largest of the three Fabaceae subfamilies, *Medicago* is closely related to the majority of crop legumes (Gepts et al. 2005). Second, it is autogamous and produces a large number of seeds on a plant of relatively small stature, making it amenable to high-density cultivation (Barker et al. 1990). Third, *Medicago* has a relatively small, diploid genome (~550 Mbp) distributed over eight chromosomes, with most of its euchromatin recently sequenced (Young et al. 2005; Young and Udvardi 2009; http://www.Medicago.org/genome). Fourth, genetic transformation is possible, allowing reverse genetics for hypotheses testing (De Sousa Araújo et al. 2004; Chabaud et al. 2003). Since its adoption by the international community as a model species, a number of useful tools and resources have been developed for *Medicago*, including high-density genetic and physical maps for genetic research (Choi et al. 2004; Thoquet et al. 2002), several mutant populations based on EMS-induced point mutations (le Signor et al. 2009), fast-neutron bombardment deletions (Rogers et al. 2009), and a large population based on *Tnt1* retrotransposon insertions (Tadege et al. 2008). Tools and protocols are available for transcriptome, proteome, and metabolome analyses (Broeckling et al. 2005; Gallardo et al. 2003, 2007; Manthey et al. 2004; Watson et al. 2003). These resources are being applied to many research fields, including seed development and storage protein biosynthesis (Benedito et al. 2008; Djemel et al. 2005; Firnhaber et al. 2005; Gallardo et al. 2007; le Signor et al. 2005; Verdier et al. 2008).

Medicago has come to age as a research model given the near completion of its genome sequence (http://www.medicagohapmap.org; Young et al. 2011). Another addition to the functional genomics toolkit for *Medicago* is the Gene Expression Atlas (Benedito et al. 2008), which provides gene expression data for most of the *Medicago* genes in all major organs of the plant and developmental series (including seeds) (http://mtgea.noble.org/v2/).

This chapter highlights how the *Medicago truncatula* Gene Expression Atlas (MtGEA) can be used as a tool to increase our understanding of legume seed biology. Here, we will describe the data available in the MtGEA that relate to seed biology, including a detailed time-series through seed development. We also present some examples of how this data set can provide molecular insights into legume seed development and differentiation.

7.1.4 Seed Development

Seed development is initiated by double fertilization of the egg and central cell, resulting in development and differentiation of the major tissues of the seed: embryo, endosperm, and seed coat (Coelho and Benedito 2008; Goldberg et al. 1994). These tissues play different roles: (i) the seed coat is a maternal tissue that protects and transfers nutrients to the developing embryo; (ii) the endosperm is a triploid tissue derived from the central cell, and nourishes the embryo before being consumed by it in legume seeds; and (iii) the embryo (cotyledons, radicle, and hypocotyl) is diploid and derived from the fertilized egg (zygote) *via* multiple rounds of cell division and differentiation, and represents the new sporophytic generation. The completely developed legume seed consists of shoot and root meristems, and the two cotyledons, which store nutrients (including proteins, lipids, starch, and minerals) used eventually during germination and the first stages of seedling development. The major metabolic pathways responsible for storage protein, lipid, and starch biosynthesis are well characterized in the model plant *Arabidopsis thaliana* and other species, although it remains unclear how partitioning between these pathways and their end products is regulated (Santos-Mendoza et al. 2008). Differences between cereals (monocots) and legumes (dicots) also limit translation of our knowledge about seed biology from cereal models to legumes.

7.2 The *Medicago truncatula* Gene Expression Atlas (MtGEA)

A comprehensive gene expression atlas based on the Affymetrix GeneChip® *Medicago* Genome Array is publicly available for *Medicago* (Benedito et al. 2008; He et al. 2009). This array contains 61,101 probe sets, which includes 50,900 probe sets based on *Medicago* transcripts (ESTs) and gene models predicted from genome sequence, as well as 1,896 probe sets based on alfalfa (*M. sativa*) transcripts and 8,305 probe sets based on predicted gene models of the nitrogen-fixing symbiotic bacteria *Sinorhizobium meliloti*. Each of the Affymetrix probe sets consists of 11, 25-mer probes, designed to perfectly match the 3' region of target transcripts in addition to a mismatching probe for each perfectly-matching probe as negative control. This is a key feature that makes Affymetrix chips highly reliable with consistent results.

Because of the rapid growth of publically available *Medicago* affymetrix gene expression data from a great range of tissues, cell types, growth conditions, and stress treatments (Benedito et al. 2008), the legume research community needed an effective and accessible bioinformatics system to store and explore data for functional genomics. The MtGEA web server was developed for this purpose (He et al. 2009). This server is a centralized platform for analyzing the *Medicago* transcriptome. It enables flexible, multifaceted analyses of gene transcription data and provides a wealth of additional information about genes, including different types of gene annotation and links to the genome sequence, which helps users to formulate hypotheses about gene function. Transcript data can be accessed using Affymetrix probe set identification number, transcript sequence *via* BLAST search, keywords, functional description in natural language, GO & KEGG annotation terms, enzyme classification number, and conserved (InterPro) domain name or number. Efforts are being made to map the latest version of the *Medicago* gene models to each of the Affymetrix probe sets.

MtGEA is updated on a regular basis to incorporate new gene expression data and genome annotation, and is accessible at http://mtgea.noble.org/v2. The first version of this web server was launched in 2008 and was composed by 60 microarrays corresponding to 20 different experiments related to organs of mature plants grown under optimal conditions, and two developmental series (root nodules and seeds). Currently, the web server is in its second version and hosts gene expression data derived from 156 microarrays of 64 distinct experiments, covering a broad range of experimental conditions in different genetic backgrounds (i.e., ecotypes, mutants, and transformed lines). A third version of the gene atlas is under development and will be launched with an increased number of experiments, improved analysis, and visualization tools.

As far as seed development in *Medicago* is concerned, MtGEA contains data from six stages, ranging from embryogenesis to physiological maturity of the seeds, in addition to data from seed coats, whole seed pods (a pool of developmental stages), and transgenic seeds.

7.3 The MtGEA and Seed Development

Seed development is a highly complex process, which involves coordinated expression and regulation of thousands of genes in different cell and tissue types. It is commonly divided into three major phases according to changes in seed weight (Fig. 7.1). The first phase of development is embryogenesis, which involves many cell divisions but little gain in seed mass. This phase is conserved among angiosperms and starts with the fertilization and the division of the zygote, which establishes the embryo polarity with the basal cell developing into the suspensor and the apical cell originating the rest of the embryo. Embryogenesis continues with cell divisions and differentiation through several stages (globular, heart, torpedo, and then bent cotyledon). Most of the increase in seed mass occurs during the second phase, loosely called seed filling. This gain is due to the accumulation of compounds stored in the cotyledons of the embryo, especially seed storage proteins (SSPs) in the case of *Medicago*. The final desiccation phase begins at physiological maturity, when seed weight starts to decrease due to water loss. At the end of this phase, seeds enter into dormancy until germination.

The MtGEA offers comprehensive and quantitative data on changes in gene expression during the aforementioned major phases of seed development, and hence allows gene expression comparisons to be made between seeds and other organs. Although a change in steady-state mRNA level does not always correspond to a

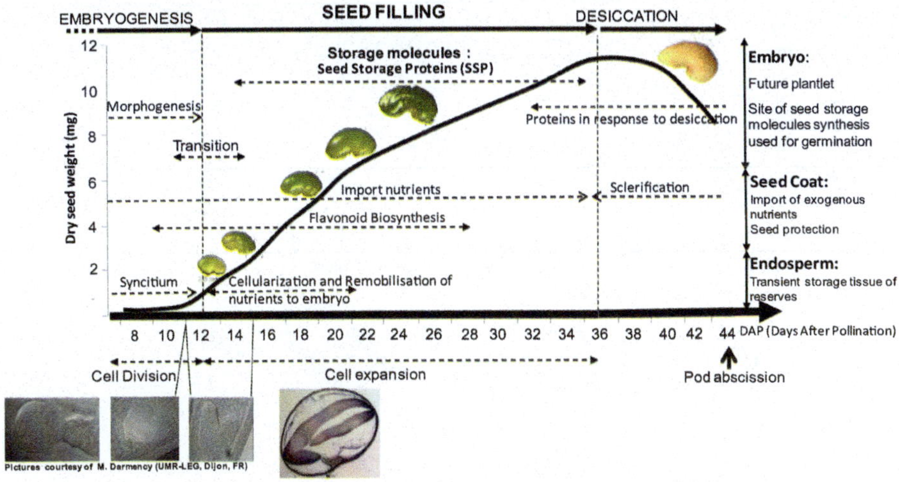

Fig. 7.1 Time course of major cellular and molecular events occurring during *Medicago* seed development. For details, refer to the text

change in protein synthesis and protein activity, transcription is a primary level of regulation in any organism and changes in transcript level generally indicate altered biological activity.

Six important stages of seed development were chosen for MtGEA, on the basis of proteomic and transcriptomic data (Gallardo et al. 2003, 2007), to determine the expression pattern of *Medicago* genes and to identify genetic mechanisms regulating seed development and maturation (Benedito et al. 2008). The samples: (i) 10 DAP corresponds to late embryogenesis; (ii) 12 DAP marks the transition from embryogenesis to seed filling, characterized by cessation of cell divisions of embryogenesis and preceding the synthesis of storage compounds; (iii) 16, 20, and 24 DAP correspond to stages of seed filling with active SSP synthesis, especially the globulins 7S (vicilins and convicilins) and 11S (legumins); and finally (iv) 36 DAP corresponds to physiological maturity and entry into desiccation phase.

Transcripts of approximately 39,000 genes, corresponding to 39,192 probe sets were detected in one or more of the seed samples. By comparison, transcripts of about 30,000 genes were detected in leaves. More than half of the seed-expressed genes were active at all stages of seed development. These results underscore the generally-accepted statement that seed development is a complex process, involving the large majority of genes and many metabolic and ontological pathways.

Comparative analysis of gene expression in seeds and other organs revealed that more than 5 % of genes are specifically or preferentially expressed in seed with transcript level ≥ 10-fold higher in seeds than any non-seed sample (Fig. 7.2). Among the seed-enhanced genes, we identified genes encoding proteins known to be involved in seed maturation, such as SSPs (e.g., vicilin gene, *Mtr.26812.1.S1_at*; legumin A gene, *Mtr.26812.1.S1_at*; legumin J gene, *Mtr.8458.1.S1_at*), and desiccation (e.g., LEA, *Mtr.2930.1.S1_at* and *Mtr.5956.1.S1_at*).

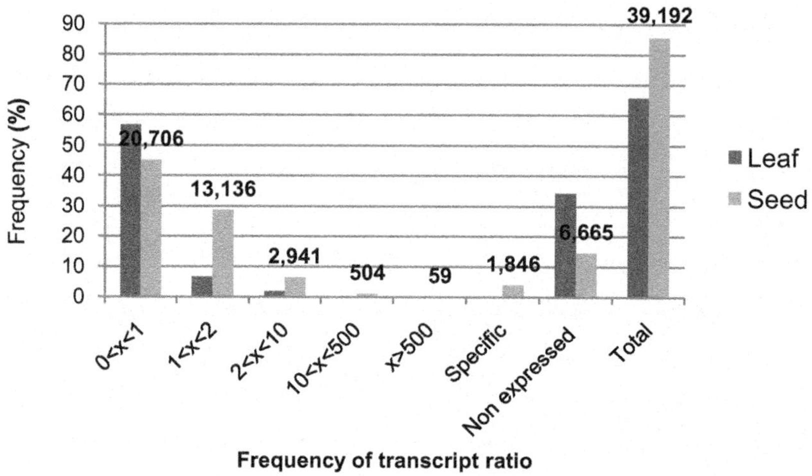

Fig. 7.2 Comparative analysis of gene expression in seeds or leaves *versus* other plant organs. Transcript ratios were calculated by dividing RNA level in seed or leaf by that of the organ with the next highest transcript level. The number of genes expressed in seeds is indicated for each category

7.3.1 Framework of Medicago Seed Development

Using one-way ANOVA, a total of 15,786 genes were found to be differentially-expressed during seed maturation (p-value < 0.01). The K-means cluster analysis of expression data derived from the six developmental time points revealed four major clusters (Fig. 7.3). The first cluster (Cluster I) contains 5,309 genes displaying a peak of expression at 10 DAP, associating them with late embryogenesis. Clusters II and III consist of genes associated with seed filling. Cluster II includes 2,545 genes peaking at 16 DAP. This cluster may correspond to genes related to the onset of expression of SSP (i.e., 14 DAP for vicilin and 16 DAP for legumin). Cluster III includes 2,620 genes with expression maxima at 20 or 24 DAP, corresponding to the expression climax of SSPs (Verdier et al. 2008). Finally, cluster IV contains 5,312 genes with the maximal expression levels at 36 DAP, implicating them in processes associated with desiccation. To obtain a global picture of the metabolic and other processes involved in different stages of seed development, we used GeneBins (Goffard and Weiller 2007a) to assign a functional class to each seed-regulated gene according to the KEGG ontology (http://bioinfoserver.rsbs.anu.edu.au/utils/GeneBins). This system is organized into six modules, four of them applicable to plants: (i) metabolic pathways; (ii) genetic information processing; (iii) environmental information processing; and (iv) cellular processes (Kanehisa et al. 2004). Although the majority of seed-regulated genes fell into a fifth class, "unclassified with or without homolog" bin, 42.4 % of the genes were assigned into functional bins (p-value ≤ 0.05 with Bonferroni correction).

More than 45 % of functionally-classified seed-regulated genes were assigned to metabolic pathways, including carbohydrate, amino acid, lipid, energy, and second-

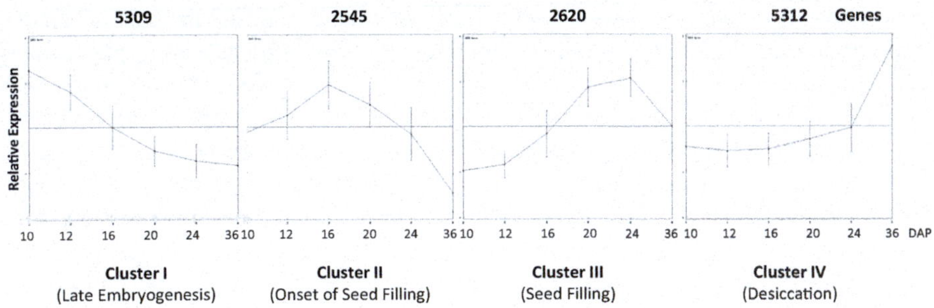

Fig. 7.3 *K*-means clustering of 15,000 genes differentially expressed during seed development

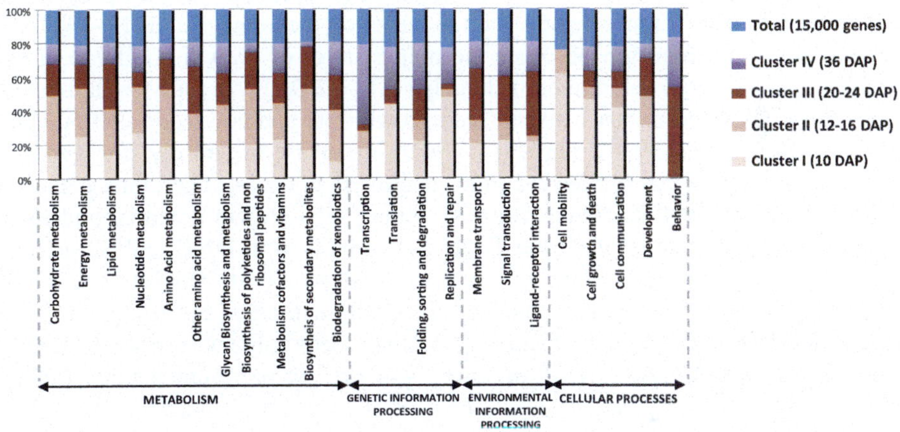

Fig. 7.4 Functional classifications of 15,000 differentially-expressed seed genes, divided by cluster (refer to Fig. 7.3). Classification was performed according to the KEGG ontology

ary metabolisms. Many metabolic genes were most active during the seed-filling phase of development (Clusters II and III), as might be expected (Fig. 7.4). In contrast, genes involved in genetic information processing were generally most active during embryogenesis (Cluster I) and desiccation (Cluster IV), presumably reflecting greater rates of cellular differentiation during these phases of development. Finally, genes involved in environmental information and cellular processes were more active at 10 DAP (Cluster I) and 20–24 DAP (Cluster III), possibly reflecting their supporting roles in cellular differentiation and partitioning of storage products.

7.3.2 Embryo Development

Embryogenesis The embryo is derived from the fertilized egg or zygote *via* multiple rounds of cell division and differentiation and represents the new sporophytic generation. It consists of a shoot and a root meristem and, in the case of dicots, two cotyledons that store nutrients, including proteins, lipids, starch, and minerals, required for seedling growth and germination. Embryogenesis is characterized by embryo histodifferentiation through the globular, heart, torpedo, then finally, bent cotyledon stages. During this period, the developing embryo acquires polarity along the shoot-root axis and meristems are established. Using the MtGEA data, we identified a cluster (Cluster I) made up of genes with maximal expression at 10 DAP, which corresponds to late embryogenesis. In this cluster, we observed that the major overrepresented functional classes are related to cell growth, replication/repair, and cell mobility (Fig. 7.4). All these processes are associated with intense cell division activity.

According to our observations and those of Wang and Grusak (2005) at 10 DAP stage, the embryo is at the heart/torpedo stage and occupies only a small proportion of the seed volume. Intense cell division in the embryo continues until 12–14 DAP, after which cell expansion becomes the major driver of growth (Gallardo et al. 2006a). Thus, embryo cell number is largely set by 12 DAP. A positive relationship has been established in pea between seed growth rate and cotyledon cell number (Munier-Jolain et al. 1998). Thus, it is during the phase of cell division and before the seed filling that the seed acquires its growth potential. Therefore, genes of cluster I are potential candidates for determining seed growth potential and final size. Amongst the genes of cluster I are many involved in DNA replication and repair, including DNA polymerases (e.g., *Mtr.1014.1.S1_at*, *Mtr.10966.1.S1_at*, and *Mtr.13328.1.S1_at*). Genes that are maximally-expressed at 10 DAP encode nucleotide biosynthesis enzymes, which produce deoxyribonucleotides (e.g., adenosine kinases, *Mtr.30425.1.S1_at*, *Mtr.17483.1.S1_at* or ribonucleotide reductases, *Mtr.45868.1.S1_s_at*, *Mtr.11322.1.S1_at*) for replication and energy metabolism. Finally, the cell mobility class encompasses genes (e.g., *Mtr.1106.1.S1_at*, *Mtr.10599.1.S1_at*) that regulate the actin cytoskeleton and determine cell shape.

Seed-Filling Phase During the filling phase, the seed grows and accumulates storage macromolecules, including carbohydrates, lipids, and proteins. Using MtGEA, we identified two distinct clusters of genes with maximal expression during this phase. Cluster II consists of genes with maximal expression at 16 DAP, which corresponds to early filling. Cluster III is composed of genes expressed maximally at late filling from 20 through 24 DAP. Using PathExpress (http://bioinfoserver.rsbs.anu.edu.au/utils/PathExpress; Goffard and Weiller 2007b), we identified four major metabolic pathways (carbohydrate, amino acid, lipid, and secondary metabolite pathways) that are regulated during these stages [False Discovery Rate (FDR)-adjusted p-value <0.1] and preferentially associated with cluster II (16 DAP) (Fig. 7.4).

Carbohydrate metabolism is central to early seed development. Although mature seeds of *Medicago* are composed of only 6–10 % soluble sugars and < 1 % starch (Djemel et al. 2005), carbohydrates represent a major carbon and energy source during development. Sucrose plays a dual role in the cell; it is central in carbohydrate metabolism and acts as a signal molecule. The balance between sucrose and hexoses regulates the direction that metabolism takes (anabolism or catabolism;

reviewed by Koch 2004; Smeekens 2000; Weber et al. 1997). For instance, the hexose/sucrose ratio is high during embryogenesis but low during seed filling, when high sucrose acts as a signal to coordinate synthesis of SSPs. In our data set, we observed a decrease of expression of SSP genes from 12 DAP to the end of seed filling (e.g., *Mtr.22018.1.S1_s_at*).

Upregulation of genes involved in amino acid metabolism during seed filling, like many amino transferase genes (e.g., *Mtr.37617.1.S1_at, Mtr.24739.1.S1_at, Mtr.37672.1.S1_at, Mtr.46230.1.S1_at,* and *Mtr.37617.1.S1_at*), is consistent with intensive amino acid synthesis at this stage, which is essential for the production of SSPs. SSPs accumulate mainly in the embryo and represent >30 % of *Medicago* seed DW (Djemel et al. 2005). SSPs belong mostly to the globulin family and consist of three major types: vicilins, legumins, and convicilins. They accumulate in a sequential fashion during seed development. This pattern is different in legumes than in *Arabidopsis* (Gallardo et al. 2003; Gatehouse et al. 1982, 1986; Walling et al. 1986). In *Medicago*, vicilin genes are first expressed at 14 DAP, followed by legumins at 16 DAP. The maximal expression occurs at 20 DAP for vicilin, 24 DAP for legumin K, and 36 DAP for legumin A (Verdier et al. 2008). MtGEA confirms this differential expression between vicilin (e.g., *Mtr.26812.1.S1_at* and *Mtr.19932.1.S1_s_at*) and legumin (e.g., *Mtr.37270.1.S1_at* and *Mtr.31203.1.S1_s_at*) and reveals the convicilin expression pattern (e.g. *Mtr.37289.1.S1_at* and *Mtr.37290.1.S1_at*) starting at 16 DAP with the maximal expression at 20–24 DAP close to that of vicilin. Expression patterns of these genes are consistent with the kinetics of protein accumulation in *Medicago* (Gallardo et al. 2007).

Cluster II also has an over representation of genes involved in lipid metabolism, especially those associated with the fatty acids (FAs) biosynthesis. At maturity, 8–10 % of *Medicago* seed DW is accounted for by FAs. Synthesis of predominantly C16:0, C18:1, C18:2, and C18:3 FAs is boosted early during seed filling *via* induction of the genes, such as FA ligases (e.g., FA ligase gene; *Mtr.46462.1.S1_at, Mtr.31183.1.S1_at,* and *Mtr.32721.1.S1_at*) or acyl carrier proteins (e.g., *Mtr.12210.1.S1_at, Mtr.35792.1.S1_at,* and *Mtr.41591.1.S1_at*).

Secondary metabolism is also over represented at early seed filling (i.e., Cluster II). Several enzymes related to flavonoid biosynthesis were identified, including chalcone synthases (e.g., *Mtr.42237.1.S1_at*), chalcone isomerases (e.g., *Mtr.40331.1.S1_at*) or flavonol synthases (e.g., *Mtr.45897.1.S1_at*). Although some enzymes may drive isoflavone accumulation in the embryo (Dhaubhadel et al. 2003), we assume that some of the genes in this category are related to the accumulation of (pro)anthocyanins in the integument (see below the subsection, The Seed Coat).

The entire seed-filling period (Clusters II and III) was characterized by a dearth of gene induction related to genetic information processing and cellular processes, including DNA replication, transcription, translation, cell growth, and cell communication (Fig. 7.4). Lack of genes involved in DNA replication is consistent with observations that cell division largely ceases prior to seed filling. Lack of transcriptional regulators amongst clusters II and III genes suggests that TFs required for transcription of genes with maximal expression during seed filling are synthesized prior to the filling phase, or at least maximal expression of their genes precedes filling. This notion is consistent with the identification of a large pool of ribosomal proteins at 12 DAP in *Medicago* seeds (Repetto et al. 2008), confirming that the transition between embryogenesis and seed filling is a key period of regulatory activity.

Genes with activity peaking at late seed filling (20–24 DAP) are grouped in cluster III. In addition to genes encoding SSPs known to accumulate during late seed filling (e.g., legumin genes; *Mtr.37270.1.S1_at* and *Mtr.31203.1.S1_s_at*), this cluster also contains genes involved in the processing of cell environment information, including several hundred kinases and membrane transporters. Active cell signaling may be required to coordinate molecule storage in specialized organelles or bodies, and is undoubtedly required to prepare the seed for desiccation.

Desiccation Phase The last phase of seed development prepares the seed for desiccation. This stage is characterized by water loss and entrance into dormancy around 40 DAP, and is coordinated with pod abscission. Cluster IV is made up of genes with peak expression at 36 DAP, making them desiccation-related genes, including genes known to be involved in this process (e.g., seed maturation proteins or LEA genes; *Mtr.43533.1.S1_at* and *Mtr.12942.1.S1_at*). Relatively few metabolic genes are in cluster IV, consistent with entry of the seed into metabolic quiescence or physiological maturity. The MtGEA data also reveals activation of transcriptional machinery at this stage, including expression of many RNA polymerases (e.g., *Mtr.1331.1.S1_at*, *Mtr.22814.1.S1_at*, and *Mtr.38548.1.S1_at*), transcription initiation factors (e.g., *Mtr.28076.1.S1_at*, *Mtr.38039.1.S1_at*, and *Mtr.32068.1.S1_at*) and TF (e.g., bHLH-type, *Mtr.11605.1.S1_s_at*; AP2-type, *Mtr.23282.1.S1_s_at*; and NAC-type, *Mtr.42696.1.S1_at*) genes. This machinery may account for accumulation of mRNA in mature seeds, which presumably facilitates rapid protein synthesis during germination (Rajjou et al. 2004; Rushton and Bray 1987).

7.3.3 Seed Coat and Endosperm Development

The MtGEA contains little data at present that are specifically related to the seed coat and endosperm. We present a brief description of these tissues as a framework for future tissue-specific transcriptome analysis, which will be incorporated into the expression atlas.

Seed Coat The seed coat (also called testa) is a diploid maternal tissue surrounding the developing embryo, which helps to nourish and ultimately protect the embryo from the external environment (Baskin and Baskin 1998). A comprehensive morphological description of seed coat development in *Medicago* was published by Wang and Grusak (2005). During seed development, the seed coat is mainly involved in transport and synthesis of nutrients, such as amino acids and flavonoids. At maturity, the dry seed coat provides a physical barrier, contributing to desiccation tolerance and seed dormancy in *Medicago* (Faria et al. 2005).

One of the most studied processes in the *Medicago* seed coat is biosynthesis of condensed tannins. Genes encoding some of the enzymes, TFs, and transporters involved in this pathway have been characterized functionally (Pang et al. 2007, 2008; Zhao and Dixon 2009; Verdier et al. 2012), where the MtGEA has played a fundamental role. The strong involvement of secondary metabolism in early filling (Cluster II), especially for flavonoid biosynthesis, is coherent with pigment biosynthesis in *Medicago* seed coat. Some of these transcripts are related to the accumulation of tannins in the integument (Pang et al. 2008).

Another major process occurring in seed coat is transportation of nutrients from maternal tissues to the embryo. Seed coat is involved in the metabolism of nutrients, such as amino acids biosynthesis (van Dongen et al. 2003), and in transient storage during the early seed development of minerals (Barnabas and Arnott 1990). Seed coat is also a transient storage tissue for unloaded sucrose from the phloem; the sucrose is transported to the endosperm and then on to the embryo after 13 DAP at the onset of seed filling (Djemel et al. 2005; Gallardo et al. 2006b). The seed coat probably determines the concentration of sucrose in the embryo *via* invertases, which affects the hexose/sucrose ratio and in turn controls the transition between embryogenesis and seed filling (for review see, Weber et al. 2005).

Endosperm The endosperm is a triploid embryo-nourishing tissue that arises from the triple fusion of a sperm nucleus with two polar nuclei. In most cereals, this tissue develops into a large portion of the mature seed and functions as a storage tissue. In most dicots (including *Arabidopsis* and *Medicago*), the endosperm plays a nourishing role during embryo formation, but it is consumed at later stages of seed development when cotyledons take over the role of storing reserves in mature seeds. Endosperm development in *Medicago* is still poorly characterized, but in *Arabidopsis*, Berger (2003) and Berger et al. (2006) have provided comprehensive descriptions of endosperm development. The endosperm development can be divided in two phases: a syncitial phase marked by intense synchronous division without cellularization leading to massive endoduplication coupled with storage of reserves in large cells; followed by a phase of cellularization that leads to differentiation and acquisition of endosperm polarity. According to microscopic observations from Wang and Grusak (2005), cellularization of *Medicago* endosperm occurs around 12 DAP, which is consistent with the high-ploidy level at 10 DAP (Gallardo et al. 2006b). During seed filling nutrients from the endosperm are translocated to the embryo, resulting in reduction of the *Medicago* endosperm to only few cell layers from 20 DAP until seed maturity.

7.4 Concluding Remarks

Medicago is arguably the premier model system for functional genomics of legume seed development, given the near complete sequence of its genome, large mutant populations for facile reverse genetics, and MtGEA described in this chapter. We have shown how temporal data in MtGEA provides a comprehensive view of genetic and molecular processes at work during seed development in *Medicago*. In future, addition of spatially-resolved expression data from isolated tissues and cell types will extend the utility of MtGEA. In the interim, however, MtGEA is already playing an instrumental role in identifying interesting regulatory and metabolic genes with potential role in determining seed composition and, therefore, seed quality. Several groups are now using available genetic resources, including *Tnt1*-insertion, fast neutron deletion, and EMS mutant populations to decipher the roles of specific genes actively expressed during seed development. We are optimistic that these available resources will lead to new approaches to modifying legume seed composition for food, feed, and industry. We also see a role for MtGEA in comparative functional genomics, which will integrate similar data from other legume and nonlegume species to shed light on the evolution of seed development. This too will have practical implications, not only for biotechnology of legume seeds, but also seeds of nonlegumes.

References

Ariza-Nieto M, Blair MW, Welch RM, Glahn RP (2007) Screening of iron bioavailability patterns in eight bean (*Phaseolus vulgaris L.*) genotypes using the caco-2 cell *in vitro* model. J Agric Food Chem 55:7950–7956

Barker DG, Bianchi S, Blondon F, Dattée Y, Duc G, Essad S, Flament P, Gallusci P, Génier G, Guy P, Muel X, Tourneur J, Dénarié J, Huguet T (1990) *Medicago truncatula*, a model plant for studying the molecular genetics of the *Rhizobium*-legume symbiosis. Plant Mol Biol Rep 8:40–49

Barnabas AD, Arnott HJ (1990) Calcium oxalate crystal formation in the bean (*Phaseolus vulgaris* L.) seed coat. Bot Gazette 151:331–341

Baskin CC, Baskin JM (1998) Seeds: ecology, biography, and evolution of dormancy and germination. In: Baskin CC, Baskin JM (eds) . Academic, San Francisco, pp 666

Beddington J (2010) Food security: contributions from science to a new and greener revolution. Philos Trans Royal Soc B: Biol Sci 365:61–71

Benedito VA, Torres-Jerez I, Murray JD, Andriankaja A, Allen S, Kakar K, Wandrey M, Verdier J, Zuber H, Ott T, Moreau S, Niebel A, Frinckey T, Weiller G, He J, Dai X, Zhao PX, Tang Y, Udvardi MK (2008) A gene expression atlas of the model legume, *Medicago truncatula*. Plant J 55:504–513

Berger F (2003) Endosperm: the crossroad of seed development. Curr Opin Plant Biol 6:42–50

Berger F, Grini PE, Schnittger A (2006) Endosperm: an integrator of seed growth and development. Curr Opin Plant Biol 9:664–670

Broeckling CD, Huhman DV, Farag MA, Smith JT, May GD, Mendes P, Dixon RA, Sumner LW (2005) Metabolic profiling of *Medicago truncatula* cell cultures reveals the effects of biotic and abiotic elicitors on metabolism. J Exp Bot 56:323–336

Campion B, Sparvoli F, Doria E, Tagliabue G, Galasso I, Fileppi M, Bollini R, Nielsen E (2009) Isolation and characterisation of an *lpa* (low phytic acid) mutant in common bean (*Phaseolus vulgaris* L.) Theor Appl Genet 118:1211–1221

Cannon SB, May GD, Jackson SA (2009) Three sequenced legume genomes and many crop species: rich opportunities for translational genomics. Plant Physiol 151:970–977

Chabaud M, de Carvalho-Niebel F, Barker DG (2003) Efficient transformation of *Medicago truncatula* cv. Jemalong using the hypervirulent *Agrobacterium tumefaciens* strain AGL1. Plant Cell Rep 22:46–51

Choi HK, Mun JH, Kim DJ, Zhu H, Baek JM, Mudge J, Roe B, Ellis N, Doyle J, Kiss GB, Young ND, Cook DR (2004) Estimating genome conservation between crop and model legume species. Proc Natl Acad Sci U S A 101:15289–15294

Coelho CMM, Benedito VA (2008) Seed development and reserve compound accumulation in common bean (*Phaseolus vulgaris* L.). Seed Sci Biotechnol 2:42–52

De Sousa Araujo S, Roldao Lopes Amaral Duque AS, Metelo Fernandes Dos Santos DM, Salema Fevereiro MP (2004) An efficient transformation method to regenerate a high number of transgenic plants using a new embryogenic line of *Medicago truncatula* cv. Jemalong. Plant Cell, Tissue Organ Culture 78:123–131

Dhaubhadel S et al (2003) Isoflavonoid biosynthesis and accumulation in developing soybean seeds. Plant Mol Biol 53:733–743. doi:10.1023/B:PLAN.0000023666.30358.ae

Dixon RA, Pasinetti GM (2010) Flavonoids and isoflavonoids: from plant biology to agriculture and neuroscience. Plant Physiol 154:453–457

Dixon RA, Sumner LW (2003) Legume natural products: understanding and manipulating complex pathways for human and animal health. Plant Physiol 131:878–885

Dixon RA, Xie DY, Sharma SB (2005) Proanthocyanidins—a final frontier in flavonoid research? New Phytol 165:9–28

Djemel N, Guedon D, Lechevalier A, Salon C, Miquel M, Prosperi JM, Rochat C, Boutin JP (2005) Development and composition of the seeds of nine genotypes of the *Medicago truncatula* species complex. Plant Physiol Biochem 43:557–566

FAO (2009) Declaration of the world summit on food security. (ftp://ftp.fao.org/docrep/fao/meeting/018/k6119e.pdf)

Faria JMR, Buitink J, van Lammeren AAM, Hilhorst HWM (2005) Changes in DNA and microtubules during loss and re-establishment of desiccation tolerance in germinating *Medicago truncatula* seeds. J Exp Bot 56:2119–2130

Firnhaber C, Pühler A, Küster H (2005) EST sequencing and time course microarray hybridizations identify more than 700 *Medicago truncatula* genes with developmental expression regulation in flowers and pods. Planta 222:269–283

Gallardo K, Le Signor C, Vandekerckhove J, Thompson RD, Burstin J (2003) Proteomics of *Medicago truncatula* seed development establishes the time frame of diverse metabolic processes related to reserve accumulation. Plant Physiol 133:664–682

Gallardo K, Kurt C, Thompson R, Ochatt S (2006a) *In vitro* culture of immature *M. truncatula* grains under conditions permitting embryo development comparable to that observed *in vivo*. Plant Sci 170:1052–1058

Gallardo K, Le Signor C, Darmency M, Burstin J, Thompson RD, Rochat C, Boutin J-P, Küster H, Buitink J, Leprince O, Limami A, Grusak MA (2006b) Seed biology of *Medicago truncatula*. In: The *Medicago truncatula* handbook. http://www.noble.org/MedicagoHandbook/pdf/SeedBiology.pdf

Gallardo K, Firnhaber C, Zuber H, Hericher D, Belghazi M, Henry C, Kuster H, Thompson RD (2007) A combined proteome and transcriptome analysis of developing *Medicago truncatula* seeds. Mol Cell Proteomics 6:2165–2179

Gatehouse JA, Evans IM, Bown D, Croy RR, Boulter D (1982) Control of storage-protein synthesis during seed development in pea (*Pisum sativum L.*). Biochem J 208:119–127

Gatehouse JA, Evans IM, Croy RRD, Boulter D (1986) Differential expression of genes during legume seed development. Phil Trans R Soc Lond B 314:367–384

Gepts P, Beavis WD, Brummer EC, Shoemaker RC, Stalker HT, Weeden NF, Young ND (2005) Legumes as a model plant family. Genomics for food and feed report of the cross-legume advances through genomics conference. Plant Physiol 137:1228–1235

Goffard N, Weiller G (2007a) GeneBins: a database for classifying gene expression data, with application to plant genome arrays. BMC Bioinformatics 8:87

Goffard N, Weiller G (2007b) PathExpress: a web-based tool to identify relevant pathways in gene expression data. Nucl Acids Res 35:176–181

Goldberg RB, de Paiva G, Yadegari R (1994) Plant embryogenesis-zygote to seed. Science 266:605–614

Grela ER, Günter KD (1995) Fatty acid composition and tocopherol content of some legume seeds. Animal Feed Sci Tech 52:325–331

Guillon F, Champ MM-J (2002) Carbohydrate fractions of legumes: uses in human nutrition and potential for health. British J Nutr 88:293–306

Hagerman AE, Butler LG (1981) The specificity of proanthocyanidin-protein interactions. J Biol Chem 256:4494–4497

He J, Benedito VA, Wang M, Murray JD, Zhao PX, Tang Y, Udvardi MK (2009) The *Medicago truncatula* gene expression atlas web server. BMC Bioinformatics 10:441

Kanehisa M, Goto S, Kawashima S, Okuno Y, Hattori M (2004) The KEGG resource for deciphering the genome. Nucl Acids Res 32:D277–D280

Koch K (2004) Sucrose metabolism: regulatory mechanisms and pivotal roles in sugar sensing and plant development. Curr Opin Plant Biol 7:235–246

Le Signor C, Gallardo K, Prosperi JM, Salon C, Quillien L, Thompson R, Duc G (2005) Genetic diversity for seed protein composition in *Medicago truncatula*. Plant Genet Res 3:59–71

Le Signor C, Savois V, Aubert G, Verdier J, Nicolas M, Pagny G, Moussy F, Sanchez M, Baker D, Clarke J, Thompson R (2009) Optimizing TILLING populations for reverse genetics in *Medicago truncatula*. Plant Biotech J 7:430–441

Manthey K, Krajinski F, Hohnjec N, Firnhaber C, Pühler A, Perlick AM, Küster H (2004) Transcriptome profiling in root nodules and arbuscular mycorrhiza identifies a collection of novel genes induced during *Medicago truncatula* root endosymbioses. Mol Plant-Microbe Interact 17:1063–1077

McPhee KE, Zemetra RS, Brown J, Myers JR (2002) Genetic analysis of the raffinose family oligosaccharides in common bean. J Amer Soc Hort Sci 127:376–382

Molvig L, Tabe LM, Eggum BO, Moore AE, Craig S, Spencer D, Higgins TJV (1997) Enhanced methionine levels and increased nutritive value of seeds of transgenic lupins (*Lupinus angustifolius* L.) expressing a sunflower seed albumin gene. Proc Natl Acad Sci U S A 94:8393–8398

Munier-Jolain NG, Munier-Jolain NM, Roche R, Ney B, Duthion C (1998) Seed growth rate in grain legumes I. Effect of photoassimilate availability on seed growth rate. J Exp Bot 49:1963–1969

Pang Y, Peel GJ, Wright E, Wang Z, Dixon RA (2007) Early steps in proanthocyanidin biosynthesis in the model legume *Medicago truncatula*. Plant Physiol 145:601–615

Pang Y, Peel GJ, Sharma SB, Tang Y, Dixon RA (2008) A transcript profiling approach reveals an epicatechin-specific glucosyltransferase expressed in the seed coat of *Medicago truncatula*. Proc Natl Acad Sci U S A 105:14210–14215

Raboy V (1990) The biochemistry and genetics of phytic acid synthesis in higher plants. In: Inositol Metabolism in Plants (Editors: Morre EJ, Boss WS, Loewus FA). Wiley, New York, pp 55–76

Raboy V (2007) The ABCs of low-phytate crops. Nat Biotechnol 25:874–875

Rajjou L, Gallardo K, Debeaujon I, Vandekerckhove J, Job C, Job D (2004) The effect of α-amanitin on the *Arabidopsis* seed proteome highlights the distinct roles of stored and neosynthesized mRNAs during germination. Plant Physiol 134:1598–1613

Repetto O, Rogniaux H, Firnhaber C, Zuber H, Küster H, Larré C, Thompson R, Gallardo K (2008) Exploring the nuclear proteome of *Medicago truncatula* at the switch towards seed filling. Plant J 56:398–410

Rogers C, Wen J, Chen R, Oldroyd G (2009) Deletion-based reverse genetics in *Medicago truncatula*. Plant Physiol 151:1077–1086

Rushton PJ, Bray CM (1987) Stored and *de novo* synthesised polyadenylated RNA and loss of vigour and viability in wheat seed. Plant Sci 51:51–59

Santos-Mendoza M, Dubreucq B, Baud S, Parcy F, Caboche M, Lepiniec L (2008) Deciphering gene regulatory networks that control seed development and maturation in *Arabidopsis*. Plant J 54:608–620

Smeekens S (2000) Sugar-induced signal transduction in plants. Annu Rev Plant Physiol Plant Mol Biol 51:49–81

Tadege M, Wen J, He J, Tu H, Kwak Y, Eschstruth A, Cayrel A, Endre G, Zhao PX, Chabaud M, Ratel P, Mysore SK (2008) Large-scale insertional mutagenesis using the Tnt1 retrotransposon in the model legume *Medicago truncatula*. Plant J 54:335–347

Thoquet P, Gherardi M, Journet E-P, Kereszt A, Ane J-M, Prosperi J-M, Huguet T (2002) The molecular genetic linkage map of the model legume *Medicago truncatula*: an essential tool for comparative legume genomics and the isolation of agronomically important genes. BMC Plant Biol 2:1

van der Mensbrugghe D, Osorio Rodarte I, Burns A, Baffes J (2009) How to feed the world in 2050: Macroeconomic environment, commodity markets—a longer term outlook. The World Bank & FAO. http://mpra.ub.uni-muenchen.de/19061/. Accessed 12 Oct 2009

van Dongen JT, Ammerlaan AMH, Wouterlood M, van Aelst AC, Borstlap AC (2003) Structure of the developing pea seed coat and the post phloem transport pathway of nutrients. Annals Bot 91:729–737

Veitch NC (2007) Isoflavonoids of the leguminosae. Nat Prod Rep 24:417–464

Verdier J, Kakar K, Gallardo K, Le Signor C, Aubert G, Schlereth A, Town CD, Udvardi MK, Thompson RD (2008) Gene expression profiling of *M. truncatula* transcription factors identifies putative regulators of grain legume seed filling. Plant Mol Biol 67:567–580

Verdier J et al (2012) MtPAR MYB transcription factor acts as an on switch for proanthocyanidin biosynthesis in Medicago truncatula. Proc Natl Acad Sci U S A 109(5):1766–1771. doi:10.1073/pnas.112091610

Walling L, Drews GN, Goldberg RB (1986) Transcriptional and post-transcriptional regulation of soybean seed protein mRNA levels. Proc Natl Acad Sci U S A 83:2123–2127

Wang HL, Grusak MA (2005) Structure and development of *Medicago truncatula* pod wall and seed coat. Annals Bot 95:737–747

Watson BS, Asirvatham VS, Wang L, Sumner LW (2003) Mapping the proteome of barrel medic (*Medicago truncatula*). Plant Physiol 131:1104–1123

Weber H, Borisjuk L, Wobus U (1997) Sugar import and metabolism during seed development. Trends Plant Sci 2:169–174

Weber H, Borisjuk L, Wobus U (2005) Molecular physiology of legume seed development. Annu Rev Plant Biol 56:253–279

Young ND, Cannon SB, Sato S, Kim D, Cook DR, Town CD, Roe BA, Tabata S (2005) Sequencing the genespaces of *Medicago truncatula* and *Lotus japonicus*. Plant Physiol 137:1174–1181

Young ND, Udvardi M (2009) Translating *Medicago truncatula* genomics to crop legumes. Curr Opin Plant Biol 12:193–201

Young ND et al (2011) The Medicago genome provides insight into the evolution of rhizobial symbioses. Nature 480:520–524. doi:10.1038/nature10625

Zhao J, Dixon RA (2009) MATE Transporters facilitate vacuolar uptake of epicatechin 3'-O-glucoside for proanthocyanidin biosynthesis in *Medicago truncatula* and *Arabidopsis*. Plant Cell 21:2323–2340

Chapter 8
Transcriptomics of Legume Seed: Soybean a Model Grain Legume

Sangeeta Dhaubhadel and Frédéric Marsolais

Abstract Grain legumes are the crop plants belonging to family Leguminosae that are cultivated for their seed yield for the purpose of human food or animal feed. They possess high metabolic activity and fluxes in seeds and are an important source of protein in vegetarian diet or when the intake from animal or fish source is not available. Soybean (*Glycine max*) L. Merr. is the world's most widely grown grain legume. In addition to a high seed protein content (approximately 40 %), soybean seeds are also a major source of vegetable oil (approximately 20 %) and many beneficial plant natural compounds, such as isoflavonoids. Recently, soybean has gained considerable attention as a major crop for biodiesel production. It has a complex genome with a large genome size (~ 1115 Mb) that has undergone at least two genome duplication events within the last 60 million years (Gill et al. Plant Physiol, 151: 1167–1174, 2009; Schlueter et al. Genome, 47: 868–876, 2004; Shoemaker et al. Genetics, 144: 329–338, 1996). The research efforts in the past decades in soybean in the area of genome mapping, molecular breeding, genomics, and whole genome sequencing have generated a vast amount of data and knowledge, providing unique opportunities to the legume community to conduct both basic and applied research in soybean and its close relatives.

Keywords Legume · Microarray · Seed development · Sequencing · Soybean · Transcriptomics

8.1 Soybean Seed Development: An Overview

Soybean seed development commences with a double fertilization process, leading to the differentiation of seed tissues, such as embryo, endosperm, and seed coat. Soybean seed is almost devoid of an endosperm and consists of a well-developed embryo and a seed coat (Goldberg et al. 1994). Seed coat surrounds the embryo and is a maternal tissue. Seed coat also serves as transient storage organs in many legume seeds, accumulating proteins and starch before the storage activity starts in

S. Dhaubhadel (✉) · F. Marsolais
Southern Crop Protection and Food Research Centre,
Agriculture and Agri-Food Canada, N5V 4T3, London, Ontario, Canada
e-mail: sangeeta.dhaubhadel@agr.gc.ca

G. K. Agrawal, R. Rakwal (eds.), *Seed Development: OMICS Technologies toward Improvement of Seed Quality and Crop Yield*,
DOI 10.1007/978-94-007-4749-4_8, © Springer Science+Business Media Dordrecht 2012

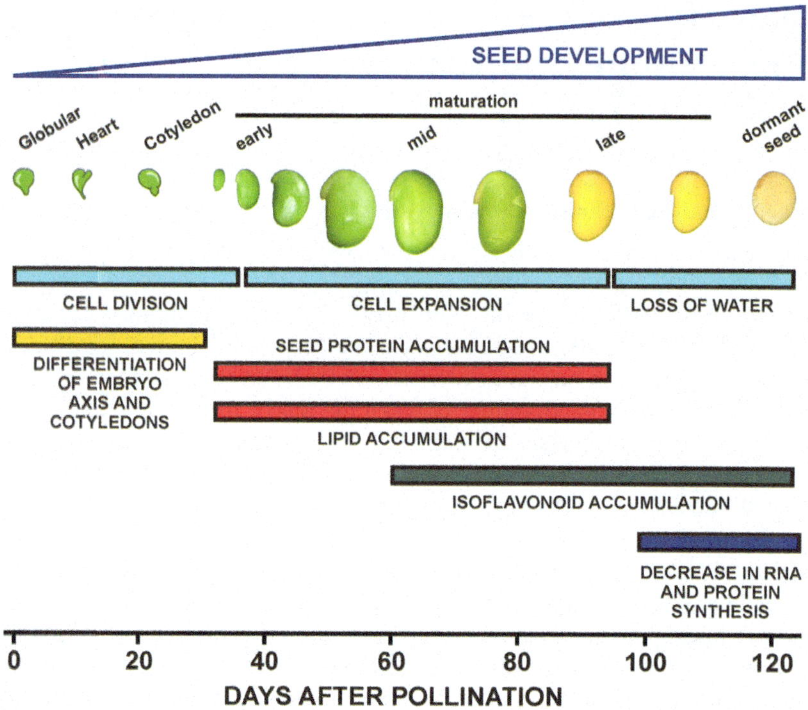

Fig. 8.1 A schematic diagram representing physiological and developmental events that occurs during soybean seed development. (Modified from Le et al. 2007)

embryo (reviewed in Weber et al. 2005). These are highly differentiated and complex seed tissues that differ structurally during the development (Miller et al. 1999) as well as among species (Corner 1951). Soybean seed coat is developmentally transitory, where a number of tissues present during early- and mid-development does not persist in the mature seed coat (Miller et al. 1999). It can be stated that the seed coat plays a critical role in soybean seed development by providing support and nourishment for the developing embryo and a protective covering for the embryo at maturity.

The embryo development starts with the differentiation of fertilized egg cell into a mature embryo containing cells with different roles. The entire process of embryogenesis can be grouped into following five phases: globular, heart, cotyledon, maturation, and dormancy (Walbot 1978). As shown in Fig. 8.1, differentiation of embryo axis and cotyledons occurs during the initial three phases of embryogenesis, while synthesis and accumulation of major seed reserve takes place during the maturation process and continues until seed reaches to the dormant state. During the late seed-maturation phase, the inner layers of seed coat tissues are compressed by the developing embryo leaving the seed coat only with a thin-walled parenchyma, hourglass cells, and palisade cells. The whole seed then undergoes desiccation, leading to reduced transcripts and protein synthesis.

8.2 Transcriptomics of Soybean Seed: Recent Developments

During the period of seed development, seed becomes the primary sink for assimilates within the plant, which is determined by the genetic factors as well as the environment and cultural practices that the maternal parent is exposed to (Copeland and McDonald 2001). The developing seed undergoes several morphological changes, which occur concurrently with the changes in cytological, chemical, and overall weight of the seed. All these changes are determined by the reprogramming of gene expression patterns reconciled by several different pathways (Gehring et al. 2004; Wobus and Weber 1999).

Even though legume seed development has been studied for more than a century, research on gene expression analysis of soybean seed started only in the early 1980s. Using RNA-excess DNA-RNA hybridization, Goldberg et al. (1989, 1981a, b) showed that a small number of mRNAs are regulated quantitatively at specific stages of seed development; however, most of the diverse transcripts continue to express throughout the seed development and later gets stored in the dormant seeds. These studies used seed protein genes as models because of their economic importance as well as transcript abundance in seeds. The recent development of genetic resources and advanced technologies has allowed the researchers to study global gene expression analysis in seeds during the entire developmental phase and in any specific seed tissue.

The public EST project was the first large-scale gene expression analysis in soybean that consisted of different tissues, including seeds under several different environmental, developmental, and stress conditions (Shoemaker et al. 2002). This project generated over 80 cDNA libraries from different tissues and conditions. Currently, there are 1,354,268 ESTs available at the DFCI gene index (http://compbio.dfci.harvard.edu/tgi/cgi-bin/tgi/gimain.pl?gudb=soybean) that includes 73,178 tentative contig (TC) sequences and 137,174 unique cDNAs with a goal to represent nonredundant view of soybean genes and the data on their expression patterns, potential functions, and evolutionary relationships. The drawback of using EST analysis for gene expression study is the relatively lower rate of novel transcript discovery as the rare transcripts are not usually represented well in EST collections.

The cDNA and oligonucleotide microarray technology is the most widely-used method of transcriptomics study in soybean. Microarray is a targeted expression profiling approach, where the expression levels of genes, for which a clone or a sequence is available, can be determined using spotted cDNAs or oligonucleotides on glass slides (Schena et al. 1995; Stears et al. 2003). Two soybean cDNA arrays are available at http://soybeangenomics.cropsci.uiuc.edu, each containing 18,432 single PCR products from low redundancy cDNA sets Gm-c1021, Gm-c1083, and Gm-c1070 (18k-A array) and Gm-r1088 and Gm-r1089 (18k-B array) (Vodkin et al. 2004). Similarly, several soybean oligonucleotide arrays are also available. The advantage of using oligoarrays over cDNA arrays is that the oligos are designed from most distant 3' region of the cDNAs, which helps to distinguish between the different members of a multigene family sharing high sequence identity at the coding region. This is particularly important for soybean whose genome has undergone at least two whole genome duplication events (Gill et al. 2009; Shoemaker et al. 1996), resulting into genes with large multigene families. An oligo array containing 38,000 unique 70-mer oligos printed on two sets of slides, each with 19,000 spots by Illumina, Inc./Invitrogen (San Diego, CA, USA), is available. Another oligo array widely used by the soybean research community is the Affymetrix array, which uses 25-mers containing 37,500 probe sets on the array.

Thibaud-Nissen et al. (2003) used a 9,280 cDNA array to study gene expression during early stage of seed development. The study identified the transcript patterns associated with somatic embryogenesis in soybean. Their results suggested that the formation of somatic globular embryos is associated with the transcription of SPs and GA synthesis. To gain understanding of the transcriptional changes during embryo development that may control seed quality traits in soybean, a global gene expression analysis was performed using an 18k-A cDNA array (Dhaubhadel et al. 2007). This study was the first large-scale gene expression analysis of soybean seeds. The experiment consisted of five different developmental stages of embryos (early to late seed maturity) in two soybean cultivars, which differed in the levels of seed oil and isoflavonoids. Samples were collected from the soybean plants grown at two different locations for two consecutive years. The analysis performed by Sangeeta Dhaubhadel and her team was aimed at measuring the followings: (i) changes in the gene expression during embryo development within a cultivar, and (ii) comparison of gene expression at specific stage of embryo development between two selected cultivars. The analysis identified 5,910 genes differentially expressed in between the cultivars in at least one of the five stages of embryo development and indicated that the gene expression changes due to environmental factors are greater than those due to cultivar specific or developmental stage specific differences. Dhaubhadel et al. (2007) found that among the seed storage product genes, majority of the cDNAs for soybean seed proteins were abundantly expressed in all the stages of development. Only 17 soybean seed protein genes were differentially expressed during development, whereas 46 of FAs-related genes and 168 of shikimic acid- and phenylpropanoid pathways-related genes were expressed differentially in soybean embryos during the development. The array contained 720 cDNAs encoding for protein kinases, where expression levels of 229 kinases changed during soybean seed development (Fig. 8.2). As shown in Fig. 8.2, soybean seed kinases are grouped into two major clusters. Cluster I consisted of the kinase transcripts expressed during early to mid seed maturity, whereas the kinase transcripts belonging to cluster II were more abundant during the later stages of seed maturation. The general patterns of kinase gene expression was similar in both the soybean cultivars; Harovinton and RCAT Angora. These two cultivars are distantly related in their pedigree. The similarity in gene expression in these two cultivars suggests that the results could be broadly applicable to other soybean cultivars or other related species. Protein kinases play a central role in signal transduction pathways and mediate many molecular responses within the cell. A detailed study of kinase genes and their potential function in the regulation of the activity of target enzymes may help identification of key genes involved in seed development or in response to stress.

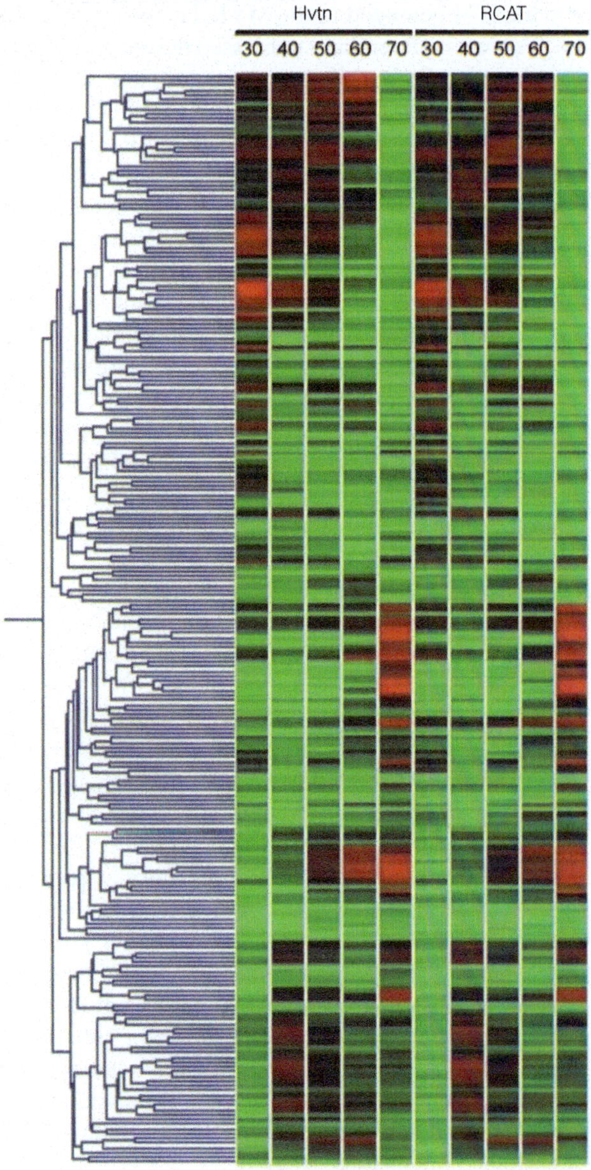

Fig. 8.2 Expression profiles of soybean protein kinase genes during embryo development as revealed by hierarchical clustering. The protein kinase genes differentially-expressed in at least one of the growth stages or expressed at all the stages during embryo development were grouped using hierarchical clustering. For each time point, normalized hybridization ratios averaged from four biological replicates are represented in color. The signals are shown in *red-green* color scale, where red represents high expression and green represents low expression. The numbers at the top represent stages of embryo development (30, 40, 50, 60, or 70) in DAP. Hvtn, Harovinton; RCAT, RCAT Angora

Comparison of gene expression patterns between two soybean cultivars, RCAT Angora (high isoflavonoid cultivar) and Harovinton (low isoflavonoid cultivar), revealed that *chalcone synthase* (*CHS*) *7* and *CHS8* genes play a critical role in isoflavonoid synthesis. RNAi silencing of these two *CHS* genes in soybean hairy roots confirmed their direct role in isoflavonoid biosynthesis (Yi et al. 2010b, c). The seed transcriptomic data set was used as a primary resource to identify the TFs that regulate expression of *CHS8* gene. The 18k-A array contained 368 TF genes;

of which, 74 belonged to the MYB TF family. Microarray analysis showed that the 39 TFs were differentially expressed during at least one of the five stages of embryo development. By correlating the expression pattern of *CHS8* with the *MYB* genes, Yi et al. (2010b) were able to make a short list of candidate MYB factors for further analysis and identified GmMYB176, an R1 MYB TFs, as a regulatory factor for expression of *CHS8* genes and isoflavonoid synthesis. These are some of the examples, demonstrating the importance of the gene expression data for the identification of candidate genes involved in valuable seed quality traits.

A similar study on transcript patterns during soybean seed development was conducted by Jones et al. (2010) using two different types of spotted arrays. This study included an additional stage of seed development, the mature dry seed, in addition to the earlier study by Dhaubhadel et al. (2007). The most interesting observation made by Jones and coworkers was that expression of many genes, such as genes for protein degradation and TFs were increased in the mature seed that are at the desiccating and dry stages of seed development. Those authors speculated that the increased levels of expression of certain genes in dry seeds may be for the preparation of pathways required in the early stages of seed imbibition and germination (Jones et al. 2010).

To investigate the global gene expression profiles in different compartments of the seed, laser capture microdissection technology was combined with Affymetrix GeneChip technology (Le et al. 2007). The results suggested that at least 22,000 diverse mRNAs are involved in fashioning the globular-stage soybean seed. Despite that the majority of diverse transcripts are expressed in different compartments of the globular embryo, small subsets of mRNAs are unique to a specific seed region. Identification of the region-specific mRNAs and other mRNAs, whose expressions are coregulated in different compartments of the seed together with the availability of soybean genome sequence (Schmutz et al. 2010), offers the researchers the opportunity to identify conserved regulatory sequences connecting seed genes with regulatory networks. The 18k-A array was also used to study gene expression profiles during soybean seed coat development. The results revealed a complex transcriptome of seed coat as compared to the embryo, emphasizing the importance of seed coat in overall seed development (Ranathunge et al. 2010). During seed development, the seed coat provides a source of nutrients to growing embryos; whereas at maturity stage it serves as a protective layer and controls seed germination by regulating the water uptake. The gene expression patterns in soybean seed coat may provide an understanding of seed coat development-related traits, such as hard seedness (Ranathunge et al. 2010).

Even though the array-based transcriptomics investigation using spotted microarrays (Dhaubhadel et al. 2007; Jones et al. 2010; Ranathunge et al. 2010) or Affymetrix GeneChip (Le et al. 2007) are very informative and are major achievements in understanding the soybean seed development, yet these data sets do not provide a complete picture of gene expression patterns. Recently, the development of high-through put DNA sequencing technologies, such as NGS have enabled the analysis of gene expression independent of information on their genomic sequence. In addition, NGS technologies are quick, inexpensive, and offer the advantage of greater sensitivity and wider range of gene expression of complex nucleic acid population (Metzker 2010). There are several NGS technology platforms available, such as 454 Life Sciences (http://www.454.com/index.asp), Illumina Genome Analyzer (http://www.illumina.com/), Applied Biosystems SOLiD (http://www.appliedbiosystems.com/absite/us/en/home/applications-technologies/solid-next-generation-sequencing.html), tSMS by Helicos Bioscience (http://www.helicosbio.com/), and SMRT by Pacific Biosciences (http://www.pacificbiosciences.com/). These platforms use different methods of template preparation and sequencing chemistries and produce sequences of different read lengths varying from 32 bp to 900 bp. The total output for single read runs varies from 0.4 Gbp (454 Life Sciences) to 30 Gbp (Applied Biosystems SOLiD).

The NGS technology can be used to study genomes, epigenomes, and transcriptomes of any organism, where one NGS platform may be more suitable than the other depending on the application. Transcriptomics study using NGS technology, also known as RNA sequencing (RNA-seq) is gaining popularity over microarrays as it does not require the knowledge of the genome sequence as prerequisite. In addition, the RNA-seq method introduces less variation between technical replicates (Marioni et al. 2008). Recently, RNA-seq was used to study seed development in soybean. This study generated 2.7–9.6 million 36 bp reads for each of seven stages of seed development (Severin et al. 2010) and created a RNA-seq atlas of soybean. The study also included a total of seven aerial and underground tissues. The hierarchical clustering gene expression profiles in 14 different tissues grouped seeds into a separate clade and created a list of genes with high expression profiles during seed development. The SoySeq-Atlas (http://soybase.org/soyseq/) is a valuable resource for soybean and other related crops of agronomic importance.

Using a deep sequencing approach with the Illumina Solexa platform, Joshi et al. (2010) sequenced four miRNA libraries, including seed library and generated over one million sequences per library. The miRNA expression levels were then compared with the predicted target genes to get the insight into the regulatory patterns. By comparing the hybridization and deep sequencing methods for the same set of experiment, Bolon et al. (2010) identified 13 genes displaying significant differences in gene expression between two near-isogenic lines contrasting in the levels of seed protein and oil. Those genes were mapped to the protein quantitative trait loci (QTL) region at linkage group 1, suggesting their potential involvement in the regulation of seed protein content.

8.3 Soybean: A Model Grain Legume for Legume Transcriptomics

As one of three legumes with substantial genomic information (Kim et al. 2010; Schmutz et al. 2010), soybean constitutes an important model for grain legumes. In the legume phylogeny, soybean belongs to the tribe *Phaseoleae* (Fig. 8.3). This tribe also includes several major pulse crops for which development of genomic

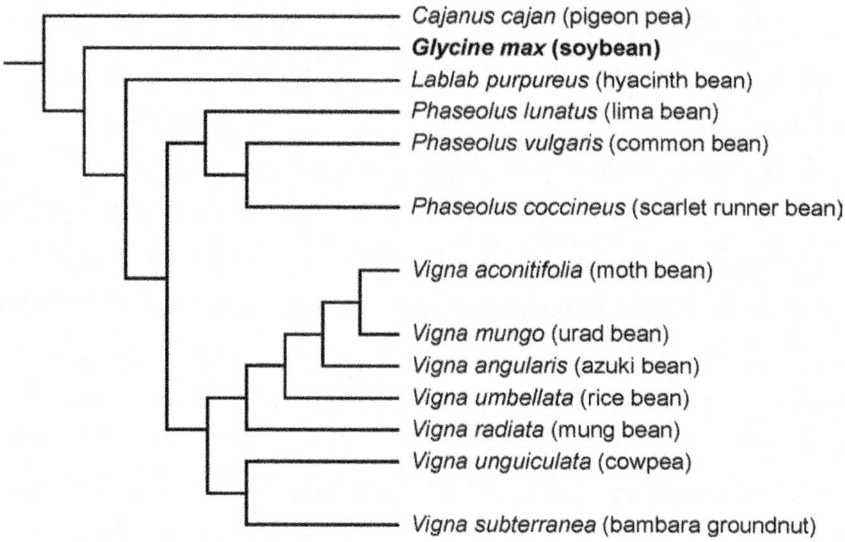

Fig. 8.3 Phylogenetic relationships between soybean and pulse crops member of the tribe Phaseoleae. The diagram is based on the phylogenetic analyses of Stefanovic et al. (2009), Wang et al. (2008), and Goel et al. (2002)

resources is not nearly as advanced as for soybean. Pigeon pea (*Cajanus cajan*) is grown in semi-arid regions of Asia, particularly in India, and exhibits significant drought tolerance (Varshney et al. 2009). Common bean (dry bean, *Phaseolus vulgaris*) is the leading pulse crop worldwide both by acreage and production (Broughton et al. 2003; McClean et al. 2004). Among *Vigna* species, cowpea (*Vigna unguiculata*) is a major crop in Africa and also exhibits drought tolerance (Ehlers and Hall 1997), while in Asia, urad bean (*V. mungo*), mung bean (*V. radiata*), and azuki bean (*V. angularis*) are of significant importance (Tomooka et al. 2002).

Given the relationships between legume species, soybean is expected to be most useful as a model for these pulse crops belonging to the *Phaseoleae*. By comparison, *Medicago truncatula* and *Lotus japonicus*, the two other models with extensive genome sequence information are more closely related to the temperate legumes, which include members of the tribe *Viciae* (lentil, faba bean and pea) and *Cicereae* (chickpea). Studies with cowpea and common bean revealed extensive macrosynteny with the soybean genome (McClean et al. 2010; Muchero et al. 2009). As expected, each syntenic block in cowpea or common bean matches two blocks in soybean, which is consistent with a genome duplication event in soybean taking place 11 million years ago, after the divergence between lineages leading to soybean and cowpea or common bean, which happened 19 million years ago (Lavin et al. 2005; Stefanovic et al. 2009). This duplication, which likely arose through outcrossing (Gill et al. 2009) has led to the present paleoploid soybean genome. Since common bean has a simpler genome, it has also been proposed as a genetic model for soybean (McClean et al. 2008).

In the context of transcriptomics and seed development, soybean is likely to be a very important model to understand the genetic bases underlying protein yield, the negative relationship between protein and oil content and high isoflavonoid content. These are key traits in soybean (Lightfoot 2008). Importantly, they display significant genetic variability. Reciprocally, research on pulse crops within the *Phaseoleae* may improve our understanding of the seed developmental and metabolic processes in soybean. For example, scarlet runner bean (*P. coccineus*), due to the large size of its seeds and embryos, has been used as a model to study gene expression in the embryo-proper and suspensor during early embryo differentiation (Le et al. 2007; Weterings et al. 2001). This research enabled the identification of the *cis*-regulatory sequences necessary for transcriptional activation in the suspensor region (Kawashima et al. 2009).

Another important trait in soybean seed is the content of sulfur amino acids (Yi et al. 2010a). Like in other grain legumes, the levels of essential sulfur amino acids are suboptimal for nutrition, and use of soybean meal as animal feed requires supplementation with methionine. The QTLs influencing cysteine and methionine contents have already been identified in a soybean population (Panthee et al. 2006), indicating that natural genetic variation could be exploited to improve this trait. Methionine varied by up to two-fold with a heritability of 0.57 in recombinant inbred lines, while a lower variability and heritability were observed for cysteine (1.4-fold and 0.15). Common bean constitutes a unique model to investigate the metabolism of sulfur amino acids in seed. Large amounts of a nonprotein sulfur amino acid (*S*-methyl-cysteine) are accumulated to 0.3 % of seed weight (Taylor et al. 2008). *S*-Methyl-cysteine constitutes a storage form of sulfur, but cannot substitute for methionine or cysteine in the diet (Padovese et al. 2001). Variation in the composition of major seed proteins in related genetic stocks is associated with an increase of cysteine and methionine content, by up to 70 % and 10 % respectively, with a corresponding decrease in *S*-methyl-cysteine (Taylor et al. 2008), and increased levels of specific sulfur-rich proteins (Marsolais et al. 2010). In the *Vigna* genus, accumulation of *S*-methyl-cysteine in seed represents a taxonomic marker, being present in cowpea and mung bean, and absent in urad and azuki bean (Baldi and Salamini 1973; Otoul et al. 1975).

A goal integral to the transcriptomics approach is the identification of transcriptional regulatory network(s) controlling the trait or process of interest. In this regard, the general resources consisting in databases of soybean TFs, SoyDB (http://casp.rnet.missouri.edu/soydb/) (Wang et al. 2010) and LegumeTFDB (http://legumetfdb.psc.riken.jp/) (Mochida et al. 2010), and plant miRNAs, PMRD (http://bioinformatics.cau.edu.cn/PMRD/) (Zhang et al. 2010) should be very helpful to the investigators, along with the already mentioned soybean tissue-specific seed expression profiles present at (http://seedgenenetwork.net/) (Le et al. 2007).

From a more technical point of view, high-density arrays developed for soybean transcriptomics research are expected to be applicable to other *Phaseoleae* crops through cross-specific hybridization, at least until similar tools become available for some of these crops. Meanwhile, the development of NGS technologies may enable to by-pass the need for microarrays in transcript profiling studies. Development of genomic information is underway for the most important *Phaseoleae* pulse crops. For example, there are currently three common bean genome sequencing projects: the American whole-genome sequence of common bean project in the United States of America (http://www.phytozome.com/commonbean_er.php), the Iberoamerican genome sequencing of the Mesoamerican bean genome (*P. vulgaris*) (http://www.langebio.cinvestav.mx/?pag=170), and Canadian *Phaseolus* Genomic for Improved Bio-Product Development (http://www.beangenomics.ca/). Information from these projects should enable the development of an efficient transcript profiling platform for common bean. Generation of ESTs from *Phaseoleae* pulses has not focused primarily on the seed, with few exceptions. A collection of approximately 30,000 Sanger ESTs was developed from four seed developmental stages of the common bean reference genotype BAT93 (Yin et al. 2011). A large number of Sanger ESTs (approximately 400,000) and random 454 cDNA sequences were obtained from suspensor and embryo-proper cDNA libraries of scarlet runner bean forming an impressive resource (http://estdb.biology.ucla.edu/PcEST/) (Kawashima et al. 2009; Le et al. 2007; Weterings et al. 2001).

Le et al. (2007) reported the use of a soybean Affymetrix GeneChip to profile RNAs from laser-captured suspensor and embryo-proper tissues of scarlet runner bean. This analysis identified small subsets of transcripts specifically expressed in either tissue, including some encoding TFs. In a recent and very interesting study, Mensack et al. (2010) used seed transcriptomics, proteomics, and metabolomics profiles to evaluate diversity among common bean lines from the Mesoamerican and Andean centers of domestication. Among the three techniques, metabolomics was the most powerful and enabled classification of cultivated varieties following principle component analysis. For transcript profiling, the authors reported a low efficiency of cross-specific hybridization to an Affymetrix soybean microarray, and therefore used the previously described cDNA array developed at the University of Illinois (Vodkin et al. 2004). The analysis identified transcripts differentially expressed between navy and white kidney beans two weeks after anthesis (WAA), related to transcription, translation or protein synthesis, and modification. Fewer differentially-expressed transcripts were observed at three WAA. However, this analysis failed to reveal differentially-expressed genes relevant to the physiological distinctions between centers of origin, which might be due in part to the use of a cross-specific platform. In a recent nonseed study, Yang et al. (2010) described a method whereby the Affymetrix GeneChip soybean genome array was successfully used for transcript profiling of common bean, by masking probes matching interspecies variable regions between the two species. Similarly, Das and coworkers successfully used the soybean Affymetrix GeneChip to profile transcripts in cowpea (Das et al. 2010).

8.4 Concluding Remarks

As legume seeds are an important source of human food and animal feed, there is a continuous effort in improving the overall nutritional quality and increasing the total seed yield. The availability of whole genome sequence of model legumes

M. truncatula, *L. japonicus*, and grain legume soybean together with the major developments in transcriptomics tools may allow us to bridge the gap between transcriptomics with seed traits of economic importance. Integration of the omics tools by treating RNAs, proteins, and metabolites as quantitative traits may offer new insights into how these traits may be regulated in the seed. Understanding the biochemistry and physiology of legume seeds at molecular level will help in manipulating metabolic routes towards increasing valuable traits in the seeds through metabolic engineering.

References

Baldi G, Salamini F (1973) Variability of essential amino acid content in seeds of 22 *Phaseolus* species. Theor Appl Genet 43:75–78

Bolon YT, Joseph B, Cannon S, Graham M, Diers B, Farmer A, May G, Muehlbauer G, Specht J, Tu Z, Weeks N, Xu W, Shoemaker R, Vance C (2010) Complementary genetic and genomic approaches help characterize the linkage group I seed protein QTL in soybean. BMC Plant Biol 10:41

Broughton WJ, Hernandez G, Blair M, Beebe S, Gepts P, Vanderleyden J (2003) Beans (*Phaseolus spp.*)—model food legumes. Plant Soil 252:55–128

Copeland LO, McDonald MB (2001) Seed formation and development. In: Copeland LO, McDonald MB (eds) Principles of seed science and technology, 4th edition. Kluwer Academic, Massachusetts, pp 17–38

Corner EJH (1951) The leguminous seed. Phytomorphology 1:117–150

Das S, Ehlers J, Close T, Roberts P (2010) Transcriptional profiling of root-knot nematode induced feeding sites in cowpea (*Vigna unguiculata* L. Walp.) using a soybean genome array. BMC Genomics 11:480

Dhaubhadel S, Gijzen M, Moy P, Farhangkhoee M (2007) Transcriptome analysis reveals a critical role of *CHS7* and *CHS8* genes for isoflavonoid synthesis in soybean seeds. Plant Physiol 143:326–338

Ehlers JD, Hall AE (1997) Cowpea (*Vigna unguiculata* L. Walp.). Field Crops Res 53:187–204

Gehring M, Choi Y, Fischer RL (2004) Imprinting and seed development. Plant Cell 16:S203–S213

Gill N, Findley S, Wallling JG, Hans C, Ma J, Doyle J, Stacey G, Jackson SA (2009) Molecular and chromosomal evidence for allopolyploidy in soybean. Plant Physiol 151:1167–1174

Goel S, Raina SN, Ogihara Y (2002) Molecular evolution and phylogenetic implications of internal transcribed spacer sequences of nuclear ribosomal DNA in the *Phaseolus-Vigna* complex. Mol Phylogenet Evol 22:1–19

Goldberg RB, Hoschek G, Ditta GS, Breidenback RW (1981a) Developmental regulation of cloned superabundant embryo mRNAs in soybean seeds. Dev Biol 83:218–231

Goldberg RB, Hoschek G, Tam SH, Ditta GS, Breidenback RW (1981b) Abundance, diversity, and regulation of mRNA sequence sets in soybean embryogenesis. Dev Biol 83:201–217

Goldberg RB, Barker SJ, Perez-Grau L (1989) Regulation of gene expression during plant embryogenesis. Cell 56:149–160

Goldberg RB, Depaiva G, Yadegari R (1994) Plant embryogenesis-zygote to seed. Science 266:605–614

Jones SI, Gonzalez DO, Vodkin LO (2010) Flux of transcript patterns during soybean seed development. BMC Genomics 11:136

Joshi T, Yan Z, Libault M, Jeong D-H, Park S, Green P, Sherrier DJ, Farmer A, May G, Meyers B, Xu D, Stacey G (2010) Prediction of novel miRNAs and associated target genes in *Glycine max*. BMC Bioinformatics 11:S14

Kawashima T, Wang X, Henry KF, Bi Y, Weterings K, Goldberg RB (2009) Identification of cis-regulatory sequences that activate transcription in the suspensor of plant embryos. Proc Natl Acad Sci U S A 106:3627–3632

Kim HK, Choi YH, Verpoorte R (2010) NMR-based metabolomic analysis of plants. Nat Protoc 5:536–549

Lavin M, Herendeen PS, Wojciechowski MF (2005) Evolutionary rates analysis of Leguminosae implicates a rapid diversification of lineages during the tertiary. Syst Biol 54:575–594

Le BH, Wagmaister JA, Kawashima T, Bui AQ, Harada JJ, Goldberg RB (2007) Using genomics to study legume seed development. Plant Physiol 144:562–574

Lightfoot DA (2008) Soybean genomics: developments through the use of cultivar "Forrest". Int J Plant Genomics 2008:793158

Marioni JC, Mason CE, Mane SM, AStephens M, Gilad Y (2008) RNA-seq: an assessment of technical reproducibility and comparison with gene expression arrays. Genome Res 18:1509–1517

Marsolais F, Pajak A, Yin F, Taylor M, Gabriel M, Merino DM, Ma V, Kameka A, Vijayan P, Pham H, Huang S, Rivoal J, Bett K, Hernández-Sebastià C, Liu Q, Bertrand A, Chapman R (2010) Proteomic analysis of common bean seed with storage protein deficiency reveals up-regulation of sulfur-rich proteins and starch and raffinose metabolic enzymes, and down-regulation of the secretory pathway. J Proteomics 73:1587–1600

McClean P, Gepts P, Jackson S, Lavin M (2008) *Phaseolus vulgaris*: a diploid model for soybean. In: Stacey G (ed) Genetics and genomics of soybean. Springer, New York, pp 55–78

McClean P, Kami J, Gepts P (2004) Genomics and genetic diversity in common bean. In: Wilson RF, Stalker HT, Brummer EC (eds) Legume crop genomics. AOCS, Illinois, pp 60–82

McClean PE, Mamidi S, McConnell M, Chikara S, Lee R (2010) Synteny mapping between common bean and soybean reveals extensive blocks of shared loci. BMC Genomics 11:184

Mensack MM, Fitzgerald VK, Ryan EP, Lewis MR, Thompson HJ, Brick MA (2010) Evaluation of diversity among common beans (*Phaseolus vulgaris* L.) from two centers of domestication using 'omics' technologies. BMC Genomics 11:686

Metzker ML (2010) Sequencing technologies—the next generation. Nat Rev Genet 11:31–46

Miller SS, Bowman LA, Gijzen M, Miki BLA (1999) Early development of the seed coat of soybean (*Glycine max*). Ann Bot 84:297–304

Mochida K, Yoshida T, Sakurai T, Yamaguchi-Shinozaki K, Shinozaki K, Tran LS (2010) LegumeTFDB: an integrative database of *Glycine max*, *Lotus japonicus* and *Medicago truncatula* transcription factors. Bioinformatics 26:290–291

Muchero W, Diop NN, Bhat PR, Fenton RD, Wanamaker S, Pottorff M, Hearne S, Cisse N, Fatokun C, Ehlers JD, Roberts PA, Close TJ (2009) A consensus genetic map of cowpea [*Vigna unguiculata* (L.) Walp.] and synteny based on EST-derived SNPs. Proc Natl Acad Sci U S A 106:18159–18164

Otoul E, Maréchal R, Dardenne G, Desmedt F (1975) Des dipeptides soufrés différencient nettement *Vigna radiata* de *Vigna mungo*. Phytochemistry 14:173–179

Padovese R, Kina SM, Barros RMC, Borelli P, Marquez UML (2001) Biological importance of gamma-glutamyl-*S*-methylcysteine of kidney bean (*Phaseolus vulgaris* L.). Food Chem 73:291–297

Panthee DR, Pantalone VR, Sams CE, Saxton AM, West DR, Orf JH, Killam AS (2006) Quantitative trait loci controlling sulfur containing amino acids, methionine and cysteine, in soybean seeds. Theor Appl Genet 112:546–553

Ranathunge K, Shao S, Qutob D, Gijzen M, Peterson CA, Bernards MA (2010) Properties of the soybean seed coat cuticle change during development. Planta 31:1171–1188

Schena M, Shalon D, Davis RW, Brown PO (1995) Quantitative monitoring of gene-expression patterns with a complementary-DNA microarray. Science 270:467–470

Schlueter JA, Dixon P, Granger C, Grant D, Clark L, Doyle JJ, Shoemaker RC (2004) Mining EST databases to resolve evolutionary events in major crop species. Genome 47:868–876

Schmutz J, Cannon SB, Schlueter J, Ma J, Mitros T, Nelson W, Hyten DL, Song Q, Thelen JJ, Cheng J, Xu D, Hellsten U, May GD, Yu Y, Sakurai T, Umezawa T, Bhattacharyya MK, Sandhu D, Valliyodan B, Lindquist E, Peto M, Grant D, Shu S, Goodstein D, Barry K, Futrell-Griggs M, Abernathy B, Du J, Tian Z, Zhu L, Gill N, Joshi T, Libault M, Sethuraman A, Zhang XC, Shinozaki K, Nguyen HT, Wing RA, Cregan P, Specht J, Grimwood J, Rokhsar D, Stacey G, Shoemaker RC, Jackson SA (2010) Genome sequence of the palaeopolyploid soybean. Nature 463:178–183

Severin A, Woody J, Bolon YT, Joseph B, Diers B, Farmer A, Muehlbauer G, Nelson R, Grant D, Specht J, Graham M, Cannon S, May G, Vance C, Shoemaker R (2010) RNA-seq atlas of *Glycine max*: a guide to the soybean transcriptome. BMC Plant Biol 10:160

Shoemaker RC, Polzin K, Labate J, Specht J, Brummer EC, Olson T, Young N, Concibido V, Wilcox J, Tamulonis JP, Kochert G, Boerma HR (1996) Genome duplication in soybean (*Glycine* subgenus *soja*). Genetics 144:329–338

Shoemaker R, Keim P, Vodkin LO, Retzel E, Clifton SW, Waterson R, Smoller D, Coryell V, Khanna A, Erpelding J, Gai X, Brendel V, Raph-Schmidt C, Shoop EG, Vielweber CJ, Schmatz M, Pape D, Bowers Y, Theising B, Martin J, Dante M, Wylie T, Granger C (2002) A compilation of soybean ESTs: generation and analysis. Genome 45:329–338

Stears RL, Martinsky T, Schena M (2003) Trends in microarray analysis. Nat Med 9:140–145

Stefanovic S, Pfeil BE, Palmer JD, Doyle JJ (2009) Relationships among Phaseoloid legumes based on sequences from eight chloroplast regions. Syst Botany 34:115–128

Taylor M, Chapman R, Beyaert R, Hernández-Sebastià C, Marsolais F (2008) Seed storage protein deficiency improves sulfur amino acid content in common bean (*Phaseolus vulgaris* L.): redirection of sulfur from gamma-glutamyl-*S*-methyl-cysteine. J Agric Food Chem 56:5647–5654.

Thibaud-Nissen F, Shealy RT, Khanna A, Vodkin LO (2003) Clustering of microarray data reveals transcript patterns associated with somatic embryogenesis in soybean. Plant Physiol 132:118–136

Tomooka N, Vaughan DA, Maxted N, Moss H (2002) The Asian *Vigna*: Genus *Vigna* subgenus *Ceratotropis* genetic resources. Academic, Dordrecht

Varshney RK, Penmetsa RV, Dutta S, Kulwal PL, Saxena RK, Datta S, Sharma TR, Rosen B, Carrasquilla-Garcia N, Farmer AD, Dubey A, Saxena KB, Gao J, Fakrudin B, Singh MN, Singh BP, Wanjari KB, Yuan M, Srivastava RK, Kilian A, Upadhyaya HD, Mallikarjuna N, Town CD, Bruening GE, He G, May GD, McCombie R, Jackson SA, Singh NK, Cook DR (2009) Pigeonpea genomics initiative (PGI): an international effort to improve crop productivity of pigeonpea (*Cajanus cajan* L.). Mol Breed 1–16

Vodkin LO, Khanna A, Shealy R, Clough SJ, Gonzalez DO, Philip R, Zabala G, Thibuad-Nissen F, Sidarous M, Stromvik MV, Shoop E, Schmidt C, Retzel E, Erpelding J, Shoemaker RC, Rodriguez-Huete AM, Polacco JC, Coryell V, Keim P, Gong G, Liu L, Pardinas J, Schweitzer P (2004) Microarrays for global expression constructed with a low redundancy set of 27,500 sequenced cDNAs representing an array of developmental stages and physiological conditions of the soybean plant. BMC Genomics 5:73

Walbot V (1978) Control mechanisms for plant embryogeny. In: Clutter M (ed) Dormancy and developmental arrest. Academic, New York, pp 113–166

Wang ML, Barkley NA, Gillaspie GA, Pederson GA (2008) Phylogenetic relationships and genetic diversity of the USDA *Vigna* germplasm collection revealed by gene-derived markers and sequencing. Genetics Res 90:467–480

Wang Z, Libault M, Joshi T, Valliyodan B, Nguyen HT, Xu D, Stacey G, Cheng J (2010) SoyDB: A knowledge database of soybean transcription factors. BMC Plant Biol 10:14

Weber H, Borisjuk L, Wobus U (2005) Molecular physiology of legume seed development. Annu Rev Plant Biol 56:253–279

Weterings K, Apuya NR, Bi Y, Fischer RL, Harada JJ, Goldberg RB (2001) Regional localization of suspensor mRNAs during early embryo development. Plant Cell 13:2409–2425

Wobus U, Weber H (1999) Seed maturation: Genetic programs and control signals. Curr. Opin. Plant Biol 2:33–38

Yang SS, Valdes-Lopez O, Xu WW, Bucciarelli B, Gronwald JW, Hernandez G, Vance CP (2010) Transcript profiling of common bean (*Phaseolus vulgaris* L.) using the GeneChip soybean genome array: Optimizing analysis by masking biased probes. BMC Plant Biol 10:85
Yi H, Ravilious GE, Galant A, Krishnan HB, Jez JM (2010a) From sulfur to homoglutathione: thiol metabolism in soybean. Amino Acids 39:963–978
Yi J, Derynck MR, Chen L, Dhaubhadel S (2010b) Differential expression of *CHS7* and *CHS8* genes in soybean. Planta 231:741–753
Yi J, Derynck MR, Li X, Telmer P, Marsolais F, Dhaubhadel S (2010c) A single repeat MYB transcription factor, GmMYB176, regulates *CHS8* gene expression and affects isoflavonoid biosynthesis in soybean. Plant J 62:1019–1034
Yin F, Pajak A, Chapman R, Sharpe A, Huang S, Marsolais F (2011) Analysis of common bean expressed sequence tags identifies sulfur metabolic pathways active in seed and sulfur-rich proteins highly expressed in the absence of phaseolin and major lectins. BMC Genomics 12:268
Zhang Z, Yu J, Li D, Liu F, Zhou X, Wang T, Ling Y, Su Z (2010) PMRD: plant microRNA database. Nucl Acids Res 38:D806–D813

Chapter 9
Peanut Seed Development: Molecular Mechanisms of Storage Reserve Mobilization and Effect of Water Deficit Stress on Seed Metabolism

Kameswara Rao Kottapalli, Pratibha Kottapalli and Paxton Payton

Abstract Seeds function as an energy storehouse, providing nourishment to the developing embryo and the young seedling during germination. In addition, due to their oil, protein, and carbohydrate composition, seeds serve as an important food source for both human and animal consumption. Legume seeds and peanuts, in particular, are an inexpensive source of plant proteins and edible oil. Owing to their importance in global food security, it is necessary to understand the genetic, biochemical, and physiological mechanisms involved in controlling seed quality and nutritive attributes. A comprehensive understanding of seed development and the environmental factors that impact the incorporation of the main storage reserves in seeds, such as proteins, fatty acids, starch, and secondary metabolites will enhance our ability to improve seed quality and yield through molecular breeding programs. Here, we discuss the current understanding of seed development in peanut and conserved developmental mechanisms between legumes and the model plant *Arabidopsis*. Primarily, the biochemistry, genetics, and transcriptional regulation during seed development are discussed with emphasis on metabolic pathways of storage reserve biosynthesis. Additionally, we discuss the effects of water deficit stress on seed metabolism, oil quality, and composition.

Keywords Fatty acids · Flavonoids · Peanut · Starch · Storage proteins · Water deficit stress

9.1 Introduction

Peanut (*Arachis hypogaea* L.) is the second most important crop legume, cultivated across the world on 23.5 million hectares with an annual production of 34.5 million tons (FAOSTAT 2009). Peanut is mainly grown for its seed in the United States of America, but represents a rich source of edible fats and a primary source of oil

P. Payton (✉) · K. R. Kottapalli · P. Kottapalli
Center for Biotechnology and Genomics, Texas Tech University and Cropping Systems Research Laboratory, USDA-ARS, 3810 4th Street, Lubbock, TX 79415, USA
e-mail: paxton.payton@ars.usda.gov

G. K. Agrawal, R. Rakwal (eds.), *Seed Development: OMICS Technologies toward Improvement of Seed Quality and Crop Yield*,
DOI 10.1007/978-94-007-4749-4_9,

in many parts of the world. Numerous studies conducted during the past several decades on peanut seed, emphasized SP composition (Cherry et al. 1973; Hinds 1995; Jones and Horn 1930; Johnson et al. 1950), nutritional value (Millerd et al. 1975), and immunological properties of the protein fractions (Kang and Gallo 2007; Koppelman et al. 2004; Viquez et al. 2003). However, metabolic pathways and mechanisms of differential accumulation of various seed storage reserves that result in differences in seed composition and quality remain largely unknown. It is believed that application of transcriptomics and proteomics approaches will provide an insight into the mechanisms involved in development and maturation and the effects of biotic and abiotic stresses on these processes. Ultimately, this knowledge should provide a means for improving peanut seed quality and yield. Transcriptional profiling of developing *Arabidopsis* seeds has provided insight into the primary transcriptional networks that coordinate the genome-wide response to seed developmental programs and lead to the distribution of carbon among oil, protein, and carbohydrate reserves (Girke et al. 2000; Ruuska et al. 2002). Cellular signaling and metabolic events are also driven by protein-protein interactions, PTMs of proteins, and enzymatic activities that cannot be predicted accurately or described by transcriptional profiling approaches alone. Since proteins are directly associated with function, proteomics approaches are being applied increasingly to address biochemical and physiological questions in many model species, such as *Arabidopsis*, rice, maize, *Medicago*, soybean, and *Brassica*.

Much of what we know today regarding the molecular aspects of seed development comes from the basic studies carried out in model legumes and *Arabidopsis*. However, several differences exist between the developmental programs of peanut seeds and those of *Arabidopsis* and even other legumes. The primary difference is that peanut seeds develop underground and have a large endosperm tissue accumulating proteins and lipids, while *Arabidopsis* and most model legume seeds develop aerially in green pods with large embryos. Since seeds harbor the nutritional reserves to support life of humans and other living organisms, it is important to understand the genetic and biochemical mechanisms favoring an enhanced accumulation of main seed reserves, such as proteins, carbohydrates, and lipids. Here, we review our current understanding of seed development in peanut with special emphasis on mechanisms of accumulation of storage compounds. Additionally, we discuss the effect of water-deficit stress on seed metabolism and gene expression.

9.2 Stages of Seed Development

The peanut plant has one of the most fascinating growth habits in the plant kingdom. It produces flowers aerially, but develops its pod (fruit) underground. Upon fertilization of the ovules in the peanut flower, a specialized organ called a gynophore or peg carries the young pod into the soil, where pod maturation occurs (Moctezuma 2003). While the peg is aerial, the embryo, having undergone three to four divisions, is in an arrested state of development (Pattee and Mohapatra 1986,

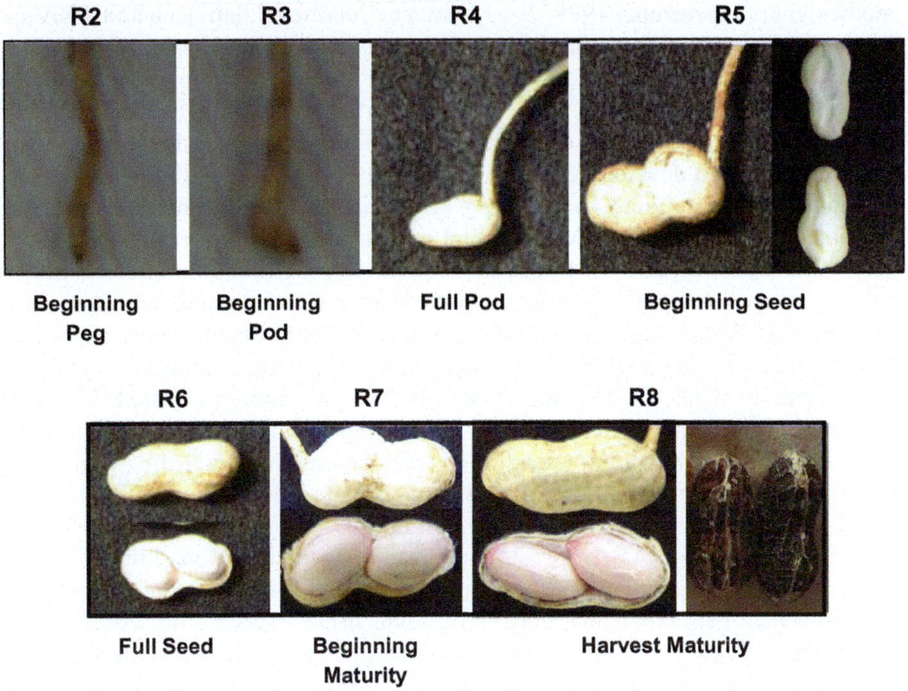

Fig. 9.1 Appearance of pod specific developmental stages (R3 to R8) in FlavRunner 458 plants. R1 (beginning bloom) and R9 (over-mature pod) stages are not shown

1987). Upon contact with the soil, development begins and eventually the ovule is pushed a few centimeters below the surface, the peanut pod begins to grow, the tip bends sideways(parallel to the soil surface) and the pod begins to develop. Development is divided into nine reproductive events (Fig. 9.1), annotated as R1–9 (Boote et al. 1982). The R1 and R2 correspond to anthesis and peg elongation, respectively. The pod-specific stages are R3 (beginning pod, with the peg in the soil and pod turned diageotropically and swollen), R4 (fully-expanded pod), R5 (beginning seed, visible cotyledon growth), R6 (full seed, pod cavity is completely filled), R7 (onset of maturity, visible coloration of pericarp), R8 (harvest maturity, majority of pods have pericarp coloration), and R9 (over-mature pod). Earlier studies on peanut pods pertained to growth and maturation of individual fruits *per se* and were limited to the visual phenological events in the physiological development of pods (Boote et al. 1982; Pattee et al. 1974; Schenk 1961; Williams and Drexler 1981).

In contrast to the wealth of information regarding the phenotypic and morphological changes that define peanut seed development, little is known about the molecular and cellular mechanisms. Peg growth and development is highly regulated by auxin (Moctezuma 1999). The mechanical stimuli provided by the soil surface, in combination with the darkness and soil moisture, trigger the necessary hormonal cues for pod development to begin and correspond with a decline in the auxin signal (Moctezuma 1999, 2003). On a cellular level, ethylene and ABA appear to be the primary hormones controlling peanut seed development, but little is known about their role, particularly during later stages of development. Ethylene appears to be a key hormone in the development of peanut pod. Shlamovitz et al. (1995) reported a release of ethylene by the pod during initial development and retardation of pod development in the presence of ethylene inhibitors. However, the exact role of ethylene in signaling of pod development is not known. ABA regulates a wide range of developmental events and mediates responses to environmental stress conditions. In peanut seed development, the hormone is necessary to attain seed maturation and acquire desiccation tolerance and dormancy (Weber et al. 2005). In *Arabidopsis*, ABA was shown to be intimately connected to the transcriptional control of seed development governed by major regulatory genes, such as *ABI3*, *FUS3*, *LEC1*, and *LEC2* (Finkelstein et al. 2002; Nambara and Marion-Poll 2003). An increase in ABA levels during maturation induces the expression of a CDK inhibitor, leading to termination of cell division (Wang et al. 1998). Similarly, the *FUS3/LEC* genes also regulate cell division as mutation in these genes causes prolonged cell division during seed maturation. Together, an integrated network of ABA, FUS3, and sugar derived signals may regulate seed maturation (Weber et al. 2005). Gibberellic acid (GA) is an ABA antagonist and is involved in cell expansion and seed growth in pea (Ozga et al. 2002; Swain et al. 1995). With current genomics and proteomics techniques, we are now able to examine the underlying molecular events, controlling seed development and impacting seed quality characteristics.

9.3 Signaling Networks and Seed Development

Protein kinases are major signaling molecules in plants. The SnRK1 is a major component in sugar sensing in legume seeds (Weber et al. 2005). The SnRK1 β- and γ-subunits in peas and SnRK1 β-subunit in tomato increase upon maturation. The regulatory γ-subunit is also responsive to GA, ABA, and stress (Bradford et al. 2003). SnRK1 proteins are closely related to a cyclin-dependent protein kinase. As evidenced in rice, cyclin-dependent protein kinase isoforms are required for storage product accumulation *via* phosphorylation of SUS (Asano et al. 2002). Thus, ABA or certain stress conditions derived from carbohydrate and/or energy depletion activate SnRK1s during seed maturation. In transgenic pea with seed specific antisense expression of SnRK1, the maturation phase is delayed resulting in green seeds. Microarray analysis of the silenced plants revealed that genes involved in cell cycle regulation and cytosolic metabolism remained high, indicating that SnRK1-defecient seeds have a prolonged prestorage phase. However, genes related to SP synthesis and ABA-mediated signal transduction were induced much later compared to wild type. These results suggest that SnRK1 and ABA interacts and mediate responses related to seed differentiation at least at the transcript level (Weber et al. 2005).

Current knowledge of the TFs involved in regulating gene expression in developing seeds has been generated from the genetic studies primarily with *Arabidopsis* and recently with soybean (Le et al. 2010). Analyses of mutant phenotypes in *Arabidopsis* indicated that ABI3, FUS3, and LEC1 are major regulators of seed maturation and synergistically control multiple elementary processes during seed development (Parcy et al. 1997). It was demonstrated that ABI3 interacts genetically with both FUS3 and LEC1 in controlling each of the elementary processes, namely, accumulation of chlorophyll and anthocyanins, sensitivity to ABA, and expression of individual members of the 12S SP gene family. In addition, it was reported that both FUS3 and LEC1 regulate positively the abundance of ABI3 protein in the seed (Parcy et al. 1997). In a recent microarray study, it was found thatFUS3-controlled genes are not confined to previously known seed maturation-related genes but also include genes that are involved in the production of secondary metabolites, photosynthesis, and FA biosynthesis (Yamamoto et al. 2010). Another B3 domain TF LEC2 directly activates seed maturation. Many LEC2-induced transcripts are directly involved in storage reserve accumulation, a hallmark of the maturation phase. These transcripts include 2S and 12S SPs, oleosin, and steroleosin (Braybrook et al. 2006). In addition, LEC2 activates seed protein genes before an increase in transcripts encoding LEC1 or FUS3 is observed. Hence, LEC2 is sufficient to initiate seed protein gene expression and likely acts in concert with ABI3, FUS3, and LEC1 to control the initiation and maintenance of the seed maturation phase (Stone et al. 2008). Recently, Le et al. (2010) identified 48 seed-specific TF genes (Table 9.1) that are active at different developmental times and may be important for controlling stage-specific biological events during seed formation. This list includes seven known regulators of seed development, includingLEC1, LEC1-like, LEC2, and FUS3. The rest represents different classes of TFs with unknown roles in seed development (Le et al. 2010). According to a recent study, many of the soybean seed-specific TFs are also seed-specific in *Arabidopsis* (Goldberg 2010). In another soybean study using cDNA microarray, it was found that genes encoding many TFs and DNA-binding proteins showed higher expression levels in the later stages of seed development compared to the green early stages (Jones et al. 2010). From our 14k microarray (Payton et al. 2009) experiments in peanut, we have detected 39 of the 48 *Arabidopsis* seed-specific TFs differentially expressed in four pod development stages. These preliminary results with peanut crop suggest a moderate level of conservation in regulation of seed development across plant species.

9.4 Storage Protein Synthesis

In higher plants seed SPs accumulate during embryogenesis and stored in the dormant seed (Higgins 1984; Walling et al. 1986). Cultivated legumes contain 20–40 % protein by DW in their seeds. In general, up to 80 % of legume seed protein is made up of two distinct types of SPs, vicilin or conglycinin and legumin or

Table 9.1 Seed-specific TF families observed in *Arabidopsis*. (Table adopted from Le et al. 2010)

Gene symbol	TF family	Gene description	Seed stages detected	Phenotype	Reference
MEA	PcG	MEA (MEDEA)	MG	EMB defective	Grossniklaus et al. 1998
LEC1	CCAAT-HAP3	LEC1 (LEAFY COTYLEDON 1)	GLOB, COT	EMB defective	West et al. 1994
LEC2	ABI3-VP1	LEC2 (LEAFY COTYLEDON 2)	GLOB, COT	EMB defective	Meinke et al. 1994
MINI3	WRKY	WRKY10 (MINISEED3)	GLOB	EMB defective	Luo et al. 2005
FUS3	ABI3-VP1	FUS3 (FUSCA 3)	GLOB, COT, MG, PMG	EMB defective	Meinke et al. 1994
PEI1	C3H	PEI1	COT, MG, PMG	EMB defective	Li and Thomas 1998
L1L	CCAAT-HAP3	CCAAT-box binding TF family protein/leafy cotyledon 1-like (L1L)	GLOB, COT, MG	EMB defective	Kwong et al. 2003
RBE	C2H2	RBE (RABBIT EARS)	24H, GLOB	Flower defects	Takeda et al. 2004
SAP	SAP	SAP (STERILE APETALA)	COT	Flower defects	Byzova et al. 1999
AGL67	MADS	TF	MG, PMG	Normal seed	Le et al. 2010
NA	C2H2	Zinc finger (C2H2 type) family protein	GLOB, COT	Normal seed	Le et al. 2010
AtMYB107	MYB	MYB107 (myb domain protein 107); DNA binding / TF	MG, PMG	Normal seed	Le et al. 2010
NA	HB	Homeobox-leucine zipper family protein	GLOB, COT	Normal seed	Le et al. 2010
ARR22	ARR-B	ARR22 (Arabidopsis response regulator 22)	24H, GLOB, COT, MG, PMG	Normal seed	Gattolin et al. 2006; Horák et al. 2008
NA	MYB	myb family TF	GLOB, COT	Normal seed	Le et al. 2010
IAA31	AUX-IAA	IAA31 (indoleacetic acid-induced protein 31)	24H, GLOB, COT, MG	Normal seed	Overvoorde et al. 2005
NA	PcG	Zinc knuckle (CCHC-type) family protein	PMG	Normal seed	Le et al. 2010
AtbZIP67; DPBF2	bZIP	DPBF2 (basic leucine zipper transcription factor 67)	COT, MG, PMG	Normal seed	Le et al. 2010
AGL91	MADS	MADS-box family protein	GLOB	Normal seed	Le et al. 2010
NA	HSF	Heat shock transcription factor family protein	GLOB, COT	Normal seed	Le et al. 2010

Table 9.1 (continued)

Gene symbol	TF family	Gene description	Seed stages detected	Phenotype	Reference
ARR21; ARR13	ARR-B	[AT5G07210, ARR21 (Arabidopsis response regulator 21)]; [AT2G27070, ARR13 (Arabidopsis response regulator 13)]	COT	Normal seed p.n.d.	Horák et al. 2003, n.d.
AGL35	MADS	MADS-box protein (AGL35)	GLOB	Normal seed	Le et al. 2010
AtbZIP15	bZIP	Basic leucine zipper TF (BZIP15)	GLOB, COT	Normal seed	Le et al. 2010
AGL69/MAF4	MADS	MADS affecting flowering 4	MG, PMG	Normal seed	Le et al. 2010
ARF21	ARF	ARF21 (auxin response factor 21)	GLOB	Normal seed	Okushima et al. 2005
NA	AP2-EREBP	DNA binding/TF	PMG	Normal seed	Le et al. 2010
HDG3	HB	Homeobox-leucine zipper family protein	GLOB, COT	Normal seed	Le et al. 2010
AGL45	MADS	MADS-box protein (AGL45)	GLOB	Normal seed	Le et al. 2010
MYB118	MYB	MYB118	COT	Normal seed	Le et al. 2010
AtbZIP72	bZIP	DNA binding/TF	COT	Normal seed	Le et al. 2010
AtHSFA9	HSF	AT-HSFA9 (Arabidopsis thaliana heat shock TF A9)	PMG	Normal seed	Le et al. 2010
NA	ABI3-VP1	Transcriptional factor B3 family protein	COT, MG	p.n.d.	p.n.d.
ARR19	ARR-B	ARR19 (Arabidopsis response regulator 19)	GLOB, COT	p.n.d.	p.n.d.
AGL33	MADS	MADS-box protein (AGL33)	GLOB, COT	p.n.d.	p.n.d.
LBD18	AS2	LOB domain protein 18/LBD18	COT, MG	p.n.d.	p.n.d.
AGL57	MADS	MADS-box family protein	GLOB	p.n.d.	p.n.d.
AtMYB67; AtY53	MYB	AtMYB67/AtY53 (myb domain protein 67)	PMG	p.n.d.	p.n.d.
NA	C3H	Zinc finger (C3HC4-type RING finger) family protein	GLOB	p.n.d.	p.n.d.
NA	NAC	No apical meristem (NAM) family protein	COT	p.n.d.	p.n.d.
NA	C2C2-Dof	Dof-type zinc finger domain-containing protein	GLOB, COT	p.n.d.	p.n.d.
NA	HB	Homeobox protein-related	GLOB, COT, MG	p.n.d.	p.n.d.
NA	MYB	myb family transcription factor	GLOB, COT	p.n.d.	p.n.d.

Table 9.1 (continued)

Gene symbol	TF family	Gene description	Seed stages detected	Phenotype	Reference
AGL90; AGL36	MADS	MADS-box protein (AGL90); DNA binding (AGL36)	GLOB	p.n.d.	p.n.d.
LBD35	AS2	LOB domain family protein/LBD35	GLOB	p.n.d.	p.n.d.
NA	C3H	Zinc finger (C3HC4-type RING finger) family protein	GLOB, COT	p.n.d.	p.n.d.
NA	C2H2	Zinc finger (C2H2 type) family protein	GLOB	p.n.d.	p.n.d.
NA	CCAAT-HAP5	CCAAT-box binding TF Hap5a, putative	GLOB	p.n.d.	p.n.d.
AIL7	AP2-EREBP	AIL7 (AINTEGUMENTA-LIKE 7)	COT, MG, PMG	p.n.d.	p.n.d.

24H 24 h post pollination seed, *GLOB* globular stage seed, *COT* cotyledon stage seed, *MG* mature-green-stage seed, *PMG* postmature green stage seed, *p.n.d.* phenotype not determined as on date, *LBD* lateral organ boundaries domain family protein glycinin (Boulter and Croy 1997). Other seed proteins may also function as primary SPs, such as pea albumin, soybean lectin and a 15 kDa protein, phaseolin and phytohaemagglutinins in beans, urease, narborin, and some trypsin inhibitors (Bailey and Boulter 1971; Boulter and Croy 1997; Goldberg et al. 1989; Richardson 1991). Seed proteins are encoded by several nonhomologs multigene families that vary in size and organization (Casey et al. 1986). Each of the seed protein gene family is independently regulated at gene level and differentially represented in mature plant organs (Goldberg et al. 1989). In soybean, these individual gene families share a common expression pattern where: (i) they are expressed at high levels during embryogenesis; (ii) their mRNAs accumulate and decay in a precise developmental timeline; and (iii) they are inactive or weakly expressed in mature plant organs (Jones et al. 2010; Walling et al. 1986). Table 9.2 lists the legume seed proteins genes that have been characterized at molecular level and studied for their transcript patterns. Initial studies with soybean seed protein genes concluded that both transcriptional and posttranscriptional processes regulate seed protein mRNA levels (Walling et al. 1986), but the exact nature of this regulation is still not understood in soybean (Golombek et al. 2001). In French bean, the seed SPs, phaseolin and phytohemagglutinin, were regulated by the TFPvAlf (*Phaseolus vulgaris* ABA-insensitive 3-like factor). This gene is homologous to the protein encoded by the *Arabidopsis* ABI3 TF (Bobb et al. 1995). The PvAlf gene interacts with the RY repeat in the promoter of SSPs (Bobb et al. 1997). Promoters of the SSP genes also interacted with ROM1 and ROM2, two leucine zipper type TFs, which repressed the PvAlf-activated transcription of the SSP genes (Chern et al. 1996a, b). To manipulate assimilate partitioning and elevated protein content in legume seeds, transgenic *Vicia* plants have been generated that over express a bacterial phosphoenolpyruvate carboxylase (PEPC) (Rolletschek et al. 2004). PEPC provides carbon skeletons for seed specific amino acid biosynthesis (Turpin and Weger 1990). Seeds of transgenic plants show a higher CO_2 uptake, a shift of metabolic flux from sugars into organic acids and free amino acids, and 20 % more protein per gram of seed DW (Rolletschek et al. 2004). In another attempt to increase nitrogen influx and seed protein content, the amino acid permease gene VfAAP1 was over expressed in pea seeds (Rolletschek et al. 2005). Overexpression of the transporter gene resulted in increased seed sink strength for nitrogenand 43 % increase in globulin seed protein. Thus, both PEPC and VfAAP1 are promising targets for molecular breeding for high seed protein content.

From peanut seed EST libraries, we identified seven different SSP gene families (Table 9.2). They include two major genes, *Ara h1* and *Ara h3/4* encoding conarachin and arachin proteins, respectively. These two proteins constitute about 87 % of peanut SSPs (Boldt et al. 2005). Other minor SSP gene families found in peanut are lectins, globulins, legumin b, phytohemagglutinin, and enzyme inhibitors (protease inhibitor/kunitz trypsin inhibitor). Unlike soybean, peanut lectin proteins do not accumulate at mid maturation and totally disappear after first undifferentiated pod stage (24 DAF) (Kottapalli et al. unpublished data). Physiological relevance of this lectin protein in early peanut pod is yet to be investigated.

Table 9.2 Expression of seed protein genes in legumes

Gene symbol	Gene family	Protein encoded	Organism	References
Cgy 1	Vicillin/conglycinin	β-Conglycinin α subunit	Soybean	Chen et al. 1988
Cgy 4	Vicillin/conglycinin	β-Conglycinin β subunit	Soybean	Barker et al. 1988; Bray et al. 1987
Le 1	Lectin	Seed lectin	Soybean	Okamuro et al. 1986
Gy 1	Legumin/glycinin	Glycinin	Soybean	Goldberg et al. (unpublished data)
Kti 1	Trypsin inhibitor	Kunitz trypsin inhibitor minor polypeptide	Soybean	Jofuku and Goldberg (unpublished data)
Kti 2	Trypsin inhibitor	Kunitz trypsin inhibitor minor polypeptide	Soybean	Jofuku and Goldberg (unpublished data)
Kti 3	Trypsin inhibitor	Kunitz trypsin inhibitor major polypeptide	Soybean	Jofuku et al. (unpublished data)
leg A	Legumin/glycinin	Legumin A polypeptide	Pea	Ellis et al. 1988
leg B	Legumin/glycinin	Legumin B polypeptide	Field bean	Baumlein et al. 1988
dlec 2	Globulin	Phytohemagglutinin L subunit	Common bean	Voelker et al. 1989
Phaseolin	Vicillin/conglycinin	Major 7S storage protein	Common bean	Sengupta-Gopalan et al. 1985; Greenwood and Chrispeels 1985
Ara h1	Vicillin/conglycinin	Conarachin	Peanut	Boldt et al. 2005; Kottapalli et al. (unpublished data)
Ara h3/4	Legumin/glycinin	Arachin	Peanut	Boldt et al. 2005; Kottapalli et al. (unpublished data)
arachin 6	Legumin/glycinin	Arachin	Peanut	Boldt et al. 2005; Kottapalli et al. (unpublished data)
Gly 1	Legumin/glycinin	Arachin	Peanut	Boldt et al. 2005; Kottapalli et al. (unpublished data); Kottapalli et al. 2008
arachin Ahy-1	Legumin/glycinin	Arachin	Peanut	Boldt et al. 2005; Kottapalli et al. (unpublished data)
arachin Ahy-4	Legumin/glycinin	Arachin	Peanut	Boldt et al. 2005; Kottapalli et al. (unpublished data)
allergen II	Vicillin/conglycinin	Conarachin	Peanut	Boldt et al. 2005; Kottapalli et al. (unpublished data)
2S protein 1	Vicillin/conglycinin	Conarachin	Peanut	Boldt et al. 2005; Kottapalli et al. (unpublished data)
SSP2	Vicillin/conglycinin	Conarachin	Peanut	Boldt et al. 2005; Kottapalli et al. (unpublished data)
Ara h6	Vicillin/conglycinin	Conglutin	Peanut	Boldt et al. 2005; Kottapalli et al. (unpublished data)
lectin	Lectin	Seed lectin	Peanut	Kottapalli et al. (unpublished data)
leg B	Legumin/glycinin	Legumin b	Peanut	Kottapalli et al. (unpublished data)
protease inhibitor	Trypsin inhibitor	Protease inhibitor	Peanut	Kottapalli et al. (unpublished data)
Kti 1	Trypsin inhibitor	Kunitz trypsin inhibitor	Peanut	Kottapalli et al. (unpublished data)
dlec 2	Globulin	Phytohemagglutinin	Peanut	Kottapalli et al. (unpublished data)

9.5 Fatty Acid Metabolism

The FAs in seeds of oilseed crops are stored in oil bodies (OBs) as TAG. In seeds, they not only act as reserves of carbon and energy, but also determine the economic value of seeds in many crops (Baud and Lepiniec 2009). Production of TAG commences during maturation phase of seed development and results in steady increase of seed DW. TAGs are esters of glycerol in which FAs are esterified to each of the three hydroxyl groups of the glycerol backbone. Once synthesized, TAGs accumulate in the subcellular organelles, the OBs, surrounded by phospholipid monolayers (Robenek et al. 2004; Yatsu and Jacks 1972). In seeds, sucrose is imported into embryo cells and cleaved either by the enzymes, invertase or SUS (Baud et al. 2004; Schwender et al. 2003). Sucrose cleavage generates hexose phosphates, which are metabolized through the oxidative pentose phosphate pathway (OPPP) and glycolytic pathway, providing precursors for FA production in the form of acetyl-CoA. In *Arabidopsis*, a multi-subunit heteromeric acetyl-CoA carboxylase (ACCase) catalyzes the first committed step of the FA biosynthetic pathway (Fig. 9.2; Ohlrogge and Browse 1995). ACCase contains four subunits with three functional domains and catalyzes the ATP dependent synthesis of malonyl-CoA from acetyl-CoA and bicarbonate. This reaction takes place in two steps. First, the carboxylation of covalently bound biotin cofactor to the biotin carboxyl carrier protein (BCCP) is catalyzed by biotin carboxylase domain. Second, the carboxyl group from carboxy-biotin is transferred to acetyl-CoA to form malonyl-CoA. This reaction is catalyzed by the carboxyltransferase domain. Malonyl-CoA is the carbon donor for FA synthesis but the malonyl group has to be transferred from CoA to a protein cofactor like ACP before entering the FA biosynthesis pathway (Ohlrogge and Browse 1995). This reaction is catalyzed by a malonyl-CoA: ACP malonyltransferase (MAT). FAs are produced from acetyl-CoA and malonyl-ACP, where acetyl-CoA is used as the starting unit and the malonyl-CoA as the elongator. The malonylthioesters formed in the above reaction enters into a series of condensation reactions with acetyl-CoA and acyl-ACP acceptors. Three separate condensing enzymes, namely 3-ketoacyl-ACP synthases (KAS) I, II, and III are necessary for the production of an 18-carbon FA. Initial condensation reaction is catalyzed by KASIII, while KASI is responsible for producing chain lengths from six to 16 carbons. KASII finally elongates 16:0-ACP to 18:0-ACP (Pidkowitch et al. 2007). Three additional reactions are required after each condensation step to obtain a saturated FA that is two carbons longer than at the start of the cycle. These reactions are catalyzed by 3-ketoacyl-ACP reductase (KAR), 3-hydroxyacyl-ACP dehydratase (HD), and enoyl-ACP reductase (ENR) (Mou et al. 2000). The enzymes involved in *de novo* FA synthesis form an easily dissociable multisubunits complex referred to as fatty acid synthase (FAS) (Brown et al. 2006). Some 16:0-ACP is released from FAS, while those elongated to 18:0-ACP are desaturated by the enzyme stearoyl-ACP desaturase (SAD). Long chain FA synthesis in the plastids is terminated, when the acyl group is removed from ACP by specific acyl-ACP thioesterases (FAT). Once activated to CoA esters and transported to the endoplasmic reticulum (ER), these FA species are modified by desaturation

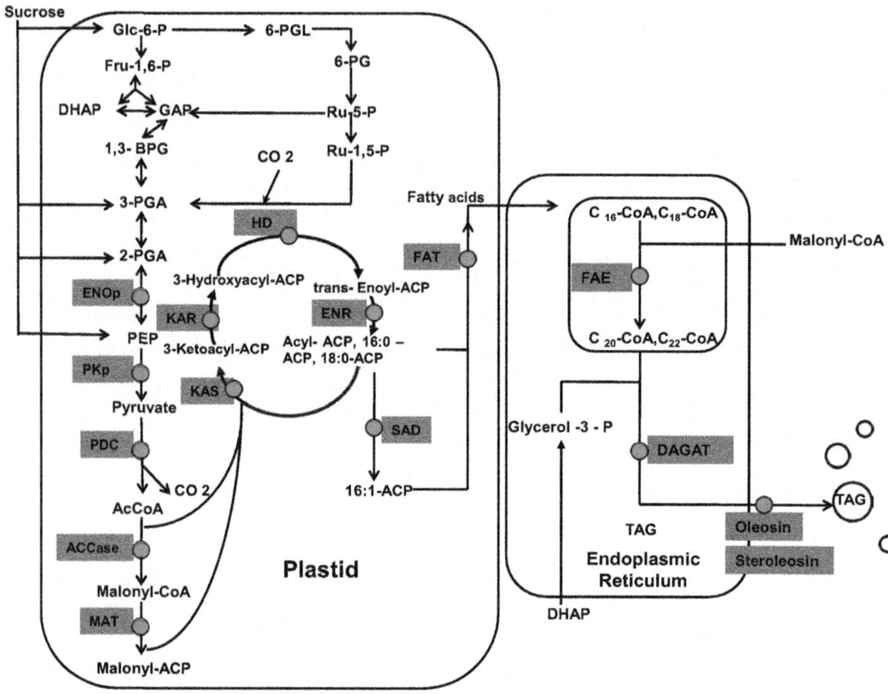

Fig. 9.2 Simplified pathway of triacylglycerol (TAG) biosynthesis in seeds. Genes for enzymatic functions or structural proteins involved in fatty acid (FA) synthesis in plastids and FA modification in the ER are indicated in gray boxes. *Glc-6-P*, glucose-6-phosphate; *Fru-1,6-P* fructose-1,6-biphosphate, *DHAP* dihydroxyacetone-3-phosphate, *GAP* glyceraldehydes-3-phosphate, *1,3-BPG* 1,3-biphosphoglycerate, *3-PGA* 3-phosphoglycerate, *2-PGA* 2-phosphoglycerate, *6-PGL* 6-phosphogluconolactone, *6-PG* 6-phosphogluconate, *Ru-5-P* ribulose-5-phosphate, *Ru-1,5-P* ribulose-1,5-biphosphate, *ENOp* plastidial enolase, *PEP* phosphoenolpyruvate, *PKp* plastidial pyruvate kinase, *PDC* pyruvate dehydrogenase complex, *Ac-CoA* acetyl-coenzyme A, *ACCase* acetyl-CoA carboxylase, *ACP* acyl carrier protein, *MAT* malonyl-CoA:ACP transacylase, *KAS* 3-ketoacyl-ACP synthase, *KAR* 3-ketoacyl-ACP reductase, *HD* 3-hydoxyacyl-ACP dehydratase, *ENR* enoyl-ACP reductase, *SAD* stearoyl-ACP desaturase, *FAT* fatty acyl-ACP thioesterases, *FAE* fatty acid elongase complex, and *DAGAT* diacylglycerol acyltransferase

or elongation. In ER, very long chain fatty acids (VLCFA) are synthesized by the multienzyme fatty acid elongase (FAE) complex. This membrane bound multienzyme complex is involved in four reactions catalyzed by 3-ketoacyl-CoA synthase (FAE1), 3-ketoacyl-CoA reductase, 3-hydroxyacyl-CoA dehydrase, and enoyl-CoA reductase. FA elongation cycles in ER are similar to the cycles of elongation for FA synthesis taking place in plastids. In rapeseed embryos, the cytosolic malonyl-CoA species used for VLCFA biosynthesis are derived from citrate through the action of ATP-citrate lyase and homomeric acetyl-CoA carboxylase (Baud et al. 2003, 2004; Fatland et al. 2005). In the ER, TAG assembly proceeds through different routes (Napier 2007). However, relatively simple route is the Kennedy pathway, which relies on the sequential acylation of glycerol-3-phosphate (G-3-P) backbone. This reaction is catalyzed by glycerol-3-phosphate acyltransferase, lysophosphatidic acid acyltransferase, and diacylglycerolacyltransferase (DAGAT) (Hobbs et al. 1999; Zou et al. 1999). Ultimately the TAGs are stored in OBs covered by a set of proteins, including steroleosin (HSD1) and five distinct oleosins (S1–5). These proteins determine the size of OBs, and prevent their coalescence when in contact with each other in the embryo cells (Li et al. 2007; Siloto et al. 2006).

Most enzymes of the FA biosynthesis pathway have been biochemically characterized and genes encoding most of the enzymes have been cloned from different plant species (Beisson et al. 2003; Mu et al. 2008). In *Arabidopsis*, the plastidial FA synthesis and ER oil synthesis and storage includes at least 26 enzymes or subunits, which are encoded by 65 genes (Beisson et al. 2003). These enzymes, number of genes, and ESTs available for each enzyme in the Lipid Gene Catalog are listed in Table 9.3, along with 22 enzymes and their respective genes identified in peanut EST libraries. In peanut, where only a fraction of genome is sequenced, we have identified >90 EST sequences annotated as genes for FA biosynthesis, including ACCase, a major regulator of the FAS pathway (Baud et al. 2003; Ohlrogge and Jaworski 1997; Sasaki and Nagano 2004). In *Arabidopsis*, the lipogenic genes encoding the plastidic enzymes of FA biosynthesis, had 2- to13-fold increased transcript expression at the onset of maturation (4 DAP). These genes display a bell-shaped pattern of expression with a decline in expression at 10 DAP. By contrast, the expression of genes encoding enzymes of TAG metabolism (e.g., FAE1, TAG1, S3, and HSD1) increases later and remains high during the maturation phase. This differential timing of expression of genes involved in FA biosynthesis and TAG metabolism suggests the existence of a global regulation system for the entire FA biosynthetic pathway (Wobus and Weber 1999). On the other hand, the enzymes of TAG metabolism and storage may be regulated by a different set of TFs (Baud and Lepiniec 2009).

Recently, three independent studies have provided evidence to support the above mentioned hypothesis and identified TFs regulating the changes in mRNA levels of lipogenic genes. Among the master regulators mentioned in the previous section (signaling of seed development), LEC2 is of particular interest since its induction tightly correlates with the onset of oil accumulation in maturing seeds. In addition, ectopic expression of LEC2 confers embryonic characters to transgenic seedlings, triggering TAG accumulation in rosette leaves (Stone et al. 2001; Santos-Mendoza et al. 2005). Using transgene induction system and qRT-PCR experiments, it was shown that LEC2 triggers the expression of genes encoding seed specific lipid body-associated oleosins and steroleosin. However, the genes encoding enzymes required for FA biosynthesis are not direct targets for LEC2 (Santos-Mendoza et al. 2005). In a separate experiment using over expressing transgenics and microarrays, it was shown that *Arabidopsis* LEC1 master regulator causes globally increased expression of the FA biosynthetic genes (Mu et al. 2008). In the plastidial FA synthetic pathway, 18 known enzyme-coding genes were reported to be upregulated in the LEC1-overexpressing plants, including those encoding three subunits of the key enzyme ACCase. Concomitantly, levels of major FAs species and lipids were substantially increased in the transgenic plants. It was also shown that LEC1

Table 9.3 Enzymes of FA synthesis in *Arabidopsis* and peanut

Cellular function	Cellular activity	*Arabidopsis*		Peanut
		Number of genes	Number of ESTs	Number of ESTs
Synthesis of FAs in plastids	F1. Plastidial homomeric acetyl-CoA carboxylase (ACCase; EC 6.4.1.2)	1	0	2
	F2a. α-Carboxyltransferase of heteromeric ACCase	1	19	2
	F2b. β-Carboxyltransferase of heteromeric ACCase (plastid encoded)	1	0	0
	F2c. Biotin carboxyl carrier protein of heteromeric ACCase	2	68	1
	F2d. Biotin carboxylase of heteromeric ACCase	1	21	4
	F3. Malonyl-CoA: ACP malonyltransferase (EC 2.3.1.39)	1	3	0
	F4a. Ketoacyl-ACP synthase I (KAS I; EC 2.3.1.41)	1	39	1
	F4b. KAS II (EC 2.3.1.41)	1	2	2
	F4c. KAS III (EC 2.3.1.41)	1	6	0
	F5. Plastidial ketoacyl-ACP reductase (EC 1.1.1.100)	5	35	3
	F6. Plastidial hydroxyacyl-ACP dehydrase (EC 4.2.1.*)	2	7	1
	F7. Plastidial enoyl-ACP reductase (EC 1.3.1.9)	1	11	1
	F8. Stearoyl-ACP desaturase (EC 1.14.19.2)	7	43	2
	F9. Plastidial ACP	5	42	2
	F10a. Acyl-ACP thioesterase FatA (EC 3.1.2.14)	2	5	0
	F10b. Acyl-ACP thioesterase FatB (EC 3.1.2.*)	1	35	1
	F11a. Plastidial pyruvate dehydrogenase E1α of pyruvate DH complex (EC 1.2.4.1)	2	46	0
	F11b. Plastidial pyruvate dehydrogenase E1β of pyruvate DH complex (EC 1.2.4.1)	2	36	2
	F11c. Plastidial dihydrolipoamide acetyltransferase of PDH complex (EC 2.3.1.12)	4	33	0
	F11d. Plastidial dihydrolipoamide dehydrogenase of PDH complex (EC 1.8.1.4)	2	20	0
	F12. Plastidial lipoate synthase	1	1	0
	F13. Plastidial lipoyltransferase	1	0	0
Synthesis and storage of oil	O1. Acyl-CoA: diacylglycerol acyltransferase (DAGAT; EC 2.3.1.20)	2	7	1
	O2. Oil body oleosin	8	132	2
	O3. Caleosin	7	42	0
	O4. Phospholipid: diacylglycerol acyltransferase (PDAT; EC 2.3.1.158)	3	9	4
Total		*65*	*662*	*31*

function is partially dependent on ABI3, FUS3, and WRINKLED1 (WRI1) TFs in the regulation of FA biosynthesis (Mu et al. 2008). In another study, it was shown that WRI1, an AP2-type TF binds the AW-box promoter sequences, which are conserved among the genes involved in FA synthesis (Maeo et al. 2009). These genes included a subunit of pyruvate kinase, pyruvate dehydrogenase (PDH), BCCP2, ACP1, and KAS1. However, the genes encoding enzymes for synthesis of TAG and genes for seed-type oleosins do not contain the typical AW-box sequences and are not regulated by WRI1. In sesamum, another seed-specific TF, basic region/helix-loop-helix (SebHLH) binds to the E-box (CANNTG) or G-box (CACGTG) *cis*-elements of ER specific genes, such as oleic acid desaturase (FAD2), DAGAT, SAD, oleosin, and caleosin and regulates their transcription (Kim et al. 2007). Together, these studies suggest that there are separate global regulators for FA biosynthesis, and TAG synthesis and storage.

Peanut oil quality depends on the polyunsaturated FA composition. Ideally genotypes with high oleic acid (monounsaturated FA) and low linoleic acid (polyunsaturated FA) are desirable. The mutation in the FAD2 gene results in the production of peanut germplasm containing 80 % oleate and as little as 2 % linoleate (Norden et al. 1987). The mutation disrupts the enzyme oleoylphosphotidyl choline desaturase enzyme and occurs either naturally or induced by chemical mutagens (Patel et al. 2004) or RNAi (Yin et al. 2007).

9.6 Starch Metabolism

In developing seeds, starch is synthesized as storage starch in heterotrophic plastids called amyloplasts. Storage starch is distinct from the transient starches, which are synthesized in leaf chloroplasts during the day and degraded at night to provide carbon for non-photosynthetic metabolism. Starch biosynthesis principally involves three committed enzymes namely, AGPase, starch synthase (SS), and SBE. In the first committed step of starch synthesis, ADP-glucose is formed from glucose-1-phosphate (Glc-1-P) and ATP in a reaction catalyzed by AGPase. This enzyme is active within the plastids and its substrates, Glc-1-P and ATP are imported into plastids through Glc-6-P/Pi antiporter and the ATP/ADP transporter, respectively. In seeds, Glc-6-P is interconverted to Glc-1-P by the plastidial isoform of phosphoglucomutase (PGM) (Hill and Smith 1991). The pyrophosphate produced by the action of AGPase is removed by a plastid-bound inorganic pyrophosphatase (iPP). This removal of pyrophosphate effectively displaces equilibrium of the AGPase reaction in favor of ADP-glucose synthesis (Weiner et al. 1987). In the next step, ADP-glucose is utilized by SS to synthesize α-(1–4) linkage between the nonreducing end of a preexisting glucan chain. Subsequently, the glucan chain is branched and debranched by SBE and debranching enzymes, respectively, to form amylopectin. Granule bound SS also utilizes ADP-glucose to synthesize the amylose fraction of storage starch.

There is a complex relationship between metabolic processes during seed development and understanding those relations can provide insights into yield-related processes (Borisjuk et al. 2003). For example, manipulation of AGPase activity alone can induce complex and pleotropic changes in seed development. In *Vicia narbonensis*, transgenic plants expressing antisense AGPase showed altered storage product composition of mature cotyledons and altered cotyledon development (Weber et al. 2000). Additionally, the seeds from these plants had elevated sugar and water levels and an extended seed-filling phase (Weber et al. 2000). While AGPase activity and ADP-glucose levels were strongly reduced, starch was only moderately reduced and seed protein content was increased due to an increase in sulfur-rich albumins.

9.7 Flavonoids in Developing Seeds

Flavonoids are aromatic amino acids derived secondary metabolites with diverse roles in plant seeds. The type, amount, and localization of these flavonoids depend on the plant species and developmental stage of the tissues. The major flavonoids found in seeds are flavonols, anthocyanins, phlobaphenes, isoflavones, and proanthocyanidins (PA; condensed tannins) (Lepiniec et al. 2006). Among legumes, anthocyanins are found in the seed coat of bean (Beninger and Hosfield 2003), while isoflavones constitutes the colorless metabolites in the embryo and seed coat (Dhaubhadel et al. 2003). In *Arabidopsis*, PAs accumulate in the seed coat and protect the embryo and endosperm against pathogen attack (Debeaujon et al. 2003; Shirley 1998). During seed desiccation the PAs oxidizes to brown pigments and confer color to the mature seed. PAs are synthesized in the endothelium of inner integument in *Arabidopsis*. Similarly, peanut seed coats are also composed of two layers and the seed color is determined by the color of inner integument (Rodriguez and Norden 1970). In peanuts, tannin (speculated as PAs) content determines the seed coat color and prevents colonization by *Aspergillus flavus*, an aflatoxin producing fungus (Carter 1970, 1973).

PAs are synthesized in the flavonoid biosynthetic pathway (Fig. 9.3; Kleindt et al. 2010). To date, at least 17 genes have been identified encoding the structural proteins, regulatory proteins, and proteins for flavonoid production (Lepiniec et al. 2006). Two additional genes encoding leucoanthocyanidin reductase and anthocyanidin reductase were identified, which act at the beginning of two major stereospecific pathways of PA biosynthesis (Dixon et al. 2005; Tanner et al. 2003; Xie et al. 2003). Various loci involved in PA biosynthesis in *Arabidopsis* seed are listed in Table 9.4 (adapted from Lepiniec et al. 2006). Most of the regulatory genes of flavonoids biosynthesis belong to MYB and bHLH families, the two largest families of regulatory genes in plants (Heim et al. 2003; Stracke et al. 2001).

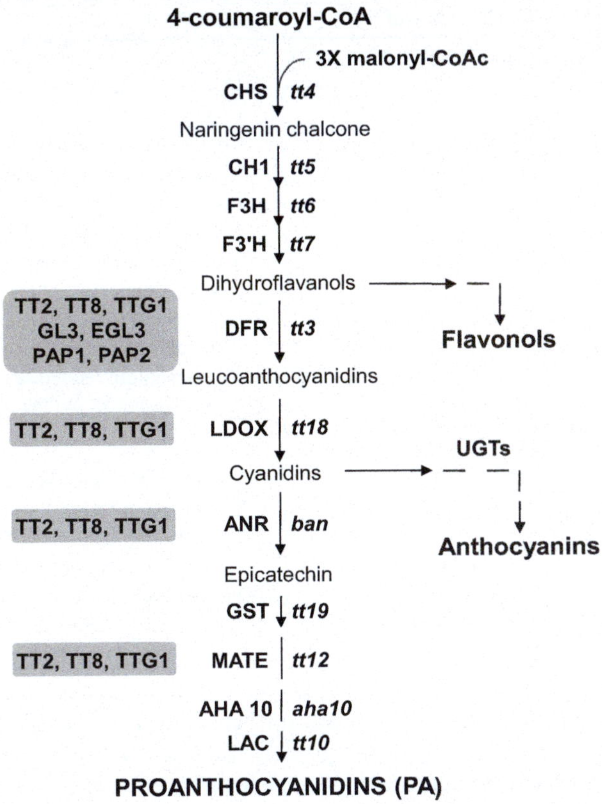

Fig. 9.3 Simplified schematic representation of seed flavonoid biosynthesis in *Arabidopsis* with focus on proanthocyanidin (PA) synthesis. Transcriptional regulators of the pathway are shaded in grey and enzymes are indicated in bold capital letters. Mutant phenotypes for the genes coding the enzymes and TFs are represented in small case letters and italics. Much of the intermediate steps leading to the synthesis of PAs are not known. *CHS* chalcone synthase, *CHI* chalcone isomerase, *F3H* flavonoid 3-hydoxylase, *DFR* dihydroflavonol 4-reductase, *LDOX* leucoanthocyanidin dioxygenase, *ANR/ban* anthocyanidin reductase, *GST* glutathione-S-transferase, *MATE* multidrug and toxic compound extrusion-type transporter, *AHA10* P-type H^+-ATPase, *LAC* laccase-like; *UGTs* UDP-dependent glycosyltransferases, *TT* transparent testa, *TTG* transparent testa glabra, *GL3* glabra 3, *EGL3* enhancer of glabra 3, and *PAP* production of anthocyanins pigments. (Modified from Kleindt et al. 2010)

9.8 Effect of Water Deficit Stress on Seed Metabolism at Molecular Level

Water-deficit during seed development causes severe losses in crop production. In peanut, late season water-deficit stress has significant effects on yield and seed quality (Stirling et al. 1989b). The detrimental effects of water deficit on pod growth are

Table 9.4 Loci involved in PA biosynthesis

Locus	Gene product
Structural genes	
tt3	Dihydroflavonol reductase (DFR)
tt4	Chalcone synthase (CHS)
tt5	Chalcone isomerase (CHI)
tt6	Flavanone-3-hydoxylase (F3H)
tt7	Flavanone-3'-hydoxylase (F3H)
tt10	Polyphenoloxydase (PPO)
tt12	MATE secondary transpoter
tt15	Glycosyltransferase (GT)
tt18/tds4/tt11	Leucocyanidin dioxygenase (LDOX)
tt19/tt14	Glutahthione S-transferase (GST)
ban	Anthocyanidin reductase (ANR)
aba10	Autoinhibited H^+-ATPase isoform10
Regualtory genes	
tt1	Transcription factor WIP-type Zn-finger
tt2	Transcription factor AtMYB123
tt8	Transcription factor AtbHLH042
tt16/abs	Transcription factor MADS AtAGL32
ttg1	Transcription factor ("WD40" or "WDR")
ttg2	Transcription factor AtWRKY44

reported to be due to an inadequate supply of assimilates (Stirling et al. 1989a) or calcium (Skelton and Shear 1971). It was further suggested that cultivars with high-harvest index achieved by rapid pod growth at the expense of canopy growth are more tolerant to early season water deficit (Chapman et al. 1993). In another study, late season water-deficit stress coinciding with pod filling, significantly reduced total oil and linoleic FA content, while total protein and oleic acid content increased (Dwivedi et al. 1996). However, to our knowledge, in peanut or in any grain legume, there are no reports of systematic evaluation of transcript and protein responses to water-deficit stress on developing seeds. Due to paucity of published data, in this section, we will review the water-deficit stress response, which is a combination of heat and water-deficit stress in cereal crop species. In maize, water deficit at 9 DAP caused significant differential transcriptional response in placenta and endosperm tissues (Yu and Setter 2003). Placenta experienced a low-tissue water status than endosperm and concomitant induction of several known stress tolerance genes. These genes mainly include heat shock proteins (HSPs), chaperonins, and major intrinsic proteins. In addition, placenta accumulated four to five-fold higher concentration of ABA and caused a decrease in sugar flux during stress. In contrast, genes for cell division and growth and cell wall degrading enzymes were downregulated in endosperm, suggesting a strong inhibition of cell proliferation during water stress (Setter and Flannigan 2001). Water deficit also upregulated a homeodomain leucine zipper TF (*ZmOCL5*), which may provide tissue-specific stress regulation in kernels (Yu and Setter 2003).

Wheat grown in warmer areas lost significant amount of grain weight due to decrease in starch synthesis (Majoul et al. 2003). In cereals, heat stress generally results in decrease in the overall synthesis of starch and SPs and ultimate reduction in grain yield (Mangelsen et al. 2011; Passarella et al. 2008). In severe cases, it also leads to grain abortion. An interesting finding is that several heat-stress responses in *Arabidopsis* shoots and drought-stress responses in barley are conserved (Mangelsen et al. 2011). In barley at cellular level, short-term heat stress consistently induced HSP-mediated protein folding, ROS scavenging, and the biosynthesis of compatible solutes (Mangelsen et al. 2011). In parallel, genes involved in embryo development, hormone biosynthesis, and cell signaling were altered, indicating rapid sensing, signal transduction, and adaptation of central processes to abiotic stress. In addition, physiology and development of barley caryopsis were negatively affected as evident from decreased starch biosynthesis, lipid metabolism, and amino acid metabolism immediately after heat stress.

Tolerant peanut genotypes grown under deficit irrigation conditions of 70 % potential evapotranspiration (PET) replacement, exhibited a 15–20 % decline in yield compared to the 100 % PET treatment (Kottapalli et al. 2009). To better understand the effect of water-deficit stress on peanut seed development, we have applied deficit irrigation equaling 50 % PET replacement. Preliminary results (Fig. 9.4) indicate that PEPC transcripts induced four-fold at stage 2 pod during stress. As PEPC provides carbon skeletons for amino acid biosynthesis, water deficit resulted in increased protein content as reported earlier(Dwivedi et al. 1996). In addition, water stress induced SUS transcripts at stage 2 and 3,which may generate higher amounts of acetyl-CoA, the precursors for FA production. However, the transcripts levels of this protein declined more than four-fold at stage 4, when the rate of FA production is at a peak. Thus, lower SUS and concomitant decrease in acetyl-CoA at stage 4, seeds might be responsible for reduced total oil content during water stress in peanut. Transcript levels of lipoxygenase, a SP in seeds (Siedow 1991),was initially induced at stage 2 and 3 but declined more than two-fold at stage 4 seeds during stress. The exact role of this protein in seed development is not known, but the protein is greatly required under stress conditions in common bean (Porta et al. 1999).

9.9 Concluding Remarks

Peanut seed is an important dietary staple and biologically, presents a unique model crop in terms of its development. However, compared to other legumes, little is known about the genetic control of development and the impact of biotic and abiotic stresses on development and quality. With limited sequence data and very few ESTs of peanut in the public domain, the challenges are significant. However, with the advent of low-cost, high-throughput sequencing, and increased computational capacity for sequence assembly, the sequencing of complex genomes and thus the transcriptomics analysis is rapidly becoming routine. An integrated approach of agronomic, genetic, and biochemical methods coupled with emerging omics data on the spatial and temporal dynamics of transcript, protein, and metabolite levels will

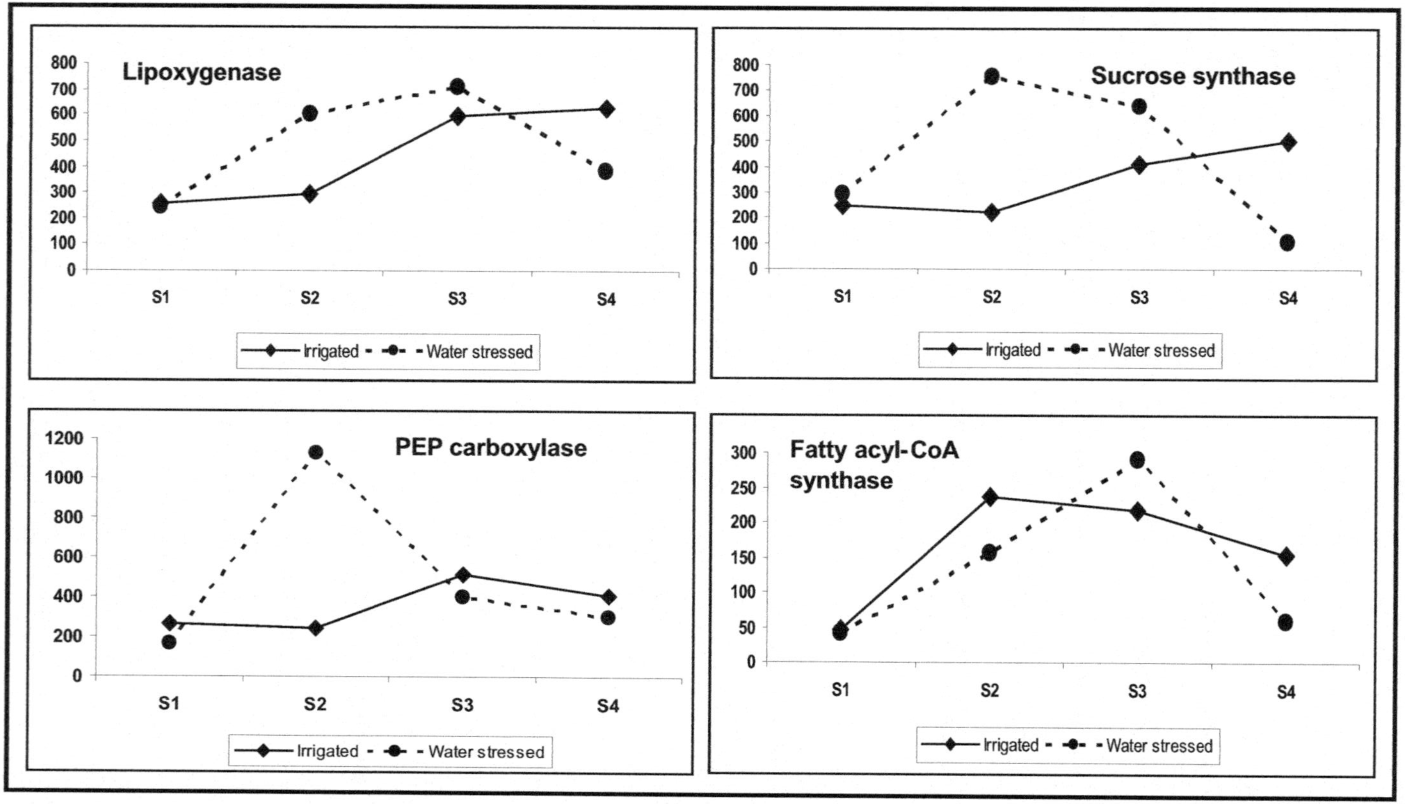

Fig. 9.4 Graphical representation of transcript levels of genes involved in fatty acid (FA) and amino acid biosynthesis during four developmental stages of peanut pod development. *Solid lines* indicate the pattern of expression under normally irrigated/control conditions. *Dashed lines* indicate the gene expression during water-stressed conditions. Values presented are the means of at least three replicates of experiments on agilent arrays

be key to our tailoring crops for production in low-input regions or for specific nutritive and quality traits.

References

Asano T, Kunieda N, Omura Y, Ibe H, Kawasaki T, Takano M, Sato M, Furuhashi H, Mujin T, Takaiwa F, Wu CY, Tada Y, Satozawa T, Sakamoto M, Shimada H (2002) Rice SPK, a calmodulin-like domain protein kinase, is required for storage product accumulation during seed development. Plant Cell 14:619–628

Bailey CJ, Boulter D (1971) Urease, a typical seed protein of the Leguminosae. In: Harbone JB, Boulter D, Turner BL (eds) Chemotaxonomy of the Leguminosae. Academic, New York, pp 485–502

Barker SJ, Harada JJ, Goldberg RB (1988) Cellular localization of soybean storage protein mRNA in transformed tobacco seeds. Proc Natl Acad Sci U S A 85:458–462

Baud S, Lepiniec L (2009) Regulation of *de novo* fatty acid synthesis in maturing oilseeds of *Arabidopsis*. Plant Physiol Biochem 47:448–455

Baud S, Guyon V, Kronenberger J, Wuilléme S, Miquel M, Caboche M, Lepiniec L, Rochat C (2003) Multifunctional acetyl-coA carboxylase1 is essential for very long chain fatty acid elongation and embryo development in *Arabidopsis*. Plant J 33:75–86

Baud S, Bellec Y, Miquel M, Bellini C, Caboche M, Lepiniec L, Faure JD, Rochat C (2004) *gurke* and *pasticcino3* mutants affected in embryo development are impaired in acetyl-CoA carboxylase. EMBO Rep 5:515–520

Baumlein H, Muller AJ, Schiemann J, Helbing D, Manteuffel R, Wobus U (1988) Expression of a *Vicia faba* legumin B gene in transgenic tobacco plants: gene dosage dependent protein accumulation. Biochem Physiol Pflanzen 183:205–210

Beisson F, Koo AJK, Ruuska S, Schwender J, Pollard M, Thelen JJ, Paddock T, Salas JJ, Savage L, Milcamps A, Mhaske VB, Cho Y, Ohlrogge JB (2003) *Arabidopsis* genes involved in acyl lipid metabolism: a 2003 census of the candidates, a study of the distribution of expressed sequence tags in organs, and a web-based database. Plant Physiol 132:681–697

Beninger CW, Hosfield GL (2003) Antioxidant activity of extracts, condensed tannin fractions and pure flavonoids from *Phaseolus vulgaris* L. seed coat color genotypes. J Agric Food Chem 51:7879–7883

Bobb AJ, Eiben HG, Bustos MM (1995) PvAIf, an embryo-specific acidic transcriptional activator enhances gene expression from phaseolin and phytohemagglutinin promoters. Plant J 8:331–343

Bobb AJ, Chern MS, Bustos MM (1997) Conserved RY-repeats mediate transactivation of seed-specific promoters by the developmental regulator PvALF. Nucl Acids Res 25:641–647

Boldt A, Fortunato D, Conti A, Petersen A, Ballmer-Weber B, Lepp U, Reese G, Becker WM (2005) Analysis of the composition of an immunoglobulin E reactive high molecular weight protein complex of peanut extract containing Ara h1 and Ara h3/4. Proteomics 5:675–686

Boote KJ, Stansell JR, Schubert AM, Stone JF (1982) Irrigation, water use and water relations. In: Patte HE, Young CT (eds) Peanut science and technology. American Peanut Research and Education Association, Yoakum, pp 164–205

Borisjuk L, Rolletschek H, Wobus U, Weber H (2003) Differentiation of legume cotyledons as related to metabolic gradients and assimilate transport into seeds. J Exp Bot 54:503–512

Boulter D, Croy RRD (1997) The structure and biosynthesis of legume seed storage proteins: a biological solution to the storage of nitrogen in seeds. Adv Botan Res 27:1–84

Bradford KJ, Downie AB, Gee OH, Alvarado V, Yang H, Dahal P (2003) Abscisic acid and gibberellin differentially regulate expression of genes of the SNF1-related kinase complex in tomato seeds. Plant Physiol 132:1560–1576

Bray EA, Naito S, Pan NS, Anderson E, Dube P, Beachy RN (1987) Expression of the *β*- subunit of β-conglycinin in seeds of transgenic plants. Planta 172:364–370
Braybrook SA, Stone SL, Park S, Bui AQ, Le BH, Fischer RL, Goldberg RB, Harada JJ (2006) Genes directly regulated by LEAFY COTYLEDON2 provide insight into the control of embryo maturation and somatic embryogenesis. Proc Natl Acad Sci U S A 103:3468–3473
Brown AP, Affleck V, Fawcett T, Slabas AR (2006) Tandem affinity purification tagging of fatty acid biosynthetic enzymes in *Synechocystis* sp. PCC6803 and *Arabidopsis thaliana*. J Exp Bot 57:1563–1571
Byzova MV, Franken J, Aarts MG, de Almeida-Engler J, Engler G, Mariani C, Van Lookeren Campagne MM, Angenent GC (1999) *Arabidopsis* STERILE APETALA, a multifunctional gene regulating inflorescence, flower, and ovule development. Genes Dev 13:1002–1014
Carter JBH (1970) Studies on the growth of *Aspergillus flavus* on groundnut kernels. Ph.D. Thesis, University of Reading, pp 170
Carter JBH (1973) The influence of testa, damage and seed dressing on emergence of groundnut *(Arachis hypogaea)*. Ann Appl Biol 74:315–323
Casey R, Domoney C, Ellis N (1986) Legume storage proteins and their genes. In: Miflin BJ (ed) Oxford surveys of plant molecular and cell biology, vol 3. Oxford University Press, Oxford, pp 1–95
Chapman SC, Ludlow MM, Blamey FPC, Fischer KS (1993) Effect of drought during early reproductive development on growth of cultivars of groundnut (*Arachis hypogaea* L.). II. Biomass production, pod development and yield. Field Crops Res 32:211–225
Chern MS, Bobb AJ, Bustos MM (1996a) The Regulator of MAT2 (ROM2) protein binds to early maturation promoters and represses PvALF-activated transcription. Plant Cell 8:305–321
Chern MS, Eiben HG, Bustos MM (1996b) The developmentally regulated bZIP factor ROM1 modulates transcription from lectin and storage protein genes in bean embryos. Plant J 10:135–148
Chen ZL, Pan NS, Beachy RN (1988) A DNA sequence element that confers seed-specific enhancement of a constitutive promoter. EMBO J 7:297–302
Cherry JP, Dechary JM, Ory RL (1973) Gel electrophoretic analysis of peanut proteins and enzymes. I. Characterization of DEAE-cellulose separated fractions. J Agric Food Chem 21:652–655
Debeaujon I, Nesi N, Perez P, Devic M, Grandjean O, Caboche M, Lepiniec L (2003) Proanthocyanidin-accumulating cells in *Arabidopsis* testa: regulation of differentiation and role in seed development. Plant Cell 15:2514–2531
Dhaubhadel S, McGarvey BD, Williams R, Gijzen M (2003) Isoflavonoid biosynthesis and accumulation in developing soybean seeds. Plant Mol Biol 53:733–743
Dixon RA, Xie DY, Sharma SB (2005) Proanthocyanidins—A final frontier in flavonoid research? New Phytol 165:9–28
Dwivedi SL, Nigam SN, Nageswara Rao RC, Singh U, Rao KVS (1996) Effect of drought on oil, fatty acids and protein contents of groundnut (*Arachis hypogaea* L.) seeds. Field Crops Res 48:125–133
Ellis JR, Shirsat AH, Hepher A, Yarwood JN, Gatehouse JA, Croy RRD, Boulter D (1988) Tissue-specific expression of a pea legumin gene in seeds of *Nicotina plumbaginifolia*. Plant Mol Biol 10:203–214
FAOSTAT (2009) http://faostat.fao.org/. Accessed 18 April 2011
Fatland BL, Nikolau BJ, Wurtele ES (2005) Reverse genetic characterization of cytosolic acetyl-CoA generation by ATP-citrate lyase in *Arabidopsis*. Plant Cell 17:182–203
Finkelstein RR, Gampala SSL, Rock CD (2002) Abscisic acid signaling in seeds and seedlings. Plant Cell 14:S15–S45
Gattolin S, Alandete-Saez M, Elliott K, Gonzalez-Carranza Z, Naomab E, Powell C, Roberts JA (2006) Spatial and temporal expression of the response regulators ARR22 and ARR24 in *Arabidopsis thaliana*. J Exp Bot 57:4225–4233
Girke T, Todd J, Ruuska S, White J, Benning C, Ohlrogge J (2000) Microarray analysis of developing *Arabidopsis* seeds. Plant Physiol 124:1570–1581

Goldberg RB (2010) Using genomics to dissect soybean seed development. 13th biennial molecular and cellular biology of the soybean conference, Durham, North Carolina.8–11 August 2010
Goldberg RB, Barker SJ, Perez-Grau L (1989) Regulation of gene expression during plant embryogenesis. Cell 56:149–160
Golombek S, Rolletschek H, Wobus U, Weber H (2001) Control of storage proteinaccumulation during legume seed development. J Plant Physiol 158:457–464
Greenwood JS, Chrispeels MJ (1985) Correct targeting of the bean storage protein phaseolin in the seeds of transformed tobacco. Plant Physiol 79:65–71
Grossniklaus U, Vielle-Calzada JP, Hoeppner MA, Gagliano WB (1998) Maternal control of embryogenesis by *MEDEA*, a polycomb group gene in *Arabidopsis*. Science 280:446–450
Heim MA, Jakoby M, Werber M, Martin C, Weisshaar B, Bailey PC (2003) The basic helix-loop-helix transcription factor family in plants: a genome-wide study of protein structure and functional diversity. Mol Biol Evol 20:735–747
Higgins TJV (1984) Synthesis and regulation of major proteins in seeds. Annual Rev Plant Physiol 35:191–221
Hill LM, Smith AM (1991) Evidence that glucose-6-phosphate is imported as the substrate for starch synthesis by the plastids of developing pea embryos. Planta 185:91–96
Hinds MJ (1995) Fatty acid composition of Caribbean-grown peanuts (*Arachis hypogaea* L.) at three maturity stages. Food Chem 53:7–14
Hobbs DH, Lu C, Hills MJ (1999) Cloning of a cDNA encoding diacylglycerol acyltransferase from *Arabidopsis thaliana* and its functional expression. FEBS Lett 452:145–149
Horák J, Brzobohaty B, Lexa M (2003) Molecular and physiological characterisation of an insertion mutant in the *ARR21* putative response regulator gene from *Arabidopsis thaliana*. Plant Biol 5:245–254
Horák J, Grefen C, Berendzen KW, Hahn A, Stierhof YD, Stadelhofer B, Stahl M, Koncz C, Harter K (2008) The *Arabidopsis thaliana* response regulator ARR22 is a putative AHP phospho-histidine phosphatase expressed in the chalaza of developing seeds. BMC Plant Biol 8:77
Johnson P, Shooter EM, Rideal EK (1950) The globulins of the ground nut. (*Arachis hypogaea*) II. Electrophoretic examination of the arachin system. Biochim Biophys Acta 5:376–396
Jones DB, Horn MJ (1930) The properties of arachin and conarachin and the proportionate occurrence of these proteins in the peanut. J Agric Res 40:673–682
Jones SI, Gonzalez DO, Vodkin LO (2010) Flux of transcript patterns during soybean seed development. BMC genomics 11:136
Kang IH, Gallo M (2007) Cloning and characterization of a novel peanut allergen Ara h3 isoform displaying potentially decreased allergenicity. Plant Sci 172:345–353
Kim MJ, Kim JK, Shin JS, Suh MC (2007) The SebHLH transcription factor mediates trans-activation of the *SeFAD2* gene promoter through binding to E- and G-box elements. Plant Mol Biol 64:453–466
Kleindt CK, Stracke R, Mehrtens F, Weisshaar B (2010) Expression analysis of flavonoid biosynthesis genes during *Arabidopsis thaliana* silique and seed development with a primary focus on the proanthocyanidin biosynthetic pathway. BMC Res Notes 3:255
Koppelman SJ, Wensing M, Ertmann M, Knulst AC, Knol EF (2004) Relevance of Ara h1, Ara h2 and Ara h3 in peanut-allergic patients, as determined by immunoglobulin E Western blotting, basophil-histamine release and intracutaneous testing: ara h2 is the most important peanut allergen. Clin Exp Allergy 34:583–590
Kottapalli KR, Payton P, Rakwal R, Agrawal GK, Shibato J, Burow M, Puppala N (2008) Proteomics analysis of mature seed of four peanut cultivars using two-dimensional gel electrophoresis reveals distinct differential expression of storage, anti-nutritional, and allergenic proteins. Plant Sci 175:321–329
Kottapalli KR, Rakwal R, Shibato J, Burow G, Tissue D, Burke J, Puppala N, Burow M, Payton P (2009) Physiology and proteomics of water-deficit stress response in three contrasting peanut genotypes. Plant Cell Environ 32:380–407

Kwong RW, Bui AQ, Lee H, Kwong LW, Fischer RL, Goldberg RB, Harada JJ (2003) LEAFY COTYLEDON1-LIKE defines a class of regulators essential for embryo development. Plant Cell 15:5–18

Le BH, Chenga C, Buia AQ, Wagmaistera JA, Henrya KF, Pelletierb J, Kwongb L, Belmonteb M, Kirkbrideb R, Horvathc S, Drewsd GN, Fischere RL, Okamurof JK, Haradab JJ, Goldberg RB (2010) Global analysis of gene activity during Arabidopsis seed development and identification of seed-specific transcription factors. PNAS 107: 8063–8070

Lepiniec L, Debeaujon I, Routaboul JM, Baudry A, Pourcel L, Nesi N, Caboche M (2006) Genetics and biochemistry of seed flavonoids. Annu Rev Plant Biol 57:405–430

Li Z, Thomas TL (1998) *PEI1*, an embryo-specific zinc finger protein gene required for heart-stage embryo formation in *Arabidopsis*. Plant Cell 10:383–398

Li L, Li Q, Rohlin L, Kim U, Salmon K, Rejtar T, Gunsalus RP, Karger BL, Ferry JG (2007) Quantitative proteomic and microarray analysis of the archaeon *Methanosarcina acetivorans* grown with acetate versus methanol. J Proteome Res 6:759–771

Luo M, Dennis ES, Berger F, Peacock WJ, Chaudhury A (2005) *MINISEED3* (*MINI3*), a *WRKY* family gene, and *HAIKU2* (*IKU2*), a leucine-rich repeat (*LRR*) *KINASE* gene, are regulators of seed size in *Arabidopsis*. Proc Natl Acad Sci U S A 102:17531–17536

Maeo K, Tokuda T, Ayame A, Mitsui N, Kawai T, Tsukagoshi H, Ishiguro S, Nakamura K (2009) An AP2-type transcription factor, WRINKLED1, of *Arabidopsis thaliana* binds to the AW-box sequence conserved among proximal upstream regions of genes involved in fatty acid synthesis. Plant J 60:476–487

Majoul T, Bancel E, Triboï E, Ben Hamida J, Branlard G (2003) Proteomic analysis of the effects of heat stress on hexaploid wheat grain: characterization of heat-responsive proteins from total endosperm. Proteomic 3:175–183

Mangelsen E, Kilian J, Harter K, Jansson C, Wanke D, Sundberg E (2011) Transcriptome analysis of high-temperature stress in developing barley caryopses: early stress responses and effects on storage compound biosynthesis. Mol Plant 4:97–115

Meinke DW, Franzmann LH, Nickle TC, Yeung EC (1994) Leafy cotyledon mutants of *Arabidopsis*. Plant Cell 6:1049–1064

Millerd A, Spencer D, Dudman WF, Stiller M (1975) Growth of immature pea cotyledons in culture. Aust J Plant Physiol 2:51–60

Moctezuma E (1999) Changes in auxin patterns in developing gynophores of the peanut plant (*Arachis hypogaea* L.). Ann Bot 83:235–242

Moctezuma E (2003) The peanut gynophore: a developmental and physiological perspective. Can J Bot 81:183–190

Mou Z, He Y, Dai Y, Liu X, Li J (2000) Deficiency in fatty acid synthase leads to premature cell death and dramatic alterations in plant morphology. Plant Cell 12:405–418

Mu J, Tan H, Zheng Q, Fu F, Liang Y, Zhang J, Yang X, Wang T, Chong K, Wang XJ, Zuo J (2008) *LEAFY COTYLEDON1* is a key regulator of fatty acid biosynthesis in *Arabidopsis*. Plant Physiol 148:1042–1054

Nambara E, Marion-Poll A (2003) ABA action and interaction in seeds. Trends Plant Sci 8:213–217

Napier JA (2007) The Production of unusual fatty acids in transgenic plants. Annu Rev Plant Biol 58:295–319

Norden AJ, Gorbet DW, Knauft DA, Young CT (1987) Variability in oil quality among peanut genotypes in the Florida breeding program. Peanut Sci 14:7–11

Ohlrogge J, Browse J (1995) Lipid biosynthesis. Plant Cell 7:957–970

Ohlrogge JB, Jaworski JG (1997) Regulation of fatty acid synthesis. Annu Rev Plant Physiol Plant Mol Biol 48:109–136

Okamuro JK, Jofuku KD, Goldberg RB (1986) Soybean seed lectin gene and flanking nonseed protein genes are developmentally regulated in transformed tobacco plants. Proc Natl Acad Sci U S A 83:8240–8244

Okushima Y, Overvoorde PJ, Arima K, Alonso JM, Chan A, Chang C, Ecker JR, Hughes B, Lui A, Nguyen D, Onodera C, Quach H, Smith A, Yu G, Theologis A (2005) Functional genomic analysis of the *AUXIN RESPONSE FACTOR* gene family members in *Arabidopsis thaliana*: unique and overlapping functions of ARF7 and ARF19. Plant Cell 17:444–463

Overvoorde PJ, Okushima Y, Alonso JM, Chan A, Chang C, Ecker JR, Hughes B, Lui A, Onodera C, Quach H, Smith A, Yu G, Theologis A (2005) Functional genomic analysis of the AUXIN/*INDOLE-3-ACETIC ACID* gene family members in *Arabidopsis thaliana*. Plant Cell 17:3282–3300

Ozga JA, van Huizen R, Reinecke DM (2002) Hormone and seed-specific regulation of pea fruit growth. Plant Physiol 128:1379–1389

Parcy F, Valon C, Kohara A, Misera S, Giraudat J (1997) The *ABSCISIC ACID-INSENSITIVE3*, *FUSCA3*, and *LEAFY COTYLEDON1* loci act in concert to control multiple aspects of *Arabidopsis* seed development. Plant Cell 9:1265–1277

Passarella VS, Savin R, Slafer GA (2008) Are temperature effects on weight and quality of barley grains modified by resource availability? Aust J Agric Res 59:510–516

Patel M, Jung S, Moore K, Powell G, Ainsworth C, Abbott A (2004) High-oleate peanut mutants result from a MITE insertion into the *FAD2* gene. Theor Appl Genet 108:1492–1502

Pattee HE, Mohapatra SC (1986) Hypanthium and style senescence in relation to ovary development in *Arachis hypogaea* L. Bot Gaz 147:302–311

Pattee HE, Mohapatra SC (1987) Anatomical changes during ontogeny of the peanut (*Arachis hypogaea* L.) fruit: Mature megagametophyte through heart shaped embryo. Bot Gaz 148: 156–164

Pattee HE, Johns EB, Singleton JA, Sanders TH (1974) Composition changes of peanut fruit parts during maturation. Peanut Sci 1:57–62

Payton P, Kottapalli KR, Rowland D, Faircloth W, Guo B, Burow M, Puppala N, Gallo M (2009) Gene expression profiling in peanut using high density oligonucleotide microarrays. BMC Genomics 10:265

Pidkowitch MS, Nguyen HT, Heilmann I, Ischebeck T, Shanklin J (2007) Modulating seed beta-ketoacyl-acyl carrier protein synthase II level converts the composition of a temperate seed oil to that of a palm-like tropical oil. Proc Natl Acad Sci U S A 104:4742–4747

Porta H, Rueda-Benítez P, Campos F, Colmenero-Flores JM, Colorado JM, Carmona MJ, Covarrubias AA, Rocha-Sosa M (1999) Analysis of lipoxygenase mRNA accumulation in the common bean (*Phaseolus vulgaris* L.) during development and under stress conditions. Plant Cell Physiol 40:850–858

Richardson MJ (1991) Seed storage proteins: the enzyme inhibitors. In: Rogers LJ (ed) Methods in Plant biochemistry, vol 5. Academic, New York, pp 259–305

Robenek MJ, Severs NJ, Schlattmann K, Plenz G, Zimmer KP, Troyer D, Robenek H (2004) Lipids partition caveolin-1 from ER membranes into lipid droplets: updating the model of lipid droplet biogenesis. FASEB J 18:866–868

Rodriguez VA, Norden AJ (1970) Inheritance of inner seed-coat color in peanuts. J Hered 61:161–163

Rolletschek H, Borisjuk L, Radchuk R, Miranda M, Heim U, Wobus U, Weber H (2004) Seed-specific expression of a bacterial phosphoenolpyruvate carboxylase in *Vicia narbonensis* increases protein content and improves carbon economy. Plant Biotechnol J 2:211–219

Rolletschek H, Hosein F, Miranda M, Heim U, Götz KP, Schlereth A, Borisjuk L, Saalbach I, Wobus U, Weber H. (2005) Ectopic expression of an amino acid transporter (VfAAP1) in seeds of *Vicia narbonensis* and pea increases storage proteins. Plant Physiol 137:1236–1249

Ruuska SA, Girke T, Benning C, Ohlrogge JB (2002) Contrapuntal networks of gene expression during *Arabidopsis* seed filling. Plant Cell 14:1191–1206

Santos-Mendoza M, Dubreucq B, Miquel M, Caboche M, Lepiniec L (2005) LEAFY COTYLEDON 2 activation is sufficient to trigger the accumulation of oil and seed specific mRNAs in *Arabidopsis* leaves. FEBS Lett 579:4666–4670

Sasaki Y, Nagano Y (2004) Plant acetyl-CoA carboxylase: structure, biosynthesis, regulation, and gene manipulation for plant breeding. Biosci Biotechnol Biochem 68:1175–1184

Schenk RU (1961) Development of the peanut fruit. Georgia Agric Exp Stn Tech Bull NS 22:5–53

Schwender J, Ohlrogge JB, Shachar-Hill Y (2003) A flux model of glycolysis and the oxidative pentosephosphate pathway in developing *Brassica napus* embryos. J Biol Chem 278:29442–29453

Sengupta-Gopalan C, Reichert NA, Barker RF, Hall TC, Kemp JD (1985) Developmentally regulated expression of the bean β-phaseolin gene in tobacco seed. Proc Natl Acad Sci U S A 82:3320–3324

Setter TL, Flannigan BA (2001) Water deficit inhibits cell division and expression of transcripts involved in cell proliferation and endoreduplication in maize endosperm. J Exp Bot 52:1401–1408

Shirley BW (1998) Flavonoids in seeds and grains: physiological function, agronomic importance and the genetics of biosynthesis. Seed Sci Res 8:415–422

Shlamovitz N, Ziv M, Zamski E (1995) Light, dark and growth regulator involvement in groundnut (*Arachis hypogaea* L.) pod development. Plant Growth Regul 16:37–42

Siedow JN (1991) Plant lipoxygenase: structure and function. Annu Rev Plant Physiol Plant Mol Biol 42:145–188

Siloto RMP, Findlay K, Lopez-Villalobos A, Yeung EC, Nykiforuk CL, Moloney MM (2006) The accumulation of oleosins determines the size of seed oilbodies in *Arabidopsis*. Plant Cell 18:1961–1974

Skelton BJ, Shear GM (1971) Calcium translocation in the peanut *(Arachis hypogaea* L.). Agronomy J 63:409–412

Stirling CM, Black CR, Ong CK (1989a) The response of groundnut (*Arachis hypogaea* L.) to timing of irrigation. II. 14C-partitioning and plant water status. J Exp Bot 40:1363–1373

Stirling CM, Ong CK, Black CR (1989b) The response of groundnut (*Arachis hypogaea* L.) to timing of irrigation. I. Development and growth. J Exp Bot 40:1145–1153

Stone SL, Kwong LW, Yee KM, Pelletier J, Lepiniec L, Fischer RL, Goldberg RB, Harada JJ (2001) *LEAFY COTYLEDON2* encodes a B3 domain transcription factor that induces embryo development. Proc Natl Acad Sci U S A 98:11806–11811

Stone SL, Braybrook SA, Paula SL, Kwong LW, Meuser J, Pelletier J, Hsieh TF, Fischer RL, Goldberg RB, Harada JJ (2008) *Arabidopsis* LEAFY COTYLEDON2 induces maturation traits and auxin activity: implications for somatic embryogenesis. Proc Natl Acad Sci U S A 105:3151–3156

Stracke R, Werber M, Weisshaar B (2001) The *R2R3-MYB* gene family in *Arabidopsisthaliana*. Curr Opin Plant Biol 4:447–456

Swain SM, Ross JJ, Reid JB, Kamiya Y (1995) Gibberellins and pea seed development. Expression of the *lhi*, *ls* and *le*5839 mutations. Planta 195:426–433

Takeda S, Matsumoto N, Okada K (2004) *RABBIT EARS*, encoding a SUPERMAN-like zinc finger protein, regulates petal development in *Arabidopsis thaliana*. Development 131:425–434

Tanner GJ, Francki KT, Abrahams S, Watson JM, Larkin PJ, Ashton AR (2003) Proanthocyanidin biosynthesis in plants: purification of legume leucoanthocyanidin reductase and molecular cloning of its cDNA. J Biol Chem 278:31647–31656

Turpin DH, Weger HG (1990) Interactions between photosynthesis, respiration and nitrogen assimilation. In: Dennis DT, Turpin DH (eds) Plant physiology, biochemistry and molecular biology. Longman Scientific, Singapore, pp 422–433

Viquez OM, Konan KN, Dodo HW (2003) Structure and organization of the genomic clone of a major peanut allergen gene, *Ara h1*. Mol Immunol 40:565–571

Voelker TA, Herman EM, Chrispeels MJ (1989) In vitro mutated phytchemagglutinin genes expressed in tobacco seeds: role of glycans in protein targeting and stability. Plant Cell 1: 95–104

Walling L, Drews GN, Goldberg RB (1986) Transcriptional and post-transcriptional regulation of soybean seed protein mRNA levels. Proc Natl Acad Sci U S A 83:2123–2127

Wang H, Qi Q, Schorr P, Cutler AJ, Crosby WL, Fowke LC (1998) ICK1, a cyclin-dependent protein kinase inhibitor from *Arabidopsis thaliana* interacts with both Cdc2a and CycD3, and its expression is induced by abscisic acid. Plant J 15:501–510

Weber H, Rolletschek H, Heim U, Golombek S, Gubatz S, Wobus U (2000) Antisense-inhibition of ADP-glucose pyrophosphorylase in developing seeds of *Vicia narbonensis* moderately decreases starch but increases protein content and affects seed maturation. Plant J 24:33–43

Weber H, Borisjuk L, Wobus U (2005) Molecular physiology of legume seed development. Annu Rev Plant Biol 56:253–279

Weiner H, Stitt M, Heldt HW (1987) Subcellular compartmentation of pyrophosphate and alkaline pyrophosphatase in leaves. Biochim Biophys Acta 893:13–21

West M, Yee KM, Danao J, Zimmerman JL, Fischer RL, Goldberg RB, Harada JJ (1994) LEAFY COTYLEDON1 is an essential regulator of late embryogenesis and cotyledon identity in *Arabidopsis*. Plant Cell 6:1731–1745

Williams EJ, Drexler JS (1981) A non-destructive method for determining peanut pod maturity. Peanut Sci 8:134–141

Wobus U, Weber H (1999) Sugars as signal molecules in plant seed development. Biol Chem 380:937–944

Xie DY, Sharma SB, Paiva NL, Ferreira D, Dixon RA (2003) Role of anthocyanidin reductase, encoded by *BANYULS* in plant flavonoid biosynthesis. Science 299:396–399

Yamamoto A, Kagaya Y, Usui H, Hobo T, Takeda S, Hattori T (2010) Diverse roles and mechanisms of gene regulation by the *Arabidopsis* seed maturation master regulator FUS3 revealed by microarray analysis. Plant Cell Physiol 51:2031–2046

Yatsu LY, Jacks TJ (1972) Spherosome membranes. Half unit-membranes. Plant Physiol 49:937–943

Yin D, Deng S, Zhan K, Cui D (2007) High-olieic peanut oils produced by HpRNA-mediated gene silencing of oleate desaturase. Plant Mol Biol Rep 25:154–163

Yu LX, Setter TL (2003) Comparative transcriptional profiling of placenta and endosperm in developing maize kernels in response to water deficit. Plant Physiol 131:568–582

Zou J, Wei Y, Jako C, Kurnar A, Selvaraj G, Taylor DC (1999) The *Arabidopsis thaliana* TAG1 mutant has a mutation in a diacylglycerol acyltransferase gene. Plant J 19:645–653

Chapter 10
Probing the Genes Expressed in Developing Seed of Oilseed Plants: *Brassica Napus* (L.) as A Case Example

Sreekala Chellamma, Bhinu Pillai and Abdelali Hannoufa

Abstract Seed development is controlled by complex regulatory interactions involving transcriptional, biochemical, and metabolic processes. In this chapter, we will briefly discuss the major changes in molecular components involved in the process of seed development using *Brassica napus* as a case example. The two major groups of storage compounds that accumulate during seed development in *B. napus* are lipids (~ 60 %) and proteins (~ 37 %). The accumulation of secondary metabolites also occurs around the same time as lipid accumulation. It is also worth noting that even though *B. napus* seeds are exalbuminous, there is considerable accumulation of carbohydrates during the initial stages of seed development. These carbohydrates contribute to fatty acid accumulation during the later stages of development. Several regulatory machineries, including transcription factors, chromatin remodeling elements, ubiquitination, and phosphoregulation by protein kinases are involved in this process. Transcripts of genes responsible for photosynthetic activities and hormonal regulation were also detected in the developing seed, but these processes are not covered in this chapter.

Keywords *Brassica* · Gene expression · Metabolism · Oilseeds · Seed development · Storage compounds

10.1 Introduction

Seed development proceeds through a series of spatiotemporal controls of gene expression. Different stages of seed development feature specific molecular and metabolic events that define the course of physiological changes. Complex regulatory interactions, involving transcriptional, biochemical, and metabolic processes define the seed development program (Xiang et al. 2008). Therefore, a genome-wide view

A. Hannoufa (✉)
Agriculture and Agri-Food Canada, Southern Crop Protection and Food Research Centre, 1391 Sandford Street, N5V 4T3, London, Ontario, Canada
e-mail: Abdelali.Hannoufa@agr.gc.ca

S. Chellamma · B. Pillai
Dow AgroSciences LLC, 9330 Zionsville Road, Indianapolis, IN 46268, USA

G. K. Agrawal, R. Rakwal (eds.), *Seed Development: OMICS Technologies toward Improvement of Seed Quality and Crop Yield*,
DOI 10.1007/978-94-007-4749-4_10, © Springer Science+Business Media Dordrecht 2012

of gene expression profiles in developing seeds is needed to understand the underlying molecular mechanisms governing seed development. Knowledge of gene expression patterns associated with a specific stage of seed development is crucial to understanding the molecular and biochemical events characteristic of that stage. A good understanding of the genomic and transcriptomic makeup of the seed will facilitate the development of suitable tools for altering seed development and seed quality traits, such as oil content and quality, using molecular genetics approaches.

Like most other dicot plants, *Brassica napus* (also called rapeseed, oilseed rape, or canola) seed is composed of a seed coat, endosperm, and embryo. The cruciferous rapeseeds are exalbuminous, where the endosperm is eventually consumed and its space is occupied by the embryo during seed development. The main constituent of the early developing seeds is the liquid endosperm, which serves as a significant carbon resource for the ensuing stages of seed development from the embryo through to the endosperm. The endosperm is depleted to a single aleuronic layer in mature rapeseeds as a result of the significant cellular and biochemical changes in the endosperm during seed development (Huang et al. 2009b).

In general, three main stages mark seed development in higher plants: pattern formation/cell proliferation, maturation (including the seed filling), and desiccation (Goldberg et al. 1989; Thomas 1993). Rapeseeds undergo active cell proliferation during the 10-12 DAP, in which they simultaneously establish metabolic networks for subsequent seed maturation (Dong et al. 2004). Seed reserves (such as carbohydrates, oils, and proteins) accumulate rapidly during the seed-filling phase of maturation stage, which is followed by seed desiccation and acquisition of dormancy (Close 1996). The *LEA* genes play an important role in establishing tolerance to desiccation during maturation, as well as tolerance to osmotic stress (Xu et al. 1996).

10.2 Differential Expression of Genes During Seed Development

Analysis of differential gene expression is a commonly used approach to determine gene expression profiles during different stages of plant development. Several studies have been conducted to identify genes that are expressed preferentially during seed development, as well as the patterns of their expression at various seed developmental stages (Dong et al. 2004; Huang et al. 2009b; Niu et al. 2009; Xiang et al. 2008; Yu et al. 2010). Some of these studies revealed conserved metabolic processes in seeds (Niu et al. 2009), while others revealed developmental stage-specific genes and some differences (Dong et al. 2004; Yu et al. 2010). The cell proliferation is an important stage in seed development owing to its critical role in determining the pattern of embryo formation. This stage also signifies establishment of a network for the biosynthesis of storage products in the embryo. A sufficient number of ESTs derived from a given tissue can provide a reasonable picture of tissue-specific gene-expression profile (Girke et al. 2000; Park et al. 1993). A comparison of gene expression profiles between rapeseed embryo and leaf tissues identified 2,561 differentially-expressed genes with changes ranging from 1.5 to 65 folds (Xiang et al. 2008). Based on SAGE and further biometric analysis ~35,000 expressed transcripts were identified during rapeseed early seed development (http://rapeseed.plantsignal.cn; Wu et al. 2008). Dong et al. (2004) analyzed a rapeseed cDNA library derived from seeds at 15 DAP and identified 104 differentially expressed and 54 unique genes. These genes indicated a diverse array of functions ranging from cell structure and development to unknown function. Twenty-five genes were found to be expressed only in seeds, of which 11 were expressed as early as 5–10 DAP. Four of the seed-specific genes were expressed only in the seed coat and another five in both the embryos and seed coats.

Yu et al. (2010) identified a large number of genes with diverse functions that were differentially regulated during seed development. These genes were involved in photosynthesis, CW biosynthesis, lipid metabolism, secondary metabolism, hormone metabolism, protein metabolism, signaling, development, and transport. In their experiments, seeds at 35 days post anthesis (DPA) showed 10-fold higher number of upregulated genes involved in photosynthesis, including those implicated in light reactions. In addition, photosynthetic genes involved in Calvin cycle were downregulated at 20 DPA (Yu et al. 2010). Fei et al. (2007) noticed a decline in expression of a substantial number of genes during seed maturation. Many of the downregulated genes were associated with energy pathways, including carbohydrate metabolism, protein metabolism, and photosynthesis. The gene encoding a 12 S globulin (cruciferin) seed SP *CRA1* was two-fold lower in seeds at 20 DPA relative to 35 DPA (Yu et al. 2010). However, *Arabidopsis* 12 S seed SP family was previously reported to be associated with embryo development with high abundance during the latter half of embryogenesis (Pang et al. 1988). Huang et al. (2009b) identified 429 unisequences across three endosperm development stages: globular, heart-shaped, and cotyledon. The most evident differential expression was detected in globular-shaped embryos relative to heart shaped and cotyledon at the early stages of endosperm development. However, they were unable to detect any obvious differences in gene expression between the heart-shaped embryo and cotyledon stages. In another study to understand gene expression during seed maturation in rapeseed in relation to the induction of secondary dormancy, the majority of genes were found to be downregulated during maturation in cultivars DH12075 and AC Excel (Fei et al. 2007). Over 300 genes each in DH12075 and AC Excel cultivars were found to have significantly altered expression during seed maturation (Fei et al. 2007). Differences in gene expression patterns between cultivars were more evident during the transition from full-size embryo to mature seed.

10.3 Genes Involved in Lipid Metabolism

Increased knowledge on the regulation of lipid metabolism in *B. napus* helps plant breeders and biotechnologists to modify seed oil composition to meet market needs. Our ability to enhance lipid quality and quantity in rapeseed could be aided by deciphering its genome. However, *B. napus*, an allotetraploid (AACC genome,

~1200 Mb), is a combination of the genomes from *B. rapa* (AA, 700 Mb) and *B. oleracea* (CC, 500 Mb) that underwent duplication during evolution. The availability of a close relative model plant *A. thaliana* (125 Mb) has made it easy to use comparative genomics approaches to understand metabolic processes. The development of high-throughput technologies, such as transcriptomics, SAGE, and next generation whole genome sequencing (Schena et al. 1995; Velculescu et al. 1995), makes it possible to perform a comprehensive gene expression profiling during *B. napus* seed development and to identify regulatory factors for FA metabolism. For example, ESTs obtained from developing *Arabidopsis* seeds served as the useful resource to investigate the conversion of carbohydrates to seed oil in higher plants (White et al. 2000). So far, approximately 700 genes encoding lipid metabolism-related proteins have been identified in *Arabidopsis* (Beisson et al. 2003) and this could serve as an important resource to understand the regulation of lipid metabolism in rapeseed.

The *B. napus* oilseed is rich in oleic and linoleic acids, and is a widely used source of oil for food and other industrial purposes. TAGs are the major storage oil in *B. napus* (Chia et al. 2005). Genes responsible for active FA metabolism were reported to express as early as nine DAF (O'Hara et al. 2002). Elaborate studies on seed development in *Arabidopsis* revealed that seed weight and lipid content were relatively low during early embryo development, whereas carbohydrate accumulation was high (Baud et al. 2002). It was also found that seed weight and FA accumulation occurred concurrently and accumulation of different types of FAs occurred at different periods of development. Moreover, 93 % of FAs were acetylated by 19 DAF. Weselake et al. (1993) reported that the appearance of DAGAT activity coincided well with the onset of lipid accumulation, and its activity was directly related to lipid accumulation. According to them, the DAGAT activity was increased during 14–33 DAF and then decreased rapidly as the lipid content stabilized in seeds.

Niu et al. (2009) classified the *B. napus* ESTs involved in lipid metabolism into four subprocesses: (i) metabolic pathways converting the photosynthate into seed oil; (ii) FA elongation and degradation; (iii) lipid metabolism in the ER and mitochondria; and (iv) proteins involved in lipid metabolism. These authors observed that most FA synthesis-related genes and some genes related to the OPPP were highly expressed at the advanced stages of seed development (21 DAF through 31 DAF) compared to sugar or starch metabolism-related genes. It has been proposed that during lipid synthesis, sugars are transported from the source tissues to the endosperm and then absorbed by the embryo (Fischer and Weber 2002). The cleavage product of sucrose by SUS, glc-6-P is utilized by both the cytosolic and plastidic glycolytic pathways. A glycolytic pathway in oilseeds helps conversion of the intermediates to pyruvate. Acetyl-CoA generated from pyruvate is then used as the source of carbon for the synthesis of FAs. This process requires combined action of acetyl-CoA carboxylase and FA synthase (Eastmond and Rawsthorne 2000). The glycoxylate cycle, β-oxidation, and gluconeogenesis enzymes are active during embryo development and their activities increase during embryo maturation and desiccation (Chia et al. 2005). The activities of these enzymes were also detected in the cotyledon, where the majority of lipid accumulation occurs (Chia et al. 2005).

Proteomics studies showed that FA synthesis-related proteins were expressed prominently at the midpoint of seed filling (bell-shaped pattern) and were highly expressed at four or five weeks after flowering (WAF; then decrease after six WAF; Hajduch et al. 2006). This finding is consistent with the findings reported by Niu et al. (2009). It was reported that types I and II genes related to FA synthesis started to show increased expression at 21 or 25 DAF and reached their maximum expression at 31 DAF. An independent transcriptomics study also reported that numerous genes including those involved in FA synthesis, FA elongation, triglyceride synthesis, and lipid transfer proteins (LTPs) were differentially expressed in seeds at 20 DPA compared to 35 DPA (Yu et al. 2010).

Numerous other factors, including plant hormones, were found to be involved in regulating lipid biosynthesis and FA accumulation. Expression of genes involved in wax FA metabolism, including *CER1*, *KCS1*, and *CER2*, were found to be regulated by WIN1, an *Arabidopsis* ethylene response factor (ERF)-type TF (Broun et al. 2004). While ABA is known to induce lipid biosynthesis in the developing rapeseed embryo (Zou et al. 1995), ABI4 served as a repressor of lipid breakdown during *Arabidopsis* seed germination (Penfield et al. 2006).

10.4 Biology of Starch and Sugar Metabolism

Embryos accumulate lipid as a major storage product and generally contain little or no starch at maturity. However, some oilseeds were reported to accumulate starch at the intermediate stages of development (da Silva et al. 1997; Kang and Rawsthorne 1994; Niu et al. 2009; Yu et al. 2010). There appears an active metabolic flux from photoassimilation to primary carbon metabolism in the endosperm. Given that starch is a carbon reserve, it could be utilized at later stages of seed development and in the synthesis of lipids (Norton and Harris 1975). Starch may also serve as the carbon source for synthesis of sugars, such as sucrose and stachyose, which have been proposed to enhance desiccation tolerance during the dry-out phase (Leprince et al. 1990). Active starch synthesis and decomposition occur during the early stages of embryo development, and typically peaks at the early to mid-stages of development (Niu et al. 2009). Expression of genes involved in carbohydrate production was found to be higher at 20 DPA compared to 35 DPA and was consistent with peak starch accumulation during 32–33 DPA (Yu et al. 2010).

Norton and Harris (1975) reported that oil content in seeds was five-fold higher than the peak starch content. King et al. (1997) suggested that even though starch was the predominant carbohydrate in seeds, it was not present in sufficient amounts to meet oil biosynthesis requirements. Starch synthesis in rapeseed embryos is known to occur in both the cotyledons and embryonic axis prior to and during the early phase of storage lipid accumulation (da Silva et al. 1997). According to that study, starch was not synthesized in the embryos by photosynthesis, but rather accumulated from carbon imported from various vegetative parts of the plant. They also showed that starch degradation starts after occurrence of one-third of the oil accumulation, and gradually disappears at maturity. Interestingly, blocking starch biosynthesis in the embryo was accompanied by a dramatic reduction in the importation of carbon source into the plastid (Vigeolas et al. 2004), and the impairment of formation of the embryo as a sink tissue. These findings indicate that imported sugars might be the main carbon source. Starch synthesis at the early stages may be essential in transforming the embryo into a sink organ prior to lipid synthesis (da Silva et al. 1997). In support of this, Niu et al. (2009) showed that the sucrose transporter (*EL626999*) gene was expressed two-fold higher at 19 DAF and later compared them to early stages (3 DAF through 9 DAF). A study on starch and FA accumulation in plastids of developing embryos of rapeseed showed that several cytosolic metabolites contribute to starch and/or FA synthesis during embryo development (Kang and Rawsthorne 1994). The gene encoding the phosphoenolpyruvate (PEP) translocator (PPT) was found to be highly expressed at all stages of seed development and was shown to be responsible for importing PEP, the main carbon source into the plastid during FA synthesis in *Arabidopsis* (Ruuska et al. 2002).

The process by which sugar and starch disappear at the later stages of seed maturity is worth noting. Imported sucrose is degraded by both SUS and invertase (Huang et al. 2009b). In developing rapeseed siliques, the switch to oil accumulation was accompanied by an increase in seed SUS activity (3.6-fold) and an approximately 76 % reduction in soluble acid invertase activity (King et al. 1997). Niu et al. (2009) also reported high expression of SUS throughout seed development. It is interesting to note that these authors also reported the isolation of a neutral invertase and found significant increase in its expression during seed development. Ruuska et al. (2002) reported increased expression of SUS genes after accumulation of storage products in the cotyledon. These independent studies also observed that the highest SUS activity coincided well with the formation of optimum seed size achieved by the accumulation of osmoticum (such as hexose), speculating that SUS accumulation might be related to seed size regulation (Huang et al. 2009b).

10.5 Storage Protein Accumulation in Rapeseed

The SPs are very useful in understanding developmental control of gene expression in seeds because of their abundance and developmental stage-specific accumulation (DeLisle and Crouch 1989). In *Arabidopsis*, protein deposition was reported to occur mainly at the seed-maturation stage following a sigmoid path (Baud et al. 2002). This report also indicated that protein synthesis continued at later maturity stages when oil content decreases on a whole seed basis, which could be due to the breakdown of lipids previously stored. Napin (2 S) and cruciferin (12 S) are the two major SPs of rapeseeds (Schwenke et al. 1983; Scofield and Crouch 1987). Napin and cruciferin were first detected in well-developed cotyledons. Cruciferin accumulation continues till the full mature seed stage, whereas napin accumulation ceases at the onset of embryo desiccation (Crouch and Sussex 1981). However, Ruuska et al. (2002) reported differential expression patterns of napin and cruciferin genes during seed development. In their study, napin genes were expressed abundantly between five DAF and 13 DAF with two-fold increase towards later stages of development. On the other hand, cruciferin genes were expressed weakly compared to napin genes and their expressions were increased by 10-fold at 8 DAF relative to 13 DAF. Napin gene, cruciferin gene *CRB*, cruciferin gene *CRC*, a putative cruciferin gene (*Arabidopsis At 1g03890*), and four seed SPs were identified by *B. napus* microarray analysis (Yu et al. 2010). This study showed that expression of *CRA1* (*Arabidopsis At 5g44120*) was two-fold lower in *B. napus* seeds at 20 DPA relative to 35 DPA, supporting earlier reports that cruciferins are expressed strongly during later stages of seed development. A SAGE analysis of winter *B. napus* seeds revealed abundant expression of three cruciferin subunits at the later stages of seed development, showing higher expression at 35 DPA compared to 20 DPA (Sjödahl et al. 1993). In winter *B. napus* seeds, however, napin and cruciferin genes showed differential expressions in seeds at 35 DAP compared to seeds at 23 DAP (Obermeier et al. 2009).

Oleosins are the most abundant proteins associated with oil bodies and play a role in the synthesis, metabolism, and stability of these bodies (Frandsen et al. 2001; Huang 1992). It is interesting to note that a positive correlation between accumulation of oleosins and oils exists in *B. napus* seeds (Huang 1992).

10.6 Carotenoid Accumulation and Gene Expression

Carotenoids are the second-most abundant pigments in nature. The carotenoid family consists of over 700 members (Britton 1998). These compounds are well-known antioxidants in higher plants. They are involved in the assembly of the photosystems and are essential components of the photosynthetic machinery. They harvest light in a broader range of the blue spectrum than chlorophyll and transfer energy to chlorophyll. Carotenoids also serve as photoprotective compounds by quenching both the triplet chlorophyll and singlet oxygen derived from excess light energy thus limiting membrane damage. In addition to their potential therapeutic properties for humans, carotenoids are also used as colorants in the food and cosmetics industries, and as supplements in livestock and fish feed formulations (Botella-Pavía and Rodríguez-Concepción 2006; Fraser and Bramley 2004; Taylor and Ramsay 2005). Carotenoids accumulate in *B. napus* seeds throughout development with the highest levels detected at approximately 35–45 DPA (Yu et al. 2008). Carotenoid accumulation drops significantly at the later stages of seed development with β-carotene and lutein accounting for more than 90 % of the total carotenoid content (Yu et al. 2008). Findings indicate that carotenoids provide oxidative stability to *B. napus* oil (Frankel 2005).

Transcriptomics of developing *B. napus* seeds have identified a large number of genes involved in secondary metabolism having differential expressions between

Table 10.1 Expression of *B. napus* Seed Genes Encoding Enzymes Involved in Carotenoid and Isoprenoid Biosynthesis at 20 DPA Relative to 35 DPA. [1]Fold change expressed as 20 DPA/35 DPA. (Reproduced from Yu et al. (2010; Plant Sci 178: 381–389) with permission from Elsevier)

Probe name	[1]Expression Ratio	Corresponding *Arabidopsis* locus name	Description
Carotenoid biosynthetic genes			
BN15353	3.67	At5g57030	lycopene ε-cyclase (*epCYC*)
BN24711	2.40	At1g08550	violaxanthin de-epoxidase precursor
BN18399	1.74	At3g10230	lycopene β-cyclase (*bLYC*)
BN12582	1.72	At5g17230	phytoene synthase (*PSY*)
BN18688	0.61	At4g25700	β-carotene hydroxylase (*HYD*)
BN12888	0.26	At5g67030	zeaxanthin epoxidase (*ZEP*)
BN15039	0.07	At4g19170	9-cis-epoxycarotenoid dioxygenase (*NCED4*)
Isoprenoid biosynthetic genes			
BN14336	9.38	At1g12900	glyceraldehyde 3-phosphate dehydrogenase A subunit 2, chloroplast (*Gapa-2*)
BN16125	4.75	At4g15560	1-deoxy-D-xylulose 5-phosphate synthase (*DXPS*)
BN18780	4.50	At3g26650	glyceraldehyde 3-phosphate dehydrogenase A subunit, chloroplast (*GAPA*-1)
BN11845	4.04	At1g42970	glyceraldehyde-3-phosphate dehydrogenase B, chloroplast (*GAPB*)
BN13608	3.16	At1g13440	glyceraldehyde 3-phosphate dehydrogenase, cytosolic (*GAPC2*)
BN24791	2.43	At2g18620	geranylgeranyl pyrophosphate synthase (*GGPS*)
BN23461	2.03	At2g23800	geranylgeranyl pyrophosphate synthase 2 (*GGPS2*)

20 DPA and 35 DPA (Yu et al. 2010). This study also showed that in developing *B. napus* seeds, expressions of carotenoid genes varied with the changes in development stages and enzymes responsible for various carotenoid biosynthetic pathways showed their expression at distinct stages of development. The same study also noted higher expressions of the upstream genes involved in isoprenoid biosynthesis in seeds at 20 DPA compared to 35 DPA (Table 10.1).

10.7 Phenolics Accumulation

Phenolics are secondary metabolites with variable phenolic structures and play important roles in protection against biotic and abiotic stresses (Auger et al. 2010; Yu and Jez 2008). These include flavonoids, phenylpropanoids, pro anthocyanins (PAs), and others. Phenolics are well known for their health benefits in humans, including antioxidant and antitumor properties (Gullett et al. 2010). Anthocyanins and PAs are important plant pigments, sharing common biosynthesis steps and intermediates with flavonoids (Li et al. 2010). Phenolics of *B. napus* seed coat consist mainly of PAs and/or their precursors (Auger et al. 2010). Among different *Brassica* species, elevated phenolics content was reported in the seed coat of a yellow-seeded *B. carinata* line compared to a genetically related brown-seeded line (Li et al. 2010; Marles et al. 2003). Furthermore, Li et al. (2010) reported that accumulation of flavonoids, phenylpropanoids, and lignans occurred in yellow seed coats, whereas the brown seed coats accumulated only phenylpropanoids and lignans. Transcriptomics analysis showed stronger expression of a gene encoding a chalcone-flavanoneisomerase-related protein (*Arabidopsis At1g53520*) involved in flavonoid biosynthetic pathway in developing seeds of *B. napus* especially at the early stages of development (20 DPA relative to 35 DPA; Yu et al. 2010).

One of the most abundant phenolics in *B. napus* seed is sinapine, also known as sinpoyl choline. Sinapine is unique to cruciferous oilseeds with its levels in *B. napus* seeds ranging from 0.7 to 4 % (Blair and Reichert 1984); most of which (approximately 90 %) is located in the embryo (non-hull) fraction (Wang et al. 1998). *B. napus* accumulates a mixture of sinapate esters, including glucose, gentiobiose, and kaempferol glycoside esters as well as sinapine (sinapoylcholine), sinapoylmalate, and an unusual cyclic spermidine amide (Baumert et al. 2005). The expression of two genes, involved in the final two steps of sinapine biosynthesis, SGT (UDP-glucose:sinapate glucosyl transferase) and SCT (sinapoylglucose:choline sinapoyl transferase), have been studied (Milkowski et al. 2004). The activities of SGT and SCT were regulated at the transcriptional level during seed and seedling development of *B. napus*. The expression of SCT was specific to developing seeds, whereas SGT expression increased during early stages of seed development until the early cotyledonary stage, and then maintained that level until later stages. The SGT expression reached its highest level in two-day-old seedlings, and then decreased significantly.

Because of the antinutritional nature of sinapine, repeated attempts have been made to reduce its levels in *Brassica* seeds (Baumert et al. 2005; Hüsken et al. 2005; Nair et al. 2000; Velasco and Möllers 1998), but the highest reduction of ~90 % was obtained by concomitantly silencing of both ferulic acid 5-hydroxylase (FAH) and SCT genes in the phenylpropanoid pathway (Bhinu et al. 2009). The sinapine reduction trait was stable over several generations and also under normal field conditions. However, silencing of FAH and SCT genes in *Arabidopsis* had pleiotropic effects beyond sinapine biosynthesis (Huang et al. 2009a). Transcriptomics of *Arabidopsis fah* and *sct* mutants documented more than 4,500 different transcripts with greater than four-fold difference in expression in developing siliques of the mutants compared to wild type control. Most of the genes involved in a variety of metabolic processes showed up to 20 % difference in their expression levels compared to wild type. Many genes of the phenylpropanoid pathway also showed changes in their expression levels (Fig. 10.1; Huang et al. 2009a).

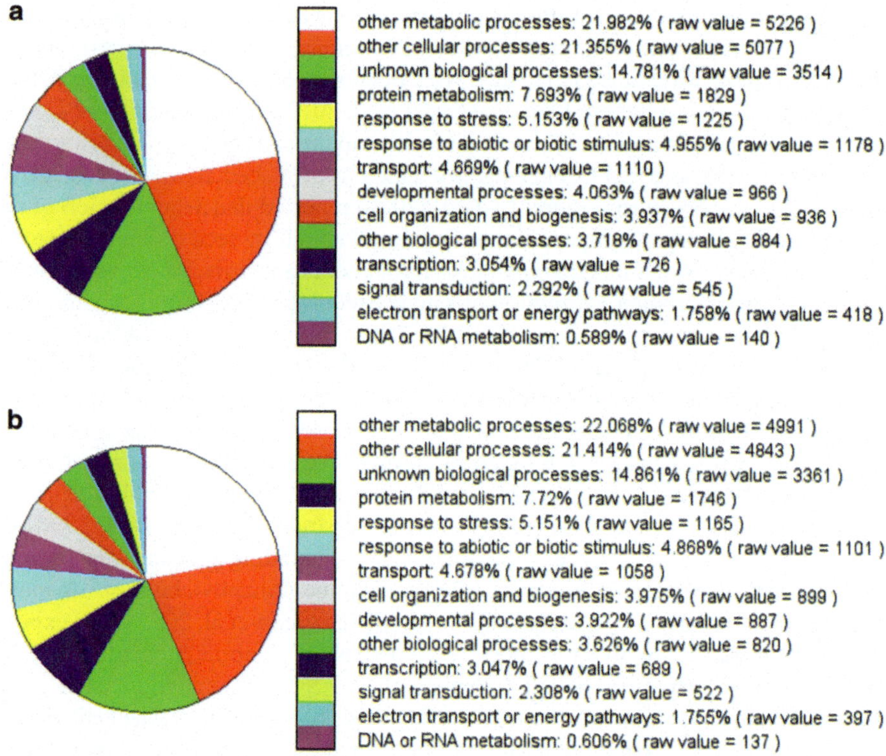

Fig. 10.1 Microarray-based identification of a large number of genes with prominent expression changes in developing siliques of **a** *f5h* and **b** *sct* plants harboring a single T-DNA insertion. The color-coded legends represent different functional groups of genes. (With kind permission from Springer Science + Business Media: Huang et al. 2009a. (Fig. 1))

10.8 Regulatory and Signaling Genes Related to Seed Development

Embryonic genes in *B. napus* are repressed in vegetative organs, and their derepression during seed development is regulated by a complex network of transcriptional and post transcriptional regulatory factors (Vicente-Carbajosa and Carbonero 2005). Changes in seed physiology and composition during seed development are regulated by regulatory networks, involving TFs and protein modifying enzymes. Such factors would play major roles in the timing of expression of storage compounds. In *Arabidopsis*, as many as 2,000 TF genes have been reported, representing more than 6 % of all total genes (Riechmann and Ratcliffe 2000; Riechmann et al. 2000). In *B. napus,* 320 TFs were reported to be expressed during seed development, and seven of them had expression patterns similar to those of genes related to FA biosynthesis (Niu et al. 2009). A number of TFs were found to be differentially expressed in seeds at 20 DPA and 35 DPA (Yu et al. 2010). A B3 family protein, similar to *Arabidopsis At3g18960*, is a good example as its expression was strongly upregulated in seeds at 20 DPA. The B3 domain-containing TFs were previously reported to regulate several aspects of seed development and accumulation of seed storage compounds in *Arabidopsis* (Meinke et al. 1994; Santos-Mendoza et al. 2005; Stone et al. 2001). Two TFs, belonging to the AP2/ERF family, which play important roles in determining seed size, weight, and regulation of storage compound biosynthesis, were reported to be expressed at a higher level at 35 DPA relative to 20 DPA (Yu et al. 2010). Of these, the AtRAP2.2 TF was shown to regulate phytoene synthase and phytoenedesaturase genes, which catalyze important steps in the carotenoid biosynthesis pathway.

Yu et al. (2010) reported a number of bZIP TFs to be differentially expressed over the course of *B. napus* DH12075 seed development. They reported that increased expression of *ABI5*, a bZIP TF involved in ABA signaling (Finkelstein and Lynch 2000), was accompanied by stimulation of 13 *LEA* genes. Abiotic stress-induced TF, *AZF2*, was also upregulated in *B. napus* seeds at 35 DPA compared to 20 DPA (Yu et al. 2010).

Analysis of differential gene expression between two cultivars of *B. napus* showed that regulation of TFs during seed development can also be cultivar specific (Fei et al. 2007). Two genes encoding TFs (*At1g50600* and *At5g10280*) were specifically downregulated in seeds of AC Excel cultivar of *B. napus* compared to DH12075. The gene *At1g50600* coded for a scarecrow-like TF has been shown to be associated with signaling and development in *Arabidopsis* (Pysh et al. 1999). The TFs, *MYB66* and the gene encoding a MYB family member (homolog to *Arabidopsis At5g14750*) was upregulated in DH12075. The MYB family of TFs is known to be involved in the regulation of secondary metabolism, control of cell shape, disease resistance, hormone, and stress response (Martin and Paz-Ares 1997). Furthermore, a gene encoding ethylene responsive MBF1 family protein, a transcriptional coactivator, was upregulated specifically in DH12075 (Matilla 2000; Yu et al. 2010).

Seed development is accompanied by large-scale chromatin remodeling and histone modifications, with chromatin modifications starting as early as the decondensation of chromosomes from pollen grains immediately after fertilization (reviewed by Baroux et al. 2007). Histones are a component of chromatin, the protein-DNA complex involved in DNA packaging and transcriptional regulation (Okada et al. 2005). Covalent histone modifications are important for chromatin remodeling and transcriptional regulation of gene expression (Jenuwein and Allis 2001). Four histone H3 genes (*Arabidopsis At1g09200*, *At5g10400*, *At5g65350*, and *At5g65360*) and four H2B genes (*Arabidopsis At1g07790*, *At2g37470*, *At3g53650*, and *At5g22880*) had higher expression in *B. napus* seeds at 20 DPA relative to 35 DPA (Yu et al. 2010).

Signal transduction through phosphorylation processes regulates many cellular events and metabolic pathways (Chevalier and Walker 2005). Six signal transduction-related genes were found to have similar expression patterns as genes involved in FA synthesis in developing *B. napus* seeds (Niu et al. 2009). These genes include MAP kinase, the phosphatidylinositol 3- and 4-kinase families, and phospholipase. Yu et al. (2010) also observed higher expression of *CDPK6* (*Arabidopsis At4g23650*) and *CDPK9* (*Arabidopsis At5g23580*) at 20 DPA, while *CDPK19* (*Arabidopsis At5g19450*) had a weaker expression in seeds at 20 DPA relative to 35 DPA. Previous reports showed that rice CDPK (SPK) played a crucial role in the accumulation of storage products during seed development (Asano et al. 2002). Other kinases differentially expressed during seed development were phosphatidylinositol-4-monophosphate 5-kinase (*PIP5K9*; *Arabidopsis At3g09920*) and pyruvate orthophosphate dikinase (*Arabidopsis At4g15530*), which were downregulated in seeds at 20 DPA compared to 35 DPA (Yu et al. 2010). In addition, cultivar-specific differential expression of protein kinases was also reported (Fei et al. 2007). Four upregulated genes encoding protein kinases (*Arabidopsis At3g09830, At1g28440, At1g75970*, and *At2g33830*) were specifically identified in seeds of cultivar DH 12075 compared to AC Excel.

Ubiquitination plays an essential role in a large number of eukaryotic cellular processes by targeting proteins for proteasome-mediated degradation (reviewed in Hershko and Ciechanover 1998). Polyubiquitin genes are transcribed and translated into polyproteins that are cleaved into monomers by specific proteases (Baker et al. 1992). Four such genes (*Arabidopsis At2g01340, At4g05050, At5g03240*, and *At5g19990*) were found to be upregulated specifically in seeds of cultivar AC Excel in comparison to DH12075 (Fei et al. 2007). These genes are related to protein metabolism, including polyubiquitins. These results reinforce the fact that differential regulation of protein metabolism and signal transduction exists between different cultivars of *B. napus*.

10.9 Concluding Remarks

Numerous efforts were concentrated on better understanding the underlying mechanisms regulating seed development in both the model plant *Arabidopsis* and the economically important oilseed crop rapeseed. Attempts so far have included conventional approaches like studying the seed composition at various stages of development, and novel approaches involving global gene expression profiling directly from the target seed compartments. These studies have contributed greatly to our understanding of the whole processes and complexities of different pathways operating simultaneously in the developing seeds. Yet, biology behind seed development remains largely a mystery and the underlying mechanisms a paradox. Recent advances in technologies, such as laser capture microdissection, will help deep profiling of gene expression from different cell types and seed compartments. Completion of the *B. napus* genome sequencing and advances in NGS will further our understanding of developmental mechanisms in seeds. Transcriptomics and proteomics approaches combined with a sequence database could complement sequence analysis and contribute to deciphering the seed development blackbox in *B. napus* and other oilseed plants.

References

Asano T, Kunieda N, Omura Y, Ibe H, Kawasaki T, Takano M, Sato M, Furuhashi H, Mujin T, Takaiwa F, Wu CY, Tada Y, Satozawa T, Sakamoto M, Shimada H (2002) Rice SPK, a calmodulin-like domain protein kinase, is required for storage product accumulation during seed development. Plant Cell 14:619–628

Auger B, Marnet N, Gautier V, Maia-Grondard A, Leprince F, Renard M, Guyot S, Nesi N, Routaboul JM (2010) A detailed survey of seed coat flavonoids in developing seeds of *Brassica napus* L. J Agric Food Chem 58:6246–6256

Baker RT, Tobias JW, Varshavsky A (1992) Ubiquitin-specific proteases of *Saccharomyces cerevisiae.* Cloning of *UBP2* and *UBP3*, and functional analysis of the *UBP* gene family. J Biol Chem 267:23364–23375

Baroux C, Pien S, Grossniklaus U (2007) Chromatin modification and remodeling during early seed development. Curr Opin Genet Dev 17:473–479

Baud S, Boutin JP, Miquel M, Lepiniec L, Rochat C (2002) An integrated overview of seed development in *Arabidopsis thaliana* ecotype WS. Plant Physiol Biochem 40:151–160

Baumert A, Milkowski C, Schmidt J, Nimtz M, Wray V, Strack D (2005) Formation of a complex pattern of sinapate esters in *Brassica napus* seeds, catalyzed by enzymes of a serine carboxypeptidase-like acyltransferase family? Phytochemistry 66:1334–1345

Beisson F, Koo AJK, Ruuska S, Schwender J, Pollard M, Thelen JJ, Paddock T, Salas JJ, Savage L, Milcamps A, Mhaske VB, Cho Y, Ohlrogge JB (2003) *Arabidopsis* genes involved in acyl lipid metabolism: a 2003 census of the candidates, a study of the distribution of expressed sequence tags in organs, and a web-based database. Plant Physiol 132:681–697

Bhinu VS, Schäfer UA, Li R, Huang J, Hannoufa A (2009) Targeted modulation of sinapine biosynthesis pathway for seed quality improvement in *Brassica napus*. Transgenic Res 18:31–44

Blair R, Reichert RD (1984) Carbohydrate and phenolic constituents in a comprehensive range of rapeseed and canola fractions: nutritional significance for animals. J Sci Food Agric 35:29–35

Botella-Pavía P, Rodríguez-Concepción M (2006) Carotenoid biotechnology in plants for nutritionally improved foods. Physiol Plant 126:369–381

Britton G (1998) Overview of carotenoid biosynthesis. In: Britton G, Liaaen Jensen S, Pfander H (eds) Carotenoids. Birkhauser, Basel, pp 13–147

Broun P, Poindexter P, Osborne E, Jiang CZ, Riechmann JL (2004) WIN1, a transcriptional activator of epidermal wax accumulation in *Arabidopsis*. Proc Natl Acad Sci U S A 101:4706–4711

Chevalier D, Walker JC (2005) Functional genomics of protein kinases in plants. Brief Funct Genomic Proteomic 3:362–371

Chia TY, Pike MJ, Rawsthorne S (2005) Storage oil breakdown during embryo development of *Brassica napus* (L.). J Exp Bot 56:1285–1296

Close TJ (1996) Dehydrins: emergence of a biochemical role of a family of plant dehydration proteins. Physiol Plant 97:795–803

Crouch ML, Sussex IM (1981) Development and storage-protein synthesis in *Brassica napus* L. embryos *in vivo* and *in vitro*. Planta 153:64–74

da Silva PMFR, Eastmond PJ, Hill LM, Smith AM, Rawthorne S (1997) Starch metabolism in developing embryos of oilseed rape. Planta 203:480–487

DeLisle AL, Crouch ML (1989) Seed storage protein transcription and mRNA levels in *Brassica napus* during development and in response to exogenous abscisic acid. Plant Physiol 91:617–623

Dong J, Keller WA, Yan W, Georges F (2004) Gene expression at early stages of *Brassica napus* seed development as revealed by transcript profiling of seed-abundant cDNAs. Planta 218:483–491

Eastmond PJ, Rawsthorne S (2000) Coordinate changes in carbon partitioning and plastidial metabolism during the development of oilseed rape embryos. Plant Physiol 122:767–774

Fei H, Tsang E, Cutler AJ (2007) Gene expression during seed maturation in *Brassica napus*in relation to the induction of secondary dormancy. Genomics 89:419–428

Finkelstein RR, Lynch TJ (2000) The *Arabidopsis* abscisic acid response gene *ABI5* encodes a basic leucine zipper transcription factor. Plant Cell 12:599–610
Fischer K, Weber A (2002) Transport of carbon in non-green plastids. Trends Plant Sci 7:345–351
Frandsen GI, Mundy J, Tzen JTC (2001) Oil bodies and their associated proteins, oleosin and caleosin. Physiol Plant 112:301–307
Frankel EN (2005) Lipid oxidation, 2nd Edition. The Oily, Bridgewater
Fraser PD, Bramley PM (2004) The biosynthesis and nutritional uses of carotenoids. Prog Lipid Res 43:228–265
Girke T, Todd J, Ruuska S, White J, Benning C, Ohlrogge J (2000) Microarray analysis of developing *Arabidopsis* seeds. Plant Physiol 124:1570–1581
Goldberg RB, Barker SJ, Perez-Grau L (1989) Regulation of gene expression during plant embryogenesis. Cell 56:149–160
Gullett NP, Ruhul Amin AR, Bayraktar S, Pezzuto JM, Shin DM, Khuri FR, Aggarwal BB, Surh YJ, Kucuk O (2010) Cancer prevention with natural compounds. Semin Oncol 37:258–281
Hajduch M, Casteel JE, Hurrelmeyer KE, Song Z, Agrawal GK, Thelen JJ (2006) Proteomic analysis of seed filling in *Brassica napus*. Developmental characterization of metabolic isozymes using high-resolution two-dimensional gel electrophoresis. Plant Physiol 141:32–46
Hershko A, Ciechanover A (1998) The ubiquitin system. Annu Rev Biochem 67:425–479
Huang AHC (1992) Oil bodies and oleosins in seeds. Annu Rev Plant Physiol Plant Mol Biol 43:177–200
Huang J, Bhinu VS, Li X, Dallal Bashi Z, Zhou R, Hannoufa A (2009a) Pleiotropic changes in *Arabidopsisf5h* and *sct* mutants revealed by large-scale gene expression and metabolite analysis. Planta 230:1057–1069
Huang Y, Chen L, Wang L, Vijayan K, Phan S, Liu Z, Wan L, Ross A, Xiang D, Datla R, Pan Y, Zou J (2009b) Probing the endosperm gene expression landscape in *Brassica napus*. BMC Genomics 10:256
Hüsken A, Baumert A, Milkowski C, Becker HC, Strack D, Möllers C (2005) Resveratrol glucoside (Piceid) synthesis in seeds of transgenic oilseed rape (Brassica napus L.). Theor Appl Genet 111:1553–1562
Jenuwein T, Allis CD (2001) Translating the histone code. Science 293:1074–1080
Kang F, Rawsthorne S (1994) Starch and fatty acid synthesis in plastids from developing embryos of oilseed rape (*Brassica napus* L.). Plant J 6:795–805
King SP, Lunn JE, Furbank RT (1997) Carbohydrate content and enzyme metabolism in developing canola siliques. Plant Physiol 114:153–160
Leprince O, Bronchart R, Deltour R (1990) Changes in starch and soluble sugars in relation to the acquisition of desiccation tolerance during maturation of *Brassica campestris* seed. Plant Cell Environ 13:539–546
Li X, Westcott N, Links, M, Gruber MY (2010) Seed coat phenolics and the developing silique transcriptome of *Brassica carinata*. J Agric Food Chem 58:10918–10928
Marles MAS, Gruber MY, Scoles GJ, Muir AD (2003) Pigmentation in the developing seed coat and seedling leaves of *Brassica carinata* is controlled at the dihydroflavonol reductase locus. Phytochemistry 62:663–672
Martin C, Paz-Ares J (1997) MYB transcription factors in plants. Trends Genet 13:67–73
Matilla AJ (2000) Ethylene in seed formation and germination. Seed Sci Res 10:111–126
Meinke DW, Franzmann LH, Nickle TC, Yeung EC (1994) Leafy cotyledon mutants of *Arabidopsis*. Plant Cell 6:1049–1064
Milkowski C, Baumert A, Schmidt D, Nehlin L, Strack D (2004) Molecular regulation of sinapate ester metabolism in *Brassica napus*: expression of genes, properties of the encoded proteins and correlation of enzyme activities with metabolite accumulation. Plant J 38:80–92
Nair RB, Joy RW 4th, Kurylo E, Shi X, Schnaider J, Datla RSS, Keller WA, Selvaraj G (2000) Identification of a CYP84 family of cytochrome P450-dependent mono-oxygenase genes in *Brassica napus* and perturbation of their expression for engineering sinapine reduction in the seeds. Plant Physiol 123:1623–1634

Niu Y, Wu GH, Ye R, Lin WH, Shi QM, Xue LJ, Xu XD, Li Y, Du YG, Xue HW (2009) Global analysis of gene expression profiles in *Brassica napus* developing seeds reveals a conserved lipid metabolism regulation with *Arabidopsis thaliana*. Mol Plant 2:1107–1122
Norton G, Harris JF (1975) Compositional changes in developing rape seed (*Brassica napus* L.). Planta 123:163–174
O'Hara P, Slabas AR, Fawcett T (2002) Fatty acid and lipid biosynthetic genes are expressed at constant molar ratios but different absolute levels during embryogenesis. Plant Physiol 129:310–320
Obermeier C, Hosseini B, Friedt W, Snowdon R (2009) Gene expression profiling via LongSAGE in a non-model plant species: a case study in seeds of *Brassica napus*. BMC Genomics 10:295
Okada T, Endo M, Singh MB, Bhalla PL (2005) Analysis of the histone H3 gene family in *Arabidopsis* and identification of the male-gamete-specific variant AtMGH3. Plant J 44:557–568
Pang PP, Pruitt RE, Meyerowitz EM (1988) Molecular cloning, genomic organization, expression and evolution of 12s seed storage protein genes of *Arabidopsis thaliana*. Plant Mol Biol 11:805–820
Park YS, Kwak JM, Kwon OY, Kim YS, Lee DS, Cho MJ, Lee HH, Nam HG (1993) Generation of expressed sequence tags of random root cDNA clones of *Brassica napus* by single-run partial sequencing. Plant Physiol 103:359–370
Penfield S, Li Y, Gilday AD, Graham S, Graham IA (2006) *Arabidopsis* ABA INSENSITIVE4 regulates lipid mobilization in the embryo and reveals repression of seed germination by the endosperm. Plant Cell 18:1887–1899
Pysh LD, Wysocka-Diller JW, Camilleri C, Bouchez D, Benfey PN (1999) The GRAS gene family in *Arabidopsis*: sequence characterization and basic expression analysis of the SCARECROW-LIKE genes. Plant J 18:111–119
Riechmann JL, Ratcliffe OJ (2000) A genomic perspective on plant transcription factors. Curr Opin Plant Biol 3:423–434
Riechmann JL, Heard J, Martin G, Reuber L, Jiang CZ, Keddie J, Adam L, Pineda O, Ratcliffe OJ, Samaha RR, Creelman R, Pilgrim M, Broun P, Zhang JZ, Ghandehari D, Sherman BK, Yu GL (2000) *Arabidopsis* transcription factors: genome-wide comparative analysis among eukaryotes. Science 290:2105–2110
Ruuska SA, Girke T, Benning C, Ohlrogge JB (2002) Contrapuntal networks of gene expression during *Arabidopsis* seed filling. Plant Cell 14:1191–1206
Santos-Mendoza M, Dubreucq B, Miquel M, Caboche M, Lepiniec L (2005) LEAFY COTYLEDON 2 activation is sufficient to trigger the accumulation of oil and seed specific mRNAs in *Arabidopsis* leaves. FEBS Lett 579:4666–4670
Schena M, Shalon D, Davis RW, Brown PO (1995) Quantitative monitoring of gene-expression patterns with a complementary-DNA microarray. Science 270:467–470
Schwenke KD, Raab B, Plietz P, Damaschun G (1983) The structure of the 12S globulin from rapeseed (*Brassica napus* L.). Die Nahrung 27:165–175
Scofield SR, Crouch ML (1987) Nucleotide sequence of a member of the napin storage protein family from *Brassica napus*. J Biol Chem 262:12202–12208
Sjödahl S, Gustavsson HO, Rödin J, Lenman M, Höglund AS, Rask L (1993) Cruciferin gene families are expressed coordinately but with tissue-specific differences during *Brassica napus* seed development. Plant Mol Biol 23:1165–1176
Stone SL, Kwong LW, Yee KM, Pelletier J, Lepiniec L, Fischer RL, Goldberg RB, Harada JJ (2001) *LEAFY COTYLEDON2* encodes a B3 domain transcription factor that induces embryo development. Proc Natl Acad Sci U S A 98:11806–11811
Taylor M, Ramsay G (2005) Carotenoid biosynthesis in plant storage organs: recent advances and prospects for improving plant food quality. Physiol Plant 124:143–151
Taylor NL, Heazlewood JL, Day DA, Millar AH (2005) Differential impact of environmental stresses on the pea mitochondrial proteome. Mol Cell Proteomics 4:1122–1133
Thomas TL (1993) Gene expression during plant embryogenesis and germination: an overview. Plant Cell 5:1401–1410

Velasco L, Möllers C (1998) Nondestructive assessment of sinapic acid esters in Brassica species: II. Evaluation of germplasm and identification of phenotypes with reduced levels. Crop Sci 38:1650–1654

Velculescu VE, Zhang L, Vogelstein B, Kinzler KW (1995) Serial analysis of gene expression. Science 270:484–487

Vigeolas H, Möhlmann T, Martini N, Neuhaus HE, Geigenberger P (2004) Embryo-specific reduction of ADP–Glc pyrophosphorylase leads to an inhibition of starch synthesis and a delay in oil accumulation in developing seeds of oilseed rape. Plant Physiol 136:2676–2686

Vicente-Carbajosa J, Carbonero P (2005) Seed maturation: developing an intrusive phase to accomplish a quiescent state. Int J Dev Biol 49:645–651

Wang SX, Oomah BD, McGregor DI, Downey RK (1998) Genetic and seasonal variation in the sinapine content of seed from *Brassica* and *Sinapis* species. Can J Plant Sci 78:395–400

Weselake RJ, Pomeroy MK, Furukawa TL, Golden JL, Little DB, Laroche A (1993) Developmental profile of diacylglycerol acyltransferase in maturing seeds of oilseed rape and safflower and microspore-derived cultures of oilseed rape. Plant Physiol 102:565–571

White JA, Todd J, Newman T, Focks N, Girke T, de Ilarduya OM, Jaworski JG, Ohlrogge JB, Benning C (2000) A new set of *Arabidopsis* expressed sequence tags form developing seeds. The metabolic pathway from carbohydrates to seed oil. Plant Physiol 124:1582–1594

Wu GZ, Shi QM, Niu Y, Xing MQ, Xue HW (2008) Shanghai RAPESEED Database: a resource for functional genomics studies of seed development and fatty acid metabolism of *Brassica*. Nucl Acids Res 36:D1044–D1047

Xiang D, Datla R, Li F, Cutler A, Malik MR, Krochko JE, Sharma N, Fobert P, Georges F, Selvaraj G, Tsang E, Klassen D, Koh C, Deneault JS, Nantel A, Nowak J, Keller W, Bekkaoui F (2008) Development of a *Brassica* seed cDNA microarray. Genome 51:236–242

Xu D, Duan X, Wang B, Hong B, Ho THD, Wu R (1996) Expression of a late embryogenesis abundant protein gene, *HVA1*, from barley confers tolerance to water deficit and salt stress in transgenic rice. Plant Physiol 110:249–257

Yu O, Jez JM (2008) Nature's assembly line: biosynthesis of simple phenylpropanoids and polyketides. Plant J 54:750–762

Yu B, Lydiate DJ, Young LW, Schäfer UA, Hannoufa A (2008) Enhancing the carotenoid content of *Brassica napus* seeds by downregulating lycopene epsilon cyclase. Transgenic Res 17:573–585

Yu B, Gruber M, Khachatourians GG, Hegedus DD, Hannoufa A (2010) Gene expression profiling of developing *Brassica napus* seed in relation to changes in major storage compounds. Plant Sci 178:381–389

Zou J, Abrams GD, Barton DL, Taylor DC, Pomeroy MK, Abrams SR (1995) Induction of lipid and oleosin biosynthesis by (+)-abscisic acid and its metabolites in microspore-derived embryos of *Brassica napus* L. cv Reston. Biological responses in the presence of 8′-hydroxyabscisic acid. Plant Physiol 108:563–571

Chapter 11
Networks of Seed Storage Protein Regulation in Cereals and Legumes at the Dawn of the Omics Era

Jerome Verdier and Richard D. Thompson

Abstract The regulation of gene expression in the developing seed has received much attention due to its economic importance but also because it represents a paradigm for plant gene regulation. The wealth of information obtained on a few well-studied systems is potentially much more widely applicable. Recent findings have revealed the participation of functionally homologous transcription factors in the cereal endosperm and cotyledon of dicots, suggesting these two regulation networks are composed of conserved or related elements.

Keywords Dicots · Monocots · Regulatory network · Regulatory elements · SSP · Transcription factor

11.1 Introduction

11.1.1 Economic Interest of Seeds

A major part of the human diet is derived from seeds. According to the FAO, 70 % of our diet is directly derived from seeds and large part of the remaining comes from animals, which mainly have seed-based diets. In developing countries, dietary dependence on seeds is accentuated by the high cost of meat production. Besides major dietary components (i.e., carbohydrates, proteins, and lipids), seeds also contain fibres, minerals, vitamins, and other secondary compounds beneficial for humans and livestock (Wang et al. 2003). Today, seeds are not only used for food production but are also a source of industrial raw materials of major economic interest (e.g., production of bioethanol, plastics, and adhesives from corn starch, and biodiesel from soybean and rapeseed oils). Finally, extensive genetic engineering advances have enhanced interest in seeds with development of nutritionally enhanced varieties [e.g., 'golden rice' technology (Ye et al. 2000)] and molecular farming, seeds

R. D. Thompson (✉)
INRA UMRLEG, 21065 Dijon, France
e-mail: thompson@dijon.inra.fr

J. Verdier
Plant Biology Division, The Samuel Roberts Noble Foundation, Ardmore, OK 73401, USA

G. K. Agrawal, R. Rakwal (eds.), *Seed Development: OMICS Technologies toward Improvement of Seed Quality and Crop Yield*,
DOI 10.1007/978-94-007-4749-4_11, © Springer Science+Business Media Dordrecht 2012

being ideal biofactories for production of high-value pharmaceutical or nutraceutical compounds (reviewed in, Stoger et al. 2005). Rice is for instance used for the production of antibodies [e.g., antibodies against a tumor-associated carcinoembryogenic antigen (Stöger et al. 2000)], hormones [e.g., glycogen-like peptide for type II diabetes treatment (Yasuda et al. 2005)], and vaccines. Advantages of producing such compounds in plants rather than in microbes or mammalian cell culture include increased safety for human consumption, extraction, and purification cost benefits and natural long-term storage capacity of dry seeds (Stöger et al. 2000; Twyman et al. 2003). However, before full exploitation, we need a deeper understanding of the regulation of seed maturation, and particularly of the dynamic expression of seed storage genes, for producing and storing high-value compounds in seeds.

11.1.2 Legumes and Cereals

Of the 19 crop plants that the FAO considers to be major suppliers for human diet, 15 belong to two different families: the Poaceae (cereals) and the Fabaceae (legumes).

Cereals Cereals are grasses and members of the monocot family. Cereal grains provide more food worldwide than any other type of crop and are the leading source of carbohydrate for animal and human consumption. The major cereals are maize, rice, and wheat, which together accounted for 87 % of all grain production worldwide and 43 % of all food calories in the year 2003. Maize is a staple food for the human diet in America and Africa but also for livestock worldwide. A large proportion of maize crops is grown for purposes other than human consumption with a production of 792×10^6 tons in the year 2007 (FAO). Rice is the primary cereal for human consumption. It accounts for more than one-fifth of calories consumed in the global human diet (Smith 1998). Its production was 659×10^6 tons in the year 2007 (FAO). Finally, wheat is the third most produced cereal worldwide with 606×10^6 tons in the year 2007 (FAO). Wheat is consumed worldwide and constitutes a staple food in North America, Europe, and Australia.

Legumes Legumes are part of the dicot family. Although their production is limited in comparison to cereals with 242×10^6 tons in the year 2007 (Association Européenne des Protéagineux, AEP), their importance lies in their ability to fix N_2 *via* symbiosis with nitrogen-fixing bacteria in root nodules. Legume seeds provide the leading source of protein for animal consumption. Three quarters of the world production of grain legumes (185×10^6 million tons) is soybean, grown mainly in North and South America.

Cereals and legumes produce contrasting seed types As mentioned previously, cereals and legumes belong to two major angiosperm plant families: the monocotyledonous and dicotyledonous plants. Monocot and dicot divergence is predicted to have occurred between 200 and 150 million years ago (Chaw et al. 2004; Laroche et al. 1995; Wolfe et al. 1989). Concerning seed morphology, monocots and dicots share the same principal seed tissues but differ in the repartition of functions between them.

Cereal seeds are made up of three major compartments, the embryo and the endosperm, which are derived from the double fertilization event, surrounded by a complex maternal structure, the coat. The cereal seed coat is composed of the pericarp, an adhering tissue derived from the ovary wall and a seed coat, deriving from the outer epidermal wall of the ovule, which forms a layer between the pericarp and endosperm. Cereal grains are also referred to as caryopses, corresponding to a seed surrounded by a fruit tissue, the pericarp. The embryo comprises shoot/root meristems and embryonic axis surrounded by a unique massive cotyledon, the scutellum. After germination, embryo will give rise to the future plantlet using reserve carbohydrates, proteins, and oils stored in the endosperm. The role of the cereal endosperm is very different to its role in dicot species. In cereals, the endosperm constitutes the bulk of the seed, and persists at seed maturity as a nutritive dead tissue.

In legume seeds and more generally in dicot seeds, the functions of seed tissues differ. The seed coat is the only maternal tissue, which plays nutritive and protective roles towards the endosperm and embryo. The embryo also comprises shoot/root meristems and an embryonic axis, but is surrounded by two cotyledons. Cotyledons constitute the main storage tissue essential for developing dicot embryo after germination. Another important difference is the role of endosperm, which is a transient storage tissue in the majority of dicots. During the maturation phase of seed development, endosperm nutrients are absorbed by the embryo *via* remobilization. At maturity, endosperm remains a thin layer in most dicot seeds or even will disappear, depending if the seed is endospermic or nonendospermic.

11.1.3 Seed Development

Seed development can be divided into three phases (Fig. 11.1). The first phase corresponds to embryogenesis. It is a period of rapid cell divisions during which the embryo acquires polarity and, the root and shoot apical meristems are established. When cell divisions cease, the embryo enters into the second or maturation phase. This phase is characterized by differentiation of a proliferating tissue into a specialized storage tissue. The cells elongate and seed weight increases mainly due to the accumulation of storage compounds. Finally, when seeds reach their maximum DW corresponding to physiological maturity, the third, desiccation phase begins. During this phase, seeds start to desiccate, resulting in a shutdown of metabolic activity in order to enter into dormancy. In parallel, gene expression directing the accumulation of diverse protective compounds occurs that confer cellular tolerance to desiccation on the mature seed.

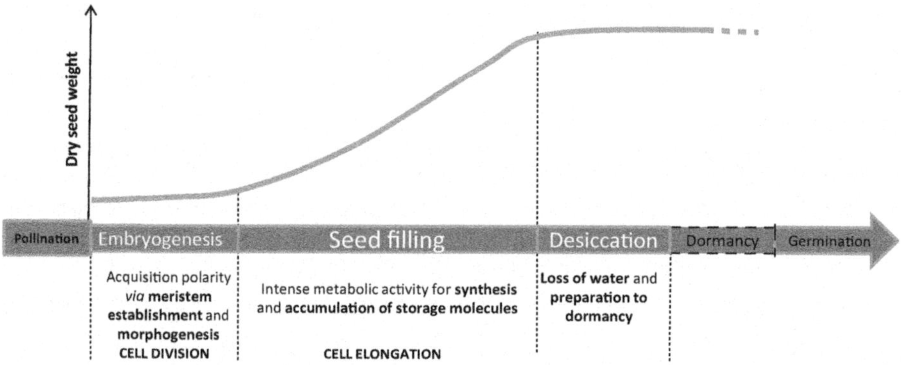

Fig. 11.1 Overview of molecular events associating with seed weight during seed development

11.1.4 Transcriptional Regulation

The TFs are master regulators of cellular processes and excellent candidates for modifying complex traits in crop plants (Century et al. 2008). Therefore, TF-based technologies are likely to be a prominent part of the next generation of successful biotech crops. An important goal of the research proposed here is to review TF genes that control seed development and differentiation, and ultimately affect the nutrient composition of mature seeds. TFs control every aspect of plant growth and development. Their importance in organizing the manifold activities of plants is highlighted by the fact that over 5 % of all plant genes encode TF proteins (Gao et al. 2006; Riechmann et al. 2000). Seed development and differentiation are governed by a network of TFs (Brocard-Gifford et al. 2003).

11.1.5 Scope of Review

In this chapter, we will focus on the maturation phase of seed development and particularly on the biosynthesis of SSPs. A complex transcriptional regulation network controls this process. Spatial and temporal expression of SSP genes requires concerted action between a genetic program and environmental signals. Classical genetics approaches have uncovered some of the TF genes, required for normal seed development and/or storage metabolism. Recent development in omics strategies has permitted to decipher the genetic regulatory networks of seed maturation associated with these TFs. Despite fundamental differences between legumes and cereals, similarity of the storage programs suggests that common or analog regulatory systems may exist. This supposition is being confirmed by recent findings of common regulatory components. Here, we draw together the possible relationships between the best-characterized regulatory networks for both plant families.

11.2 Seed Storage Proteins

Plant protein characterization is traditionally based on Thomas Bur Osborne's classification (Osborne 1924) according to their solubility: albumins (soluble in water), globulins (soluble in dilute salt solution), prolamins (soluble in 70:30 ethanol/water mixture), and glutelins (soluble in dilute alkali solution). Even if this classification is still widely used, today we prefer to classify plant proteins using their function into three different categories: structural/metabolic proteins, protective proteins, and SPs. By definition, SPs should be present in sufficient quantity and appear to have no biological role except storage during seed development. Another way to define proteins is their coefficient of sedimentation (in Svedberg, S units), which has been widely used to characterize SPs.

11.2.1 Storage Proteins and Their Location in Cereal Grain

Cereal grains contain less protein [averaging 10–12 % of DW; (Shewry and Halford 2002)] than legume seeds (20–40 % DW according to Müntz 1998). However considering the volume of cereal grain production, the quantity of cereal proteins used in feed and food exceeds that of legume seeds. Cereal grain proteins play a major role in determining technological properties of grain meal for food processing (e.g., cohesive and visco-elastic proprieties of gluten protein fraction of wheat flour). Cereal seeds contain two major SSP classes, the prolamins and globulins; both of which have been extensively studied.

Prolamin Family Except for oats and rice, prolamins represent the major SSP fraction in all cereals (Shewry et al. 1995). Prolamins are related in sequence to the less-soluble glutenins which in their native form are linked by cysteine cross-bridges to form large polymers. Prolamins are characterized by the presence of repetitive blocks of short-peptide motifs enriched in nitrogen-rich amino acid residues and embedded in nonrepeated sequences, including a few highly conserved residues reflecting a common evolutionary origin. Prolamins of the Triticeae are known under several names: gliadins in wheat, hordeins in barley, and secalins in rye. They are classified into the sulfur-rich (S-rich) (or B hordein in barley), sulfur-poor (S-poor) (or C hordein) subfamilies (Shewry and Tatham 1990). In parallel, prolamins of the Panicoideae like maize are named zeins and are divided into four groups: α, β, γ, and δ-zeins (Coleman and Larkins 1999; for review see Halford and Shewry 2007).

Globulin Family Globulins represent the most important class of SSP in rice and oats but are present in other cereal grains in smaller amounts. These proteins are soluble in dilute salt solutions and can be divided into 7S and 11S according to their sedimentation coefficients. The 11S globulins represent about 70–80 % of the total seed protein fraction in rice and oats (Casey 1999; Takaiwa et al. 1999). In rice, these proteins are not fully soluble in dilute salts and hence are also termed glutelins. In

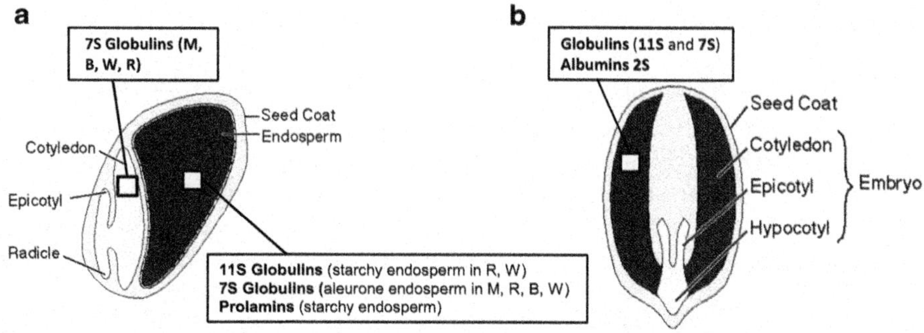

Fig. 11.2 Overview of seed storage proteins (SSP) locations in monocots (**a**), and dicots (**b**). *M* maize, *B* barley, *W* wheat, *R* rice

oats, the high content of 11S globulins contributes to the high-nutritional value of their seeds in comparison to other cereals, such as barley and wheat for livestock nutrition (Cuddeford 1995). Triticin, an 11S globulin found in wheat, accounts for 5 % of total seed protein (Singh et al. 1991). The 7S globulins are present in cereal grains in small amounts (Burgess and Shewry 1986; Heck et al. 1993; Yupsanis et al. 1990).

Location of Storage Proteins Cereal endosperm can be divided in two different domains: the aleurone, and the starchy endosperm rich in storage molecules (Fig. 11.2a). Cereal SSPs are mainly expressed and stored in the endosperm with the exception of the 7S globulins, which are both in the embryo and aleurone endosperm in maize, barley, wheat, and rice (Burgess and Shewry 1986; Heck et al. 1993; Kriz 1999; Yupsanis et al. 1990). In contrast, the 11S globulins were only identified in the starchy endosperm in rice (Casey 1999) and wheat (Singh et al. 1991).

The prolamin fraction is restricted to the starchy endosperm, where it exhibits a gradient in protein composition (Kent 1966). Using immunolocalization and *in-situ* hybridization, Lending and Larkins(1989) have shown clear differences in the locations of different prolamin classes in maize. The α-zeins are restricted to the inner part of starchy endosperm, whereas β, γ, and δ-zeins are in the subaleurone cells. A similar pattern has been reported for the prolamin accumulation in barley with the subaleurone richer in the high molecular weight fraction (or D hordein), and the inner-part richer in S-poor (C hordein), and S-rich (B hordein) fractions (Shewry and Halford 2002). Complementary studies using promoter-GUS fusions validated these results (Lamacchia et al. 2001; Stöger et al. 2001).

11.2.2 Legume Storage Proteins

Legumes are the second most important source of seed proteins after cereals. Grain legumes are usually deficient in sulfur amino acids (cysteine and methionine) in comparison to cereals. However, they are rich in lysine and provide an essential dietary complement to cereal proteins. Legume SSPs are mainly represented by two protein families: the globulins and albumins.

Globulins The globulins are the most widely distributed family of SSPs. They represent the major form of SSPs in dicot species corresponding to 70 % of the total seed nitrogen. They consist of two groups according to their coefficients of sedimentation: the 7S group (or vicilin-like, 7–8S range) and the 11S group (or legumin-like, 11–13S range). These two groups share no obvious sequence similarity but they form similar holoprotein structures and have a common evolutionary origin (Shutov et al. 2003). The globulin family was studied in detail in different legume species because of its economic interest (e.g., in pea; Tzitzikas et al. 2006).

The 11S globulins or legumin-like proteins represent the major SSP class in most dicots, including legumes. They are known under several names according to the legume species: legumin for *Pisum sativum* and *Vicia faba*; glycoprotein II for *Phaseolus vulgaris*; or glycinin for soybean. Mature legumins occur as holoproteins of 320–400 kDa composed of six nonidentical subunits (52–65 kDa). Each subunit contains a large acidic polypeptide of 33–24 kDa and a small basic polypeptide of 19–23 kDa, linked by a disulfide bond.

The 7S globulins or vicilin-like proteins are usually heterogeneous in term of amino acid composition, charge, and size due to a large variety of posttranslational processing events. They are typically present as a trimeric complex of 145–190 kDa, composed of three non-identical polypeptides of 48–83 kDa. For example in pea, vicilin subunits undergo proteolysis and glycosylation to form subunits of 12–33 kDa (for review see, Casey 1999). In soybean or *P. vulgaris,* subunits are highly glycosylated but not proteolytically cleaved (Bollini and Chrispeels 1978; Hall et al. 1977).

Albumins (2S) Albumin proteins are the second group of SSPs and are widely distributed in dicot species. They represent a minor SSP form in legumes but have been extensively analyzed in the Crucifereae family, where they are called napins. The 2S albumins are synthesized as a precursor, consisting of two polypeptides of 4–9 kDa linked by disulfide bonds (Ericson et al. 1986). They are processed by cleavage of a series of peptides from both *C*- and *N*-terminal sequences. The heterodimeric structure linked by a disulfide bond is the most common form of albumins in dicot species, such as lupin (Lilley and Inglis 1986). However, some structural variation exists, such as in pea, where the disulfide bond does not occur (Higgins et al. 1986).

Location of Storage Proteins In most dicot species, including legumes, the main seed storage tissues are the cotyledons (Fig. 11.2b). Whereas several studies confirmed a common spatial expression of SSPs in embryo, differences in temporal expression between these protein classes were reported in pea (Gatehouse et al. 1986), soybean (Ladin et al. 1987), and *M. truncatula* (Gallardo et al. 2007).

11.2.3 Comparison Legume-Cereals

Globulins from cereal and legume grains are closely related (Kriz 1999). Vicilin proteins have limited sequence similarity but are conserved between dicot and monocot species and share common properties, including low cysteine and methionine contents and similar three-subunit secondary structure (Kriz et al. 1999). Cereals also contain 11S storage globulins related to the legume 11S globulins; these constitute the main SSPs for rice and oats, for example (Casey 1999). All 11S proteins in legumes and those from rice and oat share the same structure, existing as a hexameric complex linked by disulfide bonds (Takaiwa et al. 1999; Shotwell 1999). In contrast, the wheat triticin 11S globulins share a common structure with large acidic and small basic chains but form dimers rather than the typical hexameric complex (Singh et al. 1988, 1993).

Related prolamin SSPs are found throughout cereal species, but are not reported from noncereals. Sequence conservation between prolamins from barley, rye, and wheat and, 2S albumins from castor oil, sunflower, and oil seed rape indicates however a common origin of these proteins between monocots and dicots (Häger et al. 1995; Kreis et al. 1985; Shewry et al. 1995). The 11S globulins from legumes, rice, and oat are expressed specifically in the embryo, more precisely in the cotyledons. Other SSPs, like prolamins and albumins, are mainly expressed in the endosperm (cereals) or embryo (legumes), but may share a similar control of their expression during seed maturation.

11.3 Transcriptional Regulation of SSPs

As discussed previously, SSPs are highly regulated in a spatial manner (see paragraph above). However, gene expression may also be differentially regulated between different classes of SSP during development. For instance, in *M. truncatula* seeds, sequential expression of vicilin, then K-type legumin, and finally A-type legumin is observed (Gallardo et al. 2007; Verdier et al. 2008). A series of mutants have been identified, in which the accumulation of SSPs is differentially affected, providing evidence that these processes are under discrete genetic regulation. A small set of genes controlling the synthesis and regulation of SSP genes will be described in more detail.

11.3.1 Transcriptional Regulation in Cereals

Cis- and Trans-Elements in Cereals In cereals, prolamin gene promoters from maize, wheat, and barley share a conserved sequence. This regulatory *cis*-element is located around 300-bases upstream of the start codon and called the '−300 box' or more commonly the 'prolamin box' (Forde et al. 1985). This highly-conserved sequence was found in the promoters of many monocot SSP genes, where its presence is necessary for endosperm-specific expression (Wu et al. 1998). This motif is around 30 bp long, and is composed of two-conserved motifs: the endosperm or E motif (i.e., TGTAAAG) and GCN4-like (GLM) or N motif (i.e., ATGAGTCAT) (Hammond-Kosack et al. 1993; Knudsen 1993) (Fig. 11.3a).

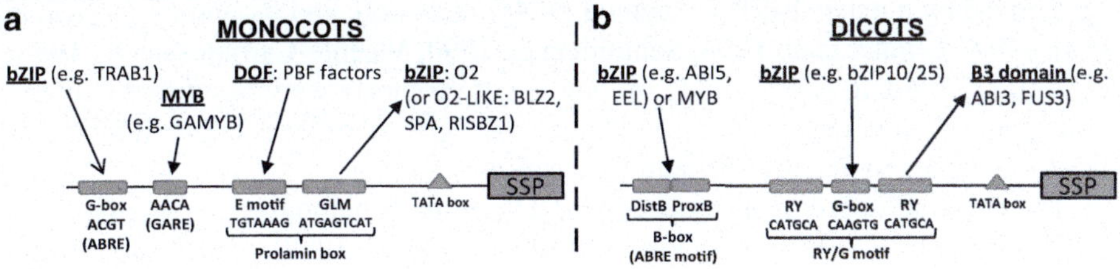

Fig. 11.3 A model illustrating *cis*- and *trans*-regulatory elements of SSP genes in monocots (**a**), and dicots (**b**)

The prolamin box is found, with small variations, in promoters of cereal SSP genes from many cereal species (Knudsen 1993; Vicente-Carbajosa et al. 1997). In maize, some minor changes have been reported. The α-zein does not have a perfectly conserved N motif. The β and γ-zeins share both motifs but they are more distant. Finally in the δ-zein promoter, neither N nor E motif was identified so far (Halford and Shewry 2007). Transient expression experiments using a reporter gene showed that the N motif acts as a negative regulator of gene expression in low-nitrogen conditions, whereas with sufficient nitrogen, interaction between N and E motifs stimulates gene expression (Knudsen 1993). These results demonstrated the importance of the N motif in the response of the prolamin box to nitrogen.

A second conserved motif in glutelin (globulin-like) and hordein (prolamin) genes from rice and barley was identified as part of a complex conferring endosperm-specific expression (Takaiwa et al. 1996; Zheng et al. 1993). This motif (AACA) confers the endosperm-specific enhancement of a truncated 35S promoter together with the GCN4 motif in glutelin genes (Yoshihara and Takaiwa 1996). This motif also constitutes a GARE (Gubler et al. 1999; Suzuki et al. 1998).

Another motif widely present in the promoters of SSP genes constitutes an ABA response element (ABRE). Many ABA-regulated genes have been identified that are associated with seed development processes. Analysis of their promoters revealed a common *cis*-regulatory element with a consensus sequence (T/G/C)ACGT(G/T) GC, also referred to by the more generic terms 'ACGT motif' or 'G-box' (Marcotte et al. 1989). Several bZIP TFs able to bind to these ABRE were identified as candidates for mediating ABA-responsive transcription (Guiltinan et al. 1990; Oeda et al. 1991; Williams and Tsang 1992).

In transgenic rice, a truncated promoter of a rice glutelin gene, containing only the GCN4, endosperm box, AACA motif, and ACGT motif, was able to confer endosperm-specific expression (Wu et al. 1998), showing the importance of these elements for spatial regulation of glutelin gene expression. Wu et al. (2000) studied the combinatorial role of these elements using substitution mutations. They found that the GCN4 box acts as the essential element conferring endosperm-specific expression, whereas AACA, ACGT, and endosperm boxes can contribute in combination with GCN4 boxes to the gene expression regulation.

Combinatorial Networks in Cereals Specific DNA-binding proteins recognize the GCN4 box, endosperm box, ACGT, and AACA motifs. The GCN4 box is recognized by a maize bZIP TF, named *OPAQUE2* (*O2*), and its homologs (Albani et al. 1997; de Pater et al. 1994; Schmidt et al. 1990; Vicente-Carbajosa et al. 1997). Several *O2*-binding sites located on one target promoter were identified (Lohmer et al. 1991; Schmidt et al. 1992). Mutation of *O2* greatly reduces expression of a set of α-zein genes in maize (Burr and Burr 1982).

The endosperm box is recognized by DOF proteins belonging to the zinc finger TF family (Mena et al. 1998; Vicente-Carbajosa et al. 1997). These proteins are called PBF. The AACA motif is recognized by proteins from the R2R3 MYB TF family (Diaz et al. 2002; Suzuki et al. 1998). The ACGT or ABRE motif is able to bind proteins from the bZIP TF family (Guiltinan et al. 1990; Oeda et al. 1991; Williams and Tsang, 1992), such as TRAB1 in rice (Hobo et al. 1999).

The O2 and PBF proteins have been shown to interact *in vivo* and to synergistically activate prolamin promoters in the transient expression experiments (Hwang et al. 2004; Vicente-Carbajosa et al. 1997). Other paralog bZIP TFs were identified in wheat, barley, and rice, termed respectively SPA, BLZ2, and RISBZ1. These proteins resemble O2 and OHP1 from maize in binding to the GCN4 box (Albani et al. 1997; Vicente-Carbajosa et al. 1997) and also by their action. For instance, in rice, RISBZ1 acted synergistically with rice (R)PBF on the GCN4 motif to activate gene expression (Yamamoto et al. 2006).

The *O2* is a key gene for regulation of the prolamin gene expression, whose activity is subject to diurnal regulation through phosphorylation (Ciceri et al. 1997). Several *O2*–related factors were identified capable of binding to the same target sequence. For instance, OHP1 can bind the same binding site as O2, either as a homodimer or heterodimer with O2, but unlike O2 it does not interact with a PBF protein (Pysh et al. 1993; Vicente-Carbajosa et al. 1997).

In maize, *VIVIPAROUS1 (VP1)*, a TF ortholog of *Arabidopsis ABI3*, plays a crucial role in several processes of seed development, including ABA signaling, activation of maturation-specific genes, and establishment of dormancy (McCarty et al. 1989; Suzuki et al. 2003). This gene is expressed specifically in the embryo and aleurone layer of endosperm, the sites of synthesis of globulin proteins. The *VP1* regulates maturation and dormancy by activating ABA-regulated genes *via* their ABRE (Hattori et al. 1995; Vasil et al. 1995). However, as no direct interaction between VP1 and this *cis*-regulatory element has been identified, it has been proposed that VP1 acts by interacting with another protein that directly binds ABRE motif (Suzuki et al. 1997). The ABRE motifs are recognized by bZIP family proteins, and a rice bZIP factor, called *TRAB1*, has been identified as a mediator of ABA-induced transcription through its interaction with VP1 and an ABRE motif (Hobo et al. 1999).

The fourth SSP promoter element, the ACAA motif, has been reported to interact with MYB proteins. Several proteins able to bind this domain have been identified in barley and called HvGAMYB. The *HvGAMYB* genes are expressed in cereal aleurone cells and have been shown to activate GA-responsive promoters by binding the ACAA motif (Gubler et al. 1999). Diaz et al. (2002) demonstrated that an interaction between HvGAMYB proteins and barley PBF proteins is essential for activating the hordein gene expression. Rubio-Somoza et al. (2006) proposed that two further MYB factors, HvMYBS3 and HvMCB1, may be a part of the complex, controlling the expression of SSP genes. Whereas HvMCB1 acts as an activator of the maturation phase and as a repressor of the germination, HvMYBS3 acts as an activator of transcription in both processes (Rubio-Somoza et al. 2006).

Recently, a B3 domain TF, closely related to *FUS3* identified in dicots, has been discovered in barley and rice. This gene is expressed in endosperm during the mid-maturation phase and is involved in transcriptional activation of the hordein genes (prolamin proteins) through RY-motifs conserved in the storage protein promoters from barley, rice, wheat, and maize (Moreno-Risueno et al. 2008). Another TF, *Zm-LEC1*, closely related to the well-known *LEC1* gene in dicots, has been identified in maize, but appears to be specifically involved in seed oil biosynthesis (Shen et al. 2010).

11.3.2 Transcriptional Regulation in Dicots

Much effort has gone into deciphering the genetic regulatory network, controlling the biosynthesis of SSP in dicots. Although we lack a comprehensive understanding of mechanisms and genes involved in this process, we have now a framework to describe this network (for review, Verdier and Thompson 2008).

Cis- and Trans-Regulatory Elements in Dicot Species In legumes and more broadly in dicots, the regulatory elements involved in SSP activation have been identified, mainly on the *napinA* promoter (2S albumin) from *Brassica napus*, on legumin B4 (11S globulin) from *V. faba*, and also on phaseolin (7S globulin) from *P. vulgaris* promoters (Bäumlein et al. 1992; Chandrasekharan et al. 2003; Ezcurra et al. 1999; Stålberg et al. 1996). Sequence comparison of these promoters revealed two highly-conserved binding sites, the RY/G motif and B-box (Fig. 11.3b).

The RY/G motif is a tripartite motif composed of two RY elements (i.e., CATGCA) (Bäumlein et al. 1992) surrounding a G-box (i.e., CACGTG) (de Pater et al. 1993). Interactions between these motifs and regulatory genes have been elucidated by several studies. The B3 domain proteins (i.e., ABI3 and FUS3) have been shown to bind RY motifs (Ezcurra et al. 2000; Reidt et al. 2000), whereas bZIP or bHLH proteins can interact with the G-box (Kawagoe and Murai 1996). This *cis*-regulatory element is the site of elicitation of the ABA response (Chandrasekharan et al. 2003).

The second essential element found in the promoters of storage protein genes is a B-box. This motif is composed of two domains: DistB (GCCACTTGTC) and ProxB (CAAACACC). This element was also identified as a mediator of ABA response in seeds *via* an ABRE domain (Ezcurra et al. 1999). The B-box element interacts with various proteins, including bZIPs, and may interact with a MYB TF due to an overlap of the sequence with a MYB-recognition site (Pla et al. 1993).

These two motifs (i.e., RY/G and B boxes) act synergistically, as mutations in either of these elements strongly reduce gene expression (Ezcurra et al. 1999; Kawagoe et al. 1994). Moreover, according to *in vitro* experiments, these elements were shown to be required for the seed-specific expression of SSPs (Ezcurra et al. 1999). Using promoter mutations, RY motifs have been identified as the major element responsible for spatial and temporal regulation of SSP gene expression (Chandrasekharan et al. 2003).

11.3.3 Combinatorial Network

Four master regulators of seed development. As a result of extensive mutant analyses, an overview was obtained of the combinatorial genetic network, involved in the regulation of seed SP synthesis (for review see, Verdier and Thompson 2008). This regulatory network is based on the four master regulators of seed development: *LEC1, LEC2, ABI3, and FUS3*. These TFs have been extensively studied in *Arabidopsis* for their roles in programming embryo development and desiccation tolerance, including the regulation of SSP expression. The four genes show pleiotropic, overlapping, and partially redundant functions (Kroj et al. 2003; Santos-Mendoza et al. 2008).

The *LEC1* (Lotan et al. 1998) belongs to the HAP3 family, which constitutes one subunit of the CCAAT-binding factor (CBF: Maity and de Crombrugghe 1998). *LEC2* (Santos Mendoza et al. 2005) is a member of plant-specific B3 domain family, closely related to *ABI3* (Parcy et al. 1997) and *FUSCA3* (*FUS3*: Bäumlein et al. 1994; Keith et al. 1994). The four genes are specifically expressed in seeds during the late embryogenesis/seed maturation processes.

LEC1 and *LEC2* control the upstream part of the regulatory cascade. They are required for the maintenance of cotyledon identity during seed filling, and thus for the completion of the maturation phase (Braybrook et al. 2006; Lotan et al. 1998). Mutations in these genes induce formation of viviparous embryos, partial transformation of cotyledons in leaves, compromise seed desiccation tolerance, and reduce accumulation of storage molecules, including SPs (Bäumlein et al. 1994; Keith et al. 1994; Meinke et al. 1994; West et al. 1994). Conversely, ectopic expression of *LEC1* or *LEC2* induces embryogenic pathways in vegetative cells (Lotan et al. 1998; Santos Mendoza et al. 2005). *LEC1* is a positive regulator of *LEC2* genes but also of *FUS3,* implying a hierarchical network (Kagaya et al. 2005). *LEC2* is also capable of activating the gene expression of *ABI3* and *FUS3,* considered downstream of it (Braybrook et al. 2006; Kroj et al. 2003; Mönke et al. 2004).

As *ABI3* and *FUS3* act downstream of the regulatory cascade, *abi3* and *fus3* mutants share similar phenotypes with those of *lec1* and *lec2* (e.g., loss of cotyledon identity, vivipary, desiccation sensitivity, and defect in storage molecule accumulation) (Bäumlein et al. 1994; Castle and Meinke 1994; Giraudat et al. 1992; Keith et al. 1994; Parcy et al. 1994). ABI3 and FUS3 were showed to be direct positive regulators of seed filling. Whilst they appear to have a similar and redundant function, genes are activated by ABI3 and FUS3 with different kinetics, suggesting the existence of an intermediate factor for the FUS3 activation pathway (Kagaya et al.

2005). Moreover, both ABI3 and FUS3 are able to bind the RY-motif present in target gene promoters (Bobb et al. 1997; Ezcurra et al. 2000; Mönke et al. 2004; Reidt et al. 2000, 2001). Direct binding to RY elements was observed using *in vitro* and *in vivo* experiments for FUS3 (Kroj et al. 2003), whereas for ABI3, only *in vitro* experiments showed this binding (Mönke et al. 2004), suggesting a putative association between ABI3 and a second partner to enhance its binding *in vivo* (Kroj et al. 2003).

A complex network of local and redundant regulations has been identified between *LEC1, LEC2, ABI3*, and *FUS3* genes forming regulatory feedbacks essential for cell-type specific expression. An additional level of complexity in this regulatory network is reflected by differences in their spatial expression within embryo, resulting in differing genetic controls in different parts of embryo (for review, Santos-Mendoza et al. 2008; To et al. 2006).

Other partners. Roles for other partners have been identified in regulation of the SSP regulatory complex. Lara et al. (2003) identified two bZIPs (i.e., *AtbZIP10* and *AtbZIP25*), which are able to bind to the G-box present in the SSP promoters and also able to interact with ABI3 in 2H experiments. These bZIP proteins may be ABI3 partners, which enhance binding between ABI3 and the RY/G-box motif, a hypothesis supported by a synergistic effect between these bZIPs and ABI3 on the activation of SSP gene expression in transient assays (Lara et al. 2003).

Two proteins have been shown to bind the B-box and mediate ABA-induced transcription (Nakamura et al. 2001). These proteins, termed *ABI5* and *ENHANCED EM LEVEL (EEL)*, from the bZIP TF family, are also able to interact with ABI3 proteins (Nakamura et al. 2001). However, differences in their timing of expression (i.e., ABI5 and EEL are expressed respectively during the mid and the late-filling phase) and in their mutant phenotypes reflect antagonistic functions. *ABI5* is necessary for expression of the SSP genes, whereas EEL, by competing for the same binding site, modulates or even represses ABI5-mediated gene expression (Bensmihen et al. 2002).

Two further bZIP TFs, named *Repressor Of Maturation* (*ROM*), have been identified in *P. vulgaris* to play an important role before and after the seed-filling phase. Both are able to bind the G-box of the RY/G-box motif *in vitro* and able to repress the expression of the phaseolin promoter *in vivo* (Chern et al. 1996a, b). The *ROM1* is expressed during embryogenesis and is downregulated at the onset of seed-filling phase (Chern et al. 1996b). In contrast, *ROM2* is expressed after the seed-filling phase and a positive correlation was demonstrated between an increase of ROM2 DNA-binding activity and the repression of maturation gene expression during the desiccation phase (Chern et al. 1996a).

Finally, the end of seed-filling phase is also regulated *via* repression of master regulator expression. A group of three B3 domain TFs, called the *VP1/ABI3-LIKE* genes (*VAL* genes), has been identified, which repress the LEC1-B3 network, possibly *via* binding the RY motif (Suzuki et al. 2007). In addition, a SWI/SNF remodelling chromatin factor, encoded by *PICKLE* (*PKL*) has also been described as a repressor of LEC1 expression upon germination (Henderson et al. 2004; Ogas et al. 1999; Rider et al. 2003; for review Zhang and Ogas 2009).

11.3.4 A Common Transcriptional Network in Dicot Cotyledons and Cereal Endosperm

The programs of seed maturation between monocots and dicots are markedly different. Different families of SSPs are synthesized and are stored in different tissues. However, numerous similarities are emerging. Although there is no evidence yet to date about the existence of *LEC2* genes in monocots, parts of the maturation gene program are common between these two families. Several *cis*-regulatory elements have been identified in the promoters of seed storage genes from dicot and monocot plants as essential for expression during seed maturation (e.g., AACA motif, GCN4-like motif, RY motif, and DOF-binding domain) (Fig. 11.3). Moreover, several regulators with similar roles in both species are functionally exchangeable.

One of the central regulators of seed maturation in cereals is the *VP1* gene, which is closely related to the dicot *ABI3* gene. They have similar functions in regulating seed filling and ABA-dependent preparation for desiccation. Both proteins interact with a bZIP partner to bind the ABRE motif [TRAB1 for VP1 and ABI5 for ABI3: Hobo et al. 1999; Nakamura et al. 2001]. Interestingly, these genes are exchangeable between both systems. For instance, Gampala et al. (2002) showed that *Arabidopsis ABI5* and maize *VP1* can be used to activate expression of ABA-inducible promoters in cereals (i.e., wheat and barley) and dicots (i.e. *Arabidopsis* and *P. vulgaris*). Moreover, *VP1* can complement *Arabidopsis abi3* mutants originally impaired in the expression of SSP genes (Suzuki et al. 2001).

Another example of this remarkable conservation between the genetic programs is the presence of *FUS3* gene in both cereals and dicots. Moreno-Risueno et al. (2008) complemented a *fus3* mutant in *Arabidopsis* with the barley *HvFUS3* gene, resulting in restoration of SSP expression. In both systems, FUS3 acts similarly by binding an RY motif with a bZIP partner (bZIP10/25 in dicots and an O2-LIKE factor in cereals) (Moreno-Risueno et al. 2008). Thus, whilst the data is incomplete, it seems probable that seed development in dicots and monocots possesses some level of conservation in regulatory programs, adapted for directing reserve accumulation in cereal and dicot seeds. In contrast, the *LEC1-type* genes were identified in both cereals and dicots, but exhibit slight functional differences. Whereas *LEC1* is well characterized in dicots, affecting all seed storage molecules including SSPs, its maize ortholog only affects seed oil biosynthesis (Shen et al. 2010).

11.3.5 Other Transcriptional Programs Operating in Developing Seeds

The previously described regulatory network represents only one of several programs of gene expression during seed development. Global transcriptomics approaches have revealed a large number of discrete seed mRNA expression profiles (Gallardo et al. 2007; Huang et al. 2009; Jones et al. 2010; Le et al. 2007), reflecting the action (and probable interaction) of several regulatory networks. Each seed-developmental phase is accompanied by expression of discrete TF populations (Verdier et al. 2008; Wang et al. 2010), and these TFs are in part specific for different seed tissues, or even cell-types within a tissue (Le et al. 2010; Verdier et al. 2008). The gene expression profile of a given seed cell is partly specified by the intrinsic-expressed TF complement, but this is also modulated by diverse environmental factors at both transcriptional and posttranscriptional levels.

11.4 Environmental Regulation

In parallel to the genetic regulation of SSPs, environmental factors influence seed protein quantity and composition. For instance, environmental factors, such as temperature, play important roles in regulating seed filling and final seed quality. In this section, we will focus on two important factors involved in the regulation of these proteins: hormones and nutrients.

11.4.1 Hormones

The ABA plays a central role during seed maturation and SP synthesis (Finkelstein et al. 2002; Nambara and Marion-Poll 2003). We know now that ABA is not the only hormone regulating this process, but its action is modulated by an antagonistic effect of GA3. The ratio of concentrations of ABA to GA3 is a key readout determining seed maturation, acquisition of desiccation tolerance, and entry into seed dormancy.

The central role of ABA in dicots is reflected by the interaction between ABA and some of the key regulators of seed maturation. For instance, gene activation implying *FUS3* and *ABI3* is strongly enhanced by ABA action (Kagaya et al. 2005; Parcy et al. 1994). Moreover, a positive correlation between *FUS3* expression and ABA level has been demonstrated (Gazzarrini et al. 2004). Conversely, *FUS3* is also a negative regulator of GA3 biosynthesis (Gazzarrini et al. 2004).

Auxin is also implicated in the regulation of seed maturation. Ectopic expression of *LEC2* induces the genes of auxin biosynthesis and the activity of auxin-responsive gene (Stone et al. 2008). Interactions between auxin and *LEC1* (Casson and Lindsey 2006), and *FUS3* (Gazzarrini et al. 2004), have also been identified, suggesting an action of these genes mediated by this hormone.

11.4.2 Nutrients

Another crucial factor acting in combination with genetic regulation is the concentration (or ratio) of certain metabolites. Among the metabolites able to regulate seed maturation, the sucrose/hexose ratio and nitrogen content appear to be the most critical. Sucrose is an important metabolite for seed development with its central role in carbohydrate metabolism. It also acts as a signal molecule, controlling regulation of seed maturation (for review see, Koch 2004; Smeekens 2000; Weber et al. 2005). *In vitro* and *in vivo* studies showed that increasing sucrose concentration induces an increase of storage activity-related transcripts (Ambrose et al. 1987; Corke et al. 1990). In *V. faba*, pea, and barley, sucrose is an inducer of seed storage gene expression, triggering the switch from embryogenesis to seed filling (Tegeder et al. 1999; Weber et al. 1997; Weschke et al. 2000). This result was confirmed for *Arabidopsis* seeds, in which sucrose induces globulin and albumin gene expression, but also that of the master regulators *LEC1, LEC2,* and *FUS3* (Tsukagoshi et al. 2007). In contrast, *apetala2* mutants, with increased seed mass, have an increased hexose/sucrose ratio resulting in a prolongation of the cell-division phase (i.e., embryogenesis phase) (Ohto et al. 2005). These results confirm that sucrose signaling, and specifically the sucrose/hexose ratio, control the switch between the differentiation and storage programs.

Nitrogen metabolites also act to regulate SSP synthesis. In pea, a positive correlation has been demonstrated between the accumulation of seed nitrogen availability and storage protein (Salon et al. 2001). Moreover, in soybean, this nitrogen signaling is mediated *via* the concentration of specific amino acids, such as asparagines (Hernández-Sebastià et al. 2005). In cereals, the response to nitrogen regime was demonstrated to act *via* the GCN4 box (i.e., in barley, wheat, or maize), where an increase in nitrogen availability induces prolamin gene expression (Knudsen 1993; Singletary et al. 1990; Weichert et al. 2010).

Other metabolic signals have been identified as showing a positive correlation between their accumulation and SSP accumulation, including ATP and oxygen levels (Borisjuk et al. 2003; Vigeolas et al. 2003; for review, Weber et al. 2005). Metabolic activities in developing seeds are sensitive to cellular oxygen levels, which are in part determined locally by photosynthetic activity of seed tissues. The transcription levels of genes associated with glycolysis and mitochondrial energy metabolism are regulated by p[O_2]. The response to p[O_2] may be detected indirectly *via* [ATP] (Borisjuk and Rolletschek 2009). These authors also propose an important role for NO in regulating metabolic activity due to p[O_2] changes, but the mechanisms of generation and transduction of these signals remain to be elucidated. This is also more generally applicable to the mechanisms of cellular response to metabolic and environmental signals.

11.5 Future Directions

11.5.1 Alternative Splicing

Alternative splicing is a mechanism of posttranscriptional control, whose significance in plants is being increasingly recognized. In the seed, there is evidence for regulation of the related TFs *ABI3* (*Arabidopsis*) and *VP1* (maize) *via* accumulation of alternatively spliced mRNAs late in seed development (Sugliani et al. 2010; Wilkinson et al. 2005).

11.5.2 Regulation by miRNA

A further emerging level of posttranscriptional gene regulation involves miRNAs. The seed, like other tissues, has its complement of small RNAs, most of which are in the process of being characterized and have not yet had functions attributed. One of the first examples to be characterized was miR159, which exists in three forms (a, b, and c); two of which repress at least two GAMYB genes in non-seed tissues. In the absence of miR159a and b, MYB33 and MYB65 induce the synthesis of a series of GA-regulated aleurone genes (Alonso-Peral et al. 2010). In an analysis of rice seed small RNA sequences, signatures matching miRNA precursors account for <1.5 % of total genome-matching signatures, which is much lower than that in rice inflorescences, seedlings, and stems (Xue et al. 2009). One interesting example of miRNA-controlled gene expression in seeds is the regulation of ARF8 by miR167 (Yang et al. 2006). The ARF8 regulates the expression of OsGH3.2, a rice IAA-conjugating enzyme, which contributes to cellular free IAA levels. Os-miR167, which is induced by auxin, acts to decrease ARF8 mRNA abundance, and may confer feedback regulation on auxin-responsive gene activity.

11.6 Concluding Remarks

The regulation study of seed development and composition will remain of fundamental socio-economic importance with increasing pressure on food supplies and the potential seeds offer as sources of renewable resources. The gene regulatory networks identified for *Arabidopsis* seed maturation, and those being developed for other species, whilst complex, are not intractable. The coming years will see increasing exploitation of omics data, combined with high-throughput phenotyping under precisely-controlled environmental conditions to define precisely the responses of developing seed to external factors, such as nutrient supply and abiotic stresses, and to provide leads for smart breeding to optimize genetically the performance of the seed factory.

References

Albani D, Hammond-Kosack MC, Smith C, Conlan S, Colot V, Holdsworth M, Bevan MW (1997) The wheat transcriptional activator SPA: a seed-specific bZIP protein that recognizes the GCN4-like motif in the bifactorial endosperm box of prolamin genes. Plant Cell 9:171–184

Alonso-Peral MM, Li J, Li Y, Allen RS, Schnippenkoetter W, Ohms S, White R, Millar AA (2010) The microRNA159-regulated *GAMYB*-like genes Inhibit growth and promote programmed cell death in *Arabidopsis*. Plant Physiol 154:757–771

Ambrose MJ, Wang TL, Cook SK, Hedley CL (1987) An analysis of seed development in *Pisum sativum*. J Exp Bot 38:1909–1920

Bäumlein H, Nagy I, Villarroel R, Inzé D, Wobus U (1992) *Cis*-analysis of a seed protein gene promoter: the conservative RY repeat CATGCATG within the legumin box is essential for tissue-specific expression of a legumin gene. Plant J 2:233–239

Bäumlein H, Miséra S, Luerßen H, Kölle K, Horstmann C, Wobus U, Müller AJ (1994) The *FUS3* gene of *Arabidopsis thaliana* is a regulator of gene expression during late embryogenesis. Plant J 6:379–387

Bensmihen S, Rippa S, Lambert G, Jublot D, Pautot V, Granier F, Giraudat J, Parcy F (2002) The homologous ABI5 and EEL transcription factors function antagonistically to fine-tune gene expression during late embryogenesis. Plant Cell 14:1391–1403

Bobb AJ, Chern MS, Bustos MM (1997) Conserved RY-repeats mediate transactivation of seed-specific promoters by the developmental regulator PvALF. Nucl Acids Res 25:641–647

Bollini R, Chrispeels MJ (1978) Characterization and subcellular localization of vicilin and phytohemagglutinin, the two major reserve proteins of *Phaseolus vulgaris*. Planta 142:291–298

Borisjuk L, Rolletschek H (2009) The oxygen status of the developing seed. New Phytol 182:17–30

Borisjuk L, Rolletschek H, Wobus U, Weber H (2003) Differentiation of legume cotyledons as related to metabolic gradients and assimilate transport into seeds. J Exp Bot 54:503–512

Braybrook SA, Stone SL, Park S, Bui AQ, Le BH, Fischer RL, Goldberg RB, Harada JJ (2006) Genes directly regulated by LEAFY COTYLEDON2 provide insight into the control of embryo maturation and somatic embryogenesis. Proc Natl Acad Sci U S A 103:3468–3473

Brocard-Gifford IM, Lynch TJ, Finkelstein RR (2003) Regulatory networks in seeds integrating developmental, abscisic acid, sugar, and light signaling. Plant Physiol 131:78–92

Burgess SR, Shewry P (1986) Identification of homologous globulins from embryos of wheat, barley, rye, and oats. J Exp Bot 37:1863–1871

Burr FA, Burr B (1982) Three mutations in *Zea mays* affecting zein accumulation: a comparison of zein polypeptides, *in-vitro* synthesis and processing, mRNA levels, and genomic organization. J Cell Biol 94:201–206

Casey R. (1999). Distribution of some properties of seed globulins. In: Shewry PR, Casey R (eds) Seed proteins. Kluwer Academic, Dordrecht, pp 617–633

Casson SA, Lindsey K (2006) The turnip mutant of *Arabidopsis* reveals that LEAFY COTYLEDON1 expression mediates the effects of auxin and sugars to promote embryonic cell identity. Plant Physiol 142:526–541

Century K, Reuber TL, Ratcliffe OJ (2008) Regulating the regulators: the future prospects for transcription-factor-based agricultural biotechnology products. Plant Physiol 147:20–29

Chandrasekharan MB, Bishop KJ, Hall TC (2003) Module-specific regulation of the β-phaseolin promoter during embryogenesis. Plant J 33:853–866

Chaw SM, Chang CC, Chen HL, Li WH (2004) Dating the monocot-dicot divergence and the origin of core eudicots using whole chloroplast genomes. J Mol Evol 58:424–441

Chern MS, Bobb AJ, Bustos MM (1996a) The Regulator of MAT2 (ROM2) protein binds to early maturation promoters and represses PvALF-activated transcription. Plant Cell 8:305–321

Chern MS, Eiben HG, Bustos MM (1996b) The developmentally regulated bZIP factor ROM1 modulates transcription from lectin and storage protein genes in bean embryos. Plant J 10:135–148

Ciceri P, Gianazza E, Lazzari B, Lippoli G, Genga A, Hoschek G, Schmidt RJ, Viotti A (1997) Phosphorylation of Opaque2 changes diurnally and impacts its DNA binding activity. Plant Cell 9:97–108

Coleman CE, Larkins BA (1999) The prolamins of maize. In: Shewry PR, Case R (eds) Seed proteins. Kluwer Academic, Dordrecht, pp 109–139

Corke FMK, Hedley CL, Wang TL (1990) An analysis of seed development *Pisum sativum*. XI. Cellular development and the deposition of storage protein in mature embryos grown *in vivo* and *in vitro*. Protoplasma 155:127–135

Cuddeford D (1995) Oats for animal feed. In: Welch RW (ed) The oat crop: production and utilization. Chapman and Hall, London, pp 321–358

de Pater S, Pham K, Chua NH, Memelink J, Kijne J (1993) A 22-bp fragment of the pea lectin promoter containing essential TGAC-like motifs confers seed-specific gene expression. Plant Cell 5:877–886

de Pater S, Katagiri F, Kijne J, Chua NH (1994) bZIP proteins bind to a palindromic sequence without an ACGT core located in a seed-specific element of the pea lectin promoter. Plant J 6:133–140

Diaz I, Vicente-Carbajosa J, Abraham Z, Martínez M, Isabel-La MI, Carbonero P (2002) The GAMYB protein from barley interacts with the DOF transcription factor BPBF and activates endosperm-specific genes during seed development. Plant J 29:453–464

Ericson ML, Rödin J, Lenman M, Glimelius K, Josefsson LG, Rask L (1986) Structure of the rapeseed 1.7S storage protein, napin, and its precursor. J Biol Chem 261:14576–14581

Ezcurra I, Ellerström M, Wycliffe P, Stålberg K, Rask L (1999) Interaction between composite elements in the *napA* promoter: both the B-box ABA-responsive complex and the RY/G complex are necessary for seed-specific expression. Plant Mol Biol 40:699–709

Ezcurra I, Wycliffe P, Nehlin L, Ellerström M, Rask L (2000) Transactivation of the *Brassica napus* napin promoter by ABI3 requires interaction of the conserved B2 and B3 domains of ABI3 with different *cis*-elements: B2 mediates activation through an ABRE, whereas B3 interacts with an RY/G-box. Plant J 24:57–66

Finkelstein RR, Gampala SSL, Rock CD (2002) Abscisic acid signaling in seeds and seedlings. Plant Cell 14:S15--S45

Forde BG, Heyworth A, Pywell J, Kreis M (1985) Nucleotide sequence of a B1 hordein gene and the identification of possible upstream regulatory elements in endosperm storage protein genes from barley, wheat, and maize. Nucl Acids Res 13:7327–7339

Gallardo K, Firnhaber C, Zuber H, Hericher D, Belghazi M, Henry C, Kuster H, Thompson RD (2007) A combined proteome and transcriptome analysis of developing *Medicago truncatula* seeds. Mol Cell Proteomics 6:2165–2179

Gampala SSL, Finkelstein RR, Sun SSM, Rock CD (2002) ABI5 interacts with abscisic acid signaling effectors in rice protoplasts. J Biol Chem 277:1689–1694

Gao G, Zhong Y, Guo A, Zhu Q, Tang W, Zheng W, Gu X, Wei L, Luo J (2006) DRTF: a database of rice transcription factors. Bioinformatics 22:1286–1287

Gazzarrini S, Tsuchiya Y, Lumba S, Okamoto M, McCourt P (2004) The transcription factor *FUSCA3* controls developmental timing in *Arabidopsis* through the hormones gibberellin and abscisic acid. Developmental Cell 7:373–385

Gubler F, Raventos D, Keys M, Watts R, Mundy J, Jacobsen J (1999) Target genes and regulatory domains of the GAMYB transcriptional activator in cereal aleurone. Plant J 17:1–9

Guiltinan MJ, Marcotte WR Jr, Quatrano RS (1990) A plant leucine zipper protein that recognizes an abscisic acid response element. Science 250:267–271

Häger KP, Braun H, Czihal A, Müller B, Bäumlein H (1995) Evolution of seed storage protein genes: legumin genes of *Ginkgo biloba*. J Mol Evol 41:457–466

Halford NG, Shewry PR (2007) The structure and expression of cereal storage protein genes. In: Olsen OA (ed) Plant cell monographs. endosperm: development and molecular biology, vol 8. Springer, Berlin, pp 196–218

Hall TC, McLeester RC, Bliss FA (1977) Equal expression of the maternal and paternal alleles for the polypeptide subunits of the major storage protein of the bean *Phaseolus vulgaris* L. Plant Physiol 59:1122–1124

Hammond-Kosack MCU, Holdsworth MJ, Bevan MW (1993) *In vivo* footprinting of a low molecular weight glutenin gene (LMWG-1D1) in wheat endosperm. EMBO J 12:545–554

Hattori T, Terada T, Hamasuna S (1995) Regulation of the *Osem* gene by abscisic acid and the transcriptional activator VP1: analysis of *cis*-acting promoter elements required for regulation by abscisic acid and VP1. Plant J 7:913–925

Heck GR, Chamberlain AK, Ho THD (1993) Barley embryo globulin 1 gene, *Beg1*: characterization of cDNA, chromosome mapping and regulation of expression. Mol Gen Genet 239:209–218

Henderson JT, Li HC, Rider SD, Mordhorst AP, Romero-Severson J, Cheng JC, Robey J, Sung ZR, de Vries SC, Ogas J (2004) *PICKLE* acts throughout the plant to repress expression of embryonic traits and may play a role in gibberellin-dependent responses. Plant Physiol 134:995–1005
Hernández-Sebastià C, Marsolais F, Saravitz C, Israel D, Dewey RE, Huber SC (2005) Free amino acid profiles suggest a possible role for asparagine in the control of storage-product accumulation in developing seeds of low- and high-protein soybean lines. J Exp Bot 56:1951–1963
Higgins TJV, Chandler PM, Randall PJ, Spencer D, Beach LR, Blagrove RJ, Kortt AA, Inglis AS (1986) Gene structure, protein structure, and regulation of the synthesis of a sulfur-rich protein in pea seeds. J Biol Chem 261:11124–11130
Hobo T, Kowyama Y, Hattori T (1999) A bZIP factor, TRAB1, interacts with VP1 and mediates abscisic acid-induced transcription. Proc Natl Acad Sci U S A 96:15348–15353
Huang J, Bhinu VS, Li X, Dallal Bashi Z, Zhou R, Hannoufa A (2009) Pleiotropic changes in *Arabidopsisf5h* and *sct* mutants revealed by large-scale gene expression and metabolite analysis. Planta 230:1057–1069
Hwang YS, Ciceri P, Parsons RL, Moose SP, Schmidt RJ, Huang N (2004) The Maize O2 and PBF proteins act additively to promote transcription from storage protein gene promoters in rice endosperm cells. Plant Cell Physiol 45:1509–1518
Jones SI, Gonzalez DO, Vodkin LO (2010) Flux of transcript patterns during soybean seed development. BMC genomics 11:136
Kagaya Y, Toyoshima R, Okuda R, Usui H, Yamamoto A, Hattori T (2005) LEAFY COTYLEDON1 controls seed storage protein genes through its regulation of *FUSCA3* and *ABSCISIC ACID INSENSITIVE3*. Plant Cell Physiol 46:399–406
Kawagoe Y, Campell BR, Murai N (1994) Synergism between CACGTG (G-box) and CACCTG *cis*-elements is required for activation of the bean seed storage protein β-*phaseolin* gene. Plant J 5:885–890
Kawagoe Y, Murai N (1996) A novel basic region/helix-loop-helix protein binds to a G-box motif CACGTG of the bean seed storage protein [beta]-phaseolin gene. Plant Sci 116:47–57
Keith K, Kraml M, Dengler NG, McCourt P (1994) fusca3: a heterochronic mutation affecting late embryo development in *Arabidopsis*. Plant Cell 6:589–600
Kent NL (1966) Subaleurone endosperm cells of high protein content. Cereal Chem 43:585–601
Knudsen S (1993) The nitrogen response of a barley C-hordein promoter is controlled by positive and negative regulation of the GCN4 and endosperm box. Plant J 4:343–355
Koch K (2004) Sucrose metabolism: regulatory mechanisms and pivotal roles in sugar sensing and plant development. Curr Opin Plant Biol 7:235–246
Kreis M, Forde BG, Rahman S, Miflin BJ, Shewry PR (1985) Molecular evolution of the seed storage proteins of barley, rye and wheat. J Mol Biol 183:499–502
Kriz AL (1999) 7S globulins of cereals. In: Shewry PR, Casey R (eds) Seed proteins. Kluwer Academic, Dordrecht, pp 477–498
Kroj T, Savino G, Valon C, Giraudat J, Parcy F (2003) Regulation of storage protein gene expression in *Arabidopsis*. Development 130:6065–6073
Ladin BF, Tierney ML, Meinke DW, Hosángadi P, Veith M, Beachy RN (1987) Developmental regulation of {beta}-conglycinin in soybean axes and cotyledons. Plant Physiol 84:35–41
Lamacchia C, Shewry PR, Di Fonzo N, Forsyth JL, Harris N, Lazzeri PA, Napier JA, Halford NG, Barcelo P (2001) Endosperm specific activity of a storage protein gene promoter in transgenic wheat seed. J Exp Bot 52:243–250
Lara P, Oñate-Sánchez L, Abraham Z, Ferrándiz C, Díaz I, Carbonero P, Vicente-Carbajosa J (2003) Synergistic activation of seed storage protein gene expression in *Arabidopsis* by ABI3 and two bZIPs related to OPAQUE2. J Biol Chem 278:21003–21011
Laroche J, Li P, Bousquet J (1995) Mitochondrial DNA and monocot-dicot divergence time. Mol Biol Evol 12:1151–1156
Le BH, Wagmaister JA, Kawashima T, Bui AQ, Harada JJ, Goldberg RB (2007) Using genomics to study legume seed development. Plant Physiol 144:562–574
Le BH, Cheng C, Bui AQ, Wagmaister JA, Henry KF, Pelletier J, Kwong L, Belmonte M, Kirkbride R, Horvath S, Drews GN, Fischer RL, Okamuro JK, Harada JJ, Goldberg RB (2010) Global analysis of gene activity during *Arabidopsis* seed development and identification of seed-specific transcription factors. Proc Natl Acad Sci U S A 107:8063–8070

Lending CR, Larkins BA (1989) Changes in the zein composition of protein bodies during maize endosperm development. Plant Cell 1:1011–1023

Lilley GG, Inglis AS (1986) Amino acid sequence of conglutin [delta], a sulfur-rich seed protein of *Lupinus angustifolius* L.: sequence homology with the C-III [alpha]-amylase inhibitor from wheat. FEBS Lett 195:235–241

Lohmer S, Maddaloni M, Motto M, Di Fonzo N, Hartings H, Salamini F, Thompson RD (1991) The maize regulatory locus Opaque-2 encodes a DNA-binding protein which activates the transcription of the *b-32* gene. EMBO J 10:617–624

Lotan T, Ohto M, Yee KM, West MAL, Lo R, Kwong RW, Yamagishi K, Fischer RL, Goldberg RB, Harada JJ (1998) *Arabidopsis* LEAFY COTYLEDON1 is sufficient to induce embryo development in vegetative cells. Cell 93:1195–1205

Maity SN, de Crombrugghe B (1998) Role of the CCAAT-binding protein CBF/NF-Y in transcription. Trends Biochem Sci 23:174–178

Marcotte WR Jr, Russell SH, Quatrano RS (1989) Abscisic acid-responsive sequences from the *em* gene of wheat. Plant Cell 1:969–976

McCarty DR, Carson CB, Stinard PS, Robertson DS (1989) Molecular analysis of viviparous-1: an abscisic acid-insensitive mutant of maize. Plant Cell 1:523–532

Meinke DW, Franzmann LH, Nickle TC, Yeung EC (1994) Leafy cotyledon mutants of *Arabidopsis*. Plant Cell 6:1049–1064

Mena M, Vicente-Carbajosa J, Schmidt RJ, Carbonero P (1998) An endosperm-specific DOF protein from barley, highly conserved in wheat, binds to and activates transcription from prolamin-box of a native B-hordein promoter in barley endosperm. Plant J 16:53–62

Mönke G, Altschmied L, Tewes A, Reidt W, Mock HP, Bäumlein H, Conrad U (2004) Seed-specific transcription factors ABI3 and FUS3: molecular interaction with DNA. Planta 219:158–166

Moreno-Risueno MA, Gonzalez N, Diaz I, Parcy F, Carbonero P, Vicente-Carbajosa J (2008) FUSCA3 from barley unveils a common transcriptional regulation of seed-specific genes between cereals and *Arabidopsis*. Plant J 53:882–894

Müntz K (1998) Deposition of storage proteins. Plant Mol Biol 38:77–99

Nakamura S, Lynch TJ, Finkelstein RR (2001) Physical interactions between ABA response loci of *Arabidopsis*. Plant J 26:627–635

Nambara E, Marion-Poll A (2003) ABA action and interaction in seeds. Trends Plant Sci 8:213–217

Oeda K, Salinas J, Chua NH (1991) A tobacco bZip transcription activator (TAF-1) binds to a G-box-like motif conserved in plant genes. EMBO J 10:1793–1802

Ogas J, Kaufmann S, Henderson J, Somerville C (1999) PICKLE is a CHD3 chromatin-remodeling factor that regulates the transition from embryonic to vegetative development in *Arabidopsis*. Proc Natl Acad Sci U S A 96:13839–13844

Ohto MA, Fischer RL, Goldberg RB, Nakamura K, Harada JJ (2005) Control of seed mass by *APETALA2*. Proc Natl Acad Sci U S A 102:3123–3128

Osborne TB (1924) The vegetable proteins (2nd edn). Longmans, Green & Co, London

Parcy F, Valon C, Kohara A, Misera S, Giraudat J (1997) The *ABSCISIC ACID-INSENSITIVE3*, *FUSCA3*, and *LEAFY COTYLEDON1* loci act in concert to control multiple aspects of *Arabidopsis* seed development. Plant Cell 9:1265–1277

Parcy F, Valon C, Raynal M, Gaubier-Comella P, Delseny M, Giraudat J (1994) Regulation of gene expression programs during *Arabidopsis* seed development: roles of the ABI3 locus and of endogenous abscisic acid. Plant Cell 6:1567–1582

Pla M, Vilardell J, Guiltinan MJ, Marcotte WR, Niogret MF, Quatrano RS, Pagès M (1993) The *cis*-regulatory element CCACGTGG is involved in ABA and water-stress responses of the maize gene *rab28*. Plant Mol Biol 21:259–266

Pysh LD, Aukerman MJ, Schmidt RJ (1993) OHP1: a maize basic domain/leucine zipper protein that interacts with Opaque2. Plant Cell 5:227–236

Reidt W, Wohlfarth T, Ellerström M, Czihal A, Tewes A, Ezcurra I, Rask L, Bäumlein H (2000) Gene regulation during late embryogenesis: the RY motif of maturation-specific gene promoters is a direct target of the *FUS3* gene product. Plant J 21:401–408

Reidt W, Ellerstrom M, Kölle K, Towes A, Tiedemann J, Altschmied L, Baumlein H (2001) *FUS3*-dependent gene regulation during late embryogenesis. J Plant Physiol 158:411–418

Rider SD Jr, Henderson JT, Jerome RE, Edenberg HJ, Romero-Severson J, Ogas J (2003) Coordinate repression of regulators of embryonic identity by PICKLE during germination in *Arabidopsis*. Plant J 35:33–43

Riechmann JL, Heard J, Martin G, Reuber L, Jiang CZ, Keddie J, Adam L, Pineda O, Ratcliffe OJ, Samaha RR, Creelman R, Pilgrim M, Broun P, Zhang JZ, Ghandehari D, Sherman BK, Yu GL (2000) *Arabidopsis* transcription factors: genome-wide comparative analysis among eukaryotes. Science 290:2105–2110

Rubio-Somoza I, Martinez M, Abraham Z, Diaz I, Carbonero P (2006) Ternary complex formation between HvMYBS3 and other factors involved in transcriptional control in barley seeds. Plant J 47:269–281

Salon C, Munier-Jolain NG, Duc G, Voisin AS, Grandgirard D, Larmure A, Emery RJN, Ney B (2001) Grain legume seed filling in relation to nitrogen acquisition: a review and prospects with particular reference to pea. Agronomie 21:539–552

Santos-Mendoza M, Dubreucq B, Miquel M, Caboche M, Lepiniec L (2005) LEAFY COTYLEDON 2 activation is sufficient to trigger the accumulation of oil and seed specific mRNAs in *Arabidopsis* leaves. FEBS Lett 579:4666–4670

Santos-Mendoza M, Dubreucq B, Baud S, Parcy F, Caboche M, Lepiniec L (2008) Deciphering gene regulatory networks that control seed development and maturation in *Arabidopsis*. Plant J 54:608–620

Schmidt RJ, Burr FA, Aukerman MJ, Burr B (1990) Maize regulatory gene *opaque-2* encodes a protein with a "leucine-zipper" motif that binds to zein DNA. Proc Natl Acad Sci U S A 87:46–50

Schmidt RJ, Ketudat M, Aukerman MJ, Hoschek G (1992) Opaque-2 is a transcriptional activator that recognizes a specific target site in 22-kD zein genes. Plant Cell 4:689–700

Shen B, Allen WB, Zheng P, Li C, Glassman K, Ranch J, Nubel D, Tarczynski MC (2010) Expression of *ZmLEC1* and *ZmWRI1* increases seed oil production in maize. Plant Physiol 153:980–987

Shewry PR, Halford NG (2002) Cereal seed storage proteins: structures, properties and role in grain utilization. J Exp Bot 53:947–958

Shewry PR, Tatham AS (1990) The prolamin storage proteins of cereal seeds: structure and evolution. Biochem J 267:1–12

Shewry PR, Napier JA, Tatham AS (1995) Seed storage proteins: structures and biosynthesis. Plant Cell 7:945–956

Shotwell MA (1999) Oat globulins. In: Shewry PR, Casey R (eds) Seed proteins. Kluwer Academic, Dordrecht, pp 389–400.

Shutov AD, Baumlein H, Blattner FR, Müntz K (2003) Storage and mobilization as antagonistic functional constraints on seed storage globulin evolution. J Exp Bot 54:1645–1654

Singh NK, Shepherd KW, Langridge P, Gruen LC (1991) Purification and biochemical characterization of triticin, a legumin-like protein in wheat endosperm. J Cereal Sci 13:207–219

Singh NK, Donovan GR, Carpenter HC, Skerritt JH, Langridge P (1993) Isolation and characterization of wheat triticin cDNA revealing a unique lysine-rich repetitive domain. Plant Mol Biol 22:227–237

Singh NK, Shepherd KW, Langridge P, Gruen LC, Skerritt JH, Wrigley CW (1988) Identification of legumin-like proteins in wheat. Plant Mol Biol 11:633–639

Singletary GW, Doehlert DC, Wilson CM, Muhitch MJ, Below FE (1990) Response of enzymes and storage proteins of maize endosperm to nitrogen supply. Plant Physiol 94:858–864

Smeekens S (2000) Sugar-induced signal transduction in plants. Annu Rev Plant Physiol Plant Mol Biol 51:49–81

Smith BD (1998) The emergence of agriculture. Scientific American Library, New York

Stålberg K, Ellerstöm M, Ezcurra I, Ablov S, Rask L (1996) Disruption of an overlapping E-box/ABRE motif abolished high transcription of the *napA* storage-protein promoter in transgenic *Brassica napus* seeds. Planta 199:515–519

Stöger E, Parker M, Christou P, Casey R (2001) Pea legumin overexpressed in wheat endosperm assembles into an ordered paracrystalline matrix. Plant Physiol 125:1732–1742

Stöger E, Ma JKC, Fischer R, Christou P (2005) Sowing the seeds of success: pharmaceutical proteins from plants. Curr Opin Biotechnol 16:167–173

Stöger E, Vaquero C, Torres E, Sack M, Nicholson L, Drossard J, Williams S, Keen D, Perrin Y, Christou P, Fischer R (2000) Cereal crops as viable production and storage systems for pharmaceutical scFv antibodies. Plant Mol Biol 42:583–590

Stone SL, Braybrook SA, Paula SL, Kwong LW, Meuser J, Pelletier J, Hsieh TF, Fischer RL, Goldberg RB, Harada JJ (2008) *Arabidopsis* LEAFY COTYLEDON2 induces maturation traits and auxin activity: implications for somatic embryogenesis. Proc Natl Acad Sci U S A 105:3151–3156

Sugliani M, Brambilla V, Clerkx EJM, Koornneef M, Soppe WJJ (2010) The conserved splicing factor SUA controls alternative splicing of the developmental regulator *ABI3* in *Arabidopsis*. Plant Cell 22:1936–1946

Suzuki M, Kao CY, McCarty DR (1997) The conserved B3 domain of VIVIPAROUS1 has a cooperative DNA binding activity. Plant Cell 9:799–807

Suzuki A, Wu CY, Washida H, Takaiwa F (1998) Rice MYB protein OSMYB5 specifically binds to the AACA motif conserved among promoters of genes for storage protein glutelin. Plant Cell Physiol 39:555–559

Suzuki M, Kao CY, Cocciolone S, McCarty DR (2001) Maize VP1 complements *Arabidopsis* abi3 and confers a novel ABA/auxin interaction in roots. Plant J 28:409–418

Suzuki M, Ketterling MG, Li QB, McCarty DR (2003) Viviparous1 alters global gene expression patterns through regulation of abscisic acid signaling. Plant Physiol 132:1664–1677

Suzuki M, Wang HHY, McCarty DR (2007) Repression of the *LEAFY COTYLEDON 1/B3* regulatory network in plant embryo development by *VP1/ABSCISIC ACID INSENSITIVE 3-LIKE B3* genes. Plant Physiol 143:902–911

Takaiwa F, Ogawa M, Okita TW (1999) Rice glutelins. In: Shewry PR, Casey R (eds) Seed proteins. Kluwer Academic, Dordrecht, pp 401–425

Tegeder M, Wang XD, Frommer WB, Offler CE, Patrick JW (1999) Sucrose transport into developing seeds of *Pisum sativum* L. Plant J 18:151–161

To A, Valon C, Savino G, Guilleminot J, Devic M, Giraudat J, Parcy F (2006) A network of local and redundant gene regulation governs *Arabidopsis* seed maturation. Plant Cell 18:1642–1651

Tsukagoshi H, Morikami A, Nakamura K (2007) Two B3 domain transcriptional repressors prevent sugar-inducible expression of seed maturation genes in *Arabidopsis* seedlings. Proc Natl Acad Sci U S A 104:2543–2547

Twyman RM, Stoger E, Schillberg S, Christou P, Fischer R (2003) Molecular farming in plants: host systems and expression technology. Trends Biotechnol 21:570–578

Tzitzikas EN, Vincken JP, de Groot J, Gruppen H, Visser RGF (2006) Genetic variation in pea seed globulin composition. J Agric Food Chem 54:425–433

Vasil V, Marcotte WR Jr, Rosenkrans L, Cocciolone SM, Vasil IK, Quatrano RS, McCarty DR (1995) Overlap of viviparous1 (VP1) and abscisic acid response elements in the *em* promoter: G-box elements are sufficient but not necessary for VP1 transactivation. Plant Cell 7:1511–1518

Verdier J, Kakar K, Gallardo K, Le Signor C, Aubert G, Schlereth A, Town CD, Udvardi MK, Thompson RD (2008) Gene expression profiling of *M. truncatula* transcription factors identifies putative regulators of grain legume seed filling. Plant Mol Biol 67:567–580

Verdier J, Kakar K, Gallardo K, Le Signor C, Aubert G, Schlereth A, Town CD, Udvardi MK, Thompson RD (2008) Gene expression profiling of *M. truncatula* transcription factors identifies putative regulators of grain legume seed filling. Plant Mol Biol 67:567–580

Verdier J, Thompson RD (2008) Transcriptional regulation of storage protein synthesis during dicotyledon seed filling. Plant Cell Physiol 49:1263–1271

Vicente-Carbajosa J, Moose SP, Parsons RL, Schmidt RJ (1997) A maize zinc-finger protein binds the prolamin box in zein gene promoters interacts with the basic leucine zipper transcriptional activator Opaque2. Proc Natl Acad Sci U S A 94:7685–7690

Vigeolas H, van Dongen JT, Waldeck P, Huhn D, Geigenberger P (2003) Lipid storage metabolism is limited by the prevailing low oxygen concentrations within developing seeds of oilseed rape. Plant Physiol 133:2048–2060

Wang GF, Wang H, Zhu J, Zhang J, Zhang XW, Wang F, Tang YP, Mei B, Xu ZK, Song RT (2010) An expression analysis of 57 transcription factors derived from ESTs of developing seeds in maize (*Zea mays*). Plant Cell Rep 29:545–559

Wang TL, Domoney C, Hedley CL, Casey R, Grusak MA (2003) Can we improve the nutritional quality of legume seeds? Plant Physiol 131:886–891

Weber H, Borisjuk L, Wobus U (1997) Sugar import and metabolism during seed development. Trends Plant Sci 2:169–174

Weber H, Borisjuk L, Wobus U (2005) Molecular physiology of legume seed development. Annu Rev Plant Biol 56:253–279

Weichert N, Saalbach I, Weichert H, Kohl S, Erban A, Kopka J, Hause B, Varshney A, Sreenivasulu N, Strickert M, Kumlehn J, Weschke W, Weber H (2010) Increasing sucrose uptake capacity of wheat grains stimulates storage protein synthesis. Plant Physiol 152:698–710

Weschke W, Panitz R, Sauer N, Wang Q, Neubohn B, Weber H, Wobus U (2000) Sucrose transport into barley seeds: molecular characterization of two transporters and implications for seed development and starch accumulation. Plant J 21:455–467

West M, Yee KM, Danao J, Zimmerman JL, Fischer RL, Goldberg RB, Harada JJ (1994) LEAFY COTYLEDON1 is an essential regulator of late embryogenesis and cotyledon identity in *Arabidopsis*. Plant Cell 6:1731–1745

Wilkinson M, Lenton J, Holdsworth M (2005) Transcripts of Vp-1 homoeologues are alternatively spliced within the *Triticeae tribe*. Euphytica 143:243–246

Williams BA, Tsang A (1992) Nucleotide sequence of an abscisic acid-responsive, embryo-specific maize gene. Plant Physiol 100:1067–1068

Wolfe KH, Gouy M, Yang YW, Sharp PM, Li WH (1989) Date of the monocot-dicot divergence estimated from chloroplast DNA sequence data. Proc Natl Acad Sci U S A 86:6201–6205

Wu CY, Suzuki A, Washida H, Takaiwa F (1998) The GCN4 motif in a rice glutelin gene is essential for endosperm-specific gene expression and is activated by Opaque2 in transgenic rice plants. Plant J 14:673–683

Wu CY, Washida H, Onodera Y, Harada K, Takaiwa F (2000) Quantitative nature of the prolamin-box, ACGT and AACA motifs in a rice glutelin gene promoter: minimal *cis*-element requirements for endosperm-specific gene expression. Plant J 23:415–421

Xue LJ, Zhang JJ, Xue HW (2009) Characterization and expression profiles of miRNAs in rice seeds. Nucl Acids Res 37:916–930

Yamamoto MP, Onodera Y, Touno SM, Takaiwa F (2006) Synergism between RPBF Dof and RISBZ1 bZIP activators in the regulation of rice seed expression genes. Plant Physiol 141:1694–1707

Yang X, Tuskan GA, Cheng MZM (2006) Divergence of the Dof gene families in poplar, *Arabidopsis*, and rice suggests multiple modes of gene evolution after duplication. Plant Physiol 142:820–830

Yasuda H, Tada Y, Hayashi Y, Jomori T, Takaiwa F (2005) Expression of the small peptide GLP-1 in transgenic plants. Transgenic Res 14:677–684

Ye X, Al-Babili S, Klöti A, Zhang J, Lucca P, Beyer P, Potrykus I (2000) Engineering the provitamin A (β-carotene) biosynthetic pathway into (carotenoid-free) rice endosperm. Science 287:303–305

Yoshihara T, Takaiwa F (1996) *Cis*-regulatory elements responsible for quantitative regulation of the rice seed storage protein glutelin *GluA-3* gene. Plant Cell Physiol 37:107–111

Yupsanis T, Burgess SR, Jackson PJ, Shewry PR (1990) Characterization of the major protein component from aleurone cells of barley (*Hordeum vulgare* L.). J Exp Bot 41:385–392

Zhang H, Ogas J (2009) An epigenetic perspective on developmental regulation of seed genes. Mol Plant 2:610–627

Zheng Z, Kawagoe Y, Xiao S, Li Z, Okita T, Hau TL, Lin A, Murai N (1993) 5' distal and proximal *cis*-acting regulator elements are required for developmental control of a rice seed storage protein *glutelin* gene. Plant J 4:357–366

Part III
Proteomics

Chapter 12
Organelle Proteomics of Developing Seeds: Comparison with Other Plant Tissue Organelles

Ombretta Repetto and Karine Gallardo

Abstract Finely regulated and orchestrated events occur during seed development and germination in different cell organelles, like mitochondria, peroxisomes, plastids, vacuoles, endoplasmic reticulum, oil bodies, and nuclei. Therefore, some of the ongoing investigations mainly concentrate on subcellular organelle purifications for a deeper comprehension of these complex developmental processes. In seed biology, recent ambitious efforts in proteomics have been directed toward organelle isolation from seeds at defined developmental stages followed by high-throughput protein separation and identification. The construction of reference maps has allowed identifying organelle-specific proteins. The variation(s) analyses of those proteins depending on both internal and external stimuli are beginning to help in deciphering the extent to which those proteins play fundamental role in seed physiology. Here, we report organelle proteomics applied to the characterization of seed physiological processes and compare these advances with studies from other plant tissues.

Keywords Mass spectrometry · Organelle · Plant tissues · Proteomics · Seed physiology · Subcellular fractionation

12.1 Introduction

Seed biology can be defined as the study of finely regulated and orchestrated events that occur during seed development and germination. One of the most ambitious goals in the seed biology field is the construction of functional models for an understanding of seed metabolism through high-throughput omics technologies (genomics, transcriptomics, proteomics, and metabolomics). Inside plant cells, metabolic pathways are confined to organelles which interact in a highly-dynamic and disci-

O. Repetto (✉)
INRA (National Institute for Agronomic Research),
UMR 102 Genetics and Ecophysiology of Grain Legumes, BP 86510, 21065 Dijon, France
e-mail: ombrettarepetto@gmail.com

K. Gallardo
INRA (National Institute for Agronomic Research), UMR1347 Agroécologie,
BP 86510, 21065 Dijon, France
e-mail: Karine.Gallardo@dijon.inra.fr

G. K. Agrawal, R. Rakwal (eds.), *Seed Development: OMICS Technologies toward Improvement of Seed Quality and Crop Yield*,
DOI 10.1007/978-94-007-4749-4_12,

Table 12.1 Reviews about the most important plant organelles proteomics

Topics/organelles	References
General	Mo et al. 2003
	Taylor et al. 2003
	Dunkley et al. 2006
	Haynes and Roberts 2007
	Lilley and Dupree 2007
	Plöscher et al. 2009a
	Agrawal et al. 2010
Mitochondria and peroxisomes	Brookes et al. 2002
	Heazlewood et al. 2003
	Millar et al. 2005
	Heazlewood and Millar 2007
Peroxisomes	Palma et al. 2009
Plastids	van Wijk 2004
	Zychlinski and Gruissem 2009
Chloroplasts	Rolland et al. 2003
	Baginksy and Gruissem 2004
	Joyard et al. 2010
Oil bodies	Shimada and Hara-Nishimura 2010
Nuclei	Erhardt et al. 2010

plined manner, under organ/tissue/cell specific controls as well as in response to developmental and environmental stimuli. Subcellular fractionation is a powerful approach to reduce sample complexity, to enrich intracellular organelles and thus, to provide information on organelle-specific molecular processes. Combination of subcellular fractionation techniques with proteomics approaches leads to the modern 'organelle proteomics', an ideal partnership whose major benefits and applications in plant biology have been exhaustively reviewed (Table 12.1).

As evidenced by Agrawal et al. (2010), organelle proteomics in plants, unlike animals, has seen limited works mainly due to: (i) inter-plant and inter-tissue complexity; (ii) difficulties in isolation of subcellular compartments; and (iii) enrichment and purity. In the last decade, organelle proteomics has been adopted to get deeper insight into the physiology of plant tissues (leaves, seeds, seedlings, and roots) or cells (Table 12.2).

Subcellular fractionation techniques were either combined with high-resolution 2-DGE, 1-DGE, or gel-independent analytical proteomics approaches to dissect the proteome of organelles. Several methods are currently available for shotgun proteomics in plants. Recent quantitative proteomics has succeeded in comparing two or more physiological states using labeling or label-free based strategies. Sample preparation is the most critical step, which determines proteome coverage, especially for low-abundance proteins (Fig. 12.1). The method consists in organelle purification, assessment of their degree of purity by microscopic observation and/or immunodetection, protein extraction with specific buffers, protein digestion, protein or peptide separation, MS analysis [generally MALDI for peptide mass fingerprint (PMF) data or nESI-MS/MS to fragment more than one precursor and to get putative sequence information or *in silico* analyses], and sequence database interrogation (Fig. 12.1).

Table 12.2 List of plant organelles purified and analyzed by proteomics in the last 10 years

Organelle	Tissue-cell	Plant	Proteomic approach	Major objectives and findings	References
Mitochondrion	Seed	*Pisum sativum*	cDNA cloning; Northern blot; transient expression of GFP fusion protein and submitochondrial localization; 1-DGE and WB; dehydration protection assays	Characterization of a LEA protein of group 3 localized within the matrix space: its expression during late seed development, in dry seeds and throughout germination	Grelet et al. 2005
		P. sativum	Measurements of respiratory enzyme activities and carbohydrate contents; Cd and mineral analyses	Investigations of citric acid cycle enzymes and electron transport after Cd stress in embrionic axis	Smiri et al. 2009
		Glycine max	2-DGE; ESI-MS/MS [Micromass Q-TOF]; respiration and enzymes activity; TEM; malondialdehyde contents; electric conductivity; SOD, CAT, and APX enzyme activity assays	Analyses of the effects induced by imbibition temperature towards mitochondria: identifications of eight differentially-expressed proteins at chilling	Yin et al. 2009
		Oryza sativa	1-DGE and WB; MS analyses on digested bands: C18 RP-LC and ESI-MS/MS [Applied Biosystems QStar Pulsar i] 1-DGE and WB; MS analyses on on whole extracts: Q-TOF MS [Agilent 6510]; qRT-PCR	Investigations of membrane transport capacities during germination under anaerobic or aerobic conditions: characterization of a basic amino acid carrier	Taylor et al. 2010
		Lupinus luteus, Brassica oleracea, Arabidopsis thaliana	Purity assays by enzymes activities; thermostability tests; 1-DGE and WB	Proteomic analyses of mitochondrial membrane and matrix to characterize the level of dehydrin-like protein accumulation under cold and heat stress	Rurek 2010
	Coleoptile	*O. sativa*	1-DGE and WB; 2-DGE; MS analyses on digested spots: Q-TOF [Applied Biosystems QStar Pulsar] 1-DGE and WB; 2-DGE; MS analyses on on whole extracts: Zorbax C18 RP [Agilent] LC and MS/MS	Comparison of mitochondria under anoxia following 1 day of air adaptation with those isolated from 7 days aerobically grown seedlings	Millar et al. 2004
	Seedling	*Triticum aestivum, Secale cereal, Zea mays*	1-DGE and WB	Accumulation of dehydrins in mitochondria: differential expression depending on plant cryosensitivity	Borovskii et al. 2002

Table 12.2 (continued)

Organelle	Tissue-cell	Plant	Proteomic approach	Major objectives and findings	References
	Leaf	*P. sativum*	2-DGE, spot blotting, CBB staining, N-terminal sequencing [Applied Biosystem ABI 492 Procise sequence], MALDI-TOF [Bruker-Daltonics Biflex] or ESI MS/MS [Micromass Q-TOF]	A mitochondrial proteome survey in relation to development: detection of 433 spots and identification of 73 %, some of which were specific of roots or seeds	Bardel et al. 2002
		P. sativum	Stress treatment; CO_2 consumption and synthesis; 2-DGE and BN/SDS-PAGE; C18 RP LC and Q-TOF MS/MS [Applied Biosystems QStar Pulsar i]	Investigation of the impact of environmental stresses on mitochondrial proteome: differential expression of 33 spots in a least one stress treatment	Taylor et al. 2005
		P. sativum, Solanum tuberosum, Spinacia olearia	1-DGE and WB	Simultaneous isolation of intact mitochondria and chloroplasts from a single pulping	Rődiger et al. 2010
	Shoot	*O. sativa*	2-DGE and BN/SDS-PAGE; MS analyses by digested spots: Q-TOF MS/MS [Applied Biosystems QStar Pulsar] 2-DGE and BN/SDS-PAGE; MS analyses by digested whole extracts: microbore C18 RP (Agilent) LC and MS/MS	Analyses of rice mitochondrial proteome: identification of rice proteins and comparison with *Arabidopsis* and pea mitochondrial proteomes	Heazlewood et al. 2003
		O. sativa	1-DGE and WB; 2-DGE; DIGE 2-DGE; MS analyses [Agilent Technologies XCT Ultra Ion Trap]; microarrays	Identification of 322 NR proteins in the context of mitochondrial biogenesis	Huang et al. 2009
	Root	*O. sativa*	1-DGE and WB; 2D IEF/SDS-PAGE; BN-PAGE; MALDI-TOF/TOF MS [Bruker-Daltonics AutoFlex TOF/TOF LIFT]	Protein changes towards the salt stress-induced programmed cell death	Chen et al. 2009
	Tuber	*S. tuberosum*	Phosphorylation assay; BN/Tricine SDS-PAGE; 2-DGE; RP C18 LC and nESI-MS/MS [Finnigan LCQ quadrupole ion storage]; MALDI-TOF MS [Bruker-Daltonics Reflex IV]	Phosphoprotein mitochondrial patterns: identifications of 14 phosphoproteins as involved in the signal integration of mitochondrial complexes	Bykova et al. 2003
	Tuber and stem	*S. tuberosum*	2D BN-SDS-PAGE; 2D BN/BN-PAGE; in gel activity stains for cytochrome c oxidase; O_2 electrode measurements	Characterization of respirasomes: identification of novel supercomplexes as comparatively low abundant	Eubel et al. 2004

Table 12.2 (continued)

Organelle	Tissue-cell	Plant	Proteomic approach	Major objectives and findings	References
	Cell	*A. thaliana*	Enzyme and respiratory measurements; 2-DGE; MALDI-TOF MS [Applied Biosystem Voyager-DE/Micromass ToFSpec 2E]	Analysis of mitochondrial proteome: excision of 170 spots, of which 81 had defined functions	Millar et al. 2001
		A. thaliana	2-DGE; RP C18 LC and MS with either a Qstar Pulsar i [Applied Biosystems] or a 6510 Q-TOF [Agilent Technologies]	Characterization of phosphorylation sites of 7 from the 18 affinity-enriched mitochondrial phosphoproteins	Ito et al. 2009
	Cell, stem and leaf	*A. thaliana*	2-DGE; BN-SDS-PAGE; WB and N-terminal sequencing by Edman degradation; nESI-MS/MS [Micromass Q-TOF]	Proteomic approach to identify novel mitochondrial proteins: identification of 52 proteins	Kruft et al. 2001
	Cell and shoot	*A. thaliana*	WB; DIGE 2-DGE; BN-SDS-PAGE; RP C18 LC and ESI-MS [Agilent Technologies XCT Ultra Ion Trap]; O_2 electrode and spectrophotometric measurements; microarrays	Quantitative comparison between the mitochondrial proteomes isolated from a nonphotosynthetic cell culture model and photosynthetic shoots	Lee et al. 2008
	Ear	*Z. mays*	2-DGE; MALDI-TOF MS [PerSeptive Biosystems Voyager-DE PRO]	Analysis of mitochondrial proteome from unpollinated ears: identification of 100 spots from 197 analyzed	Hochholdinger et al. 2004
Peroxisome	Seedling	*G. hirsutum*, *H. annuus*, *O. sativa*, *R. communis*	Gradient fraction analysis; 1-DGE and WB; NO detection in mutants expressing GFP-PTS1; ONOO- detection with fluorescent reagent APF; CNBr cleavage and sequencing	Characterization of a 36 kDa polypeptide as an outer membrane porin, typical of oilseed glyoxysomes and also detected in leaf-type peroxisomes and mitochondria	Corpas et al. 2000
		A. thaliana	2-DGE; WB; MALDI-TOF MS [Bruker]	Establishment of a method to isolate peroxisomes at high purity and concentration, and proteome characterization	Fukao et al. 2002
		A. thaliana	2-DGE and WB; MALDI-TOF MS [Bruker]; phosphorylation assay; RT-PCR for GPK1 mRNA	2-D map of glyoxysomal proteins from etiolated cotyledons: identification of 19 proteins, including 13 novel	Fukao et al. 2003
		Pumpkin	1-DGE; EM	Isolation of glyoxysomes from pumkin cotyledons: protein profiles were used to check the entity of organelle isolation	Harrison-Lowe and Olsen 2006

Table 12.2 (continued)

Organelle	Tissue-cell	Plant	Proteomic approach	Major objectives and findings	References
		G. max	1-DGE and WB; 2-DGE and MALDI-TOF MS [Bruker Daltonics Reflex III]	Proteomic analysis of highly-purified peroxisomes: identification of 92 NR proteins	Arai et al. 2008a
		G. max, *A. thaliana*	1-DGE and WB; 2D BN-PAGE and MALDI-TOF MS [Bruker Daltonics Reflex III]	Identification of the Gm PNC1 protein and characterization in the context of germination and post-germinative growth	Arai et al. 2008b
	Green and red fruit	*Capsicum annuum*	EM; enzyme activities (ascorbate and glutathione); BN-SDS-PAGE and WB	Characterization of antioxidant enzymes and corresponding metabolites during fruit maturation	Mateos et al. 2003
	Leaf	*A. thaliana*	2-DGE and MALDI-TOF MS [Bruker Daltonics Ultraflex I]; shotgun analyses; C18 RP LC and nanoLC MS [Thermo Electron LTQ]; subcellular localization	Proteome analysis of leaf peroxisomes by complementary gel-based and gel-free approaches: identification of 78 NR proteins	Reumann et al. 2007
		A. thaliana	1-DGE and WB; LC-ESI-MS/MS [Waters BEH C18 peptide trap and nanoAcquity UPLC coupled to a ThermoFisher LTQ-FTICR]; verification of peroxisome targeting	In-depth analysis of leaf peroxisomes combined with *in vivo* subcellular targeting to search for new peroxisomal metabolic and regulatory functions	Reumann et al. 2009
		S. oleracea	2-DGE and MALDI-MS [Bruker Biflex]; LC-MS/MS [Micromass Q-Tof 2]; RT-PCR and subcellular localization	Proteomic approach extended by *in vivo* subcellular localization analyses: finding of around 100 spots	Babujee et al. 2010
	Cell	*A. thaliana*	DIGE 2-DGE; 1-D Tricine SDS-PAGE; WB; RP C18 LC and ESI-MS/MS [Agilent XCT Ultra IonTrap]; phosphopetide isolation and MS [Agilent 6510 Q-TOF]; GFP localization	Quantitative peroxisomal proteomic analysis: identification of 89 proteins, mostly showing novel peroxisomal metabolic functions	Eubel et al. 2008
Plastid	Seed	*B. napus*	Chlorophyll determination; TEM; 1-DGE and WB; copolymerization of embryoplast proteins with acrylamide, in-gel digestion and peptide analyses by semicontinuous MudPIT [Thermo Fisher ProteomeX LTQ workstation]	Preparation of highly pure embryoplasts and proteome analyses: identification of 80 NR proteins, 53 % of which being components of photosystem, light harvesting, cytochrome b/f and ATP synthase	Jain et al. 2008

Table 12.2 (continued)

Organelle	Tissue-cell	Plant	Proteomic approach	Major objectives and findings	References
	Root	*Medicago truncatula*	Ezyme assays; 1-DGE and WB; C18 RP-LC and nESI-MS/MS [Thermo Electron LCQ Deca XP+ion trap]	Identification of 266 proteins, mostly implicated in nucleic-acid and carbohydrate/nitrogen/sulfur metabolisms	Daher et al. 2010
	Cell	*Nicotiana tabacum*	1-DGE; RP-LC and ESI-MS/MS [Thermo Finnigan LCQ DecaXP]	Identification of proteins especially involved in amino acid biosynthetic pathway	Baginsky et al. 2004
Chloroplast	Seedling	*A. thaliana*	1-DGE; LC/ESI-MS/MS [Thermo Finnigan LCQ-DecaXP ion trap]; RNA expression analysis by RNA hybridization to the full *Arabidopsis* Affymetrix GeneChip	Development of a map of all metabolic and regulatory pathways by shotgun proteomics and RNA expression analysis: identification of 690 different proteins	Kleffmann et al. 2004
	Leaf	*A. thaliana*	EM; import assay using RuBisCO SSU preprotein at-pSS	Development of a simple and cost-saving method to isolate yields of integer and functional chloroplasts	Aronsson and Jarvis 2002
		A. thaliana	1-DGE and WB; C18 RP-LC and MS/MS [Waters Q-TOF Ultima]; transient expression of GFP-fusion proteins	Analyses of highly-purified chloroplast envelope membranes: identification of more than 100 proteins	Ferro et al. 2003
		A. thaliana	1-DGE; SEC; IEC; 'top-down' approach by Fourier Transform MS	Characterization of chloroplast proteins by a *top-down* FTMS approach without a careful plastid purification	Zabrouskov et al. 2003
		Z. mays	2-DGE and MALDI-TOF MS [Perspective Biosystems Voyager-DE Pro]	A reproducible method to analyze changes in protein composition of plastids during greening	Lonosky et al. 2004
		A. thaliana	1-DGE and WB; 2-DGE; MS/MS [Waters nanoACQUITY nUPLC SYNAPT HDMS System]	A novel fast and high-yield isolation procedure for chloroplasts from small samples	Kley et al. 2010
		Sinapis alba	HS CL-6B chromatography; 2D BN-SDS-PAGE; WB; LC and ESI-MS/MS [Thermo Finnigan LCQ-DecaXP ion trap]; T-DNA know-out experiment of PEP-associated thioredoxin	Investigation of accumulation and structure of nucleic acid-binding oligomeric protein complexes, potentially involved in plastid development	Schröter et al. 2010

Table 12.2 (continued)

Organelle	Tissue-cell	Plant	Proteomic approach	Major objectives and findings	References
	Leaf and stem	*Salicornia europaea*, *A. thaliana*	Enzyme assays; 2D IEF/SDS-PAGE and MALDI-TOF MS [Bruker Daltonics Autoflex]	Development of an efficient MS-compatible method to extract chloroplast proteins and separate them by 2-DGE	Fan et al. 2009
	Fruit	*C. annuum*	Enzyme assays of the ASC-GSH cycle; lipid peroxidation and protein oxidation; TEM; immunolabelling	Characterization of the antioxidant system and the chloroplast to chromoplast conversion during ripening	Martí et al. 2009
Chromoplast	Fruit	*C. annuum*	1-DGE; C18 RP-LC and nESI-MS/MS [Finnigan LCQ-DecaXP ion trap]	Deciphering chromoplast-specific metabolic pathways: identification of 151 proteins	Siddique et al. 2006
		S. lycorpersicon	1-DGE and WB; nLC-MS/MS [Thermo-Fischer Scientific LTQ-Orbitrap]	Proteomic analysis of the non-photosynthetic plastids: 988 proteins were detected	Barsan et al. 2010
Etioplast	Dark-grown seedling	*Hordeum vulgare*	BN SDS-PAGE and WB; enzyme activity measurements	Characterization of membrane integral LMW proteins from photosystem II	Plöscher et al. 2009b
		O. sativa	1-DGE; C18 RP-LC and nESI-MS/MS [Thermo Finnigan LCQ-DecaXP ion trap]	A systematic analysis of etioplast proteomics: identification of 240 unique proteins	von Zychlinski et al. 2005
Amyloplast	Seed	*T. aestivum*	Marker enzyme assays; 1-DGE; 2-DGE; C18 RP-LC and ESI-MS/MS [Thermo Finnigan LCQ- Deca ion trap]	Characterization of 171 proteins: identification of 108 unique proteins, categorized into seven functional classes	Andon et al. 2002
		T. aestivum	2-DGE and LC-MS/MS [Applied Biosystems/MDS Sciex QSTAR PULSAR i TOF]	Amyloplasts from developing endosperm: identification of 289 proteins	Balmer et al. 2006
		Z. mays and *T. aestivum*	1-DGE and WB; GPC, purification by anion-exchange chromatography, immunoprecipitation; ESI-MS/MS	Purification and characterization of HMW complexes containing starch biosynthetic enzymes	Hennen-Bierwagen et al. 2009
		T. aestivum	2-DGE and MALDI-TOF [Perspective BioSystems Voyager DE-Pro] or LC-MS/MS [Micromass/Waters Q-TOF Global]	Development of an efficient extraction method for starch granules and quantitative proteomic analyses	Bancel et al. 2010
	Fruit	*S. tuberosum*	1-DGE & WB; MALDI-TOF MS [Bruker-daltonics Reflex III], MALDI-TOF/TOF [Applied Biosystems MS 4700] and LC-ESI MS/MS [Waters Micromass Micro]	Amyloplast proteomics from mini- and micro-tubers: identification of the major starch granule-bound and soluble proteins	Stensballe et al. 2008

Table 12.2 (continued)

Organelle	Tissue-cell	Plant	Proteomic approach	Major objectives and findings	References
Vacuole	Leaf	*H. vulgare*	1-DGE; WB; C18 RP-LC and MS [Thermo Finnigan LCQ-Deca XP ion trap]; RT-PCR; subcellular localization	Tonoplast proteomic approach to characterize the *Arabidopsis* homolog AtSUT4	Endler et al. 2006
		M. crystallinum, *A. thaliana*, *Ananas comosus*	1-DGE & WB; DIGE 2-DGE; C18 RP-LC & nLC-MS/MS [Thermo LTQ-Orbitrap]; analyses of glycolityc and V-ATPase hydrolitic enzyme activities and immunoprecipitation	Quantitative proteomics of the tonoplast in salt tolerance, with particular emphasis to the membrane association of two glycolytic enzymes	Barkla et al. 2009
		H. vulgare	IMAC; TiO_2 chromatography; C18 RP-LC and nESI-MS/MS [Thermo Fischer LTQ-Orbitrap]	A targeted phosphoproteomics approach to analyze the regulation of tonoplast proteins	Endler et al. 2009
		H. vulgare	iTRAQ™ labelling; SCX-LC, C18 RP-LC and MALDI-TOF/TOF MS [Applied Biosystems]; 1-DGE and WB	Quantitative changes in mesophyll tonoplast proteome in response to Cd: identification of 56 transporters	Schneider et al. 2009
	Leaf and stem	*A. thaliana*	1-DGE and WB; enzyme activities; RP-LC and Edman sequencing [Applied Biosystems 491cLC protein sequencer]	Characterization of vacuolar transmembrane and membrane-bound proteins: identification of 34 proteins	Sazuka et al. 2004
	Cell	*A. thaliana*	Multidimensional LC-MS/MS; 1-DGE, C18 RP-LC and nESI-MS/MS [Waters Q-TOF API-US]	Characterization of vegetative vacuolar proteome: identification of 402 proteins	Carter et al. 2004
		A. thaliana	Enzyme activities; 1-DGE and WB; C18 RP-LC and MS [Waters Q-TOF Ultima]	Proteomic analysis to identify new tonoplast proteins: identification of 163 proteins	Shimaoka et al. 2004
		A. thaliana	Gel filtration and anion-exchange chromatography; SDS-PAGE and MALDI-TOF MS [Bruker Daltonics BiFlex III]	Identification of 70 integral proteins from vacuolar membrane	Szponarski et al. 2004

Table 12.2 (continued)

Organelle	Tissue-cell	Plant	Proteomic approach	Major objectives and findings	References
		A. thaliana	1-DGE and WB; enzymatic assays; C18 RP-LC and nESI-MS/MS [Micromas/Waters Q-TOF Ultima]; subcellular localization by transient expression of GFP fusion constructs	Targeted proteomic approach to identify proteins from membrane and soluble vacuolar fractions: identification of more than 650 proteins, 2/3 being membrane hydrophobic	Jaquinod et al. 2007
	Cell and bud	*B. oleracea* and *A. thaliana*	1-DGE and WB; C18 RP-LC nESI MS/MS [Thermo Finnigan LCQDeca XP ion trap]; subcellular localization	Proteomics of purified vacuoles from growing tissues: identification of 102 integral and 214 peripheral proteins	Schmidt et al. 2007
Endoplasmic reticulum	Seed	*R. communis*	N-terminal sequencing [Applied Biosystems 477A protein sequencer]; 2-DGE and MALDI-TOF MS [PE Biosystems Voyager-DE] or C18 RP-LC Q-TOF MS/MS [Micromass]	Proteomic analysis of ER from developing and germinating seeds: identification of proteins playing roles in protein processing and storage, as well as lipid metabolism	Maltman et al. 2002
		R. communis	1-DGE; DIGE; MALDI-TOF MS [Applied Biosystems Voyager DE-STR] and C18 RP-LC and nESI-MS/MS [Applied Biosystems QSTAR Q-TOF & Ettan MDLC™ nHPLC workstation]	Differential proteomic analysis of ER from developing and germinating seeds: 91 proteins were found up-regulated during seed germination, 19 of which being identified	Maltman et al. 2007
Oil body	Mature seed	*A. thaliana*	1-DGE; C18 RP-LC nESI MS/MS [Thermo Finnigan LCQ Deca xp]	Characterization of OB protein composition: identification of eight proteins	Jolivet et al. 2004
		B. napus	1- and 2-DGE; C18 RP-LC and nESI MS/MS; multidimensional LC and nanoESI-MS/MS [Thermo Finnigan LTQ linear Ion Trap]; TEM	Characterization of OB protein and lipid composition: identification of 91 spots with a very similar overall composition	Katavic et al. 2006
		G. max	Immunoblotting of oleosin 24-kDa suppressed seeds; 2-DGE; DNA array analysis; ESI-MS/MS [Applied Biosystems QSTAR hybrid Q-TOF]; MO and TEM	Biological interpretation for the role of the 24 kDa oleosin by its knockdown using RNAi: OBs formed large OB-ER complexes with an interior dominated by micro-OBs	Schmidt and Herman 2008

Table 12.2 (continued)

Organelle	Tissue-cell	Plant	Proteomic approach	Major objectives and findings	References
		B. napus	1-DGE and WB; 2-DGE; C18 RP-LC nESI MS/MS [Thermo Electron LCQ Deca ion-trap]; SCX followed by nano-scale C18 RP, nESI-MS/MS [Waters Global Q-TOF]	OB protein composition: identification and post-translational modifications of some proteins with high level of sequence conservation with *Arabidopsis* counterparts	Jolivet et al. 2009
Nucleus	Seed	*P. sativum*	1-DGE and WB; Edman degradation [Waters Pro-sequencer 6625]; construction and screening of a cDNA library from pea embryonic axes and DNA sequencing	Characterization of the nuclear dehydration-induced p16 protein as a putative member of the vicilin superfamily and playing roles in chromatin protection against desiccation	Castillo et al. 2000
		Z. mays	1-DGE and WB	Analyses at several stages of endosperm development: visualization of several differentially expressed proteins	Ferreira et al. 2006
		O. sativa	MS analyses on spots: MALDI-TOF MS [Applied Biosystems ABI 4700 Proteomics Analyzer] MS analyses on whole extracts: SCX, C18 RP-LC and nESI-MS/MS [Thermo Finnigan LCQ-Deca XP ion trap]	Shot-gun proteomics to recover the low abundant endosperm nuclear proteins and to analyze protein acetylation and methylation: identifications of 468 proteins, 44 % of which being hypothetical proteins	Li et al. 2008a
		M. truncatula	1-DGE and WB; C18 RP-LC nESI MS/MS [Water Q-TOF Global]; qRT-PCR and microarray	Exploring the nuclear proteome at the switch towards seed filling: identification of 143 proteins, highlighting new nuclear proteins with roles in ribosome biogenesis/nucleocytoplasmic trafficking	Repetto et al. 2008
	Seedling	*Cicer arietinum*	2-DGE; 1-DGE and WB; LC and TOF-MS [Applied Biosystems Q-Trap 4000]; enzyme assays	Comparative nuclear proteome: differential regulation of 205 spots	Pandey et al. 2008
		O. sativa	1-DGE; WB; enzyme assays; C18 RP-LC and Q-TOF ESI-MS [Applied Biosystems Q-TRAP 4000]; subcellular localization	Investigation of dehydration-responsive proteome: differential expression of 150 spots in the tolerant variety	Choudhary et al. 2009

Table 12.2 (continued)

Organelle	Tissue-cell	Plant	Proteomic approach	Major objectives and findings	References
	Seedling and cell	*A. thaliana*	SCX and phosphopeptide enrichment; C18 RP-LC and nESI-MS/MS [Thermo Fisher LTQ]	Nuclear phosphoproteins; identification of 416 phosphopetides with phosphorylation sites	Jones et al. 2009
	Leaf	*A. thaliana*	1-DGE and WB; 2-DGE; MALDI-TOF MS [Applied Biosystems Voyager-DE STR]; RNA blotting; transient expression of GFP-fusion proteins for subcellular localization	Characterization of nuclear proteome and its changes in response cold treatment: around 500–700 spots were detected, 184 of which were identified	Bae et al. 2003
		Xerophyta viscosa	1-DGE and WB; 2-DGE; LC and TOF MS [Applied Biosystems Q-Trap 4000]	Development of a protocol to isolate nuclei and extract their proteins from plant samples	Abdalla et al. 2009
		O. sativa	DNA and heparin affinity chromatography; 1-DGE; C18 RP-LC and nESI-MS/MS [Agilent Technologies LC/MSD Trap XCT]; subcellular localization; qRT-PCR	Identification of more than 600 proteins by a shot-gun approach, some of which being transcription/splicing factors or sugar-responsive nuclear factors	Aki and Yanagisawa 2009
	Root	*Lepidium sativum*	2-DGE; WB; immunogold EM	Analyses of distribution and levels of fibrillarin and NopA64 in nucleoli under altered gravity conditions	Sobol et al. 2007
		Z. mays	Acetic acid-urea SDS-PAGE; acetic acid-urea-Triton X-100/acetic acid-urea-CTAB 2-DGE; WB and densitometry	Analyses of alterations in core histone variant ratios depending on differentiation *versus* de-differentiation and/or physiology	Alatzas and Foundouli 2009
	Cell	*O. sativa*	2-DGE; N-terminal sequencing [Applied Biosystems Procise 494 protein sequencer] and MALDI-TOF MS [Applied Biosystems Voyager DE-STR]	Analysis of nuclear proteomics: resolution of 549 proteins, 257 proteins of which being analyzed by Edman sequencing/MS	Khan and Komatsu 2004
		O. sativa	2-DGE; WB; Pro-Q Diamond in gel stain; MS analyses on spots: MALDI-TOF MS [Applied Biosystems ABI 4700 Proteomics Analyzer] 2-DGE; WB; Pro-Q Diamond in gel stain; MS analyses on whole extracts: SCX, C18 RP-LC and nESI-MS/MS [Thermo LCQDeca XP]	Shot-gun analysis of chromatin proteome and phosphoproteome: resolution of 972 proteins, 509 of which being identified as 269 unique proteins, and some phosphoproteins being copurified with chromatin	Tan et al. 2007

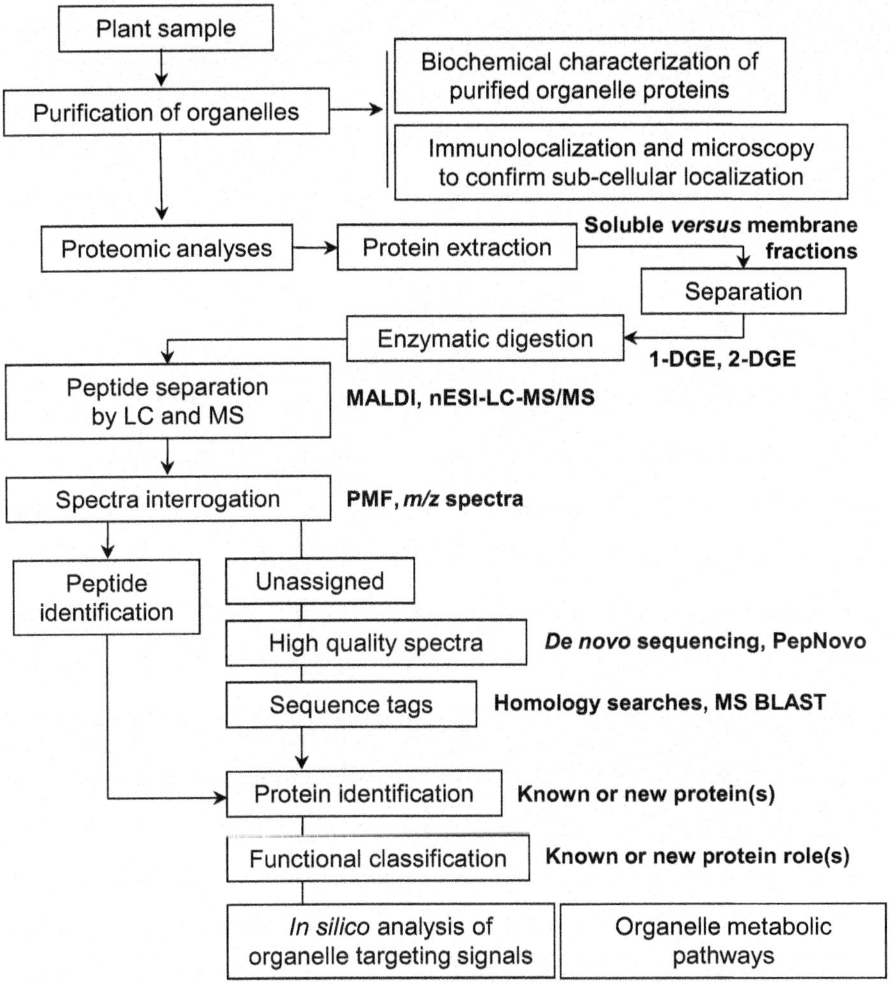

Fig. 12.1 Typical workflow of a plant subproteomic approach, from organelle isolation to MS-based protein identification and data interpretation in the context of organelle metabolic pathways. Organelles are purified followed by preparation of proteins. In-gel or liquid-digested peptides are analyzed by MS and, subsequently, data are subjected to a standard database search. Unidentified spectra are scored for their quality. High-quality spectra are subjected to *de novo* peptide sequencing. Sequences extracted from MS spectra are submitted to MS BLAST for peptide identification by homology. Identified proteins can be either known or novel, not previously described. Their functional classification as well as searches for organelle peptide signals are generally performed in order to better understand their contribution to organelle metabolic pathways

Some plant subcellular protein databases have been created that can be interrogated to identify organelle proteins (Table 12.3). The identified peptides are usually searched for putative subcellular localization using publicly available prediction algorithms (Table 12.4). These techniques have been applied to plant/seed organelles, establishing proteome maps for mitochondria, peroxisomes, plastids, vacuoles, ER,

Table 12.3 Plant subcellular protein database

Subcellular compartment	Name	References
Subcellular proteomic database	SUBA	Heazlewood et al. 2006
Membrane-spanning regions	HMMTOP	Tusnady and Simon 2001
	ARAMEMNON	Schwacke et al. 2003
Plastids protein database	plprot	Kleffmann et al. 2006
Mitocondria	AMPDB	Heazlewood and Millar 2005
	MITOP2	Elstner et al. 2009
Peroxisomes	AraPerox	Reumann et al. 2004
	PeroxisomeDB2.0	Schlüter et al. 2010
Nucleoli	AtNoPDB	Brown et al. 2005

Table 12.4 Softwares used for prediction of protein subcellular localization

Subcellular compartment	Name	References
Mitochondria	Mitopred	Guda et al. 2004
	MitoProt II	Elstner et al. 2009
N-terminal targeting sequences	Predotar	Small et al. 2004
Subcellular localization	PSORT	Nakai and Horton 1999
	pTARGET	Guda 2006
	eSLDB	Pierleoni et al. 2007
	Plant-mPloc	Chou and Shen 2010

oil bodies (OBs), and nuclei. These investigations have significantly increased our knowledge on the processes occurring in specific organelles during seed development and germination (Table 12.2), most of which will be discussed below in comparison with studies from other plant tissues.

12.2 Mitochondria: Strategic Roles in Development and Upon Abiotic Stresses

Mitochondria are essential organelles, possessing their own genome and playing pivotal roles in many cellular processes as sites of tricarboxylic acid cycle (TCA cycle; also known as the citric acid cycle, *Krebs cycle*, or Szent-Györgyi-*Krebs cycle*), oxidative phosphorylation, and energy production. Moreover, a strategic role in eukaryotic cell energetics scenario has been recently attributed to these organelles (Lane and Martin 2010). Since many years, respiratory chain super complexes have been characterized in plant mitochondria at the intricate levels of membrane structural organization, photosynthesis *versus* photorespiration interactions, amino acid/lipid/cofactor biosynthesis, and redox regulation (Dudkina et al. 2006; Davy de Virville et al. 2010; Fernie et al. 2004). It has been shown that the number, position, and respiratory capacity of mitochondrial populations (called 'chondriosomes') dramatically vary between tissues/cells. In 2005, the plant mitochondrial proteome was proposed to be composed of ~3,000 proteins and described as highly responsive to developmental, genetic, and environmental stimuli and, varying in composition depending on both plant organs and cell types, where it may play specialized functions (Millar et al. 2005). Recent advances in mitochondria proteome characterization after organelle purification and high-throughput identifications by MS allowed deciphering complex mechanisms taking place in this organelle, such as PTMs (Table 12.2).

The first 2-D reference map of mitochondria proteome came from *Arabidopsis* cell cultures, representing 100 abundant proteins and 250 low-abundance proteins in the 3–10 *pI* range. Among the 170 proteins analyzed by MALDI-TOF, only 81 have known functions (Millar et al. 2001). In the same species, Kruft et al. (2001) separated 800 mitochondria proteins from green tissues (leaves and stems) and/or cell suspension cultures by 2-D IEF or blue native (BN)/Tricine SDS-PAGE. Fifty-two protein spots were identified by immunoblotting, Edman sequencing, MALDI-TOF, or nESI-MS/MS. Of 52 proteins, 20 % were newly associated to mitochondria. The other proteins belonged to eight functional classes illustrating the multiple activities taking place in the mitochondrion. These classes are: (i) respiratory chain and ATP synthase complex; (ii) pyruvate decarboxylation and citric cycle; (iii) amino acid, nitrogen and sulfur metabolism; (iv) nucleotide metabolism; (v) transport across membranes; (vi) chaperones; (vii) DNA replication, transcription, translation, DNA- and RNA-binding proteins; and (viii) protection against oxygen.

One of the first targeted approaches studying mitochondrial proteome variations between plant organs was performed by Bardel et al. (2002) in *P. sativum*. They combined 2-DGE with gel filtration as third dimension and adopted three methods of protein identification (Edman degradation, MALDI-, and ESI-MS/MS) to characterize the soluble proteome of pea leaf mitochondria. Then, they compared the leaf soluble mitochondrial proteome to those of etiolated leaves, roots, or seeds. Among the proteins detected in seeds are thiosulfate sulfurtransferase (TST, rhodanese) with role in detoxification processes, HSP22, and proteins related to late seed maturation (maturation protein PM32 and late embryogenesis abundant 'LEA' III protein). Several mitochondria seed proteins were also present in mitochondria from others plant tissues, such as HSP22 and LEA III. The remarkable tolerance of seed mitochondria to adverse temperatures (both heat and cold) was related to the high amounts of mitochondrial HSPs (Stupnikova et al. 2006). High levels of LEA proteins were found within the matrix space of mitochondria isolated from pea seeds, where a beneficial role in respect to the protection of cell components during dehydration was proposed (Grelet et al. 2005).

During early stages of seed germination, respiration is anaerobic. The imbibition of water during germination allows aerobic respiration to occur through hydrolysis of the stored reserves providing substances for respiration. The pivotal role of mitochondria in germinated seeds subjected to anaerobic conditions has been recently investigated by Taylor et al. (2010). Anaerobic-germinated seeds were shown to contain large amounts of arginine, ornithine, and proline. Moreover, seven distinct carrier proteins belonging to the *O. sativa* mitochondrial carrier family were identified and quantified by MS in aerobic- *versus* anaerobic-germinated seeds. Anaerobiosis was shown to induce a mitochondrial basic amino acid carrier1 (BAC1) and an arginase involved in the ornithine-urea cycle, concomitantly with the activation of ornithine-arginine-citrate metabolism. The authors proposed that during anaerobic respiration rice germination BAC1 plays a specific role in the ornithine-arginine-citrate metabolism through arginine and ornithine transport. This transport may facilitate the mitochondria-localized arginase reaction and may be needed for specific aspects of basic amino acid metabolism of anaerobic rice, leading to putrescine-stimulated coleoptile elongation and early anaerobic proline synthesis in the rice embryo.

During seed imbibition (i.e., when respiration becomes aerobic), the mitochondrion appears highly sensitive to environmental conditions, as shown in soybean seed axis subjected to different imbibition temperatures (22, 10, and 4 °C: Yin et al. 2009). The mitochondria isolated from soybean seed axis were divided into light and heavy, and proteins differentially accumulated between the different imbibition temperatures were analyzed by ESI-Q-TOF-MS/MS. The axes imbibed at 4 °C mainly contained light mitochondria, which showed poorly developed internal membrane system and few cristae, lower levels of oxidative phosphorylation activity, and specific mitochondrial enzymes, including malate dehydrogenase (MDH), putative ATP synthase subunit, chaperonin-60, arginase, and elongation factor Tu. In contrast, the axes imbibed at 22 °C mainly contained heavy mitochondria, which exhibited higher metabolism (Yin et al. 2009). During the development of seed embryonic axis, respiratory enzymes and related carbohydrate metabolism were found to be responsive to other abiotic stresses. In mitochondria purified from *Pisum sativum* seeds exposed to the highly toxic heavy metal cadmium, the activities of MDH, succinate dehydrogenase (SDH), NADH cytochrome c reductase, succinate cytochrome c reductase, and invertase were rapidly inhibited, while the activities of NADPH-generating enzymes (glc-6-P dehydrogenase and 6-phosphogluconate dehydrogenase) and fermentation enzymes [alcohol dehydrogenase (ADH)] were stimulated (Smiri et al. 2009). Interestingly, differential effects of cold and heat stress towards the synthesis and accumulation of 'dlps', which are hydrophilic proteins similar to 'dehydrins', were recently observed in mitochondria isolated from imbibed seeds and hypocotyls (Rurek 2010). These proteins belong to the LEA protein family discovered first in the mitochondria (Grelet et al. 2005), where a role was shown in protecting enzymes from desiccation. The entity of thermostable dlp accumulation was evaluated after immunoprecipitation and identification of dlp-cross reactive proteins from 1-D (SDS-PAGE) gels. The pool of dlps from different plant organs highly varied; their molecular mass ranging from 10 to 100 kDa and 20 to 90 kDa in *Arabidopsis*, cauliflower and lupin, respectively. Interestingly, cold stress selectively induced accumulation of some small dlps (Rurek 2010). The effect of heat stress was less significant in lupin and *Arabidopsis* than in cauliflower inflorescence mitochondria, leading to the preferential accumulation of 100 kDa dlps. Such dehydrin-like proteins were also detected in nucleus, cytoplasm, and plasma membrane in response to cold, frost, and drought, respectively (Borovskii et al. 2002). Altogether, the data suggest dehydrins might protect the components of various organelles during late seed development, where an intense desiccation occurs and later during the plant life cycle when conditions become unfavorable.

12.3 Peroxisomes Are Key Players in Carbon Metabolism (Breakdown of Storage Lipids) During Germination of Oilseeds and Seedling Growth

Peroxisomes are ubiquitously found eukaryotic organelles that compartmentalize a variety of oxidative metabolic reactions by a single lipid bilayer membrane. They were the focus of many studies this last decade, probably because they were the latest organelles discovered. The name peroxisome comes from the presence of several enzymes, involved in production/degradation of hydrogen peroxide (H_2O_2). An interesting feature of peroxisomes is their lack of genome. The peroxisomal proteins are imported from the cytoplasm or *via* the ER into peroxisomes. The import of peroxisomal matrix proteins occurs through two specific receptor-mediated import pathways: each pathway is defined by one of two conserved 'peroxisome targeting signals (PTSs)', types 1 and 2 (PTS1, PTS2) on the cargo proteins that specific cytosolic receptors (Pex5p, Pex7p) recognize and transport (Brown and Baker 2003). In plants, there are several classes of peroxisomes. Plant specific glyoxysomes are found in young seedlings and contain enzymes specific of the glyoxylate cycle, which plays a role in the conversion of stored FA to carbohydrates during the heterotrophic-growth phase. When the young plant becomes photosynthetic (i.e., autotrophic), leaf peroxisomes allows photorespiration to occur through the activity of key enzymes, such as glycolate oxidase and hydroxypyruvate reductase. Peroxisomes are transformed back into glyoxysomes in occurrence of leaf senescence (Kaur and Hu 2009). These organelles are extremely dynamic and their enzymatic and metabolite contents vary depending on plant species, tissue, and environmental stimuli (Aung et al. 2010; Baker et al. 2006; Gabaldón 2010). Computational analyses of protein sequences helped in identifying a wide range of their biochemical pathways. The 'AraPerox' database lists 400 candidate proteins endowed of the PTS and thus, potentially targeted to *Arabidopsis* peroxisomes (Reumann et al. 2004). Novel plant PTS peptides have been discovered in the postgenomic era, thus increasing the number of putative proteins targeted to peroxisomal membrane or matrix (Reumann et al. 2009).

In plants, these organelles contain enzymes for the oxidative photosynthetic carbon cycle of photorespiration, glyoxylate cycle, β-oxidation of indole butyric acid, biosynthesis of JA and auxin, ureid metabolism, and cell signaling by ROS and nitrogen species (Kaur and Hu 2009). In oilseeds, the presence of two pathways of FA β-oxidation and glyoxylate during postgerminative growth of seedlings allows this organelle to convert the seed reserve lipids into sugars necessary for seed germination and early plant growth (Cornah and Smith 2002; Linka et al. 2008). This pivotal role in carbon metabolism has been recently described in germinating *Arabidopsis* seeds (Pracharoenwattana et al. 2010), where a peroxisomal hydroxypyruvate reductase (HPR) can oxidize NADH at sufficient rate and contribute to FA β-oxidation in the absence of peroxysomal MDH, which normally serves to reoxidise NADH for β-oxidation. The peroxisomal NADH has been thought not to be exported to the cytosol for oxidation but oxidized by resident oxidoreductases. In spite of 40 years of research, peroxisomes are still considered as a 'mysterious organelle'. As evidenced by Palma et al. (2009) in their review about the proteome of plant peroxisomes, less proteomics data are available in comparison with animals and yeast because of the difficulties in purifying peroxisomes from plant samples, in term of both yield and absence of interfering compounds. These organelles are poorly abundant under standard growth conditions and physically associated with mitochondria and chloroplasts. Despite these difficulties in isolating peroxisomes, some remarkable proteomics investigations of this organelle have been performed in plants (Table 12.2).

Analyses on seedlings mostly concern peroxisome/glyoxisome enzyme characterization. In *Helianthus annuus* cotyledons, Jiang et al. (1994) characterized HPR and malate synthase activities of peroxisome membrane proteins during the conversion of oilseed glyoxisomes into leaf-type peroxisomes. Bunkelmann and Trelease (1996) investigated a glyoxysomal membrane ascorbate peroxidase (gmAPX) and a monodehydroascorbate reductase in dark-grown cotyledons of growing cotton seedlings with respect to H_2O_2 production during the β-oxidation of lipids stored in cotyledons. One of the first works reported for seeds is from Reumann et al. (1997), who isolated glyoxysomes from endosperm of *Ricinus communis* seedlings and demonstrated the presence in their membranes of an integral protein with porin activity, very similar to porins of leaf peroxisomes, and putatively involved in the translocation of small organic and amino acids (e.g., succinate, citrate, malate, and aspartate). The existence of porin-like polypeptides was further confirmed by immunoblots and electron microscopy (EM) immunogold analyses in the boundary membranes of seedling glyoxysomes of oilseeds species (sunflower, castor, bean, and cotton; Corpas et al. 2000).

Fukao et al. (2002) analyzed peroxisomes purified from glyoxysomes of greening *Arabidopsis* cotyledons. Apart from the identification of nine typical proteins of leaf peroxisomes (five related to glycolate pathway and four related to H_2O_2 scavengers), they identified 20 novel proteins, such as protein kinases and phosphatases, thus hypothesizing that phosphorylation may be a regulatory mechanism in leaf peroxisomes. Among phosphorylated glyoxysomal proteins, a serine/threonine (S/T) protein kinase 1 was further characterized. The serine/threonine protein kinase 1 was found in peripheral membranes of *Arabidopsis* etiolated cotyledons. Its putative kinase domain being located inside the glyoxisomes appears different from the four protein kinases previously identified in leaf peroxisomal proteome (Fukao et al. 2002), thus supporting the evidence of distinct peroxisome-specific phosphorylation events.

Recent developments in methods of isolating peroxisomes suitable for proteome analyses were performed. In pumpkin seedlings, Harrison-Lowe and Olsen (2006) developed a rapid procedure and proposed catalase and isocitrate lyase assays on 1-D protein bands to verify the presence of intact glyoxysomes, together with morphological observation by EM. By combining Percoll and iodixanol density-gradient centrifugations, Arai et al. (2008a) succeeded in optimizing a protocol for peroxisome isolation from etiolated soybean cotyledons. This method allowed the identification of 92 polypeptides assigned to 70 sequences. Of which, 30 were predicted as located in peroxisomes, including proteins related to FA β-oxidation, glyoxylate and glycolate cycles, stress response, metabolic transport, a voltage-dependent anion-selective channel protein, proteins from the short-chain dehydrogenase/reductase, the enoyl-CoA hydratase/isomerase, and the 3-hydroxyacyl-CoA dehydrogenase-like families. Interestingly, of the 30 reported peroxisomal proteins, 18 contained PTS1, 4 contained PTS2, but 8 of them, which were orthologs of MDH, citrate synthase, glycolate oxidase, APX, and voltage-dependent anion-selective channel protein from other species, had no PTS sequence.

A proteomic analysis of leaf peroxisomes from mature *Arabidopsis* leaves has also been recently reported (Reumann et al. 2007). A total of 68 nonredundant (NR) proteins were identified, 42 of which were newly associated to peroxisomes. Seventeen of the novel proteins carried predicted signals PTS1 or PTS2, and 11 proteins contained PTS-related peptides. In fact, the investigation on the targeting function of predicted and unpredicted signals shed light on three novel functional PTS1 peptides (SSL, SSI, and ASL), which were not previously characterized. A combination of Percoll and sucrose-density gradient centrifugations was also performed to obtain highly pure peroxisomes from mature *Arabidopsis* leaves. Among the 42 novel proteins, a large number of them were associated to FA β-oxidation. Among the other proteins identified, some are related to detoxification (such as glutathione reductase and glutathione-S-transferase, GST), pathogen, and herbivore defense (β-glucosidases 1 and PYK10 among others), supporting an important role of leaf peroxisomes in defense processes.

The comparative analysis of the proteome of peroxisomes between plant tissues reported thus far showed significant discrepancies. Only 11 proteins were in common between *Arabidopsis* cotyledons and mature leaves. In a novel approach for the investigation of the peroxisomal proteome from *Arabidopsis*, the classical density gradient centrifugation was followed by free-flow electrophoresis and quantitative proteomics (Eubel et al. 2007, 2008). These authors provided evidence for the peroxisomal localization of 89 proteins, 35 of which had not previously been identified. Twenty-one of these proteins lack a recognizable PTS1 or PTS2 signals and most of them are involved in and/or related to FA oxidation. Other reported proteins were associated to processes involved in hormone biosynthesis/activation, valine catabolism, carbon/nitrogen/sulfur metabolisms, antioxidant defense, and reducing metabolism. The functional roles of many putative peroxisomal proteins and their interactions with metabolic pathways localized in other cell organelles are still unknown. Interestingly, an emerging role played by peroxisomes was recently shown regarding vitamin K1 biosynthesis in spinach leaves (Babujee et al. 2010), salinity stress in *Arabidopsis* seedlings (Corpas et al. 2009), cadmium oxidative alterations in *Arabidopsis* leaves (Rodríguez-Serrano et al. 2009), and during cytokinesis (Collings and Harper 2008). Innovative subproteomics approaches and analyses of PTMs (e.g., glycosylation, phosphorylation, S-nitrosylation, or oxidation) on purified peroxisomes may thus offer a good complementary strategy to classical proteomic approaches to better decipher, together with genomics information, the biology of this dynamic organelle, the functionality of several enzymes, as well as their interactions with other cellular components, including metabolite exchanges with other cell compartments.

12.4 Plastids: Photosynthetic Carbon Fixation, Synthesis of Amino- and Fatty-Acids, Starch, and Secondary Metabolites

Plastids are characteristic plant cell organelles performing essential biosynthetic and metabolic functions (photosynthetic carbon fixation, synthesis of amino and fatty acids, starch, and secondary metabolites). They develop from proplastids and differentiate in a tissue/cell specific and signal-dependent manner into highly-specialized plastid types with specific structure, pigment composition, cell/tissue localization, and functional properties, such as chloroplasts (photosynthetically active leaf tissues), chromoplasts (fruits and petals), and amyloplasts (roots and storage tissues). Plastids can be distinguished as photosynthetic or nonphotosynthetic, depending on their autotrophic or heterotrophic metabolism (Lopez-Juez and Pyke 2005). A comparison of proteins identified from different plastid types revealed that proteomes of heterotrophic and autotrophic plastids highly differ, especially in their energy metabolism. Although plastids are of significant biological interest, current knowledge of their different metabolic functions is still limited. In this context, subcellular proteomics of purified plastids represents a powerful approach for high-throughput protein identification and elucidation of plastid-type specific functions (Table 12.2).

A first overall plastid proteome characterization was performed in *Brassica napus* seeds (Jain et al. 2008), whose embryos store up to 50 % TAG, to characterize the so-called 'embryoplasts' containing chlorophyll, but heterotrophic and similar to the nonphotosynthetic leucoplasts. Highly-pure plastids were successfully isolated from developing embryos and isolated proteins were analyzed by MudPIT. Interestingly, around 53 % of the 80 NR identified proteins were components of photosystem, light harvesting, cytochrome b/f, and ATP synthase complexes. These data supported the term often used to describe embryoplasts as photoheterotrophic, since these organelles harvest light for ATP and NADPH production but obtain carbon principally from the source organs. The authors also showed a close similarity of embryoplasts to chloroplasts than leucoplasts or amyloplasts. Baginsky et al. (2004) compared the proteome of undifferentiated proplastids of tobacco BY-2 cell culture with that of differentiated heterotrophic amyloplasts by using MudPIT. This comparative shotgun proteome analysis revealed significant differences between these two plastid types, suggesting functional adaptations of the plastid proteome during differentiation.

Proteomics also proved to be an extremely useful technology to get new insights into the root plastidome. The proteome of plastids isolated from *Medicago truncatula* roots was analyzed by nESI-LC-MS/MS (Daher et al. 2010). A total of 266 proteins were identified that belong to different classes, including nucleic acid-related processes (16 %), carbohydrate (15 %), and nitrogen/sulfur (12 %) metabolisms. Interestingly, a new role for root plastids was hypothesized in the context of plant cell responses towards biotic/abiotic stresses, since 10 % of the identified proteins were found to be involved in stress response processes.

12.4.1 Chloroplasts: Autotrophic Plastids Specialized in Photosynthesis Processes

Chloroplasts are typical photosynthetic organelles of cyanobacterial origin (Goksøyr 1967). Chloroplasts collect light energy and synthesize sugar phosphates, which release NADPH and ATP through oxidative metabolism. Mature chloroplasts are thought to contain about 3,000 proteins (Leister 2003). They are composed of several compartments, each of them being constituted of a subset of proteins or subproteomes, thus implying the establishment of a variety of purification and fractionation techniques for chloroplast subproteomics studies in different plant species (Leister 2003; van Wijk 2004) (Table 12.2). Several functional proteomics approaches have been reported for purified chloroplasts in *Arabidopsis* seedlings (Kleffmann et al. 2004), in leaves of *Arabidopsis* (Aronsson and Jarvis 2002; Ferro et al. 2003; Kley et al. 2010; Zabrouskov et al. 2003), *Sinapis alba* (Schröter et al. 2010), and *Zea mays* (Lonosky et al. 2004), in stems and leaves of *Salicornia europaea* and *Arabidopsis* (Fan et al. 2009), and in *Capsicum annuum* fruits (Martí et al. 2009). All these works, usually combined with transcript profiling, have succeeded in characterizing the proteome of photosynthetically active chloroplasts, especially after overcoming the constraint of the highly-abundant proteins to obtain new insight into chloroplast biochemical pathways and regulatory networks (Table 12.2) (for a review, see Joyard et al. 2010). Moreover, mechanisms regulating chloroplast biogenesis have been studied during maize de-etiolation ('greening'). When exposed to light, photosynthetically incompetent etioplasts in dark-grown seedlings develop into photosynthetically competent chloroplasts, which involve the synthesis of photosynthetic apparatus components and pronounced ultrastructure alterations. In this context, Lonosky et al. (2004) proposed a valid and reproducible protocol based on 2-D and MS to identify changes in the *Z. mays* plastid proteome during the greening process and to set up a clustering technique for discriminating protein expression classes during greening.

12.4.2 Chromoplasts: Heterotrophic Plastids Synthesizing and Accumulating Caroteinoids

Flower development and fruit ripening are usually accompanied by several biochemical and physiological events, which include the loss of chlorophyll and the synthesis of colored compounds, such as carotenoids accumulating in chromoplasts. Chromoplasts are special heterotrophic plastids. They play a crucial role in the generation of major metabolites essential for sensory and nutritional quality of fruits. These chromoplasts are derived from photosynthetic (autotrophic) chloroplasts. Chloroplast to chromoplast conversion is accompanied by the disassembly of the chloroplast thylakoid membrane system and by the massive synthesis and accumulation of carotenoid pigments. Evidences of carotenoid biosynthesis as primary metabolic pathway come from a comprehensive database-independent proteome analysis of chromoplasts from *Capsicum annum* (Siddique et al. 2006). A total of 151 proteins were identified with a high-level of confidence. All of them were classified into 16 functional classes, 60 % being enzymes involved in carotenoid biosynthesis. The other proteins were associated to carbohydrate and amino acid metabolism, or were not previously reported. Recent advances in the elucidation of mechanisms governing fruit ripening were achieved for red fruit tomato chromoplasts. Both soluble and insoluble proteins extracted from purified chromoplasts were sequenced using the LC-MS/MS LTQ-Orbitrap technology of high-mass accuracy and sensitivity. A total of 988 proteins corresponded to 802 *Arabidopsis* unigenes, among which 209 had not been listed so far in plastidial databanks. These data substantially revealed several new features of chromoplasts and thus enriched our knowledge of nonphotosynthetic plastids (Barsan et al. 2010) (Table 12.2). An example of new feature is the presence in chromoplasts of the entire set of Calvin cycle proteins, including ribulose 1,5-bisphosphate carboxylase/oxygenase (RuBisCO), as well as the OPPP.

12.4.3 Etioplasts and Amyloplasts: Heterotrophic Plastids Storing Starch Lipid and Proteins

Etioplasts and amyloplasts are both heterotrophic organelles, acting as chloroplast precursors and starch storage, respectively. In particular, etioplasts develop in dark-grown angiosperm seedlings. Upon light, their extensive inner prolamellar and prothylakoid membranes are converted into the chloroplast lamellar thylakoid membrane system, and photosystem (PS) I and II are assembled in a few hours. A first extensive proteome analysis of purified etioplasts was performed by von Zychlinski and Gruissem (2009) from dark-grown rice leaves. A total of 240 unique proteins were identified, revealing complex etioplast-specific metabolic pathways and novel regulatory functions. In etioplasts isolated from dark-grown *Hordeum vulgare* seedlings, Plöscher et al. (2009b) investigated how plants grown in the darkness prepare for the induction of chlorophyll dependent photosystem assembly upon light perception. All low molecular weight proteins of PSII accumulate in the etioplast membrane fraction in the darkness, whereas PsaI and PsaJ of PSI were the only low molecular weight proteins which do not accumulate in the darkness.

The amyloplast, like the chloroplast, is surrounded by a double membrane containing the stroma in which starch granules accumulate. Because this organelle is a major sink for starch in many types of plant cells, proteomics has been applied to purified amyloplast in order to characterize the starch biosynthetic enzymes and the regulatory networks underlying starch accumulation during grain filling in *Z. mays* and *Triticum aestivum* (Andon et al. 2002; Balmer et al. 2006; Bancel et al. 2010; Hennen-Bierwagen et al. 2009) and during fruit development in *Solanum tuberosum* (Stensballe et al. 2008) (Table 12.2). Starch granules are the major carbohydrate storage organites for many types of plant cells. In seeds of cereals, starch is the major storage compound of the wheat endosperm (i.e., the filial tissue). It is composed of amylose [a linear chain of R(1-4) D-glucan polymer] and amylopectin [a branched chain of a R(1-4) and (1-6) D-glucan polymer]. These components are synthesized in amyloplasts in the form of distinct and developmentally regulated granules. Large-type A granules (diameters ≥10 μm) are formed in early stages of seed development (Bechtel et al. 1990). Smaller type B granules (diameters from 5 to 10 μm) are initiated during mid-development, and much smaller type C granules (diameters <5 μm) appear late in development (Bechtel et al. 1990). As evidenced by Andon et al. (2002), among the most important factors determining final protein yield of cereal grains, the rate of grain filling is largely controlled by endosperm-located amyloplasts. The authors identified 108 proteins from whole amyloplasts and 63 from amyloplast membranes isolated from wheat endosperm, among which several candidates were putatively involved in regulating grain filling (e.g., hexose phosphate transporter homologs, putative import associated proteins, proteins with potential ADPG binding or mitocarrier motifs, and proteins containing transmembrane domains). Seed amyloplasts possess a broad metabolic versatility. Identification of 289 proteins in wheat seed amyloplasts has evidenced, besides starch biosynthesis and degradation, a wide range of metabolic pathways (e.g., cytoskeleton/plastid division, energetics, nitrogen/sulfur metabolism, nucleic acid- or protein- related reactions, synthesis of various building blocks, transport, signaling, and stress), emphasizing the importance of amyloplasts in fulfilling essential biosynthetic functions for the parent cell (Balmer et al. 2006).

Recently, Bancel et al. (2010) set up an efficient and selective purification method of wheat seed starch granules of types A and B by minimizing the yield of contaminations by SPs. Proteome 2-D maps were established for both types of granules and compared. Around 85 % of the identified spots were related to carbohydrate (hexose or starch) metabolism and stress defense, reflecting similar metabolic activities. Few spots also showed significant variation in abundance between A-type and B-type granules. Some enzymes involved in seed starch biosynthesis (starch synthase IIa 'SSIIa', SSIII, starch branching enzyme IIb 'SBEIIb', and 'SBEIIa') were purified by anion-exchange chromatography and immunoprecipitation together with additional proteins, and their identification by MS provided insights into multisubunit complexes and carbon partitioning between metabolic pathways in developing seeds (Hennen-Bierwagen et al. 2009).

12.5 Endoplasmic Reticulum: An Inticulum Endomembrane System for SSP Processing, Trafficking, and Lipid Biosynthesis

Seed development is accompanied by high rates of SP and/or lipid synthesis, which are degraded during germination to provide nutrients for the young heterotrophic seedling. The ER plays an important role in seed filling since it is the site of SP maturation and lipid synthesis. Therefore, there is strong interest in identifying the proteins related to this endomembrane system in view of stimulating SP or lipid synthesis/maturation in developing seeds (Table 12.2). Subproteomics approaches were employed to better decipher the complex events occurring in ER during seed development and germination. ER was purified from developing and germinating seeds of castor bean (*R. communis*), whose endosperm is liquid and thus highly suitable to intact organelle isolation (Maltman et al. 2002, 2007). Proteins present in the soluble ER proteome from developing seeds were involved in SP trafficking/processing and those present in the soluble ER proteome from germinating seeds were involved in folding processes. Putative components of lipid biosynthesis were unidentified in urea soluble ER fractions; these proteins were hypothesized as either insoluble in the 2-D extraction buffer or located in ER integral membranes (Maltman et al. 2007). Some luminal proteins involved in protein folding were identified, such as protein disulfide isomerase (PDI) and putative chaperonin of the binding protein (BiP) class. PDI catalyzes the formation and rearrangement of disulfide bonds of the newly synthesized proteins, whereas BiP, a molecular chaperone related to the HSPs, is involved in the assembling of the nascent proteins by preventing their denaturation or aggregation and in the recognition and disposal of misfolded polypeptides. In *M. truncatula* seeds, the temporal induction of the proteins identified as BiP and PDI coincided with the onset of SP accumulation (Gallardo et al. 2003), which makes these chaperones good candidates for controlling folding and assembly of SP during seed development.

12.6 Oil Bodies: A Reservoir of Lipids As Carbon and Energy Source

In oleoproteaginous seeds, lipids are deposited in the form of TAG during seed development and are mobilized upon germination to provide carbon and energy for the developing seedling. After trafficking in the ER, TAGs accumulate in embryo or endosperm in small spherical organelles named OBs, composed of a core of TAG surrounded by a monolayer of phospholipids embedded with proteins (Shimada and Hara-Nishimura 2010). The exact mechanisms of OBs biogenesis and degradation still remain to be elucidated. Because of the central role played by proteins in OB stability and resistance, several works focused on their characterization in oil-rich mature seeds of *Arabidopsis* (Jolivet et al. 2004), rapeseed (Jolivet et al. 2009; Katavic et al. 2006), and soybean (Schmidt and Herman 2008) (Table 12.2).

First information insights into OB-protein complement were achieved through a subproteomic approach, which combined protein extraction from purified OBs derived from *Arabidopsis* mature seeds with 1-DGE and MS. Eight proteins were identified; oleosins represented up to 79 % of OB protein content and the 18.5 kDa polypeptides were being the most abundant (Jolivet et al. 2004). Among these proteins, the putative embryo specific protein ATS1 (a novel *Arabidopsis thaliana* seed protein) and the 11-β-hydroxysteroid dehydrogenase-like protein were also detected in mature seed OBs of two *B. napus* crop cultivars. Among the novel OB-proteins identified in *B. napus*, a short chain dehydrogenase/reductase was found to be similar to a TAG-associated factor from narrow-leafed lupin, while a myrosinase showed a high similarity to the lipase/hydrolase family of enzymes with GDSL motifs. Both proteins were suggested to act in OB degradation. The 24 kDa oleosin was investigated by a functional proteomics approach combining ultrastructural observation and protein analyses in transgenic soybean lines unable to synthesize this protein. Transgenic seeds showed large OB-ER complexes with the interior dominated by micro-OBs, which were abundant in caleosin and lipoxygenase but, as expected, absent in oleosin, for which a surfactant function was thus proposed (Schmidt and Herman 2008). Consistent methodological improvements of OB proteomics were gained through the use of highly efficient OB purification steps to remove non-specifically OB-trapped proteins. Genomics and conventional/shotgun proteomics approaches were combined and allowed, together with searches for PTMs and immune detections, a precise characterization of the major seed OB-proteins (e.g., 19 oleosins, 5 steroleosins, and 9 caleosins), for some of which, PTMs were detected (e.g., acetylation of some oleosins, low level of phosphorylation of steroleosins and caleosins) (Jolivet et al. 2009). Further analyses about OB proteins during seed development and their possible role(s) in OB biogenesis will help to decipher part of the processes governing oil accumulation in seeds.

12.7 Nuclei: Site of Gene Expression and Regulatory Processes for Seed Development

The eukaryotic nucleus is an organelle particularly complex both in structure and function. It is surrounded by a double membrane, and an intense trafficking of proteins and RNAs occurs through its nuclear pore complex. It hosts mostly of the DNA-coded genetic information but regulatory networks still need to be identified. Its structure is highly dynamic. Both euchromatin and heterochromatin move continuously through the fibrillar network of the nuclear matrix, depending on cell transcriptional status (Erhardt et al. 2010). The cardinal role of nucleus in eukaryotic gene expression and regulation poses this organelle as preferential site of proteins (e.g., transcription factors/regulators) controlling plant developmental programs or physiological states, including seed development (Table 12.2).

The nuclear proteome was explored at the switch towards seed filling in order to understand molecular mechanisms governing the transition from embryogenesis to reserve accumulation in the legume model plant *M. truncatula*. This study revealed an unexpected high pool of ribosomal proteins, in preparation for the subsequent intense synthesis activity for SPs in embryos (Repetto et al. 2008). In comparison with *Arabidopsis* leaf nuclei (Bae et al. 2003) and chickpea (Pandey et al. 2008), the *M. truncatula* seed nuclear proteome was enriched in three classes (DNA metabolism, RNA processing, and ribosome biogenesis; Repetto et al. 2008), the proportion of ribosomal proteins exceeding that in nuclear proteomes from vegetative tissues or cell cultures. In particular, among the 143 proteins identified by 1-DGE and nLC ES-MS/MS, several new nuclear proteins were highlighted with possible roles in the biogenesis of ribosomal subunits (PESCADILLO-like) or in nucleocytoplasmic trafficking (dynamin-like GTPase), together with others involved in transcriptional regulation (MADS-box protein AGL11), RNA processing, silencing, or transport. Some of these proteins were not previously reported in seeds, such as the DIP2 and ES43 transcriptional regulators or RNA silencing-related ARGONAUTE proteins. Interestingly, this study also revealed a possible control of gene expression through modification of genome architecture/accessibility during seed filling and maturation, as suggested by the timing of expression of genes encoding chromatin-modifying enzymes (C-5 MTase, HAT, HDA) (Repetto et al. 2008). In the year 2008, another approach was used by Li et al. (2008) to elucidate regulatory mechanisms of endosperm development in rice and, indirectly, grain yield improvement in cereals. The authors developed a method for the enrichment of endosperm nuclei and the recovery of relatively low-abundance proteins for shotgun proteomics. Among the 468 identified proteins, 44 % were hypothetical, supporting the importance of nuclear proteome characterization for genome annotation, while a large proportion was putatively acetylated (e.g., putative H2B, retrotansposon, transposon, and 14 function unknown nuclear proteins) or methylated (bZIP transcription factor, H2A, putative HSF, TRAB1, transposon protein, retrotransposon protein, and 13 unknown and hypothetical nuclear proteins). Acetylation and methylation of these proteins may affect their interactions with the nuclear genome, particularly in the imprinted endosperm genome. Similarly to Repetto et al. (2008), many ribosomal proteins were identified by Li et al. (2008) but the authors did not include them in the list of nuclear proteins as they can be both nuclear and cytoplasm. A differential expression of nuclear proteins during seed endosperm development (from 8 to 35 DAP) was reported in *Z. mays* (Ferreira et al. 2006). Some unidentified proteins appeared by 1-DGE as specific to early *versus* late developmental stages. Among the nuclear proteins that may play a role during late seed development is the p16 protein identified by Castillo et al. (2000). This protein may play a dehydrin like-function in the protection of chromatin structure against water deficit during seed development.

Several nuclear proteomics investigations have been performed in other tissues than seeds and many nucleus-specific proteome reference maps have been created to understand various plant physiological processes: dehydration responses in seedlings of *Oryza sativa* (Pandey et al. 2008) and *Cicer arietinum* (Choudhary et al. 2009), and phosphorylation in those of *Arabidopsis* (Jones et al. 2009); cold stress and sugar responses in leaves of *Arabidopsis* and *O. sativa*, respectively (Bae et al. 2003; Aki and Yanagisawa 2009); and microgravity and differentiation/callus formation in roots of *Lepidium sativum* (Sobol et al. 2007) and *Z. mays*, respectively (Alatzas and Foundouli 2009). All these works illustrate the interest of plant biologists in nucleus and show the usefulness of combining high-purity and yield nuclei purification techniques with 1- or 2-DGE protein separations and/or high-throughput identifications by MS and bioinformatics, to decipher the regulatory network governing plant development or physiological states.

12.8 Concluding Remarks

In plants, proteomics investigations of organelles have undoubtedly increased our knowledge of their metabolic and regulatory pathways in better understanding seed biology. In the last decade, advancements in proteomics technology (i.e., MS instruments, software development) along with access to the genome sequence of several plants have helped in digging dipper into organelle proteome, including the identification of low-abundance proteins and new organelle-specific proteins. The findings of tissue-specific metabolic roles for some organelles underline the importance to exploit organelle function(s) in distinct plant compartments. The reconstruction of how organelle metabolic network interact/change during plant development is of particular interest in the systems biology area (e.g., switches of leaf chloroplast and seed amyloplast/OB proteomics during leaf photosynthetate remobilization and seed formation). The integration of proteomics with the other high-throughput omics technologies (genomics, transcriptomics, and metabolomics) could offer ambitious perspectives to reconstruct the seed metabolic networks. The validation of hypothetical role(s) for selected protein(s) and their importance in specific physiological processes necessarily pass through gene depletion.

References

Abdalla KO, Thomson JA, Rafudeen MS (2009) Protocols for nuclei isolation and nuclear protein extraction from the resurrection plant *Xerophyta viscosa* for proteomic studies. Anal Biochem 384:365–367

Agrawal GK, Bourguignon J, Rolland N, Ephritikhine G, Ferro M, Jaquinod M, Alexiou KG, Chardot T, Chakraborty N, Jolivet P, Doonan JH, Rakwal R (2010) Plant organelle proteomics: collaborating for optimal cell function. Mass Spectrom Rev 30:772–853

Aki T, Yanagisawa S (2009) Application of rice nuclear proteome analysis to the identification of evolutionarily conserved and glucose-responsive nuclear proteins. J Proteome Res 8:3912–3924

Alatzas A, Foundouli A (2009) Alterations in core histone variant ratios during maize root differentiation, callus formation and in response to plant hormone treatment. Biol Res 42:445–460

Andon NL, Hollingworth S, Koller A, Greenland AJ, Yates JR 3rd, Haynes PA (2002) Proteomic characterization of wheat amyloplasts using identification of proteins by tandem mass spectrometry. Proteomics 2:1156–1168

Arai Y, Hayashi M, Nishimura M (2008a) Proteomic analysis of highly purified peroxisomes from etiolated soybean cotyledons. Plant Cell Physiol 49:526–539

Arai Y, Hayashi M, Nishimura M (2008b) Proteomic identification and characterization of a novel peroxisomal adenine nucleotide transporter supplying ATP for fatty acid beta-oxidation in soybean and *Arabidopsis*. Plant Cell 20:3227–3240

Aronsson H, Jarvis P (2002) A simple method for isolating import-competent *Arabidopsis* chloroplasts. FEBS Lett 529:215–220

Aung K, Zhang X, Hu J (2010) Peroxisome division and proliferation in plants. Biochem Soc Trans 38:817–822

Babujee L, Wurtz V, Ma C, Lueder F, Soni P, van Dorsselaer A, Reumann S (2010) The proteome map of spinach leaf peroxisomes indicates partial compartmentalization of phylloquinone (vitamin K1) biosynthesis in plant peroxisomes. J Exp Bot 61:1441–1453

Bae MS, Cho EJ, Choi EY, Park OK (2003) Analysis of the *Arabidopsis* nuclear proteome and its response to cold stress. Plant J 36:652–663

Baginsky S, Gruissem W (2004) Chloroplast proteomics: potentials and challenges. J Exp Bot 55:1213–1220

Baginsky S, Siddique A, Gruissem W (2004) Proteome analysis of tobacco bright yellow-2 (BY-2) cell culture plastids as a model for undifferentiated heterotrophic plastids. J Proteome Res 3:1128–1137

Baker JM, Hawkins ND, Ward JL, Lovegrove A, Napier JA, Shewry PR, Beale MH (2006) A metabolomic study of substantial equivalence of field-grown genetically modified wheat. Plant Biotech J 4:381–392

Balmer Y, Vensel WH, DuPont FM, Buchanan BB, Hurkman WJ (2006) Proteome of amyloplasts isolated from developing wheat endosperm presents evidence of broad metabolic capability. J Exp Bot 57:1591–1602

Bancel E, Rogniaux H, Debiton C, Chambon C, Branlard G (2010) Extraction and proteome analysis of starch granule-associated proteins in mature wheat kernel (*Triticum aestivum* L.). J Proteome Res 9:3299–3310

Bardel J, Louwagie M, Jaquinod M, Jourdain A, Luche S, Rabilloud T, Macherel D, Garin J, Bourguignon J (2002) A survey of the plant mitochondrial proteome in relation to development. Proteomics 2:880–898

Barkla BJ, Vera-Estrella R, Hernández-Coronado M, Pantoja O (2009) Quantitative proteomics of the tonoplast reveals a role for glycolytic enzymes in salt tolerance. Plant Cell 21:4044–4058

Barsan C, Sanchez-Bel P, Rombaldi C, Egea I, Rossignol M, Kuntz M, Zouine M, Latché A, Bouzayen M, Pech JC (2010) Characteristics of the tomato chromoplast revealed by proteomic analysis. J Exp Bot 61:2413–2431

Bechtel DB, Zayas I, Kaleikau L, Pomeranz Y (1990) Size-distribution of wheat-starch granules during endosperm development. Cereal Chem 67:59–63

Borovskii GB, Stupnikova IV, Antipina AI, Vladimirova SV, Voinikov VK (2002) Accumulation of dehydrin-like proteins in the mitochondria of cereals in response to cold, freezing, drought and ABA treatment. BMC Plant Biol 2:5

Brookes PS, Pinner A, Ramachandran A, Coward L, Barnes S, Kim H, Darley-Usmar VM (2002) High throughput two-dimensional blue-native electrophoresis: a tool for functional proteomics of mitochondria and signaling complexes. Proteomics 2:969–977

Brown JW, Shaw PJ, Shaw P, Marshall DF (2005) *Arabidopsis* nucleolar protein database (AtNoPDB). Nucl Acids Res 33:633–636

Brown LA, Baker A (2003) Peroxisome biogenesis and the role of protein import. J Cell Mol Med 7:388–400

Bunkelmann JR, Trelease RN (1996) Ascorbate peroxidase. A prominent membrane protein in oilseed glyoxysomes. Plant Physiol 110:589–598

Bykova N, Egsgaard H, Møller I (2003) Identification of 14 new phosphoproteins involved in important plant mitochondrial processes. FEBS Lett 540:141–146

Carter C, Pan S, Zouhar J, Avila EL, Girke T, Raikhel NV (2004) The vegetative vacuole proteome of *Arabidopsis thaliana* reveals predicted and unexpected proteins. Plant Cell 16:3285–3303

Castillo J, Rodrigo MI, Márquez JA, Zúñiga A, Franco L (2000) A pea nuclear protein that is induced by dehydration belongs to the vicilin superfamily. Eur J Biochem 267:2156–2165

Chen X, Wang Y, Li J, Jiang A, Cheng Y, Zhang W (2009) Mitochondrial proteome during salt stress-induced programmed cell death in rice. Plant Physiol Biochem 47:407–415

Chou KC, Shen HB (2010) Plant-mPLoc: a top-down strategy to augment the power for predicting plant protein subcellular localization. PLoS One 5:e11335

Choudhary MK, Basu D, Datta A, Chakraborty N, Chakraborty S (2009) Dehydration-responsive nuclear proteome of rice (*Oryza sativa* L.) illustrates protein network, novel regulators of cellular adaptation, and evolutionary perspective. Mol Cell Proteomics 8:1579–1598

Collings DA, Harper JDI (2008) Peroxisome aggregation during cytokinesis in different angiosperm taxa. Int J Plant Sci 169:241–252

Cornah JE, Smith SM (2002) Plant Peroxisomes. In: Baker A, Graham IA (eds) Biochemistry, cell biology and biotechnological applications. Kluwer Academic, Dordrecht, pp 57–101

Corpas FJ, Hayashi M, Mano S, Nishimura M, Barroso JB (2009) Peroxisomes are required for in *vivo* nitric oxide accumulation in the cytosol following salinity stress of *Arabidopsis* plants. Plant Physiol 151:2083–2094

Corpas FJ, Sandalio LM, Brown MJ, del Río LA, Trelease RN (2000) Identification of porin-like polypeptide(s) in the boundary membrane of oilseed glyoxysomes. Plant Cell Physiol 41:1218–1228

Daher Z, Recorbet G, Valot B, Robert F, Balliau T, Potin S, Schoefs B, Dumas-Gaudot E (2010) Proteomic analysis of *Medicago truncatula* root plastids. Proteomics 10:2123–2137

Davy de Virville J, Brown S, Cochet F, Soler MN, Hoffelt M, Ruelland E, Zachowski A, Collin S (2010) Assessment of mitochondria as a compartment for phosphatidylinositol synthesis in *Solanum tuberosum*. Plant Physiol Biochem 48:952–960

Dudkina NV, Heinemeyer J, Sunderhaus S, Boekema EJ, Braun HP (2006) Respiratory chain supercomplexes in the plant mitochondrial membrane. Trends Plant Sci 11:232–240

Dunkley TPJ, Hester S, Shadforth IP, Runions J, Weimar T, Hanton SL, Griffin JL, Bessant C, Brandizzi F, Hawes C, Watson RB, Dupree P, Lilley KS (2006) Mapping the *Arabidopsis* organelle proteome. Proc Natl Acad Sci U S A 103:6518–6523

Elstner M, Andreoli C, Klopstock T, Meitinger T, Prokisch H (2009) The mitochondrial proteome database: MitoP2. Methods Enzymol 457:3–20

Endler A, Meyer S, Schelbert S, Schneider T, Weschke W, Peters SW, Keller F, Baginsky S, Martinoia E, Schmidt UG (2006) Identification of a vacuolar sucrose transporter in barley and *Arabidopsis* mesophyll cells by a tonoplast proteomic approach. Plant Physiol 141:196–207

Endler A, Reiland S, Gerrits B, Schmidt UG, Baginsky S, Martinoia E (2009) *In vivo* phosphorylation sites of barley tonoplast proteins identified by a phosphoproteomic approach. Proteomics 9:310–321

Erhardt M, Adamska I, Franco OL (2010) Plant nuclear proteomics—inside the cell maestro. FEBS J 277:3295–3307

Eubel H, Heinemeyer J, Braun HP (2004) Identification and characterization of respirasomes in potato mitochondria. Plant Physiol 134:1450–1459

Eubel H, Lee CP, Kuo J, Meyer EH, Taylor NL, Millar AH (2007) Free-flow electrophoresis for purification of plant mitochondria by surface charge. Plant J 52:583–594

Eubel H, Meyer EH, Taylor NL, Bussell JD, O'Toole N, Heazlewood JL, Castleden I, Small ID, Smith SM, Millar AH (2008) Novel proteins, putative membrane transporters, and an integrated metabolic network are revealed by quantitative proteomic analysis of *Arabidopsis* cell culture peroxisomes. Plant Physiol 148:1809–1829

Fan P, Wang X, Kuang T, Li Y (2009) An efficient method for the extraction of chloroplast proteins compatible for 2-DE and MS analysis. Electrophoresis 30:3024–3033

Fernie AR, Carrari F, Sweetlove LJ (2004) Respiratory metabolism: glycolysis, the TCA cycle and mitochondrial electron transport. Curr Opin Plant Biol 7:254–261

Ferreira BS, Branco AT, de Oliveira MA, Pereira MG, de Souza Filho GA (2006) Methodological improvements on extraction of nuclear proteins and its preliminary analysis during the maize (*Zea mays* L.) endosperm development. Protein Pept Lett 13:981–984

Ferro M, Salvi D, Brugiere S, Miras S, Kowalski S, Louwagie M, Garin J, Joyard J, Rolland N (2003) Proteomics of the chloroplast envelope membranes from *Arabidopsis thaliana*. Mol Cell Proteomics 2:325–345

Fukao Y, Hayashi M, Hara-Nishimura I, Nishimura M (2003) Novel glyoxysomal protein kinase, GPK1, identified by proteomic analysis of glyglyoxysomes in etiolated cotyledons of *Arabidopsis thaliana*. Plant Cell Physiol 44:1002–1012

Fukao Y, Hayashi M, Nishimura M (2002) Proteomic analysis of leaf peroxisomal proteins in greening cotyledons of *Arabidopsis thaliana*. Plant Cell Physiol 43:689–696

Gabaldón T (2010) Peroxisome diversity and evolution. Philos Trans R Soc Lond B Biol Sci 365:765–773

Gallardo K, Le Signor C, Vandekerckhove J, Thompson RD, Burstin J (2003) Proteomics of *Medicago truncatula* seed development establishes the time frame of diverse metabolic processes related to reserve accumulation. Plant Physiol 133:664–682

Goksøyr J (1967) Evolution of eucaryotic cells. Nature 214:1161

Grelet J, Benamar A, Teyssier E, Avelange-Macherel MH, Grunwald D, Macherel D (2005) Identification in pea seed mitochondria of a late-embryogenesis abundant protein able to protect enzymes from drying. Plant Physiol 137:157–167

Guda C (2006) pTARGET: a web server for predicting protein subcellular localization. Nucl Acids Res 34:210–213

Guda C, Fahy E, Subramaniam S (2004) MITOPRED: a genome-scale method for prediction of nucleus-encoded mitochondrial proteins. Bioinformatics 20:1785–1794

Harrison-Lowe N, Olsen LJ (2006) Isolation of glyoxysomes from pumpkin cotyledons. Curr Protoc Cell Biol (Chapter 3: Unit 3.19)

Haynes PA, Roberts TH (2007) Subcellular shotgun proteomics in plants: looking beyond the usual suspects.Proteomics 7:2963–2975

Heazlewood JL, Howell KA, Whelan J, Millar AH (2003) Towards an analysis of the rice mitochondrial proteome. Plant Physiol 132:230–242

Heazlewood JL, Millar AH (2005) AMPDB: the *Arabidopsis* Mitochondrial Protein Database. Nucl Acids Res 1:605–610

Heazlewood JL, Millar AH (2007) *Arabidopsis* mitochondrial proteomics. Methods Mol Biol 372:559–571

Heazlewood JL, Verboom RE, Tonti-Filippini J, Small I, Millar AH (2006) SUBA: the *Arabidopsis* Subcellular Database. Nucl Acids Res 35:213–218

Hennen-Bierwagen TA, Lin Q, Grimaud F, Planchot V, Keeling PL, James MG, Myers AM (2009) Proteins from multiple metabolic pathways associate with starch biosynthetic enzymes in high molecular weight complexes: a model for regulation of carbon allocation in maize amyloplasts. Plant Physiol 149:1541–1559

Hochholdinger F, Guo L, Schnable PS (2004) Cytoplasmic regulation of the accumulation of nuclear-encoded proteins in the mitochondrial proteome of maize. Plant J 37:199–208

Huang Y, Chen L, Wang L, Vijayan K, Phan S, Liu Z, Wan L, Ross A, Xiang D, Datla R, Pan Y, Zou J (2009) Probing the endosperm gene expression landscape in *Brassica napus*. BMC Genomics 10:256

Ito J, Taylor NL, Castleden I, Weckwerth W, Millar AH, Heazlewood JL (2009) A survey of the *Arabidopsis thaliana* mitochondrial phosphoproteome. Proteomics 9:4229–4240

Jain R, Katavic V, Agrawal GK, Guzov VM, Thelen JJ (2008) Purification and proteomic characterization of plastids from *Brassica napus* developing embryos. Proteomics 8:3397–3405

Jaquinod M, Villiers F, Kieffer-Jaquinod S, Hugouvieux V, Bruley C, Garin J, Bourguignon J (2007) A proteomics dissection of *Arabidopsis thaliana* vacuoles isolated from cell culture. Mol Cell Proteomics 6:394–412

Jiang LW, Bunkelmann J, Towill L, Kleff S, Trelease RN (1994) Identification of peroxisome membrane proteins (PMPs) in sunflower (*Helianthus annuus* L.) cotyledons and influence of light on the PMP developmental pattern. Plant Physiol 106:293–302

Jolivet P, Boulard C, Bellamy A, Larré C, Barre M, Rogniaux H, d'Andréa S, Chardot T, Nesi N (2009) Protein composition of oil bodies from mature *Brassica napus* seeds. Proteomics 9:3268–3284

Jolivet P, Roux E, D'Andrea S, Davanture M, Negroni L, Zivy M, Chardot T (2004) Protein composition of oil bodies in *Arabidopsis thaliana* ecotype WS. Plant Physiol Biochem 42:501–509

Jones AM, MacLean D, Studholme DJ, Serna-Sanz A, Andreasson E, Rathjen JP, Peck SC (2009) Phosphoproteomic analysis of nuclei-enriched fractions from *Arabidopsis thaliana*. J Proteomics 72:439–451

Joyard J, Ferro M, Masselon C, Seigneurin-Berny D, Salvi D, Garin J, Rolland N (2010) Chloroplast proteomics highlights the subcellular compartmentation of lipid metabolism. Prog Lipid Res 49:128–158

Katavic V, Agrawal GK, Hajduch M, Harris SL, Thelen JJ (2006) Protein and lipid composition analysis of oil bodies from two *Brassica napus* cultivars. Proteomics 6:4586–4598

Kaur N, Hu J (2009) Dynamics of peroxisome abundance: a tale of division and proliferation. Curr Opin Plant Biol 12:781–788

Khan M, Komatsu S (2004) Rice proteomics: recent developments and analysis of nuclear proteins. Phytochemistry 65:1671–1681

Kleffmann T, Hirsch-Hoffmann M, Gruissem W, Baginsky S (2006) plprot: a comprehensive proteome database for different plastid types. Plant Cell Physiol 47:432–436

Kleffmann T, Russenberger D, von Zychlinski A, Christopher W, Sjolander K, Gruissem W, Baginsky S (2004) The *Arabidopsis thaliana* chloroplast proteome reveals pathway abundance and novel protein functions. Curr Biol 14:354–362

Kley J, Heil M, Muck A, Svatos A, Boland W (2010) Isolating intact chloroplasts from small *Arabidopsis* samples for proteomic studies. Anal Biochem 398:198–202

Kruft V, Eubel H, Jänsch L, Werhahn W, Braun HP (2001) Proteomic approach to identify novel mitochondrial proteins in *Arabidopsis*. Plant Physiol 127:1694–1710

Lane N, Martin W (2010) The energetics of genome complexity. Nature 467:929–934

Lee CP, Eubel H, O'Toole N, Millar AH (2008) Heterogeneity of the mitochondrial proteome for photosynthetic and non-photosynthetic *Arabidopsis* metabolism. Mol Cell Proteomics 7:1297–1316

Leister D (2003) Chloroplast research in the genomic age. Trends Genet 19:47–56

Li G, Babi R, Reddy N, Nallamilli BR, Tan F, Peng Z (2008) Removal of high-abundance proteins for nuclear subproteome studies in rice (*Oryza sativa*) endosperm. Electrophoresis 29:604–617

Lilley KS, Dupree P (2007) Plant organelle proteomics. Curr Opin Plant Biol 10:594–599

Linka N, Theodoulou FL, Haslam RP, Linka M, Napier JA, Neuhaus HE, Weber APM (2008) Peroxisomal ATP import is essential for seedling development in *Arabidopsis thaliana*. Plant Cell 20:3241–3257

Lonosky PM, Zhang X, Honavar VG, Dobbs DL, Fu A, Rodermel SR (2004) A proteomic analysis of maize chloroplast biogenesis. Plant Physiol 134:560–574

Lopez-Juez E, Pyke KA (2005) Plastids unleashed: their development and their integration in plant development. Int J Dev Biol 49:557–577

Maltman DJ, Gadd SM, Simon WJ, Slabas AR (2007) Differential proteomic analysis of the endoplasmic reticulum from developing and germinating seeds of castor (*Ricinus communis*) identifies seed protein precursors as significant components of the endoplasmic reticulum. Proteomics 7:1513–1528

Maltman DJ, Simon WJ, Wheeler CH, Dunn MJ, Wait R, Slabas AR (2002) Proteomic analysis of the endoplasmic reticulum from developing and germinating seed of castor (*Ricinus communis*). Electrophoresis 23:626–639

Martí MC, Camejo D, Olmos E, Sandalio LM, Fernández-García N, Jiménez A, Sevilla F (2009) Characterisation and changes in the antioxidant system of chloroplasts and chromoplasts isolated from green and mature pepper fruits. Plant Biol (Stuttg) 11:613–624

Mateos RM, León AM, Sandalio LM, Gómez M, del Río LA, Palma JM (2003) Peroxisomes from pepper fruits (*Capsicum annuum* L.): purification, characterization and antioxidant activity. J Plant Physiol 160:1507–1516

Millar AH, Heazlewood JL, Kristensen BK, Braun HP, Møller IM (2005) The plant mitochondrial proteome. Trends Plant Sci 10:36–43

Millar AH, Sweetlove LJ, Giegé P, Leaver CJ (2001) Analysis of the *Arabidopsis* mitochondrial proteome. Plant Physiol 127:1711–1727

Millar AH, Trend AE, Heazlewood JL (2004) Changes in the mitochondrial proteome during the anoxia to air transition in rice focus around cytochrome-containing respiratory complexes. J Biol Chem 279:39471–39478

Mo B, Tse YC, Jiang L (2003) Organelle identification and proteomics in plant cells. Trends Biotechnol 21:331–332

Nakai K, Horton P (1999) PSORT: a program for detecting sorting signals in proteins and predicting their subcellular localization. Trends Biochem Sci 24:34–36

Palma JM, Corpas FJ, del Río LA (2009) Proteome of plant peroxisomes: New perspectives on the role of these organelles in cell biology. Proteomics 9:2301–2312

Pandey A, Chakraborty S, Datta A, Chakraborty N (2008) Proteomics approach to identify dehydration responsive nuclear proteins from chickpea (*Cicer arietinum* L.). Mol Cell Proteomics 7:88–107

Pierleoni A, Martelli PL, Fariselli P, Casadio R (2007) eSLDB: eukaryotic subcellular localization database. Nucl Acids Res 35:208–212

Plöscher M, Granvogl B, Reisinger V, Masanek A, Eichacker LA (2009a) Organelle proteomics. Methods Mol Biol 519:65–82

Plöscher M, Granvogl B, Zoryan M, Reisinger V, Eichacker LA (2009b) Mass spectrometric characterization of membrane integral low molecular weight proteins from photosystem II in barley etioplasts. Proteomics 9:625–635

Pracharoenwattana I, Zhou W, Keech O, Francisco PB, Udomchalothorn T, Tschoep H, Stitt M, Gibon Y, Smith SM (2010) *Arabidopsis* has a cytosolic fumarase required for the massive allocation of photosynthate into fumaric acid and for rapid plant growth on high nitrogen. Plant J 62:785–795

Repetto O, Rogniaux H, Firnhaber C, Zuber H, Küster H, Larré C, Thompson R, Gallardo K (2008) Exploring the nuclear proteome of *Medicago truncatula* at the switch towards seed filling. Plant J 56:398–410

Reumann S, Babujee L, Ma C, Wienkoop S, Siemsen T, Antonicelli GE, Rasche N, Lüder F, Weckwerth W, Jahn O (2007) Proteome analysis of *Arabidopsis* leaf peroxisomes reveals novel targeting peptides, metabolic pathways, and defense mechanisms. Plant Cell 19:3170–3193

Reumann S, Bettermann M, Benz R, Heldt HW (1997) Evidence for the presence of a porin in the membrane of glyoxysomes of castor bean. Plant Physiol 115:891–899

Reumann S, Ma C, Lemke S, Babujee L (2004) AraPerox. A database of putative *Arabidopsis* proteins from plant peroxisomes. Plant Physiol 136:2587–2608

Reumann S, Quan S, Aung K, Yang P, Manandhar-Shrestha K, Holbrook D, Linka N, Switzenberg R, Wilkerson CG, Weber AP, Olsen LJ, Hu J (2009) In-depth proteome analysis of *Arabidopsis* leaf peroxisomes combined with *in vivo* subcellular targeting verification indicates novel metabolic and regulatory functions of peroxisomes. Plant Physiol 150:125–143

Rödiger A, Baudisch B, Klösgen RB (2010) Simultaneous isolation of intact mitochondria and chloroplasts from a single pulping of plant tissue. J Plant Physiol 167:620–624

Rodríguez-Serrano M, Romero-Puertas MC, Sparkes I, Hawes C, del Río LA, Sandalio LM (2009) Peroxisome dynamics in *Arabidopsis* plants under oxidative stress induced by cadmium. Free Radic Biol Med 47:1632–1639

Rolland N, Ferro M, Seigneurin-Berny D, Garin J, Douce R, Joyard J (2003) Proteomics of chloroplast envelope membranes. Photosynth Res 78:205–230

Rurek M (2010) Diverse accumulation of several dehydrin-like proteins in cauliflower (*Brassica oleracea* var. *botrytis*), *Arabidopsis thaliana* and yellow lupin (*Lupinus luteus*) mitochondria under cold and heat stress. BMC Plant Biol 10:181

Sazuka T, Keta S, Shiratake K, Yamaki S, Shibata D (2004) A proteomic approach to identification of transmembrane proteins and membrane-anchored proteins of *Arabidopsis thaliana* by peptide sequencing. DNA Res 11:101–113

Schlüter A, Real-Chicharro A, Gabaldón T, Sánchez-Jiménez F, Pujol A (2010) PeroxisomeDB 2.0: an integrative view of the global peroxisomal metabolome. Nucl Acids Res 38:800–805

Schmidt MA, Herman EM (2008) Suppression of soybean oleosin produces micro-oil bodies that aggregate into oil body/ER complexes. Mol Plant 1:910–924
Schmidt UG, Endler A, Schelbert S, Brunner A, Schnell M, Neuhaus HE, Marty-Mazars D, Marty F, Baginsky S, Martinoia E (2007) Novel tonoplast transporters identified using a proteomic approach with vacuoles isolated from cauliflower buds. Plant Physiol 145:216–229
Schneider T, Schellenberg M, Meyer S, Keller F, Gehrig P, Riedel K, Lee Y, Eberl L, Martinoia E (2009) Quantitative detection of changes in the leaf-mesophyll tonoplast proteome in dependency of a cadmium exposure of barley *(Hordeum vulgare* L.) plants. Proteomics 9:2668–2677
Schröter Y, Steiner S, Matthäi K, Pfannschmidt T (2010) Analysis of oligomeric protein complexes in the chloroplast sub-proteome of nucleic acid-binding proteins from mustard reveals potential redox regulators of plastid gene expression. Proteomics 10:2191–2204
Schwacke R, Schneider A, Van Der Graaff E, Fischer K, Catoni E, Desimone M, Frommer WB, Flügge UI, Kunze R (2003) ARAMEMNON, a novel database for *Arabidopsis* integral membrane proteins. Plant Physiol 131:16–26
Shimada TL, Hara-Nishimura I (2010) Oil-body-membrane proteins and their physiological functions in plants. Biol Pharm Bull 33:360–363
Shimaoka T, Ohnishi M, Sazuka T, Mitsuhashi N, Hara-Nishimura I, Shimazaki KI, Maeshima M, Yokota A, Tomizawa KI, Mimura T (2004) Isolation of intact vacuoles and proteomic analysis of tonoplast from suspension-cultured cells of *Arabidopsis thaliana*. Plant Cell Physiol 45:672–683
Siddique MA, Grossmann J, Gruissem W, Baginsky S (2006) Proteome analysis of bell pepper (*Capsicum annuum* L.) chromoplasts. Plant Cell Physiol 47:1663–1673
Small I, Peeters N, Legeai F, Lurin C (2004) Predotar: a tool for rapidly screening proteomes for *N*-terminal targeting sequences. Proteomics 4:1581–1590
Smiri M, Chaoui A, El Ferjani E (2009) Respiratory metabolism in the embryonic axis of germinating pea seed exposed to cadmium. J Plant Physiol 166:259–269
Sobol MA, González-Camacho F, Kordyum EL, Medina FJ (2007) Changes in the two-dimensional proteome of the soluble fraction of nuclear proteins from *Lepidium sativum* root meristematic cells grown under clinorotation. J Gravit Physiol 14:109–110
Stensballe A, Hald S, Bauw G, Blennow A, Welinder KG (2008) The amyloplast proteome of potato tuber. FEBS J 275:1723–1741
Stupnikova I, Benamar A, Tolleter D, Grelet J, Borovskii G, Dorne AJ, Macherel D (2006) Pea seed mitochondria are endowed with a remarkable tolerance to extreme physiological temperatures. Plant Physiol 140:326–335
Szponarski W, Sommerer N, Boyer JC, Rossignol M, Gibrat R (2004) Large-scale characterization of integral proteins from *Arabidopsis* vacuolar membrane by two-dimensional liquid chromatography. Proteomics 4:397–406
Tan F, Li G, Chitteti BR, Peng Z (2007) Proteome and phosphoproteome analysis of chromatin associated proteins in rice (*Oryza sativa*). Proteomics 7:4511–4527
Taylor NL, Heazlewood JL, Day DA, Millar AH (2005) Differential impact of environmental stresses on the pea mitochondrial proteome. Mol Cell Proteomics 4:1122–1133
Taylor NL, Howell KA, Heazlewood JL, Tan TY, Narsai R, Huang S, Whelan J, Millar AH (2010) Analysis of the rice mitochondrial carrier family reveals anaerobic accumulation of a basic amino acid carrier involved in arginine metabolism during seed germination. Plant Physiol 154:691–704
Taylor SW, Fahy E, Ghosh SS (2003) Global organellar proteomics. Trends Biotechnol 21:82–88
Tusnady GE, Simon I (2001) The HMMTOP transmembrane topology prediction server. Bioinformatics 17:849–850
van Wijk KJ (2004) Plastid proteomics. Plant Physiol Biochem 42:963–977
von Zychlinski A, Kleffmann T, Krishnamurthy N, Sjölander K, Baginsky S, Gruissem W (2005) Proteome analysis of the rice etioplast: Metabolic and regulatory networks and novel protein functions. Mol Cell Proteomics 4:1072–1084

Yin G, Sun H, Xin X, Qin G, Liang Z, Jing X (2009) Mitochondrial damage in the soybean seed axis during imbibition at chilling temperatures. Plant Cell Physiol 50:1305–1318

Zabrouskov V, Giacomelli L, van Wijk KJ, McLafferty FW (2003) A new approach for plant proteomics: Characterization of chloroplast proteins of *Arabidopsis thaliana* by top-down mass spectrometry. Mol Cell Proteomics 2:1253–1260

Zychlinski A, Gruissem W (2009) Preparation and analysis of plant and plastid proteomes by 2DE. Methods Mol Biol 519:205–220

Chapter 13
Proteomics in Identifying New Regulatory Mechanisms Involved in Seed Development and Ultimately Seed Quality

Jean-Louis Prioul, Domenica Manicacci, Catherine Damerval and Valérie Méchin

Abstract Many changes in gene expression occur during maize kernel development. Most changes are associated with the three main developmental stages: lag phase, grain filling, and grain maturation. Proteomics approaches dedicated to non storage proteins coupled with measurements of enzyme activities and metabolite levels provide a way to get an overview of the main metabolic changes and to look at the coordinated variation of the enzymes involved in the main biosynthetic pathways. It also enables to single out specific variation, which may have been overlooked by targeted approaches. Identification of proteins during maize kernel development not only uncovers physiologically consistent protein patterns associated with each stage but also reveals the unexpected importance of some pathways. A major modification occurs at the transition from lag phase (establishment of potential grain size) to grain-filling phase, where starch and proteins are accumulated in the endosperm storage tissue. Although the expression of enzymes involved in the biosynthetic pathway for storage product is dominant in the accumulation phase, the proportion of protein destination (mainly, chaperonins) and protein synthesis machinery is still important. Detailed proteomics analysis of metabolism shows a surprising upsurge of the pyruvate-Pi-dikinase (PPDK) at the late grain-filling period (21 DAP onwards), and that is interpreted as a switch in the starch/protein balance. This hypothesis is based on biochemical arguments involving the negative effect of PPi generated by PPDK activity on the cytosolic ADP-glucose pyrophosphorylase, a key-enzyme in starch synthesis, and the role of phosphoenolpyruvate in aromatic amino acid synthesis. It is substantiated by the molecular genetic data on the *O2* gene, which encodes a transcription factor with pleiotropic

J.-L. Prioul (✉)
Institut de Biologie des Plantes, Université Paris-Sud,
Bât. 630, UMR 8618, CNRS, 91405, Orsay, Cedex, France
e-mail: jean-louis.prioul@u-psud.fr

D. Manicacci · C. Damerval
Université Paris-Sud, UMR 0320/UMR 8120 Génétique Végétale,
91190, Gif sur Yvette, France

V. Méchin
Institut Jean-Pierre Bourgin, INRA, UMR 1318 INRA-AgroParisTech,
78000, Versailles, France

G. K. Agrawal, R. Rakwal (eds.), *Seed Development: OMICS Technologies toward Improvement of Seed Quality and Crop Yield*,
DOI 10.1007/978-94-007-4749-4_13,

effects on lysine content, carbohydrate metabolism, and expression of the *cyPPDK1* gene. One way to test this hypothesis is to use association genetics on a large panel covering most of the genetic variability in tropical, American, and European maize. After genotyping of 375 lines for *O2* and *cyPPDK1* genes, a polymorphism in the *O2* coding sequence and several polymorphisms along the *cyPPDK1* promoter have been identified which are associated in anon-additive way to both the lysine content and the protein *versus* starch balance. These and other findings are the subject of discussion of this chapter.

Keywords cyPPDK · Lysine · Maize kernel · Opaque-2 · Proteomics · Starch/protein

13.1 Introduction

Cereals represent a major part of human diet and animal feeding as they provide both carbohydrate (starch) and proteins, although unbalanced in some essential amino acids as lysine, especially in the case of maize (Young et al. 1998). Since the origin of agriculture, human populations have tried to favor grains with larger endosperm, the main storage organ. Maize (*Zea mays* L.), one of the major crops, accumulates large amount of starch (75–80 % of endosperm dry matter at maturity) and SPs (10–12 %; mainly, zeins) in its kernels. Such resources are primarily devoted to seed germination and seedling initial growth. These reserve components constitute major traits in both agronomical and natural conditions. The starch/protein balance is of major importance in the use of maize for animal feeding. However, the attempts to increase the protein complement by selecting for that trait irremediably resulted in lower starch content as illustrated by the well-known Illinois Long-Term Selection Experiment, which is spanning over more than 100 generations of classical breeding (Moose et al. 2004).

A physiological explanation of the apparent antagonism between starch and protein accumulation is likely to originate from different source-sink relationships for each reserve types (Prioul 1996). Indeed, the carbon source for grain filling mainly originates from current photosynthesis especially in maize, where carbon remobilization from vegetative parts is poor, especially under stress conditions (Tollenaar and Daynard 1978; Prioul and Schwebel-Dugué 1992). Conversely, most nitrogen (amino acids) entering kernels comes from protein remobilization in the source organs of previously assimilated nitrogen (Cliquet et al. 1990). Given that protein reserve in leaves is largely in the form of carboxylating enzymes [RuBisCO (35 %) and PEPC (15 %)] (Rocher et al. 1989), it is clear that an active protein remobilization allowing high kernel protein content would be detrimental for photosynthesis and thus starch synthesis. This mechanism may be overcome in some stay-green genotypes, where late nitrogen uptake and assimilation from the root system reduce the drain in leaf proteins, thus the photosynthetic decline (Pommel et al. 2006).

The biochemical and molecular mechanisms controlling starch *versus* protein partitioning in the endosperm are yet to be uncovered. Many kernel mutants are available in maize, suggesting that endosperm development is driven by the coordinate expression of several hundreds of genes (Liu et al. 2008; Neuffer and Sheridan 1980; Scanlon et al. 1994; Verza et al. 2005). Mutant analysis has been of great help in dissecting the biosynthetic pathway for reserves synthesis. Accordingly, each enzyme involved in starch synthesis from phloem sucrose import to starch accumulation in amyloplast granules has been identified, as well as were some of the regulatory mechanisms (Hannah 2005; Hannah and James 2008). For protein, *o2* mutation was recognized in the mid-twentieth century not only as affecting kernel aspect but more importantly as providing high-lysine grains, thus increasing the nutritional value for monogastric animals. Numerous efforts have been made to understand the *o2*-caused pleiotropic effects and its associated drawbacks. A major step in this direction was the identification of the *O2* gene as a transcriptional activator of the bZIP protein family (Hartings et al. 1989; Schmidt et al. 1990) and its expression in the subaleurone layer from 10 DAP (Gallusci et al. 1994). Direct evidence on *O2* pleiotropic effects were provided from the *o2* mutant which presents reduced transcript level of the 22 kD zein gene (Schmidt et al. 1992), increased accumulation of transcripts for the albumin class genes (Cord Neto et al. 1995), and also a decreased level of the gene encoding the bifunctional enzyme LKR/SDH (lysine-ketoglutarate reductase/saccharopine dehydrogenase). LKR/SDH is the first enzyme of lysine catabolism. The LKR/SDH activity temporally coincides with SP synthesis and accumulation in seeds and is strongly decreased in the *o2* mutants compared to the wild type (Azevedo et al. 2003; Brochetto-Braga et al. 1992). The coordinated and opposite regulation of zein synthesis and lysine degradation pathways provide a basis for two important explanations. Firstly, the potential excess of lysine in the wild type seeds, if not utilized for the zein synthesis, would be degraded and, in turn, could act as a limiting factor for the synthesis of lysine-containing proteins. Secondly, the higher kernel lysine content in the *o2* mutant indicates increase in the lysine-containing proteins (Landry et al. 2002). However, a demonstrated proof for limitation of lysine accumulation in seeds is still lacking.

In addition to the effect of *O2* on SP expression, altered transcript expressions were found for the genes encoding enzymes of the starch biosynthetic pathway in the *o2* mutant seeds as compared to wild type (Giroux et al. 1994), suggesting that the *O2* gene also affects enzymes involved in the carbon metabolism. Accordingly, expression of the cytosolic form of the pyruvate-Pi-dikinase (*cyPPDK1*) is reduced at the mRNA, protein, and enzymatic activity levels due to the *o2* mutation (Maddaloni et al. 1996). Furthermore, a comparative proteomics analysis of the seven *o2* mutants along with their corresponding isogenic wild type provided a full picture of the extensive changes in the protein complements, including zeins and PPDK (Damerval and Le Guilloux 1998). Hence, the *O2* and *cyPPDK1* genes are excellent candidates for controlling the starch/protein ratio in seeds (Prioul et al. 2008b).

Among the natural population of maize, genetic variation has largely been observed for traits, such as yield, endosperm content in lipids, proteins, starch, or protein quality. For example, quantitative variation of zeins in various genotypes was demonstrated using an improved extraction method and2-DGE-based proteomics approach (Consoli and Damerval 2001). To date, genes and alleles have been seldom identified that control natural variation among cultivated maize. Omics-identified molecular markers and their use inbreeding programs might help in understanding the mechanisms and regulations involved in seed quality and yield. However, a rigorous validation of the candidate genes is necessary before utilizing them for genetic breeding. In this respect, association genetics provides a powerful tool. A similar approach is largely being used for human and animal genetics. Since the pioneer work of Thornsberry et al. (2001), this method has received an increasing interest from plant geneticists, especially with the development of high-throughput genotyping methods. The principle is to analyze a core population of large genetic diversity for molecular polymorphisms of candidate genes, and to test the correlation between variation at phenotypic traits of interest and genetic diversity, such as SNP.

Cereal endosperm development is quite well characterized from its physiological point of view. Three main key phases are already described that include a total of five successive phenomena. The 'lag phase' begins from within a few hours after pollination to 12 DAP and consists of three stages, coenocytic, cellularization, and differentiation, resulting in rapid expansion of the endosperm. The subsequent period ranging from 12 to 40 DAP is dedicated to kernel filling. Biochemically, the filling involves the conversion of imported sucrose and amino acids into starch and SPs. Most starch present in the endosperm is thus synthesized within a 28-day period after the lag phase. Afterwards, the maturation-desiccation phase occurs, leading to the mature kernel at 70 DAP. A proteomic analysis of non SPs performed at seven successive stages (4 through 40 DAP) of developmental process revealed the main shifts in individual protein amount (Méchin et al. 2007). These proteins were mostly enzymes involved in starch biosynthesis and provided information on regulation, when comparing to the variation in key enzyme activities and carbohydrate levels and also upon complementary transcriptomics analyses (Prioul et al. 2008a).

This chapter will mainly present the contribution of proteomics and some comparison with transcriptomics results towards understanding of the main changes in protein expression during seed development using maize as a model system. The potential role of two candidate genes *O2* and *cyPPDK1* in controlling starch/protein ratio and lysine content will be tested using association genetics.

13.2 Functional View of Maize Endosperm Development from Proteomics Analysis and Complementary Transcriptomics Measurements

Grains were sampled from the middle part of each ear at 4, 7, 10, 14, 21, 30, and 40 DAP to cover the main phases of endosperm development. Starting from 14 DAP (as soon as kernels developed enough to allow easy dissection), kernels were dissected to collect embryo and pericarp. The 2-DGE was performed using the IPG strip of 4–7 pH gradient in order to eliminate the zeins, basic proteins ($pI > 7$) and to enrich low-abundance proteins. A reference map was established for the 14-DAP

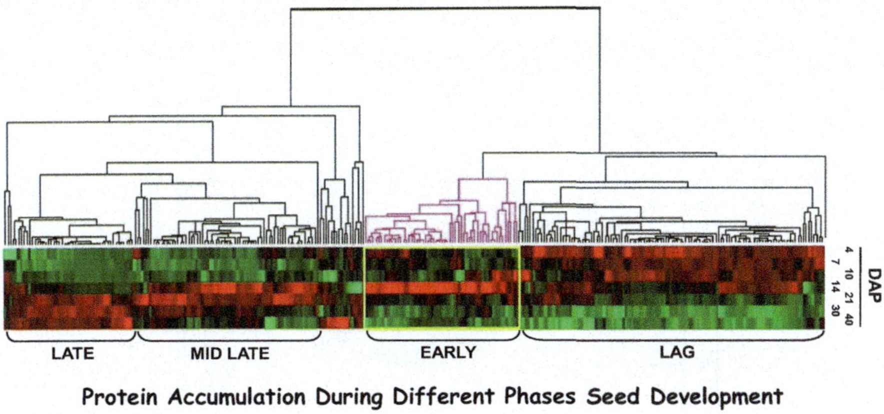

Fig. 13.1 Hierarchical clustering analysis of 302 proteins quantified from 2-DGE-based analysis of kernel (4–10 DAP) and endosperm (14–40 DAP). The time course of maximum protein quantities (*red* color) can be grouped into four clusters corresponding to the accumulation of proteins (excluding the storage proteins) at the lag, early, mid-late, and late phases

stage, which is likely to contain proteins from both the lag and filling phases. Of the 632 excised protein spots, 504 spots were unambiguous identified by MS/MS analysis (Méchin et al. 2004). A total of 423 out of 504 spots were identified and quantified at four through 40 DAP (Méchin et al. 2007). Of 423 identified spots, the function of 121 spots could not be assigned. Hence, hierarchical clustering analysis was performed on the 302 remaining protein spots. The cluster analysis revealed four main developmental profiles, presenting a maximum protein accumulation at 4–10 DAP, 10–14 DAP, 14–30 DAP, and 30–40 DAP, respectively (Fig. 13.1). Protein compositions in each cluster were physiologically meaningful.

Cluster I (4–10 DAP) covers the lag phase, including cellularization, cell division, and CW deposition (Fig. 13.2). Proteins within this cluster belonged to functional categories metabolism, protein destination, cell rescue (defense and anti-ROS), organization, cell division, and protein synthesis. While the relative importance of respiration metabolism (glycolysis and Krebs cycle) is not surprising, the large contribution of cell rescue proteins and notably protection against ROS is rather unexpected. Presence of actin, tubulin, and cell organization proteins might be related to the active division and cellularization at this early stage. The predicted involvement of a large number of proteins in proteolysis is consistent with an important protein turnover process, which is likely to be associated to the switch from growth and differentiation to storage.

Cluster II corresponds to the early accumulation phase with minimum diversity in protein categories. Metabolism and protein destination categories appears to be equally important at early phase, but the protein destination category largely consist of proteins dedicated to protein degradation which is consistent with complete renewal of the enzyme set when shifting from lag to accumulation phase. In clusters III and IV, the relative representation of metabolism-related enzymes increases from 37 to 47 %, which is in agreement with endosperm storage filling. Nevertheless,

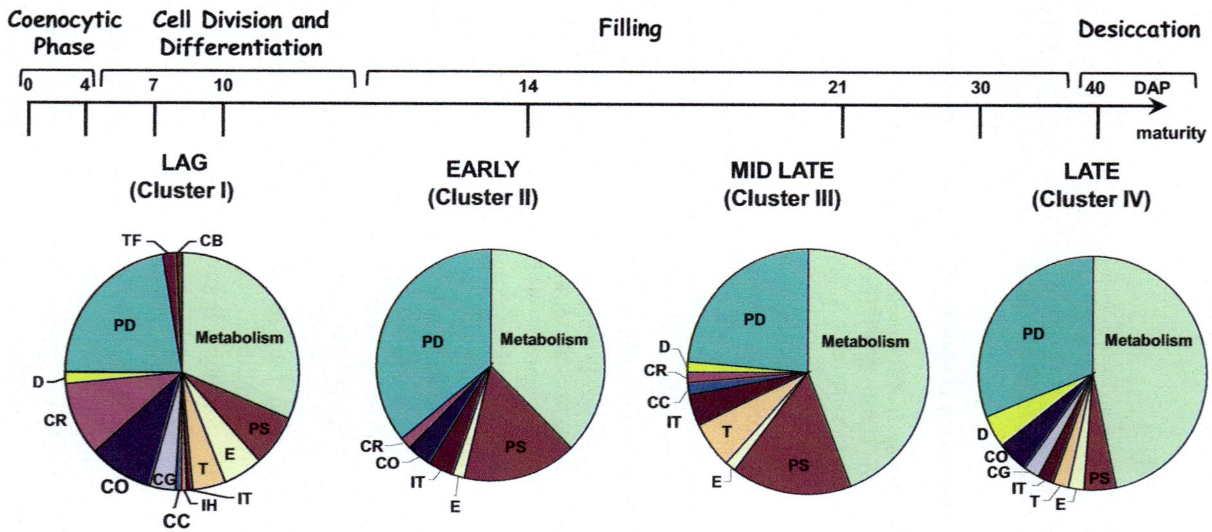

Fig. 13.2 Changes in the composition of the four clusters for the proteins quantified from 2-D electrophoresis from maize endosperm from 4 to 40 DAP. Storage proteins were excluded. *PS* protein synthesis, *E* energy, *T* transcription, *IT* intracellular transport, *IH* ionic homeostasis, *CC* cellular communication and signal transduction, *CG* cell growth, cell division, and DNA synthesis, *CO* cellular organization, *CR* cell rescue, defence, cell death, and ageing, *D* development, *PD* protein destination, *TF* transport facilitation, *CB* cell biogenesis. Adapted from Méchin et al. (2007)

proteins frequently overlooked at this stage are also in fair amount. The 'protein destination' category is a good example but its composition changed from previous clusters in favor of proteins involved in protein folding (i.e., chaperones, BiP, and PDI); this change in protein composition is consistent with the timing of SP accumulation. In contrast, the diversity in the category of protein synthesis-RNA processing decreased (Fig. 13.2).

The complementary transcriptomics analysis at 10 (end of lag phase), 14 (beginning of reserves accumulation), and 21 (maximum rate of reserves accumulation) DAP typical stages led to the identification of 711, 757, and 384 ESTs, respectively (Prioul et al. 2008a). A comparative analysis of transcriptome and proteome data showed a good agreement between category rank in the 10 DAP transcriptome and 14 DAP proteome, as expected because of the likely lag between transcription and mature protein accumulation. The only notoriously missing categories in the proteome were kinases and proteins for RNA processing, which are likely to be expressed at a level below the detection limit of 2-DGE. Transcriptomics also provides information of the time course of expression of SPs, which were not considered by proteomics for technical reasons. The SP sequences increased dramatically from 5 to 38 % between 10 to 21 DAP (Prioul et al. 2008a).

13.3 Interpretation of Protein Expression Associated to Metabolism in Relation with Evolution of Endosperm Physiology During Development

A close examination of the 'metabolism' category in all the four clusters (Fig. 13.3) revealed several trends, which could be interpreted in accordance to both changes in metabolism orientation and endosperm microenvironment. In addition, PPDK abundance at the late phase provides new insights into seed development.

13.3.1 Glycolysis, TCA Cycle, and Anoxia

A major change was found in the balance between the oxygenic and nonoxygenic components of the respiratory pathway during seed development. The expressions of Krebs-TCA cycle enzymes were at a maximum during the lag phase (4–10 DAP), whereas glycolytic enzymes reached their maximum expressions during early and mid-phases (Fig. 13.3). It is important to note that TCA cycle requires oxygen for its proper function but glycolysis does not. Hence, the metabolic shift between two processes of the respiratory pathway could be interpreted in the light of the recent demonstration that starch accumulation in the endosperm (i.e., after 12 DAP) proceeds under very hypoxic conditions (Rolletschek et al. 2005). The lack of oxygen preventing function of the Krebs cycle is likely to influence the overall expression

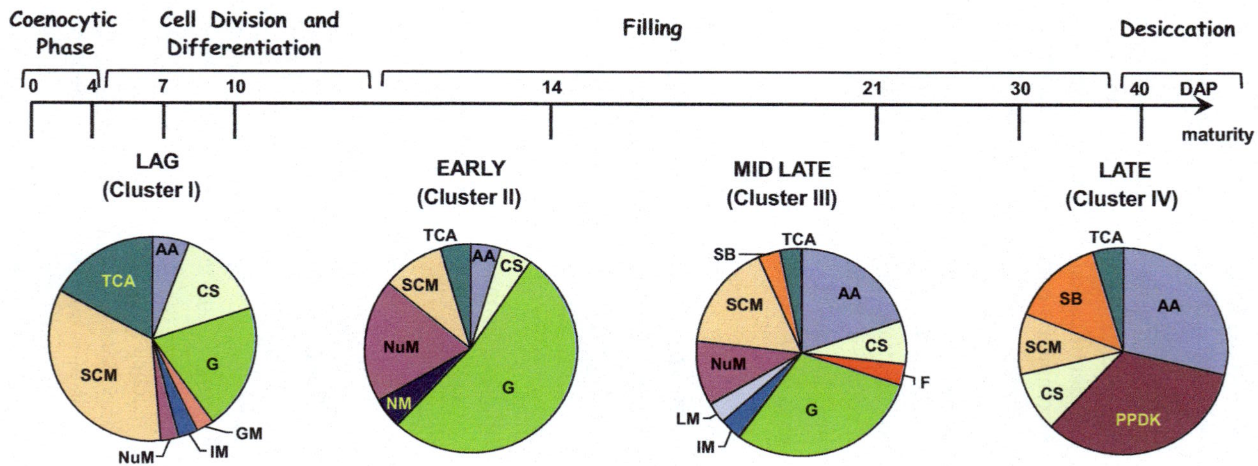

Fig. 13.3 Changes in the composition of each cluster for the enzymes involved in primary and secondary metabolism. Proteins dedicated to metabolism: *AA* amino acid biosynthesis, *PPDK* pyruvate-Pi dikinase isoforms, *CS* carbohydrate synthesis, *F* fermentation, *G* glycolysis, *GM* glycerol metabolism, *IM* inositol metabolism, *NM* nitrogen metabolism, *LM* lipid metabolism, *NuM* nucleotide metabolism, *SCM* secondary compound metabolism, *SB* starch biosynthesis, *TCA* TCA cycle. Adapted from Méchin et al. (2007)

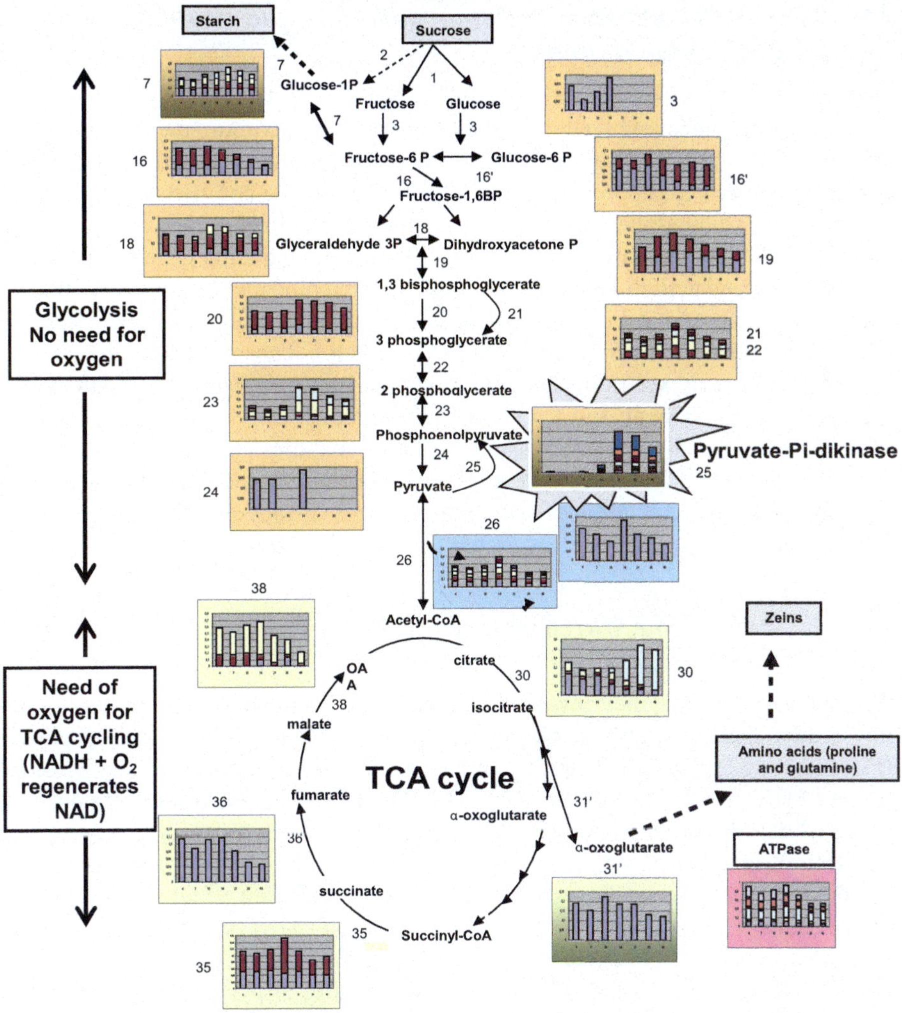

Fig. 13.4 Variation in the content of the enzymes of the respiratory metabolism (glycolysis and Krebs cycle) from 4 to 40 DAP as measured by quantitative 2-D electrophoresis. *1* invertase, *2* sucrose synthase, *3* hexokinase, *7* phosphoglucomutase, *16* fructokinase-II, *16'* diphosphate-fructose-6-phosphate-1-phosphotransferase, *18* triosephosphate-isomerase, *19* phosphoglycerate-dehydrogenase, *20* phosphoglycerate-kinase-3, *21* phosphoglycerate-mutase (bis), *22* phosphoglycerate-mutase, *23* enolase, *24* pyruvate-kinase, *25* pyruvate-orthophosphate-dikinase, *26* A, pyruvate-dehydrogenase, *26* B, dihydrolipoamide S-acetyltransferase, *30* aconitate-hydratase = citrate-hydro-lyase, *31'* NADP-specific-isocitrate-dehydrogenase, *35* succinyl-CoA-ligase (synthetase), *36* succinate-dehydrogenase, and *38* malate-dehydrogenase. Adapted from Prioul et al. (2008)

of its corresponding enzymes. The hypoxic condition fails to affect starch synthesis as ATP needed for the synthesis of ADP-glucose (the starch precursor) is produced without oxygen by the SUS UDP-glucose pyrophosphorylase/AGPase (UGPase/AGPase) cycle and by the first steps of glycolysis (Fig. 13.4).

13.3.2 Starch Synthesis: From Transcripts and Protein to Enzyme Activities

Most of the enzymes involved in starch synthesis had maximum expression during the seed-filling phase (Prioul et al. 2008a). As few enzymes could not be identified using the 2-DGE approach, focus was directed on the comparison of enzymes involved in carbohydrate metabolism over the time course period of 10 through 40 DAP with the quantitative information of their corresponding gene transcripts as per qRT-PCR and activities of eight key enzymes and nine metabolites. Two distinct patterns were observed. First, the invertases and their resulting products (hexoses) reached their maximum from 10 to 20 DAP, which is consistent with the need of hexose during early seed development as shown by the CW invertase mutant *miniature-1* analysis (Cheng and Chourey 1999). Second, as observed previously for AGPase (Prioul et al. 1994), the starch pathway enzymes appeared from 14 DAP (Prioul et al. 2008a). In most cases, it appears that transcriptional control is responsible for the global regulation of starch biosynthesis (Prioul et al. 2008a).

13.3.3 PPDK and Starch/Protein Balance

Accumulation of enzymes involved in metabolite production, namely amino acids, carbon skeletons, and starch, tended to increase during the two 'late' phases (Figs. 13.1–13.3; Clusters III & IV). But the most prominent change is sudden onset of the PPDK protein starting from 21 DAP. The PPDK is known as a key enzyme of the C4 photosynthetic pathway as it allows the C4 cycle to proceed. Given this, its location in nonphotosynthetic organs (like endosperm) would seem puzzling, but the literature survey shows several reports on abundant PPDK activity in the cereal endosperm (Meyer et al. 1982), such as wheat (Aoyagi and Bassham 1984), rice (Imaizumi et al. 1997), and maize (Gallusci et al. 1996). In nonphotosynthetic tissue, it was suggested that PPDK functions parallel to pyruvate kinase (PK) in the glycolytic pathway, but due to its reversibility it may function in the opposite direction, thus producing a so-called futile cycle. However, this cycle provides the interesting property of regulating the PEP/pyruvate ratio, two organic acids being at the entry point for aromatic and alanine-aspartate amino acid families, respectively.

The PPDK-catalyzed reversible reaction also has a consequence on the levels of ATP and PPi. Indeed, when PPDK acts in the PEP-to-pyruvate direction, one more ATP is produced to catalyze the same reaction by PK. In the reverse direction, PPDK produces one AMP and one PPi, which is potentially important for starch synthesis in cereal seeds because in these plants the synthesis of ADP-glucose takes place in the cytosol. Under these conditions, PPi accumulation is pushing the reversible activity of AGPase toward ADP-glucose degradation rather than ADP-glucose synthesis, a mandatory step for starch synthesis. Thus, PPDK may play a central role in storage product composition, since PPi production will reduce starch accumulation in favor of amino acid synthesis, especially aspartate and aromatic acids. Genetic data provide several interesting arguments, which support above-mentioned hypothesis of a possible PPDK role in starch/protein balance. First, the QTL for starch/protein ratio was colocalized on the chromosome 6 (bin 6.05) with the locus of *cyPPDK1* gene, expressed specifically in the endosperm cytosol (Aoyagi and Bassham 1984). Conversely, no such colocation has been observed for the other reported PPDK gene on the chromosome 8 (http://www.maizegdb.org/). Second, from the *o2* mutant analysis it is inferred that *cyPPDK1* is upregulated by the *O2* TF (Maddaloni et al. 1996).

13.4 Epistatic Relationship Between *cyPPDK1* and *O2* Genes in Relation with Starch/Protein Balance and Lysine Content

Association genetics approach has been applied to test the possible causal relationship between DNA polymorphism at *cyPPDK1* and *O2* genes and natural variation in the seed contents of starch and protein. Association genetics is a powerful method to track gene polymorphisms responsible for phenotypic variation since it takes advantage of existing collections and historical recombination to study the correlation between large genetic diversity and phenotypic variation. When simultaneously examining two genes, as in the case of *cyPPDK1* and *O2* genes, it is also possible to test for epistatic effect. The epistatic effect means that pairs of polymorphism in each gene result in combined effect on the associated phenotypic traits. For this purpose, a panel of 375 inbred lines, representative of the natural diversity in tropical, American, and European maize has been characterized for genome-wide neutral markers and population structure (Fig. 13.5). In this panel, SNPs were identified at 14 and 10 positionsin the *cyPPDK1* and *O2* genes, respectively, either in promoter, intronic, or exonic regions (Fig. 13.6). Only two SNPs had effect on amino acids encoding while the other SNPs were silent (Manicacci et al. 2009). Phenotypic traits, such as kernel starch/protein ratio and individual amino acid content, were measured in each line by NIRS (near-infrared spectrometry) in a high-throughput manner. Individual SNPs and epistatic interactions between *O2* and *cyPPDK1* polymorphisms were found to be associated to variation in starch/protein ratio and lysine content using the statistical models, which correct for population structure and individual kinship (Manicacci et al. 2009) (Fig. 13.7).

13.4.1 Kernel Protein/Starch Balance

The *cyPPDK1* gene polymorphism is significantly linked to diverse phenotypic traits: amino acid and protein contents as well as the protein/starch ratio. All these

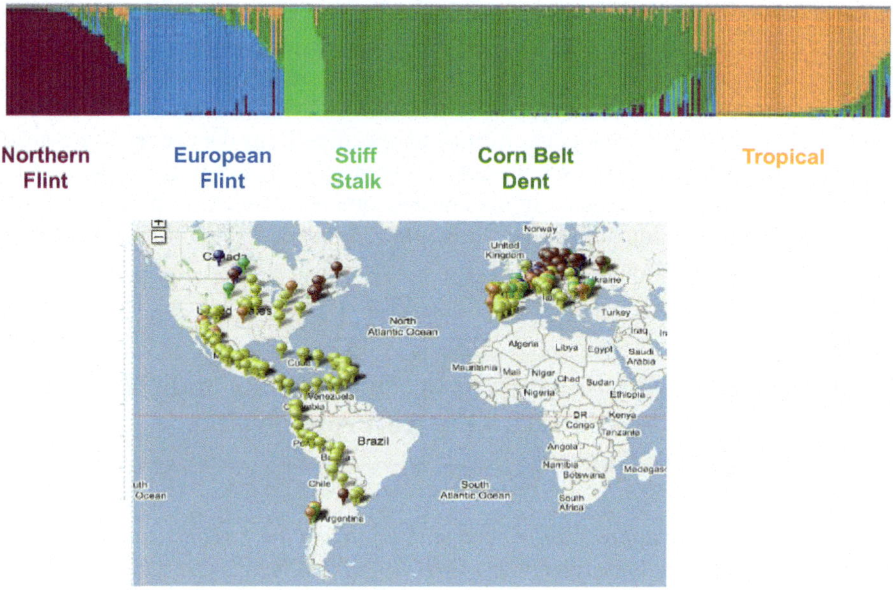

Fig. 13.5 Structure of the 375 inbred line panel covering most of the genetic variability in American and European maize. This highly recombined population with large diversity was characterized for population structure and individual kinship before searching for associations between SNPs in *cyPPDK* and *O2* genes and the phenotypic traits

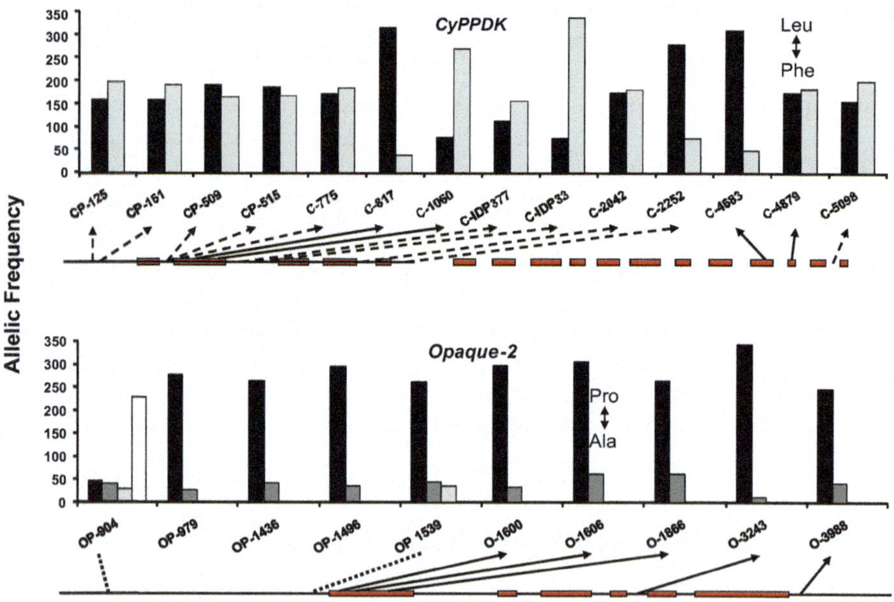

Fig. 13.6 Polymorphisms in the *cyPPDK1* and *O2* genes among 375 inbred lines, either in promoters, introns, or exons

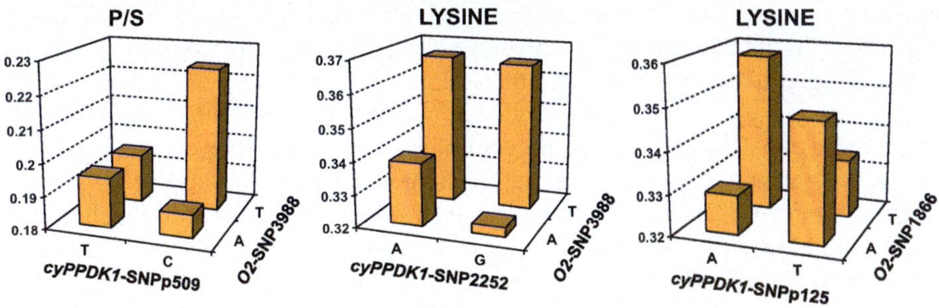

Fig. 13.7 Epistatic interactions between *cyPPDK1* and *O2* genes. Changes in the protein starch ratio (P/S) and Lysine content as a function of single nucleotide changes at pairs of SNPs. *A* adenine, *C* cytosine, *T* thymine

associations involve a SNP polymorphism located in the coding region (C817), suggesting that this gene has a direct effect on amino acid synthesis. This is consistent with the fact that *cyPPDK1* activity reversibly catalyzes the conversion of pyruvate into PEP, which is the initial substrate for aromatic amino acid biosynthesis, namely phenylalanine. PEP is also indirectly involved *via* oxaloacetate in the synthesis of aspartate-derived amino acids, such as threonine, isoleucine, methionine, and lysine. Interestingly, SNP C817 is associated with the contents of all these amino acids, except lysine. Although the C817 polymorphism itself might not be the causative element for the variation in amino acid content because it is partly linked with other polymorphisms in the promoter, it is likely that the promoter plays an important role in aromatic and aspartate-derived amino acid synthesis, except for the lysine accumulation, which is likely the subject of a more complex regulation.

Additionally, strong associations were observed between many phenotypic traits related to kernel starch and protein contents and the combination of *O2* SNP O3988 and one of the SNPs in the *cyPPDK1* promoter (CP125, CP161, CP509, or CP515; Fig 13.7). As all of these phenotypic traits being strongly correlated with each other and all the four SNPs in the *cyPPDK1* promoter being in linkage disequilibrium (more or less linked), these associations are likely due to the same cause. Specific analysis of all significant interactions between these SNP pairs in the *O2* and *cyPPDK1* alleles showed that low starch content, high protein content, and high protein/starch ratio are only obtained for the simultaneous change of allele A to T at O3988 and allele T to C at CP509 (or G to A at CP125, A to G at CP161, or TG to a 2-bp deletion at CP515) (Manicacci et al. 2009). These observations reinforce the afore-mentioned hypothesis as raised from the proteomics study, stating that the cyPPDK protein has a critical role in protein/starch balance in the kernel (Méchin et al. 2007). However, the natural polymorphism in the *cyPPDK1* gene alone shows only slight effect on endosperm protein content and no association with starch content, whereas the protein and starch contents as well as protein/starch balance are affected by epistatic interactions between the *O2* coding sequence and the *cyPPDK1* promoter. This finding suggests that an increase in protein/starch ratio is unlikely to be achieved through selection of more efficient *cyPPDK1* alleles but rather through specific combinations of compatible alleles that allow an increased activation of *cyPPDK1* by its transcriptional activator *O2*. Maddaloni et al. (1996) have shown that *O2* upregulates the *cyPPDK1* transcription through DNA binding of the O2 protein on two specific domains located at positions 163–172 and 295–304. However, these two domains show no polymorphism among the 30 inbred lines initially sequenced; thus, no SNP has been defined in these regions. This result leads to the assumption that other sequences are involved in the ability of *O2* to regulate *cyPPDK1* transcription. They could be located in the *cyPPDK1* promoter and 3′-untranslated region (UTR) in *O2*, as discussed by Manicacci et al. (2009).

13.4.2 Complex Control of Lysine Content by O2 and cyPPDK1

The strongest association for a single SNP was found between *O2* polymorphism O3988 and lysine content (FDR $< 10^{-4}$) (Manicacci et al. 2009). Although such association is expected from the well-known effect of the mutant allele *o2* on kernel lysine content because of the 50–70 % reduction in the lysine-free zeins (Landry et al. 2002) and an increase in the lysine-rich proteins (Habben et al. 1993), it was not detected by an QTL approach, demonstrating the strength of the association genetics within a large collection of inbred lines. In the present example, it is possible to show how to use *O2* natural polymorphism to control lysine content. No single SNP association was significant between *cyPPDK1* polymorphism and lysine content but the role of *cyPPDK1* could be revealed when combined with *O2* polymorphism. Such epistatic effect is observed with the combination of SNP C2252 and SNP O3988 polymorphisms on lysine content (Fig. 13.7). Even more striking example of epistasis between both genes can be obtained by O1866 and CP125 SNPs. While none of these SNPs is individually associated with lysine content, their interaction strongly correlates with its variation. Two combinations of alleles (i.e., allele A at CP125 and allele T at O1866 or allele T at CP125 and allele A at O1866) produced high lysine, while the two remaining combinations lead to low lysine.

The strong association of O3988 with lysine in the absence of interaction with C2252 suggests that there is an additional effect of *O2* on lysine content, independent from *cyPPDK1*. This may be achieved through *O2* transcriptional control of lysine degradation by the LKR/SDH enzyme (Arruda et al. 2000). Further studies show that the activities of the first two enzymes (AK, aspartate kinase; HSDH, homoserine dehydrogenase) of the aspartate pathway producing lysine, threonine, and methionine are also controlled by *O2* (Azevedo et al. 2003). The rationale for this effect is based on the fact that very little lysine is required in maize endosperm since the main SP zeins do not contain this amino acid. Thus, excess of free lysine has to be continuously catabolized through the saccharopine pathway. The association genetics approach further suggests that the natural diversity in lysine degradation controlled by *O2* plays a central role in grain nutritional value.

13.5 Concluding Remarks

Recent nontargeted approaches using omics methods provide an overview of the molecular regulation and the metabolic reorientation occurring during seed development. Proteomics data are central as they yield access to posttranscriptional regulation when compared to transcriptomics, and to corresponding enzymes activities and substrate or product contents. Looking at the data from a physiologist point of view may allow uncovering phenomena, the importance of which have been overlooked by targeted approach, enabling to raise new functional hypotheses.

In the present example of maize kernel development, the likely importance of some overlooked phenomena is revealed. During the lag phase, a large number of proteins and genes involved in protection against biotic and abiotic stresses were found, in addition to the expected proteins involved in division, cellularization, cell organization, and protein synthesis. The transition to the storage phase is marked by a maximum expression of proteases, which may be explained by resetting of the protein pool, thus enabling the reorientation of the metabolism towards the synthesis of the storage products (i.e., accumulation of starch and protein in the endosperm). However, among the proteins having their maximum expression during grain filling, the proportion of proteins related to storage metabolism was not higher than 40–48 %. Contribution of the protein folding (15–16 %) and protein synthesis (10–12 %) categories remains important. Furthermore, a meaningful important change was also observed in respiratory metabolism since the enzymes of the oxygenic pathway (Krebs cycle) become less expressed as compared to nonoxygenic enzymes (glycolysis). This is consistent with the reported deeply hypoxic environment in the endosperm at that stage because of the low permeability to gas of the kernel pericarp. Another dramatic change related to respiration and carbon metabolism is the sudden upsurge of cyPPDK, an enzyme generating a futile cycle with PK in glycolysis. The result of this cycle provides a way to regulate PEP and pyruvate levels, two key metabolites for the synthesis of aromatic amino acids and aspartate-derived amino acids, respectively. In addition, the generation of PPi by cyPPDK in endosperm cytosol could slow down the synthesis of ADP-glucose, the substrate for starch synthesis. The combined action of cyPPDK products, namely PEP and PPi, on starch and amino acid synthesis raises the hypothesis of a role of this enzyme in regulating the starch/protein ratio at the end of the rapid grain-filling period.

In order to provide support for this hypothesis, genetic tools were used to make progress towards the establishment of causative relationships. An attracting genetic observation came from reported data on the *O2* gene, a pleiotropic TF known to act on lysine content but also on the expression of *cyPPDK1*. Taking advantage of the now available high-throughput sequencing and phenotyping techniques, association genetics was performed on a 375 inbred line panel covering the maize genetic diversity. The sequence polymorphism of *cyPPDK1* and *O2* genes was analyzed in parallel to measurement of individual amino acid content and starch/protein ratio. Highly significant associations were found between the phenotypical traits and SNPs in both genes, thus validating their likely role in controlling the variation in these traits. In addition, a prominent result is that the maximum effect is observed from specific pair combinations of SNPs in the two genes, suggesting the major role of epistatic interaction between the TF *O2* and its target gene *cyPPDK1*.

References

Aoyagi K, Bassham JA (1984) Pyruvate orthophosphate dikinase of C3 seeds and leaves as compared to the enzyme from maize. Plant Physiol 75:387–392

Arruda P, Kemper EL, Papes F, Leite A (2000) Regulation of lysine catabolism in higher plants. Trends Plant Sci 5:324–330

Azevedo RA, Damerval C, Landry J, Lea PJ, Bellato CM, Meinhardt LW, Le Guilloux M, Delhaye S, Toro AA, Gaziola SA, Berdejo BDA (2003) Regulation of maize lysine metabolism and endosperm protein synthesis by opaque and floury mutations. Eur J Biochem 270:4898–4908

Brochetto-Braga MR, Leite A, Arruda P (1992) Partial purification and characterization of lysine-ketoglutarate reductase in normal and Opaque-2 maize endosperms. Plant Physiol 98:1139–1147

Cheng WH, Chourey PS (1999) Genetic evidence that invertase-mediated release of hexoses is critical for appropriate carbon partitioning and normal seed development in maize. Theor Appl Genet 98:485–495

Cliquet JB, Deléens E, Bousser A, Martin M, Lescure JC, Prioul JL, Mariotti A, Morot-Gaudry JF (1990) Estimation of carbon and nitrogen allocation during stalk elongation by 13C and 15N tracing in *Zea mays* L. Plant Physiol 92:79–87

Consoli L, Damerval C (2001) Quantification of individual zein isoforms resolved by two-dimensional electrophoresis: genetic variability in 45 maize inbred lines. Electrophoresis 22:2983–2989

Cord Neto G, Yunes JA, da Silva MJ, Vettore AL, Arruda P, Leite A (1995) The involvement of Opaque 2 on *β-prolamin* gene regulation in maize and Coix suggests a more general role for this transcriptional activator. Plant Mol Biol 27:1015–1029

Damerval C, Le Guilloux M (1998) Characterization of novel proteins affected by the o2 mutation and expressed during maize endosperm development. Mol Gen Genet 257:354–361

Gallusci P, Salamini F, Thompson RD (1994) Differences in cell type-specific expression of the gene *Opaque 2* in maize and transgenic tobacco. Mol Gen Genet 244:391–400

Gallusci P, Varotto S, Matsuoko M, Maddaloni M, Thompson RD (1996) Regulation of cytosolic pyruvate, orthophosphate dikinase expression in developing maize endosperm. Plant Mol Biol 31:45–55

Giroux MJ, Boyer C, Feix G, Hannah LC (1994) Coordinated transcriptional regulation of storage product genes in the maize endosperm. Plant Physiol 106:713–722

Habben JE, Kirleis AW, Larkins BA (1993) The origin of lysine-containing proteins in opaque-2 maize endosperm. Plant Mol Biol 23:825–838

Hannah LC (2005) Starch synthesis in the maize endosperm. Maydica 50:497–506

Hannah LC, James M (2008) The complexities of starch biosynthesis in cereal endosperms. Curr Opin Biotechnol 19:160–165

Hartings H, Maddaloni M, Lazzaroni N, Di Fonzo N, Motto M, Salamini F, Thompson R (1989) The *O2* gene which regulates zein deposition in maize endosperm encodes a protein with structural homologies to transcriptional activators. EMBO J 8:2795–2801

Imaizumi N, Ku MSB, Ishihara K, Samejima M, Kaneko S, Matsuoka M (1997) Characterization of the gene for pyruvate, orthophosphate dikinase from rice, a C3 plant, and a comparison of structure and expression between C3 and C4 genes for this protein. Plant Mol Biol 34:701–716

Landry J, Delhaye S, Damerval C (2002) Effect of the *Opaque-2* gene on accumulation of protein fractions in maize endosperm. Maydica 47:59–66

Liu X, Fu J, Gu D, Liu W, Liu T, Peng Y, Wang J, Wang G (2008) Genome-wide analysis of gene expression profiles during the kernel development of maize (*Zea mays* L.). Genomics 91: 378–387

Maddaloni M, Donini G, Balconi C, Rizzi E, Gallusci P, Forlani F, Lohmer S, Thompson R, Salamini F, Motto M (1996) The transcriptional activator Opaque-2 controls the expression of a cytosolic form of pyruvate orthophosphate dikinase-1 in maize endosperms. Mol Gen Genet 250:647–654

Manicacci D, Camus-Kulandaivelu L, Fourmann M, Arar C, Barrault S, Rousselet A, Feminias N, Consoli L, Frances L, Méchin V, Murigneux A, Prioul JL, Charcosset A, Damerval C (2009) Epistatic interactions between Opaque2 transcriptional activator and its target gene CyPPDK1 control kernel trait variation in maize. Plant Physiol 150:506–520

Méchin V, Balliau T, Château-Joubert S, Davanture M, Langella O, Négroni L, Prioul JL, Thévenot C, Zivy M, Damerval C (2004) A two-dimensional proteome map of maize endosperm. Phytochemistry 65:1609–1618

Méchin V, Thévenot C, Le Guilloux M, Prioul JL, Damerval C (2007) Developmental analysis of maize endosperm proteome suggests a pivotal role for pyruvate orthophosphate dikinase. Plant Physiol 143:1203–1219

Meyer AO, Kelly GJ, Latzko E (1982) Pyruvate orthophosphate dikinase from the immature grains of cereal grasses. Plant Physiol 69:7–10

Moose SP, Dudley JW, Rocheford TR (2004) Maize selection passes the century mark: a unique resource for 21st century genomics. Trends Plant Sci 9:358–364

Neuffer MG, Sheridan WF (1980) Defective kernel mutants of maize. I. Genetic and lethality studies. Genetics 95:929–944

Pommel B, Gallais A, Coque M, Quillere I, Hirel B, Prioul JL, Andrieu B, Floriot M (2006) Carbon and nitrogen allocation and grain filling in three maize hybrids differing in leaf senescence. Eur J Agro 24:203–211

Prioul JL (1996) Corn. In: Zamski E, Schaffer AA (eds) Photoassimilate distribution in plants and crops. Marcel Decker Inc., New York, pp 549–553

Prioul JL, Schwebel-Dugué N (1992) Source-sink manipulations and carbohydrate metabolism in maize. Crop Sci 32:751–756

Prioul JL, Jeannette E, Reyss A, Gregory N, Giroux M, Hannah LC, Causse M (1994) Expression of ADP-glucose pyrophosphorylase in maize (*Zea mays* L.) grain and source leaf during grain filling. Plant Physiol 104:179–187

Prioul JL, Méchin V, Lessard P, Thévenot C, Grimmer M, Chateau-Joubert S, Coates S, Hartings H, Kloiber-Maitz M, Murigneux A, Sarda X, Damerval C, Edwards KJ (2008a) A joint transcriptomic, proteomic and metabolic analysis of maize endosperm development and starch filling. Plant Biotechnol J 6:855–869

Prioul JL, Méchin V, Damerval C (2008b) Molecular and biochemical mechanisms in maize endosperm development: the role of pyruvate-Pi-dikinase and Opaque-2 in the control of C/N ratio. C R Biol 331:772–779

Rocher JP, Prioul JL, Lecharny A, Reyss A, Joussaume M (1989) Genetic variability in carbon fixation, sucrose-P-synthase and ADP glucose pyrophosphorylase in maize plants of differing growth rate. Plant Physiol 89:416–420

Rolletschek H, Koch K, Wobus U, Borisjuk L (2005) Positional cues for the starch/lipid balance in maize kernels and resource partitioning to the embryo. Plant J 42:69–83

Scanlon MJ, Stinard PS, James MG, Myers AM, Robertson DS (1994) Genetic analysis of 63 mutations affecting maize kernel development isolated from Mutator stocks. Genetics 136:281–294

Schmidt RJ, Burr FA, Aukerman MJ, Burr B (1990) Maize regulatory gene *opaque-2* encodes a protein with a "leucine-zipper" motif that binds to zein DNA. Proc Natl Acad Sci USA 87: 46–50

Schmidt RJ, Ketudat M, Aukerman MJ, Hoschek G (1992) Opaque-2 is a transcriptional activator that recognizes a specific target site in 22-kD zein genes. Plant Cell 4:689–700

Thornsberry JM, Goodman MM, Doebley J, Kresovich S, Nielsen D, Buckler ES (2001) Dwarf8 polymorphisms associate with variation in flowering time. Nat Genet 28:286–289

Tollenaar M, Daynard TB (1978) Effect of defoliation on kernel developmemt in maize. Can J Plant Sci 58:207–212

Verza NC, E Silva TR, Neto GC, Nogueira FT, Fisch PH, de Rosa VE Jr, Rebello MM, Vettore AL, da Silva FR, Arruda P (2005) Endosperm-preferred expression of maize genes as revealed by transcriptome-wide analysis of expressed sequence tags. Plant Mol Biol 59:363–374

Young VR, Scrimshaw NS, Pellett PL (1998) Significance of dietary protein source in human nutrition: animal and/or plant proteins? In: Waterlow JC, Armstrong DG, Fowden L, Riley R (eds) Feeding a world population of more than eight billion people. Oxford University Press, Oxford, pp 205–221

Chapter 14
Digging Deeper into the Seed Proteome: Prefractionation of Total Proteins

Ján A. Miernyk and Mark L. Johnston

Abstract Seeds are a common experimental system for many reasons. Among these: (i) they occupy a major niche in agriculture and human nutrition; (ii) they are a rich source of critical genetic information; and (iii) they are a near-ideal system for the study of phytohormone action or the transition from either dormancy or quiescence to active growth and development. One important component of all of these considerations is occurrence of the highly-abundant seed storage proteins (SSP). While on the one hand the high levels of proteins present in seeds make them attractive subjects, the SSP themselves are anathema to proteomics analyses. Without some sort of pretreatment removal of SSP, they will be virtually the only proteins identified in shotgun proteomics analyses. Here in, we describe and compare several methods commonly used to deplete samples of SSP, present the relatively recent application of combinatorial-ligand random-peptide libraries to seed proteomics studies, and speculate briefly on the short-term future.

Keywords Affinity chromarography · Dynamic range · Polyacrylamide gel electrophoresis · Proteomics · Solubility · Storage proteins

14.1 Background

Anonymous proteomics profiling is the cornerstone of contemporary systems biology (Cox and Mann 2011). The advances in high performance ion trap (IT) MS employing ESI and nanoflow LC have significantly improved the ability to analyze proteins from biological samples (Chait 2011). However, despite these advances there remain serious inherent limitations. Simply stated, the range of protein concentrations in biological samples is very large (up to 10^{12} in serum; Hortin and Sviridov 2010; Wu and Han 2006) and the dynamic range of the analytical ap-

J. A. Miernyk (✉) · M. L. Johnston
USDA, Agricultural Research Service, Plant Genetics Research Unit, University of Missouri, Columbia, MO 65211, USA
e-mail: MiernykJ@missouri.edu

J. A. Miernyk
Department of Biochemistry, University of Missouri, Columbia, MO 65211, USA

Interdisciplinary Plant Group, University of Missouri, Columbia, MO 65211, USA

G. K. Agrawal, R. Rakwal (eds.), *Seed Development: OMICS Technologies toward Improvement of Seed Quality and Crop Yield*,
DOI 10.1007/978-94-007-4749-4_14, © Springer Science+Business Media Dordrecht 2012

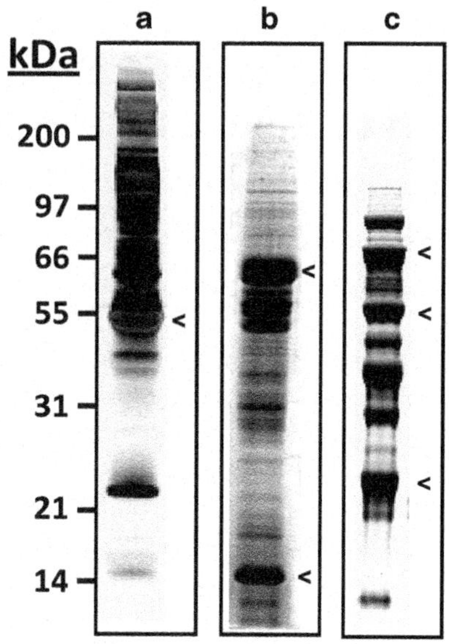

Fig. 14.1 The villains of dynamic range. Lane **a**, total human serum proteins. The position of serum albumin is indicated with a carat to the *right*. This is an image of a silver-stained gel that over represents the levels of the non-albumin proteins. Lane **b**, total soybean (*G. max*) leaf proteins. The positions of RuBisCO large and small subunits are indicated by the carats to the *right*. The image is of a Coomassie brilliant blue (CBB) stained gel. Lane **c**, total proteins from developing soybean seeds. The positions of, from top to bottom, the β-conglycinin acidic subunits, glycinin, and the β-conglycinin basic subunits are indicated by carats to the right

proaches employed is small (less than 10^3 in most cases; Vestal 2011). The only practical way to overcome the protein dynamic range problem is by adding some sort of prefractionation/depletion step to the workflow.

14.1.1 Mammalian Serum

Prefractionation of proteins prior to proteolysis, chromatography, and MS analyses will decrease the dynamic range and increase the number of confident identifications. A number of methods were evaluated for prefractionating serum samples as part of the human Plasma Proteome Project (Barnea et al. 2005). These methods include strong cation exchange (SCX) chromatography, preparative SDS-PAGE, liquid-phase IEF (LP-IEF), and serial immunoremoval. These methods have been effective in removing or greatly reducing abundant proteins, such as albumin (Fig. 14.1a), IgG, and transferrin (Ogata et al. 2005; Steel et al. 2003; Zolotarjova et al. 2005). The efficacy of the different prefractionation methods was evaluated as an increase in the number of proteins identified, and higher confidence levels and sequence coverage of the identified peptides. Prefractionation based on serial immunoremoval gave the best results, followed by SCX, preparative electrophoresis, and LP-IEF. It is important to note that each of the methods revealed a distinct set of proteins, so that determining a comprehensive overview of the serum proteome required the use of several different prefractionation approaches in parallel. It was the prefractionation research on serum that ultimately gave rise to application of this strategy to plant proteomics analyses.

14.1.2 Leaves

Proteomics analyses of plant tissues and organs are often impeded by the very high levels of a relatively small number of individual proteins. The Calvin cycle enzyme RuBisCO comprises 40–80 % of total leaf protein (Parry et al. 2003; Fig. 14.1b). Indeed, RuBisCO is considered the most abundant protein in the biosphere (Ellis 1979). Luckily, the RuBisCO primary sequence is relatively conserved and the use of an immunoremoval strategy improved analyses of the leaf proteome (Cellar et al. 2008). Subsequently a chemical-precipitation method for RuBisCO depletion has been described that is simple and inexpensive (Krishnan and Natarajan 2009). Use of either method resulted in a significant increase in the number of proteins identified.

14.1.3 Seeds

Unfortunately, the situation with seeds is comparably extreme to those of serum or leaves (Fig. 14.1c). The SSPs can comprise as much as 70 % of total protein fraction from seed storage organs, such as cotyledons or endosperm (Miernyk et al. 2011). The extremes in dynamic range of seed proteins make it difficult, if not impossible, to achieve resolution by 2-DE in amounts sufficient for identification of less-abundant proteins by MS. Reduction/depletion of the SSP from total seed protein fractions is necessary in order to reduce the dynamic range problems and effectively increase the levels of the less-abundant components. There are several different strategies by which reduction in dynamic range can be achieved.

14.2 Strategies for Addressing the Dynamic Range Problem Inherent to Seed Biology

14.2.1 Immunoremoval of Abundant Seed Proteins

Removal of supra-abundant proteins with antibodies is the most specific of all of the depletion methods, and has successfully been used to remove albumin from

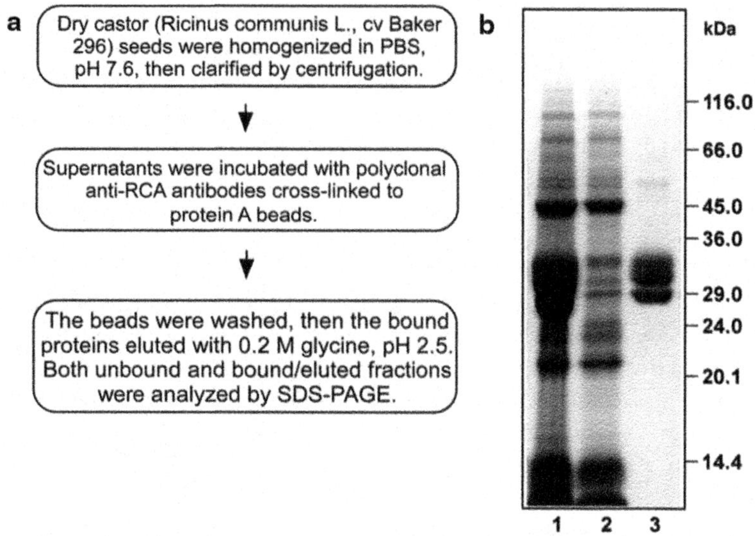

Fig. 14.2 Pretreatment of a clarified homogenate prepared from developing castor seeds by immunoremoval of the supra-abundant *Ricinus communis* agglutinin subunits. The protocol used is briefly described in panel (**a**), and the results obtained is shown in panel (**b**). Gel *lane 1*, total castor seed proteins detected by staining with CBB; *lane 2*, the non-binding fraction after incubation with the anti-RCA antibodies; and *lane 3*, proteins eluted from the immobilized anti-RCA antibodies at pH 2.5

serum (Cellar et al. 2009; Steel et al. 2003) and RuBisCO from leaf preparations (Cellar et al. 2008). The efficacy of immunoremoval depends upon the avidity of the antibodies used (Schneider et al. 1982). In general, antibodies prepared against a native antigen are more useful in immunoremoval than antibodies prepared against a denatured antigen. This can be problematic with SSP, some of which are only slightly soluble in aqueous solutions (Osborne 1908). However, at least some of the difficulties in antigen solubility can be overcome through the use of synthetic peptides (Houston et al. 2009) (Fig. 14.2).

There are several potential difficulties that must be considered before committing to an immunoremoval strategy. Antibody preparation can be expensive, and is not always efficient. Polyclonal antibodies are generally available in limited quantity. This is not a problem with monoclonal antibodies. However, storage and growth of hybridomas can be quite expensive. Finally, sometimes immunoremoval simply does not work as the result of unanticipated technical problems. For example, monoclonal antibodies against the ricin castor seed lectin (Brandon and Hernlem 2009) gave excellent results when used to detect the protein on immunoblots but surprisingly were inefficient at binding and removal of ricin from aqueous solutions (MLJ unpublished data). Nonetheless, the typical selectivity and efficiency of immunoremoval make it an excellent first choice when designing an experimental strategy.

14.2.2 Depletion of Abundant Seed Glycoproteins

Many abundant SSP are *N*-glycosylated (Doyle et al. 1986; Lauer et al. 2004; Lerouge et al. 1998; Maruyama et al. 1998; Wilson 2002), making lectin-affinity chromatography a potentially useful removal method in this context (Hirabayashi 2004; Van Damme 2011). Concanavalin A (con A), which recognizes and binds high-mannose type glycans (those with terminal α-D-mannopyranoside or α-D-glucopyranoside residues), is the best understood and most commonly used lectin (Rüdiger and Gabius 2001). Many additional lectins are available with glycan-specificities that encompass most of the typical complex-type glycans (those with terminal galactopyranoside residues, branching xylose, or fucose residues).

In soybean seeds, the abundant SSPs are the 7S (β-conglycinin) and 11S (glycinin) globulins (Krishnan 2000; Miernyk and Hajduch 2011). Together, these two fractions comprise up to 70 % of total seed protein. The β-conglycinin SSPs are known to be *N*-glycosylated (Maruyama et al. 1998) and, at least, one glycan is of the high-mannose type, which should allow the use of con A in a lectin-based affinity-depletion step. Our initial attempts at using con A-Sepharose for glycoprotein-depletion resulted in the appearance of an abundant band at approximately Mr 26,000 in all fractions (Miernyk and Johnston 2006). This is close to the deduced molecular weight of the con A monomer (22.5 kDa). At pH values above 5.6, con A is a homo-tetramer of 22.5 kDa subunits (Lis and Sharon 1973), which could dissociate in the urea-containing buffer systems commonly used in electrophoresis-based proteomic analyses. The Mr 26,000 protein band was subsequently verified as con A by MALDI-TOF-MS peptide mass fingerprint analysis (Miernyk and Johnston 2006).

Using a combination of dimethyl pimelimidate and dimethyl suberimidate, we were able to chemically cross-link the subunits of the con A-immobilized on Sepharose 4B. After cross-linking, the immobilized lectin appeared to remain fully active, and there was no evidence of the con A subunit bleeding from the column even in buffer containing 4 M urea. Use of the cross-linked con A-Sepharose effectively reduced the amount of the β-conglycinin storage proteins (Fig. 14.3). To validate the improvement in 2-D spot definition resulting from glycoprotein-depletion, six proteins were selected for MS analysis. Prior to con A-Sepharose glycoprotein affinity-depletion, the selected protein spots were faintly visible, and no ions useful for identification by MALDI-TOF PMF were obtained. After glycoprotein depletion, each was identified with high confidence (Miernyk and Johnston 2006). Mild chemical cross-linking of proteins has potential for application to proteomics analysis tools beyond immobilized-con A (e.g., it is additionally possible to chemically cross-link antibodies; MLJ unpublished data).

Con A-affinity chromatography removes all high-mannose type glycoproteins from a complex mixture, including low-abundance non-SSP proteins. These can be eluted from the lectin with the hapten α-methyl-mannoside (Lis and Sharon 1973), separated from the β-conglycinin by electrophoresis, digested, and analyzed by LC-MS.

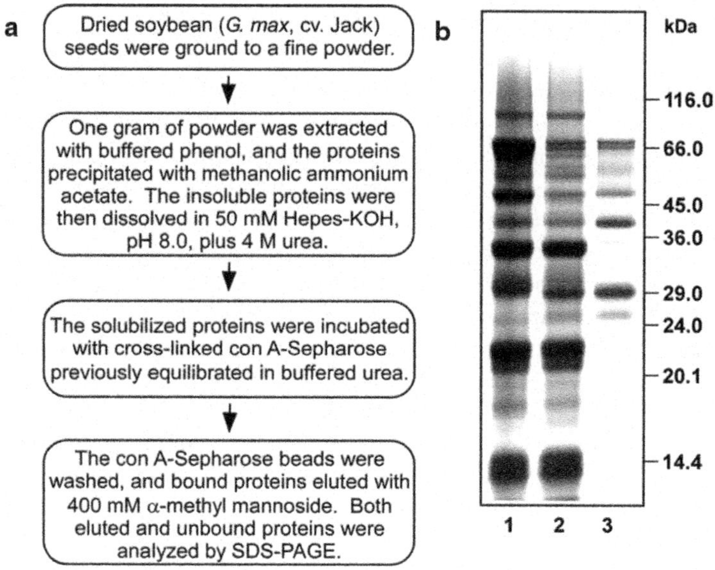

Fig. 14.3 Pretreatment of a clarified homogenate prepared from developing soybean seeds by immobilized-lectin affinity chromatography. Cross-linked con A-Sepharose was used to selectively remove abundant high-mannose type glycoproteins, including the abundant β-conglycinin SSP. The protocol used is briefly described in panel (**a**), and the results obtained is shown in panel (**b**). Gel *lane 1*, total soybean seed proteins detected by staining with CBB; *lane 2*, the non-binding fraction after incubation with the con A-Sepharose beads; and *lane 3*, proteins eluted from the lectin beads with 400 mM α-methyl-mannoside

14.2.3 Liquid-Phase Isoelectric Focusing

The first practical, usable, and commercially available device for LP-IEF (Bier 1998) was dubbed the Rotofor. These devices are available for separations on three different scales (preparative, as much as 1 g of protein in a total volume of up to 60 mL; "Mini," up to 18 mL; and "Micro", up to 2.5 mL). A typical instrumental setup is assembled from 20 sample chambers, separated by liquid-permeable nylon screens. Upon completion of a separation, the 20 focused fractions are collected simultaneously by piercing a septum at the bottom of each chambers with 20 needles connected to a vacuum source. The proteins in each of the individual, narrow-pH range fractions can then be further separated by SDS-PAGE (Hochstrasser et al. 1991). This method has been used extensively in both shotgun proteomics analyses (Righetti et al. 2003) and in individual, specific applications (Ayala et al. 1998; Petrash et al. 1991; Vincent et al. 2009). Separation of *Glycine max* seed proteins can be seen in Fig. 14.4.

More recently, Agilent Technologies has marketed a LP-IEF system called the 3100 OFFGEL fractionator. In this system, the pH gradient results from use of

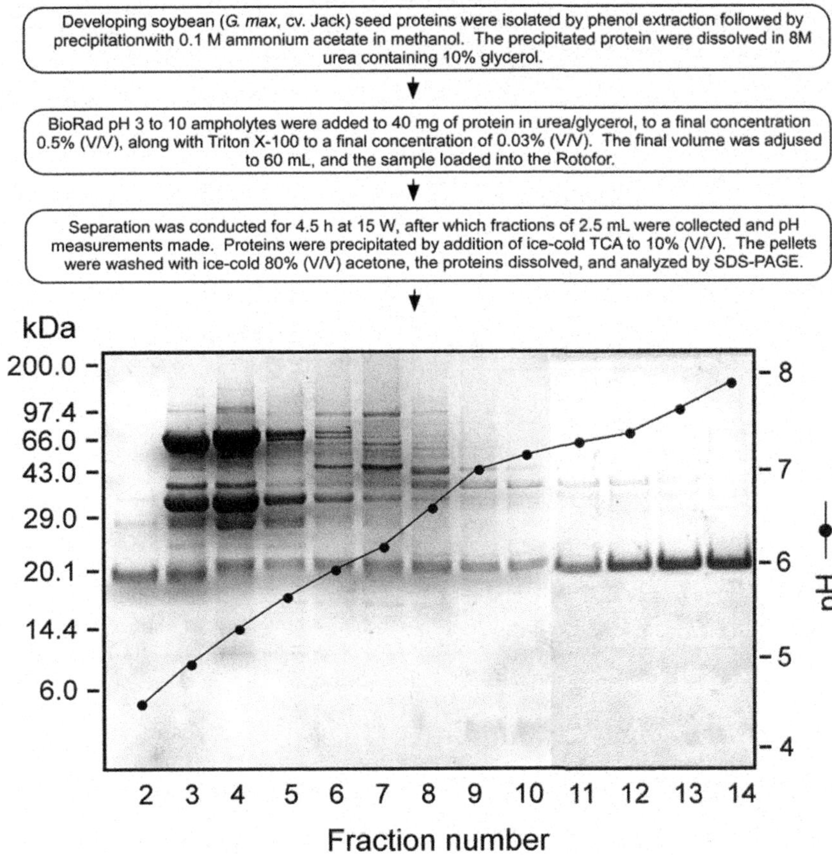

Fig. 14.4 Fractionation of proteins from developing soybean seeds by Rotofor preparative liquid phase isoelectric focusing (IEF). The protocol used is briefly described in the upper panel. The results obtained is shown in the lower panel. The supra-abundant β-conglycinin and glycinin SSP are concentrated in the low *pI* fractions three and four

an immobilized pH gradient strip, rather than pH limit ampholytes. The OFFGEL fractionator apparently works equally well with proteins and peptides, allowing flexibility with regard to when the proteolytic digestion is done (Chenau et al. 2008; Hörth et al. 2006). The Rotofor and OFFGEL, and similar devices, such as the ZOOM IEF Fractionator (Richardson et al. 2008), allow the possibility of 3- or even 4-D experimental separations with exceptional resolution. Typically these separations would include LP-IEF, gel electrophoresis, or RP-HPLC, and then nano-flow LC-MS analysis (Elschenbroich et al. 2009; Scruggs et al. 2010; Thorsell et al. 2007; Wagner 2011). These strategies have seldom been applied to depletion of SSP but, as can be seen in Fig. 14.4, they can be effective and deserve more attention.

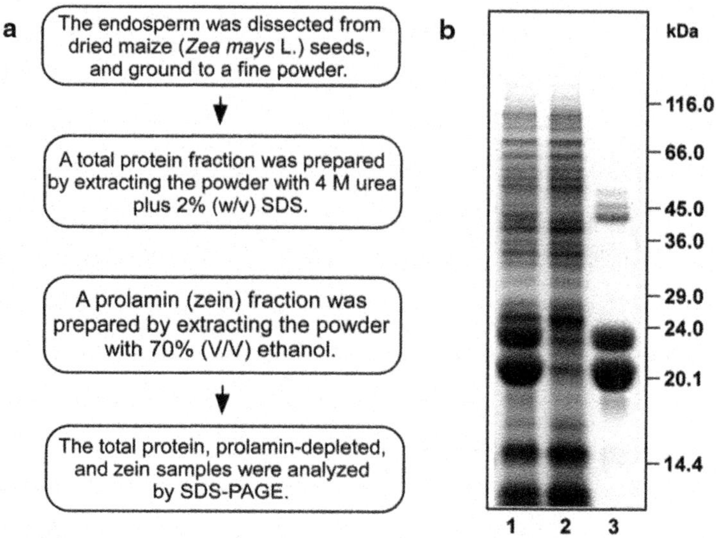

Fig. 14.5 Prefractionation of maize seed proteins by differential solubility. The protocol used is briefly described in panel (**a**). The results obtained is shown in panel (**b**). Gel *lane 1*, total maize seed proteins detected by staining with CBB; *lane 2*, maize proteins after extraction with 70 % ethanol; and *lane 3*, the zein seed storage prolamins extracted with 70 % ethanol

14.2.4 *Differential Solubility*

The first systematic classification system for proteins was developed by Osborne (1908), and it remains in use today (Miernyk and Hajduch 2011; Miernyk et al. 2011). The SSP are grouped on the basis of solubility in H_2O (albumins), dilute saline solutions (globulins), alcohol:water mixtures (prolamins), and dilute alkali or acid (glutelins). The majority of agriculturally important SSP are albumins, globulins, or prolamins. While albumins are found in all seeds, prolamins and glutelins are most abundant in monocotyledon seeds and globulins are prevalent in dicotyledon seeds (Krishnan 2000; Shewry et al. 1995).

In principle, it should be possible to deplete a sample of SSP by either differential solubility or isoelectric precipitation. In fact, these simple strategies seldom work satisfactorily. For example, when a total protein fraction prepared from developing soybean seeds was dialyzed exhaustively against deionized water, it should have yielded a globulin-depleted albumin fraction and an insoluble globulin fraction. Both fractions, however, were heavily cross-contaminated. Results were better with rice seed glutelins, and best with maize prolamins (Fig. 14.5). Extraction of dry, mature maize endosperm did yield a prolamin (zein)-enriched fraction, but some of the zeins remained associated with the ethanol-insoluble material. This was probably because of oxidative cross-linking of the zein subunits (Field et al. 1983), and might have been improved by including 2 % (v/v) 2-mercaptoethanol (2-ME) in the 70 % ethanol solvent.

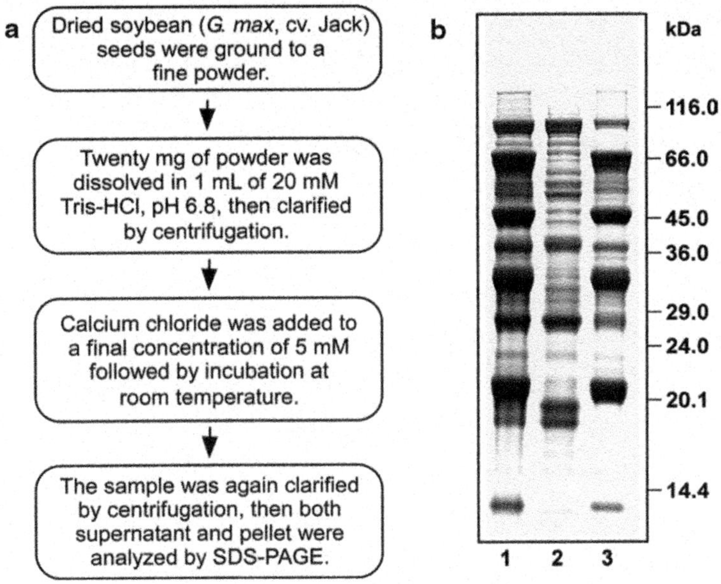

Fig. 14.6 Prefractionation of a clarified homogenate prepared from developing soybean seeds by calcium-mediated precipitation. The protocol used is briefly described in panel (**a**). The results obtained is shown in panel (**b**). Gel *lane 1*, total soybean seed proteins detected by staining with CBB; *lane 2*, proteins remaining soluble after treatment with 5 mM $CaCl_2$; and *lane 3*, proteins precipitated by $CaCl_2$ treatment

Krishnan et al. (2009) adopted a somewhat different strategy for depleting total soybean seed proteins of the very abundant 7S and 11S globulins. They found that incubation of a total protein fraction with 10 mM Ca^{2+} led to precipitation of the seed globulins glycinin and β-conglycinin (Fig. 14.6). This simple and inexpensive method removed 87±4 % of the SSP from the sample and allowed identification of 541 previously inconspicuous proteins. In preliminary experiments, this method appeared to work equally well with seeds from other legumes (e.g., peanut, bean, pea, and alfalfa). While it would be difficult to improve the ease or cost of the calcium-precipitation strategy, it does not always work perfectly and might be more productively combined with one of the other methods described herein.

14.2.5 Combinatorial-Ligand Random-Peptide Beads

Amino acids in a protein primary sequence that are complementary to an immobilized ligand allow that protein to be removed from a complex mixture, up to the point of saturation of the ligand (Righetti et al. 2007). With sufficient diversity, it is theoretically possible to have an immobilized ligand complementary to each protein in a complex mixture, ensuring that they would all be adsorbed. When a

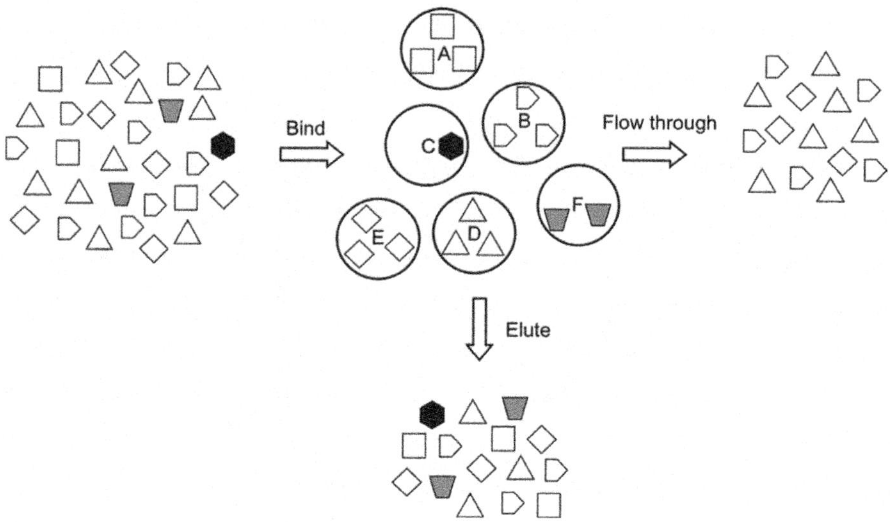

Fig. 14.7 A schematic representation of the use of combinatorial-ligand random-peptide beads to reduce sample protein dynamic range. Each bead (*A–F*) is decorated with a unique linear hexapeptide that interacts with a single protein species. A portion of the proteins species that are in excess of bead capacity will flow through. All of the proteins from the input sample are represented in the eluted fraction, but the overall dynamic range has been reduced. Adapted from Boschetti and Righetti 2009

biological sample is incubated with such a ligand-library under capacity-restrained conditions, abundant proteins will saturate all available high-affinity ligands and the remaining non-binding majority of the protein will remain in solution (Boschetti and Righetti 2009). In contrast, a low-abundance protein will not saturate the corresponding high-affinity ligand, and most of this protein will be removed from solution (Fig. 14.7). Based on this saturation-overload principle, use of a combinatorial library should enrich for low-abundance proteins relative to those of high abundance (Righetti et al. 2007). Elution of the entire population of proteins adsorbed to the beads should result in a solution with a smaller dynamic range than the starting material but still including representatives of all of the original proteins.

Results from preliminary experiments indicated that hexapeptides were the best length to capture the largest population of proteins (Boschetti and Righetti 2008a). Libraries of hexapeptide ligands can be synthesized on resin beads. Each bead is covered with millions of copies of a unique-sequence hexapeptide, and each bead potentially has a hexapeptide of a different sequence. Using the 20 naturally occurring amino acids, it is feasible to prepare a library of linear hexapeptides with 64 million different sequences (Boschetti and Righetti 2008a). Relatively recently, combinatorial-ligand libraries composed of millions of hexapeptides have been use to "amplify" the low-abundance proteins that comprise ~50 % of any proteome (Righetti et al. 2010). An example is the ability to explore the low-abundance proteome in chicken egg white and yolk. Using the peptide library method, it was possible to detect at least twice as many protein species compared to the best results obtained with un-fractionated samples (Righetti et al. 2007, 2010).

The combinatorial-peptide beads, in the format of the ProteoMiner kit (Boschetti and Righetti 2008b), have been productively used in analyses of *Hevea brasiliensis* latex (Boschetti et al. 2009) and the cytoplasmic proteome from maize (Boschetti et al. 2009), and more recently spinach leaves (Fasoli et al. 2011). At the time of writing, there were no publications describing the use of ProteoMiner for analysis of seed proteins. Our attempts to use ProteoMiner kit, under the conditions developed for spinach leaves, to either deplete samples from developing soybeans of the abundant SSP or to prepare samples that yielded an increased number of protein ID's have been without success to date (MLJ unpublished data). The method is known to be sensitive to pH and solvent characteristics, and will likely have to be customized for individual applications. Furthermore, we speculate that the often unusual solubility of SSP might be problematic in terms of nonspecific adsorption. The possible use of a two-stage strategy might overcome these problems, e.g., starting with a calcium-precipitation step followed by use of the combinatorial-peptide beads.

14.3 Concluding Remarks

The short-term future of seed proteomics will feature more and better protein identifications. This goal will be achieved through combined advances in instrumentation (Chait 2011), bioinformatic analyses and methods for protein identification (Bandeira et al. 2011), improved databases (e.g., phytozome.net/), and, of course, improvements in the methods used to deplete the supra-abundant SSPs from input samples. In the latter case, the strategy of using combinatorial-ligand random-peptide beads has substantial potential. One obvious improvement would be to remove the "random" component. In part because of their abundance, genes and cDNAs for SSPs were among the first sequenced at the beginning of the genomics era (Miernyk and Hajduch 2011). Because the sequences of many SSP are extant (Esen 1990; Feeney et al. 2003; Gibbs et al. 1989), it should be relatively simple to prepare "designer sequence"-beads that would have both the capacity and specificity to efficiently remove these supra-abundant components of the seed proteome. This strategy should additionally be amenable to incorporation into high-throughput experimental designs.

References

Ayala A, Parrado J, Machado A (1998) Use of Rotofor preparative isoelectrofocusing cell in protein purification procedure. Appl Biochem Biotechnol 69:11–16

Bandeira N, Nesvizhskii A, McIntosh M (2011) Advancing next-generation proteomics through computational research. J Proteome Res 10:2895–2895

Barnea E, Sorkin R, Ziv T, Beer I, Admon A (2005) Evaluation of prefractionation methods as a preparatory step for multidimensional based chromatography of serum proteins. Proteomics 5:3367–3375
Bier M (1998) Recycling isoelectric focusing and isotachophoresis. Electrophoresis 19:1057–1063
Boschetti E, Righetti PG (2008a) Hexapeptide combinatorial ligand libraries: the march for the detection of the low-abundance proteome continues. BioTechniques 44:663–665
Boschetti E, Righetti PG (2008b) The ProteoMiner in the proteomic arena: a non-depleting tool for discovering low-abundance species. J Proteomics 71:255–264
Boschetti E, Righetti PG (2009) The art of observing rare protein species in proteomes with peptide ligand libraries. Proteomics 9:1492–1510
Boschetti E, Bindschedler LV, Tang C, Fasoli E, Righetti PG (2009) Combinatorial peptide ligand libraries and plant proteomics: a winning strategy at a price. J Chromatogr A 1216:1215–1222
Brandon DL, Hernlem BJ (2009) Development of monoclonal antibodies specific for *Ricinus* agglutinins. Food Agric Immunol 20:11–22
Cellar NA, Kuppannan K, Langhorst ML, Ni W, Xu P, Young SA (2008) Cross species applicability of abundant protein depletion columns for ribulose-1,5-bisphosphate carboxylase/oxygenase. J Chromatogr B Analyt Technol Biomed Life Sci 861:29–39
Cellar NA, Karnoup AS, Albers DR, Langhorst ML, Young SA (2009) Immunodepletion of high abundance proteins coupled on-line with reversed-phase liquid chromatography: a two-dimensional LC sample enrichment and fractionation technique for mammalian proteomics. J Chromatogr B Analyt Technol Biomed Life Sci 877:79–85
Chait BT (2011) Mass spectrometry in the postgenomic era. Annu Rev Biochem 80:239–246
Chenau J, Michelland S, Sidibe J, Seve M (2008) Peptides OFFGEL electrophoresis: a suitable pre-analytical step for complex eukaryotic samples fractionation compatible with quantitative iTRAQ labeling. Proteome Sci 6:9
Cox J, Mann M (2011) Quantitative, high-resolution proteomics for data-driven systems biology. Annu Rev Biochem 80:273–299
Doyle JJ, Schuler MA, Godette WD, Zenger V, Beachy RN, Slightom JL (1986) The glycosylated seed storage proteins of *Glycine max* and *Phaseolus vulgaris*. Structural homologies of genes and proteins. J Biol Chem 261:9228–9238
Ellis RJ (1979) The most abundant protein in the world. Trends Biochem Sci 4:241–244
Elschenbroich S, Ignatchenko V, Sharma P, Schmitt-Ulms G, Gramolini AO, Kislinger T (2009) Peptide separations by on-line MudPIT compared to isoelectric focusing in an off-gel format: application to a membrane-enriched fraction from C2C12 mouse skeletal muscle cells. J Proteome Res 8:4860–4869
Esen A (1990) An immunodominant site of gamma-zein1 is in the region of tandem hexapeptide repeats. J Protein Chem 9:453–460
Fasoli E, D'Amato A, Kravchuk AV, Boschetti E, Bachi A, Righetti PG (2011) Popeye strikes again: the deep proteome of spinach leaves. J Proteomics 74:127–136
Feeney KA, Wellner N, Gilbert SM, Halford NG, Tatham AS, Shewry PR, Belton PS (2003) Molecular structures and interactions of repetitive peptides based on wheat glutenin subunits depend on chain length. Biopolymers 72:123–131
Field JM, Shewry PR, Miflin BJ (1983) Aggregation states of alcohol-soluble storage proteins of barley, rye, wheat and maize. J Sci Food Agric 34:362–369
Gibbs PE, Strongin KB, McPherson A (1989) Evolution of legume seed storage proteins—a domain common to legumins and vicilins is duplicated in vicilins. Mol Biol Evol 6:614–623
Hirabayashi J (2004) Lectin-based structural glycomics: glycoproteomics and glycan profiling. Glycoconj J 21:35–40
Hochstrasser AC, James RW, Pometta D, Hochstrasser D (1991) Preparative isoelectrofocusing and high resolution 2-dimensional gel electrophoresis for concentration and purification of proteins. Appl Theor Electrophor 1:333–337
Hortin GL, Sviridov D (2010) The dynamic range problem in the analysis of the plasma proteome. J Proteomics 73:629–636

Houston NL, Hajduch M, Thelen JJ (2009) Quantitative proteomics of seed filling in castor: comparison with soybean and rapeseed reveals differences between photosynthetic and nonphotosynthetic seed metabolism. Plant Physiol 151:857–868
Hörth P, Miller CA, Preckel T, Wenz C (2006) Efficient fractionation and improved protein identification by peptide OFFGEL electrophoresis. Mol Cell Proteomics 5:1968–1974
Krishnan HB, Natarajan SS (2009) A rapid method for depletion of rubisco from soybean (*Glycine max*) leaf for proteomic analysis of lower abundance proteins. Phytochemistry 70:1958–1964
Krishnan HB, Oehrle NW, Natarajan SS (2009) A rapid and simple procedure for the depletion of abundant storage proteins from legume seeds to advance proteome analysis: a case study using *Glycine max*. Proteomics 9:3174–3188
Krishnan HB (2000) Biochemistry and molecular biology of soybean seed storage proteins. J New Seeds 2:1–25
Lauer I, Foetisch K, Kolarich D, Ballmer-Weber BK, Conti A, Altmann F, Vieths S, Scheurer S (2004) Hazelnut (*Corylus avellana*) vicilin Cor a11: molecular characterization of a glycoprotein and its allergenic activity. Biochem J 383:327–334
Lerouge P, Cabanes-Macheteau M, Rayon C, Fischette-Lainé AC, Gomord V, Faye L. (1998) N-glycoprotein biosynthesis in plants: recent developments and future trends. Plant Mol Biol 38:31–48
Lis H, Sharon N (1973) The biochemistry of plant lectins (phytohemagglutinins). Annu Rev Biochem 42:541–574
Maruyama N, Katsube T, Wada Y, Oh MH, Barba De La Rosa AP, Okuda E, Nakagawa S, Utsumi S (1998) The roles of the N-linked glycans and extension regions of soybean beta-conglycinin in folding, assembly and structural features. Eur J Biochem 258:854–862
Miernyk JA, Hajduch M (2011) Seed proteomics. J Proteomics 74:389–400
Miernyk JA, Johnston ML (2006) Chemical cross-linking immobilized concanavalin A for use in proteomic analyses. Prep Biochem Biotechnol 36:203–214
Miernyk JA, Preťová A, Olmedilla A, Klubicova K, Obert B, Hajduch M (2011) Using proteomics to study sexual reproduction in angiosperms. Sexual Plant Reprod 24:9–22
Ogata Y, Charlesworth MC, Muddiman DC (2005) Evaluation of protein depletion methods for the analysis of total-, phospho- and glycoproteins in lumbar cerebrospinal fluid. J Proteome Res 4:837–845
Osborne TB (1908) Our present knowledge of plant proteins. Science 28:417–427
Parry MA, Andralojc PJ, Mitchell RA, Madgwick PJ, Keys AJ (2003) Manipulation of rubisco: the amount, activity, function and regulation. J Exp Bot 54:1321–1333
Petrash JM, DeLucas LJ, Bowling E, Egen N (1991) Resolving isoforms of aldose reductase by preparative isoelectric focusing in the Rotofor. Electrophoresis 12:84–90
Richardson MR, Liu S, Ringham HN, Chan V, Witzmann FA (2008) Sample complexity reduction for two-dimensional electrophoresis using solution isoelectric focusing prefractionation. Electrophoresis 29:2637–2644
Righetti PG, Boschetti E, Monsarrat B (2007) The "invisible proteome": how to capture the low abundance proteins via combinatorial ligand libraries. Curr Proteomics 4:198–208
Righetti PG, Boschetti E, Zanella A, Fasoli E, Citterio A (2010) Plucking, pillaging and plundering proteomes with combinatorial peptide ligand libraries. J Chromatogr A 1217:893–900
Righetti PG, Castagna A, Herbert B, Reymond F, Rossier JS (2003) Prefractionation techniques in proteome analysis. Proteomics 3:1397–1407
Rüdiger H, Gabius HJ (2001) Plant lectins: occurrence, biochemistry, functions and applications. Glycoconj J 18:589–613
Schneider C, Newman RA, Sutherland DR, Asser U, Greaves MF (1982) A one-step purification of membrane proteins using a high efficiency immunomatrix. J Biol Chem 257:10766–10769
Scruggs SB, Reisdorph R, Armstrong ML, Warren CM, Reisdorph N, Solaro RJ, Buttrick PM (2010) A novel, in-solution separation of endogenous cardiac sarcomeric proteins and identification of distinct charged variants of regulatory light chain. Mol Cell Proteomics 9:1804–1818
Shewry PR, Napier JA, Tatham AS (1995) Seed storage proteins: structures and biosynthesis. Plant Cell 7:945–956

Steel LF, Trotter MG, Nakajima PB, Mattu TS, Gonye G, Block T (2003) Efficient and specific removal of albumin from human serum samples. Mol Cell Proteomics 2:262–270

Thorsell A, Portelius E, Blennow K, Westman-Brinkmalm A (2007) Evaluation of sample fractionation using micro-scale liquid-phase isoelectric focusing on mass spectrometric identification and quantitation of proteins in a SILAC experiment. Rapid Commun Mass Spectrom 21;771–778

Van Damme EJ (2011) Lectins as tools to select for glycosylated proteins. Methods Mol Biol 753:289–297

Vestal ML (2011) The future of biological mass spectrometry. J Am Soc Mass Spectrom 22:953–959

Vincent D, Balesdent MH, Gibon J, Claverol S, Lapaillerie D, Lomenech AM, Blaise F, Rouxel T, Martin F, Bonneu M, Amselem J, Dominguez V, Howlett BJ, Wincker P, Joets J, Lebrun MH, Plomion C (2009) Hunting down fungal secretomes using liquid-phase IEF prior to high resolution 2-DE. Electrophoresis 32:4118–4136

Wagner L, Wermann M, Rosche F, Rahfeld JU, Hoffmann T, Demuth HU (2011) Isolation of dipeptidyl peptidase IV (DP 4) isoforms from porcine kidney by preparative isoelectric focusing to improve crystallization. Biol Chem 392:665–677

Wilson IB (2002) Glycosylation of proteins in plants and invertebrates. Curr Opin Struct Biol 12:569–577

Wu L, Han DK (2006) Overcoming the dynamic range problem in mass spectrometry-based shotgun proteomics. Expert Rev Proteomics 3:611–619

Zolotarjova N, Martosella J, Nicol G, Bailey J, Boyes BE, Barrett WC (2005) Differences among techniques for high-abundant protein depletion. Proteomics 5:3304–3313

Chapter 15
The Central Role of Phosphoenolpyruvate Metabolism in Developing Oilseeds

William C. Plaxton and Brendan O'Leary

Abstract Metabolite interconversion at the phosphoenolpyruvate (PEP)-pyruvate-oxaloacetate node involves a complex set of cytosolic and plastidial reactions that interconnect the major pathways of carbohydrate metabolism, thereby making a crucial contribution to the distribution of carbon flux among catabolism, anabolism, and ATP and NAD(P)H supply to plant cells. Enzymes involved in plant PEP metabolism catalyze a diverse array of reactions, including a major metabolic branchpoint between primary and secondary (shikimate pathway) metabolism. Carbon partitioning at the PEP branchpoint is complicated by an intricate network of posttranslational enzyme controls, including allosteric effectors and protein kinase mediated phosphorylation. Experiments on transgenic or mutant plants possessing altered amounts of PEP metabolizing enzymes or transporters are enhancing our understanding of the functional organization and control of oilseed PEP metabolism. Such experiments illustrate the highly flexible nature of plant PEP metabolism and the crucial biosynthetic function played by glycolysis and respiration beyond their role in catabolic ATP generation. This chapter summarizes what is known about the key PEP metabolizing enzymes and corresponding metabolic fluxes during the reserve deposition stage of oilseed development. PK and PEPC are of particular interest since they play an essential role in controlling the provision of: (i) pyruvate for mitochondrial ATP production *via* oxidative phosphorylation; (ii) tricarboxylic acid cycle intermediates needed for nitrogen assimilation and amino acid biosynthesis; and (iii) precursors and cofactors (e.g., pyruvate, acetyl-CoA, malate, ATP, and NAD(P)H) needed for plastidial fatty acid synthesis. Novel insights into the functions, and molecular and regulatory characteristics of oilseed PK and PEPC isozymes have arisen through their purification and detailed biochemical and molecular characterization, as well as advances in functional genomics, proteomics, and metabolic flux analysis. PK and PEPC are becoming important targets for metabolic engineering of the PEP branchpoint to modify levels of agronomically important end products, such as storage proteins and lipids in oilseeds.

Keywords Metabolic control · Oil seed metabolism · Phosphoenolpyruvate carboxylase · Posttranslational modifications · Protein-protein interactions · Pyruvate kinase

---

W. C. Plaxton (✉) · B. O'Leary
Department of Biology, Queen's University, K7L 3N6, Kingston, ON, Canada
e-mail: plaxton@queensu.ca

G. K. Agrawal, R. Rakwal (eds.), *Seed Development: OMICS Technologies toward Improvement of Seed Quality and Crop Yield*,
DOI 10.1007/978-94-007-4749-4_15,

15.1 Introduction

In developing seeds, the partitioning of imported photosynthate between starch, storage lipid, and SP biosynthesis is of major agronomic concern. Metabolism of the glycolytic intermediate PEP (Fig. 15.1a) plays a central role in controlling the flux of seed carbohydrates towards plastidic FA biosynthesis *versus* the mitochondrial production of ATP and organic acids needed for amino acid interconversion in support of SP biosynthesis (Baud and Lepiniec 2010; Rawsthorne 2002; Weber et al. 2005). PEP's universal metabolic importance largely stems from the fact that it occupies the highest position on the thermodynamic scale of known phosphorylated metabolites (Davies 1979). The equilibrium position between the keto and enol forms of pyruvate greatly favors the keto form. The enol form is at a much higher energy level, and thus PEP is referred to as a 'high energy' compound owing to its stabilization of the enol form. PEP is a versatile metabolite that: (i) participates in a wide range of reactions by enzymatic cleavage on either side of its enol oxygen atom; C-O cleavage producing an active pyruvyl group and O-P cleavage producing an active phosphoryl group (Fig. 15.1a); (ii) is situated at a major metabolic branchpoint leading into a variety of primary and secondary metabolism pathways (Fig. 15.1b); and (iii) is a potent allosteric effector of several cytosolic and plastidic enzymes (Fig. 15.2a) (Davies 1979; Plaxton and Podestá 2006). A comprehensive analysis of PEP metabolism will greatly benefit our overall understanding of the control and integration of photosynthate partitioning during seed filling.

The major storage compounds accumulated in developing oilseeds are oil (TAGs) and SPs, derived from imported sugars and amino acids. However, attempts to engineer seed oils have met with limited success. In order to improve oil yield and quality, a more detailed understanding of the underlying metabolism during oilseed development is needed, as well as the degree to which cytosolic, mitochondrial, and plastidial metabolic pathways are integrated and regulated to control the flow of carbon from maternal sources into seed storage products (Baud and Lepiniec 2010). In plants, FA synthesis occurs predominantly in the plastids and requires carbon in the form of acetyl-CoA, ATP, and reducing power in the form of NADH and NADPH (Fig. 15.2b) (Rawsthorne 2002). Precursors for acetyl-CoA synthesis must be generated within the plastid or imported from the cytosol. In plastids of green oilseeds, such as *Brassica napus* (also called oilseed rape, rapeseed, or canola), the photosynthetic electron transport chain makes an important contribution to the generation of plastidic ATP and NADPH (Ruuska et al. 2004; Schwender et al. 2004). By contrast, leucoplasts from heterotrophic nongreen oilseeds, such as castor endosperm or sunflower and maize embryo's, must generate these compounds without relying on a photosynthetic electron transport chain. In both green and nongreen developing oilseeds, however, PEP metabolism *via* PEPC, and the cytosolic and plastidic isozymes of PK_c and PK_p, respectively, makes a crucial contribution to the control of photosynthate partitioning towards leucoplast FA biosynthesis *versus* the mitochondrial production of C-skeletons and ATP needed for amino acid interconversion in support of SP biosynthesis (Baud and Lepiniec 2010; O'Leary et al.

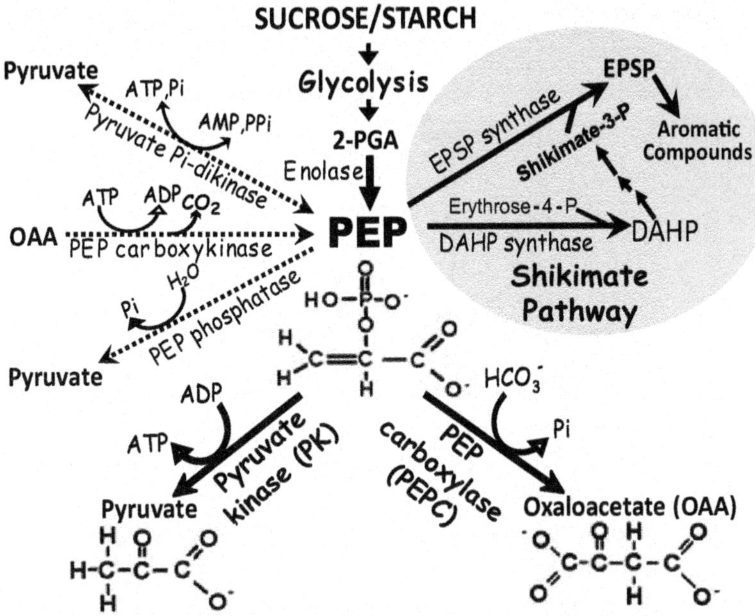

Fig. 15.1 The central role of PEP in plant metabolism. At least eight different enzymes are involved in plant PEP metabolism. This includes a major metabolic branch point between primary and secondary (shikimate) metabolism. *2-PGA* 2-P-glycerate, *DAHP* 3-deoxy-D-arabino-heptulosonate-7-P, *EPSP* 5-enolpyruvyl shikimate-3-P

2011b; Rawsthorne 2002). A number of reviews concerning various aspects of the organization and control of photosynthate partitioning to storage end products by developing seeds have appeared over the past decade (Baud and Lepiniec 2010; Borisjuk et al. 2004; Gallardo et al. 2008; Hills 2004; Rawsthorne 2002; Santos-Mendoza et al. 2008; Weber et al. 2005; Weselake et al. 2009). The aim of this chapter is to provide an overview of the central role played by PEP metabolism in supporting storage oil and protein biosynthesis during the reserve deposition phase of oilseed development. Our focus is on the molecular and biochemical properties, and posttranslational controls of the key regulatory enzymes that coordinate flux through the PEP branch point to storage end products, namely PK_c, PK_p, and PEPC.

15.2 PK and PEPC Exert a Major Influence on the Metabolic Control of Oilseed Glycolysis and Respiration

In nonplant systems such as mammalian liver, primary control of glycolytic flux of hexose-phosphates to pyruvate is mediated by ATP-dependent phosphofructokinase (ATP-PFK), with secondary control at PK (Plaxton 1996; Plaxton and Podestá 2006). Activation of ATP-PFK increases the level of its product, fructose-1,6-P_2,

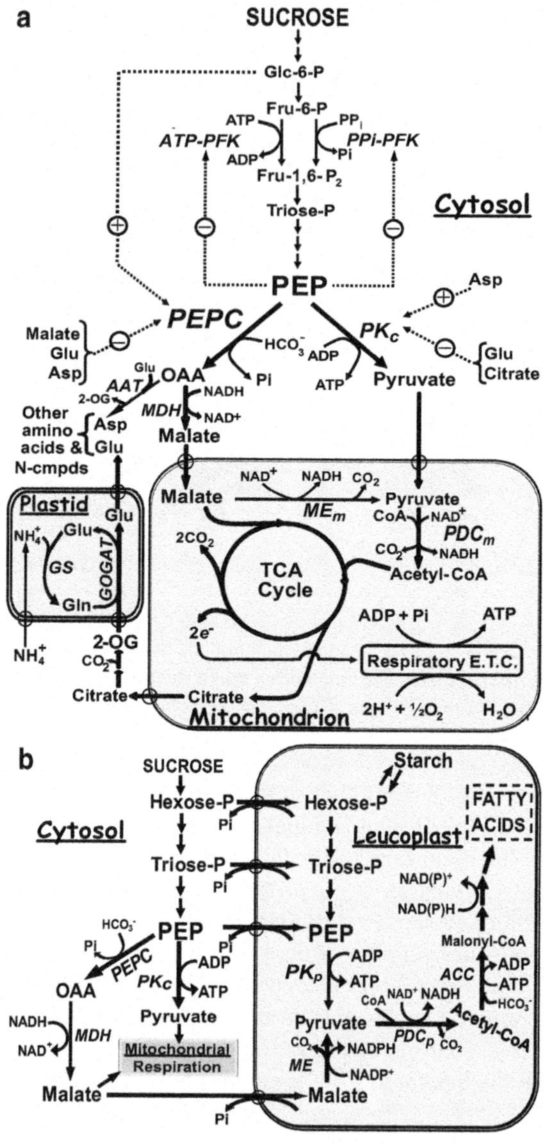

Fig. 15.2 Glycolytic control and PEP partitioning in primary plant metabolism is largely mediated by PK and PEPC. **a** This scheme highlights the essential role of the two terminal enzymes of plant cytosolic glycolysis, PK_c and PEPC, in controlling the provision of mitochondria with respiratory substrates (pyruvate and malate), as well as for generating the 2-OG and OAA respectively required for NH_4^+-assimilation *via* GS/GOGAT in plastids and Asp aminotransferase in the cytosol. The coordinate control of PEPC and PK_c by allosteric effectors, particularly Glu and Asp, provides a mechanism for the regulation of cytosolic glycolytic flux and PEP partitioning during and following NH_4^+-assimilation as discussed in the text. PEP's role in providing 'bottom up' allosteric control of glycolysis is also illustrated. *Dotted arrows* with a *circled plus* and *minus* sign indicate enzyme activation and inhibition, respectively, by allosteric effectors. **b** Alternative

which is a potent feed-forward allosteric activator of the majority of nonplant PKs examined to date. By contrast, a wide variety of studies in vascular plants and green algae consistently demonstrated that plant glycolysis is controlled from the 'bottom up' with primary and secondary regulation exerted at the levels of PEP and fructose-6-P utilization, respectively (Fig. 15.2a) (Plaxton 1996; Plaxton and Podestá 2006). These findings are compatible with enzymological studies demonstrating that plant ATP- and PPi-dependent PFKs demonstrate potent allosteric (feedback) inhibition by PEP (Fig. 15.2a) (Plaxton and Podestá 2006). It follows that any enhancement in the activity of PK or PEPC will relieve PEP's inhibition of ATP-PFK and PPi-PFK, thereby elevating glycolytic flux from hexose-P. In this way, PEPC and/or PK_c play a central role in the overall control of plant respiration, since the control of their activities *in vivo* will ultimately dictate the rate of mobilization of starch or sucrose for respiration, while simultaneously controlling the provision of: (i) pyruvate for ATP production *via* oxidative phosphorylation and (ii) TCA cycle C-skeletons needed for nitrogen-assimilation or as biosynthetic precursors. PEPC and PK_c represent important targets for metabolic engineering of plant respiration and carbon partitioning in developing seeds.

15.3 Cytosolic *Versus* Plastidic PEP Metabolism of Developing Oilseeds

One prominent feature of primary plant metabolism is the presence of the complete or nearly complete sequence of glycolytic enzymes in plastids, distinct and spatially separated from glycolytic enzymes located in the cytosol (Fig. 15.2b). The integration of cellular metabolism necessitates controlled interactions between pathways sequestered in various subcellular organelles. Thus, cytosolic and plastidic glycolytic pathways of developing oilseeds interact through the action of highly selective metabolite transporters of the inner plastid envelope (Weber and Linka 2011). These transporters include the PEP/Pi translocator, which mediates the import of cytosolic PEP in counterexchange with Pi. Analysis of the *Arabidopsis cue1* mutant, which lacks PEP/Pi translocator activity, demonstrated its essential role in providing PEP to the plastid in support of storage oil biosynthesis by the developing seeds (Prabhakar et al. 2010). The prime function of glycolysis in nonphotosynthetic plastids is to generate C-skeletons, reductant, and ATP for key anabolic pathways, particularly FA synthesis (Fig. 15.2b). A complete glycolytic pathway appears to be present in plastids isolated from developing pea

metabolic routes for the conversion of sucrose into FAs in developing oilseeds. PK_p generates pyruvate and ATP for plastidial FA synthesis. In nongreen oilseeds PEPC controls cytosolic PEP partitioning to malate as a source of pyruvate and reducing power (NADPH and NADH) for leucoplast FA synthesis. *AAT* Asp aminotransferase, *ETC* electron transport chain, ME_m/PDC_m and ME_p/PDC_p, mitochondrial and plastidial isozymes of ME and PDC, respectively and *B. napus* embryos, and castor seed endosperm (Rawsthorne 2002). However, leucoplasts from other developing oilseeds (such as sunflower) appear to lack one or several enzymes of the lower half of glycolysis (e.g., enolase) (Troncoso-Ponce et al. 2009). In this case, the import of cytosolic PEP (or one of its derivatives, such as pyruvate or malate) is the only mechanism to provide primary carbon sources for lipid biosynthesis.

Plants, unlike animals and fungi, have additional flexibility at the PEP node because they express PEPC, a tightly controlled cytosolic enzyme that catalyzes the irreversible β-carboxylation of PEP to oxaloacetate (OAA) using Mg^{2+} as a cofactor. Besides its fundamental and widely studied function in the initial fixation of atmospheric CO_2 during C_4 and crassulacean acid metabolism (CAM) photosynthesis, PEPC has many nonphotosynthetic roles, including supporting seed formation and germination (O'Leary et al. 2011b). A ubiquitousand pivotal function for plant PEPC is to anaplerotically replenish TCA cycle intermediates consumed during biosynthesis and NH_4^+-assimilation (Fig. 15.2a). In conjunction with cytosolic MDH and mitochondrial NAD-dependent malic enzyme (ME), PEPC can also indirectly bypass the reaction catalyzed by PK_c (Fig. 15.2a). Indeed, *in vivo* metabolic flux studies of developing *B. napus* embryos using ^{13}C-labeled precursors demonstrated that approximately 50 % of mitochondrial pyruvate was derived *via* NAD-ME and 50 % *via* PK_c (Schwender et al. 2004). Experiments with transgenic tobacco plants lacking PK_c highlighted the importance of this enzyme in the control and integration of plant carbon and energy metabolism while confirming the exceptional flexibility with respect to the manner by which cytosolic PEP can be metabolized into pyruvate (Grodzinski et al. 1999). However, in contrast to the cytosol, PK_p represents the only known route by which PEP can be converted into pyruvate in the plastid (Fig. 15.2b). As discussed below, profound differences have been documented in the respective physical, immunological, and kinetic/regulatory features of oilseed PK_c and PK_p isozymes. Experiments with mutant *B. napus* plants lacking expression of seed-specific PK_p subunits have confirmed PK_p's central role in generating pyruvate and ATP needed for long-chain fatty acid biosynthesis within the leucoplast (Andre et al. 2007).

15.4 PEPC Supports Oilseed Storage Protein and Fatty Acid Biosynthesis

Seed SP synthesis has been traditionally linked to PEPC activity owing to the requirement for OAA and 2-oxoglutarate (2-OG) needed to assimilate NH_4^+ during formation of amino acids (Fig. 15.2a). Seeds import nitrogen in the form of amino acids, but still require additional C-skeletons because: (i) glutamine, a major component of the supplied amino acids contains two amino groups per 2-OG skeleton and (ii) amino acids such as alanine can be deaminated and respired through the TCA cycle, yielding NH_4^+ that must be rapidly assimilated into glutamine and glutamate *via* glutamine synthetase (GS) and glutamate-oxoglutarate amino transferase (GOGAT) (Fig. 15.2a). PEPC's role in seed amino acid metabolism has been corroborated by a study of transgenic bean plants overexpressing *Cornybacterium* PEPC, which is not feedback inhibited by malate, in a seed specific manner; repartitioning of carbon from sucrose/starch into organic and free amino acids was observed during seed development (Rolletschek et al. 2004). These changes were consistent with enhanced PEP partitioning through the anaplerotic pathway catalyzed by PEPC. Consequently, the transgenic seeds were up to 30 % larger, and accumulated up to 20 % more SP (Rolletschek et al. 2004).

De novo FA synthesis occurs within the plastid and requires stoichiometric amounts of acetyl-CoA and ATP, and NADPH and NADH for each C2 addition to the growing acyl chain in the reactions catalyzed by ACCase and FA synthase (Rawsthorne 2002). However, the pathway leading from sucrose-derived cytosolic hexose-phosphates to plastidial acetyl-CoA (and the corresponding metabolism generates plastidic ATP and NAD(P)H) is highly flexible (Fig. 15.2b) and certainly varies within and between species. Studies of isolated leucoplasts from nongreen developing castor and sunflower seeds showed that relative to other precursors exogenous malate supported maximal rates of FA synthesis (Pleite et al. 2005; Smith et al. 1992). As a consequence of malate's oxidation into acetyl-CoA *via* the plastidialisozymes of NADP-ME and the pyruvate dehydrogenase complex (PDC_p) all of the NADPH and NADH required for carbon incorporation into FAs is produced (Fig. 15.2b). Both PEPC and plastidial NADP-ME are abundant in developing castor bean endosperm (Sangwan et al. 1992b; Shearer et al. 2004), and a novel malate-Pi antiporter exists in the castor and sunflower leucoplast envelope (Fig. 15.2b) (Eastmond et al. 1997; Pleite et al. 2005). Furthermore, the most significant increase in the activity and concentration of castor endosperm PEPC and plastidial NADP-ME coincides with the most active phase of storage oil accumulation (Blonde and Plaxton 2003; Gennidakis et al. 2007; Sangwan et al. 1992b; Shearer et al. 2004). These results are compatible with the hypothesis that PEPC has an additional function to control sucrose partitioning at the cytosolic PEP branchpoint for generation of precursors and reducing power needed for FA biosynthesis, particularly in nongreen oilseeds such as castor bean (Fig. 15.2b). This hypothesis was corroborated by a quantitative proteomic analysis of castor bean reserve deposition that concluded that cytosolic glycolysis and malate synthesis *via* PEPC and MDH are important contributors of both carbon and reducing equivalents for leucoplast FA synthesis (Houston et al. 2009). PEPC also refixes CO_2 released by ME, PDC, and the TCA cycle to improve overall seed carbon economy. However, FA synthesis also requires a source of ATP for ACCase, and this may be an indispensible function for PK_p (Fig. 15.2b). PK_p is also highly expressed during the oil synthesis phase of developing castor, tobacco, and *B. napus* seeds (Ireland et al. 1980; McHugh et al. 1995; Plaxton et al. 1990; Sangwan et al. 1992a). The respective contributions of PEPC-MDH-ME *versus* PK_p pathways for supporting oilseed FA synthesis require further clarification. It has been suggested that the relative importance of each route in developing castor beans may vary diurnally according to the rate of photosynthate provision from source leaves (Sangwan et al. 1992b).

Metabolic flux analysis has reported a spectrum of values regarding anaplerotic PEP carboxylation to support biosynthetic pathways of developing oilseeds. The flux from malate to plastidial pyruvate (and hence, acetyl-CoA) appears to be negligible in developing *B. napus* embryos, whereas about 20 % of total pyruvate that feeds into FA biosynthesis in developing soybean seeds appears to be generated from malate *via* plastidial NADP-ME (Allen et al. 2009; Ruuska et al. 2004). However, FA synthesis by green *B. napus* embryos is absolutely light dependent, since much of the NADPH and ATP they need to assemble FAs from acetyl-CoA is generated *via* the light reactions of photosynthesis (Ruuska et al. 2004; Schwender et al. 2004). In the nongreen developing maize embryo, however, 30 % of carbon flux into FAs was derived from PEPC-generated malate (Alonso et al. 2010). One caveat of metabolic flux analyses is that they require artificially static conditions, such as constant light and sucrose supply, and non-physiological levels of oxygen (O_2). Conversely, the activities of the enzymes surrounding the PEP node and their support of FA synthesis are likely to be tightly controlled to be in tune with changing environmental conditions, including the diurnal cycle.

15.5 Molecular and Biochemical Properties of Cytosolic and Plastidic PK Isozymes of Developing Oilseeds

The PK catalyzes the irreversible conversion of PEP and ADP into pyruvate and ATP (Fig. 15.2a), thus providing one of the only ways that cells can generate ATP *via* substrate-level phosphorylation. Plant PK biochemistry is quite complex owing to the existence of multiple PK_c and PK_p isozymes with different biochemical properties depending on the tissue and plant source. *Arabidopsis*, for instance, has 14-annotated PK genes encoding eight PK_c and four PK_p isozymes and that likely exhibit a large degree of variation with respect to regulation of gene expression and enzymatic activity (by contrast, the human genome only encodes four different PK isozymes) (Andre et al. 2007). Plant PK_c and PK_p isozymes are immunologically distinct owing to their amino acid sequence dissimilarity (Plaxton 1989; Plaxton et al. 1990, 2002). PK_c and PK_p from developing oilseeds have thus far been subjected to concurrent biochemical characterization in two systems—heterotrophic *B. napus* suspension cell cultures[1] and the oil-rich endosperm of developing castor beans (Blonde and Plaxton 2003; Ireland et al. 1980; Moraes and Plaxton 2000; Negm et al. 1995; Plaxton 1991; Plaxton et al. 1990, 2002; Smith et al. 2000; Turner et al. 2005).

[1] Heterotrophic cell suspension cultures of embryos derived *in vitro* from pollen grains of *B. napus* closely resemble their developing zygotic (seed embryo) counterpart with respect to FA and storage lipid accumulation, and expression of PK_c and PK_p (Sangwan et al. 1992a; Weselake et al. 2009).

15.5.1 Cytosolic PK

Antibodies against *B. napus* PK_c were used to examine PK_c subunit composition and developmental profiles in developing castor and soybean seeds (Turner et al. 2005). A 56 kDa immunoreactive PK_c polypeptide was uniformly detected on immunoblots of clarified extracts from developing castor endosperm or soybean embryos. Maximal PK_c activities occurred early in castor and soybean development and were up to 25-fold greater than those of fully mature seeds. Time-course studies revealed a close correlation between extractable PK_c activity and the relative amount of immunoreactive PK_c subunits. The results imply that oilseed PK_c synthesis *versus* proteolytic turnover is tightly controlled such that net synthesis predominates earlier in seed development, whereas proteolytic turnover prevails during seed maturation (Turner et al. 2005). *In vivo* PK_c phosphorylation appears to tag the enzyme for subsequent polyubiquitination and proteolytic degradation in maturing soybean seeds (Tang et al. 2003). To evaluate the physical and kinetic/regulatory properties of oilseed PK_c, the developing castor bean and *B. napus* suspension cell enzymes were purified to homogeneity as a ~230 kDa homotetramers composed of 56 kDa subunits (Smith et al. 2000; Turner et al. 2005). Their activities exhibited: (i) a broad symmetrical pH/activity profile with maximal activity centered around pH 6.5; (ii) hyperbolic substrate saturation kinetics, with similarly high affinities for PEP and ADP; (iii) absolute dependence on a univalent and bivalent metal cation cofactor (K^+ and Mg^{2+}); and (iv) greater catalytic efficiency (V_{max}/K_m) at pH 7.4 (physiological pH) than pH 6.4 (Smith et al. 2000; Turner et al. 2005). These characteristics are similar to PK_c preparations from several other plant sources, but contrast with PK_p isolated from *B. napus* suspension cells and developing castor beans, which uniformly displayed sharp pH/activity profiles centered at pH 8.0 (Ireland et al. 1980; Plaxton et al. 2002). Light-dependent alkalinization of the chloroplast stroma is believed to contribute to the metabolic control of several Calvin cycle enzymes, which display similar pH-activity profiles as the castor bean and *B. napus* PK_p. However, we are unaware of any stromal pH estimates for leucoplasts from nongreen oilseeds. Thus, the potential role of pH in the control of leucoplast PEP metabolism awaits further research.

The PK_c of developing castor and soybeans also demonstrated potent allosteric (mixed-competitive) inhibition by glutamate (Tang et al. 2003; Turner et al. 2005). Glutamate is thus expected to have major significance in the *in vivo* allosteric control of oilseed PK_c. It is intriguing that aspartate functioned as an activator of purified castor bean, soybean, and *B. napus* PK_cs only in the presence of glutamate (Smith et al. 2000; Tang et al. 2003; Turner et al. 2005). The inhibition of these PK_cs by glutamate and its reversal by aspartate was specific for these particular amino acids. This reciprocal allosteric control is remarkable considering that glutamate differs from aspartate only by having an extra carbon atom in its functional group. The data indicate that the cytosolic [aspartate]:[glutamate] ratio is a major determinant of *in vivo* PK_c activity in developing oilseeds. Both glutamate and aspartate have been implicated in the allosteric control of PK_c *versus* PEPC in additional vascular plant tissues and green algae (Plaxton 1996; Plaxton and Podestá 2006).

Allosteric control of plant PK_cs and PEPCs by glutamate provides a tight feedback control that is hypothesized to balance their overall activity with the production of C-skeletons required for NH_4^+-assimilation and transamination reactions in tissues active in amino acid and protein synthesis. By contrast, the activity of the hetero-tetrameric PK_c isozyme isolated from endosperm of 5-day-germinated castor seedlings was insensitive to glutamate or aspartate (Podestá and Plaxton 1991). These allosteric differences reflect tissue-specific PK_c isozymes that have evolved to suit the distinctive metabolic requirements of developing *versus* germinating oilseeds (i.e., storage lipid and protein accumulation *versus* mobilization).

15.5.2 Plastidial PK

The PK_p polypeptide and activity levels correlate with the most active phase of storage lipid biosynthesis in developing castor bean endosperm, as well as tobacco, *B. napus*, and *Arabidopsis* embryos (Andre et al. 2007; Blakeley et al. 1991; McHugh et al. 1995; Sangwan et al. 1992a). This correlation supports the hypothesis that an essential function for PK_p is to generate pyruvate and ATP needed for long chain FA biosynthesis in leucoplasts. This was corroborated by the observation that T-DNA insertional *Arabidopsis* mutants lacking expression of seed-specific PK_p subunits showed a drastic reduction in seed FA and oil content (Andre et al. 2007). All seed PK_ps examined to date exist as unusual heteromeric complexes composed of equivalent amounts of α- and β-subunits having molecular masses of about 64 kDa and 57 kDa, respectively, and that are encoded by different PK_p genes within the nucleus. In developing castor beans, transit peptides of 44 and 60 amino acids are respectively cleaved from PK_p α- and β-subunit preproteins following their import from the cytosol into leucoplast (Negm et al. 1995). The α- and β-subunits of castor bean PK_p are extremely susceptible to partial *in vitro* proteolysis by an endogenous asparaginyl endopeptidase (Cornel and Plaxton 1994; Negm et al. 1995). However, nonproteolyzed native PK_p from *B. napus* cell cultures was purified to near homogeneity and characterized (Plaxton et al. 2002). A novel kinetic feature of the *B. napus* PK_p (subsequently observed with heterologously expressed PK_p from *Arabidopsis* embryos), was its allosteric activation by 6-P-gluconate (Andre et al. 2007; Plaxton et al. 2002), the product of glucose-6-P dehydrogenase, which catalyzes the first committed step of the plastid's OPPP. Feedforward activation of leucoplast PK_p by 6-P-gluconate was hypothesized to coordinate the stromal production of reductant (NADPH) by the OPPP with pyruvate and ATP generation by PK_p for leucoplast FA biosynthesis (Andre et al. 2007; Plaxton et al. 2002).

15.6 The Incredible Complexity of Oilseed PEPC

Plant PEPCs belong to a small multigene family encoding several plant-type PEPCs (PTPCs), along with at a distantly related bacterial-type PEPC (BTPC). PTPC genesencode similar 105–110 kDa polypeptides having a conserved *N*-terminal seryl-phosphorylation domain (acidic-basic–XX-**pSIDAQLR**) and that typically exist as homotetrameric class-1 PEPCs (O'Leary et al. 2011b). PTPCs have been categorized as being either the photosynthetic (C_4 and CAM PEPCs) or nonphotosynthetic (C_3) isozymes. PTPCs evolved from a common ancestral gene and display a high degree of conservation at the genetic level. All plant genomes sequenced to date, including that of ancestral green algae, also contain an enigmatic BTPC gene encoding a larger 116–118 kDa polypeptides that exhibits low (<40 %) sequence identity with PTPCs. BTPCs appear to function as catalytic and regulatory subunits of extraordinary class-2 PEPC heteromeric complexes, whose novel oligomeric-structure, control, expression patterns, and putative functions are discussed below.

Our current understanding of oilseed PEPC biochemistry was built upon a foundation of integrating classical enzyme biochemistry with modern tools of MS and associated bioinformatic databases. In particular, low and highmolecular masses PEPC isoforms of approximately equal abundance were purified, identified, and characterized from the endosperm of developing castor beans; their respective physical and kinetic/regulatory properties were determined to be remarkably analogs to the highly distinctive class-1 *versus* class-2 PEPC isoforms of unicellular green algae (Blonde and Plaxton 2003; O'Leary et al. 2011b; Rivoal et al. 2001; Uhrig et al. 2008a). Thus, class-1 PEPC of developing castor beans exists as a classic 410 kDa PEPC homotetramer composed of 107 kDa PTPC subunits (encoded by *RcPpc3*). By contrast, the corresponding castor bean class-2 PEPC 910 kDa heterooctameric complex consists of the same class-1 PEPC (RcPPC3) homotetrameric core tightly associated with four 118 kDa BTPC subunits (encoded by *RcPpc4*) (Fig. 15.3) (Gennidakis et al. 2007; O'Leary et al. 2009).

15.7 Castor Bean Class-1 PEPC Is Controlled by Allosteric Effectors, Reversible Phosphorylation, and Regulatory Monoubiquitination

15.7.1 Allosteric Effectors

PTPC subunits of class-1 PEPCs from various plant tissues show a range of sensitivities to allosteric effectors, typically activation by hexose-Ps and inhibition by malate (Fig. 15.2a). C_3 PEPCs (e.g., class-1 PEPC from castor beansor *B. napus* embryos) are typically very sensitive to allosteric effectors relative to C_4 PEPCs, which function during C_4 photosynthesis under higher malate concentrations (O'Leary et al. 2011b). As discussed above, aspartate and glutamate are also important feedback effectors of class-1 PEPC and PK_c isozymes in green algae and plant tissues active in nitrogen-assimilation and/or transamination reactions, including developing castor beans and *B. napus* embryos (Blonde and Plaxton 2003; Moraes and Plaxton 2000; Smith et al. 2000; Turner et al. 2005). These effectors provide a regulatory link between nitrogen metabolism and the control of respiratory carbohydrate

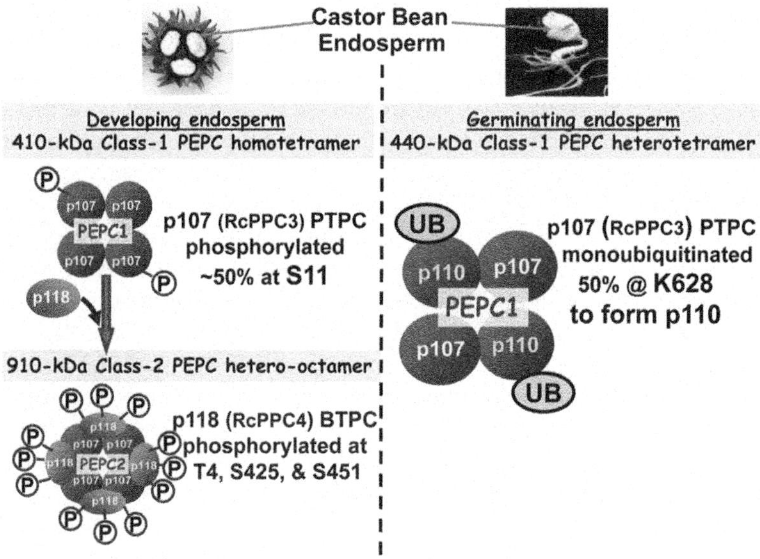

Fig. 15.3 Model illustrating the amazing biochemical complexity of castor bean PEPC. The PTPC (RcPPC3) of developing castor bean endosperm exists: (i) as a typical class-1 PEPC homotetramer (PEPC1) which is activated *in vivo* by sucrose-dependent phosphorylation of its 107 kDa subunit (p107) at Ser^{11} (Tripodi et al. 2005), and (ii) tightly associated with 118 kDa BTPC (RcPPC4) subunits (p118) to form the allosterically-desensitized Class-2 PEPC hetero-octameric complex (PEPC2) (Blonde and Plaxton 2001; Gennidakis et al. 2007; Uhrig et al. 2008a; O'Leary et al. 2009). The BTPC subunits are subject to *in vivo* multisite phosphorylation at Thr^{4}, Ser^{425}, and Ser^{451} (Dalziel et al. 2012; Uhrig et al. 2008a; O'Leary et al. 2011b, c). Castor bean maturation is accompanied by disappearance of the class-2 PEPC complex and BTPC polypeptides and transcripts, and a marked reduction in the amount of class-1 PEPC coupled with dephosphorylation of its 107 kDa PTPC subunits (Tripodi et al. 2005). Imbibition and germination is accompanied by increased PTPC (*RcPpc3*) gene expression, class-1 PEPC activity and amount, and ubiquitination of 50 % of its PTPC subunits at Lys^{628} to form a novel monoubiquitinated class-1 PEPC heterotetramer (Gennidakis et al. 2007; Uhrig et al. 2008b). *UB* ubiquitin

metabolism (Fig. 15.2a); they provide a mechanism for decreasing flux from PEP to aspartate (*via* PEPC and aspartate aminotransferase) while promoting PK_c activity whenever cytosolic aspartate levels become elevated. This elevation will occur when the cell's demands for nitrogen are satisfied, and the overall rate of protein synthesis becomes more dependent upon ATP availability, rather than the supply of amino acids. In this instance, PK_c and respiratory carbon flux may assume a more significant role, in terms of satisfying a large ATP demand, than the anaplerotic generation of biosynthetic precursors via PEPC. Metabolomic analyses of organic and amino acid developmental profiles is required to achieve a better understanding of the coordinate control of oilseed PEPC *versus* PK_c by allosteric effectors.

15.7.2 Reversible Phosphorylation

A cDNA microarray study identified numerous putative protein kinase and phosphoprotein phosphatase genes, whose expressions were significantly enhanced during *Arabidopsis* embryogenesis (Ruuska et al. 2002). Similarly, seed filling in developing *B. napus* embryos is accompanied by dynamic phosphorylation changes in a large number of functionally diverse phosphoproteins (Agrawal and Thelen 2006). Nevertheless, very little is known about the specific roles of enzyme phosphorylation and the identification and characterization of the responsible protein kinases and phosphatases in the integration and control of carbon and nitrogen metabolism and photosynthate partitioning to FAs during oilseed development. It is well established that class-1 PEPC from C_4- and CAM leaves is controlled by reversible phosphorylation at a conserved *N*-terminal seryl residue, which activates the enzyme by decreasing inhibition by malate while increasing activation by glucose-6-P. Tripodi et al. (2005) produced antiphosphorylation-site specific antibodies to the conserved *N*-terminal regulatory phosphorylation-site (pSer^{11}) of class-1 PEPC from developing castor beans. The anti-pSer^{11}-IgG helped to provide the first definitive evidence for: (i) regulatory PEPC phosphorylation-activation in maturing seeds and (ii) reversible *in vivo* dephosphorylation of a developing seed PEPC in response to disruptions in photosynthate supply (Tripodi et al. 2005). These results implied that a PEPC protein kinase (PPCK) is functional during castor bean development. The presence of PPCKwas confirmed in a follow-up study in which the enzyme was purified ~ 1,500 folds from developing castor beans (Murmu and Plaxton 2007). Size exclusion chromatography and immunoblotting with anti-(rice PPCK)-immune serum established that this Ca^{2+}-insensitive PPCK exists as a 31 kDa monomer (Murmu and Plaxton 2007). PPCK-mediated rephosphorylation of castor bean class-1 PEPC activated the enzyme by ~ 80 % when assayed under suboptimal conditions (pH 7.3, 0.2 mM PEP, 0.125 mM malate). Castor bean PPCK displayed remarkable selectivity for phosphorylating castor class-1 PEPC (relative to tobacco, sorghum, or maize PEPCs), exhibited a broad pH-activity optima of pH 8.5, and at pH 7.3 was activated 40–65 % by 1 mM PEP, or 10 mM glutamine or asparagine, but inhibited 65 % by 10 mM malate (Murmu and Plaxton 2007). The possible control of castor bean PPCK by disulfide-dithiolinterconversion was suggested by its rapid inactivation and subsequent reactivation when incubated with oxidized glutathione and then dithiothreitol (DTT). *In vitro* PPCK activity showed good correlation with *in vivo* class-1 PEPC phosphorylation status, with both peaking during the mid-cotyledon to full-cotyledon stages of castor bean development (Murmu and Plaxton 2007; Tripodi et al. 2005). Notably, PPCK activity and class-1 PEPC phosphorylation were both eliminated, following pod excision or prolonged darkness of intact castor plants; both effects were fully reversed 24 h following reillumination of darkened plants (Murmu and Plaxton 2007; Tripodi et al. 2005). These results implicate a direct relationship between the upregulation of PPCK and consequent activation of class-1 PEPC by phosphorylation during the recommencement of photosynthate delivery from illuminated leaves to the nonphotosynthetic castor beans. Overall, the results support the hypothesis that class-1 PEPC and its PPCK participate in the control of photosynthate partitioning into C-skeletons needed as precursors for key biosynthetic pathways of developing oilseeds. An important area for future studies will be to establish signaling pathway(s) that link sucrose supply to enhanced PPCK expression and class-1 PEPC phosphorylation-activation in developing oilseeds.

15.7.3 Regulatory Monoubiquitination

Characterization of native class-1 PEPC purified from germinated castor seed endosperm provided the first example of regulatory monoubiquitination[2] of a metabolic enzyme in nature (Uhrig et al. 2008b). A 440 kDa class-1 PEPC heterotetramer, composed of an equivalent ratio of nonphosphorylated 110 and 107 kDa subunits, was purified from three-day-old germinated COS endosperm. MS and immunoblotting established that both subunits arose from the same PTPC gene (*RcPpc3*) encoding the phosphorylated class-1 PEPC homotetramer of developing COS, but that the 110 kDa subunit is a monoubiquitinated form of the 107 kDa subunit. MS sequencing of tryptic peptides identified Lys^{628} as PEPC's monoubiquitination site (Uhrig et al. 2008b). Lys^{628} is absolutely conserved in all PTPCs and BTPCs, and is proximal to a PEP binding/catalytic domain (O'Leary et al. 2011b). Monoubiquitination was shown to be inhibitory in nature by increasing the class-1 PEPC's K_m (PEP) and sensitivity to allosteric effectors (Uhrig et al. 2008b). It is notable that elimination of photosynthate supply to developing castor beans caused by detaching intact developing pods from the plant caused class-1 PEPC's 107 kDa PTPC subunits in the endosperm and cotyledon to become dephosphorylated, and then subsequently monoubiquitinated *in vivo* (O'Leary et al. 2011a; Tripodi et al. 2005; Uhrig et al. 2008a). Because depodding abolishes photosynthate delivery, it is conceivable that the carbon metabolism of depodded developing castor beans is reorganized to parallel that of the gluconeogenic germinating endosperm. Moreover, the distinctive developmental patterns of PTPC phosphorylation *versus* monoubiquitination

[2] Ubiquitin is a highly conserved globular protein of eukaryotic cells that modifies target proteins *via* its covalent attachment through an isopeptide bond between the C-terminal Gly residue of ubiquitin and the ε-amino group of a Lys residue on a target protein. A multienzyme system consisting of activating (E1), conjugating (E2), and ligating (E3) enzymes attach ubiquitin to cellular proteins. Polyubiquitination is a well known PTM that tags many proteins for their proteolytic elimination by the 26S proteasome. However, protein monoubiquitination has been demonstrated to play a variety of crucial, nondestructive functions in yeast and mammalian cells. Monoubiquitination mediates protein-protein interactions (by recruiting ubiquitin-binding domain client proteins) and localization to help control processes, such as endocytosis, DNA repair, transcription and translation, and signal transduction (Uhrig et al. 2008b). Ubiquitin-related pathways are believed to be of widespread importance in the plant kingdom. Genome annotation indicates that the ubiquitin-related pathway alone comprises over 6 % of the *Arabidopsis* or rice proteomes with thousands of different proteins being probable targets.

throughout various green and nongreen tissues of the castor plant indicated that these PTMs are mutually exclusive (O'Leary et al. 2011a). As discussed above, class-1 PEPC phosphorylation-activation has been observed in tissues in which a high and tightly controlled flux of PEP to malate has an obvious metabolic role; e.g., as occurs during photosynthate partitioning to storage endproducts by developing castor beans. Clearly, additional research is warranted to assess the interplay between, and metabolic functions of class-1 PEPC phosphorylation versus monoubiquitination. Future studies also need to characterize the: (i) ubiquitin-binding domain proteins that might interact with the ubiquitin 'docking site' of monoubiquitinated-class-1 PEPCs; (ii) possible influence of this PTM on their subcellular location; and (iii) signaling pathways and specific E3 ligase that mediate tissue-specific class-1 PEPC monoubiquitination.

15.8 Castor Bean Class-1 PEPC May Interact with a Cytosolic-Targeted PDC

A surprising outcome of the recent coIP proteomics study of the class-1 PEPC interactome of developing castor bean endosperm was the identification of PDC_p as a putative interactor (Uhrig et al. 2008b). Immunoblotting using monospecific antibodies raised against E1α, E1β, E2, and E3 subunits of *Arabidopsis* PDC_p verified the presence of all four PDC_p subunits in eluates from the anti-(COS class-1 PEPC [RcPPC3]) immunoaffinity column. Although castor class-1 PEPC and PDC_p are believed to be localized in different metabolic compartments, there are only two enzymatic steps between them (MDH and ME). A previous report indicated that PDH_p may not be exclusively plastidic in developing castor bean endosperm (Reid et al. 1977). In this study, the plastid marker enzyme ACCase demonstrated 98 % of its total activity to be leucoplast-localized, while only 62 % of total PDC activity was in the same fraction. Conversely, 2 % and 38 % of total ACCase and PDC activity were respectively measured in the corresponding cytosolic fraction (Reid et al. 1977). Further studies are required to establish PDH_p's distribution in the plastid *versus* cytosol. Nevertheless, the aforementioned findings indicate that a specific class-1 PEPC-PDC_p interaction might exist in developing castor beans. This interaction could facilitate CO_2 recycling from PDC to PEPC and/or occur as part of a metabolon that channels PEP to cytosolic acetyl-CoA required for the biosynthesis of isoprenoids, flavonoids, and malonated derivatives, in addition to the elongation of C_{16} and C_{18} FAs. Although cytosolic acetyl-CoA is generally believed to be generated from citrate and CoA by ATP-citrate lyase (Rawsthorne 2002), PDC_p could provide an alternative metabolic route for acetyl-CoA production within the castor bean cytosol. It will be interesting to determine whether the observed *in vitro* interaction between a class-1 PEPC and PDC_p exists *in vivo*, and if so, the role that it plays in carbohydrate partitioning and CO_2 recycling.

15.9 The Class-2 PEPC Complex of Developing Castor Beans Exhibits Remarkable Structural and Kinetic/Regulatory Properties

15.9.1 BTPC Subunits of Class-2 PEPC

Vascular plant BTPCs constitute a monophyletic group, separate from either the PTPCs or bacterial and archaeal PEPCs, and appear to have evolved in green algae (O'Leary et al. 2011b). All deduced BTPC sequences contain residues critical for PEPC catalysis, and heterologous expression of green algal (*Chlamydomonas reinhardtii*) and castor bean BTPCs yielded active PEPCs (O'Leary et al. 2009, 2011b). Deduced PEPC polypeptides are readily classified as a BTPC or PTPC by three main criteria: (i) their C-terminal tetrapeptide is either (R/K)NTG for BTPCs or QNTG for PTPCs; (ii) BTPCs lack the distinctive *N*-terminal seryl phosphorylation motif characteristic of PTPCs; and (iii) BTPCs contain a unique and highly divergent insertion of approximately 10 kDa that exists in a largely unstructured and highly flexible conformation, known as an intrinsically disordered region (O'Leary et al. 2011b, c). Disordered regions typically exist as a flexible linker that connects two globular domains to mediate protein-protein interactions. Indeed, a recent imaging study of transiently-expressed wild type and truncation mutants of developing castor bean BTPC (RcPPC4) and PTPC (RcPPC3) fluorescent fusion proteins in tobacco BY2 suspension cells demonstrated that BTPC's disordered region mediates its *in vivo* interaction with the PTPC as a class-2 PEPC complex (Park et al. 2012). Several lines of evidence indicate that PTPC is an essential binding partner for the corresponding BTPC. For example: (i) native vascular plant and green algal BTPC subunits tightly associate with corresponding PTPC subunits during purification; (ii) heterologously-expressed castor BTPC subunits aggregate and precipitate in the soluble fraction of *Escherichia coli* lysates unless combined with a PTPC; and (iii) fluorescent protein-tagged castor bean PTPC and BTPC subunits interact *in vivo* during their transient coexpression in tobacco BY2 cells (O'Leary et al. 2009, 2011b; Park et al. 2012; Rivoal et al. 2001). PTPCs have thus been hypothesized to be compulsory BTPC binding partners that maintain BTPCs in their proper structural and functional state in class-2 PEPCs (O'Leary et al. 2009, 2011c; Park et al. 2012).

The tissue-specific expression of BTPC transcripts has been investigated in several species. Expression of BTPC transcripts tends to be quite low relative to PTPC transcripts, and is dependent upon the developmental stage and metabolic status of the tissue being examined (O'Leary et al. 2011a, b). An emerging pattern is the increased expression of BTPC in rapidly dividing and/or biosynthetically very active cells. For example, BTPC transcripts and polypeptides are coordinately and highly expressed during castor bean filling; their developmental profiles coincide with class-2 PEPC expression during rapid endosperm growth and oil accumulation (Gennidakis et al. 2007). The class-2 PEPC complex of developing castor beans has been independently documented by a variety of *in vitro* techniques, including coIP and nondenaturing PAGE of clarified endosperm extracts coupled with in-gel PEPC activity staining and parallel immunoblotting using BTPC- and PTPC-specific antibodies (Gennidakis et al. 2007; Uhrig et al. 2008a). These approaches were used to demonstrate comparable class-1 and class-2 PEPC isoforms in extracts of developing lily pollen (Igawa et al. 2010). BTPC transcripts and polypeptides (and thus class-2 PEPC) are also expressed in developing (rapidly expanding) castor leaves, but disappear at leaf maturity (O'Leary et al. 2011a).

15.9.2 Kinetic Properties and Allosteric Effectors of Class-2 PEPC

Upon extraction, the BTPC subunits of green algal and castor bean class-2 PEPCs are extremely susceptible to rapid proteolytic cleavage by an endogenous thiolendopeptidase at a specific site within their disordered region (Gennidakis et al. 2007; Rivoal et al. 2001). This *in vitro* proteolysis has prevented purification of native castor bean class-2 PEPC containing nontruncated BTPC subunits, even in the presence of a wide assortment of protease inhibitors and cocktails. To circumvent this problem, *E. coli* lysates containing heterologously-expressed castor BTPC (RcPPC4) and *Arabidopsis* PTPC (AtPPC3) subunits were mixed together *in vitro* to create intact, stable chimeric class-2 PEPCs, which were then purified and characterized (O'Leary et al. 2009, 2011c). The PTPC and BTPC subunits were both catalytically active and consequently the chimeric class-2 PEPC displayed biphasic PEP saturation kinetics, as previously documented for the purified, nonproteolyzed, native class-2 PEPC from the green alga *Selenastrum minutum* (O'Leary et al. 2009, 2011c; Rivoal et al. 2001). BTPC is a high V_{max}, low affinity, allosterically desensitized class-2 PEPC subunit that exhibits K_m (PEP) and I_{50} (malate) values about 10- and 15-fold higher, respectively, than those of the PTPC subunits. The BTPC is also unaffected by glucose-6-P, a characteristic allosteric activator of class-1 PEPCs (Fig. 15.2a). The resulting class-2 PEPC complex as a whole was far less sensitive to allosteric effectors compared to the corresponding class-1 PEPC of developing castor seeds and thus appears to be constitutively active (O'Leary et al. 2009, 2011c).

Significant levels of class-2 PEPC and its constituent BTPC and PTPC subunits occur in developing castor beans and lily pollen (Blonde and Plaxton 2003; Gennidakis et al. 2007; Igawa et al. 2010; Uhrig et al. 2008a). Both are reproductive sink tissues in which mitosis has been completed and the cells are rapidly expanding while simultaneously metabolizing large amounts of imported photosynthate into storage lipids and proteins. BTPC and thus class-2 PEPC upregulation appears to be a distinctive feature of rapidly growing and/or biosynthetically active tissues that require a large anaplerotic flux from PEP to replenish TCA cycle C-skeletons being withdrawn for anabolism. The unique kinetic and regulatory properties of class-2 PEPC have been hypothesized to function as a 'metabolic overflow' mechanism capable of sustaining significant flux from PEP to malate under *in vivo* conditions, where the corresponding class-1 PEPC activity would become largely suppressed by feedback allosteric inhibitors, such as malate, glutamate, and aspartate (O'Leary et al. 2009). For example, malate levels in developing castor bean endosperm have been measured at 5 mM, a value ~80-fold higher than the I_{50} (malate) of the castor class-1 PEPC at physiological pH, but within the range of the I_{50} (malate) of the BTPC subunit of castor class-2 PEPC (O'Leary et al. 2009, 2011c; Smith et al. 1992).

15.9.3 Regulatory Phosphorylation of Castor Bean BTPC

Extensive MS analysis of native BTPC coIP'd from developing castor bean endosperm led to the discovery of three novel *in vivo* phosphorylation sites: Thr^4 at the *N*-terminus, and Ser^{425} and Ser^{451} within its intrinsically disordered region (Dalziel et al. 2012; O'Leary et al. 2011b, c; Uhrig et al. 2008a). The Thr^4 and Ser^{451} phosphorylation sites are conserved in other BTPC orthologs, but the Ser^{425} site is only partially conserved. Phosphomimetic mutagensis of the Ser^{425} and Ser^{451} sites have shown them to be regulatory in nature, with both causing inhibition of the BTPC subunits within a class-2 PEPC by increasing their K_m (PEP) and sensitivity to feedback inhibition by malate and aspartate. The developmental patterns of BTPC phosphorylation at Ser^{425} and Ser^{451} were examined by immunoblotting clarified castor seed extracts with the respective phosphosite specific antibodies, and shown to be very distinct from the *in vivo* phosphorylation-activation of castor bean class-1 PEPC at Ser^{11} of its PTPC subunits, implying control by distinct PPCKs and signaling pathways (Dalziel et al. 2012; O'Leary et al. 2011b, c; Tripodi et al. 2005). The function of the third phosphorylation site ($pThr^4$) of castor bean BTPC is unknown. Kinetic analysis of a T4D phosphomimetic RcPPC4 mutant indicated that it is not regulatory in nature. However, it is intriguing that the Thr^4 of plant BTPC exists in a conserved forkhead-associated (FHA) binding domain (**pTXXD**). As FHA domains have gained considerable prominence as phospho-Thr dependent protein interaction modules (O'Leary et al. 2011b), it will be necessary to establish the role that BTPC phosphorylation at Thr^4 might play in mediating the interaction of class-2 PEPCs with FHA domain-containing proteins. Characterization of BTPC from additional oilseed species, alongside identification of the responsible protein kinases and related signaling pathways, will also be essential to further validate and extend the role of BTPC and its *in vivo* multisite phosphorylation in class-2 PEPCs.

15.10 Class-2 PEPC of Developing Castor Beans Appears to Associate with the Mitochondrial Outer Envelop

The subcellular location and *in vivo* interaction of castor bean PTPC (RcPPC3) and BTPC (RcPPC4) were recently assessed by imaging PEPC-fluorescent protein fusions (using confocal laser scanning microscopy; CLSM) that had been transiently expressed in heterotrophic tobacco BY2 suspension cells (Park et al. 2012). This was complemented by transmission electron microscopy (TEM) of immunogold-labeled developing castor bean endosperm and cotyledons using monospecific castor BTPC and PTPC antibodies. The overall results indicated that: (i) castor bean BTPC and PTPC interact *in vivo* as a class-2 PEPC complex; (ii) BTPC's intrinsically disordered region mediates its tight interaction with PTPC; (iii) the BTPC-containing class-2 PEPC appears to be located on the mitochondrial outer membrane; whereas (iv) the PTPC-containing class-1 PEPC is uniformly distributed throughout the cytosol. Mitochondrial-associated cytosolic glycolytic isozymes have also been reported in several studies. Sweetlove and colleagues showed that seven different glycolytic enzymes formed a metabolon on the mitochondrial surface of *Arabidopsis* suspension cells during periods of increased respiration so as to channel carbon from cytosolic metabolite pools into the mitochondria while restricting substrate use by competing metabolic pathways (Graham et al. 2007). We therefore need to determine the prevalence, mechanism, and metabolic role(s) of mitochondrial associated class-2 PEPC complexes of developing oilseeds. In addition, it will be important to identify any class-2 PEPC interacting proteins that may form a metabolon to facilitate respiratory CO_2 refixation (e.g., *via* its possible association with carbonic anhydrase and/or bicarbonate transporter of the mitochondrial inner envelop) and/or anaplerotic PEP partitioning to metabolic end products, such as storage lipids and proteins.

15.11 Concluding Remarks

While the central importance of PEP in plant metabolism and physiology has been appreciated for many years (Davies 1979), recent research has provided numerous insights into the complex biochemical and molecular mechanisms that underpin the organization andcontrol of PEP metabolism in developing oilseeds. Integrative analyses at the genomics, transcriptomics, proteomics, enzymological/biochemical, and cellular levels have revealed the biochemical complexity the key PEP metabolizing enzymes: PK_c, PK_p, and class-1 *versus* class-2 PEPC. As documented in Part II of this book, important insights into oilseed development have been acquired through extensive transcriptomics analyses that have measured the expression of a wide variety of genes encoding enzymes involved in the metabolism of imported photosynthate into storage end products. However, a change in transcript abundance does not automatically result ina corresponding change in cognate protein levels (and even less so in a change to *in vivo* enzyme activityor pathway flux). This is reflected by integrated proteomic and transcriptomics studies of developing *Arabidopsis* and *Medicago trunculata* seeds that only observed a ~50 % correlation between transcript levels and amounts of the encoded enzymes (Gallardo et al. 2003; Hajduch et al. 2010). Likewise, the differences in protein accumulation and mRNA expression patterns of tobacco seed PK_c *versus* PK_p indicated that the developmental expression of these isozymes is controlled by independent transcriptional and posttranscriptional mechanisms (McHugh et al. 1995). This corroborates theoretical and experimental studies indicating that the majority of metabolic control occurs at the posttranscriptional level (Sweetlove and Fernie 2005; Plaxton and Podestá 2006). Moreover, transcript profiling provides no information about either the subcellular location of the encoded enzymes, their protein-protein interactions, nor crucial allosteric effectors and PTMs that may be essential for their *in vivo* function and activity. The applications of metabolic flux analysis and functional genomics tools have consistently implicated the importance of posttranslational enzyme control in ensuring the optimal regulation and plasticity of cytosolic versus plastidic PEP metabolism. This undoubtedly applies to many additional regulatory enzymes involved in the control of photosynthate partitioning and reserve deposition by developing oilseeds. For example, a phosphoproteomics study of developing *B. napus* embryosidentified 80 novel phosphoproteins, of which 45 % were enzymes involved in metabolism and energy production (Agrawal and Thelen 2006). However, the biological relevance of this dynamic phosphoproteome will remain unknown until future studies determine the impact of reversible phosphorylation on the biological function of each protein.

Many metabolic roles have been attributed to malate and other TCA cycle intermediates within different plant cell types and the versatility of these metabolites is undoubtedly linked to the diverse functions and posttranslational controls of plant PEPC (O'Leary et al. 2011b). A comprehensive understanding of the linkages between the expression of individual PEPC isozymes, their specific physiological/metabolic functions, and the synchronous regulation of PEPC with PK_c, PK_p, and other enzymes of the PEP branchpoint (e.g., PEP carboxykinase, pyruvate Pi-dikinase, the shikimate pathway, etc.) represents a major challenge. The refinement and integration of enzymology, genomics/bioinformatics, proteomics, transcriptomics, metabolomics, cell biology, and metabolite transporter and *in vivo* flux analyses will enhance our understanding of the cell-specific functions and control of cytosolic *versus* plastidial PEP metabolism of developing oilseeds. It is anticipated that the characterization of oilseed PK_c, PK_p, and class-1 and class-2 PEPCs, their *in vivo* PTMs (including the requisite interconverting enzymes, such as PPCK or E3-ubiquitin ligases, and relevant signaling pathways), and protein-protein interactions will remain a fruitful research area for the foreseeable future. Patterns of PK_c, PK_p, and class-1 *versus* class-2 PEPC (and PTPC versus BTPC) expression need to be reevaluated together with the continued application of functional genomics tools to assess the impact of altering their expression on anaplerotic PEP-partitioning to storage endproducts in developing oilseeds.

Acknowledgments W.C.P. is greatly indebted to past and present members of his laboratory, who made interesting discoveries concerning the molecular and biochemical characteristics of key PEP metabolizing enzymes of developing and germinating oilseeds. W.C.P. is also very grateful to collaborators, who have contributed to this research, particularly Drs. David Dennis, David Turpin, Brian Miki, Florencio Podestá, Wayne Snedden, Chao-fu Lu, Yi-min She, Craig Leach, and Robert Mullen.

References

Agrawal GK, Thelen JJ (2006) Large scale identification and quantitative profiling of phosphoproteins expressed during seed filling in oilseed rape. Mol Cell Proteomics 5:2044–2059

Allen DK, Ohlrogge JB, Shachar-Hill Y (2009) The role of light in soybean seed filling metabolism. Plant J 58:220–234

Alonso AP, Dale VL, Shachar-Hill Y (2010) Understanding fatty acid synthesis in developing maize embryos using metabolic flux analysis. Metab Eng 12:488–497

Andre C, Froehlich JE, Moll MR, Benning C (2007) A heteromeric plastidic pyruvate kinase complex involved in seed oil biosynthesis in Arabidopsis. Plant Cell 19:2006–2022

Baud S, Lepiniec L (2010) Physiological and developmental regulation of seed oil production. Prog Lipid Res 49:235–249

Blakeley SD, Plaxton WC, Dennis DT (1991) Relationship between the subunits of leukoplast pyruvate kinase from *Ricinus communis* and a comparison with the enzyme from other sources. Plant Physiol 96:1283–1288

Blonde JD, Plaxton WC (2003) Structural and kinetic properties of high and low molecular mass phosphoenolpyruvate carboxylase isoforms from the endosperm of developing castor oilseeds. J Biol Chem 278:11867–11873

Borisjuk L, Rolletschek H, Radchuk R, Weschke W, Wobus U, Weber H (2004) Seed development and differentiation: a role for metabolic regulation. Plant Biol 6:375–386

Cornel FA, Plaxton WC (1994) Characterization of asparaginyl endopeptidase activity in endosperm of developing and germinating castor oil seeds. Physiol Plant 91:599–604

Dalziel KJ, O'Leary B, Brikis C, Rao SK, She Y-M, Cyr T, Plaxton WC (2012) The bacterial-type phosphoenolpyruvate carboxylase isozyme from developing castor oil seeds is subject to *in vivo* regulatory phosphorylation at serine-451. FEBS Lett 586:1049–1054

Davies DD (1979) Central role of phosphoenolpyruvate in plant metabolism. Annu Rev Plant Physiol 30:131–158

Eastmond PJ, Dennis DT, Rawsthorne S (1997) Evidence that a malate/inorganic phosphate exchange translocator imports carbon across the leucoplast envelope for fatty acid synthesis in developing castor seed endosperm. Plant Physiol 114:851–856

Gallardo K, Le Signor C, Vandekerckhove J, Thompson RD, Burstin J (2003) Proteomics of *Medicago truncatula* seed development establishes the time frame of diverse metabolic processes related to reserve accumulation. Plant Physiol 133:664–682

Gallardo K, Thompson R, Burstin J (2008) Reserve accumulation in legume seeds. Comptes Rendus Biol 331:755–762

Gennidakis S, Rao S, Greenham K, Uhrig RG, O'Leary B, Snedden WA, Lu C, Plaxton WC (2007) Bacterial- and plant-type phospho*enol*pyruvate carboxylase polypeptides interact in the hetero-oligomeric Class-2 PEPC complex of developing castor oil seeds. Plant J 52:839–849

Graham JWA, Williams TCR, Morgan M, Fernie AR, Ratcliffe RG, Sweetlove LJ (2007) Glycolytic enzymes associate dynamically with mitochondria in response to respiratory demand and support substrate channeling. Plant Cell 19:3723–3738

Grodzinski B, Jiao J, Knowles VL, Plaxton WC (1999) Photosynthesis and carbon partitioning in transgenic tobacco plants deficient in leaf cytosolic pyruvate kinase. Plant Physiol 120:887–896

Hajduch M, Hearne LB, Miernyk JA, Casteel JE, Joshi T, Agrawal GK, Song Z, Zhou M, Xu D, Thelen JJ (2010) Systems analysis of seed filling in *Arabidopsis thaliana:* using general linear modeling to assess concordance of transcript and protein expression. Plant Physiol 152:2078–2087

Hills MJ (2004) Control of storage-product synthesis in seeds. Curr Opin Plant Biol 7:302–308

Houston NL, Hajduch M, Thelen JJ (2009) Quantitative proteomics of seed filling in castor: comparison with soybean and rapeseed reveals differences between photosynthetic and nonphotosynthetic seed metabolism. Plant Physiol 151:857–868

Igawa T, Fujiwara M, Tanaka I, Fukao Y, Yanagawa Y (2010) Characterization of bacterial-type phospho*enol*pyruvate carboxylase expressed in male gametophyte of higher plants. BMC Plant Biol 10:200

Ireland RJ, De Luca V, Dennis DT (1980) Characterization and kinetics of isoenzymes of pyruvate kinase from developing castor bean endosperm. Plant Physiol 65:1188–1193

McHugh SG, Knowles VL, Blakeley SD, Sangwan RS, Miki BL, Dennis DT, Plaxton WC (1995) Differential expression of cytosolic and plastid pyruvate kinase isozymes in tobacco. Physiol Plant 95:507–514

Moraes TF, Plaxton WC (2000) Purification and characterization of phospho*enol*pyruvate carboxylase from *Brassica napus* (rapeseed) suspension cell cultures. Implications for phospho*enol*pyruvate carboxylase regulation during phosphate starvation, and the integration of glycolysis with nitrogen assimilation. Eur J Biochem 267:4465–4476

Murmu J, Plaxton WC (2007) Phosphoenolpyruvate carboxylase protein kinase from developing castor oil seeds: partial purification, characterization, and reversible control by photosynthate supply. Planta 226:1299–1310

Negm FB, Cornel FA, Plaxton WC (1995) Suborganellar localization and molecular characterization of nonproteolytic degraded leukoplast pyruvate kinase from developing castor oil seeds. Plant Physiol 109:1461–1469

O'Leary B, Rao SK, Kim J, Plaxton WC (2009) Bacterial-type phosphoenolpyruvate carboxylase (PEPC) functions as a catalytic and regulatory subunit of the novel class-2 PEPC complex of vascular plants. J Biol Chem 284:24797–24805

O'Leary B, Fedosejevs ET, Hill AT, Bettridge J, Park J, Rao SK, Leach CA, Plaxton WC (2011a) Tissue-specific expression and post-translational modifications of plant- and bacterial-type phosphoenolpyruvate carboxylase isozymes of the castor oil plant, Ricinus communis L. J Exp Bot 62:5485–5495

O'Leary B, Park J, Plaxton WC (2011b) The remarkable diversity of plant PEPC (phosphoenolpyruvate carboxylase): recent insights into the physiological functions and post-translational controls of non-photosynthetic PEPCs. Biochem J 436:15–34

O'Leary B, Rao SK, Plaxton WC (2011c) Phosphorylation of bacterial-type phosphoenolpyruvate carboxylase at Ser425 provides a further tier of enzyme control in developing castor oil seeds. Biochem J 433:65–74

Plaxton WC (1989) Molecular and immunological characterization of plastid and cytosolic pyruvate kinase isozymes from castor-oil-plant endosperm and leaf. Eur J Biochem 181:443–451

Plaxton WC (1991) Leukoplast pyruvate kinase from developing castor oil seeds: characterization of the enzyme's degradation by a cysteine endopeptidase. Plant Physiol 97:1334–1338

Plaxton WC (1996) The organization and regulation of plant glycolysis. Annu Rev Plant Physiol Plant Mol Biol 47:185–214

Plaxton WC, Podestá FE (2006) The functional organization and control of plant respiration. Crit Rev Plant Sci 25:159–198

Plaxton WC, Dennis DT, Knowles VL (1990) Purification of leucoplast pyruvate kinase from developing castor bean endosperm. Plant Physiol 94:1528–1534

Plaxton WC, Smith CR, Knowles VL (2002) Molecular and regulatory properties of leucoplast pyruvate kinase from *Brassica napus* (rapeseed) suspension cells. Arch Biochem Biophys 400:54–62

Pleite R, Pike MJ, Garcés R, Martínez-Force E, Rawsthorne S (2005) The sources of carbon and reducing power for fatty acid synthesis in the heterotrophic plastids of developing sunflower (*Helianthus annuus* L.) embryos. J Exp Bot 56:1297–1303

Podestá FE, Plaxton WC (1991) Kinetic and regulatory properties of cytosolic pyruvate-kinase from germinating castor oil seeds. Biochem J 279:495–501

Prabhakar V, Lottgert T, Geimer S, Dormann P, Kruger S, Vijayakumar V, Schreiber L, Gobel C, Feussner K, Feussner I, Marin K, Staehr P, Bell K, Flügge UI, Häusler RE (2010) Phosphoenolpyruvate provision to plastids is essential for gametophyte and sporophyte development in *Arabidopsis thaliana*. Plant Cell 22:2594–2617

Rawsthorne S (2002) Carbon flux and fatty acid synthesis in plants. Prog Lipid Res 41:182–196

Reid EE, Thompson P, Lyttle CR, Dennis DT (1977) Pyruvate dehydrogenase complex from higher plant mitochondria and proplastids. Plant Physiol 59:842–848

Rivoal J, Trzos S, Gage DA, Plaxton WC, Turpin DH (2001) Two unrelated phospho*enol*pyruvate carboxylase polypeptides physically interact in the high molecular mass isoforms of this enzyme in the unicellular green alga *Selenastrum minutum*. J Biol Chem 276:12588–12597
Rolletschek H, Borisjuk L, Radchuk R, Miranda M, Heim U, Wobus U, Weber H (2004) Seed-specific expression of a bacterial phosphoenolpyruvate carboxylase in Vicia narbonensis increases protein content and improves carbon economy. Plant Biotech J 3:211–219
Ruuska SA, Girke T, Benning C, Ohlrogge JB (2002) Contrapuntal networks of gene expression during *Arabidopsis* seed filling. Plant Cell 14:1191–1206
Ruuska SA, Schwender J, Ohlrogge JB (2004) The capacity of green oilseeds to utilize photosynthesis to drive biosynthetic processes. Plant Physiol 136:2700–2709
Sangwan RS, Gauthier DA, Turpin DH, Pomeroy MK, Plaxton WC (1992a) Pyruvate-kinase isoenzymes from zygotic and microspore-derived embryos of *Brassica napus*. Developmental profiles and subunit composition. Planta 187:198–202
Sangwan RS, Singh N, Plaxton WC (1992b) Phospho*enol*pyruvate carboxylase activity and concentration in the endosperm of developing and germinating castor oil seeds. Plant Physiol 99:445–449
Santos-Mendoza M, Dubreucq B, Baud S, Parcy F, Caboche M, Lepiniec L (2008) Deciphering gene regulatory networks that control seed development and maturation in *Arabidopsis*. Plant J 54:608–620
Schwender J, Ohlrogge JB, Shachar-Hill Y (2004) Understanding flux in plant metabolic networks. Curr Opin Plant Biol 7:309–317
Shearer HL, Turpin DH, Dennis DT (2004) Characterization of NADP-dependent malic enzyme from developing castor oil seed endosperm. Arch Biochem Biophys 429:134–144
Smith CR, Knowles VL, Plaxton WC (2000) Purification and characterization of cytosolic pyruvate kinase from *Brassica napus* (rapeseed) suspension cell cultures. Implications for the integration of glycolysis with nitrogen assimilation. Eur J Biochem 267:4477–4485
Smith RG, Gauthier DA, Dennis DT, Turpin DH (1992) Malate- and pyruvate-dependent fatty acid synthesis in leucoplasts from developing castor endosperm. Plant Physiol 98:1233–1238
Sweetlove LJ, Fernie AR (2005) Regulation of metabolic networks: understanding metabolic complexity in the systems biology era. New Phytol 168:9–23
Tang GQ, Hardin SC, Dewey R, Huber SC (2003) A novel C-terminal proteolytic processing of cytosolic pyruvate kinase, its phosphorylation and degradation by the proteasome in developing soybean seeds. Plant J 34:77–93
Tripodi KE, Turner WL, Gennidakis S, Plaxton WC (2005) *In vivo* regulatory phosphorylation of novel phosphoenolpyruvate carboxylase isoforms in endosperm of developing castor oil seeds. Plant Physiol 139:969–978
Troncoso-Ponce MA, Kruger NJ, Ratcliffe G, Garcés R, Martínez-Force E (2009) Characterization of glycolytic initial metabolites and enzyme activities in developing sunflower (*Helianthus annuus* L.) seeds. Phytochemistry 70:1117–1122
Turner WL, Knowles VL, Plaxton WC (2005) Cytosolic pyruvate kinase: subunit composition, activity, and amount in developing castor and soybean seeds, and biochemical characterization of the purified castor seed enzyme. Planta 222:1051–1062
Uhrig RG, O'Leary B, Spang HE, MacDonald JA, She YM, Plaxton WC (2008a) Coimmunopurification of phosphorylated bacterial- and plant-type phospho*enol*pyruvate carboxylases with the plastidial pyruvate dehydrogenase complex from developing castor oil seeds. Plant Physiol 146:1346–1357
Uhrig RG, She YM, Leach CA, Plaxton WC (2008b) Regulatory monoubiquitination of phosphoenolpyruvate carboxylase in germinating castor oil seeds. J Biol Chem 283:29650–29657
Weber H, Borisjuk L, Wobus U (2005) Molecular physiology of legume seed development. Annu Rev Plant Biol 56:253–279
Weber PM, Linka N (2011) Connecting the plastid: transporters of the plastid envelope and their role in linking plastidial with cytosolic metabolism. Annu Rev Plant Biol 62:53–77
Weselake RJ, Taylor DC, Rahman MH, Shah S, Laroche A, McVetty PB, Harwood JL (2009) Increasing the flow of carbon into seed oil. Biotechnol Adv 27:866–878

Part IV
Metabolomics

Chapter 16
Search for Low-Molecular-Weight Biomarkers in Plant Tissues and Seeds Using Metabolomics: Tools, Strategies, and Applications

Guillaume Marti, Matthias Erb, Serge Rudaz, Ted Turlings and Jean-Luc Wolfender

Abstract This chapter summarises the metabolomic strategies currently in force in plant science and describes the methods used. The metabolite profiling and fingerprinting of plant tissues through MS- and/or NMR-based approaches and the subsequent identification of biomarkers is detailed. Strategies for the microisolation and *de novo* identification of unknown biomarkers are also discussed. The various approaches are illustrated by a metabolomic study of the maize response to herbivory. A review of recent metabolomic studies performed on seed and crop plant tissues involving various analytical strategies is provided.

Keywords Metabolomics · Biomarker identification · Liquid chromatography · Mass spectrometry · Nuclear magnetic resonance · Crop plants

16.1 Introduction

Metabolomics aims at the comprehensive and quantitative analysis of wide arrays of metabolites in biological samples. It is playing an increasingly important role in plant science, particularly in the study of significant metabolome variations in crop plants, such as the intended and unintended effects of genetic modifications, stress-related changes or fruit, and seed development. Metabolomics is a part of systems biology. The latter comprises a number of other omics technologies, such as transcrip-

J.-L. Wolfender (✉) · G. Marti · S. Rudaz
School of Pharmaceutical Sciences, EPGL, University of Geneva, University of Lausanne,
Quai Ernest-Ansermet 30, 1211 Geneva, Switzerland
e-mail: Jean-Luc.Wolfender@unige.ch

G. Marti
e-mail: guillaume.marti@unige.ch

M. Erb
Root-Herbivore Interactions Group, Max Planck Institute for Chemical Ecology,
Hans-Knoell-Str. 807745 Jena, Germany

T. Turlings
Laboratory for Fundamental and Applied Research in Chemical Ecology,
University of Neuchâtel, 2000 Neuchâtel, Switzerland

G. K. Agrawal, R. Rakwal (eds.), *Seed Development: OMICS Technologies toward Improvement of Seed Quality and Crop Yield*,
DOI 10.1007/978-94-007-4749-4_16, © Springer Science+Business Media Dordrecht 2012

tomics (gene expression) and proteomics (protein expression). Metabolomics can be considered to be a large-scale analysis of metabolites in a given organism in different physiological states (Sumner et al. 2003). Profiling the metabolome may actually provide the most 'functional' information of all of the omics technologies (Fiehn et al. 2000) by giving a broad view of the biochemical status of an organism that can be used to monitor significant metabolite variations. Indeed, because metabolites are the end products of cellular regulatory processes, their levels can be regarded as the ultimate response of biological systems to genetic or environmental changes. Finally, this information can be related to other systems biology approaches to assess gene function and provide a holistic view of a living system for its in-depth study (Rochfort 2005; Saito and Matsuda 2010; Yonekura-Sakakibara and Saito 2009).

With the recent developments in analytical methods and data mining (Dunn et al. 2011), metabolomics has rapidly evolved to provide a global picture of molecular plant organization at the metabolite level. Various biological issues have been successfully studied using this holistic approach. Growing number of examples indicate that profiling approaches can be used to expose significant sources of variation in the composition of crop and model plants caused by their genetic background, breeding method, growing environment, genotype-environment interactions, and crop cultural practices (Davies et al. 2010). In plant science, a very important increase in studies related to plant metabolomics has been recorded in recent years (Guy et al. 2008b). As an example, this approach was found to be very effective for investigating plant system problematics, such as plant stress responses (Shulaev et al. 2008) and plant-host interactions (Allwood et al. 2008; Vinale et al. 2008).

Metabolomics has rarely been applied to crop seeds, and most omics studies have mainly been focused on proteomics and transcriptomics (Thompson et al. 2009). In crop science, breeders have long been aware of composition variation through the application of targeted analytical approaches. The usage of broad-scale, 'unbiased' analysis of the metabolome now offers major upside to our understanding of the true extent of variation in a plethora of metabolites relevant to many biological issues. Metabolomics is helping to provide targets for plant breeding by linking gene expression and allelic variations to variations in metabolites (functional genomics) (Davies et al. 2010).

Plant metabolomics has benefited from a wide array of preexisting methodological approaches and bioanalytical knowledge for the characterization of many chemically diverse classes of metabolites. While the field has pushed the implementation of unbiased and generally applicable strategies for metabolite extraction, fractionation, and detection, significant challenges in fundamental activities, such as compound identification and quantification, still exist (Hegeman 2010). In this chapter, the focus will be related to the applications and methodologies used for the metabolomic study of plant tissue from crop seeds. Indeed, direct metabolomics methods have also been used on seeds themselves but these applications are more infrequent (Thompson et al. 2009) (see, Table 16.1). Because our background is more based on phytochemical analysis, a special focus will be given to methods that enable the identification of low molecular weight stress biomarkers. The metabolomic strategies discussed below will be illustrated by several examples, extracted from a comprehensive study of the maize response to biotic stress. The methods discussed are generic and can be used for the study of other plant tissues and seeds.

16.1.1 Profiling and Fingerprinting of Plant Tissues

Despite the impressive development of analytical methods in plant science, the comprehensive characterization of a plant metabolome remains challenging because its size is still unknown (estimated to have a few thousand constituents), and large differences in physicochemical properties exist between plant metabolites. Metabolites are generally of low molecular weight and are issued either from primary and secondary plant metabolism; they occupy a broad chemical space and cover a much extended dynamic range. Some compounds, such as sugars and lipids, have important nutritive value and can occur in very large amounts, while secondary metabolites are more involved in defence and resistance mechanisms. Some of these metabolites can be present at very low levels while displaying key hormonal effects related to growth or defence. In addition, secondary metabolites can be important to the beneficial health properties of some plants.

From a metabolomic perspective, one would ideally expect to be able to monitor, identify, and quantitate all of the compounds in very complex biological matrices in a single analysis with minimal sample and in an unbiased manner. Furthermore, mining of the data should provide a powerful way to interrogate datasets from various viewpoints according the biological question at hand. Metabolomics has not reached this level of advancement, mainly because of the limits of the analytical methods used. In all approaches developed, compromises have to be made, but nevertheless, a broad survey of metabolome variations can be recorded that will ultimately provide new biological knowledge.

Currently, two complementary approaches are used for metabolomic investigations: metabolic profiling and metabolic fingerprinting. The intention of the latter is not to identify each metabolite, but to compare patterns or 'fingerprints' of metabolites that change in response to various stimuli or genetic alterations. In contrast, metabolic profiling focuses on the analysis of a group of metabolites, either related to a specific metabolic pathway or class of compounds. In most cases, metabolic profiling is a hypothesis-driven approach rather than a hypothesis-generating one. Based on the questions asked, metabolites are selected for analysis, and specific analytical methods are developed for their determination. Both methods can be used in the search for new biomarkers (Dettmer et al. 2007) and are complementary. Once a biomarker has been determined without direct matching in a database, a dedicated procedure (target analysis) can be performed for identification. Furthermore, target analysis can be used when accurate quantification is required.

16.2 Metabolomics Approaches and Strategies

In metabolomics, unlike genomics and proteomics, a single analytical technique capable of profiling all of the low molecular weight metabolites of the cell does not exist (Dunn 2008). Among the different techniques enlisted for metabolome analysis, both MS and Nuclear Magnetic Resonance (NMR) represent key methods, each with advantages and disadvantages. Both detection methods can also be hyphenated with chromatography; however, only MS, mainly in gas chromatography (GC)-MS or LC-MS hyphenation, is used to profile plant tissues. LC-NMR is powerful for providing structural information on single constituents of complex mixtures (Wolfender 2010). However, profiling data obtained by LC-NMR has not been directly used for differential metabolomics, mainly because of the lack of sensitivity, cost, and low throughput of the method. The use of microNMR methods in direct relation with HPLC, mainly by at-line hyphenation, is however, a powerful way to identify *de novo* biomarkers highlighted by metabolomics.

16.2.1 NMR-Based Metabolomics

NMR fingerprinting is used to directly analyze crude plant extracts without the need for prior chromatographic separation (Choi et al. 2006). The method is simple and has a high-through put that does not require specific sample preparation and provides detection of all protons (Colquhoun 2007; Dunn 2008). Another advantage of NMR analysis is the possibility to obtain quantitative information, as the proton signal intensity is only determined by the molar concentration (Wishart 2008). Standard protocols exist for the extraction and analysis of various plant tissues that can be shared between laboratories (Kim et al. 2010). These methods are extremely useful when comparisons have to be made based on the main constituents of a plant extract and when samples have to be compared in long-term studies. Indeed, NMR fingerprints are very reproducible independent of the period of analysis. In this respect, the method is very useful for grouping samples in crop products, such as olive oil or wine, based on origin (Son et al. 2009; Vlahov et al. 2003). The main drawback is that NMR lacks sensitivity, although constant improvement is being made with new generations of magnets and probes. Further, the identification of single compounds in extracts maybe hindered by overlapping signals (Kim et al. 2010). The identification of biomarkers relies mainly on a comparison of NMR shifts of plant metabolites acquired under the same solvent conditions, while *de novo* identification can be partly based on the acquisition of complementary 2D-NMR data of a mixture (Verpoorte et al. 2007) (see below, Fig. 16.5c).

16.2.2 MS-Based Metabolomics

MS provides sensitive detection of most plant metabolites (Hall 2006), although MS ionization is compound-dependent and LC-MS approaches in both positive and negative ion modes have to be used for comprehensive metabolite surveys (De Vos et al. 2007). The ability to identify metabolites is based on MS/MS spectra, when libraries are available or on the determination of molecular formulae when spectra are recorded on high-resolution MS analyzers.

For LC-MS applications, many types of mass spectrometers can be used. Those that are low-resolution, such as single Q mass spectrometers, are the most commonly used and least expensive. Those with high-resolution and exact-mass capabilities, such as TOF instruments, are becoming increasingly popular. TOF-MS provides resolution of up to 20,000 and very good mass accuracy (<5 ppm). They can be operated with a high acquisition frequency over a broad mass range and are compatible with rapid chromatographic methods without compromising sensitivity. Other instruments, such as Fourier transform-ion cyclotoron resonance (FT-ICR) MS, provide even higher resolution (>100,000). Thanks to their extreme resolution, the see expensive instruments can be used for direct infusion MS (DIMS) analysis of complex mixtures without the need for chromatographic hyphenation (Feng and Siegel 2007).

To obtain additional structure information, molecular ions generated by soft ionization methods can be fragmented in tandem mass spectrometers using MS/MS or MS^n experiments. In this respect, triple-quadrupole (QQQ) MS/MS systems are the most commonly used. The IT mass spectrometers have the unique capability of producing multiple-stage MS-MS (MS^n) data that may be essential for structural elucidation. For high-resolution measurements in MS/MS, the Orbitrap FTMS instrument provides high quality spectra for metabolite identification (van der Hooft et al. 2011), but these IT methods do not have a high acquisition frequency compatible with rapid chromatography methods. In addition to these types of mass spectrometers, there is a growing number of additional varieties, including hybrid systems that combine low-resolution and high-resolution analyzers for specific applications (Korfmacher 2005).

In the absence of standard or instrument-dedicated databases, peak annotation remains sometimes putative (Brown et al. 2011). Because of its ability to analyze multiple compounds with a high sensitivity, MS is however playing an increasingly important role in the progression of proteomics and metabolomics. Nonhyphenated MS methods (DIMS) enable the rapid and high-throughput screening of hundreds of samples, mainly for metabolite fingerprinting, but have limited quantification and metabolite identification capabilities. The coupling of MS with separation techniques (GC, LC, or CE) ('hyphenated methods') is extremely powerful in terms of detection, quantification, and identification of a wide range of metabolites. Another advantage of hyphenated methods concerns the decrease in matrix effects, which are detrimental to the determination of analytes present in low concentrations. Hyphenation enables bidimensional detection, where each metabolite is resolved in both chromatographic [retention time (RT)] and mass spectrometric (*m/z*) dimensions. These approaches require a rather long analysis time for the chromatographic separation (typically 10–60 min). The throughput can, however, be increased using faster techniques, such as fast GC or ultra-high-pressure liquid chromatography (UHPLC) (Grata et al. 2008). These methods can be used for both fingerprinting and profiling of metabolites.

Because the response in MS is compound-dependent, absolute quantification in metabolic profiling studies is currently not feasible. However, for the analysis of many related samples in series, MS represents a good alternative, especially if the biomarkers of interest require a high sensitivity of detection.

GC-MS has fewer problems than LC-MS in terms of ion suppression and compound-dependent MS response. Furthermore, the electron ionization MS (EI-MS) spectra that can be recorded with this technique are very reproducible. This has the important advantage in that large EI-MS databases can be directly searched for peak annotation, which is not the case for LC-MS. Despite the high resolution of GC separation, the coelution of metabolites may occur and deconvolution of these spectra represents an important step. The latest advance in this area involves GC coupled to a TOF-MS instrument to take advantage of the high data-acquisition rate and spectral continuity of TOF mass spectrometers to build libraries that take into account spectra/retention-index datasets (Schauer et al. 2005). Despite these advances, very little progress has been reported for the identification of unknown peaks (Boroczky et al. 2006) in GC-MS because the method does not easily allow isolation of the biomarker of interest for *de novo* characterization, as is the case when LC-MS is used. GC-MS is, however, one of the most frequently used tools for profiling primary metabolites after adequate sample preparation and derivatization. Instruments are mature enough to run large sequences of samples; novel advancements increase the breadth of compounds that can be analyzed, and improved algorithms and databases are employed to capture and utilize biologically relevant information (Fiehn 2008). GC-MS is limited for the analysis of large secondary metabolites, and thus, comprehensive metabolomic studies require multiple orthogonal and complementary analytical methods. Today, the most powerful studies combine the advantages of MS and NMR spectroscopy (Crockford et al. 2008; Moing et al. 2011).

16.3 Metabolomic Study of the Maize Response to Herbivory as an Example

The analytical strategies in metabolomics currently used in our group will be illustrated by a comprehensive study of the maize response to herbivore attack. Maize is one of the most important crops of agronomic interest and is a model plant to evaluate environmental, physical, and chemical effects. Maize is also the second most wide spread genetically modified (GM) crop (James 2009), and many studies have focussed on the assessment of the unintended effects of genetic modifications at the transcriptome, proteome, and metabolome levels. The most recent applications in metabolomics include a GM assessment of *Bacillus thuringiensis* (Bt) maize by GC-MS profiling (Barros et al. 2010) and high-resolution-MS fingerprinting (Leon et al. 2009). The adaptation of maize seedlings to a high salinity environment has been monitored by NMR fingerprinting (Gavaghan et al. 2011). A GC-MS-based approach was used to assess the seed content in relation to the environment, season, and genetic background (Rohlig et al. 2009).

In this chapter, our investigation on the main metabolome variation that occurs in maize seedlings upon herbivore *Spodoptera littoralis* attack will be discussed, mainly to illustrate how the different metabolomic strategies provide information and how biomarkers can be characterized.

This study was motivated by the fact that maize reacts strongly to herbivore attack. Upon the perception of specific elicitors in the saliva of *Spodoptera* larvae

(Alborn et al. 1997), it begins producing high amounts of a complex, volatile blend including oxylipin breakdown products, aromatic compounds, and sesquiterpenes (Turlings and Tumlinson 1992). Recently, we found that plants attacked by *Spodoptera* also become more resistant to subsequent infestation by the same species (Erb et al. 2009). This induced immunity is positively correlated with volatile emissions (Erb et al. 2011), but the volatiles themselves do not have any direct toxic effect on the herbivores (Turlings and Veyrat unpublished data), suggesting that nonvolatile changes in the metabolome may account for the increased defensive capacity of the plant. Although several toxic secondary metabolites are known to be produced by maize plants (Byrne et al. 1996; Frey et al. 1997; Nuessly et al. 2007), little is known about induced metabolites, and we hypothesized that untargeted metabolomics may lead to the discovery of novel induced defence compounds.

To test the above idea, 12-day-old maize seedlings were infected with *S. littoralis* larvae or left herbivore-free. After 48 h, the leaf tissue was harvested and immediately frozen in liquid nitrogen. Each leaf sample was ground into a powder prior to extraction with isopropanol (IPA). The crude extracts were dried and resuspended in a mixture of methanol/water (85/15) for rapid solid-phase extraction using C18 cartridges in order to remove highly a polar compounds (chlorophylls, waxes, etc.). This sample preparation restricts the number of analytes but is important for analysis by LC-MS on a RP column to preserve instrumental repeatability. The analytical variability and biological variation of the samples was assessed. The analytical variability should always be much lower than the biological variability. Moreover, the number of replicates has to be chosen, such that enough analyses can be carried out to determine metabolome variations that are significant. The raw data provided by NMR, DIMS, and LC-MS fingerprinting for a given maize leave extract are displayed in Fig. 16.1. This shows the complementary views of metabolite composition that can be obtained with these orthogonal analytical approaches.

In the NMR fingerprint of the maize extract (Fig. 16.1a), a few tens of constituents were observed. Typically, 30–150 metabolites can be identified in an NMR fingerprint (Kim et al. 2011). DIMS analysis of the same extract obtained on a TOF-MS was much more sensitive and revealed more than 200 metabolites (Fig. 16.1c). The intensity of the different ions recorded in MS does not reflect their relative abundance, contrary to those registered in the NMR spectra. DIMS analysis does not allow the separation of isomers and many minor compounds were detected that were barely discernable from the instrumental noise. LC-MS analysis of the same extract performed on a UHPLC-TOF-MS revealed more than 1,000 features (Fig. 16.1b) because additional resolution was provided by hyphenation with a chromatographic separation.

LC-MS analysis provides3-Ddata (RT × *m/z* × intensity) (Fig. 16.1e). LC-MS has the advantage over DIMS of decreasing signal overlap and providing useful information on metabolite structure because MS spectra of a given LC peak or a given biomarker can directly be extracted from the raw data (Fig. 16.1f). However, the mining of data from hyphenated methods is more complex than that obtained with direct spectroscopic methods (see below). A 3-D plot of the LC-MS profiling of maize extract also indicated the very important dynamic range in which observation

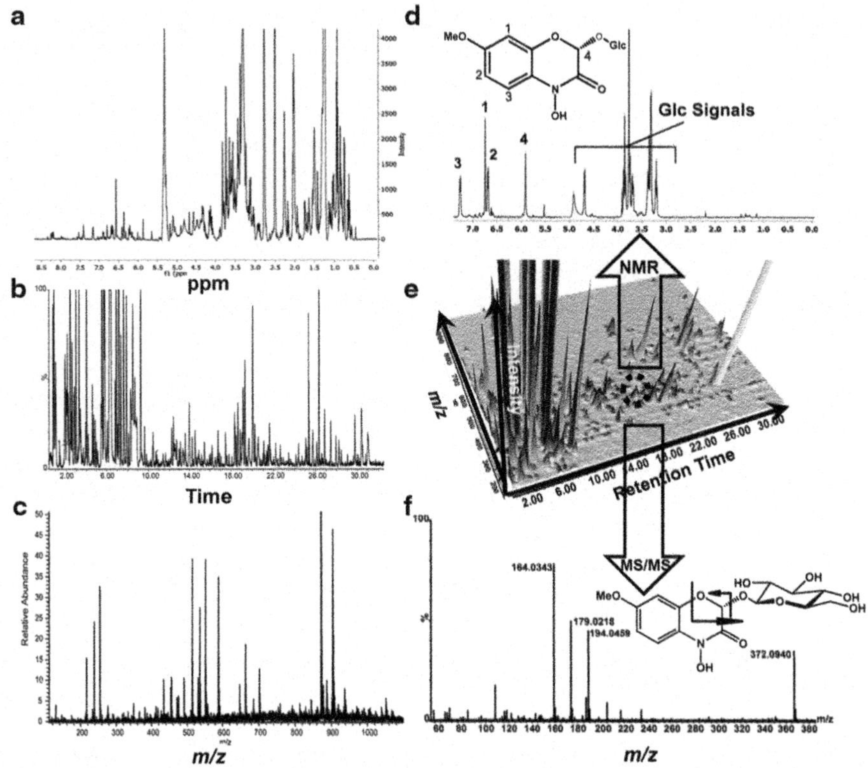

Fig. 16.1 **a** ^{1}H-NMR (500 MHz) of *Zea mays* leaves crude isopropanol extract. **b** UHPLC-TOF-MS (PI mode) profiling of the same extract on a C18 column (150 × 2.1 mm; i.d., 1.7 μm) with gradient of Water/ACN + 0.1 % formic acid in 35 min. **c** DIMS of the same extract by infusion in 5 min on a QMS. **d** ^{1}H-NMR of DIMBOA-Glc. **e** 3D visualization of UHPLC-TOF-MS profiling showed in **b**. **f** QTOF-MS/MS of DIMBOA-Glc. ACN, acetonitrile

can be made and the complementarity of both chromatographic and MS scales to resolve metabolites in both dimensions (Fig. 16.1e). Figure 16.1d shows that NMR data can be obtained on a given biomarker when the metabolomic study is performed based on LC-MS. Indeed the technique enables localization of the LC peak of interest, which can be isolated and characterized at-line by NMR after scaling up of the LC separation (see below).

Further focus will mainly be given to HPLC-related methods because of their multiple uses for fingerprinting, profiling, and the targeted isolation of biomarkers.

16.4 Multiple Facets of LC-MS in Plant Metabolomics

LC-MS represents the most robust and versatile technique for the separation of plant metabolites because it allows for the analysis of metabolites without the need for derivatization of the crude extracts. Derivatization procedures represent a limitation for GC-MS metabolomic approaches (t'Kindt et al. 2009), and relatively large and polar secondary metabolites cannot be profiled in this case (see above).

For LC-MS fingerprinting and profiling, crude extracts are usually obtained by maceration, ultra-sonication, ball milling, or ASE (accelerated solvent extraction) extraction of dried material or fresh plant tissues. In the last case, enzymatic activity has to be stopped by flash freezing of the tissue samples just after harvest. Frozen samples can be homogenized and directly extracted with solvents. Most of the extracts prepared for metabolomics are obtained with MeOH or MeOH-H_2Omixtures. However, if fewer polar constituents are targeted, fewer polar solvents (such IPA or dichloromethane) can be used. For other organisms, different extraction methods have been used, but in most cases, the final extracts are either polar, nonpolar, or cover an intermediate polarity range. Crude extracts can be injected directly in HPLC without sample pretreatment. In order to ensure good instrument repeatability and avoid column clogging or contamination with compounds that would stick on the HPLC stationary phase, simple sample preparation procedures, such as liquid-liquid extraction or solid-phase extraction, are used. The same procedures may help to enrich the samples in classes of given constituents for more targeted analyses. This type of clean-up represents the compromise needed to ensure good instrument reproducibility, which is mandatory for any metabolomic study, but it also implies that the number of metabolites observed will be reduced.

HPLC has significantly developed in terms of convenience, speed, choice of stationary column phases, sensitivity applicability to a broad variety of sample matrices, and the ability to hyphenate with spectroscopic methods (such as MS). From a chromatography viewpoint, the development of columns with different phase chemistries (especially RP) enables the separation of almost any type of plant metabolite. The latest developments in HPLC, including the recent introduction of highly pH-stable phases, sub-2-μm particles (Nguyen et al. 2006) and monolith columns, have considerably improved the performance of HPLC systems in terms of resolution, speed, and reproducibility. Efficiencies exceeding 100,000 plates and peak capacities over 900 can be attained by coupling columns together (David et al. 2007).

The separations are mostly performed on C18 material with MeCN-H_2O or MeOH-H_2O solvent systems in gradient elution mode. To improve the separation efficiency, various modifiers are added to the mobile phase that can strongly influence the sensitivity of MS detection. Because the polarity of the constituents analyzed extends over a large domain, the separations are carried out in the gradient mode in all fingerprint and profiling studies. In the RP mode on C18 columns, such profiling is performed by gradually increasing the proportion of organic solvent during the analysis. In stationary phases that behave differently, such as hydrophilic interaction liquid chromatography (HILIC), the proportion of water is increased during the analytical run. The gradient mode produces LC peak focusing during separation that is beneficial for sensitivity because the peaks are sharper than in isocratic mode.

In plant metabolomics, LC-MS is the most favourable analytical method when minor key biomarkers have to be identified. Indeed, hyphenation of MS with HPLC provides a good means to reducing ionization suppression effects and resolving multiple components in a mixture. Furthermore, use of the chromatographic dimension gives additional resolution to overlapping features in data mining that is orthogonal to MS. With such an approach, the coordinates of each feature that will subsequently be used for data mining are characterized by chromatographic retention (dependent on the polarity or lipophilicity of the metabolites) and the *m/z* value of the ions. A strategic advantage of such an approach is that biomarkers that will be highlighted by differential metabolomics with this kind of data will easily be localized in the chromatographic dimension. Such information will be useful for further isolation of biomarkers of interest for either complete *de novo* identification by complementary spectroscopic methods, such as NMR or the determination of bioactivity in a specific model (Wolfender et al. 2011) (see, Fig. 16.1d–f).

16.4.1 MS Detection for HPLC

For LC-MS detection, different interfaces are available that allow transfer of the metabolites and their ionization for further separation in various type mass analyzers. They allow the ionizations of almost any nonvolatile and thermally labile compound to be converted into a gas-phase ion. In LC-MS-based metabolomics, a large majority of applications are mainly performed using ESI sources.

An ESI source operates by passing the LC eluent through a capillary held at high voltages (2–5 kV). Ion formation occurs in the capillary followed by nebulization and desolvation to provide a transfer of ions from the liquid to gas phase and into the mass spectrometer vacuum. The source acts as an electrochemical cell. Minimal fragmentation of the molecular ion is observed, although fragmentation can be produced in-source or by tandem MS. The technique is described as soft-ionization. This technique works well for relatively large and polar metabolites. For compounds of medium to low polarity and of smaller molecular weight, other soft-ionization techniques area good alternative, namely atmospheric pressure chemical ionization (APCI) and atmospheric pressure photoionization (APPI).

APCI operates by passing the LC eluent through a heated glass tube (up to 500 °C), which produces explosive vaporization to generate a gaseous collection of solvent and metabolite molecules. A corona discharge needle provides ionization of the solvent molecules (which are in a large excess), followed by ion or charge transfer to the metabolite molecules. Degradation of molecules can occur at high temperature, though minimal fragmentation of molecular ions is observed. APPI is complementary to ESI and APCI. In this case, electron ejection from molecules is produced by photons emitted from discharge UV lamps. APPI, however, is not commonly available from many instrument manufacturers (Dunn 2008).

All of these different ionization methods can be operated in both positive (PI) and negative ion (NI) modes. For complete coverage of most of the metabolites present

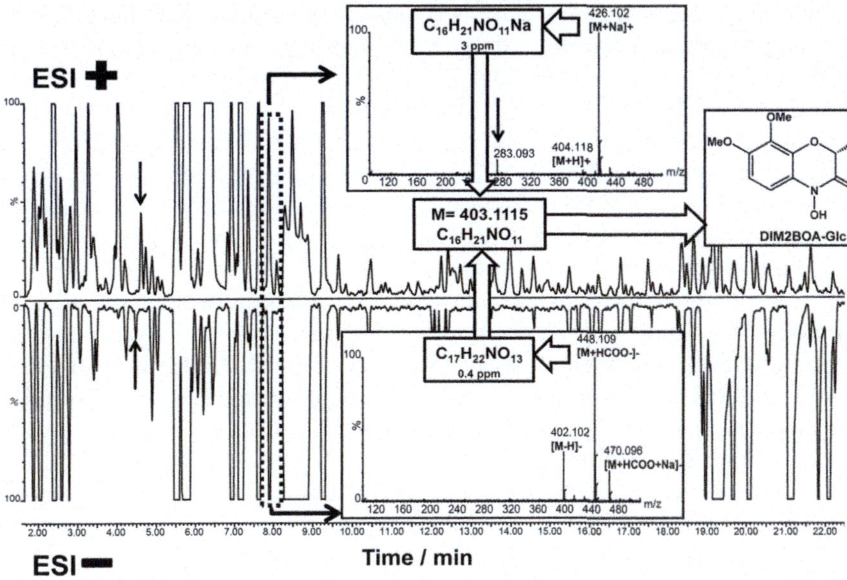

Fig. 16.2 Enlargement of the UHPLC-TOF-MS total ion chromatogram (TIC) of crude IPA maize leaves extract in PI (*upper side*) and NI (*lower side*) mode using a C18 column (150 × 2.1 mm; i.d., 1.7 μm) with gradient of Water/ACN + 0.1 % formic acid in 45 min. *Arrows* highlight features detected in a single mode. The NI and PI mass spectra extracted from the peak in *dashed line* at RT = 8 min highlight several adducts encountered for DIM2BOA-Glc

in a sample (which is the goal of a metabolomics study), the two ionization modes are complementary. With these soft-ionization methods, the ionization is mainly governed by proton affinity, and acidic molecules will tend to lose a proton to form a deprotonated molecular ion $[M\text{-}H]^-$, while compounds containing nitrogen (with alkaline properties) will mostly produce protonated molecular species $[M+H]^+$.

Another aspect of these soft-ionization methods is that most compounds produce not only protonated and deprotonated molecular ion species but also different adducts. The nature of these adducts is strongly dependent on the LC-MS conditions used. For example, if formic acid is used as an HPLC modifier, some constituents will tend to produce formate adducts in the NI mode. As an example, Fig. 16.2 displays a typical ESI PI/NI profile obtained for the crude extract of maize leaves. As shown, most compounds were ionized in both the PI and NI modes, but some were only detected with one polarity. The arrows highlight features detected only in one ionization mode (Fig. 16.2). The mass spectra extracted from the peak at retention time (8 min) in ESI PI and NI displayed a typical pattern for each mode. In NI, the ion at *m/z* 402.102was present in its $[M\text{-}H]^-$ form and was also detected as its formate and sodium formate adducts. In contrast, the same ion was detected as a sodium adduct (*m/z* 426.102) in PI, while the protonated form was almost undetectable. The compound at *m/z* 283.093, coeluted under these conditions, was detected only in PI. The complementary PI/NI analysis is, thus, very useful to determine the molecular weight when an unknown is detected. Furthermore, it can be used to deduce a putative molecular formula when high-resolution MS data is recorded using dedicated software (Draper et al. 2009). In this example, $C_{17}H_{22}NO_{13}$ was calculated from the ion at *m/z* 448.109 in NI and $C_{16}H_{21}NO_{11}Na$ was deduced for*m/z* 426.114 in PI. The exact molecular formula that fit both ionization modes was $C_{16}H_{21}NO_{11}$, which corresponded to(*2R*)-2-β-D-glucopyranosyloxy-4-hydroxy-7,8-dimethoxy-2*H*-1,4-benzoxazin-3(4*H*)-one), DIM2BOA-Glc a benzoxazinone derivative well known in maize leaves (Sicker and Schulz 2002).

This is an important factor to take into account for the identification of biomarkers as well as for data mining. A given metabolite might generate several features (*m/z*) at a given retention time if many adducts are detected. Furthermore, if a feature is highlighted by data mining, its corresponding *m/z* ion has to be correctly deconvoluted and analyzed to determine which type of adduct it corresponds to and, thus, correctly extract its molecular formula.

For LC-MS-based metabolomics, many MS analyzers can be used. Contrary to DIMS, analytes are separated by chromatography, so the technique does not requires MS instruments with extreme resolution, such as FT-ICR-MS, and simpler MS analyzers, such as QITs and TOF-MS, can be used (see above).

16.4.2 UHPLC-MS Profiling and Fingerprinting

One of the most important breakthroughs over the last few years has been the introduction of UHPLC for both the high-resolution profiling and rapid fingerprinting of crude extracts (Grata et al. 2009). For metabolic applications, UHPLC is ideally coupled to TOF-MS to provide high resolution in both the chromatographic and mass spectrometric dimensions (Guillarme et al. 2010).

From a theoretical point of view, the introduction of sub-2-μm particle supports induces a concomitant increase in efficiency and optimal velocity due to improved mass transfer (Nguyen et al. 2006). Therefore, in metabolomics, high-throughput separations can be obtained with short column lengths for metabolite fingerprinting, whereas highly efficient separations are achieved with relatively long columns for detailed metabolite profiling (Grata et al. 2008). However, the use of such small particles requires dedicated systems that are able to work at pressures higher than 600 bars. Because this approach provides very narrow LC peaks, MS detectors with very fast response times are generally mandatory, and in recent years, UHPLC-TOF-MS has been recognized to be very efficient for profiling studies in metabolomics (Grata et al. 2009; Wilson et al. 2005).

By selecting an adequate UHPLC column of short length, it is possible, from a theoretical point of view, to increase the throughput by a factor of nine compared to conventional HPLC. Such ultra-fast separations have been experimentally demonstrated and analysis times in the range of 1–5 min can be expected (Nováková et al. 2006; Wren and Tchelitcheff 2006). On the other hand, by keeping strictly identical

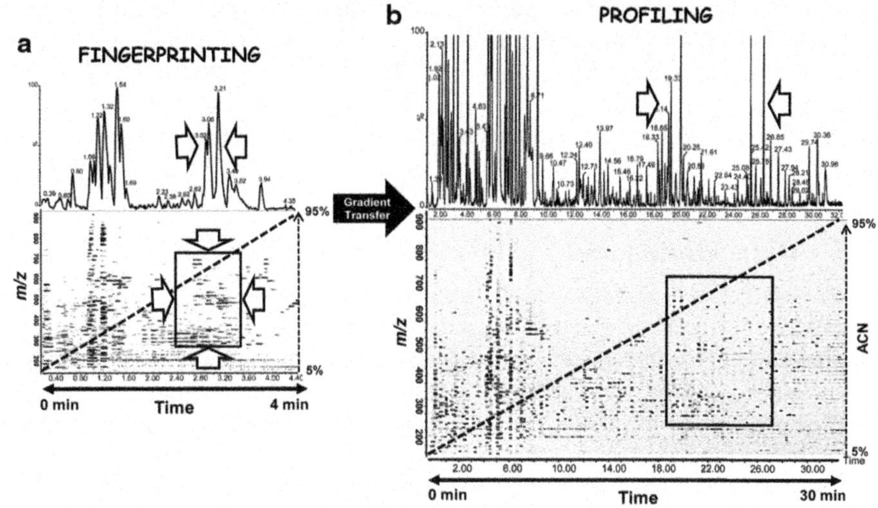

Fig. 16.3 **a** Fingerprinting: TIC (PI mode) of maize leaves IPA extract on a short C18 column (50 × 1 mm; i.d., 1.7 µm) and the corresponding 2D map with % of ACN in dashed line. **b** Profiling: UHPLC-TOF-MS (PI mode) profiling of the same extract on long C18 column (150 × 2.1 mm; i.d., 1.7 µm) and the corresponding 2D map resulting from the gradient transfer using HPLC calculator (see text). *Square* and *arrows* highlight the gain in resolution from fingerprinting to profiling

column lengths in both HPLC and UHPLC, it is hypothetically possible to increase the plate count by a factor of three. Some separations involving 150-mm, or even longer, UHPLC columns have been reported in the literature and show very high efficiency (Eugster et al. 2011).

For differential metabolomic studies in our group, we use both the high-throughput capabilities of UHPLC on short columns for the acquisition of rapid UHPLC-TOF-MS fingerprints of numerous replicates and long high-resolution profiling on long columns of pool representative samples for the localization and determination of biomarkers. The high-throughput analysis of many biological replicates improves the reproducibility of the LC-MS detection, allowing a large series of samples to be analyzed over a short time period, thereby avoiding drift of the MS detection. The increase in number of biological replicates gives more significant weight to metabolome variations in relation to a given physiological change *versus* natural biological variation of the samples. The data mining of such fingerprinting data is thus notably improved.

In the case of our maize study, a rapid metabolite fingerprint was achieved by UHPLC-TOF-MS with a short column (1.7 µm, 50 × 1 mm i.d.) using IPA extracts in less than 5 min (Fig. 16.3a). The PI and NITOF-MS data were recorded for comprehensive metabolome coverage. Only the PI data are shown in Fig. 16.3. With this fingerprinting method, more than 850 features were detected (Fig. 16.3a). The UH-PLC conditions of the fingerprinting method were transferred to a detailed profiling gradient (40 min) for profiling on a long column (1.7 µm, 150 × 2.1 mm i.d.) by applying conventional chromatographic equations(HPLC calculator; Goga-Remont et al. 2000). The transfer retained the same selectivity for both fingerprinting and profiling, while a baseline resolution of many metabolites was obtained in the long profiling run. The latter allowed the detection of approximately one thousand features in PI mode (Fig. 16.2b) and the chromatographic resolution of many isomers.

16.5 Data Mining and Localization of Biomarkers

The power of differential metabolomics in comparison with more classical metabolite profiling approaches resides in the possibility of analyzing a high number of replicates derived from various biological states and highlight biomarkers that are significantly related to the situation. This process is performed by applying adequate data mining to spectroscopic raw data that have been recorded in a reproducible manner.

Raw data from any of the omics fields are an eventual source of information and, in turn, a source of knowledge (Goodacre 2005). However, to make the leap from one to the next requires considerable data processing and statistical analysis, as well as suitable data storage formats (Allwood et al. 2008; Issaq et al. 2009). In the case of NMR or direct infusion MS fingerprinting, raw data are recorded as a single file corresponding to a specific sample, and hyphenated data files (LC-MS or GC-MS) are constituted by a set of mass spectra recorded in sequence. For the latter, each spectrum displays all of the ions detected by the mass spectrometer during a given time lapse by associating an *m/z* with their intensities. Because the data structure is more complex, the processing of such raw data constitutes the initial step of data handling. Its main goal is to extract the relevant information and recapitulate the multiple files in a single table. In this table, each sample should ideally be defined by the same number of variables, and every variable must correspond to the same metabolite measured in all samples of the dataset. With the development of metabolomic methods for the analysis of complex biological systems, values for concentrations of thousands of compounds are available in a single experiment, and comparing samples become a problem of high dimensionality. Because a high number of variables is measured, datasets not only become larger in size but are also more complex and, therefore, preprocessing of the data is required (Katajamaa and Oresic 2007; Trygg et al. 2007). This methodology could include numerous processing steps, such as noise filtering, data binning, peak detection, and/or chromatographic alignment. This data-processing procedure aims to generate homogeneous information to allow an appropriate comparison of multiple samples by statistical methods. Common successive data processing procedures are usually available in commercial solutions, including data filtering, peak detection, normalization, and alignment. Several solutions exist in the form of freely available software, where the simplest tools are tailor-made for a specific task, such as chromatographic alignment, while other software suites combine everything from instrument control to data analysis into a single package.

Extracting knowledge from these tables, revealing patterns among samples, and identifying critical or discriminatory variables are not a straight-forward task (Boccard et al. 2010). Because changes in metabolite levels may be drastic or subtle, a careful statistical processing is mandatory to determine the relevance of an observed change. Data scaling and centring are classical pretreatment procedures intended to provide relevant data for further statistical analysis. Univariate statistical tests, such as the Student's *t*-test, are used to identify relevant variables but are rather limited. Common chemometric tools, such as principal component analysis (PCA), used for projecting multivariate data to a low-dimensional plot, are generally proposed for display and exploratory analysis purposes. This method provides an overview of the preprocessed data. For further investigation, supervised approaches remain very attractive because of the strong impact of their use in human metabolomics. In this context, techniques based partial least squares (PLS) analysis, including OSC preprocessing (Li et al. 2002), are of the utmost interest in plant metabolomics (Ducruix et al. 2008). Finally, because the feature number is a key aspect that determines the data space, dimensionality reduction is often very useful when mining large datasets. Selection of a subset of representative features retaining the salient characteristics of the data is, therefore, a fundamental issue in building potent models, including variables with highly predictive abilities, and avoiding the addition of uninformative or superfluous data.

As an example, the data mining performed on the UHPLC-TOF-MS fingerprints of the IPA maize extracts is displayed in Fig. 16.4. As mentioned above, the goal of the study was to assess the metabolic changes of maize leaves upon herbivory attack; six seedlings were infested with twenty *S. littoralis* larvae, and six were left herbivore-free. The UHPLC-TOF-MS fingerprints were recorded, as shown in Fig. 16.3, and 2-D ion maps (*m/z versus* RT) were generated for the six biological replicates in both the control and stress situations (Fig. 16.4a).

Prior to data mining, the raw UHPLC-TOF-MS data were preprocessed by filtering, feature detection, chromatogram alignment, and data normalization by dedicated software (Katajamaa and Oresic 2007) to generate a peak list. In this case, a simple PCA applied to the preprocessed LC-MS fingerprinting data provided a clear separation between control and attacked leaves (Fig. 16.4b). The score plot of this PCA demonstrated that 46 % of the total variance was explained using the first PC (PC1). The corresponding loading plots, which display the variables (*m/z versus* RT) responsible for the sample clustering by PCA, were ranked according to their PC1 score (Fig. 16.4; Table 16.1). The most significant features (1–6) are highlighted on the 2-D ion maps of the attacked leaves (Fig. 16.4a). The corresponding ions belong to the biomarkers mainly induced upon herbivory attack of the maize leaf. The *m/z* and RT coordinates of these biomarkers for their localization in the long metabolite profiling method are shown in Fig. 16.3b. The long profiling data provided a deconvolution of the biomarkers form the coeluting compounds. Following a similar process to that described in Fig. 16.2, the molecular formula of each of these biomarkers was obtained. The molecular formula of compounds (1, 2, 3, and 6) matched with the reported phytoalexin of maize (benzoxazinones derivatives), according to a literature search. Benzoxazinones are well-known allelochemicals

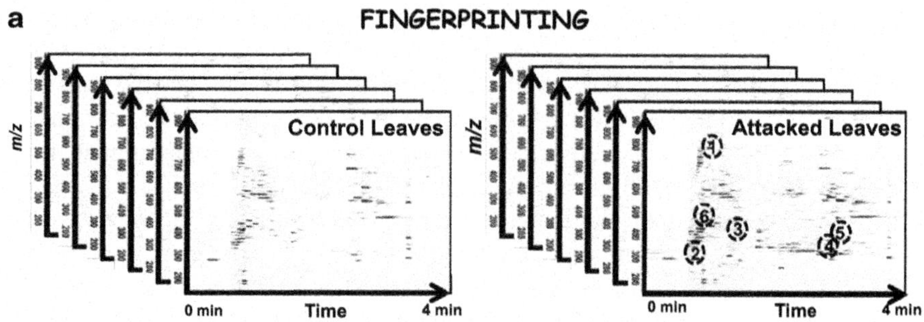

b DATA MINING

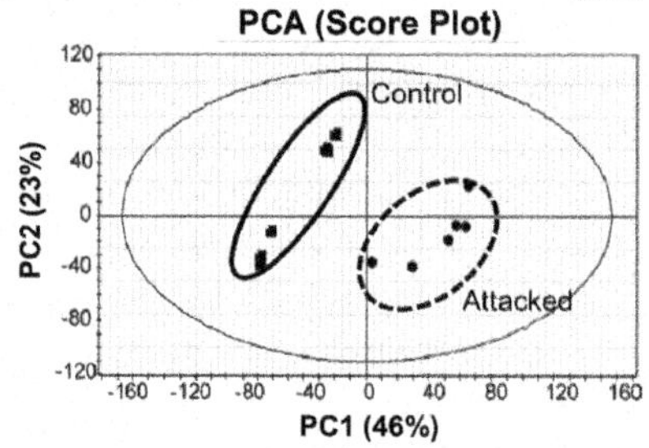

PCA (Loadings Plot)

Rank	*m/z*	RT	PC1 score
1	819.228	1.4327	-0.21585
2	282.113	0.9137	-0.144067
3	462.119	1.7468	-0.122646
4	327.216	3.2519	-0.121952
5	291.197	3.4378	-0.117186
6	470.09	1.2749	-0.106836

Fig. 16.4 **a** Fingerprinting: 2D map (*m/z* vs RT) of UHPLC-TOF-MS fingerprints of six control and six attacked leaves samples IPA extracts using a short C18 column (50 × 1 mm; i.d., 1.7 µm); **b** Data mining: Principal components analysis (PCA) of the UHPLC-TOF-MS data (*m/z* × RT × intensity), score plot on the left and the six first loadings plots on the *right* with their respective localization on the 2D map of attacked leaves in *dashed circles*

from herbaceous plants. These derivatives present significant toxicity against the larvae in their deglycosylated form (Sicker and Schulz 2002). The apolar products (4 and 5) were unknown, but their molecular formulae matched those of FAs. However, all compounds needed to be isolated for complete identification. Characteristic patterns of these biomarkers were also highlighted using an orthogonal NMR metabolomic approach (see below).

16.6 Strategies for the *de novo* Identifications of Unknown Biomarkers

A central role of metabolomics is to collect analytical data and convert them to biological knowledge. Identification of the metabolites of interest (biomarkers) represents a key step in this process. However, this identification also constitutes one of the main bottlenecks of all metabolomic studies because databases are not always available for a given instrument. As peak annotation is often putative, unknown identification requires complete *de novo* structure determination based on MS and NMR data interpretation.

16.6.1 Biomarker Identification in NMR-Based Approaches

In NMR-based approaches, structural information on the biomarker of interest can directly be extracted from the ^{1}H-NMR fingerprints recorded if no overlapping signals from the mixture are present. In this case, the first step is the assignment of resonances and comparison of the NMR chemical shifts and coupling constants with those of authentic samples measured under the same conditions (Kim et al. 2010). This represents a powerful approach because NMR chemical shifts can be compared in large databases built with data obtained under well-established conditions.

For *de novo* identification, it is necessary to apply 2-D measurements. Among the various 2D-NMR techniques, the ^{1}H-^{1}H J-resolved spectrum (a plot of the chemical shifts versus the coupling constants) has proven to be very useful for the resolution of signals in crowded areas of NMR fingerprints (Tiziani et al. 2008).

As this is the case for pure natural products, 2D-NMR techniques can be directly applied to crude extracts for biomarker identification. This includes experiments, such as correlation spectroscopy (COSY), total correlation spectroscopy (TOCSY), heteronuclear multiple bond correlation (HMBC), and hetero nuclear single quantum coherence spectroscopy (HSQC). Although COSY shows the correlation of protons, which have mutual spin-spin couplings, TOCSY shows spin systems of coupled protons in the same molecule. TOCSY is very useful for carbohydrate or amino acid identification because once typical signals are assigned (e.g., the anomeric proton in a sugar), the remaining signals, even those in crowded areas, can be assigned using this correlation. In HMBC spectra, long-range correlations between carbon and protons can be assigned, which helps to identify resonances from the same molecules (Kim et al. 2010).

However, this approach is efficient only if the most abundant biomarker in a mixture must be identified. For minor compounds, the signals are often weak, they appear in crowded areas, and their detection and identification may be hampered, even when 2D-NMR detection is used.

Following our example, the IPA extract of the maize leaves previously studied by MS metabolomics was investigated by direct NMR fingerprinting. The samples were obtained after enrichment of the IPA extracts on solid-phase extraction and analyzed in deuterated DMSO. Figure 16.5a shows the proton NMR spectra of control leaves and attacked leaves after 48 h. As this was observed by LC-MS metabolomics, signals characteristic of glycosylated benzoxazinone derivatives (highlighted in the dashed square of Fig. 16.5a and (*) for the anomeric proton of the glucose moiety) in the herbivory leaves appeared in the aromatic area of the spectra (see magnified region in Fig. 16.5b). As for the MS metabolomics approach, the data mining of all NMR fingerprints revealed the features that were most significantly induced as a result of the attack. In this case, preprocessing of the NMR raw data was performed by bucketing the spectra in small integrated portions (0.04 ppm). The corresponding loading plot was a list of NMR chemical shifts and intensities. Most of the features detected fit well with the benzoxazinone structures revealed by MS.

Finally, 2D-NMR experiments, such as HSQC (Fig. 16.5c), enabled the attribution of some proton and carbon signals directly in the crude extract. A doublet at

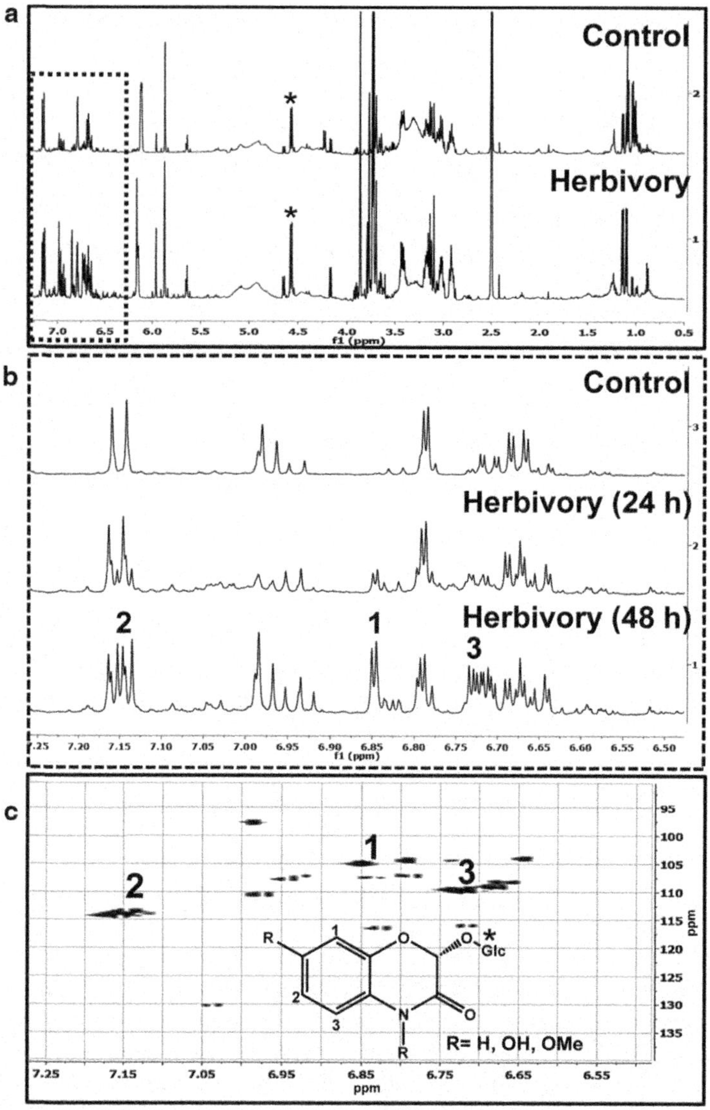

Fig. 16.5 **a** Total ^{1}H-NMR spectra from IPA extract after SPE enrichment of maize control leaves and 48 h after *S. littoralis* attack in DMSO-d6, asterisk underline the anomeric proton of glucoside derivatives. **b** Magnified region from 6.50 to 7.25 ppm of ^{1}H-NMR spectra of control leaves and 24 h and 48 h after *S. littoralis* attack. **c** Magnified region of HSQC experiment on IPA extract of leaves 48 h after attack. Aromatic protons of hydroxamic acid derivatives have been assigned on ^{1}H and 2D spectra

6.85 ppm was assigned to the proton in the *ortho* position of the aromatic region of the benzoxazinone derivatives, while two other doublets of doublets at 7.15 and 6.72 ppm, highlighted by HMBC experiments, confirmed the identification of glycosylated hydroxamic acids derivatives. The increase signal of anomeric proton at 4.57 ppm confirmed this hypothesis. This example illustrates the complementary responses that can be obtained by MS- and NMR-based metabolomic approaches.

In NMR, the most abundant biomarkers were shown and the change in ratio of their corresponding signals could be directly linked to their change in concentration because NMR is an absolute quantitative method. On the other hand, the MS-based metabolomic approach (Fig. 16.3) highlighted additional minor but significant biomarkers with higher resolution/selectivity and on a wider dynamic range. In this case, however, direct correlation between variations in the MS signal intensities and quantities could not be directly performed. Further, LC-MS quantification of a given biomarker requires targeted methods with the use of appropriate pure standards.

The complete identification of minor biomarkers by NMR alone in a mixture is not possible. In this respect, NMR hyphenation with HPLC may help in further identification, as it provides NMR data for individual metabolites after separation (Wolfender et al. 2005). This type of approach can be efficient when a given biomarker is detected by LC-MS-based metabolomics, localized in the extract, isolated and identified at the microgram scale by micro-flow NMR. This approach will be discussed below.

16.6.2 Biomarker Identification in MS-Based Approaches

MS-based approaches are very sensitive compared with those based on NMR, but the structure information that can be extracted from MS data is often not sufficient for *de novo* structure identification. Therefore, access to databases is mandatory, and in this respect, the different MS approaches are not comparable.

In GC-MS metabolomics studies, most applications are performed using electron ionization. This ionization method is always performed with the same energy (70 eV) and, thus, provides very reproducible MS spectra. Furthermore, electronic ionization MS produces an important number of fragments having characteristic patterns for each analyte. Metabolites can thus be efficiently identified based on their fragmentation patterns after an adapted deconvolution procedure (Schauer et al. 2005). However, the method is not well adapted for *de novo* structure determination, and molecular ion species are difficult to detect and need complementary information, such as that provided by chemical ionization in GC-MS.

In LC-MS, soft-ionization techniques are employed (mainly ESI or APCI), and very little structural information is obtained, with the exception of the molecular ion. Using high-resolution instruments, molecular formulas can be deduced, but this is often not sufficient for an unambiguous structure determination. On the other hand, most of these analyzers generally have a mass accuracy below 5 ppm. Even with this precision, several molecular formulae may match a given exact mass. In order to automatically constrain the thousands of possible candidate structures, rules need to be developed to select the most likely and chemically correct molecular formulae. For this purpose, algorithms for filtering molecular formulae have been derived from heuristic rules that enable determination of the most probable elemental compositions (Kind and Fiehn 2007). This approach provides the correct molecular formula with a probability of 98 % if the compound exists in a database.

For the dereplication of secondary metabolite biomarkers in plants, if the molecular formula is correctly assigned, a cross-search based on chemotaxonomic information can drastically limit the hit possibilities. Retention time information can also be exploited, but this is much more difficult than in GC because retention factors are dependent on the type of column used and the mobile phase composition (organic modifier nature, content, and pH). LC-MS/MS provides important complementary information that can be compared with libraries. Unlike electron ionization, this type of MS/MS spectra will, however, strongly depend on the instrument used and is hardly universally applicable. An example of the differences in MS/MS spectra that can be recorded for a common flavonoid C-glycoside (isovitexin) between an IT analyzer and a QQQ instrument is highlighted in Fig. 16.6 (Waridel et al. 2001). As shown, a direct matching of the two spectra recorded does not allow a correct matching in the database.

MS/MS spectra recorded on an Orbitrap FT-MS have been shown to be very stable, and MS^n fragmentation was tested for the differentiation and identification of metabolites using a series of 121 polyphenolic molecules. The results obtained were stable over a five-month time period at concentration range of 100-fold, with small changes in normalized collision energy, which is key to metabolite annotation and helpful in structure and substructure elucidation (van der Hooft et al. 2011).

In DIMS, because there is no physical separation of the metabolites, as with LC-MS, structure elucidation is limited. In this case, the structure assignment strongly relies on molecular formula assignment obtained with high-resolution mass measurements and on complementary MS/MS experiments. For example, improvements in the identification of metabolites by searching a species-metabolite relationship database, such as KNApSAcK, and structural analyses by an MS/MS method have been shown to be very efficient (Sawada et al. 2009).

Another possibility is to interpret some MS/MS characteristic fragmentation of given class of compounds, such as glycosides to help the *de novo* identification process. As an example in the case of maize, the fragmentation of DIMBOA-Glc(*2R*)-2-β-D-glucopyranosyloxy-4-hydroxy-7-methoxy-2*H*-1,4-benzoxazin-3(4*H*)-one), one of the biomarkers detected in the crude extract profiling, confirmed the structure of the benzoxazinone skeleton based on its characteristic fragmentation (Fig. 16.1f) (Bonnington et al. 2003). However, for the *de novo* identification of a given biomarker complementary NMR data are necessary as illustrated in Fig. 16.1d (see below).

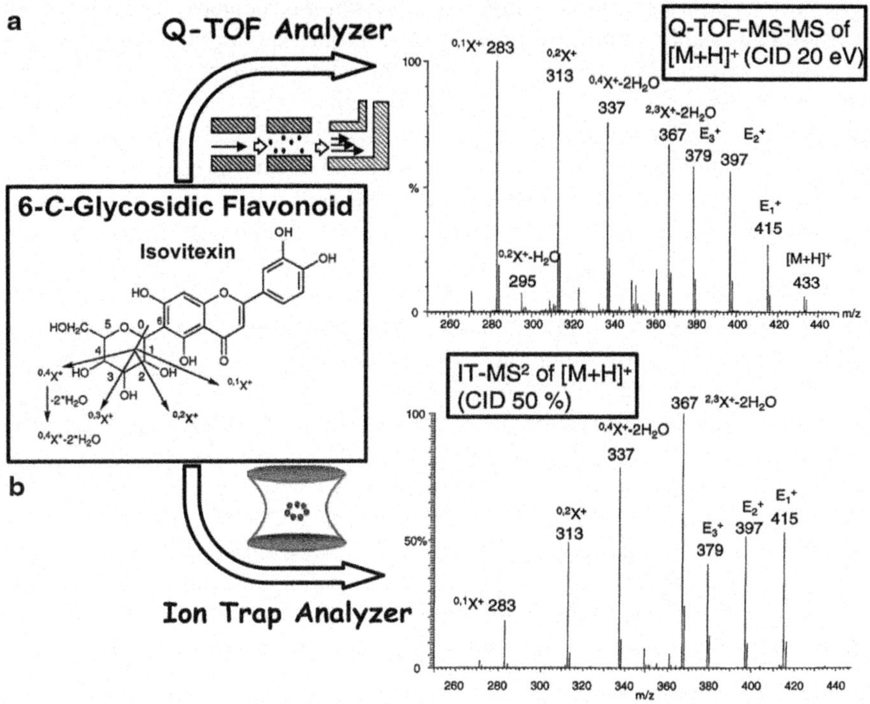

Fig. 16.6 MS/MS fragmentation pattern of isovitexin with **a** Q-TOF analyzer and **b** Ion trap (IT) analyzer. The pattern recorded in this two MS/MS regimes are not superimposable and render a search in databases difficult. (Adapted from (Waridel et al. 2001) with permission from Elsevier)

16.6.3 Stategy for the de novo Biomarker Identification by High Resolution UHPLC-TOF-MS and Subsequent Micro NMR Analyis

As discussed above, most of the MS-based metabolomics strategies provide mainly putative peak annotations for given biomarkers unless a high quality match in a given database is obtained. For most new or rare biomarkers, *de novo* structure determination represents a critical step that is often the bottleneck in many metabolomics studies. Thus, for a definitive identification, biomarkers either need to be synthesized based on the putative structures determined after peak annotation and compared to the compound detected, or the metabolites have to be isolated and their structure fully assigned by NMR. In this latter case, one advantage of LC-MS based metabolomics approaches is that because biomarkers of interest can be efficiently localized in the LC chromatogram, and provided that enough sample is available, they can be isolated using a targeted procedure and subsequently identified by sensitive microNMR methods.

In this respect, we have developed micro-isolation methods based on LC-MS profiling that enable a rapid isolation of plant biomarkers at the microgram scale for NMR analysis. Due to the complexity of plant extracts, the purification of metabolites present in low concentrations is critical. The strategy used relies on optimization of the chromatographic analysis using UPLC-TOF-MS with modeling software. The optimized method is then transferred to semi-preparative LC conditions with MS detection. Complete characterization of the isolated metabolites, often obtained in the low-microgram range from a few milligrams of plant extract, is then performed using micro-flow-NMR methods (Webb 2005), such as Cap NMR, which provide excellent sensitivity. With such NMR probes, the biomarkers need only to be dissolved in a minimal amount of deuterated solvent (5 µL). With such a small volume, high concentrations can be obtained with an optimum filling factor, and high-quality 1D- and 2D-NMR spectra can be measured (Wolfender et al. 2010). Such an approach has permitted the full characterization of new phytohormones induced by wounding in the leaf of *Arabidopsis thaliana*. In this case, even minor key biomarkers were fully characterized by Cap NMR after a two-step targeted LC-MS micro-isolation procedure (Glauser et al. 2008).

In the case of the maize example, the same strategy was applied and enabled complete *de novo* structure elucidation of several biomarkers induced upon herbivory attack. Fig. 16.7 shows the method that was applied to isolate compound 4, one of the putative biomarkers highlighted by the data mining results (see, Fig. 16.4).

To find the optimum chromatographic conditions for the targeted isolation of 4, the crude IPA extract was profiled at the analytical scale by UHPLC-TOF-MS on a C18 column (1.7 µm, 150 × 2.1 mm i.d.) using two generic gradients differing only in gradient slope (Fig. 16.7a). The retention times of the desired compound and those surrounding this peak were then introduced into chromatographic optimization software (Osiris™) (Goga-Remont et al. 2000). In a first step, the software performed modeling of the chromatographic behavior of all of the compounds selected. The second step involved an optimization of the scale-up process, which takes into account the parameters chosen for the purification. The optimized theoretical separation obtained for the purification of 4 is shown in Fig. 16.7b. The scale-up was performed by separating 30 mg of the IPA extract on a semi-preparative column with the same phase chemistry (5 µm, 250 × 10 mm i.d.). Detection was ensured by using an IT mass spectrometer (Fig. 16.7c). With the help of Cap-NMR, 100 µg of purified compound was sufficient to acquire 1D- and 2D-NMR spectra at good resolution. These data combined with the molecular formula deduced from the TOF-MS data ($C_{18}H_{31}O_5$, Fig. 16.4) enabled *de novo* identification of biomarker 4 (Fig. 16.7d).

16.7 Applications in Plant Metabolomics

Metabolomics has been applied in several domains related to plant biology and chemistry (Sumner et al. 2003). This approach was found to be particularly useful for issues related to: (i) fingerprinting of species, genotypes or ecotypes for

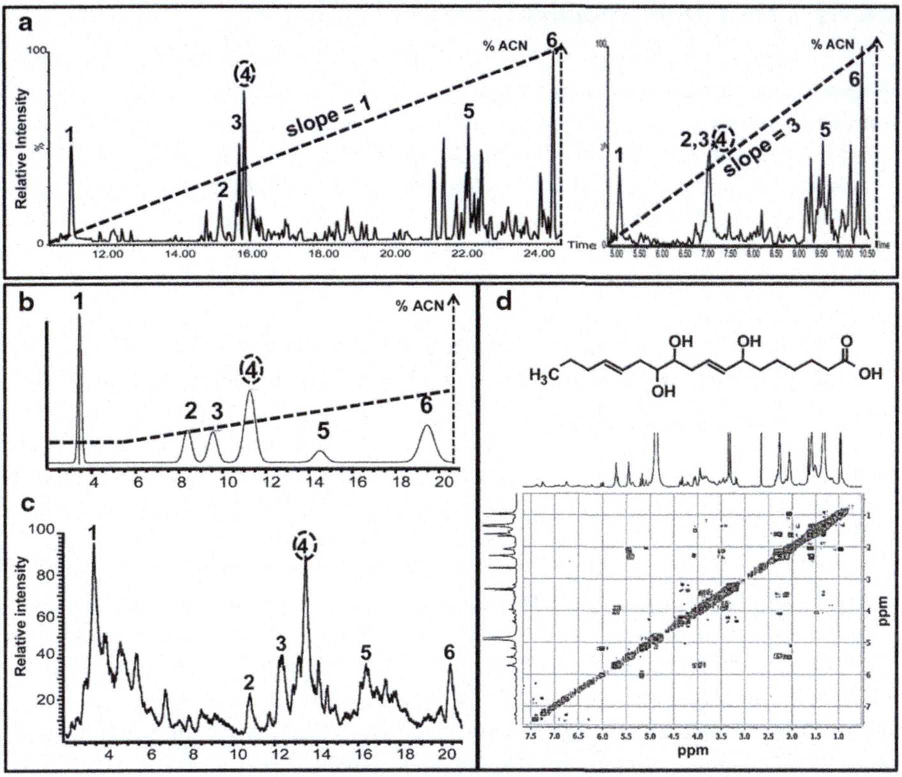

Fig. 16.7 **a** UHPLC-TOF-MS chromatogram on a C18 column (150 × 2.1 mm; i.d., 1.7 μm) performed with two gradients (1 %/min and 3 %/min slopes). **b** optimized theoretical chromatogram by Osiris™ software. **c** the same fraction injected on a semi-preparative C18 column (250 × 10 mm; i.d, 5 μm) and on-line detection using Ion-Trap MS in NI mode. **d** 2D-CapNMR-COSY spectra of *peak 4* and its structure

taxonomic, or biochemical (gene discovery) purposes (Böttcher et al. 2008); (ii) comparing and contrasting the metabolite content of mutant or transgenic plants with that of their wild type counterparts (Hall 2006); (iii) monitoring the behavior of specific classes of metabolites in relation to applied exogenous chemical and/or physical stimuli (Shulaev et al. 2008); (iv) interaction of plants and the environment (Guy et al. 2008a) or herbivore/pathogens (Jansen et al. 2009); (v) studying developmental processes, such as the establishment of symbiotic associations, fruit ripening, or germination (Shu et al. 2008); (vi) quality control of medicinal herbs and phytopharmaceuticals (Chan et al. 2007); and (vii) and determining the activity of medicinal plants (Urich-Merzenich et al. 2007) and health-affecting compounds in food (Hall et al. 2008).

The aim here is not to give a comprehensive review of all of these applications. Therefore, we will focus on a few recent studies related to crops and seeds that give an overview of the various analytical strategies involved (Table 16.1).

16.7.1 Seed Metabolomics

Legume seeds are an important source of food and feed and have been principally studied to optimize the composition of reserve substances like starch, oligosaccharides, oil, and protein. The availability of the genome and transcribed genome sequences for some legume model species has stimulated the application of omics approaches to understanding legume seed development in these and closely related crop species. To date, however, there have been relatively few published analyses at the metabolomic levels for legume seeds (Thompson et al. 2009).

For example, the seed metabolome of pea lines possessing and lacking a major reserve protein, pea albumin 2, produced by introgression of a naturally occurring PA2 locus deletion into a standard genetic background, has been studied (Vigeolas et al. 2008). In that study, the absence of pea albumin 2 was associated with differences in amino acids and polyamine content in the seed. This was attributed in part to the decreased in activity of two enzymes in polyamine synthesis, spermidine synthase and arginine decarboxylase. The metabolite profiling in this case was carried out based on a validated GC-MS method (Roessner-Tunali et al. 2003).

Another interesting study performed on seeds was an investigation of the mung bean (*Vigna radiate*). GC-MS was employed to investigate the metabolic changes in the course of sprouting in a time-dependent manner. The GC method enabled a comprehensive survey of lipophilic (e.g., FA methyl esters, hydrocarbons, fatty alcohols, and sterols) and hydrophilic (e.g., sugars, acids, amino acids, and amines) low molecular weight constituents. Investigation of the obtained fractions resulted in the detection of more than 450 distinct peaks, of which 146 were identified by electron ionization MS. Statistical assessment of the metabolite profiling data *via* PCA demonstrated that the metabolic changes during the sprouting of mung beans were reflected by time-dependent shifts of the scores, which were comparable for two sprouting processes independently conducted under the same conditions. Analysis of the loadings showed that polar metabolites were major contributors to the separation along the first principal component. The dynamic changes of single metabolites revealed significantly increased levels of monosaccharides, organic acids, and amino acids, and a decrease in FA methyl esters (Jom et al. 2011).

GC-MS has also been employed to follow naturally occurring germination processes in rice (Shu et al. 2008). Another GC-MS-based approach on several maize grains from different accessions and growing locations has been undertaken in order to assess the effects and relationship between the environment, season, and genotype on the metabolome pattern of maize kernels (Rohlig et al. 2009). This strategy was based on two successive extractions followed by crude fractionation with solid-phase extraction and liquid-liquid extraction prior to derivatization and led to the characterization of 176 compounds among the 300 features detected. The other metabolomics studies in legumes have been largely restricted to the model species and to non seed tissues, such as soybeans (Frank et al. 2009).

Metabolomic studies can also be helpful for quality control purposes or to help in breeding programs (Fernie and Schauer 2009). Targeted metabolite profiling (LC-

PDA and LC-MS/MS) of typical compounds or those having health claims, such as phenolics in the wheat grain (Dinelli et al. 2009) or capsinoids and carotenoids in the genus *Capsicum* (Wahyuni et al. 2011), allowed the detection of potential cultivars for breeding programs with improved consumer quality traits.

16.7.2 Metabolomic Studies Related to Other Plant Tissues

With the exception of the cases described above, very few studies have been performed on seed metabolomes; thus, we have mainly described here metabolomics studies on other tissues from important crop plants. Various applications to maize, tomato, and rice using different metabolomic strategies have been selected in order to obtain a larger view of how various tissues can be profiled in view of potential applications in seed research. The metabolomic applications were classified according to the following sections: (i) substantial equivalence; (ii) assessment of stress effects; and (iii) monitoring of metabolites during life cycles of a crop plant (Table 16.1).

Substantial Equivalence In order to assess the substantial equivalence of GM maize, an extensive profiling study of transcripts, proteins, and metabolites of transgenic and control maize from field experiments was carried out (Barros et al. 2010). More than 100 metabolites were detected by GC-MS, and fourteen were identified by NMR fingerprinting. Finally, sample differentiation was mainly based on environmental factors rather than on the genotypes studied. Another team used high-resolution MS fingerprinting to assess the substantial equivalence of GM maize with Bt protein (Leon et al. 2009). The use of FT-ICR-MS in PI and NI modes allowed the determination of more than 1000 putative molecular formulae per extract. The resulting raw data was submitted to dedicated software, which enabled the identification of compounds and their associated biochemical pathways based on an organism-specific database (Okuda et al. 2008). From these results, several pathways, such as amino acid metabolism, folate biosynthesis, and purine metabolism, were found to be clearly elicited in GM lines compared to their isogenic control lines. These results were in agreement with those of Levandi and coworkers, who highlighted the quantitative differences between control and GM lines based on CE-TOF-MS profiling (Levandi et al. 2008). In contrast, a targeted approach based on the quantification of major constituents (fat, protein, ash, carbohydrate, and fibre) and key nutrients (amino acids, FAs, vitamins, and minerals) found in maize highlighted an important natural variation among samples that could not be linked to a specific trait (Harrigan et al. 2010). Because of its robustness and repeatability, the compositional analysis remained the reference method for evaluating GM crop substantial equivalence and food-safety assessment (Doerrer et al. 2010). Although metabolomics studies have highlighted important changes in GM crop metabolism, the lack of standardization between studies, fragmentary knowledge on crop variation and influence of the environment, which are not well established, make omics unsuitable for regulatory purposes. However, these limitations could be overcome

Table 16.1 An overview of recent metabolomics analyses in crop science

Method(s)	Instrumentation	Species/organ(s)	Factor(s)	Application	Metabolites detected/identified	References
Substantial equivalence						
HR-MS	CE-ESI-TOF-MS (+); FT-ICR-MS ESI (±)	Maize/grain	Bt protein	Metabolite fingerprinting	1,000/450	Leon et al. 2009
^{1}H-NMR and FTIR	NMR-400 MHz	Rice/grain	Bt protein	Metabolite fingerprinting	–/–	Keymanesh et al. 2009
^{1}H-NMR, GC-MS	NMR-400 MHz; EI-Q-MS	Maize/kernels	Bt protein	Transcript, protein and metabolite fingerprinting	120/26	Barros et al. 2010
LC-MS; GC-MS; CE-MS	ESI-TOF-MS (+)	Tomato/fruit	Miraculin protein	Metabolite profiling	1,639/175	Kusano et al. 2011
LC-UV; LC-MS; LC-MS/MS	APCI; ESI-IT-MS (±)	Wheat/leaves	Antifungal genes (RIP; KP4)	Flavonoid profiling	–/14	Ioset et al. 2007
GC-MS	EI-Q-MS	Wheat/caryopses	D x 5 protein	Metabolite profiling	–/109	Stamova et al. 2009
GC-FID; GC-MS	EI-Q-MS	Rice/grain	Bt protein and CpTI	Metabolite profiling	250/25	Zhou et al. 2009
LC-UV; LC-ESI-MS	ESI-QLIT-MS (–)	Barley/leaves	cThEn and Beta-1,3-Glu	Transcript, targeted and untargeted metabolite profiling	307/72	Kogel et al. 2010
CE-UV; CE-MS	ESI-TOF-MS(+)	Maize/flour	Bt protein	Metabolite profiling	27/27	Levandi et al. 2008
NIR; LC-UV; GC-MS; ICP-AES	EI-Q-MS	Rice/grain	Bt, beta-1,3-Glu, hpt, RCH10, RAC22, beta-Glu, B-RIP	Protein, elements, and primary metabolites profiling	–	Jiao et al. 2010
CE-UV; CE-MS	ESI-TOF-MS(+)	Soybean/seeds	Glyphosate resistant (CP4-EPSPS)	Metabolite profiling	45/45	García-Villalba et al. 2008
HPLC-UV; GC-MS	EI-Q-MS	Tomato/fruits	Down-regulation of DET1 gene (light signal transduction pathway)	Transcript and metabolite profiling	–/120	Enfissi et al. 2010

Table 16.1 (continued)

Method(s)	Instrumentation	Species/organ(s)	Factor(s)	Application	Metabolites detected/identified	References
Assessment of stress effects						
^{1}H-NMR	NMR-600 MHz	Maize/roots and shoots	Salt stress	Metabolite fingerprinting	27/28	Gavaghan et al. 2011
^{1}H-NMR and HR-MAS	NMR-500 and 600 MHz	Rice/seeds	Biotic and abiotic stress	Metabolite fingerprinting	20/15	Fumagalli et al. 2009
^{1}H-NMR	NMR-400 MHz	Wheat/leaves and stem	Fusarium head blight disease	Metabolite fingerprinting	–/10	Browne and Brindle 2007
GC-MS; 1D- and 2D-NMR	NMR-400 MHz; EI-Q-MS	Rice/roots	Chromium stress	Transcript and metabolite profiling	–/42	Dubey et al. 2010
GC-MS	EI-Q-MS	Barley/roots and leaves	Salt stress	Metabolite profiling	102/72	Widodo et al. 2009
1D- and 2D-NMR	NMR-500 MHz	Pea/leaves	Drought stress	Metabolite fingerprinting	–/12	Charlton et al. 2008
ICP-AES; GC-MS	EI-TOF-MS	Lotus/shoots	Salt stress	Elements, transcript and metabolite profiling	123/82	Sanchez et al. 2011
GC-MS; LC-MS	EI-TOF-MS; ESI-TOF-MS (±)	Rice/leaves	Bacterial leaf blight disease	Metabolite profiling	796/202	Sana et al. 2010
Seeds and fruits development						
GC-MS	EI-Q-MS	Maize/grain	Environment, season and genotype impact on maize grain content	Metabolite profiling	300/167	Rohlig et al. 2009
GC-MS	EI-Q-MS	Mung beans	Sprouting process	Metabolite profiling	450/146	Jom et al. 2011
GC-MS; LC-MS; DIMS	EI-Q-MS; TQMS	Potato tuber	Life cycle development	Metabolite profiling	238/161	Shepherd et al. 2010
^{1}H-NMR; LC-PDA; GC-MS; LC-MS; ICP-MS	NMR-500 MHz; EI-TOF-MS; ESI-QTOF-MS(–)	Melon/fruit	Melon fruit development	Elements and metabolite profiling	1,932/197	Moing et al. 2011

Table 16.1 (continued)

Method(s)	Instrumentation	Species/organ(s)	Factor(s)	Application	Metabolites detected/identified	References
LC-MS; GC-MS; SPME-GC-MS	EI-Q-MS; ESI-QIT-MS (+)	Tomato/fruit	Relation between tomato metabolites and sensory attributes	Metabolite profiling	–/29	Thissen et al. 2011
LC-MS	ESI-TOF-MS (–)	Wheat/grain	Phenolic compounds profiling of several wheat varieties	Targeted metabolite profiling	–/70	Dinelli et al. 2009
GC-MS; LC-MS	ESI-QTOF-MS (±)	Strawberry/receptacle and achene	Strawberry fruit development	Metabolite profiling	2,200/160	Fait et al. 2008
GC-MS; LC-MS	EI-QMS; ESI-QTOF-MS (–)	Soybean/roots	Symbiotic response assessment	Metabolite profiling	2610/634	Brechenmacher et al. 2010
LC-PDA; LC-MS	ESI-QTOF-MS (–)	Capsicum/pericarp; seeds	Biochemical variation between pepper accessions	Targeted metabolite profiling	–/25	Wahyuni et al. 2011

APCI atmospheric pressure chemical ionization, *CE* capillary electrophoresis, *EI* electronic ionization, *ESI* electrospray ionization, *FT-ICR* fourier transform-ion cyclotron resonance, *FTIR* fourrier transform infrared spectroscopy, *GC* gas chromatography, *HPLC* high performance liquid chromatography, *HR-MS* high resolution-mass spectrometry, *HR-MAS* high-resolution magic angle spinning NMR, *ICP-AES* inductively coupled plasma-optical emission spectrometry, *IT* ion trap, *NIR* near-infrared reflectance, *NMR* nuclear magnetic resonance, *PDA* photodiode array, *Q* quadrupole, *QLIT* quadrupole-linear ion trap, *SPME* solid phase micro extraction, *TQ* triple quadrupole, *TOF* time of flight, *UV* ultra-violet, + positive ionization mode, – negative ionization mode with the rapid improvement of dedicated databases (Kind et al. 2009) and standardization of metabolomic data (Fiehn et al. 2007).

Assessment of Stress Effects Another topic that has been investigated in depth in crop fitness research is the influence of various biotic or abiotic stresses (Shulaev et al. 2008). For example, the salt-stress response of maize shoots and roots has been studied by direct NMR fingerprinting of aqueous extracts. A clear separation of the growth and saline effects was highlighted by supervised data mining methods. Based on proton chemical shifts, 28 compounds were characterized. Among them, increased levels of alanine, glutamate, asparagines, and glycine-betaine upon salt stress were detected (Gavaghan et al. 2011). The drought-stress response of pea leaves has been monitored with the help of 1D- and 2D-NMR by Charlton and coworkers (Charlton et al. 2008). The fast acquisition of several ^{1}H-NMR spectra allowed the determination of the most significant areas of changes upon drought stress after application of a supervised data mining method. Finally, a dozen compounds, mainly amino acids, were unambiguously identified in crude extracts by 2D-NMR experiments are believed to characterize the drought-stress response of pea leaves.

Monitoring of Metabolites During Life Cycles of Crop Plants The development of fruits or other plant organs can also be accurately monitored by metabolomics. In the frame of monitoring the development of organs, a more comprehensive MS-based metabolomic analysis was applied to study the potato tuber life cycle in potatoes (Shepherd et al. 2010). In this case, changes in the metabolome were assessed by LC-MS, GC-MS, and DIMS. The data was subjected to PCA and HCA (hierarchical cluster analysis) to assess the potential for separating the life cycle stages, to define the major profiles of metabolite changes during the life cycle stages, and to determine which metabolites underpinned these profiles. The study provided new insight into the temporal changes in metabolites related to acrylamide-forming potential. More than 1,000 metabolites were detected by coupling GC-MS and LC-MS data acquisition for the study of strawberry fruit development (Moing et al. 2011). Among them, 160 of the most characteristic metabolites of primary and secondary metabolism were identified. Based on these results, global networks of the putative biochemical events that trigger strawberry fruit development were developed. The same type of study was carried out to monitor melon fruit development at the metabolome level, including volatile and nonvolatile primary and secondary metabolites and mineral element profiling by using six different analytical platforms (Moing et al. 2011). Thirty-seven polar metabolites were detected by ^{1}H-NMR and more than 100 were determined through GC-MS after derivatization. The use of LC-QTOF-MS for semi-polar compounds allowed the detection of 1,200 compounds, and GC-MS analysis of the volatile fractions detected 500 components. The analytical methods were highly complementary, with fewer than 20 compounds being detected by more than one technology.

These different applications, and the others summarised in Table 16.1, demonstrate that metabolomics plays an increasingly important role in various aspects of crop plant research, including seed investigation, but also has consequences on seed modifications in whole plant tissue-specific organ germination and organ life cycle monitoring. The effect of stress at the metabolite level can also be monitored or compared as a consequence of seed selection to improve plant fitness. While the combination of several instruments enabled the detection of more than one thousand compounds in the most comprehensive studies, the identification of putative compounds remains a major drawback in metabolomics, with only one-third to one-half of the features identified.

16.8 Concluding Remarks

The characterization of natural diversity in plant metabolites using unbiased metabolite profiling approaches is already providing deeper knowledge of plant and crop composition and its variable nature, both within and between species. While genomics techniques are in the process of revolutionizing plant biology, there is a need for complementarity in other studies that can confirm the functional aspects of these changes (Thompson et al. 2009); in this respect, metabolomics plays an increasingly important role. Metabolomics was still in its infancy at the beginning of the millennium; however, approximately10 years later, this approach is mature and represents an important complement to the other omics used in systems biology.

For crop and particularly seed research, investigations have been mainly performed using a single analytical profiling approach in combination with advanced data mining methods. These work provided very interesting results. However, only a partial view of the metabolome has been obtained in this way. The most comprehensive coverage of metabolome composition at both the level of primary or secondary metabolites can only be obtained by combining analytical methods. For this, MS-based and NMR-based metabolomic approaches represent an essential complement. The use of MS in plant metabolomics will continue to grow massively and should be accompanied by further improvements in raw data filtering, deconvolution, and metabolite identification. NMR approaches will remain very important for reasons of reproducibility, throughput and identification, and both methods are very complementary (van der Kooy et al. 2009). When applied to well-defined biological issues, metabolomic data can be efficiently transformed into new biological knowledge, provided that correct experimental design, analytics, data mining, and biomarker identification are performed.

While metabolomics is garnering significant interest among plant scientists, the approach suffers from limitations that are not to be under estimated. In the case of MS-based strategies, there is still a need to increase metabolome coverage and reproducibility. For NMR-based metabolomics, the sensitivity and the number of metabolites observed remains a problem. In all approaches used, complete metabolite identification remains a challenging task, because often, only partial peak annotation is obtained. As has been shown in this chapter, innovative approaches based on efficient LC separation and subsequent *de novo* structural determinations are needed. In this respect, efforts aimed at rapid and efficient strategies for the dereplication of natural products based on on-line or at-line hyphenated methods should be strengthened. Although substantial improvements have been made in the field of metabolomics, the uniform annotation of metabolite signals in databases and informatics through international standardization efforts also remains a challenge.

Important technological advances in the biological sciences have notably strengthened the emerging field of systems biology. Although a complete understanding of living organisms at the molecular system level is far from reality, comprehensive investigations of living organisms, such as the study of plants with different omics technologies, represents an important step forward. Many important findings will continue to emerge from metabolomics in various fields of plant science, and the investigation of seeds by such an unbiased approach still has important potential and may open new avenues in crop science research.

Acknowledgments Financial support was provided by the Swiss National Science Foundation (Grant no. 205320_135190/1 to J-L.W.). The work was also supported by the National Centre of Competence in Research (NCCR) Plant Survival, a research program of the Swiss National Science Foundation, as well as by an IPP project from the SystemsX.ch initiative.

References

Alborn HT, Turlings TCJ, Jones TH, Stenhagen G, Loughrin JH, Tumlinson JH (1997) An elicitor of plant volatiles from beet armyworm oral secretion. Science 276:945–949

Allwood JW, Ellis DI, Goodacre R (2008) Metabolomic technologies and their application to the study of plants and plant-host interactions. Physiol Plant 132:117–135

Barros E, Lezar S, Anttonen MJ, Van Dijk JP, Röhlig RM, Kok EJ, Engel KH (2010) Comparison of two GM maize varieties with a near-isogenic non-GM variety using transcriptomics, proteomics and metabolomics. Plant Biotechnol J 8:436–451

Boccard J, Veuthey JL, Rudaz S (2010) Knowledge discovery in metabolomics: an overview of MS data handling. J Sep Sci 33:290–304

Bonnington LS, Barcelò D, Knepper TP (2003) Utilisation of electrospray time-of-flight mass spectrometry for solving complex fragmentation patterns: application to benzoxazinone derivatives. J Mass Spectrom 38:1054–1066

Boroczky K, Laatsch H, Wagner-Dobler I, Stritzke K, Schulz S (2006) Cluster analysis as selection and dereplication tool for the identification of new natural compounds from large sample sets. Chem Biodiversity 3:622–634

Böttcher C, von Roepenack-Lahaye E, Schmidt J, Schmotz C, Neumann S, Scheel D, Clemens S (2008) Metabolome analysis of biosynthetic mutants reveals a diversity of metabolic changes and allows identification of a large number of new compounds in *Arabidopsis*. Plant Physiol 147:2107–2120

Brechenmacher L, Lei Z, Libault M, Findley S, Sugawara M, Sadowsky MJ, Sumner LW, Stacey G (2010) Soybean metabolites regulated in root hairs in response to the symbiotic bacterium *Bradyrhizobium japonicum*. Plant Physiol 153:1808–1822

Brown WV (1960) The morphology of the grass embryo. Phytomorphology 10:215–223

Brown SJ, Asai Y, Cordell HJ, Campbell LE, Zhao Y, Liao H, Northstone K, Henderson J, Alizadehfar R, Ben-Shoshan M, Morgan K, Roberts G, Masthoff LJ, Pasmans SG, van den Akker PC, Wijmenga C, Hourihane JO, Palmer CN, Lack G, Clarke A, Hull PR, Irvine AD, McLean WH

(2011) Loss-of-function variants in the filaggrin gene are a significant risk factor for peanut allergy. J Allergy Clin Immunol 127:661–667
Browne RA, Brindle KM (2007) ^{1}H NMR-based metabolite profiling as a potential selection tool for breeding passive resistance against *Fusarium* head blight (FHB) in wheat. Mol Plant Pathol 8:401–410
Byrne PF, McMullen MD, Snook ME, Musket TA, Theuri JM, Widstrom NW, Wiseman BR, Coe EH (1996) Quantitative trait loci and metabolic pathways: genetic control of the concentration of maysin, a corn earworm resistance factor, in maize silks. Proc Natl Acad Sci U S A 93:8820–8825
Chan ECY, Yap SL, Lau AJ, Leow PC, Toh DF, Koh HL (2007) Ultra-performance liquid chromatography/time-of-flight mass spectrometry based metabolomics of raw and steamed *Panax notoginseng*. Rapid Commun Mass Spectrom 21:519–528
Charlton A, Donarski J, Harrison M, Jones S, Godward J, Oehlschlager S, Arques J, Ambrose M, Chinoy C, Mullineaux P, Domoney C (2008) Responses of the pea (*Pisum sativum* L.) leaf metabolome to drought stress assessed by nuclear magnetic resonance spectroscopy. Metabolomics 4:312–327
Choi YH, Kim HK, Linthorst HJM, Hollander JG, Lefeber AWM, Erkelens C, Nuzillard JM, Verpoorte R (2006) NMR metabolomics to revisit the tobacco mosaic virus infection in *Nicotiana tabacum* leaves. J Nat Prod 69:742–748
Colquhoun IJ (2007) Use of NMR for metabolic profiling in plant systems. J Pest Sci 32:200–212
Crockford DJ, Maher AD, Ahmadi KR, Barrett A, Plumb RS, Wilson ID, Nicholson JK (2008) ^{1}H NMR and UPLC-MSE statistical heterospectroscopy: characterization of drug metabolites (Xenometabolome) in epidemiological studies. Anal Chem 80:6835–6844
David F, Vanhoenacker G, Tienpont B, Francois I, Sandra P (2007) Coupling columns and multidimensional configurations to increase peak capacity in liquid chromatography. LCGC Europe 20:154–158
Davies HV, Shepherd LVT, Stewart D, Frank T, Röhlig RM, Engel KH (2010) Metabolome variability in crop plant species—when, where, how much and so what? Regul Toxicol Pharmacol 58:S54–S61
De Vos RCH, Moco S, Lommen A, Keurentjes JJB, Bino RJ, Hall RD (2007) Untargeted large-scale plant metabolomics using liquid chromatography coupled to mass spectrometry. Nat Protoc 2:778–791
Dettmer K, Aronov PA, Hammock BD (2007) Mass spectrometry-based metabolomics. Mass Spectrom Rev 26:51–78
Dinelli G, Segura Carretero A, Di Silvestro R, Marotti I, Fu S, Benedettelli S, Ghiselli L, Fernández Gutiérrez A (2009) Determination of phenolic compounds in modern and old varieties of durum wheat using liquid chromatography coupled with time-of-flight mass spectrometry. J Chromatogr A 1216:7229–7240
Doerrer N, Ladics G, McClain S, Herouet-Guicheney C, Poulsen LK, Privalle L, Stagg N (2010) Evaluating biological variation in non-transgenic crops: executive summary from the ILSI Health and Environmental Sciences Institute workshop, November 16–17, 2009, Paris, France. Regul Toxicol Pharmacol 58:S2–S7
Draper J, Enot DP, Parker D, Beckmann M, Snowdon S, Lin W, Zubair H (2009) Metabolite signal identification in accurate mass metabolomics data with MZedDB, an interactive *m/z* annotation tool utilising predicted ionisation behaviour 'rules'. BMC Bioinformatics 10:227
Dubey S, Misra P, Dwivedi S, Chatterjee S, Bag S, Mantri S, Asif M, Rai A, Kumar S, Shri M, Tripathi P, Tripathi R, Trivedi P, Chakrabarty D, Tuli R (2010) Transcriptomic and metabolomic shifts in rice roots in response to Cr (VI) stress. BMC Genomics 11:648
Ducruix C, Vailhen D, Werner E, Fievet JB, Bourguignon J, Tabet JC, Ezan E, Junot C (2008) Metabolomic investigation of the response of the model plant *Arabidopsis* thaliana to cadmium exposure: evaluation of data pretreatment methods for further statistical analyses. Chemom Intell Lab Syst 91:67–77
Dunn WB (2008) Current trends and future requirements for the mass spectrometric investigation of microbial, mammalian and plant metabolomes. Phys Biol 5:011001

Dunn WB, Broadhurst DI, Atherton HJ, Goodacre R, Griffin JL (2011) Systems level studies of mammalian metabolomes: the roles of mass spectrometry and nuclear magnetic resonance spectroscopy. Chem Soc Rev 40:387–426

Enfissi EMA, Barneche F, Ahmed I, Lichtlé C, Gerrish C, McQuinn RP, Giovannoni JJ, Lopez-Juez E, Bowler C, Bramley PM, Fraser PD (2010) Integrative transcript and metabolite analysis of nutritionally enhanced *DE-ETIOLATED1* downregulated tomato fruit. Plant Cell 22:1190–1215

Erb M, Flors V, Karlen D, de Lange E, Planchamp C, D'Alessandro M, Turlings TCJ, Ton J (2009) Signal signature of above ground-induced resistance upon below ground herbivory in maize. Plant J 59:292–302

Erb M, Balmer D, De Lange ES, Von Merey G, Planchamp C, Robert CAM, Röder G, Sobhy I, Zwahlen C, Mauch-Mani B, Turlings TCJ (2011) Synergies and trade-offs between insect and pathogen resistance in maize leaves and roots. Plant Cell Environ 34:1088–1103

Eugster PJ, Guillarme D, Rudaz S, Veuthey JL, Carrupt PA, Wolfender JL (2011) Ultra high pressure liquid chromatography for crude plant extracts profiling. J AOAC Int 94:51–70

Fait A, Hanhineva K, Beleggia R, Dai N, Rogachev I, Nikiforova VJ, Fernie AR, Aharoni A (2008) Reconfiguration of the achene and receptacle networks during strawberry fruit development. Plant Physiol 148:730–750

Feng X, Siegel MM (2007) FTICR-MS applications for the structure determination of natural products. Anal Bioanal Chem 389:1341–1363

Fernie AR, Schauer N (2009) Metabolomics-assisted breeding: a viable option for crop improvement? Trends Genet 25:39–48

Fiehn O (2008) Extending the breadth of metabolite profiling by gas chromatography coupled to mass spectrometry. Trends Analyt Chem 27:261–269

Fiehn O, Kopka J, Dörmann P, Altmann T, Trethewey RN (2000) Metabolite profiling for plant functional genomics. Nat Biotechnol 18:1157–1161

Fiehn O, Sumner LW, Rhee SY, Ward J, Dickerson J, Lange BM, Lane G, Roessner U, Last R, Nikolau B (2007) Minimum reporting standards for plant biology context information in metabolomic studies. Metabolomics 3:195–201

Frank T, Nörenberg S, Engel KH (2009) Metabolite profiling of two novel low phytic acid (*lpa*) soybean mutants. J Agric Food Chem 57:6408–6416

Frey M, Chomet P, Glawischnig E, Stettner C, Grun S, Winklmair A, Eisenreich W, Bacher A, Meeley RB, Briggs SP, Simcox K, Gierl A (1997) Analysis of a chemical plant defense mechanism in grasses. Science 277:696–699

Fumagalli E, Baldoni E, Abbruscato P, Piffanelli P, Genga A, Lamanna R, Consonni R (2009) NMR techniques coupled with multivariate statistical analysis: tools to analyse *Oryza sativa* metabolic content under stress conditions. J Agr Crop Sci 195:77–88

García-Villalba R, León C, Dinelli G, Segura-Carretero A, Fernández-Gutiérrez A, Garcia-Cañas V, Cifuentes A (2008) Comparative metabolomic study of transgenic versus conventional soybean using capillary electrophoresis-time-of-flight spectrometry. J Chromatogr A 1195:164–173

Gavaghan CL, Li JV, Hadfield ST, Hole S, Nicholson JK, Wilson ID, Howe PWA, Stanley PD, Holmes E (2011) Application of NMR-based metabolomics to the investigation of salt stress in maize (*Zea mays*). Phytochem Anal 22:214–224

Gavaghan CL, Li JV, Hadfield ST, Hole S, Nicholson JK, Wilson ID, Howe PWA, Stanley PD, Holmes E (2011) Application of NMR-based metabolomics to the investigation of salt stress in maize (*Zea mays*). Phytochem Anal 22:214–224

Glauser G, Guillarme D, Grata E, Boccard J, Thiocone A, Carrupt PA, Veuthey JL, Rudaz S, Wolfender JL (2008) Optimized liquid chromatography-mass spectrometry approach for the isolation of minor stress biomarkers in plant extracts and their identification by capillary nuclear magnetic resonance. J Chromatogr A 1180:90–98

Goga-Remont S, Heinisch S, Rocca JL (2000) Use of optimization software to determine rugged analysis conditions in high-performance liquid chromatography. J Chromatogr A 868:13–29

Goodacre R (2005) Making sense of the metabolome using evolutionary computation: seeing the wood with the trees. J Exp Bot 56:245–254
Grata E, Boccard J, Guillarme D, Glauser G, Carrupt PA, Farmer E, Wolfender JL, Rudaz S (2008) UPLC–TOF-MS for plant metabolomics: a sequential approach for wound marker analysis in *Arabidopsis thaliana*. J Chromatogr B 871 261–270
Grata E, Guillarme D, Glauser G, Boccard J, Carrupt PA, Veuthey JL, Rudaz S, Wolfender JL (2009) Metabolite profiling of plant extracts by ultra-high-pressure liquid chromatography at elevated temperature coupled to time-of-flight mass spectrometry. J Chromatogr A 1216:5660–5668
Guillarme D, Schappler J, Rudaz S, Veuthey JL (2010) Coupling ultra-high-pressure liquid chromatography with mass spectrometry. Trends Analyt Chem 29:15–27
Guy C, Kaplan F, Kopka J, Selbig J, Hincha DK (2008a) Metabolomics of temperature stress. Physiol Plant 132:220–235
Guy C, Kopka J, Moritz T (2008b) Plant metabolomics coming of age. Physiol Plant 132:113–116
Hall RD (2006) Plant metabolomics: from holistic hope, to hype, to hot topic. New Phytol 169:453–468
Hall RD, Brouwer ID, Fitzgerald MA (2008) Plant metabolomics and its potential application for human nutrition. Physiol Plant 132:162–175
Harrigan GG, Glenn KC, Ridley WP (2010) Assessing the natural variability in crop composition. Regul Toxicol Pharmacol 58:S13–S20
Hegeman AD (2010) Plant metabolomics-meeting the analytical challenges of comprehensive metabolite analysis. Brief Funct Genomics 9:139–148
Ioset JR, Urbaniak B, Ndjoko-Ioset K, Wirth J, Martin F, Gruissem W, Hostettmann K, Sautter C (2007) Flavonoid profiling among wild type and related GM wheat varieties. Plant Mol Biol 65:645–654
Issaq HJ, Van QN, Waybright TJ, Muschik GM, Veenstra TD (2009) Analytical and statistical approaches to metabolomics research. J Sep Sci 32:2183–2199
James C (2009) Global status of commercialized biotech/GM crops: 2009. ISAAA Brief No. 41. ISAAA, Ithaca
Jansen JJ, Allwood JW, Marsden-Edwards E, Van Der Putten WH, Goodacre R, van Dam NM (2009) Metabolomic analysis of the interaction between plants and herbivores. Metabolomics 5:150–161
Jiao Z, Si XX, Li GK, Zhang ZM, Xu XP (2010) Unintended compositional changes in transgenic rice seeds (*Oryza sativa* L.) studied by spectral and chromatographic analysis coupled with chemometrics methods. J Agric Food Chem 58:1746–1754
Jom KN, Frank T, Engel KH (2011) A metabolite profiling approach to follow the sprouting process of mung beans (*Vigna radiata*). Metabolomics 7:102–117
Katajamaa M, Oresic M (2007) Data processing for mass spectrometry-based metabolomics. J Chromatogr A 1158:318–328
Keymanesh K, Darvishi MH, Sardari S (2009) Metabolome comparison of transgenic and non-transgenic rice by statistical analysis of FTIR and NMR Spectra. Rice Science 16:119–123
Kim HK, Choi YH, Verpoorte R (2010) NMR-based metabolomic analysis of plants. Nat Protoc 5:536–549
Kim HK, Choi YH, Verpoorte R (2011) NMR-based plant metabolomics: where do we stand, where do we go? Trends Biotechnol 29:267–275
Kind T, Fiehn O (2007) Seven golden rules for heuristic filtering of molecular formulas obtained by accurate mass spectrometry. BMC Bioinformatics 8:105
Kind T, Scholz M, Fiehn O (2009) How large is the metabolome? A critical analysis of data exchange practices in chemistry. PLoS One 4:e5440
Kogel KH, Voll LM, Schäfer P, Jansen C, Wu Y, Langen G, Imani J, Hofmann J, Schmiedl A, Sonnewald S, von Wettstein D, Cook RJ, Sonnewald U (2010) Transcriptome and metabolome profiling of field-grown transgenic barley lack induced differences but show cultivar-specific variances. Proc Nat Acad Sci U S A 107:6198–6203

Korfmacher WA (2005) Principles and applications of LC-MS in new drug discovery. Drug Discov Today 10:1357–1367
Kusano M, Redestig H, Hirai T, Oikawa A, Matsuda F, Fukushima A, Arita M, Watanabe S, Yano M, Hiwasa-Tanase K, Ezura H, Saito K (2011) Covering chemical diversity of genetically-modified tomatoes using metabolomics for objective substantial equivalence assessment. PLoS One 6:e16989
Leon C, Rodriguez-Meizoso I, Lucio M, Garcia-Cañas V, Ibañez E, Schmitt-Kopplin P, Cifuentes A (2009) Metabolomics of transgenic maize combining Fourier transform-ion cyclotron resonance-mass spectrometry, capillary electrophoresis-mass spectrometry and pressurized liquid extraction. J Chromatogr A 1216:7314–7323
Levandi T, Leon C, Kaljurand M, Garcia-Canas V, Cifuentes A (2008) Capillary electrophoresis time-of-flight mass spectrometry for comparative metabolomics of transgenic versus conventional maize. Anal Chem 80:6329–6335
Li B, Morris AJ, Martin EB (2002) Orthogonal signal correction: algorithmic aspects and properties. J Chemometrics 16:556–561
Moing A, Aharoni A, Biais B, Rogachev I, Meir S, Brodsky L, Allwood JW, Erban A, Dunn WB, Kay L, de Koning S, de Vos RCH, Jonker H, Mumm R, Deborde C, Maucourt M, Bernillon S, Gibon Y, Hansen TH, Husted S, Goodacre R, Kopka J, Schjoerring JK, Rolin D, Hall RD (2011) Extensive metabolic cross-talk in melon fruit revealed by spatial and developmental combinatorial metabolomics. New Phytol 190:683–696
Nguyen DTT, Guillarme D, Rudaz S, Veuthey JL (2006) Fast analysis in liquid chromatography using small particle size and high pressure. J Sep Sci 29:1836–1848
Nováková L, Matysová L, Solich P (2006) Advantages of application of UPLC in pharmaceutical analysis. Talanta 68:908–918
Nuessly GS, Scully BT, Hentz MG, Beiriger R, Snook ME, Widstrom NW (2007) Resistance to *Spodoptera frugiperda* (Lepidoptera: Noctuidae) and *Euxesta stigmatias* (Diptera: Ulididae) in sweet corn derived from exogenous and endogenous genetic systems. J Econ Entomol 100:1887–1895
Okuda S, Yamada T, Hamajima M, Itoh M, Katayama T, Bork P, Goto S, Kanehisa M (2008) KEGG Atlas mapping for global analysis of metabolic pathways. Nucl Acids Res 36:W423–W426
Rochfort S (2005) Metabolomics reviewed: a new "Omics" platform technology for systems biology and implications for natural products research. J Nat Prod 68:1813–1820
Roessner-Tunali U, Hegemann B, Lytovchenko A, Carrari F, Bruedigam C, Granot D, Fernie AR (2003) Metabolic profiling of transgenic tomato plants overexpressing hexokinase reveals that the influence of hexose phosphorylation diminishes during fruit development. Plant Physiol 133:84–99
Rohlig RM, Eder J, Engel KH (2009) Metabolite profiling of maize grain: differentiation due to genetics and environment. Metabolomics 5:459–477
Saito K, Matsuda F (2010) Metabolomics for functional genomics, systems biology, and biotechnology. Annu Rev Plant Biol 61:463–489
Sana TR, Fischer S, Wohlgemuth G, Katrekar A, Jung KH, Ronald PC, Fiehn O (2010) Metabolomic and transcriptomic analysis of the rice response to the bacterial blight pathogen *Xanthomonas oryzae* pv. *oryzae*. Metabolomics 6:451–465
Sanchez DH, Pieckenstain FL, Escaray F, Erban A, Kraemer U, Udvardi MK, Kopka J (2011) Comparative ionomics and metabolomics in extremophile and glycophytic *Lotus* species under salt stress challenge the metabolic pre-adaptation hypothesis. Plant Cell Environ 34:605–617
Sawada Y, Akiyama K, Sakata A, Kuwahara A, Otsuki H, Sakurai T, Saito K, Hirai MY (2009) Widely targeted metabolomics based on large-scale MS/MS data for elucidating metabolite accumulation patterns in plants. Plant Cell Physiol 50:37–47
Schauer N, Steinhauser D, Strelkov S, Schomburg D, Allison G, Moritz T, Lundgren K, Roessner-Tunali U, Forbes MG, Willmitzer L, Fernie AR, Kopka J (2005) GC-MS libraries for the rapid identification of metabolites in complex biological samples. FEBS Lett 579:1332–1337

Shepherd L, Alexander C, Sungurtas J, McNicol J, Stewart D, Davies H (2010) Metabolomic analysis of the potato tuber life cycle. Metabolomics 6:274–291

Shu XL, Frank T, Shu QY, Engel KR (2008) Metabolite profiling of germinating rice seeds. J Agric Food Chem 56:11612–11620

Shulaev V, Cortes D, Miller G, Mittler R (2008) Metabolomics for plant stress response. Physiol Plant 132:199–208

Sicker D, Schulz M (2002) Benzoxazinones in plants: occurrence, synthetic access, and biological activity. In: Rahman Au (ed) Studies in natural products chemistry 27. Elsevier, Amsterdam, pp 185–232

Son HS, Hwang GS, Kim KM, Ahn HJ, Park WM, Van Den Berg F, Hong YS, Lee CH (2009) Metabolomic studies on geographical grapes and their wines using H-1 NMR analysis coupled with multivariate statistics. J Agric Food Chem 57:1481–1490

Stamova B, Roessner U, Suren S, Laudencia-Chingcuanco D, Bacic A, Beckles D (2009) Metabolic profiling of transgenic wheat over-expressing the high-molecular-weight Dx5 glutenin subunit. Metabolomics 5:239–252

Sumner LW, Mendes P, Dixon RA (2003) Plant metabolomics:large-scale phytochemistry in the functional genomics era. Phytochemistry 62:817–836

t'Kindt R, Morreel K, Deforce D, Boerjan W, Van Bocxlaer J (2009) Joint GC-MS and LC-MS platforms for comprehensive plant metabolomics: repeatability and sample pre-treatment. J Chromatogr B 877:3572–3580

Thissen U, Coulier L, Overkamp KM, Jetten J, Van Der Werff BJC, van de Ven T, Van Der Werf MJ (2011) A proper metabolomics strategy supports efficient food quality improvement: a case study on tomato sensory properties. Food Qual Prefer 22:499–506

Thompson R, Burstin J, Gallardo K (2009) Post-genomics studies of developmental processes in legume seeds. Plant Physiol 151:1023–1029

Tiziani S, Lodi A, Ludwig C, Parsons HM, Viant MR (2008) Effects of the application of different window functions and projection methods on processing of 1H J-resolved nuclear magnetic resonance spectra for metabolomics. Anal Chim Acta 610:80–88

Trygg J, Holmes E, Lundstedt T (2007) Chemometrics in metabonomics. J Proteome Res 6:469–479

Turlings TCJ, Tumlinson JH (1992) Systemic release of chemical signals by herbivore-injured corn. Proc Natl Acad Sci USA 89:8399–8402

Urich-Merzenich G, Zeitler H, Jobst D, Panek D, Vetter H, Wagner H (2007) Application of the "-omic-" technologies in phytomedicine. Phytomedicine 14:70–82

Van Der Hooft JJJ, Vervoort J, Bino RJ, Beekwilder J, de Vos RCH (2011) Polyphenol identification based on systematic and robust high-resolution accurate mass spectrometry fragmentation. Anal Chem 83:409–416

Van Der Kooy F, Maltese F, Hae Choi Y, Kyong Kim H, Verpoorte R (2009) Quality control of herbal material and phytopharmaceuticals with MS and NMR based metabolic fingerprinting. Planta Med 75:763–775

Verpoorte R, Choi Y, Kim H 2007 NMR-based metabolomics at work in phytochemistry. Phytochem Rev 6:3–14

Vigeolas H, Chinoy C, Zuther E, Blessington B, Geigenberger P, Domoney C 2008 Combined metabolomic and genetic approaches reveal a link between the polyamine pathway and albumin 2 in developing pea seeds. Plant Physiol 146:74–82

Vinale F, Sivasithamparam K, Ghisalberti EL, Marra R, Woo SL, Lorito M. 2008 *Trichoderma*-plant-pathogen interactions. Soil Biol Biochem 40:1–10

Vlahov G, Del Re P, Simone N (2003) Determination of geographical origin of olive oils using 13C nuclear magnetic resonance spectroscopy. I—Classification of olive oils of the Puglia region with denomination of protected origin. J Agric Food Chem 51:5612–5615

Wahyuni Y, Ballester AR, Sudarmonowati E, Bino RJ, Bovy AG (2011) Metabolite biodiversity in pepper (*Capsicum*) fruits of thirty-two diverse accessions: variation in health-related compounds and implications for breeding. Phytochemistry 72:1358–1370

Waridel P, Wolfender J-L, Ndjoko K, Hobby KR, Major HJ, Hostettmann K (2001) Evaluation of quadrupole time-of-flight tandem mass spectrometry and ion-trap multiple-stage mass spectrometry for the differentiation of C-glycosidic flavonoid isomers. J Chromatogr A 926:29–41

Webb AG (2005) Microcoil nuclear magnetic resonance spectroscopy. J Pharm Biomed Anal 38:892–903

Widodo, Patterson JH, Newbigin E, Tester M, Bacic A, Roessner U (2009) Metabolic responses to salt stress of barley (*Hordeum vulgare* L.) cultivars, Sahara and Clipper, which differ in salinity tolerance. J Exp Bot 60:4089–4103

Wilson ID, Nicholson JK, Castro-Perez J, Granger JH, Johnson KA, Smith BW, Plumb RS (2005) High resolution "ultra performance" liquid chromatography coupled to TOF mass spectrometry as a tool for differential metabolic pathway profiling in functional genomic studies. J Proteome Res 4:591–598

Wishart DS. (2008) Quantitative metabolomics using NMR. Trends Analyt Chem 27:228–237

Wolfender JL (2010) LC-NMR and related techniques for the rapid identification of plant metabolites. In: Waksmundzka-Hajnos M, Sherma J (eds) High performance liquid chromatography in phytochemical analysis. CRC Press, Taylor and Francis, Boca Raton, pp 287–330

Wolfender JL, Queiroz EF, Hostettmann K (2005) Phytochemistry in the microgram domain—a LC-NMR perspective. Magn Reson Chem 43:697–709

Wolfender JL, Marti G, Queiroz EF (2010) Advances in techniques for profiling crude extracts and for the rapid identification of natural products: dereplication, quality control and metabolomics. Curr Org Chem 14:1808–1832

Wolfender JL, Eugster PJ, Bohni N, Cuendet M (2011) Advanced methods for natural product drug discovery in the field of nutraceuticals. CHIMIA Intl J Chem 65:400–406

Wren SAC, Tchelitcheff P (2006) Use of ultra-performance liquid chromatography in pharmaceutical development. J Chromatogr A 1119:140–146

Yonekura-Sakakibara K, Saito K (2009) Functional genomics for plant natural product biosynthesis. Nat Prod Rep 26:1466–1487

Zhou J, Ma C, Xu H, Yuan K, Lu X, Zhu Z, Wu Y, Xu G (2009) Metabolic profiling of transgenic rice with *cryIAc* and *sck* genes: an evaluation of unintended effects at metabolic level by using GC-FID and GC-MS. J Chromatogr B 877:725–732

Part V
Towards Systems Biology: Organization, Integration and Modelization of Data

Chapter 17
Plant Metabolic Pathways: Databases and Pipeline for Stoichiometric Analysis

Eva Grafahrend-Belau, Björn H. Junker and Falk Schreiber

Abstract Mathematical modeling of plant metabolism offers new approaches to improve the understanding of complex biological processes. In this chapter an overview of resources and tools available for the reconstruction of stoichiometric models and their constraint-based analysis is given, focusing on plant metabolic pathways. To facilitate and support the modeling of metabolism, a pipeline for the constraint-based analysis of crop plant metabolic models is described and the proposed framework is applied in a case study of storage metabolism in developing barley seeds.

Keywords Constraint-based analysis (CBA) · CBA software · Crop plant metabolism · Flux balance analysis · Modeling pipeline · Plant metabolic pathway databases

17.1 Introduction

Bioinformatics has become an integral part of biological research, including basic and applied plant science. To understand the complexity inherent in the vast amount of data gathered by large-scale profiling of biological features (e.g., transcriptomics, proteomics, and metabolomics data), new approaches need to be developed with focus on the management, visualization, integration, analysis, modelization, and prediction of these data. Mathematical modeling of metabolism provides new concepts and methods to elucidate the structure, dynamics, and behavior of complex biological processes. *In silico* models of biological systems can be applied to verify and extend the understanding of complex metabolic processes, to generate and test hypothesis, and to explore *in silico* scenarios, thus supplementing or even replacing experiments.

In plant research, the issue of modeling metabolism is constantly gaining attention. Several mathematical modeling approaches applied to plant metabolism exist

---

F. Schreiber (✉)
Clayton School of Information Technology, Monash University, Victoria 3800, Australia
e-mail: schreibe@ipk-gatersleben.de

F. Schreiber · E. Grafahrend-Belau · B. H. Junker
Leibniz Institute of Plant Genetics and Crop Plant Research (IPK),
Correnssstr. 3, 06466 Gatersleben, Germany

G. K. Agrawal, R. Rakwal (eds.), *Seed Development: OMICS Technologies toward Improvement of Seed Quality and Crop Yield*,
DOI 10.1007/978-94-007-4749-4_17, © Springer Science+Business Media Dordrecht 2012

ranging from highly detailed quantitative to less complex qualitative approaches (for reviews, see Giersch 2000; Morgan and Rhodes 2002; Poolman et al. 2004; Rios-Estepa and Lange 2007). Although these approaches offer the most detailed model predictions, the application of quantitative kinetic modeling is usually restricted to small-scale biological systems due to the lack of kinetic information. In contrast, qualitative modeling approaches, such as the constraint-based approach, require less data and thereby, having the advantage of permitting an in-depth analysis of large-scale systems.

Flux balance analysis (FBA) is a constraint-based modeling approach that allows the prediction of metabolic steady-state fluxes by applying mass-balance constraints to a stoichiometric model (Edwards et al. 1999). One advantage of FBA is that it requires a relatively low prior knowledge about the metabolic system, such as the knowledge of reaction stoichiometry and directionality, biomass composition, and maximum uptake/excretion rates of nutrients and products. FBA has been applied to a variety of different biological systems, such as bacteria (Schilling et al. 2002; Reed and Palsson 2003), fungi (David et al. 2003; Famili et al. 2003), algae (Shastri and Morgan 2005), plants (de Oliveira Dal'Molin et al. 2010; Grafahrend-Belau et al. 2009b; Poolman et al. 2009), and animals (Cakir et al. 2007; Duarte et al. 2007), to study different aspects of metabolism, including the prediction of optimal metabolic yields and flux distributions (Varma et al. 1993a, b), gene deletion lethality (Edwards and Palsson 2000; Förster et al. 2003), and pathway redundancies (Van Dien and Lidstrom 2002). In plant science, metabolic flux determination is acknowledged to be an important part of plant metabolic engineering, which makes FBA a valuable tool for extracting new insights from complex plant metabolic networks. The workflow used to perform FBA comprises a three-step procedure encompassing the steps of (i) model reconstruction, (ii) model analysis, and (iii) model validation, with the evaluated model eventually representing a valuable platform to generate or test hypothesis and to explore plant metabolism *in silico*.

The reconstruction of plant metabolic networks requires detailed metabolic information. The process of manually extracting this information from literature requires huge effort and can be accelerated by accessing information stored in databases. A variety of metabolic pathway databases has been developed in conjunction with methods to access and analyze the data. Metabolic pathway databases are repositories of biochemical pathways and reactions with data related to organism-specific information about genes, gene products, enzymes, and compounds organized into pathways and maps (Karp 1998a, b; Karp et al. 1999; Wittig and De Beuckelaer 2001). The databases vary largely in coverage and representation of metabolic processes with the database content either being curated from literature and online resources or from computationally predicted pathways that are manually refined. Despite the high number of currently available 325 online pathway databases, only 14 multispecies and species-specific metabolic pathway databases contain plant-specific information.

Once an *in silico* model is constructed, it can be converted into a mathematical format for computational analysis by including parameters required for computational simulation, such as maximum uptake/excretion rates of nutrients and products. Several software tools have been developed to perform FBA, including packages freely available for academic research, such as CellNetAnalyzer (Klamt et al. 2007), Constraint-Based Reconstruction and Analysis (COBRA) toolbox (Becker et al. 2007), and FBA-SimVis (Grafahrend-Belau et al. 2009a). Most of these tools offer a variety of constraint-based analysis techniques that can be used to study different biological questions and *in silico* scenarios.

This chapter will describe available resources, a pipeline for metabolic model analysis, and an application to developing seed in barley. The first part of this chapter deals with the resources available for the constraint-based analysis of plant metabolic models, where the first section provides an overview of metabolic pathway databases containing plant-specific information and the second section provides an overview of software tools available for FBA. The second part of the chapter explains a pipeline for the integrated constraint-based analysis of plant metabolic models, focusing on the database and tools developed for constraint-based model reconstruction, analysis, and visualization. Finally, the application of the proposed pipeline is presented as a case study of storage metabolism in developing barley seeds.

17.2 Plant Metabolic Pathway Databases: Organization and Integration of Plant Metabolic Information

Metabolic pathway databases aim to provide the current knowledge of biochemical processes involved in metabolism, thereby facilitating the understanding of complex metabolic processes by integrating large and/or complex data sets of several types of metabolic pathway information (e.g., data on compounds, reactions, enzymes, and genes) in the context of the graphical representation of pathways. In addition to provide an encyclopedic reference on pathway information, metabolic pathway databases facilitate metabolic engineering, support global analysis of omics datasets, and facilitate a variety of analysis and simulation techniques used in mathematical modeling by supporting network reconstruction.

In the following section, widely-used databases are discussed that could serve as a starting point for plant metabolic network reconstruction. For a complete list of available databases, readers are advised to visit the Path guide resource (Bader et al. 2006).

17.2.1 Metabolic Pathway Databases

17.2.1.1 Multispecies Metabolic Pathway Databases

The **Kyoto Encyclopedia of Genes and Genomes** (KEGG; Kanehisa and Goto 2000; Kanehisa et al. 2010) is an integrated database resource that consists of 16 main databases containing chemical, genomic, and systems information with cross-references to a wide range of external biological databases. The KEGG pathway database contains metabolic pathways, represented as curated, manually drawn pathway maps consisting of links to information about compounds, enzymes, reactions, and genes. The KEGG database currently contains more than 140,000 species-specific pathways generated from 398 reference pathways (as of July 12, 2011). The plenitude of pathway information (e.g., metabolic, genetic, environmental, and cellular information) makes KEGG one of the most widely-used pathway databases.

With respect to plant research, KEGG offers different resources, such as the EGENE database (Masoudi-Nejad et al. 2007a, b). The EGENE comprises plant genomic information (EST contigs) of 64 species and the GENE database (Masoudi-Nejad et al. 2007b), which comprises gene catalogs for all complete genomes (12 plant species). The pathway-based integration of the plant genomic information allows generating plant species-specific pathways along with enzymes (genes) expressed in the specific organism (found by homology search) in the KEGG reference pathway map, thereby offering an overview of fundamental biological processes in plants. In addition, the KEGG plant interface especially focuses on the understanding of relationships between genomic and chemical information of natural products from plants, and offers different plant biosynthetic pathway maps for plant secondary metabolism.

MetaCyc MetaCyc (Caspi et al. 2010) belongs to the BioCyc collection (Karp et al. 2005) of pathway/genome databases (PGDBs). MetaCyc is a curated metabolic pathway database containing pathways from a wide range of organisms representing all domains of life. The database contains experimentally determined metabolic pathways curated from the literature, lists of compounds, enzymes, reactions, genes, and proteins associated with the pathways, and links to external biological databases. With more than 1,400 pathways from over 1800 different species, MetaCyc (V14.0) is one of the largest collections of metabolic pathways currently available with the majority of pathways occurring in microorganisms and plants. Relating to its role as encyclopedic reference on metabolism, MetaCyc is additionally used as a reference metabolic pathway resource for the computational prediction of the metabolic network of any organism having sequenced and annotated genome using the Pathway Tools software (Karp et al. 2002).

Reactome Reactome (Matthews et al. 2009) is a curated metabolic pathway database initially developed to represent human biology. But, Reactome has been extended to include computationally inferred pathways and reactions from 22 non-human species, including the two plant species *Arabidopsis thaliana* and *Oryza sativa.* The non-human species-specific pathways are inferred *via* orthology relationships from the respective human pathways. Pathway information includes very detailed information on equivalent events in other organisms, descriptions, and subcellular compartment information that is cross referenced to a wide range of external biological databases. The current release of Reactome (V32.0) is comprised of over 10,000 pathways from 23 different species, representing a well curated pathway database. Database contents can be viewed and downloaded in several formats (Table 17.1) and, queried *via* a search interface with a variety of options. In addition, Reactome

Table 17.1 Overview of metabolic pathway databases providing information on plant metabolism

Database	URL	Organisms	Data sources	Data quality	Level of detail	Data access
Multi-species metabolic pathway databases						
KEGG	http://www.genome.ad.jp/kegg/	A, B, F, Pt, Pz	P	M, P[b]	C[a], L, T[a]	BioPAX
MetaCyc	http://metacyc.org/	A, B, F, Pt, Pz	P	M	C[a], O[a], L, T[a]	SBML, BioPAX
Reactome	http://www.reactome.org	A, B, F, Pt, Pz	P	M	C[a], O[a], T[a]	SBML, BioPAX
WikiPathways	http://wikipathways.org/	A, B, F, Pt	P	M	C[a], L[a], T[a]	
PathCase	http://nashua.cwru.edu/PathwaysWeb/	A, B, F, Pt, Pz	P, S	I, M		BioPAX
PlantCyc	http://www.plantcyc.org/	Pt	P, S	I, M, P	C[a], O[a], L, T[a]	
Arabidopsis Reactome	http://www.arabidopsisreactome.org	Pt	P, S	I, M	C[a], O[a], T[a]	SBML, BioPAX
MetaCrop	http://metacrop.ipk-gatersleben.de	Pt	P	M	C, D, K[a], L, O, T	SBML
Species-specific plant metabolic pathway databases						
AraCyc	http://www.arabidopsis.org/biocyc/index.jsp	Pt	P	M, P	C[a], O[a], L, T[a]	
MetNetDB	http://www.metnetdb.org/MetNet_db.htm	Pt	P, S	I, M	C[a], T[a]	
Metabolic pathway database systems						
NCBI biosystems	http://www.ncbi.nlm.nih.gov/biosystems	A, B, F, Pt, Pz	S	I	C[a], O[a], L, T[a]	
Pathway commons	http://www.pathwaycommons.org/pc/	A, B, F, Pt, Pz	S	I	C[a], O[a], T[a]	BioPAX, PSI-MI

Organisms: *A* animals, *B* bacteria, *F* fungi, *Pt* plants, *Pz* protozoa. Data sources: *P* primary, *S* secondary. Data quality: *I* imported, *M* manually curated, *P* predicted. Level of detail: *C* subcellular compartment, *D* developmental stage, *K* kinetics, *L* literature references, *O* organ/tissue, *T* transport processes

[a] Information restricted to single pathways and reactions

[b] Manually curated general pathways, predicted species-specific pathways

provides several tools for pathway analysis, such as Skypainter. Skypainter allows users to visualize and analyze their own data sets in relation to the reaction maps.

17.2.1.2 Plant Metabolic Pathway Databases

PlantCyc PlantCyc is a comprehensive multispecies plant metabolic pathway database hosted by PMN (Plant Metabolic Network), a collaborative project with an aim to develop and curate plant metabolic pathway and enzyme databases. The database contains curated information on genes, enzymes, compounds, reactions, and pathways involved in primary and secondary plant metabolism. In addition to the in-house curated pathways, PlantCyc contains all pathways from AraCyc, all plant-specific pathways from MetaCyc, and a number of pathways from other plant pathway databases, including RiceCyc and MedicCyc. Furthermore, hypothetical pathways published in journals as well as computationally predicted pathways manually validated by PMN curators are provided. Together, PlantCyc is a comprehensive plant metabolic pathway database containing about 714 pathways from 329 plant species as per its current release (V3.0).

Arabidopsis Reactome *Arabidopsis* Reactome (Tsesmetzis et al. 2008) is a curated pathways database initially developed to represent biological processes in the model plant *Arabidopsis*, but which has been extended to include information on about 11 other plant species. The database contains in-house curated pathways as well as pathways imported from KEGG and AraCyc with cross-references to other biological databases. In addition to experimental evidence and literature citations, detailed pathway information is provided, including subcellular location and pathway descriptions. In its current version (V3.0), *Arabidopsis* Reactome comprises between 990 and 1,100 pathways for each of the 12 plant species. Database contents can be viewed and downloaded in several formats (Table 17.1). Software tools are also available for *Arabidopsis* Reactome.

MetaCrop MetaCrop (Grafahrend-Belau et al. 2008) is a manually curated metabolic pathway database focusing on crop plant species with high agronomical importance. A detailed description of the database is given below under the subtitle 'Model Reconstruction' and within the main section 'Pipeline for the integrated constraint-based model reconstruction, analysis, and visualization.'

17.2.1.3 Species-Specific Plant Metabolic Pathway Databases

Species-specific databases contain predicted and/or verified pathways for a single species. The majority of these databases are PGDBs. These databases are mostly computationally generated from the annotated genome of the organism using the Pathway Tools software, developed by the SRI Bioinformatics Research Group (Karp et al. 2002). Computationally generated PGDBs can be subdivided into three groups based on their quality: (i) intensively-curated; (ii) moderately-curated; and

(iii) non-curated databases. A comprehensive list of plant-specific PGDBs as well as the associated links to the databases is provided by the PMN website (http://www.plantcyc.org/) and the BioCyc website (http://biocyc.org/). In the following section, only the most comprehensive and best curated plant PGDB, AraCyc, is described due to the limitation of space.

AraCyc AraCyc (Mueller et al. 2003) is a species-specific plant metabolic pathway database that provides information about metabolic pathways in the model plant *Arabidopsis*. The database was the first computationally predicted plant PGDB using MetaCyc as a reference database. Since its automatic generation, AraCyc has been under continued manual curation with the predicted pathways being literature validated. The database now contains a mixture of computationally predicted, manually validated, and newly curated pathways extracted from the literature, with an evidence code clearly indicating the source of information. In addition, pathways and enzymes are annotated with pathway descriptions and to some extent location information, such as organ-/tissue-specificity. AraCyc (V6.0) now contains about 408 pathways, enriched by 4,803 literature references. It provides several tools for pathway analysis, such the OMICS Viewer. The OMICS Viewer is a user data visualization and analysis tool, allowing mapping different kind of user-specific high-throughput data onto the overview pathway map of AraCyc.

MetNetDB MetNetDB (Metabolic Networking Database) (Yang et al. 2005) is a single-species pathway database containing information on metabolic and regulatory networks in the model plant *Arabidopsis*. The database contains in-house curated pathways as well as pathways imported from MetaCyc and AraCyc. Pathway-associated information, such as component information (e.g., metabolites, proteins, DNA, and RNA), interaction information (e.g., translocation, transcription, and assembly), and metadata (confidence, evidence, and synonymous) are partly cross-referenced with external biological databases or, in some cases, additionally annotated by experts. The current release of MetNetDB (V3.0) contains about 334 pathways combined into a metabolic and regulatory map of *Arabidopsis*. The MetNet suite of visualization and analysis tools (http://www.metnetdb.org/MetNet_overview.htm) allows an in-depth analysis of the pathways maps provided by MetNetDB.

17.2.2 Metabolic Pathway Database Systems

In addition to metabolic pathway databases providing primary data or at least to some part in-house curated metabolic pathway information, there exist a number of different metabolic pathway database systems providing pathway data of multiple external metabolic pathway databases (secondary data). With respect to plant science, the following metabolic pathway database systems contain plant-specific metabolic pathway information.

NCBI Biosystems NCBI BioSystems (Geer et al. 2010) is a centralized repository for metabolic pathway information. The NCBI BioSystems contains biological pathways from four sources: KEGG, Human Reactome, BioCyc (including EcoCyc and MetaCyc), and the National Cancer Institute's Pathway Interaction database. The system integrates these database records into the existing NCBI Entrez databases (e.g., Gene, Protein, PubMed, and PubChem), thereby connecting the records with associated literature, molecular, and chemical data. In addition, detailed pathway maps and annotations are available on the web sites of the source databases. With respect to plant metabolism, plant-specific pathway information is provided by the source databases KEGG and MetaCyc.

Pathway Commons Pathway Commons provides access to multiple public metabolic pathway databases (e.g., Human Reactome, HumanCyc, IMID, etc.), allowing to browse, query, and download pathway information *via* the web portal. The database entries are cross-linked to the source databases, and links to external databases rely on the source databases. Therefore, the quality of the pathways available in Pathway Commons depends on the quality of the pathway source databases, which provide different levels of pathway-associated information including subcellular location, experimental evidence, and literature citations. The current release of Pathway Commons contains about 1,450 pathways from about 440 organisms. The database supports different exchange formats (Table 17.1) and can be queried using search and a filter to restrict the results to organism and source.

17.2.3 Additional Resources and Tools

In addition to metabolic pathway databases, a wide variety of different databases exits, providing valuable data for plant metabolic modeling, such as enzyme-related information [e.g., Brenda (Chang et al. 2009)], [ExPASy-ENZYME (Gasteiger et al. 2003)], protein-related information [Swiss-Prot/TrEMBL (Boeckmann et al. 2003)], reaction kinetics-related information [SABIO-RK (Rojas et al. 2007)], metabolite-related information [PubChem (Wang et al. 2010a) and KEGG LIGAND (Goto et al. 2002)], transporter-related information [ARAMEMNON (Schwacke et al. 2003)], and model-related information [BioModels (Li et al. 2010)]. In addition, WikiPathways (Pico et al. 2008) is a collaborative platform for curating, editing, and sharing biological pathways in different species. BiNCO-wiki (Telgkamp et al. 2007) is a wiki system that stores information related to biological networks and allows community-based extension of the information. Furthermore, in addition to pathway analysis and visualization tools provided by the metabolic pathway databases themselves, availability of different tools allow to browse pathways and to map omics data onto plant metabolic pathway maps, supporting the functional analysis of data for plant pathways [e.g., KaPPA-View4 (Tokimatsu et al. 2005; Sakurai et al. 2011), VANTED (visualization and analysis of networks containing experimental data; Junker et al. 2006), MapMan (Thimm et al. 2004)].

17.3 Resources for the Constraint-Based Analysis of Plant Metabolic Models: Modelization of Plant Metabolism

In biological research, including plant science, several approaches have been developed for quantitative *in silico* modeling and simulation of metabolic systems. One increasingly used method in the metabolic modeling is FBA, which is a constraint-based modeling approach having the ability to predict optimal metabolic yield and steady-state flux distributions under different environmental conditions and genetic backgrounds (Varma and Palsson 1994). FBA uses the principle of linear programming to determine the steady-state flux distribution in a metabolic model by defining an objective function and searching the allowable solution space for an optimal flux distribution that maximizes or minimizes the objective. Due to its predictive power and the advantage of not requiring the knowledge of kinetic parameters, FBA has been applied successfully to a variety of biological systems to study different aspects of metabolism (Kauffman et al. 2003; Reed and Palsson 2003). For a more detailed description of the basic concepts of FBA, readers are referred to the respective literature (Edwards et al. 2001; Rios-Estepa and Lange 2007). There are also excellent and comprehensive reviews covering in detail about FBA and its applications (Feist and Palsson 2008; Lee et al. 2006; Orth et al. 2010). In the following section, an overview of currently available resources for the constraint-based analysis of plant metabolic models is given (Table 17.2) and the most widely-used tools are discussed.

17.3.1 Standalone Applications

OptFlux OptFlux (Rocha et al. 2010) is an open-source and plug-in based software for *in silico* metabolic engineering. Focusing on the analysis of stoichiometric metabolic models, the software provides: (i) methods for phenotype simulations (e.g., FBA, MOMA, and ROOM); (ii) methods for metabolic flux analysis (MFA); (iii) elementary modes analysis; and (iv) strain optimization algorithms (OptKnock, EAs, and SA). These features are aimed at facilitating the rational design of microbial strains (i.e., aimed to identify potential targets for metabolic engineering). In addition, OptFlux comprises a plug-in for the dynamic visualization of the simulation results, which are superimposed on the model graph. Model handling is supported by allowing the import/export to several flat file formats including SBML.

Systems Biology Research Tool The Systems Biology Research Tool (SBRT) (Wright and Wagner 2008) is an open-source, integrated software platform focusing on the management, simulation, and analysis of stoichiometric models. The extendable, plug-in based software provides 35 processes for stoichiometric model analysis including FBA, flux variability analysis, and optimizing objective functions, which can be controlled from the command line or a simple graphical user interface (GUI). In addition, SBRT is capable of interacting with external software (such as

Table 17.2 Overview of software tools available for flux balance analysis (FBA)

Name	URL	Usability		Capabilities							Output			Depen-dend on		Nec. skills	Frame-works		Platform-dependencies			SBML		Avail-ability		
		GUI	Command-line	Model reconstruction	FBA	Additional CBA methods	EMA, EPA	MFA	Built-in visualization	External vis-software	Other utility functions	Flux values	Flux maps	Commercial software	Free software	Programming abilities	Linear programming	Non-linear programming	Linux	Windows	Mac	Import	Export	Open source	Acadamic use	Commercial use
Standalone applications																										
OptFlux 2.0	http://www.optflux.org	+	–	–	+	+	+	+	+	+	+	+	+	–	–	–	+	+	+	+	+	+	–	+	f	f
SBRT 2.0	http://www.bioc.uzh.ch/wagner/software/SBRT	+[a]	+	–	+	+	–	–	–	–	+	+	–	–	–	–	+	–	+	+	+	+	+	+	f	f
MetaFluxNet 1.86	http://mbel.kaist.ac.kr/lab/mfn/	+	-	+	+	+	-	+	+	–	+	+	+	–	+	–	+	?	–	+	–	+	+	–	f[g]	$
PathwayAna-lyzer 1.0	http://sourceforge.net/projects/pathwayanalyser/	–	+	–	+	+	–	–	–	–	–	+	–	–	+	–	+	+	+	–	–	+	–	+	f	f
	http://sourceforge.net/projects/fluxor	–	+	–	+	–	–	–	–	–	–	+	-	–	+	–	+	–	+	–	–	+	–	+	f	f
FBA-SImVis 1.9	http://fbasimvis.ipk-gatersleben.de/	+	–	+	+	+	–	–	+	–	+	+	+	–	+	–	+	+	–	+	–	+	+	+	f	f
Dependend on commercial platforms																										
CellNetAna-lyzer 9.0	http://www.mpi-magdeburg.mpg.de/projects/cna/cna.html	+	+	+	+	–	+	+	+	+[d]	+	+	+	+	+	–	+	–	+	+	+	+	+	–	f	$
COBRA Tool-box 1.3.1	http://gcrg.ucsd.edu/Downloads/Cobra_Toolbox	–	+	–	+	+	–	–	–	+	+	+	–	+	+	+	+	–	+	+	+	+	+	+	f	f
SNA 1.0	http://www.bioinformatics.org/project/?group_id=546	–	+	–	+	–	+	–	–	–	–	+	–	+	+	+	+	–	+	–	–	+[e]	–	+	f	f
Web-based tools																										
BioMet Toolbox	www.sysbio.se/BioMet/	+[b]	+[c]	–	+	+	+[c]	–	+	–	+	+	–	–	–	–	+	–	–	+[c]	–	+	–	–	f	f
CycSim	http://www.genoscope.cns.fr/cycsim	+[b]	–	–	+	+	–	–	–	–	+	+	+	–	–	–	+	–	–	–	–	+	+	–	f	f

Feature comparison of software tools focusing on FBA. ± denote the existence/absence of the feature

CBA constrained-based analysis, *EMA* elementary mode analysis, *EPA* extreme pathway analysis, *$* licence fee, *?* not known

[a] Very simple; [b] Web; [c] Standalone program; [d] Import of maps; [e] MATHSBML; [f] free; [g] Limited version

Metatool, Mathematica, R, GLPK, and CPLEX) and providing a high degree of flexibility.

MataFluxNet MetaFluxNet (Lee et al. 2003) is a standalone software package for quantitative metabolic flux calculation, that provides: (i) a model reconstruction and management environment allowing to set up a model in a user-defined way or in a reference retrieved way by integrating data from different online databases; (ii) an interface for quantitative and comparative flux analysis (e.g., MFA, FBA, and *in silico* gene deletion/addition); and (iii) a GUI supporting dynamic visualization of metabolic flux maps for interactive and comparative analysis of the resulting flux profiles and automated pathway layout creation. The SBML-supporting software uses lp_solve and QSopt as default linear programming (LP) solvers, and provides model conversion into a variety of LP formats, including CPLEX, LINDO, GAMS, AMPL, and MATLAB LP. MetaFluxNet offers a free academic version with only limited functionality.

Pathway Analyzer Pathway Analyzer is an open-source software tool for flux analysis of metabolic models running on the Linux platform. Accepting SBML files as input, the tool performs FBA and MOMA using the GLPK solver for linear programming and the OOQP solver for quadratic programming. The command-line tool does not provide built-in routines for dynamic flux visualization. Instead, a flux distribution output is provided in plain text format for the wild type and *in silico* deletion mutants.

Fluxor Fluxor is a simple phyton command line tool for performing FBA. Taking a SBML file as input, the tool performs FBA using the GLPK solver for linear programming. Fluxor is part of the Bio-Spice (Biological Simulation Program for Intra- and Inter-Cellular Evaluation) tool set (http://biospice.sourceforge.net/), an open-source framework for systems biology research.

FBA-SimVis FBA-SimVis (Grafahrend-Belau et al. 2009a) is an open-source software tool for integrated constraint-based model analysis and visualization. It extends the functionality of VANTED (Junker et al. 2006). A comprehensive and detailed description of the software is given below under the subtitle ‘Constraint-Based Model Analysis and Visualization’ and within the main section ‘Pipeline for the integrated constraint-based model reconstruction, analysis, and visualization.’

17.3.2 Add-Ons to Commercial Platforms

CellNetAnalyzer CellNetAnalyzer (CNA) (Klamt et al. 2007), formerly FluxAnalyzer, is a software tool for the modeling and analysis of metabolic, regulatory, and signal transduction networks that runs on the commercial MATLAB platform. In terms of constraint-based metabolic modeling, the software provides a variety of stoichiometric analysis techniques, such as FBA, MFA, elementary mode analysis, and extreme pathway analysis, as well as a static graphical representation of the simulation results. Model handling is supported by a GUI-based reconstruction environment and import/export functionalities for several flat file formats, including SBML. CNA provides a GUI as well as a command line-based running of the program.

Constraint-Based Reconstruction and Analysis The COBRA toolbox (Becker et al. 2007) for MATLAB is a software package that focuses on the modeling and analysis of stoichiometric metabolic models using constraint-based approaches. The software provides a comprehensive set of techniques, such as FBA, dynamic FBA, MOMA, flux variability analysis, robustness analysis, and gene deletion analysis, together with an interface to several free and commercial linear programming solvers (e.g., lp_solve, GLPK, LINDO, CEPLEX, and Mosek) and tools to read and manipulate constraint-based models. Although COBRA requires some programming ability to use it, the COBRA toolbox provides a high degree of flexibility as it can be extended by user-written MATLAB routines. The toolbox neither provides a GUI nor supports built-in visualization routines, although a network and flux distribution output for Cytoscape is provided (Shannon et al. 2003).

Stoichiometric Network Analysis Stoichiometric Network Analysis (SNA) (Urbanczik 2006) is a MATHEMATICA toolbox for stoichiometric analysis of metabolic networks focusing on the computation of elementary flux modes of a network, which form the basis for the analysis of its metabolic capabilities. In addition, the toolbox offers FBA. Although SNA supports SBML for the ease of model handling, the toolbox is restricted to a command line interface, requires programming abilities to use, and only runs under Linux.

17.3.3 Web-Based Tools

Only very recently, web-based tools have been released to enable the online constraint-based analysis on stoichiometric metabolic models.

BioMet BioMet Toolboxis a web-based platform for stoichiometric analysis and for integration and analysis of transcriptomics data. BioMet is aimed to provide insight into the metabolic capabilities of genome-scale metabolic models. In terms of constrain-based modeling, the toolbox provides the BioOpt tool, offering several analysis software, including FBA, reduced cost and shadow price analysis, sensitivity analysis, and gene deletion analysis. The suite of analysis is restricted to genome-scale models, which are either provided by the in-house model library containing GSMMs of different organisms or which can be submitted using SBML. BioOpt is available as a web-based application as well as in a standalone version running on the Windows platform.

CycSim CycSim (Fèvre et al. 2009) is a web application for the *in silico* analysis of genome-scale metabolic models. Based on different constraint-based modeling techniques offered by the system (e.g., FBA, metabolite producibility analysis, and gene deletion analysis), CycSim supports simulation of growth phenotypes under varying genetic and/or environmental background. The comparison of phenotype predictions and experimental results is provided by the direct visualization of both metabolic pathway maps obtained from BioCyc or KEGG. The analysis is possible for genome-scale metabolic models provided by CycSim, SBML models, and/or experimental results.

17.3.4 Commercial Tools

Different commercial software packages for FBA and stoichiometric analysis of metabolic models offering a variety of methods for (i) model reconstruction, (ii) constraint-based model analysis, and (iii) model and flux visualization have been developed, including SimPheny (Genomatica Inc., San Diego, CA, USA) and Insilico Discovery (Insilico Biotechnology Inc., Stuttgart, Germany). The reader is referred to the respective websites for further information about the tools, which are of minor interest for most academic researchers due to cost requirements.

17.3.5 Mathematical Software Platforms and Solvers

As the mathematical framework for constraint-based analysis relies primarily on linear algebra and linear programming, performing FBA does not require the use of specialized software tools. Instead, general purpose mathematical software platforms [such as MATLAB (http://www.mathworks.com/) and MATHEMATICA [(http://www.wolfram.com/)], optimization modeling packages [such as GAMS [(http://www.gams.com/)], or any other software package that allows for fast and efficient LP computation, are well suited. For greater efficiency, specialized optimization packages can be added (e.g., the Optimization Toolbox for MATLAB). In addition, several free and commercial solvers for linear and quadratic programming problems are available, including GLPK (http://www.gnu.org/software/glpk/), OOQP (http://pages.cs.wisc.edu/swright/ooqp/), CPLEX (http://www.ilog.com/products/cplex/), and LINDO (http://www.lindo.com/).

17.4 Pipeline for the Integrated Constraint-Based Model Reconstruction, Analysis, and Visualization

Numerous studies in the recent years have shown a rapidly growing interest in metabolic modeling as such modeling can be deduced by a number of resources and tools available for metabolic reconstruction, analysis, and simulation. Nevertheless, modeling and analysis of plant metabolic networks is still hampered by the fact of a reduced number of databases providing plant-specific information. Most of the heterogeneous and fine grained information required for model reconstruction are either scattered over numerous data sources or not provided by most of these databases. In addition, differences in data access methods and file formats create difficulties for users attempting to gather data from multiple sources. Such difficulties also hold true for the numerous tools and methods available for constraint-based analysis.

To facilitate and support the modeling and analysis of plant metabolic networks, a pipeline is described for the constraint-based analysis of crop plant metabolic models in the following section. The pipeline is comprised of the following steps of the modeling approach: (i) model reconstruction, management, and export; (ii) model simulation and analysis (FBA); and (iii) model and flux visualization. Focusing on the developed tools and methods, each of the modeling steps is described in this section. The application of the proposed pipeline is shown below for a case study of storage metabolism in developing barley seeds within the main section 'Application Example'. The workflow has been used for the constraint-based analysis of crop plant metabolic models and is summarized in Fig. 17.1.

17.4.1 Model Reconstruction

The reconstruction of plant metabolic models requires detailed metabolic information, which is often lacking in many data resources or scattered over multiple sources. To facilitate the construction of crop plant metabolic models, we developed MetaCrop (Grafahrend-Belau et al. 2008). MetaCrop is a curated multispecies plant metabolic pathway database focusing on crop plant metabolism. The database provides about 40 manually curated metabolic pathways of primary crop plant metabolism with emphasis on metabolism of agronomically important organs, such as seeds and tubers. In addition, the database contains detailed pathway-associated information, including clickable pathway maps and comprehensive information about metabolites (e.g., CAS number, molecular weight, and chemical formula), enzymes (e.g., EC and CAS number), stoichiometry, transport processes, location (species, organ, tissue, and compartment), and plant developmental stage as well as kinetic information for central metabolic pathways (sucrose breakdown, glycolysis, and TCA cycle). All the information was extracted from the literature and database (e.g., KEGG Pathways, EGENES, Brenda, ExPASy-Enzyme, PubChem, and ARAMEMNON). For quality assurance, information inferred from databases has been checked against literature. The information is cross-referenced to external biological databases and literature references, and the corresponding PubMed IDs are provided where available. Controlled vocabulary from ontologies (e.g., PO and GO) was used to ensure consistency and to allow the comparison of data from different sources.

The current release of MetaCrop comprises species-specific information on four monocot crop species (*Hordeum vulgare, Triticum aestivum, Oryza sativa,* and

Fig. 17.1 Pipeline used for constraint-based analysis of crop plant metabolic models

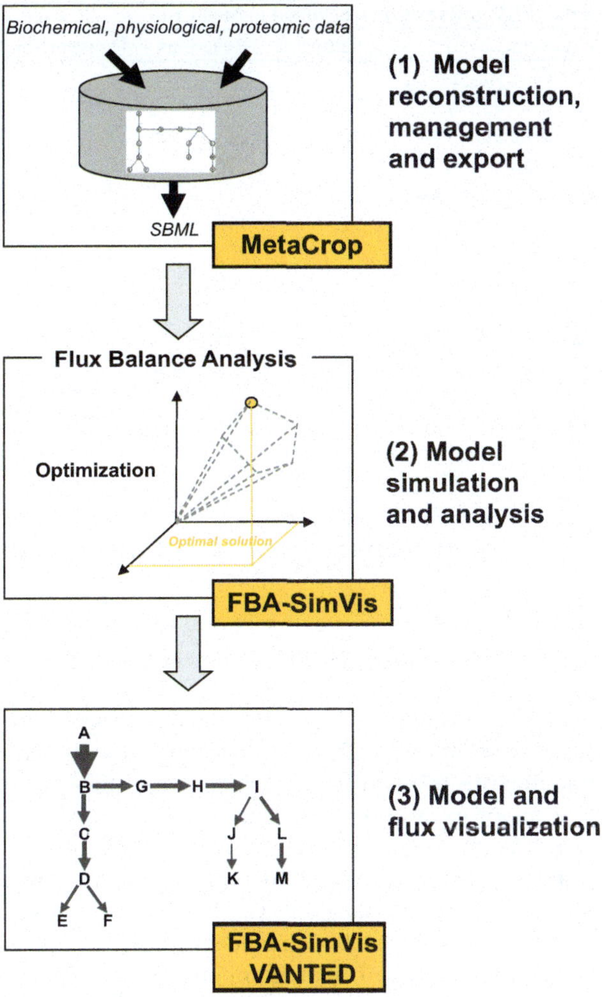

Zea mays), five dicot species, including three crop species (*Solanum tuberosum, Brassica napus*, and *Beta vulgaris*), and two model plants (*Arabidopsis thaliana* and *Medicago truncatula*). In total, currently MetaCrop contains about 400 enzymatic reactions, 60 transport processes, five compartments, and 1,100 references (Table 17.3). Due to the database focus on metabolic model reconstruction, most data in Metacrop correspond to biochemical data (e.g., taxon-specific enzymatic information). In the case of missing biochemical information, proteomics and genetics information were provided for a given enzymatic reaction or transport process. The web interface of MetaCrop (http://metacrop.ipk-gatersleben.de) supports detailed browsing and searching of data *via* searchable data tables and exploration of information from overview pathways to single reactions *via* clickable pathway image maps. The MetaCrop SBML exporter allows model creation and automatic

Table 17.3 Information Contained in MetaCrop

Organism	Pathways	Reactions	Transporters	Compartments	References
Hordeum vulgare	36	293	8	5	391
Triticum aestivum	34	273	6	5	365
Oryza sativa	35	281	9	4	355
Zea mays	35	276	27	5	392
Solanum tuberosum	34	211	14	3	272
Brassica napus	32	170	7	4	206
Beta vulgaris	35	228	–	4	366
Arabidopsis thaliana	35	289	15	4	525
Medicago truncatula	34	239	–	4	337
Total	38	394	60	5	1,100

data export *via* SBML, supporting modeling and analysis of crop plant metabolic models.

17.4.2 Constraint-Based Model Analysis and Visualization

As described above, several software tools have been developed to perform constraint-based analysis, including packages freely available for academic research, such as CellNetAnalyzer, COBRA toolbox, and OptFlux. Although most of these tools offer a variety of constraint-based analysis techniques, they have been mainly focused on model analysis, while comparative data examination and interpretation in terms of interactive visualization of FBA results has only been considered to a minor extent. To provide a visual analysis of FBA results in an interactive way with an aim to facilitate the analysis and interpretation of metabolic fluxes in response to genetic and/or environmental conditions, we developed FBA-SimVis (Grafahrend-Belau et al. 2009a). FBA-SimVis is a software tool for the constraint-based analysis of metabolic models with special focus on the visual exploration of metabolic flux data. The program provides a user-friendly environment for model reconstruction, constraint-based model analysis, and interactive visualization of the simulation results.

FBA-SimVis is implemented as a plug-in for the open source program VANTED (Junker et al. 2006), an analysis and visualization software for biological networks containing experimental data. The plug-in extends the Java-based VANTED system by integrating methods for constraint-based model analysis and interactive flux visualization. The analysis methods are implemented in MATLAB using the CLP solver (COIN-OR Linear Program Solver) for linear and nonlinear optimization. The MATLAB routines are based on the free library of the COBRA toolbox and are integrated as standalone executable, providing a MATLAB environment-independent application of the software.

FBA-SimVis supports model reconstruction by providing a GUI, which allows the user to create, edit, and store a metabolic model by using a simple drag-and-drop mechanism for network creation and a text menu for network refinement. In addition, models originating from external model repositories and/or reconstruction environments such as MetaCrop can be imported into the software *via* SBML.

FBA-SimVis supports an in-depth constraint-based analysis of metabolic models by integrating various methods for the constraint-based analysis of metabolic models, such as FBA, knock-out analysis, robustness analysis, and flux variability analysis. A major advantage of FBA-SimVis is that it integrates a nonlinear optimization procedure in addition to the linear optimization provided by most FBA toolboxes. Due to its underlying quadratic optimization, the nonlinear optimization routine does not produce multiple optima, thus having the advantage of reducing redundant scenarios and allowing to handle the problem of alternate optimal solutions.

To facilitate the analysis and interpretation of metabolic fluxes resulting from model analysis, FBA-SimVis provides a visual analysis of FBA results in a highly interactive way by offering an interactive GUI, where user interactions directly affect the visualization. Automatic mapping of the computed flux onto the network map is done by scaling the width of the reaction edges according to the flux and displaying the flux values next to the corresponding reaction edges. FBA-SimVis allows users to compare metabolic fluxes in response to varying genetic and/or environmental conditions in an interactive and comparative way by providing different visualizations (e.g., dynamic parameter analysis) for the various constraint-based analysis results. Examples of visualizations are shown in Fig. 17.2. Due to the integration of constraint-based analysis techniques with interactive visualization routines, FBA-SimVis offers a comprehensive analysis of stoichiometric models of metabolism.

17.4.3 Additional Methods and Tool for Model Analysis and Visualization

In addition to the constraint-based analysis techniques provided by FBA-SimVis, the VANTED system provides a wide variety of network-based analysis and visualization routines, allowing an in-depth analysis of metabolic networks. Focusing on the contextualization of high-throughput datasets from genomics, proteomics, and metabolomics, VANTED offers different visualization routines (e.g., line, bar, and pie charts) allowing to map/overlay experimental and computed datasets onto a given network. The network-based integration of omics data allows organizing, sorting, interrelating, and analyzing large and/or complex experimental data sets in the context of a metabolic network, thereby facilitating the global analysis of omics datasets as well as the understanding of complex metabolic processes. Furthermore, VANTED provides a variety of graph theoretical analysis techniques (e.g., shortest path length, connectivity analysis, and cycle detection) and statistical analysis techniques (e.g., *t*-test, correlation analysis, and cluster analysis), supporting a detailed topological and statistical analysis of experimental and simulation results.

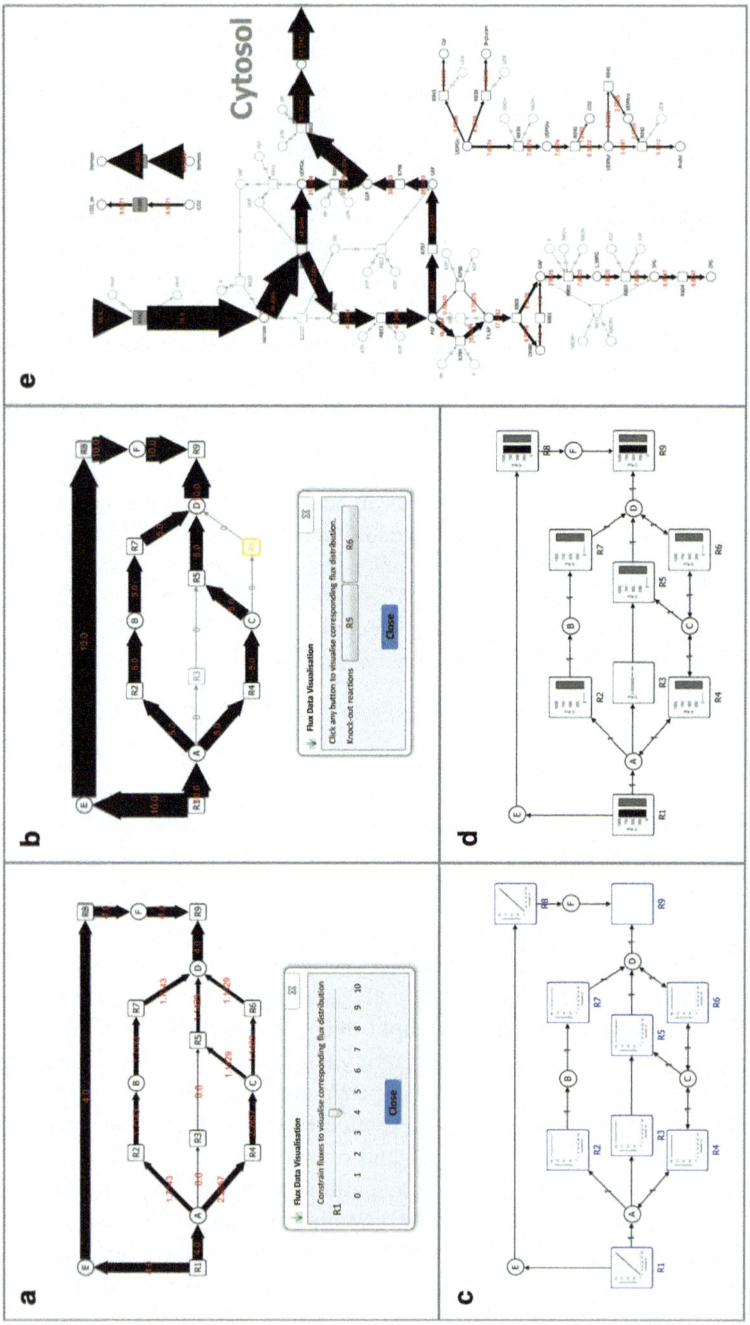

Fig. 17.2 This picture shows example visualizations for **a** flux balance analysis, **b** knock-out analysis, **c** robustness analysis, **d** flux variability analysis for a small example network, and **e** visualization of a partial flux map of a large model of barley seed metabolism

17.5 Application Example

The application of the proposed pipeline is shown by a case study of storage metabolism in developing barley seeds.

17.5.1 Model Reconstruction

To provide a framework for the *in silico* analysis of cereal seed storage metabolism in response to environmental and genetic perturbation, a stoichiometric model of primary metabolism in developing endosperm of barley (*H. vulgare*) during starch accumulation was reconstructed. The information required for network reconstruction (biochemical, physiological, and proteomic data) was collected through an extensive survey of scientific literature and online databases. The data were integrated into MetaCrop and the model was reconstructed in a stepwise manner. The resulting compartmented stoichiometric model comprises 234 metabolites and 257 reactions across four different compartments.

17.5.2 Model Analysis and Visualization

FBA was applied to the model using FBA-SimVis to study: (i) the effect of oxygen depletion on grain yield and metabolic flux distribution; and (ii) grain growth in response to enzyme deletion. In general, the simulation results were found to be in good agreement with the main biochemical properties of barley seed storage metabolism. The simulated growth rate was in the range of experimental observations. Moreover, the metabolic pathway pattern predicted by the model was in accordance with literature-based findings.

17.6 Concluding Remarks

Modeling of metabolism is increasingly important for a detailed understanding of complex biological processes. With the growing scope of applications of metabolic modeling, the range of methods for stoichiometric analysis of metabolism is constantly rising and modeling software tools for both expert and nonexpert users are commonly available. However, despite an increasing number of resources and software tools which aim to assist and facilitate the reconstruction and analysis of metabolic models, the reconstruction of high-quality metabolic models still requires manual evaluation and remains a laborious and time intensive process. There is still a long way to go before model reconstruction tools will provide high-quality models automatically.

References

Bader GD, Cary MP, Sander C (2006) Pathguide: a pathway resource list. Nucl Acids Res 34:D504–D506

Becker SA, Feist AM, Mo ML, Hannum G, Palsson BØ, Herrgard MJ (2007) Quantitative prediction of cellular metabolism with constraint-based models: the COBRA Toolbox. Nat Protoc 2:727–738

Boeckmann B, Bairoch A, Apweiler R, Blatter M-C, Estreicher A, Gasteiger E, Martin MJ, Michoud K, O'Donovan C, Phan IS, Pilbout S, Schneider M (2003) The SWISS-PROT protein knowledge-base and its supplement TrEMBL in 2003. Nucl Acids Res 31:365–370

Cakir T, Alsan S, Saybas¸ili H, Akin A, Ulgen KO (2007) Reconstruction and flux analysis of coupling between metabolic pathways of astrocytes and neurons: application to cerebral hypoxia. Theor Biol Med Model 4:e48

Caspi R, Altman T, Dale JM, Dreher K, Fulcher CA, Gilham F, Kaipa P, Karthikeyan AS, Kothari A, Krummenacker M, Latendresse M, Mueller LA, Paley S, Popescu L, Pujar A, Shearer AG, Zhang P, Karp PD (2010) The MetaCyc database of metabolic pathways and enzymes and the BioCyc collection of pathway/genome databases. Nucl Acids Res 38:D473–D479

Chang A, Scheer M, Grote A, Schomburg I, Schomburg D (2009) BRENDA, AMENDA and FRENDA the enzyme information system: new content and tools in 2009. Nucl Acids Res 37:D588–D592

David H, Akesson M, Nielsen J (2003) Reconstruction of the central carbon metabolism of Aspergillus niger. Eur J Biochem 270:4243–4253

de Oliveira Dal'Molin CG, Quek LE, Palfreyman RW, Brumbley SM, Nielsen LK (2010) AraGEM, a genome-scale reconstruction of the primary metabolic network in Arabidopsis. Plant Physiol 152:579–589

Duarte NC, Becker SA, Jamshidi N, Thiele I, Mo ML, Vo TD, Srivas R, Palsson BØ (2007) Global reconstruction of the human metabolic network based on genomic and bibliomic data. Proc Natl Acad Sci U S A 104:1777–1782

Edwards JS, Palsson BØ (2000) The Escherichia coli MG1655 in silico metabolic genotype: its definition characteristics, and capabilities. Proc Natl Acad Sci U S A 97:5528–5533

Edwards JS, Ramakrishna R, Schilling CH, Palsson BØ (1999) Metabolic flux balance analysis. In: Lee SSY, Papoutsakis ET (eds) Metabolic engineering. Marcel Dekker, New York, pp 13–57

Edwards JS, Ibarra RU, Palsson BØ (2001) In silico predictions of Escherichia coli metabolic capabilities are consistent with experimental data. Nat Biotechnol 19:125–130

Famili I, Forster J, Nielsen J, Palsson BØ (2003) Saccharomyces cerevisiae phenotypes can be predicted by using constraint-based analysis of a genome-scale reconstructed metabolic network. Proc Natl Acad Sci U S A 100:13134–13139

Feist AM, Palsson BØ (2008) The growing scope of applications of genome-scale metabolic reconstructions using Escherichia coli. Nat Biotechnol 26:659–667

Fèvre FL, Smidtas S, Combe C, Durot M, d'Alché-Buc F, Schachter V (2009) CycSim—an online tool for exploring and experimenting with genome-scale metabolic models. Bioinformatics 25:1987–1988

Förster J, Famili I, Palsson BØ, Nielsen J (2003) Large-scale evaluation of in silico gene deletions in Saccharomyces cerevisiae. OMICS 7:193–202

Gasteiger E, Gattiker A, Hoogland C, Ivanyi I, Appel RD, Bairoch A (2003) ExPASy: the proteomics server for in-depth protein knowledge and analysis. Nucl Acids Res 31:3784–3788

Geer LY, Marchler-Bauer A, Geer RC, Han L, He J, He S, Liu C, Shi W, Bryant SH (2010) The NCBI BioSystems database. Nucl Acids Res 38:D492–D496

Giersch C (2000) Mathematical modelling of metabolism. Curr Opin Plant Biol 3:249–253

Goto S, Okuno Y, Hattori M, Nishioka T, Kanehisa M (2002) Ligand: database of chemical compounds and reactions in biological pathways. Nucl Acids Res 30:402–404

Grafahrend-Belau E, Weise S, Koschützki D, Scholz U, Junker BH, Schreiber F (2008) MetaCrop: a detailed database of crop plant metabolism. Nucl Acids Res 36:D954–D958

Grafahrend-Belau E, Klukas C, Junker BH, Schreiber F (2009a) FBA-SimVis: interactive visualization of constraint-based metabolic models. Bioinformatics 25:2755–2757

Grafahrend-Belau E, Schreiber F, Koschützki D, Junker BH (2009b) Flux balance analysis of barley seeds: a computational approach to study systemic properties of central metabolism. Plant Physiol 149:585–598

Junker BH, Klukas C, Schreiber F (2006) VANTED: a system for advanced data analysis and visualization in the context of biological networks. BMC Bioinformatics 7:109

Kanehisa M, Goto S (2000) KEGG: Kyoto encyclopedia of genes and genomes. Nucl Acids Res 28:27–30

Kanehisa M, Goto S, Furumichi M, Tanabe M, Hirakawa M (2010) KEGG for representation and analysis of molecular networks involving diseases and drugs. Nucl Acids Res 38:D355–D360

Karp PD (1998a) Metabolic databases. Trends Biochem Sci. 23:114–116.

Karp PD (1998b) What we do not know about sequence analysis and sequence databases. Bioinformatics 14:753–754.

Karp PD, Krummenacker M, Paley S, Wagg J (1999) Integrated pathway-genome databases and their role in drug discovery. Trends Biotechnol 17:275–281

Karp PD, Paley S, Romero P (2002) The Pathway Tools software. Bioinformatics 18;S225–S232

Karp PD, Ouzounis CA, Moore-Kochlacs C, Goldovsky L, Kaipa P, Ahrén D, Tsoka S, Darzentas N, Kunin V, López-Bigas N (2005) Expansion of the BioCyc collection of pathway/genome databases to 160 genomes. Nucl Acids Res 33:6083–6089

Kauffman KJ, Prakash P, Edwards JS (2003) Advances in flux balance analysis. Curr Opin Biotechnol 14:491–496

Klamt S, Saez-Rodriguez J, Gilles ED (2007) Structural and functional analysis of cellular networks with CellNetAnalyzer. BMC Syst Biol 1:2

Lee JM, Gianchandani EP, Papin JA (2006) Flux balance analysis in the era of metabolomics. Brief Bioinformatics 7:140–150

Lee SY, Lee DY, Hong SH, Kim TY, Yun H, Oh YG, Park S (2003) MetaFluxNet, a program package for metabolic pathway construction and analysis and its use in large-scale metabolic flux analysis of Escherichia coli. Genome Inform 14:23–33

Li C, Donizelli M, Rodriguez N, Dharuri H, Endler L, Chelliah V, Li L, He E, Henry A, Stefan MI, Snoep JL, Hucka M, Le Novère N, Laibe C (2010) BioModels Database: an enhanced, curated and annotated resource for published quantitative kinetic models. BMC Syst Biol 4:92

Masoudi-Nejad A, Goto S, Jauregui R, Ito M, Kawashima S, Moriya Y, Endo TR, Kanehisa M (2007a) EGENES: transcriptome-based plant database of genes with metabolic pathway information and expressed sequence tag indices in KEGG. Plant Physiol 144:857–866

Masoudi-Nejad A, Goto S, Endo TR, Kanehisa M (2007b) KEGG bioinformatics resource for plant genomics research. Methods Mol Biol 406:437–458

Matthews L, Gopinath G, Gillespie M, Caudy M, Croft D, de Bono B, Garapati P, Hemish J, Hermjakob H, Jassal B, Kanapin A, Lewis S, Mahajan S, May B, Schmidt E, Vastrik I, Wu G, Birney E, Stein L, D'Eustachio P (2009) Reactome knowledgebase of human biological pathways and processes. Nucl Acids Res 37:D619–622

Morgan JA, Rhodes D (2002) Mathematical modeling of plant metabolic pathways. Metab Eng 4:80–89

Mueller LA, Zhang P, Rhee SY (2003) AraCyc: a biochemical pathway database for Arabidopsis. Plant Physiol 132:453–460

Orth JD, Thiele I, Palsson BØ (2010) What is flux balance analysis? Nat Biotechnol 28:245–248

Pico AR, Kelder T, Iersel MP, Hanspers K, Conklin BR, Evelo C (2008) WikiPathways: pathway editing for the people. PLoS Biol 6:e184

Poolman MG, Assmus HE, Fell DA (2004) Applications of metabolic modelling to plant metabolism. J Exp Bot 55:1177–1186

Poolman MG, Miguet L, Sweetlove LJ, Fell DA (2009) A genome-scale metabolic model of Arabidopsis and some of its properties. Plant Physiol 151:1570–1581

Reed JL, Palsson BØ (2003) Thirteen years of building constraint-based in silico models of Escherichia coli. J Bacteriol 185(9):2692–2699

Rios-Estepa R, Lange BM (2007) Experimental and mathematical approaches to modeling plant metabolic networks. Phytochemistry 68:2351–2374

Rocha I, Maia P, Evangelista P, Vilaça P, Soares S, Pinto JP, Nielsen J, Patil KR, Ferreira EC, Rocha M (2010) OptFlux: an open-source software platform for in silico metabolic engineering. BMC Syst Biol 4:45

Rojas I, Golebiewski M, Kania R, Krebs O, Mir S, Weidemann A, Wittig U (2007) Storing and annotating of kinetic data. In Silico Biol 7:S37–S44

Sakurai N, Ara T, Ogata Y, Sano R, Ohno T, Sugiyama K, Hiruta A, Yamazaki K, Yano K, Aoki K, Aharoni A, Hamada K, Yokoyama K, Kawamura S, Otsuka H, Tokimatsu T, Kanehisa M, Suzuki H, Saito K, Shibata D (2011) KaPPA-View4: a metabolic pathway database for representation and analysis of correlation networks of gene co-expression and metabolite co-accumulation and omics data. Nucl Acids Res 39:D677–D684

Schilling CH, Covert MW, Famili I, Church GM, Edwards JS, Palsson BØ (2002) Genome-scale metabolic model of Helicobacter pylori 26695. J Bacteriol 184:4582–4593

Schwacke R, Schneider A, Van Der Graaff E, Fischer K, Catoni E, Desimone M, Frommer WB, Flügge UI, Kunze R (2003) ARAMEMNON, a novel database for Arabidopsis integral membrane proteins. Plant Physiol 131:16–26

Shannon P, Markiel A, Ozier O, Baliga NS, Wang JT, Ramage D, Amin N, Schwikowski B, Ideker T (2003) Cytoscape: a software environment for integrated models of biomolecular interaction networks. Genome Research 13(11):2498–2504

Shastri AA, Morgan JA (2005) Flux balance analysis of photoautotrophic metabolism. Biotechnol Prog 21:1617–1626

Telgkamp M, Koschützki D, Schwöbbermeyer H, Schreiber F (2007) Community-based linking of biological network resources: databases, formats and tools. J Integr Bioinform 4:71

Thimm O, Bläsing O, Gibon Y, Nagel A, Meyer S, Krüger P, Selbig J, Müller LA, Rhee SY, Stitt M (2004) MAPMAN: a user-driven tool to display genomics data sets onto diagrams of metabolic pathways and other biological processes. Plant J 37:914–939

Tokimatsu T, Sakurai N, Suzuki H, Ohta H, Nishitani K, Koyama T, Umezawa T, Misawa N, Saito K, Shibata D (2005) KaPPA-view: a web-based analysis tool for integration of transcript and metabolite data on plant metabolic pathway maps. Plant Physiol. 138:1289–300

Tsesmetzis N, Couchman M, Higgins J, Smith A, Doonan JH, Seifert GJ, Schmidt EE, Vastrik I, Birney E, Wu G, D'Eustachio P, Stein LD, Morris RJ, Bevan MW, Walsh SV (2008) Arabidopsis reactome: a foundation knowledgebase for plant systems biology. Plant Cell 20:1426–1436

Urbanczik R (2006) SNA—a toolbox for the stoichiometric analysis of metabolic networks. BMC Bioinformatics 7:129

Van Dien SJ, Lidstrom ME (2002) Stoichiometric model for evaluating the metabolic capabilities of the facultative methylotroph Methylobacterium extorquens AM1, with application to reconstruction of C(3) and C(4) metabolism. Biotechnol Bioeng 78:296–312

Varma A, Palsson BO (1994) Metabolic flux balancing: basic concepts, scientific and practical use. Nat Biotechnol 12:994–998

Varma A, Boesch BW, Palsson BØ (1993a) Biochemical production capabilities of Escherichia coli. Biotechnol Bioeng 42:59–73

Varma A, Boesch BW, Palsson BØ (1993b) Stoichiometric interpretation of Escherichia coli glucose catabolism under various oxygenation rates. Appl Environ Microbiol 59:2465–2473

Wang Y, Bolton E, Dracheva S, Karapetyan K, Shoemaker BA, Suzek TO, Wang J, Xiao Z, Zhang J, Bryant SH (2010) An overview of the PubChem BioAssay resource. Nucl Acids Res 38:D255–D266

Wittig U, De Beuckelaer A (2001) Analysis and comparison of metabolic pathway databases.Brief Bioinform 2:126–142

Wright J, Wagner A (2008) The systems biology research tool: evolvable open-source software. BMC Syst Biol 2:55

Yang Y, Engin L, Wurtele ES, Cruz-Neira C, Dickerson JA (2005) Integration of metabolic networks and gene expression in virtual reality. Bioinformatics 21:3645–3650

Chapter 18
Coupled Transcript-Metabolite Profiling: Towards Systems Biology Approaches to Unravel Regulation of Seed Secondary Metabolism

Thierry Joët, Eleanore T. Wurtzel, Fumio Matsuda, Kazuki Saito and Stéphane Dussert

Abstract Seeds are an attractive model to decipher the regulation of secondary metabolite biosynthesis. Furthermore, given that seeds are a major source of food for humans and that numerous secondary metabolites play a role in the nutritional value and organoleptic quality of food products, a better understanding of seed secondary metabolic networks may be a key step for seed crop breeding. Network genomics offers outstanding opportunities to unravel the underlying regulation of metabolism in seeds. For now, only a few recent studies have coupled transcriptomics and metabolomics to investigate metabolism in developing seeds, either at the genome-wide level or at the quantitative level using pathway-guided approaches and multigenotype or multienvironment designs. Using three distinct examples, the present chapter aims to show how coupled transcript-metabolite profiling can help decipher the regulation of secondary metabolite synthesis in seeds. This chapter first summarizes the latest advances made in the model plant *Arabidopsis* at the genome-wide level. The second example is the metabolism of carotenoids in maize, which is a model system for provitamin A biofortification. Finally, this chapter also describes the coffee seed, whose endosperm stores spectacular amounts of alkaloids and phenylpropanoid-derived compounds. These three studies concern the biosynthesis of secondary metabolites, but similar approaches may undoubtedly be used for other seed metabolic pathways.

Keywords Carotenoid · Chlorogenic acid · Coexpression network · Flavonoid · Phenylpropanoid · Quantitative trait transcript

---

T. Joët (✉) · S. Dussert
IRD, UMR DIADE, Diversity, Adaptation and Development of Cultivated Crops,
BP 64501, 34394, Montpellier, France
e-mail: thierry.joet@ird.fr

E. T. Wurtzel
Department of Biological Sciences, Lehman College and The Graduate School, The City University of New York, 250 Bedford Park Blvd. West, Bronx, New York, NY 10468, USA

F. Matsuda · K. Saito
RIKEN Plant Science Center,
Tsurumi-ku, 230–0045, Yokohama, Japan

G. K. Agrawal, R. Rakwal (eds.), *Seed Development: OMICS Technologies toward Improvement of Seed Quality and Crop Yield*,
DOI 10.1007/978-94-007-4749-4_18,

18.1 Introduction

Collectively, plants synthesize several tens of thousands of secondary metabolites, i.e. highly diverse chemicals with no primary physiological function in the tissue or organ that produces them (D'Auria and Gershenzon 2005; Lewinsohn and Gijzen 2009; Schwab 2003). Secondary metabolites include a wide range of chemical classes, such as phenylpropanoids, flavonoids, terpenoids, glucosinolates, and alkaloids. Their functions are often linked with plant interactions with herbivores, foliar pathogens, pollinators, and seed predators. For example, many secondary metabolites act as chemical protection to ward off a wide variety of pathogens and herbivores. They may also suppress the growth of neighboring plants, while others attract pollinating or seed dispersing animals. Plant secondary metabolites are therefore essential for the plant's survival and reproductive fitness, but also represent unique sources for pharmaceuticals, food additives, flavors, and other industrial materials.

In addition to secondary metabolites induced in response to environmental stimuli, plants also constitutively produce secondary metabolites whose accumulation is often associated with specific developmental programs. In this regard, the seed is the specific site of biosynthesis and storage of diverse secondary metabolites. Actually, many secondary metabolites play a defensive role during the development and quiescent state of seeds, and also serve as reserves of energy, carbon, and nitrogen, which are rapidly mobilized during germination. Although secondary metabolites generally exist in a state of dynamic equilibrium between synthesis/translocation and degradation, catabolic activity is reduced in storage organs, such as seeds (e.g., for caffeine in coffee seeds; Vitoria and Mazzafera 1999). This means seeds are an attractive model to decipher the regulation of secondary metabolite biosynthesis. Furthermore, given that seeds are a major source of food for humans and that numerous secondary metabolites play a role in the nutritional value and organoleptic quality of food products, a better understanding of seed secondary metabolic networks may be a key step for seed crop breeding.

Using three distinct examples, the present chapter aims to show how coupled transcript-metabolite profiling can help decipher the regulation of secondary metabolite synthesis in seeds. The chapter first summarizes the latest advances made in the model plant *Arabidopsis* at the genome-wide level. The second example is the metabolism of carotenoids in maize endosperm, which is a model system for provitamin A biofortification. Finally, this chapter also describes the coffee seed, whose endosperm stores spectacular amounts of alkaloids and phenylpropanoid-derived compounds.

18.2 Different Ways of Integrating Transcript and Metabolite Data to Unravel Seed Secondary Metabolism

The biosynthetic pathways of many secondary metabolites have been successfully characterized in recent decades and are the subject of an abundant literature. Regardless of whether they are triggered by environmental or developmental stimuli, most of the secondary metabolic pathways studied to date were shown to be mainly regulated at the transcriptional level (Broun 2005; Davies and Schwinn 2003; Gachon et al. 2005). As a result, transcriptomics approaches appeared to be necessary to investigate the steps which modulate the synthesis and accumulation of secondary metabolites. Accordingly, many studies dealing with seed secondary metabolism collected data on molecules belonging to a single chemical family and the corresponding repertoire of biosynthetic genes (i.e., the pathway-guided approach). In this case, the limited number of molecules and metabolic steps under study enables the combination of highly quantitative methods (i.e., RT-PCR and conventional analytical chemistry techniques). Data are usually collected during the time-course of seed development and then mapped onto known metabolic pathways, using manual expert curation or visualization software, such as AraCyc (Mueller et al. 2003), MapMan (Thimm et al. 2004), and KaPPA-view (Tokimatsu et al. 2005). On the basis of the different biosynthetic routes described in other organs or species, this approach may help reconstructing seed-specific metabolic pathways in a given species (e.g., van Baarlen et al. 2008). Indeed, this approach enables better characterization of timing of gene expression associated with biosynthetic steps and identification of the critical associated gene paralogs. The transcript and metabolite patterns have been successfully overlaid, for instance, to characterize the metabolism of the main storage compounds in the dicot albuminous coffee seed, including the phenylpropanoid pathway, which leads to chlorogenic acid accumulation (Joët et al. 2009).

However, pathway-guided studies which couple transcript-metabolite profiling in seed systems usually have several limitations. First, as they focus on metabolites and genes that have been previously characterized elsewhere, they are not appropriate for the discovery of novel secondary metabolism. Moreover, because such studies generally use limited time-course experiments in one or very few genotype/environment combinations, causal relationships between transcriptional activity and metabolite synthesis generally remain unidentified. Many factors can impede the identification of transcriptional control points. Their identification is not facilitated by the difference in patterns generally observed between the transient expression of genes (bell-shape pattern) and the long-lived accumulation of metabolites (S-shaped). Weak correlations between transcript and metabolite levels may also be due to the time lag between gene expression and the resulting metabolite accumulation. Finally, pathway-guided studies using simple experimental designs cannot reveal whether levels of gene transcription quantitatively modulate metabolite accumulation.

Up-scaling coupled transcript-metabolite profiling by completing genome-wide transcriptomics studies in association with large metabolome surveys is the best way to bypass the first limitation described above; this is indeed a relatively unbiased method to discover novel genes regulating secondary metabolism. On the basis of such large data sets, groups of coexpressed genes and coaccumulated metabolites are usually identified by HCA, self-organizing maps (SOM), K-mean, and other clustering methods. Analysis of genomic data generated by high-throughput transcript profiling has demonstrated that genes with an overall similar expression pattern are often enriched for similar functions (Saito et al. 2008). This guilt-by-association principle can be applied to determine modular gene programs, identify *cis*-regulatory elements, or predict functions for unknown genes based on their co-expression neighborhood or their tight association with metabolites (Fukushima et al. 2009; Saito et al. 2008). Surprisingly, while many studies have used coupled transcript-metabolite approaches to investigate specific metabolic pathways in higher plants, very few have focused specifically on seed secondary metabolism. Since *Arabidopsis* has become a model plant for molecular and genetic analyses, its seed development has been largely characterized using transcriptomics approaches (Day et al. 2008; Le et al. 2010; Ruuska et al. 2002; Spencer et al. 2007). Separate metabolomic studies were performed to characterize the diversity and timing of flavonoid accumulation in *Arabidopsis* seeds (Kitamura et al. 2010; Routaboul et al. 2006). Few studies integrated transcriptome and metabolome of developing *Arabidopsis* seeds but were not specifically focused on secondary metabolites. A first step in characterizing gene expression associated with *Arabidopsis* seed-specific secondary metabolism was recently achieved with the AtMetExpress initiative (Matsuda et al. 2010), which aims to assemble an Atlas of phytochemicals accumulated in various organs at different developmental stages in a manner compatible with previous transcriptome data sets (Schmid et al. 2005). The most significant seed-specific relationships between transcripts and metabolites obtained from the AtMetExpress initiative are described in this chapter.

Understanding the genetic basis of variation for quantitative traits is a major challenge in biology. Systems approaches can help elucidating the biology that lies between genotype and phenotype in terms of causal networks of interacting genes. Molecular networks are starting to be widely used to characterize and predict biosystem behavior, giving rise to a new branch of biological knowledge (i.e., network genomics). Network genomics approaches introduce small perturbations in a system to generate subtle variation in gene expression and metabolite content. Regarding identification of regulatory mechanisms in a given tissue, one way to circumvent the limitations of simple designs consists of introducing a source of variation at the transcript and metabolite levels, by either increasing the number of genotypes studied or the number of environments tested. Gene coexpression networks and gene-metabolite networks are indeed better approached at a quantitative level by regression analysis of large data sets (Stitt et al. 2010). The detection of significant correlations between transcript abundance of genes of a given module is a further step in the demonstration that these genes are coregulated (Usadel et al. 2009). The analysis of Pearson correlations between transcript and metabolites is an original way of identifying quantitative transcript–phenotypic trait associations. Such quantitative associations are referred to as quantitative trait transcripts (QTTs; Passador-Gurgel et al. 2007) and correspond to key transcriptional control points for metabolite synthesis. This strategy is explained in the present chapter using two examples. Natural genetic variation has been recently used in maize as a tool to identify QTT for carotenoid accumulation (Vallabhaneni et al. 2009; Vallabhaneni and Wurtzel 2009), while different growing environments have been shown to provide enough variation in transcript and metabolite data sets to reconstruct metabolic networks for chlorogenic acid accumulation in the coffee seed (Joët et al. 2009).

18.3 Atlas of Gene Expression and Secondary Metabolism in *Arabidopsis* Seeds

Flavonoids (derived from the phenylpropanoid pathway) constitute the major secondary metabolite component of *Arabidopsis* seeds. These compounds are involved both in seed longevity and coat-imposed dormancy (Lepiniec et al. 2006). The two major classes of flavonoids in the *Arabidopsis* seed, flavonols and PAs, accumulate in the whole seed and in the seed coat, respectively. In the past decade, considerable progress has been made in the characterization of the *Arabidopsis* flavonoid biosynthetic pathway (Lepiniec et al. 2006) and several R2R3-MYBs TFs regulating their biosynthesis have been identified (Hirai et al. 2007; Stracke et al. 2007; Yonekura-Sakakibara et al. 2008). However, outside of flavonoids, the number of genomic studies devoted to other classes of secondary metabolites in *Arabidopsis* is relatively small (North et al. 2010). This is surprising since *Arabidopsis* has as a large and diverse range of secondary metabolites as other plant taxa, and the number of genes dedicated to secondary metabolism represents a vast part of its genome (D'Auria and Gershenzon 2005).

To fill this gap, the AtMetExpress Metabolome Atlas (Matsuda et al. 2010) was recently developed with an experimental design compatible with that of the AtGenExpress transcriptome data set (Schmid et al. 2005). This metabolome atlas was developed to enable large-scale integrated transcript-metabolite analyses. Phytochemical accumulation during development, determined using LC-MS (Matsuda et al. 2009), represents many growth stages and organs, including four stages of developing seeds.

The PCA of the entire metabolome data set revealed that seeds had distinct and characteristic metabolic profiles (Matsuda et al. 2010). For example, the seed is the main accumulation site of glucosinolates, procyanidins, lignans, and sinapoylcholine derivatives. When compared to other organs, approximately 13 % of the metabolites detected were shown to be specifically accumulated in seeds. Sinapoylcholine derivatives represented a large fraction of the metabolites and formed a class of seed-specific phenylpropanoids.

Transcript-metabolite analysis by Matsuda et al. (2010) was conducted on a single matrix containing 1,589 metabolite profiles together with 10,147 metabolism-related gene expression profiles, using the batch-learning self-organizing method (BL-SOM). Genes and metabolites were classified into lattices according to their relative expression levels across the different tissues analyzed (Fig. 18.1). By mapping organ-specific markers (Fig. 18.1a), the metabolites and genes specifically expressed and accumulated in seeds were easy to identify because they were located near the seed marker. The resulting classification could also be used to map targeted pathways, such as those involved in phenylpropanoid (Fig. 18.1b) and flavonoid (Fig. 18.1c) biosynthesis; Table 18.1 shows the positions of compounds and related genes on the BL-SOM map. For example, this approach revealed that 11 genes of the core phenylpropanoid pathway were coordinately expressed (high similarity of transcript patterns). The approach also highlighted upregulation of *SNG2* in

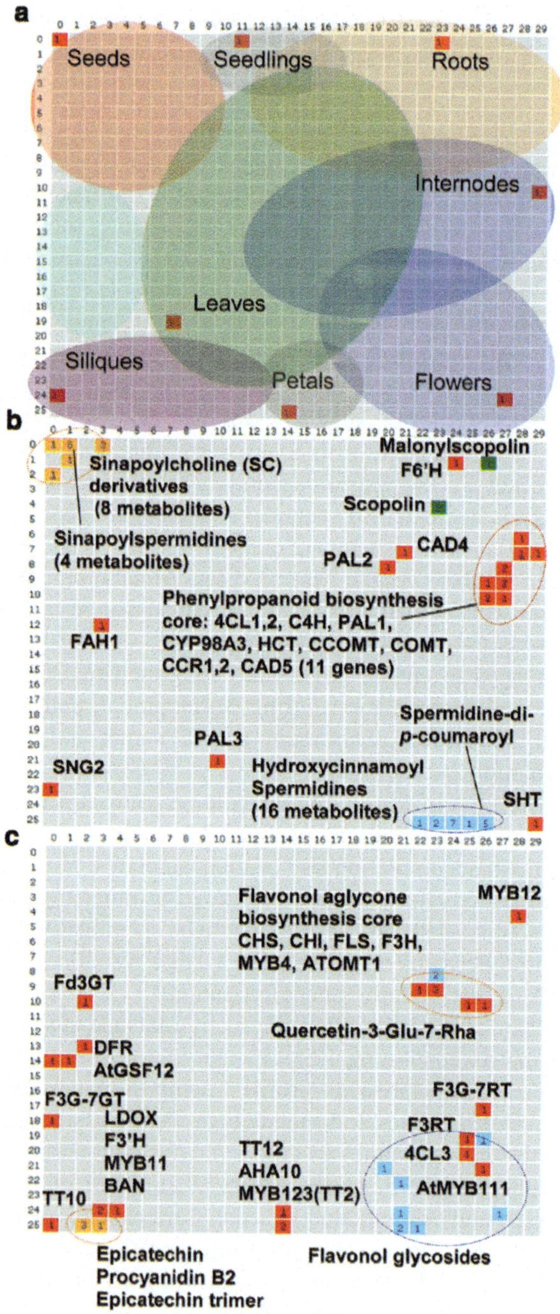

Fig. 18.1 Integrated analysis of transcriptome and metabolome data in *Arabidopsis*. **a** Batch learning self-organizing map (BL-SOM) clustering of 10,147 metabolism-related genes and 1,589 metabolite signals according to their expression or accumulation patterns across 36 tissues. Positions of tissue markers: filled circles roughly represent dominant tissues in each cell. **b** BL-SOM clustering of phenylpropanoid metabolites and their related genes. **c** BL-SOM clustering of flavonoid metabolites and their related genes. In the BL-SOM analysis, the genes and metabolites with similar expression or accumulation profiles are clustered in neighboring cells. Positions of genes are indicated by red, and other colors represent positions of metabolites. Adapted from Matsuda et al. (2010)

Table 18.1 Seed-specific secondary metabolites and their corresponding biosynthesis-related genes. Their position on batch-learning self-organizing map is indicated. Adapted from Matsuda et al. (2010)

Identifier	Map		Description (from TAIR8 and AtMetExpress database)
	X	Y	
Phenylpropanoid pathway			
At4g36220	3	12	*FAH1*; ferulate 5-hydroxylase
At5g04230	10	21	*PAL3*; phenylalanine ammonia-lyase
adp015590	3	0	Sinapoylcholine 4-*O*-hexoside
adp007692	1	0	Sinapoylcholine dehydrodimer-isomer#2
adp007728	3	0	Sinapoylcholine dimer-isomer#1
adp007736	1	0	Sinapoylcholine dimer-isomer#2
adp007735	1	0	Sinapoylcholine dimer-isomer#3
adp017396	0	0	Sinapoylcholine (4-*O*-b)G
adp024326	0	2	Sinapoylcholine (4-*O*-b)G 4′-*O*-hexoside
adp026007	1	1	Sinapoylcholine (4-*O*-b)G 4′-*O*-hexoside #2
At5g09640	0	23	*SNG2* (Sinapoylglucose accumulator 2); serine carboxypeptidase
adp029084	1	0	Spermidine *N*-Sinapoyl-*N*'-(3,5-dimethoxy-4-hexosyloxycinnamoyl)
adp020090	1	0	Spermidine *N*-Sinapoyl-*N*'-(3,5-dimethoxy-4-hexosyloxycinnamoyl)
adp020095	3	0	Spermidine-disinapyl
Flavonol biosynthesis			
At5g17220	1	14	*ATGSTF12* (Glutathione S-transferase 26)
At3g62610	3	24	*AtMYB11* (myb domain protein 11); Transcription factor
At1g61720	2	25	*BAN* (Banyuls)
At5g42800	2	13	*DFR* (Dihydroflavonol 4-reductase)
adp007051	2	25	Epicatechin
adp037126	3	25	Epicatechin trimer
At2g36790	0	18	*F3G-7GT* UDP-glycosyltransferase/transferase
At5g07990	4	24	*F3'H TT7* (Transparent testa 7); flavonoid 3′-monooxygenase
At5g17050	2	10	*Fd3GT* UDP-glucoronosyl/UDP-glucosyl transferase
At4g22880	3	24	*LDOX* (Tannin Deficient Seed 4)
adp021117	2	25	Procyanidin B2

siliques, a gene encoding sinapoylglucose:choline sinapoyltransferase, indicating that seed-specific sinapoylcholine derivatives are actively biosynthesized during seed development. Similar results were observed for the biosynthesis of procyanidin and flavonol glycoside, which accumulate specifically in the seed (Fig. 18.1c). These results indicated that the functional differentiation of the phenylpropanoid pathway among tissues was achieved by controlling expression of a small number of key biosynthetic genes. Regarding the occurrence of lignin- and neolignan-like metabolites in *Arabidopsis* seeds, previous studies suggested the participation of dirigent proteins (DPs) in lignin biosynthesis (Burlat et al. 2001). AtMetExpress was used to examine coexpression/accumulation analysis of lignin- and neolignan-like metabolites and 10 putative DP genes. The results showed that *At4g11180* (*DP1*) was expressed only in siliques, suggesting a role for neolignan accumulation in seeds. It is worth noting that the knock-out T-DNA insertion mutant for this gene lacks seed-specific neolignans (Böttcher et al. 2008). Although the function of the *DP1* gene still needs to be unequivocally characterized, this example clearly shows that the AtMetExpress approach enables hypothesis generation regarding the function of orphan genes.

18.4 Using Natural Genetic Diversity as a Source of Variation to Identify Quantitative Trait Transcripts: The Example of Carotenoid Metabolism in the Starchy Maize Endosperm

Carotenoids, a complex class of C40 isoprenoid pigments located in plastids, accumulate abundantly in some fleshy fruits (e.g., oil palm), where they play an important role as visible signals to attract animals for seed dispersal (Tanaka et al. 2008). Carotenoids are also direct precursors of apocarotenoids, including the phytohormone ABA and many volatile compounds. Among them β-ionone and geranylacetone have animal attracting characteristics with very low odor thresholds and play crucial roles in fruit dispersion. Carotenoids also accumulate in some seeds, in particular in the endosperm of maize, wheat, and sorghum, but in rather low quantities compared with fruits and leafy vegetables. These major cereal crops are generally deficient in adequate levels of nutritionally essential carotenoids. Carotenoids act as antioxidants in humans, and are also precursors of provitamin A (mainly β-carotene and α-carotene). A sustainable approach to alleviate worldwide vitamin A deficiency associated with consumption of foods poor in provitamin A carotenoids is to improve the level and composition of carotenoids in the endosperm of these taxonomically related crops. However, engineering high levels of specific carotenoids requires both increasing total carotenoid levels (enhancing pathway flux, minimizing degradation, and optimizing sequestration) and controlling routing to obtain the desired provitamin A carotenes.

The carotenoid biosynthetic pathway requires the activity of phytoene synthase (PSY), phytoene desaturase (PDS), ζ-carotene isomerase (Z-ISO), ζ-carotene desaturase (ZDS), and carotenoid isomerase (CRTISO) to produce and convert 15-ζ-phytoene into all-ε-lycopene (Fig. 18.2) (Chen et al. 2010). With the introduction of rings by the cyclase enzymes, the pathway diverges into two alternative routes to produce β-carotene or α-carotene, which require the activity of lycopene β-cyclase (LCYB) and LCYB in combination with lycopene ε-cyclase (LCYE), respectively. Finally, hydroxylation of carotenes to the nonprovitamin A xanthophylls, such as zeaxanthin and lutein, requires the activity of hydroxylase enzymes (HYD). Carotenogenic genes have been described in maize and other cereal crops (Buckner et al. 1990; Chen et al. 2010a; Gallagher et al. 2004; Harjes et al. 2008; Li et al. 1996, 2007, 2008, 2009; Matthews et al. 2003; Quinlan et al. 2007, 2012; Shumskaya et al. 2012; Singh et al. 2003; Vallabhaneni et al. 2009, 2010; Vallabhaneni and Wurtzel 2009).

The literature, which covers a wide range of organisms including model plants, indicates that various biosynthetic genes and enzymes may act as control points

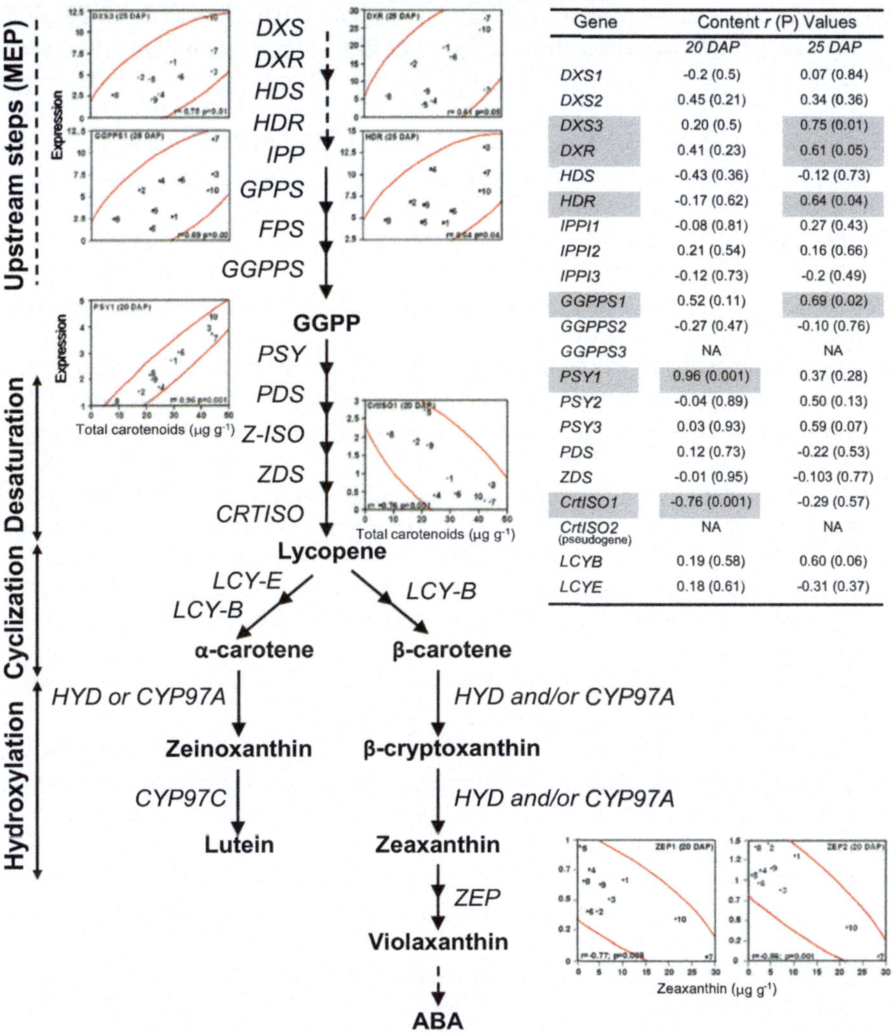

Gene	Content *r* (P) Values	
	20 DAP	*25 DAP*
DXS1	-0.2 (0.5)	0.07 (0.84)
DXS2	0.45 (0.21)	0.34 (0.36)
DXS3	0.20 (0.5)	0.75 (0.01)
DXR	0.41 (0.23)	0.61 (0.05)
HDS	-0.43 (0.36)	-0.12 (0.73)
HDR	-0.17 (0.62)	0.64 (0.04)
IPPI1	-0.08 (0.81)	0.27 (0.43)
IPPI2	0.21 (0.54)	0.16 (0.66)
IPPI3	-0.12 (0.73)	-0.2 (0.49)
GGPPS1	0.52 (0.11)	0.69 (0.02)
GGPPS2	-0.27 (0.47)	-0.10 (0.76)
GGPPS3	NA	NA
PSY1	0.96 (0.001)	0.37 (0.28)
PSY2	-0.04 (0.89)	0.50 (0.13)
PSY3	0.03 (0.93)	0.59 (0.07)
PDS	0.12 (0.73)	-0.22 (0.53)
ZDS	-0.01 (0.95)	-0.103 (0.77)
CrtISO1	-0.76 (0.001)	-0.29 (0.57)
CrtISO2 (pseudogene)	NA	NA
LCYB	0.19 (0.58)	0.60 (0.06)
LCYE	0.18 (0.61)	-0.31 (0.37)

Fig. 18.2 Identification of quantitative trait transcripts for carotenoid content in the mature maize grain. Correlation analyses showing that endosperm carotenoid content positively or negatively correlated with mRNA levels of specific pathway enzymes. Pearson correlation (*r*) and statistical significance (*P*) for correlation between transcript level and kernel carotenoid content in maize inbred lines were determined to test the statistical significance ($p<0.05$) of the relationship at a 95 % confidence interval (*red lines*). *CRTISO* carotenoid isomerase, *HYD* nonheme di-iron β-carotene hydroxylase, *CYP97A* P450 β-ring carotene hydroxylase, *CYP97C* P450 ε-ring carotene hydroxylase, *DXR* 1-deoxy-xylulose 5-phosphate reductoisomerase, *DXS* 1-deoxy-xylulose 5-phosphate synthase, *FPS* farnesyl pyrophosphate synthase, *GPPS* geranyl diphosphate synthase, *GGPPS* geranyl geranyl pyrophosphate synthase, *HDS* 4-hydroxy-3-methylbut-2-en-1-yl diphosphate synthase, *HDR* 4-hydroxy-3-methylbut-2-enyl diphosphate reductase, *IPP* isopentenyl diphosphate, *LCY-B* lycopene β-cyclase, *LCY-E* lycopene ε-cyclase, *MEP* methylerythritol phosphate, *PDS* phytoene desaturase, *PSY* phytoene synthase, *ZDS* ζ-carotene desaturase, *ZEP* zeaxanthin epoxidase, *Z-ISO* 15-cis-ζ-carotene isomerase. As described in Quinlan et al. (2012), production of lutein requires co-expression of interacting *CYP97A* and *CYP97C*. If alpha-carotene is hydroxylated by *HYD,* the product zeinoxanthin is a pathway dead-end and cannot be further converted to lutein. Adapted from Vallabhaneni and Wurtzel (2009)

for carotenoid accumulation (Cazzonelli and Pogson 2010; Cuttriss et al. 2011). In the carotenoid biosynthetic pathway, PSY and PDS, the first two committed steps, have often been suggested to play this role (Cazzonelli and Pogson 2010). However, upstream pathways for precursor biosynthesis, which include 1-deoxy-D-xylulose-5-phosphate synthase (DXS), 1-deoxy-D-xylulose-5-phosphate reductoisomerase (DXR), isopentenyl pyrophosphate isomerase (IPPI), and geranylgeranyl pyrophosphate synthase (GGPPS), may also influence carotenoid accumulation positively, while downstream degradative pathways may deplete the carotenoid pool (Cazzonelli and Pogson 2010). In this context, for a given crop, system biology approaches may help identifying specific key transcriptional control points and hence appropriate targets for breeding.

In maize, recent surveys of diverse germplasm and F1 hybrids revealed the existence of a very broad variation in endosperm carotenoid content and composition, suggesting that there is room for enhancement of these traits (Harjes et al. 2008; Li et al. 2008; Wurtzel et al. 2012). Studying correlations between transcript and metabolite levels among selected maize germplasm were therefore recently proposed as a simple way to identify the transcriptional control points that affect carotenoid synthesis in the maize endosperm (Li et al. 2008; Vallabhaneni and Wurtzel 2009). Transcript levels were quantified in developing seeds from 10 to 25 DAP, i.e., the stages when carotenoids accumulate in the endosperm. Quantitative transcript profiling was conducted in a core subset of 10 maize lines (spanning four major genetic diversity groups), which represent existing diversity regarding carotenoid content and composition (Harjes et al. 2008; Liu et al. 2003).

As a proof-of-concept for the effectiveness of this approach, the collection was first used to test the PSY gene family, which contains three paralogs, one of which (PSY1) was suspected of directing the isoprenoid flux towards carotenogenesis (Li et al. 2008). As expected, among the three paralogs, only PSY1 showed a significant correlation between transcript levels at 20 DAP and carotenoid content in the mature endosperm (Fig. 18.2). The validated germplasm subset was then used to assess the influence of other genes of the pathway using paralog-specific primers, enabling detection of QTT at the paralog level. Pearson correlation analysis revealed multiple pathway bottlenecks for carotenoid biosynthesis in specific temporal windows of endosperm development. Transcript levels of paralogs encoding isoprenoid isopentenyl diphosphate and geranylgeranyl diphosphate-producing enzymes, DXS3, DXR, HDR, and GGPPS1, were found to positively correlate with endosperm carotenoid content (Fig. 18.2). Carotenoid accumulation was inversely associated with the transcript levels for CrtISO1 and ZEP1 and ZEP2. The ZEP enzymes deplete the carotenoid pool by conversion to ABA (Fig. 18.2). The negative correlation with CrtISO was unexpected. Recent studies have shown this gene in *Arabidopsis* to be under epigenetic control (Cazzonelli et al. 2010). More recently, high transcript levels of the cleavage enzyme CCD1 were found to be associated with reduced endosperm carotenoids in maize endosperm (Vallabhaneni et al. 2010). In summary, the germplasm diversity collection revealed a number of genes for which expression affects pathway flux and total carotenoid levels. It was also observed that the timing of expression for isoprenoid pathway genes fell into a different temporal window compared to the carotenoid pathway-specific genes. This collection was also used to identify two genes, *HYD3* and *LCYE*, for which expression affects carotenoid composition (Harjes et al. 2008; Vallabhaneni et al. 2009). The *HYD3* locus encodes the enzyme that converts provitamin A carotenoids to nonprovitamin A xanthophylls and the *LCYE* affects pathway branching towards carotenoids with lower provitamin A value. As a result of these and later studies (Yan et al. 2010), useful alleles are now being used for breeding high provitamin A maize.

The combined use of a genetically diverse germplasm and coupled transcript-metabolite profiling thus appeared to be a reliable way to identify potential targets for metabolic engineering/breeding. A complementary approach to the use of natural genetic variation may be the use of reverse genetics to artificially modify a particular metabolic pathway and subsequently analyze transcript-metabolite relationships at the quantitative level. This can be achieved by increasing or decreasing the expression of individual enzymes in a pathway or by ectopic expression of heterolog genes. Recently, this was carried out in potato tubers using 24 lines carrying six different transgene combinations (Diretto et al. 2010). The assessment of perturbations in transcript and metabolite levels in transgenic tubers enabled identification of several endogenous genes likely to play a key regulatory role in carotenogenesis, showing that multiplex genetic transformation may be a reliable way of studying targeted metabolism and identifying control points. The transfer and combination of five carotenogenic genes were recently accomplished in a white seeded maize variety, indicating that this strategy could also be used to elucidate the biosynthetic steps, which control carotenoid synthesis in this crop (Zhu et al. 2008).

18.5 Use of Growth Environment as a Source of Variation to Build Metabolic Networks: The Metabolism of Chlorogenic Acids in Coffee Seeds as a Case Study

Phenylpropanoid-derived compounds are ubiquitous plant secondary metabolites. Among them, esters formed between hydroxycinnamic and quinic acids, collectively known as chlorogenic acids (CGA), are a major family of soluble plant phenolics (Clifford 1999). Since the transferase reaction that couples quinic to cinnamic acids derivatives is reversible, CGA are considered as a storage form of cinnamic acid derivatives for lignin biosynthesis (Aerts and Baumann 1994; Schoch et al. 2001). CGA are commonly found in Asteraceae, Solanaceae, and Rubiaceae (Molgaard and Ravn 1988). Species of the genus *Coffea*, which belongs to the Rubiaceae family, all accumulate CGA in their seeds, which are made up of a copious endosperm (approximate 99 % of the seed mass) surrounding a tiny rudimentary embryo (Campa et al. 2005). In *C. arabica*, CGA transiently accumulate to a spectacular extent (15–20 % of the dry mass, DM) in developing seeds during the early stage of endosperm expansion (Joët et al. 2009) and represent up to 8 % DM of the mature seed (Farah and Donangelo 2006). Coffee is therefore a valuable model to investigate the modes of regulation involved in the accumulation of these widespread plant phenolics.

5-caffeoyl quinic acid (5-CQA) is the precursor of all other CGAs and is often the most abundant CGA isomer (Fig. 18.3). The first steps of 5-CQA biosynthesis involve the well-characterized enzymes of the 'core phenylpropanoid pathway', namely phenylalanine ammonia lyase (PAL), cinnamate 4-hydroxylase (C4H), and 4-coumarate CoA ligase (4CL). Then, from p-coumaroyl-CoA, the shortest route involves only two enzymatic steps: esterification of quinic acid on p-coumaroyl-CoA by HQT (hydroxycinnamoyl-CoA shikimate/quinate hydroxycinnamoyl transferase), then hydroxylation of p-coumaroyl quinate by C3′H (p-coumaroyl ester 3′-hydroxylase) to form 5-CQA (Niggeweg et al. 2004; Schoch et al. 2001). Based on this knowledge, several pathway-guided studies using coupled transcript-metabolite profiling during seed development were performed in coffee to decipher CGA metabolism (Joët et al. 2009; Koshiro et al. 2007; Lepelley et al. 2007). By providing broad timing of 5-CQA biosynthetic transcript accumulation, these studies suggested that 5CQA synthesis is controlled at the transcriptional level, with key phenylpropanoid genes being coordinately expressed during the developmental program (Joët et al. 2009). However, the studies did not demonstrate whether transcriptional activity quantitatively controls 5-CQA biosynthesis in the developing coffee endosperm.

An approach to investigate whether 5-CQA accumulation is quantitatively modulated at the transcriptional level consists in profiling CGA and transcripts of related biosynthetic genes under various conditions affecting seed CGA content. This strategy was achieved, thanks to the dramatic influence of the environment on the CGA composition of mature coffee seeds (Joët et al. 2010b). CGA composition was shown to be influenced by the mean air temperature during seed development. For instance, the amount of minor CGA isomers [e.g., 3- and 4-caffeoyl quinate (3-CQA and 4-CQA)] was positively correlated with temperature, while the reverse trend was observed for 5-CQA. Roughly, a 10 °C decrease in mean daily temperature led to a two-fold increase in 5-CQA content. This strong effect of temperature enabled the investigation of CGA metabolism using a systems biology approach with the environment as the source of variation (Joët et al. 2010a). In this study, the transcript levels of 23 selected phenylpropanoid genes, together with the accumulation of eight CGA compounds were monitored throughout seed development across 16 locations throughout Reunion Island (S.W. Indian Ocean) having high climactic variation.

The variability of seed CGA composition induced by temperature is a valuable way to identify causative relationships between gene expression and subsequent metabolite accumulation. Highly significant correlations (p-value<0.01) were found with four genes, namely *4CL8*, *HQT*, *F5H1* and *POD* (Fig. 18.3). The final amount of 5-CQA was positively correlated with early expression of *HQT* and *4CL8*, suggesting these genes represent major QTT for CGA biosynthesis (Fig. 18.3a). By contrast, early *F5H1* and *POD* transcript levels were negatively correlated with amounts of 5-CQA at mid development, showing that transcriptional control of CGA content could also operate at the level of phenylpropanoid gene encoding

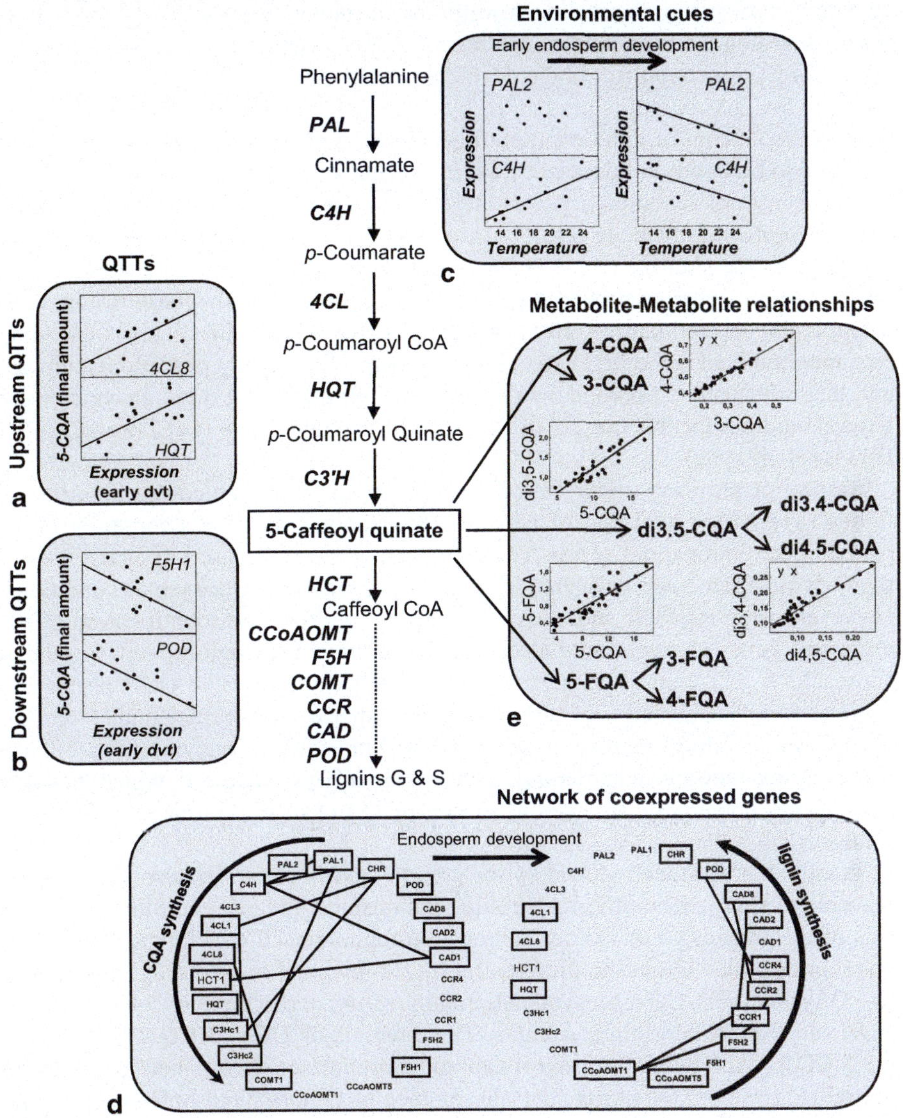

Fig. 18.3 System analysis of phenylpropanoid metabolism and chlorogenic acid accumulation in the developing coffee seed. The different boxes highlight highly significant correlations obtained between climatic, genomic, and metabolite datasets. Significant correlations between transcript abundances are shown by a straight line within the transcriptional network (*P* value cut-off of 0.001, corresponding to a Pearson's correlation coefficient >0.80). *C3'H* p-coumaroyl CoA 3-hydroxylase, *C4H* trans-cinnamate 4-hydroxylase, *CAD* cinnamyl alcohol dehydrogenase, *CCoAOMT* caffeoyl-CoA 3-O-methyltransferase, *4CL* 4-coumarate:CoA ligase, *CCR* cinnamoyl-CoA reductase, *COMT* caffeic acid O-methyltransferase, *F5H* ferulate 5-hydroxylase, *FQT* feruloyl-CoA quinate feruloyl transferase, *HCT* hydroxycinnamoyl-CoA:shikimate/quinate hydroxycinnamoyl transferase, *HQT* hydroxycinnamoyl-CoA quinate hydroxycinnamoyl transferase, *PAL* phenylalanine ammonia lyase, *POD* secretory peroxidase. Adapted from Joët et al. (2010)

enzymes involved in channeling towards lignin biosynthesis (Fig. 18.3b). These results revealed the existence of upstream and downstream transcriptional control points for CGA metabolism in the developing coffee seed.

Since 5-CQA content is modulated by temperature and CGA synthesis is under transcriptional control, transcription of genes of the phenylpropanoid pathway was expected to be tightly influenced by environmental cues. The influence of temperature on transcript accumulation was therefore investigated using linear regression. One of the most noticeable results of these analyses was the modulation of *PAL2* and *C4H* expression by temperature, where cool climates caused a delay in the upregulation of phenylpropanoid gene transcripts (Fig. 18.3c). These results indicate that the environment exerts strong control over expression of genes encoding enzymes involved in the first two committed steps of the phenylpropanoid pathway, and thus substantiate previous work which hypothesized that these enzymes were good candidates for flux control of the phenylpropanoid pathway (Bate et al. 1994; Howles et al. 1996).

Transcript profiling under various environments also enabled the description, through regression analysis, of networks of quantitatively and temporally coexpressed phenylpropanoid genes. This approach is generally used to analyze compilations of microarray data that cover a wide range of experimental conditions. However, using RT-PCR and a restricted gene repertoire, the identification of co-expressed genes clusters could also be easily achieved through a pathway-guided approach. A first module of genes, which matched the massive CGA accumulation stage, was identified very early during endosperm development and comprised most genes encoding upstream phenylpropanoid biosynthetic enzymes (Fig. 18.3d). Later, during endosperm hardening, a second module was detected, which included many genes potentially involved in partial 5-CQA channeling towards lignin biosynthesis.

Finally, the variance induced by the growth environment was also shown to be valuable to study metabolite-metabolite relationships. Indeed, combinatorial analysis of the various CGA isomer contents throughout seed development provided novel information about the preferential routes involved in 5-feruloyl quinic acid (5-FQA) and di3,5-CQA biosynthesis and in *trans*-esterification of 5-CQA, di3,5-CQA and 5-FQA into minor isomers. The three major CGA (5-CQA, 5-FQA and di3,5-CQA) exhibited highly significant intercorrelations during their biosynthetic period (Fig. 18.3e), showing that environmental factors have no impact on the dynamic properties of the isomer conversion processes involved. In other terms, while the environment significantly influenced the magnitude of 5-CQA biosynthesis, for the downstream isomers concomitantly synthesized from 5-CQA, the rate of conversion at each step was constant. The highly significant correlation found between intermediate amounts of 5-FQA and 5-CQA confirmed that feruloyl quinate is derived from 5-CQA remobilization *via* direct methylation of 5-CQA and not from a parallel shikimate ester route (Lepelley et al. 2007). Similarly, the observed biosynthetic ratio of 1:1 for 3-CQA and 4-CQA suggested that the isomerization process of 5-CQA does not present any specificity toward 3- or 4- positions for acylation.

18.6 Concluding Remarks

The use of the growth environment as a source of variation is a reliable approach to identify QTT and to decipher metabolic networks. This approach should be applicable to other metabolic pathways whose key components are under environmental control. In coffee, for instance, an effect of mean daily temperature on seed FA and free sugar composition was also observed (Joët et al. 2010b). In addition to the detection of QTT, this approach should also enable its corollary, i.e. the identification of metabolites that regulate transcription. A systems approach used on *Arabidopsis* leaves grown in highly variable growth conditions recently led to the identification of a few metabolites, including sucrose, whose content was robustly correlated with transcript abundance of hundreds of genes not directly related to their biosynthesis (Hannah et al. 2010). A systems approach could be a promising way of identifying metabolites that mediate transcription in seeds. For example, the ratio between hexoses and sucrose (H/S) has been proposed as a key determinant of metabolic regulation in developing seeds, with a high H/S ratio activating cell division, and a low H/S ratio triggering deposition of reserves (Hill et al. 2003; Wobus and Weber 1999). However, up to now, the number of metabolites identified as playing such a role in seeds is limited.

18.6.1 Perspectives

Network genomics is a new field in plant biology that offers outstanding opportunities to unravel the underlying regulation of metabolism in seeds. For now, only a few recent studies have coupled transcriptomics and metabolomics to investigate metabolism in developing seeds, either at the genome-wide level or at the quantitative level using pathway-guided approaches and multigenotype or multienvironment designs. The three examples described in this chapter are drawn from this recent literature. They all concern the biosynthesis of secondary metabolites, but similar approaches may undoubtedly be used for other seed metabolic pathways.

At the transcriptome level, analysis of coexpression networks does not only allow the prediction of gene functions, but also facilitates the characterization of pathways as a whole (Fukushima et al. 2009; Saito et al. 2008; Usadel et al. 2009). After identifying modules of coexpressed genes, the next step will be the search for TFs regulating the expression of these functional clusters of genes. Again, coexpression analysis, either by clustering or regression, has proved to be a powerful tool to identify regulatory elements of a particular metabolic pathway (Hirai et al. 2007).

With regard to coupled transcript-metabolite profiling in seed tissues, another logical extension of the studies described in this chapter is the search for correlations between transcript abundance and values of a quantitative trait using multigenotype or multienvironment designs at the genome-wide scale. To date, most studies that searched for QTTs focused on specific pathways with targeted gene expression profiling using RT-PCR. However, as the cost of microarrays and RNA-seq technologies continues to go down, simultaneous deep transcript profiling in various environments or genotypes should soon be within reach. As a preliminary step towards this goal, a recent publication reported the comparison of transcript profiles in six wheat genotypes grown in two different environments (Wan et al. 2009). Using the 55-k wheat Affymetrix chip, this transcriptomics study was carried out on whole caryopsis samples at two stages of grain filling. This work enabled identification of more than 400 transcripts, stable over many years and across environments, which were correlated with variations in yield and quality traits. The widespread use of network genomics should therefore pave the way for the identification of the QTTs involved in seed organoleptic and nutritional qualities, an important step that will highlight specific targets in breeding programs.

Acknowledgments The Wurtzel laboratory is supported by grants from the United States National Institutes of Health (GM081160) and New York State. The research performed by S. Dussert and T. Joët is financially supported by the European Union, Région Réunion, and IRD (DOCUP Réunion 2007–2013).

References

Aerts RJ, Baumann TW (1994) Distribution and utilization of chlorogenic acid in coffeaseedlings. J Exp Bot 45:497–503

Bate NJ, Orr J, Ni W, Meromi A, Nadler-hassar T, Doerner PW, Dixon RA, Lamb CJ, Elkind Y (1994) Quantitative relationship between phenylalanine ammonia-lyase levels and phenylpropanoid accumulation in transgenic tobacco identifies a rate-determining step in natural product synthesis. Proc Natl Acad Sci U S A 91:7608–7612

Böttcher C, von Roepenack-Lahaye E, Schmidt J, Schmotz C, Neumann S, Scheel D, Clemens S (2008) Metabolome analysis of biosynthetic mutants reveals a diversity of metabolic changes and allows identification of a large number of new compounds in *Arabidopsis*. Plant Physiol 147:2107–2120

Broun P (2005) Transcriptional control of flavonoid biosynthesis: a complex network of conserved regulators involved in multiple aspects of differentiation in *Arabidopsis*. Curr Opin Plant Biol 8:272–279

Buckner B, Kelson TL, Robertson DS (1990) Cloning of the y1 locus of maize, a gene involved in the biosynthesis of carotenoids. Plant Cell 2:867–876

Burlat V, Kwon M, Davin LB, Lewis NG (2001) Dirigent proteins and dirigent sites in lignifying tissues. Phytochemistry 57:883–897

Campa C, Doulbeau S, Dussert S, Hamon S, Noirot M (2005) Qualitative relationship between caffeine and chlorogenic acid contents among wild *Coffea* species. Food Chem 93:135–139

Cazzonelli CI, Pogson BJ (2010) Source to sink: regulation of carotenoid biosynthesis in plants. Trends Plant Sci 15:266–274

Cazzonelli CI, Roberts AC, Carmody ME, Pogson BJ (2010) Transcriptional control of set domain group 8 and carotenoid isomerase during *Arabidopsis* development. Mol Plant 3:174–191

Chen Y, Li F, Wurtzel ET (2010) Isolation and characterization of the Z-ISO gene encoding a missing component of carotenoid biosynthesis in plants. Plant Physiol 153:66–79

Cuttriss AJ, Cazzonelli CI, Wurtzel ET, Pogson BJ (2011) Carotenoids. In: Rébeillé F, Douce R (eds) Biosynthesis of vitamins in plants (advances in botanical research, part A), vol 58. Elsevier, Amsterdam, pp 1–36

Clifford MN (1999) Chlorogenic acids and other cinnamates—nature, occurrence and dietary burden. J Sci Food Agric 79:362–372

D'Auria JC, Gershenzon J (2005) The secondary metabolism of *Arabidopsis thaliana*: growing like a weed. Curr Opin Plant Biol 8:308–316

Davies KM, Schwinn KE (2003) Transcriptional regulation of secondary metabolism. Funct Plant Biol 30:913–925

Day RC, Herridge RP, Ambrose BA, Macknight RC (2008) Transcriptome analysis of proliferating *Arabidopsis* endosperm reveals biological implications for the control of syncytial division, cytokinin signaling, and gene expression regulation. Plant Physiol 148:1964–1984

Diretto G, Al-Babili S, Tavazza R, Scossa F, Papacchioli V, Migliore M, Beyer P, Giuliano G (2010) Transcriptional-metabolic networks in beta-carotene-enriched potato tubers: the long and winding road to the golden phenotype. Plant Physiol 154:899–912

Farah A, Donangelo CM (2006) Phenolic compounds in coffee. Braz J Plant Physiol 18:23–36

Fukushima A, Kusano M, Redestig H, Arita M, Saito K (2009) Integrated omics approaches in plant systems biology. Curr Opin Chem Biol 13:532–538

Gachon CM, Langlois-Meurinne M, Henry Y, Saindrenan P (2005) Transcriptional co-regulation of secondary metabolism enzymes in *Arabidopsis*: functional and evolutionary implications. Plant Mol Biol 58:229–245

Gallagher CE, Matthews PD, Li F, Wurtzel ET (2004) Gene duplication in the carotenoid biosynthetic pathway preceded evolution of the grasses. Plant Physiol 135:1776–1783

Hannah MA, Caldana C, Steinhauser D, Balbo I, Fernie AR, Willmitzer L (2010) Combined transcript and metabolite profiling of *Arabidopsis* grown under widely variant growth conditions facilitates the identification of novel metabolite-mediated regulation of gene expression. Plant Physiol 152:2120–2129

Harjes CE, Rocheford TR, Bai L, Brutnell TP, Kandianis CB, Sowinski SG, Stapleton AE, Vallabhaneni R, Williams M, Wurtzel ET, Yan JB, Buckler ES (2008) Natural genetic variation in lycopene epsilon cyclase tapped for maize biofortification. Science 319:330–333

Hill LM, Morley-Smith ER, Rawsthorne S (2003) Metabolism of sugars in the endosperm of developing seeds of oilseed rape. Plant Physiol 131:228–236

Hirai MY, Sugiyama K, Sawada Y, Tohge T, Obayashi T, Suzuki A, Araki R, Sakurai N, Suzuki H, Aoki K, Goda H, Nishizawa OI, Shibata D, Saito K (2007) Omics-based identification of *Arabidopsis* Myb transcription factors regulating aliphatic glucosinolate biosynthesis. Proc Natl Acad Sci U S A 104:6478–6483

Howles PA, Sewalt VJH, Paiva NL, Elkind Y, Bate NJ, Lamb C, Dixon RA (1996) Overexpression of L-phenylalanine ammonia-lyase in transgenic tobacco plants reveals control points for flux into phenylpropanoid biosynthesis. Plant Physiol 112:1617–1624

Joët T, Laffargue A, Salmona J, Doulbeau S, Descroix F, Bertrand B, de Kochko A, Dussert S (2009) Metabolic pathways in tropical dicotyledonous albuminous seeds: *Coffea arabica* as a case study. New Phytol 182:146–162

Joët T, Salmona J, Laffargue A, Descroix F, Dussert S (2010a) Use of the growing environment as a source of variation to identify the quantitative trait transcripts and modules of co-expressed genes that determine chlorogenic acid accumulation. Plant Cell Environ 33:1220–1233

Joët T, Laffargue A, Descroix F, Doulbeau S, Bertrand B, de Kochko A, Dussert S (2010b) Influence of environmental factors, wet processing and their interactions on the biochemical composition of green Arabica coffee beans. Food Chem 118:693–701

Kitamura S, Matsuda F, Tohge T, Yonekura-Sakakibara K, Yamazaki M, Saito K, Narumi I (2010) Metabolic profiling and cytological analysis of proanthocyanidins in immature seeds of *Arabidopsis thaliana* flavonoid accumulation mutants. Plant J 62:549–559

Koshiro Y, Jackson MC, Katahira R, Wang ML, Nagai C, Ashihara H (2007) Biosynthesis of chlorogenic acids in growing and ripening fruits of *Coffea arabica* and *Coffea canephora* plants. Z Naturforsch C 62:731–742

Le BH, Cheng C, Bui AQ, Wagmaister JA, Henry KF, Pelletier J, Kwong L, Belmonte M, Kirkbride R, Horvath S, Drews GN, Fischer RL, Okamuro JK, Harada JJ, Goldberg RB (2010) Global analysis of gene activity during *Arabidopsis* seed development and identification of seed-specific transcription factors. Proc Natl Acad Sci U S A 107:8063–8070

Lepelley M, Cheminade G, Tremillon N, Simkin A, Caillet V, McCarthy J (2007) Chlorogenic acid synthesis in coffee: an analysis of CGA content and real-time RT-PCR expression of HCT, HQT, C3H1, and CCoAOMT1 genes during grain development in *C. canephora.* Plant Sci 172:978–996

Lepiniec L, Debeaujon I, Routaboul JM, Baudry A, Pourcel L, Nesi N, Caboche M (2006) Genetics and biochemistry of seed flavonoids. Annu Rev Plant Biol 57:405–430

Lewinsohn E, Gijzen M (2009) Phytochemical diversity: the sounds of silent metabolism. Plant Sci 176:161–169

Li ZH, Matthews PD, Burr B, Wurtzel ET (1996) Cloning and characterization of a maize cDNA encoding phytoene desaturase, an enzyme of the carotenoid biosynthetic pathway. Plant Mol Biol 30:269–279

Li F, Murillo C, Wurtzel ET (2007) Maize *Y9* encodes a product essential for 15-*cis* zetacarotene isomerization. Plant Physiol 144:1181–1189

Li F, Vallabhaneni R, Yu J, Rocheford T, Wurtzel ET (2008) The maize phytoene synthase gene family: overlapping roles for carotenogenesis in endosperm, photomorphogenesis, and thermal stress tolerance. Plant Physiol 147:1334–1346

Li F, Tzfadia O, Wurtzel ET (2009) The phytoene synthase gene family in the Grasses: subfunctionalization provides tissue-specific control of carotenogenesis. Plant Signal Behav 4:208–211

Liu K, Goodman M, Muse S, Smith JS, Buckler E, Doebley J (2003) Genetic structure and diversity among maize inbred lines as inferred from DNA microsatellites. Genetics 165:2117–2128

Matsuda F, Yonekura-Sakakibara K, Niida R, Kuromori T, Shinozaki K, Saito K (2009) MS/MS spectral tag-based annotation of non-targeted profile of plant secondary metabolites. Plant J 57:555–577

Matsuda F, Hirai MY, Sasaki E, Akiyama K, Yonekura-Sakakibara K, Provart NJ, Sakurai T, Shimada Y, Saito K (2010) AtMetExpress development: a phytochemical atlas of *Arabidopsis* development. Plant Physiol 152:566–578

Matthews PD, Luo R, Wurtzel ET (2003) Maize phytoene desaturase and zeta-carotene desaturase catalyse a poly-Z desaturation pathway: implications for genetic engineering of carotenoid content among cereal crops. J Exp Bot 54:2215–2230

Molgaard P, Ravn H (1988) Evolutionary aspects of caffeoyl ester distribution in dicotyledons. Phytochemistry 27:2411–2421

Mueller LA, Zhang P, Rhee SY (2003) AraCyc: a biochemical pathway database for *Arabidopsis*. Plant Physiol 132:453–460

Niggeweg R, Michael AJ, Martin C (2004) Engineering plants with increased levels of the antioxidant chlorogenic acid. Nat Biotechnol 22:746–754

North H, Baud S, Debeaujon I, Dubos C, Dubreucq B, Grappin P, Jullien M, Lepiniec L, Marion-Poll A, Miquel M, Rajjou L, Routaboul JM, Caboche M (2010) *Arabidopsis* seed secrets unravelled after a decade of genetic and omics-driven research. Plant J 61:971–981

Passador-Gurgel G, Hsieh WP, Hunt P, Deighton N, Gibson G (2007) Quantitative trait transcripts for nicotine resistance in *Drosophila melanogaster*. Nat Genet 39:264–268

Quinlan RF, Jaradat TT, Wurtzel ET (2007) *Escherichia coli* as a platform for functional expression of plant P450 carotene hydroxylases. Arch Biochem Biophys 458:146–157

Quinlan RF, Shumskaya M, Bradbury LMT, Beltrán J, Ma C, Kennelly EJ, Wurtzel ET (2012) Synergistic interactions between carotene ring hydroxylases drive lutein formation in plant carotenoid biosynthesis. Plant Physiol 160:204–214

Routaboul JM, Kerhoas L, Debeaujon I, Pourcel L, Caboche M, Einhorn J, Lepiniec L (2006) Flavonoid diversity and biosynthesis in seed of *Arabidopsis thaliana*. Planta 224:96–107

Ruuska SA, Girke T, Benning C, Ohlrogge JB (2002) Contrapuntal networks of gene expression during *Arabidopsis* seed filling. Plant Cell 14:1191–1206

Saito K, Hirai MY, Yonekura-Sakakibara K (2008) Decoding genes with coexpression networks and metabolomics—'majority report by precogs'. Trends Plant Sci 13:36–43

Schmid M, Davison TS, Henz SR, Pape UJ, Demar M, Vingron M, Scholkopf B, Weigel D, Lohmann JU (2005) A gene expression map of *Arabidopsis thaliana* development. Nat Genet 37:501–506

Schoch G, Goepfert S, Morant M, Hehn A, Meyer D, Ullmann P, Werck-Reichhart D (2001) CYP98A3 from *Arabidopsis thaliana* is a 3'-hydroxylase of phenolic esters, a missing link in the phenylpropanoid pathway. J Biol Chem 276:36566–36574

Schwab W (2003) Metabolome diversity: too few genes, too many metabolites? Phytochemistry 62:837–849

Shumskaya M, Bradbury LMT, Monaco RR, Wurtzel ET (2012) Plastid localization of the key carotenoid enzyme phytoene synthase is altered by isozyme, allelic variation, and activity. Plant Cell 24:3725–3741. doi:10.1105/tpc.112.104174 (first published on Sept 28)

Singh M, Lewis PE, Hardeman K, Bai L, Rose JK, Mazourek M, Chomet P, Brutnell TP (2003) Activator mutagenesis of the pink *scutellum1/viviparous7* locus of maize. Plant Cell 15:874–884

Spencer MWB, Casson SA, Lindsey K (2007) Transcriptional profiling of the *Arabidopsis* embryo. Plant Physiol. 143:924–940 (Erratum in: Plant Physiol 143 (2007) 1982)

Stitt M, Sulpice R, Keurentjes J (2010) Metabolic networks: how to identify key components in the regulation of metabolism and growth. Plant Physiol 152:428–444

Stracke R, Ishihara H, Huep G, Barsch A, Mehrtens F, Niehaus K, Weisshaar B (2007) Differential regulation of closely related R2R3-MYB transcription factors controls flavonol accumulation in different parts of the *Arabidopsis thaliana* seedling. Plant J 50:660–677

Tanaka T, Antonio BA, Kikuchi S, Matsumoto T, Nagamura Y, Numa H, Sakai H, Wu J, Itoh T, Sasaki T, Aono R, Fujii Y, Habara T, Harada E, Kanno M, Kawahara Y, Kawashima H, Kubooka H, Matsuya A, Nakaoka H, Saichi N, Sanbonmatsu R, Sato Y, Shinso Y, Suzuki M, Takeda JI, Tanino M, Todokoro F, Yamaguchi K, Yamamoto N, Yamasaki C, Imanishi T, Okido T, Tada M, Ikeo K, Tateno Y, Gojobori T, Lin YC, Wei FJ, Hsing YI, Zhao Q, Han B, Kramer MR, McCombie RW, Lonsdale D, O'Donovan CC, Whitfield EJ, Apweiler R, Koyanagi KO, Khurana JP, Raghuvanshi S, Singh NK, Tyagi AK, Haberer G, Fujisawa M, Hosokawa S, Ito Y, Ikawa H, Shibata M, Yamamoto M, Bruskiewich RM, Hoen DR, Bureau TE, Namiki N, Ohyanagi H, Sakai Y, Nobushima S, Sakata K, Barrero RA, Sato Y, Souvorov A, Smith-White B, Tatusova T, An S, An G, OOta S, Fuks G, Messing J, Christie KR, Lieberherr D, Kim H, Zuccolo A, Wing RA, Nobuta K, Green PJ, Lu C, Meyers BC, Chaparro C, Piegu B, Panaud O, Echeverria M (2008) The Rice Annotation Project Database (RAP-DB): 2008 update. Nucl Acids Res 36:D1028–D1033

Thimm O, Bläsing O, Gibon Y, Nagel A, Meyer S, Krüger P, Selbig J, Müller LA, Rhee SY, Stitt M (2004) MAPMAN: a user-driven tool to display genomics data sets onto diagrams of metabolic pathways and other biological processes. Plant J 37:914–939

Tokimatsu T, Sakurai N, Suzuki H, Ohta H, Nishitani K, Koyama T, Umezawa T, Misawa N, Saito K, Shibata D (2005) KaPPA-view: a web-based analysis tool for integration of transcript and metabolite data on plant metabolic pathway maps. Plant Physiol. 138:1289–300

Usadel B, Obayashi T, Mutwil M, Giorgi FM, Bassel GW, Tanimoto M, Chow A, Steinhauser D, Persson S, Provart NJ (2009) Co-expression tools for plant biology: opportunities for hypothesis generation and caveats. Plant Cell Environ 32:1633–1651

Vallabhaneni R, Wurtzel ET (2009) Timing and biosynthetic potential for carotenoid accumulation in genetically diverse germplasm of maize. Plant Physiol 150:562–572

Vallabhaneni R, Gallagher CE, Licciardello N, Cuttriss AJ, Quinlan RF, Wurtzel ET (2009) Metabolite sorting of a germplasm collection reveals the hydroxylase3 locus as a new target for maize provitamin A biofortification. Plant Physiol 151:1635–1645

Vallabhaneni R, Bradbury LMT, Wurtzel ET (2010) The carotenoid dioxygenase gene family in maize, sorghum, and rice. Arch Biochem Biophys 504:104–111

van Baarlen P, van Esse HP, Siezen RJ, Thomma BP (2008) Challenges in plant cellular pathway reconstruction based on gene expression profiling. Trends Plant Sci 13:44–50

Vitoria AP, Mazzafera P (1999) Xanthine degradation and related enzyme activities in leaves and fruits of two coffea species differing in caffeine catabolism. J Agric Food Chem 47:1851–1855

Wan Y, Underwood C, Toole G, Skeggs P, Zhu T, Leverington M, Griffiths S, Wheeler T, Gooding M, Poole R, Edwards KJ, Gezan S, Welham S, Snape J, Mills EN, Mitchell RA, Shewry PR (2009) A novel transcriptomic approach to identify candidate genes for grain quality traits in wheat. Plant Biotechnol J 7:401–410

Wurtzel ET, Cuttriss A, Vallabhaneni R (2012) Maize provitamin A carotenoids, current resources and future metabolic engineering challenges. Front Plant Sci 3:29

Yan JB, Kandianis CB, Harjes CE, Bai L, Kim EH, Yang XH, Skinner DJ, Fu ZY, Mitchell S, Li Q, Fernandez MGS, Zaharieva M, Babu R, Fu Y, Palacios N, Li JS, DellaPenna D, Brutnell T, Buckler ES, Warburton ML, Rocheford T (2010) Rare genetic variation at *Zea mays* crtRB1 increases beta-carotene in maize grain. Nature Genet 42:322–374

Yonekura-Sakakibara K, Tohge T, Matsuda F, Nakabayashi R, Takayama H, Niida R, Watanabe-Takahashi A, Inoue E, Saito K (2008) Comprehensive flavonol profiling and transcriptome coexpression analysis leading to decoding gene-metabolite correlations in *Arabidopsis*. Plant Cell 20:2160–2176

Zhu C, Naqvi S, Breitenbach J, Sandmann G, Christou P, Capell T (2008) Combinatorial genetic transformation generates a library of metabolic phenotypes for the carotenoid pathway in maize. Proc Natl Acad Sci U S A 105:18232–18237

Chapter 19
Using Systems Approaches to Analyze Metabolic Networks Involved in Storage Reserve Synthesis in Developing Seeds

Christian Krach, Eva Grafahrend-Belau, Hart Poskar, Kai Schallau, Falk Schreiber and Björn H. Junker

Abstract In contrast to transcripts, proteins, and metabolites, intracellular metabolic fluxes cannot be measured directly, but rather rely on prediction *via* computational simulation or on indirect measurements. In this chapter, we describe different approaches to quantify and simulate fluxes in plant metabolic pathways. Metabolic behavior can be simulated with models ranging from basic stoichiometry to fully detailed kinetic models, and fluxes can be quantified via a combination of computer modeling and stable isotope labeling experiments. We especially emphasize the usefulness of these approaches for improving the understanding of storage compound synthesis in developing crop seeds.

Keywords Flux balance analysis (FBA) · Kinetic modeling · Metabolic flux analysis (MFA) · Metabolism · Steady state

19.1 Introduction

In recent years novel approaches have found their way into plant research, which allow researchers to quickly obtain large molecular datasets about the system under study. NSG, transcriptomics, proteomics, metabolomics, and similar types of analyses have revolutionized the way that research is conducted in plant biology. Albeit these types of experiments are indispensable to modern biological studies, they all have in common that they do not consider the underlying network and systems properties. Instead, in an approach sometimes termed "top-down systems biology", statistical data reduction methods are used to condense the dataset into an interpretable format. In contrast, "bottom-up systems biology" is where models of biological processes are built step-by-step *in silico*, including only those components that are necessary to answer predefined questions.

B. H. Junker (✉) · C. Krach · E. Grafahrend-Belau · H. Poskar · K. Schallau · F. Schreiber
Leibniz-Institute of Plant Genetics and Crop Plant Research (IPK),
Correns str.3, Gatersleben, Germany
e-mail: junker@ipk-gatersleben.de

F. Schreiber
Martin Luther University Halle-Wittenberg, 06099 Halle, Germany

G. K. Agrawal, R. Rakwal (eds.), *Seed Development: OMICS Technologies toward Improvement of Seed Quality and Crop Yield*,
DOI 10.1007/978-94-007-4749-4_19, © Springer Science+Business Media Dordrecht 2012

In (bottom-up) systems biology research on plant metabolism, various methods are used to quantify, simulate, and understand metabolic fluxes. The first method described is stoichiometric modeling. While just a surface view of the system, the description of the metabolic network is mainly built on the stoichiometries of enzymatic reactions. Nevertheless this method allows for the elucidation of arising systems properties that, in many cases, are far-reaching. These stoichiometric models can be used to find alternate routes from one metabolite to another and to calculate the yield of these routes. Enriched by substrate uptake rates and composition of the produced biomass, flux balance analysis can be performed on these models, which predicts the flux through each reaction that is optimal for a certain criterion, *e.g.*, to maximize growth. In mathematical terms this is an underdetermined problem, which means that there are not enough data points to calculate the flux distribution actually occurring *in vivo*. It is to overcome this lack of data that the optimization criterion, mentioned above, is required. Thus, the result of the prediction, even though very useful in many cases, sometimes has to be taken with caution with respect to its validity.

The underdetermined problem can be transformed into an overdetermined one by adding more experimental data. In the second method that we describe, ^{13}C metabolic flux analysis, the models are enriched by information about the position of each C atom of the substrate(s) and product(s) comprising each reaction, in conjunction with the measurements from stable isotope labeling experiments. The resulting overdetermined mathematical problem is solved numerically by minimizing the difference between the simulated and the measured distribution of label in various metabolites. Thus, provided that the underlying network is correct, the result can be regarded as the actual flux distribution occurring *in vivo*.

The most detailed types of models are kinetic models, in which each enzymatic reaction is described by a rate equation that is often motivated by the mechanism of the enzyme (e.g., Michaelis-Menten rate law). Together with the stoichiometry of the network, the rate laws form a system of differential equations. This system can be used to simulate a time course of metabolite concentration and fluxes, to calculate a steady state and assess its stability. Furthermore, control analysis may be used to determine the contribution of any enzyme to the limitation of flux through a pathway. However, due to their complexity, kinetic models are usually limited to single pathways or subnetworks with a maximum of some 10s of reactions.

All these methods contribute to the understanding of metabolism. This knowledge can be used to improve seed yield and quality. For example, with FBA, suitable knockout strategies can be predicted to enhance the flux into specific compounds. Flux analysis can be used as an analytical tool to understand the effects of modifications in central metabolism far beyond textbook knowledge, and kinetic models can be used to predict targets for tailored engineering of metabolic pathways.

19.2 Stoichiometric Modeling

Stoichiometric modeling aims at elucidating the systemic properties and capabilities of cellular metabolism by analyzing the steady state characteristics of the metabolic network under investigation (Llaneras and Picó 2008). Based on the fundamental assumption of pseudo steady state, the modeling approach is focused on the analysis of time invariant network behaviors.

19.2.1 Principles of Stoichiometric Modeling

The structure of a metabolic network with m metabolites and n reactions can mathematically be described by a *stoichiometric matrix* S of size $m \times n$, in which rows correspond to metabolites, columns to reactions, and matrix entries to the stoichiometric coefficients. Using the stoichiometric matrix S, a mathematical description of the dynamic behavior of the metabolic model is given by the *dynamic mass balance equation:*

$$\frac{dx}{dt} = S \cdot v \qquad (19.1)$$

where $x = [x_1, x_2, \cdots, x_m]$ signifies the vector of intracellular metabolite concentrations and $v = [v_1, v_2, \cdots, v_n]$ the vector of fluxes, specifying a flux distribution in the network. Under the assumption of *pseudo steady state*, expressed by a zero time derivative of the concentration $dx/dt = 0$, Eq. (19.1) simplifies to

$$S \cdot v = 0 \qquad (19.2)$$

This homogeneous set of linear equations defines the set of all possible steady state flux distributions in the metabolic network. Since the number of reactions is greater than the number of metabolites ($n > m$) in most biological systems, Eq. (19.2) is underdetermined and contains an infinite number of possible steady state flux distributions. Using stoichiometric modeling, the reduction of the feasible flux space is obtained by the application of different constraints (e.g., stoichiometric, thermodynamic and capacity constraints) which creates a bounded flux cone corresponding to the metabolic capacities of the modeled organism (Kauffman et al. 2003). The application of constraints provides a link between stoichiometric and constraint-based modeling, a term used interchangeably in the literature.

19.2.2 Methods Used in Stoichiometric Modeling

In general, methods used in stoichiometric/constraint-based modeling can be subdivided into two major approaches: (i) pathway-based network analysis, aimed at assessing the *whole* capabilities of a metabolic network, and (ii) optimization-based network analysis, aimed at identifying *optimal* network behaviors.

Pathway-based network analysis is focused on the elucidation of the properties and capabilities of a metabolic network as a whole by analyzing the set of pathways through the network. In general, pathway-based network analysis refers to two main approaches: (i) elementary modes analysis (EMA), with an elementary mode (EMM) corresponding to a steady-state flux distribution involving a minimal set of reactions, and (ii) extreme pathway analysis (EPA), with the set of EPs corresponding to a subset of EMMs.

Optimization-based network analysis, i.e., FBA is based on the assumption that cellular metabolism behaves optimally with respect to a given objective such as the maximization of biomass. FBA uses the principle of linear programming to solve Eq. (19.2) by defining an objective function and searching the feasible solution space for an optimal flux distribution that maximizes or minimizes the objective. A detailed review of stoichiometric modeling is given in Llaneras and Picó (2008). For a general introduction into EMA, EPA, and FBA, readers are referred to Lee et al. (2006), Kauffman et al. (2003), Schuster and Hilgetag (1994), Schilling et al. (2000a,b), and Schuster et al. (2002).

19.2.3 Applications of Stoichiometric Modeling to Plant Metabolism

Despite the vast number of applications resulting from stoichiometric modeling, applications to plant metabolism are, so far, sparse. Steuer et al. (2007) applied EMA to investigate the structural properties of the mitochondrial TCA cycle of plants, forming the basis for subsequent dynamic analysis of the pathway. To determine feasible pathways for the futile cycling of sucrose in sugarcane, Rohwer and Botha (2001) performed EMA and combined the resulting information with kinetic modeling, allowing the authors to formulate hypothesis that can now be tested experimentally. Poolman et al. (2003) applied EMA to outline viable pathways of primary metabolism in the chloroplast under different light conditions. Based upon these *in silico* studies, the authors proposed the OPPP to operate cyclically in the dark and, in contrast to the traditional view, the Calvin cycle and OPPP to act as complementary and overlapping components of the same system. Schwender et al. (2004) used EMA to evaluate the contribution of different alternative pathways to oil biosynthesis in developing embryos of rapeseed (*Brassica napus*). In combination with enzyme activity measurements and isotope labeling experiments, these studies demonstrated that carbon fixation operates without the Calvin cycle, which was shown to improve the carbon efficiency of developing green seeds by combining the activity of RuBisCO with the nonoxidative reactions of the pentose phosphate pathway.

Although FBA has been applied to various biological systems (e.g., bacteria, fungi, algae, and animals) to study different aspects of metabolism, only a few applications to plant metabolism have been published so far. Poolman et al. (2009) applied FBA to a genome-scale model of *Arabidopsis* to get insight into the metabolic properties of the modeled heterotrophic cell suspension system. The validated model, which showed to be capable of producing biomass components in the proportion observed experimentally, was used (i) to identify potential metabolic modules, and (ii) to infer total ATP demand for growth and maintenance, which proved to be consistent with estimates in prokaryotes and yeast. Using the *Arabidopsis* model as a basis, Williams et al. (2010) performed FBA to compare the predicted flux solution with that estimated from ^{13}C-based steady state MFA. With the model predictions being in good agreement with most of the experimental results, these studies demonstrated that constraints-based genome-scale modeling is a valuable tool to predict flux changes in central carbon metabolism under stress conditions. De Oliveira Dal'Molin et al. (2010) applied FBA to a compartmentalized genome-scale reconstruction of the primary metabolic network in *Arabidopsis*. The model was validated against various classical physiological scenarios (photosynthesis, photorespiration, and respiration) and can now be used as a viable framework to derive new hypotheses for exploring plant metabolism.

FBA is, so far, the only stoichiometric modeling approach applied to develop predictive models of crop seed metabolism. In the following a more detailed description of the respective FBA application (Grafahrend-Belau et al. 2009b) focusing on cereal seed storage metabolism is given. In this study, a compartmentalized model of primary metabolism in the developing endosperm of barley (*Hordeum vulgare*) was constructed that includes 257 biochemical and transport reactions across four different compartments. FBA was applied to the model to get insight into storage patterning in developing cereal seed in response to environmental and genetic perturbation. To study the effect of oxygen depletion on storage patterning in developing barley endosperm, *in silico* growth simulations under varying oxygen conditions were performed (Fig. 19.1).The model was validated by comparing the simulation results with published experimental results. In addition, the model predictions were used to test published controversial hypotheses on the role of pyrophosphate metabolism in oxygen depleted tissues.

Under fully anaerobic conditions (Fig. 19.1a), the model depicted characteristic anaerobic metabolic behavior (Geigenberger 2003; Kennedy et al. 1992): (i) inhibition of respiration, (ii) induction of fermentation, and (iii) stimulation of glycolysis (Pasteur effect), resulting in a decrease in the cellular energy state and an increase in the redox state. In response to the surplus of reducing equivalents, the operation of NAD(P)H generating pathways was limited, as indicated by an incomplete citrate cycle and the inactivity of the oxidative part of the pentose phosphate pathway. The simulation results under hypoxic conditions(Fig. 19.1b) were generally consistent with the main qualitative physiological characteristics reported for hypoxic cereal seed storage metabolism (Rolletschek et al. 2004, 2005; van Dongen et al. 2004). With respect to starch metabolism, the model correctly predicted the sucrose-to-starch pathway experimentally proven for barley seed metabolism (Beckles et al. 2001; Emes et al. 2003; Thorbjørnsen et al. 1996; Weschke et al. 2000) by predicting that (i) sucrose degradation is restricted to the sucrose synthase pathway and (ii) synthesis of ADP-Glc, which is the main precursor for starch synthesis, is predominantly catalyzed by the cytosolic isoform of AGPase. Under fully aerobic conditions (Fig. 19.1c), the model showed characteristic aerobic metabolic behavior of cereal seed metabolism (van Dongen et al. 2004): (i) upregulation of respiratory energy production, resulting in an increase in the cellular energy state; (ii) subsequent increase of storage metabolism; and (iii) increase of seed dry weight due to extensive storage metabolism. In general, these results indicated that the

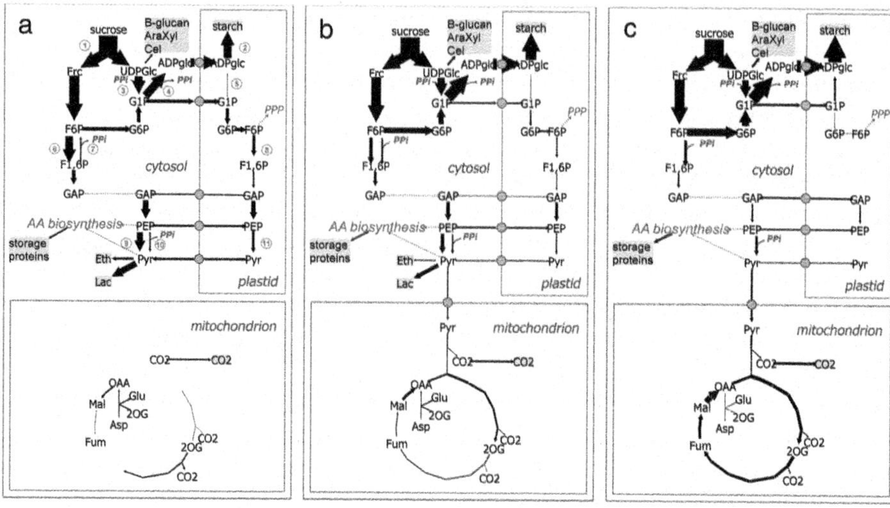

Fig. 19.1 Carbon flux maps depicting the key uptake/excretion rates and fluxes within primary metabolism of developing barley endosperm under anoxic conditions (**a**), hypoxic conditions (**b**), and aerobic conditions (**c**). Simulations were performed using the growth conditions outlined in Grafahrend-Belau et al. (2009). Metabolites excreted by the model (CO_2, ethanol, and lactate) or incorporated into biomass (starch, storage proteins, β-glucan, arabinoxylan, and cellulose) are highlighted in gray rectangles. Reactions are as follows: *1* sucrose synthase, *2* starch synthase, *3* UDP-Glc pyrophosphorylase, *4/5* cytosolic/plastidic ADP-Glc pyrophosphorylase, *6/8* cytosolic/plastidic ATP: Fru-6-P 1-phosphotransferase, *7* pyrophosphate: Fru-6-P 1-phosphotransferase, *9/11* cytosolic/plastidic pyruvate kinase, *10* pyruvate, phosphate dikinase. Metabolites are as follows: *Asp* aspartate, *ADPglc* ADP-Glc, *AraXyl* arabinoxylan, *B-glucan* β-glucan, *Cel* cellulose, *Eth* ethanol, *F1,* 6P Fru-1, 6-P, *Frc* Fru, *F6P* Fru-6-P, *Fum* fumarate, *GAP* glycerinaldehyd-*3*-phosphat, *G1P* Glc-1-P, *G6P* Glc-6-P, *Glu* glutamate, *Lac* lactate, *Mal* malate, *OAA* oxaloacetate, *2OG* 2-oxoglutarate, *PEP* phosphoenolpyruvate, *Pyr* pyruvate, *UDPglc* UDP-Glc

reconstructed model has the potential to simulate cereal seed metabolism and thus can be used as a viable framework to generate or test hypothesis and to explore cereal seed metabolism *in silico*.

With respect to the potential role of PPi metabolism in seed storage metabolism under varying oxygen supply, the simulation results supported the hypothesis of PPi availability determining the net flux through PFP and PPDK, as proposed by previous reports (Gibbs et al. 2000; Stitt 1990): (i) under anoxic conditions (Fig. 19.1a), a substantial requirement for PPi results in PFP operating in the direction of PPi synthesis to compensate for PPi deficiency; (ii) under hypoxic conditions (Fig. 19.1b), increasing PPi availability results in a net flux of PFP and PPDK operating in the direction of PPi consumption; and (iii) under aerobic conditions (Fig. 19.1c), sufficient PPi availability allows the energy-saving mode of PPi-dependent glycolysis to be the only form of glycolysis.This interpretation was supported by simulation studies, which showed that *in silico*-generated PPi deficiency leads to PFP and PPDK, operating in the direction of PPi synthesis. Although further experiments are required to evaluate the proposed hypothesis, the given examples show the potential of FBA to verify or extend the understanding of controversially-discussed biological processes by looking at systemic stoichiometric constraints alone.

19.3 Metabolic Flux Analysis

Metabolic flux is the rate at which molecules are exchanged between metabolic pools, sources, and sinks. These fluxes are regulated by the enzymes that are involved in the respective pathway.

MFA is a method for determining fluxes that combines the metabolic network, represented by the stoichiometric model as used in the dynamic mass balance equation (Eq. 19.1), together with measurement data from carbon labeling experiments (CLEs) (Blum and Stein 1982; Wiechert and de Graaf 1996). The measurement data are used to optimize the fluxes within the given model. In this way, MFA differs from FBA using the measurement data from CLEs to derive the objective function for optimization.

CLEs are experiments in which the biological organism is fed a media that incorporates one or more substrates with a known ^{13}C labeling state. By "a known labeling state", it is meant that the position of the ^{13}C atom(s) in the C backbone of each substrate compound is known. For example [1-^{13}C] glucose, in which the first[1] carbon is a ^{13}C atom, and the remaining carbons, position 2 through 6, are ^{12}C atoms. This labeled substrate is introduced into the media once the biological organism has achieved metabolic steady state[2]. The CLE then continues to distribute ^{13}C labels throughout the central metabolic network until a second steady state is reached, the isotopic steady state[3]. Once the fractional labeling or labeling profile has achieved equilibrium the isotopic enrichment is representative of the C reaction network in the central metabolism (Wiechert and de Graaf 1996).

Measurements of isotopically enriched intermediates and extracellular products from the CLEs are used to provide the positional labeling information that, when combined with the model, can be used to deduce individual pathway fluxes which cannot be directly measured. In order to use this positional information it is not enough to consider only metabolic paths. Instead, it is necessary to account for the individual path taken by each carbon atom in a given reaction.

MFA is typically characterized as either steady state ($S \cdot v = 0$) or kinetic ($S \cdot v \neq 0$). Steady state MFA refers to when the concentration of the metabolic pools has achieved both metabolic and isotopic steady state. Under these conditions the fluxes are constant, and may be directly determined. In the kinetic case, the

[1] By convention the positional numbering of the carbon backbone begins with the carbon that is the most oxidized.

[2] Metabolic steady state is when the concentration of plant metabolites becomes constant over the time frame of the study.

[3] Similar to the metabolic steady state, isotopic steady state is when the labeling profile of each metabolite becomes constant over the same time frame.

fluxes are continuously varying, and are subsequently modeled using enzyme kinetics. In this section, MFA will be used to refer specifically to steady state metabolic flux analysis.

19.3.1 MFA Theory

The theory behind MFA is that it is possible to deduce intracellular fluxes by analysis of the positional redistribution of stable carbon isotopes in the carbon backbone of intracellular metabolites. This requires three components, the stoichiometric model, the atom transition between each reaction in the model, and metabolite measurements that provide positional isotopic information.

The atom transition data is used to extend the stoichiometry to a full atom transition matrix. This allows for each individual carbon atom to be traced through the metabolic network. When combined with the measurements from CLEs this results in the augmented steady state model:

$$P \cdot v \cdot f = 0 \tag{19.3}$$

Where, P is the atom transition matrix, replacing the stoichiometric matrix in the model (Wiechert and de Graaf 1997; Zupke and Stephanopoulos 1994). In addition, we now have a second state vector f, the fractional labeling of each carbon atom, in addition to the flux vector v.

19.3.2 MFA Procedure

MFA begins with one or more CLEs. Measurement data from the CLE are then combined with a model of the metabolic network. This model is solved and iteratively optimized in such a way as to minimize an objective function based on the measurement data. The resulting 'optimized fluxes' are used to build a highly detailed flux map of the system under investigation.

Figure 19.2 Illustrates this basic workflow of MFA following two parallel streams, experimental (*in vivo*/*in vitro*) and computational (*in silico*). The experimental stream begins with the CLEs. Once the CLEs have reached steady state, the biological material must be prepared for measurement (for example, washing, biochemical extraction of desired metabolites, hydrolyzing of polymers, lyophilizing, and so on). The isotopic enrichment of desired metabolites is then measured, using one or more forms of mass spectrometry[4] (MS), such as GC-MS (Allen and Ratcliffe 2009).The last step in the experimental stream is to correct the labeled measurements. To be accurate, the isotopic distribution should only reflect the controlled distribution of the known labeled substrate. In fact, all substrates contain

[4] NMR spectroscopy may also be used to derive information on the fractional enrichment.

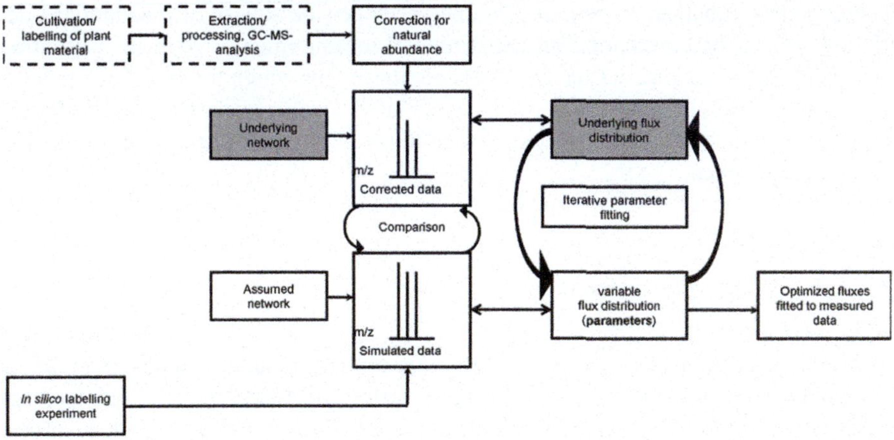

Fig. 19.2 Workflow of a steady state MFA. *Dashed boxes* wet-lab experiments; *shaded boxes* biological properties of the system, depending on the experimental conditions and the genetic background; *solid boxes in silico* calculations

a naturally occurring abundance of labeling; for example, the probability of an individual carbon atom being ^{13}C is 1.1 %. Correction for natural abundance in both the carbon backbone as well as any derivatizing agent used can make a difference on the order of 20 % in label measurements.

The computational stream in the workflow begins with the model (Eq. 19.3) and an assumed set of metabolic fluxes. This model is inherently nonlinear, and solving this model can be achieved by iterative mass balancing of all metabolite isotopic isomers, or isotopomers. An alternative approach was introduced in (Wiechert and de Graaf 1997) based on an alternative labeling description to isotopomers called cumomers[5], or cumulated isotopomer fractions. This approach eliminates the nonlinearity in the forward problem, and allows for an analytical solution based on linear programming. More recently, another labeling description has emerged called elementary metabolite units (EMU[6]) (Antoniewicz et al. 2007). Similar to cumomers, this approach allows for an analytical solution, however, it does so with a minimum number of system variables. By solving the metabolic model, a simulated set of all possible isotopomers for each metabolite is computed. From this set of isotopomers a simulated set of mass isotopomer data can be calculated.

A least squares based optimization is then performed comparing the experimental and computed sets of measurements, with a weighted objective function:

$$\Phi = \sum_{i=1}^{n} W_{ii} r_i^2 = \sum_{i=1}^{n} \frac{(\exp_i - calc_i)^2}{\sigma_i^2} \qquad (19.4)$$

[5] There is a one to one relationship between isotopomers and cumomers.

[6] An EMU is a moiety comprising any distinct subset of the desired atoms in a given metabolite. In general for a metabolite with N carbon atoms, there are 2^N isotopomers, 2^N cumomers and 2^{N-1} EMUs.

The objective function is weighted by the variance (the square of the standard deviation, σ) of the corresponding experimental measurements. With each iteration of the optimization, a new flux vector is calculated, for which the *in silico* model is solved and the simulated measurements calculated. Each new set of simulated measurements is compared with the experimental measurements until the optimization criteria are reached.

19.3.3 MFA Applications

In the recent past, steady state MFA was successfully applied to analyze plant seeds at different stages of development using a variety of plant species, *e.g.*, isolated rapeseed embryos (*B. napus*: Junker et al. 2007; Schwender et al. 2004, 2006), germinating maize kernels/isolated embryos (*Zea mays*: Alonso et al. 2005, 2007b, 2010), isolated sunflower embryos (*Helianthus annuus*; Alonso et al. 2007a) and isolated soybean embryos (*Glycine max*; Allen et al. 2009; Sriram et al. 2004). Several important discoveries concerning the central metabolism have been made using MFA. Some of them shall be further explained.

The application of stable labeled substrates together with mass balancing helped to establish a previously undescribed function of RuBisCO lacking the involvement of the Calvin cycle in developing embryos of *B. napus* (Schwender et al. 2004). In this species more than 60 % of all carbon is stored as oil. According to biochemistry textbooks oil synthesis occurs *via* sucrose hydrolysis, glycolytic hexose breakdown to pyruvate and further oxidation through PDH to acetyl-CoA and CO_2. One third of all hexose carbons are lost as CO_2 giving rise to a 2:1 ratio of acetyl-CoA:CO_2. On the other hand the measured ratio was rather 3:1; less CO_2 was released indicating a refixation of the carbon. By analyzing the labeling pattern of amino acids/FAs after feeding with $^{13}CO_2$ the authors suggested that the refixation was achieved *via* RuBisCO, as only C_1 of 3-phosphoglyceric acid (PGA) showed labeling. This is the position where RuBisCO-fixated carbon is located. A new 3-step pathway was suggested: (1) hexoseP→ Ru-1,5-P_2via the non-oxidative reaction of the OPPP together with phosphoribulokinase; (2) Ru-1,5-P_2+CO_2 (mostly from PDH)→ 2 PGA via RuBisCO; and (3) PGA →→ acetyl-CoA. The net carbon stoichiometry is:

$$5 \text{ hexoseP} \rightarrow 6 \text{ pentoseP} \rightarrow 12 \text{ acetyl-CoA} + 6\ CO_2.$$

Compared to the glycolytic pathway alone:

$$5 \text{ hexoseP} \rightarrow 10 \text{ acetyl-CoA} + 10\ CO_2$$

This results in a decrease of C loss by 40 % and a rise of available acetyl-CoA by 20 %.

Steady state MFA experiments using positionally/uniformly labeled substrate could help to further clarify the mitochondrial metabolism in *B. napus* (Schwender et al. 2006). Different labeling experiments were summed up into one data set leading to an increased reliability of the calculated fluxes. Compared to other heterotrophic/autotrophic plant tissues, the resulting mitochondrial flux map exhibits some distinctive differences. In this scenario the overall flux around the TCA is nearly absent, most of the carbon leaves the cycle via the ATP citrate lyase, the produced acetyl-CoA is then at disposal for cytosolic FA elongation. The needed pyruvate is rather produced from "recycled" malate *via* the mitochondrial malic enzyme then imported from the cytoplasm. Finally, the needed carbon backbones for SP synthesis originate in greater proportion from the uptake of amino acids then via the metabolites produced from PEP carboxylase. For more comprehensive reading about analysis of the TCA *via* MFA Sweetlove et al. (2010) is highly recommended.

Furthermore the influence of different nitrogen sources on the metabolism of developing *Brassica* embryos was analyzed by MFA (Junker et al. 2007). Depending on the chemical nature of the nitrogen source (organic/inorganic), two different flux maps could be constructed. On the other hand, the analysis of selected enzyme activities did not support the two models. In both cases, a surplus of enzymes was measured, more than enough to achieve the measured fluxes. This study once more proved the central metabolism to be very stable with regard to environmental perturbations. It could easily adopt within the experimental time frame without long term regulatory reprogramming of the enzymatic apparatus.

MFA studies on soybean, one of the most important crops on earth due to the high levels of protein and oil could help to quantify the influence of light on developing green seeds (Allen et al. 2009). In this study several labeling experiments under different light conditions were conducted. A detailed analysis of labeling pattern, uptake fluxes, fluxes into biomass, and respiratory fluxes resulted in over 750 measurements of metabolism. This large dataset was then used to validate a detailed flux map of green soy embryos. The resulting flux map could show that light is rather responsible for ATP generation but not net reductant production. RuBisCO again is involved in the efficient refixation of CO_2 through formation of 3-phosphoglycerate (see above). Also amino acids could be identified as a major carbon source for FA synthesis. Major methodological and theoretical progresses about MFA were made using the soy system (NMR2Flux: Sriram et al. 2004).

In another model system, developing sunflower embryos, the role of malate in the context of seed filling was further analyzed by MFA (Alonso et al. 2007a). One decisive difference compared to *Brassica* is the absence of the before discussed RuBisCO bypass pathway. The average carbon conversion efficiency is comparably low (only about 50 % in contrast to over 80 %). The application of [1-$^{13}C_1$] glucose, [2-$^{13}C_1$] glucose, or [U-$^{13}C_5$] glutamine leads to labeled sugars, starch, FA, and amino acids, which were analyzed *via* GC-MS and NMR. Evaluation of the calculated map revealed a good agreement to the experimental data. Only small amounts of malate found their way from the cytoplasm into biomass, especially oil. Excessive futile cycles known from other plant tissues, which are wasting high amounts of ATP (see next paragraph) could not be identified in sunflower embryos.

Similar to the RuBisCO bypass discovered in *Brassica*, MFA experiments performed with growing maize root tips lead to another previously unknown substrate cycle: the glucoseP-to-glucose cycle (Alonso et al. 2005). Steady state labeling results in a similar labeling distribution comparing C_1 and C_6 of glucose and cytosolic hexoseP. This is different to the pattern of storage derived hexosePs, indicating a high connecting flux between glucose and hexoseP, and thus, predicting a one-step conversion of the two metabolites. The presence of high amounts of corresponding glucose-6-phosphatase activity could be shown *via* enzyme assays. The ATP loss by this futile cycle is estimated to be around 40 % in contrast of the sucrose cycling with only 3–6 %. Different breeding lines exhibit very different values for this cycle (up to factor 3), as carbohydrate starvation significantly decreases it. On the other hand, the reduction of respiration *via* oxygen depletion (down to 3 %) leads to an onset of fermentation, decreased total ATP formation, low glucose consumption, small glycolytic fluxes, and slow substrate cycles. Meanwhile the proportion of consumed ATP by the slowed down cycles stays constant letting the authors conclude that substrate cycles are "… not a luxury but an integral part of the organization of the plant central metabolism" (Alonso et al. 2007b).

A recent study elucidates the synthesis of FAs in developing maize embryos in more detail (Alonso et al. 2010). The C conversion efficiency in this species is moderate (57–71 % compared to 50 % for sunflower and over 80 % in rapeseed, see above). Even though about 36 % of the entering carbon is oxidized *via* the OPPP, the demand for NADPH is much higher than this pathway alone, could provide to fuel FA synthesis. To compensate for this shortage the NADP-dependent malic enzyme kicks in. Evaluation of the flux map proposes a contribution of this enzyme for about one third of the cell's carbon and NADPH requirements for FA synthesis.

This collection of recent studies applying ^{13}C steady state MFA to different plant species and different aspects of the metabolism should show the usefulness of this technique to confirm known or identify new metabolic pathways, which would have been missed by conventional mass balancing experiments. For further reading on MFA, the following reviews are highly recommended: Libourel and Shachar-Hill 2008; Ratcliffe and Shachar-Hill 2006 and Rios-Estepa and Lange 2007.

19.4 Dynamic Behavior in Enzymatic Reaction Networks: Kinetic Modeling

Kinetic modeling represents the most complex form of all metabolic modeling approaches. While a stoichiometric model can be generated from a fully sequenced genome and a flux distribution can be set up according to biochemical constraints and optimization functions, such as biomass production or metabolite yield, a kinetic model requires detailed knowledge about metabolite concentrations, kinetic parameters, and also the kinetic rate laws of each reaction involved. This intense data coverage is necessary to predict the dynamic behavior of the metabolic system under varying conditions, such as changes in enzyme activity, substrate uptake, or light exposure.

19.4.1 From Experimental Data to Differential Equations

Like all modeling approaches, kinetic modeling describes biological processes by mathematical means. In the case of kinetic models, all metabolite concentrations

(S) and the fluxes (v) between them, which constitute the change of concentrations over time, are defined as system variables. In comparison to the metabolites and fluxes, the kinetic parameters, such as binding constants (K_m) and maximal reaction velocities (v_{max}) are not changed for a given model. Therefore, a change in a parameter would result in a new model. As the v_{max}-values are associated with the enzyme concentration, a change in enzyme concentration, either by further enzyme expression or by decay, would change the v_{max}-value of the respective enzyme. One example for such a variation in enzyme concentration is the difference between a wild type plant and a knock-out mutant. The K_m-values on the other hand are dependent on the structure of the respective enzyme and can change *via* a mutation at the binding site.

The kinetic rate law, which combines the parameters and variables, can be as simple as a Michaelis-Menten-Kinetic (see Eq. 19.5). Some rate laws even leave out details about the parameters, such as mass action kinetic, which connects both variables with a single factor, but this carries the problem of oversimplification.

$$v = \frac{v_{max} \cdot S}{K_m + S} \tag{19.5}$$

All of the fluxes in a kinetic model have an impact on the concentration of one or several metabolites, either increasing or decreasing the respective concentration. Therefore, all fluxes with an impact on one specific metabolite are unified in one differential equation, describing the change of the metabolite concentration over time. Thus, the kinetic model consists of a number of ordinary differential equations equal to the number of simulated metabolites. The status of interest in kinetic modeling is the metabolic steady state or, to be more precise, the path from one steady state to the next. The condition of a stable steady state according to fluxes and metabolite concentration is depicted in Fig. 19.3. At a given steady state with a metabolite concentration S_0, both consumption and production of the respective metabolite are equal and therefore the net flux becomes zero, which in turn will cause no further change in the metabolite concentration.

The simplest solution that achieves this steady state condition is the trivial solution with all fluxes set to zero. However in biological terms, this "solution" is equivalent to a dead system in biological terms. In a stable steady state, any concentration change away from S_0 will lead back to S_0. So, if the concentration drops below S_0, the net flux will assume a positive value and increase the concentration. If the concentration rises above S_0, the net flux will drop below zero, therefore reducing the metabolite concentration back to S_0.

A more detailed insight into the basics of kinetic modeling in plants was given in recent reviews (Rios-Estepa and Lange 2007; Schallau and Junker 2010). The challenge of the necessary data integration and model set up was addressed in various publications, creating several possibilities for an automated approach. Most of these workflows utilize simplified kinetic rate laws, such as convenience kinetics (Borger et al. 2007) to ensure the ease of computing of complex networks. An overview of

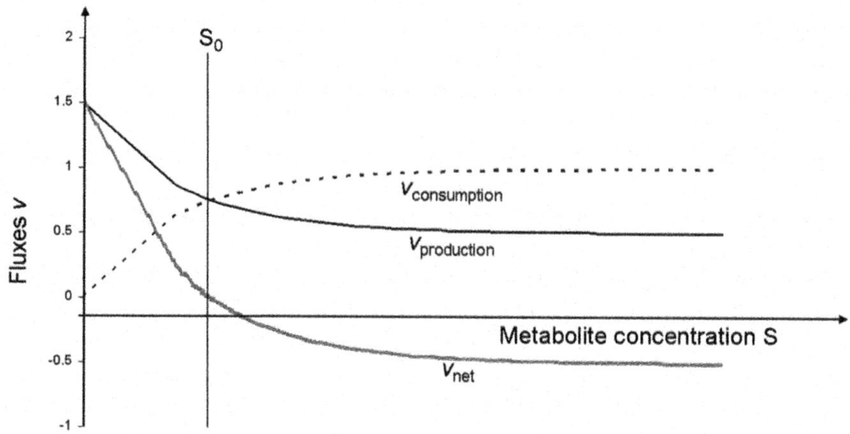

Fig. 19.3 Dynamic changes of metabolite concentrations and fluxes around a steady state (S_0). All fluxes consuming ($v_{consumption}$) and producing ($v_{production}$) a specific metabolite add up to the net flux (v_{net}). See the main text for details

existing models including but not limited to plant systems, can be found in the Java Web Simulation Online Cellular Systems Modeling (Snoep and Olivier 2002) or the Biomodels Database (Li et al. 2010a).

19.4.2 Applications of Kinetic Modeling in the Plant Kingdom

Several modeling approaches, such as network analysis or FBA have been applied to many plant species and compartments, while kinetic modeling, due to the aforementioned complexity, has been applied to a very limited number of plants. So far, there is no explicit kinetic model for the metabolism in crop seeds. While metabolic pathways of several model plants and cyanobacteria have been simulated, the number of models in crop plants is limited. The existing kinetic models for crop plants describe different parts of the metabolism of potato tubers and sugarcane (Rohwer and Botha 2001).Because of the intense research performed on model plants like *Arabidopsis thaliana* and *Nicotiana tabacum*, the data coverage predestines these model plants as candidates for kinetic modeling.

Another approach is the generalized modeling of a common mechanism, such as the photosynthetic apparatus in C3 plants (Zhu et al. 2007). In the given example, the model was used to predict an optimized photosynthesis system under changing atmospheric CO_2 concentrations.

Kinetic Models in Model Plants Because the computational efforts increase exponentially with the size of a model, the starting point is often a very small set of reactions. Such a model was set up to describe the branch point between the methionine and threonine biosynthesis in *Arabidopsis* chloroplasts (Curien et al. 2003). The simulated data of enzymatic regulation was validated with the *in vitro* data of the purified enzymes and predicted correctly the *in vivo* behavior, especially the input of effector molecules on the partition of steady state fluxes. A recent minimal model compared the wildtype and mutant accessions of *Arabidopsis* on the enzymatic level and correctly predicted the respective sucrose levels, while differences in other metabolites indicate the necessity of further subsequent changes of enzymatic parameters (Nägele et al. 2010). Another example for a small scope kinetic model describes the impact of oscillatory behavior on the concentration of ROS and the subsequent enzyme inactivation (Olsen et al. 2003).The verification of *in silico* data with *in planta* data shows the quality of the respective model. The kinetic modeling of the Calvin cycle in the leaves of *N. tabacum* predicted a bistable behavior, which was experimentally verified (Poolman et al. 2001).

Metabolic Engineering for Higher Yields: Kinetic Models of Crop Plants To predict the effect of changes on the metabolism of plants is not only an interesting scientific endeavor, but also a promising way to improve the yield of crop plants. Being an important crop plant for food and industrial use, the potato tuber is an interesting system for this approach. The aims for such a modification are versatile: increase of starch levels, prevention of cold-sweetening or acrylamide formation. The controlled change of a metabolic flux is quite complicated, because multiple feedback loops and regulation mechanisms can prevent the desired effect. An early example for a potato model could show the low impact of phosphofructokinase on the respiratory flux (Thomas et al. 1997). Although the approach primarily utilized Metabolic Control Analysis, the kinetic modeling was used to verify the outcome. A more recent kinetic model simulating the central carbon metabolism in potato tuber included data for cell compartment-specific metabolite pools (Assmus 2005), an important feature for eukaryotic models. The scope of the model was the identification of candidate genes for the control of sucrose concentration. Another kinetic model of sucrose breakdown in potato tubers was used to simulate changes in sucrose cycling in mutants overexpressing enzymes of the sucrose to starch pathway (Junker 2005).

The accumulation of sugar in developing sugarcane was analyzed with the perspective to optimize the sugar yield (Rohwer and Botha 2001). After validating the model with experimental data, candidate enzymes for the increase of sugar accumulation were identified. This model has been updated and expanded (Uys et al. 2007). The updated model included several enzymatic isoforms involved in sucrose accumulation and predicted the impact of fructose and sucrose transport on the futile cycling of sucrose. Another crop plant model simulates essential oil composition in peppermint (Rios-Estepa et al. 2008). The model correctly predicted the effect of low-light growth conditions on menthofuran levels and the predicted mechanism of regulation was further validated using a transgenic approach.

While kinetic modeling has been applied to identify key enzymes and the associated candidate genes for metabolic engineering in several plant systems, including crop plants, no kinetic model of crop plant seeds has been published. One big obstacle is the change of seed compartments during maturation, further increasing the necessary complexity of the kinetic model. But due to the fact that metabolic engineering without kinetic modeling is rather impractical because of the counterintuitive nature of metabolic networks, the necessity to increase the crop yields worldwide and the adaption to environmental stresses will facilitate the development of kinetic seed models.

19.5 Concluding Remarks

We have presented different methods for simulation and quantitative analysis of metabolic networks. It is foreseeable that these methods will gain a lot of importance in the near future, as it will be more and more important to elucidate and understand the quantitative relation between cellular compounds, rather than to determine the cellular inventory alone. Biology is currently in the middle of a transition from a qualitative and descriptive science towards a quantitative and predictive science, just as it happened to chemistry about a century ago. For this transition, as has been clearly pointed out in this chapter, it is increasingly important to train scientists in an interdisciplinary way between biology, physics, mathematics, engineering, and computer science, so that the distances still existing between these areas can be overcome.

References

Allen DK, Ohlrogge JB, Shachar-Hill Y (2009) The role of light in soybean seed filling metabolism. Plant J 58:220–234

Allen DK, Ratcliffe RG (2009) Quantification of isotope label. In: Schwender J (ed) Plant metabolic networks. Springer, New York, pp 105–150

Alonso AP, Vigeolas H, Raymond P, Rolin D, Dieuaide-Noubhani M (2005) A new substrate cycle in plants. Evidence for a high glucose-phosphate-to-glucose turnover from *in vivo* steady-state and pulse-labeling experiments with [13C] glucose and [14C] glucose. Plant Physiol 138:2220–2232

Alonso AP, Goffman F, Ohlrogge JB, Shachar-Hill Y (2007a) Carbon conversion efficiency and central metabolic fluxes in developing sunflower (*Helianthus annuus* L.) embryos. Plant J 52:296–308

Alonso AP, Raymond P, Rolin D, Dieuaide-Noubhani M (2007b) Substrate cycles in the central metabolism of maize root tips under hypoxia. Phytochemistry 68:2222–2231

Alonso AP, Dale VL, Shachar-Hill Y (2010) Understanding fatty acid synthesis in developing maize embryos using metabolic flux analysis. Metab Eng 12:488–497

Antoniewicz MR, Kelleher JK, Stephanopoulos G (2007) Elementary metabolite units (EMU): a novel framework for modeling isotopic distributions. Metab Eng 9:68–86

Assmus H (2005) Modelling carbohydrate metabolism in potato tuber cells. Thesis, Oxford Brookes University

Beckles DM, Smith AM, ap Rees T (2001) A cytosolic ADP-glucose pyrophosphorylase is a feature of graminaceous endosperms, but not of other starch-storing organs. Plant Physiol 125:818–827

Blum JJ, Stein RB (1982) On the analysis of metabolic networks. In: Goldberger RF (ed) Biological regulation and development. Plenum, New York, pp 99–124

Borger S, Uhlendorf J, Helbig A, Liebermeister W (2007) Integration of enzyme kinetic data from various sources. In Silico Biol 7:S73–S79

Curien G, Ravanel S, Dumas R (2003) A kinetic model of the branch-point between the methionine and threonine biosynthesis pathways in *Arabidopsis thaliana*. Eur J Biochem 270:4615–4627

de Oliveira Dal'Molin CG, Quek LE, Palfreyman RW, Brumbley SM, Nielsen LK (2010) AraGEM, a genome-scale reconstruction of the primary metabolic network in *Arabidopsis*. Plant Physiol 152:579–589

Emes MJ, Bowsher CG, Hedley C, Burrell MM, Scrase-Field ES, Tetlow IJ (2003) Starch synthesis and carbon partitioning in developing endosperm. J Exp Bot 54:569–575

Geigenberger P (2003) Regulation of sucrose to starch conversion in growing potato tubers. J Exp Bot 54:457–465

Gibbs J, Morrell S, Valdez A, Setter TL, Greenway H (2000) Regulation of alcoholic fermentation in coleoptiles of two rice cultivars differing in tolerance to anoxia. J Exp Bot 51:785–796

Grafahrend-Belau E, Schreiber F, Koschützki D, Junker BH (2009b) Flux balance analysis of barley seeds: a computational approach to study systemic properties of central metabolism. Plant Physiol 149:585–598

Junker BH (2005) Sucrose breakdown in the potato tuber. Thesis. http://opus.kobv.de/ubp/volltexte/2005/176/

Junker BH, Lonien J, Heady LE, Rogers A, Schwender J (2007) Parallel determination of enzyme activities and *in vivo* fluxes in *Brassica napus* embryos grown on organic or inorganic nitrogen source. Phytochemistry 68:2232–2242

Kauffman KJ, Prakash P, Edwards JS (2003) Advances in flux balance analysis. Curr Opin Biotechnol 14:491–496

Kennedy RA, Rumpho ME, Fox TC (1992) Anaerobic metabolism in plants. Plant Physiol 100:1–6

Lee JM, Gianchandani EP, Papin JA (2006) Flux balance analysis in the era of metabolomics. Brief Bioinformatics 7:140–150

Li C, Donizelli M, Rodriguez N, Dharuri H, Endler L, Chelliah V, Li L, He E, Henry A, Stefan MI, Snoep JL, Hucka M, Le Novère N, Laibe C. 2010a. BioModels Database: an enhanced, curated and annotated resource for published quantitative kinetic models. BMC Syst Biol 4:92

Libourel IGL, Shachar-Hill Y (2008) Metabolic flux analysis in plants: from intelligent design to rational engineering. Annu Rev Plant Biol 59:625–650

Llaneras F, Picó J (2008) Stoichiometric modelling of cell metabolism. J Biosci Bioeng 105:1–11

Nägele T, Henkel S, Hörmiller I, Sauter T, Sawodny O, Ederer M, Heyer AG (2010) Mathematical modeling of the central carbohydrate metabolism in *Arabidopsis* reveals a substantial regulatory influence of vacuolar invertase on whole plant carbon metabolism. Plant Physiol 153:260–273

Olsen LF, Hauser MJB, Kummer U (2003) Mechanism of protection of peroxidase activity by oscillatory dynamics. Eur J Biochem 270:2796–2804

Poolman MG, Ölçer H, Lloyd JC, Raines CA, Fell DA (2001) Computer modelling and experimental evidence for two steady states in the photosynthetic Calvin cycle. Eur J Biochem 268:2810–2816

Poolman MG, Fell DA, Raines CA (2003) Elementary modes analysis of photosynthate metabolism in the chloroplast stroma. Eur J Biochem 270:430–439

Poolman MG, Miguet L, Sweetlove LJ, Fell DA (2009) A genome-scale metabolic model of *Arabidopsis* and some of its properties. Plant Physiol 151:1570–1581

Ratcliffe RG, Shachar-Hill Y (2006) Measuring multiple fluxes through plant metabolic networks. Plant J 45:490–511

Rios-Estepa R, Lange BM (2007) Experimental and mathematical approaches to modeling plant metabolic networks. Phytochemistry 68:2351–2374

Rios-Estepa R, Turner GW, Lee JM, Croteau RB, Lange BM (2008) A systems biology approach identifies the biochemical mechanisms regulating monoterpenoid essential oil composition in peppermint. Proc Natl Acad Sci U S A 105:2818–2823

Rohwer JM, Botha FC (2001) Analysis of sucrose accumulation in the sugar cane culm on the basis of *in vitro* kinetic data. Biochem J 358:437–445

Rolletschek H, Weschke W, Weber H, Wobus U, Borisjuk L (2004) Energy state and its control on seed development: starch accumulation is associated with high ATP and steep oxygen gradients within barley grains. J Exp Bot 55:1351–1359

Rolletschek H, Koch K, Wobus U, Borisjuk L (2005) Positional cues for the starch/lipid balance in maize kernels and resource partitioning to the embryo. Plant J 42:69–83

Schallau K, Junker BH (2010) Simulating plant metabolic pathways with enzyme-kinetic models. Plant Physiol 152:1763–1771

Schilling CH, Letscher D, Palsson BØ. 2000a. Theory for the systemic definition of metabolic pathways and their use in interpreting metabolic function from a pathway-oriented perspective. J Theor Biol 203:229–248

Schilling CH, Schuster S, Palsson BØ, Heinrich R. 2000b. Metabolic pathway analysis: basic concepts and scientific applications in the post-genomic era. Biotechnol Prog 15:296–303

Schuster S, Hilgetag C (1994) On elementary flux modes in biochemical reaction systems at steady state. J Biol Syst 2:165–182

Schuster S, Hilgetag C, Woods JH, Fell DA (2002) Reaction routes in biochemical reaction systems: algebraic properties, validated calculation procedure and example from nucleotide metabolism. J Math Biol 45:153–181

Schwender J, Goffman F, Ohlrogge JB, Shachar-Hill Y (2004) RubisCO without the Calvin cycle improves the carbon efficiency of developing green seeds. Nature 432:779–782

Schwender J, Shachar-Hill Y, Ohlrogge JB (2006) Mitochondrial metabolism in developing embryos of *Brassica napus*. J Biol Chem 281:34040–34047

Snoep JL, Olivier BG (2002) Java Web Simulation (JWS); a web based database of kinetic models. Mol Biol Reports 29:259–263

Sriram G, Fulton DB, Iyer VV, Peterson JM, Zhou R, Westgate ME, Spalding MH, Shanks JV (2004) Quantification of compartmented metabolic fluxes in developing soybean embryos by employing biosynthetically directed fractional 13C labeling, two-dimensional [13C, 1H] nuclear magnetic resonance, and comprehensive isotopomer balancing. Plant Physiol 136:3043–3057

Steuer R, Nesi AN, Fernie AR, Gross T, Blasius B, Selbig J (2007) From structure to dynamics of metabolic pathways: application to the plant mitochondrial TCA cycle. Bioinformatics 23:1378–1385

Stitt M (1990) Fructose-2,6-bisphosphate as a regulatory molecule in plants. Annu Rev Plant Physiol Plant Mol Biol 41:153–185

Sweetlove LJ, Beard KF, Nunes-Nesi A, Fernie AR, Ratcliffe RG (2010) Not just a cycle: flux modes in the plant TCA cycle. Trends Plant Sci 15:462–470

Thomas S, Mooney PJ, Burrell MM and Fell DA (1997) Metabolic Control Analysis of glycolysis in tuber tissue of potato (*Solanum tuberosum*): explanation for the low control coefficient of phosphofructokinase over respiratory flux. Biochem J 322:119–127

Thorbjørnsen T, Villand P, Denyer K, Olsen OA, Smith AM (1996) Distinct isoforms of ADPglucose pyrophosphorylase occur inside and outside the amyloplasts in barley endosperm. Plant J 10:243–250

Uys L, Botha FC, Hofmeyr JHS, Rohwer JM (2007) Kinetic model of sucrose accumulation in maturing sugarcane culm tissue. Phytochemistry 68:2375–2392

van Dongen JT, Roeb GW, Dautzenberg M, Froehlich A, Vigeolas H, Minchin PE, Geigenberger P (2004) Phloem import and storage metabolism are highly coordinated by the low oxygen concentrations within developing wheat seeds. Plant Physiol 135:1809–1821

Weschke W, Panitz R, Sauer N, Wang Q, Neubohn B, Weber H, Wobus U (2000) Sucrose transport into barley seeds: molecular characterization of two transporters and implications for seed development and starch accumulation. Plant J 21:455–467

Wiechert W, de Graaf AA (1996) *In vivo* stationary flux analysis by 13C labelling experiments. Adv Biochem Eng Biotechnol 54:109–154

Wiechert W, de Graaf AA (1997) Bidirectional reaction steps in metabolic networks: I. Modeling and simulation of carbon isotope labeling experiments. Biotechnol Bioeng 55:101–117

Williams TCR, Poolman MG, Howden AJM, Schwarzlander M, Fell DA, Ratcliffe RG, Sweetlove LJ (2010) A genome-scale metabolic model accurately predicts fluxes in central carbon metabolism under stress conditions. Plant Physiol 154:311–323

Zhu XG, de Sturler E, Long SP (2007) Optimizing the distribution of resources between enzymes of carbon metabolism can dramatically increase photosynthetic rate: a numerical simulation using an evolutionary algorithm. Plant Physiol 145:513–526

Zupke C, Stephanopoulos G (1994) Modeling of isotope distributions and intracellular fluxes in metabolic networks using atom mapping matrices. Biotechnol Prog 10:489–498

Chapter 20
Metabolic Specialization of Maternal and Filial Tissues

Hélène Zuber, Mélanie Noguero, Christine Le Signor, Richard Thompson and Karine Gallardo

Abstract The seed consists of three principal components of maternal (seed coats) or zygotic (embryo and endosperm) origin with distinct functions, but that interplay throughout their development to ensure the accumulation of storage compounds for successful germination and early seedling growth. The reserves stored in mature seeds represent major human and livestock food sources. Therefore, much research and breeding efforts are concentrated on optimizing seed quality and yield. The principal filial storage organ differs between species. For example, it is the endosperm for cereal grains accumulating high amount of starch, and the embryo for protein-rich legume seeds. These organs are surrounded by tissues of maternal and/or zygotic origin, depending on the species, which represent a protective barrier and play a role in furnishing the filial organ with nutrients and oxygen. Seed tissues and cell types have been individually studied by the omics approaches with a view to dissecting the molecular processes underlying reserve accumulation. The most comprehensive analyses have been performed at the transcriptome and/or proteome levels in various species, including *Medicago*, soybean, *Arabidopsis*, sugar beet, barley, wheat, maize, rice, and tomato. Here, we report the division of metabolic activities between seed tissues, based on the identification and ontological classification of gene products differentially accumulated between seed tissues. The work allowed metabolic networks to be proposed in specific tissue-types and regulatory factors to be identified, two fundamental tasks in systems biology, with an ultimate goal to undertake a computational reconstruction of tissue-specific metabolic models.

Keywords Development · Metabolism · Omics · Regulation · Seed tissues · Transport processes

20.1 Introduction

The seed is the unit of angiosperm plant dispersal arising from a double fertilization event, which gives rise to the triploid endosperm and diploid zygotic embryo (the future plant). These entities are accommodated within a capsule of maternal

K. Gallardo (✉) · H. Zuber · M. Noguero · C. Le Signor · R. Thompson
INRA (National Institute for Agronomic Research), UMR 1347 Agroécologie,
BP 86510, 21065 Dijon, France
e-mail: Karine.Gallardo@dijon.inra.fr

G. K. Agrawal, R. Rakwal (eds.), *Seed Development: OMICS Technologies toward Improvement of Seed Quality and Crop Yield*,
DOI 10.1007/978-94-007-4749-4_20,

tissue, the seed coat (i.e., pericarp in cereals). The endosperm in monocot plants typically comprises the majority of the mature seed, serving as the major filial storage tissue, rich in starch but possessing a protein content of less than 16 %. In dicot plants, the endosperm is a more variable fraction and often reduced to a single cell layer adjacent to the embryo sac wall at maturity after degeneration of most of the tissue to remobilize nutrients for the maturing embryo. The embryo is generally the major filial organ of dicot seeds, and the reserves stored in this tissue differ among species. Whereas seeds of oilseed rape (*Brassica napus L.*) exhibit approximately 40 % oil and 15 % protein as per dry seed weight, seeds of legumes, such as pea (*Pisum sativum*) and soybean (*Glycine max*) possess 20–40 % protein and 2–20 % oil. Since seeds are major human and livestock food sources, much research and breeding efforts are concentrated on optimizing their nutritional content. Efforts often focus on FA and amino acid balance, level of essential minerals, protein content, and digestibility. Regarding amino acid balance, seeds of legumes and of some cereals (such as barley and maize) are low in tryptophan and in the sulfur-containing amino acids cysteine and methionine. These amino acids are important in the nutrition of humans and animals that are unable to synthesize methionine. Genetic manipulations have been used in attempts to improve amino acid balance. One strategy consisted in modifying the activity of biosynthetic pathways for essential amino acids through the introduction of genes encoding enzymes of these pathways insensitive to feed-back inhibition by other amino acids (Ufaz and Galili 2008). This strategy increased the level of free essential amino acids but reduced germination ability, possibly due to the toxicity of free amino acids and/or to modifications in the synthesis of compounds derived from these amino acids. Another strategy employed was to modify SP composition by expressing genes encoding sulfur-rich SPs (Molvig et al. 1997), but the accumulation of such foreign proteins in seeds occurred at the expense of other sulfur-containing compounds, such as free sulfur amino acids and glutathione (Tabe and Droux 2002). These and other investigations demonstrate the complexity of seed metabolism, emphasizing the importance of understanding its regulation in order to modify seed composition without adverse metabolic consequences. Part of the complexity of this regulation is the compartmentalization of biosynthetic pathways between seed tissues that may regulate their activity. The omics technologies are powerful tools for determining the metabolic specificity of the different seed tissues and identifying the regulatory factors. Moreover, such high-throughput technologies are able to provide an exhaustive overview of the gene products present in different seed tissues. In this chapter, we describe the developmental program followed by the maternal and filial tissues. We also report main characteristics of the seed tissues of maternal or zygotic origins revealed by the omics technologies. Focus of discussion is on the partitioning of sulfur metabolism and transport between seed tissues, highlighting several transcriptional regulators likely to determine metabolic specialization of seed tissues. The integration of these tissue-specific omics data will help to undertake a computational reconstruction of tissue-specific metabolic models in the systems biology area.

20.2 Development and Main Functions of Seed Tissues

Early seed development follows a similar program in both monocot and dicot plants, with endosperm and embryo development proceeding in parallel (Goldberg et al. 1994; Lopes and Larkins 1993; Olsen 2004). The fertilized egg cell undergoes a series of divisions to produce a globular embryo structure. This structure is attached to the embryo sac wall by a file of cells termed the suspensor, which provides nutrients during early embryo development (Kawashima and Goldberg 2010). In dicot seeds, the embryo becomes asymmetric and the cotyledons start to differentiate at 3–4 DAF. In the mature embryo, the root and SAM are separated by the axis, the shoot meristem or epicotyl being surrounded by the much larger cotyledons. The end of the embryogenic phase is marked by a progressive arrest of cell divisions. The embryo then enters the seed-filling phase, during which the storage reserves accumulate, mainly in the cotyledons and to a lesser extent in the axis cells. By this stage, the suspensor has ceased to be functional and degenerates. The seed-filling phase is followed by the maturation phase, during which the embryo acquires desiccation tolerance and is prepared for quiescence. In monocot seeds, the embryo developmental events are broadly similar to those in the dicot seed, except that the proportion of reserves deposited in the cotyledons is less, these being mainly stored in the endosperm in the form of starch.

The triploid endosperm initial undergoes a series of free nuclear divisions to give rise to a coenocyte containing several hundred nuclei. After 3 DAF, cellularization of the endosperm begins. In dicots, this starts at the micropylar end of the embryo cavity and progressively spreads to the remainder of the endosperm (Berger et al. 2006). During the seed-filling period, the main described role of the endosperm is to provide nutrients to the developing embryo. This process is achieved by transfer of nutrients from the surrounding maternal tissue, sometimes involving their further metabolism, and in later stages, by the progressive degeneration of the endosperm to mobilize nutrients. In monocot seeds, the process of endosperm cellularization results in a single peripheral cell layer, the aleurone surrounding the remaining cells that constitute the starchy endosperm. The starch and SPs accumulate during the seed-filling phase in a gradient of decreasing concentration from the distal (crown) region to the basal region of the starchy endosperm. Within the starchy endosperm, further domains of specialization can be identified, most importantly the basal endosperm transfer layer and the embryo-surrounding region (Opsahl-Ferstad et al. 1997; Thompson et al. 2001). The basal endosperm transfer layer cells are specialized for promoting solute transfer into the endosperm cavity. They are located adjacent to the site of phloem sap unloading in the placentochalaza, and putatively accelerate solute transfer across the maternal plant-zygote junction, which lacks plasmodesmata. The role of the embryo surrounding region is less well established, but it may provide nutrients or growth factors to the developing embryo. The aleurone cells also acquire desiccation tolerance during the maturation phase of monocot seed development, and they are metabolically active during seed germination. The remaining endosperm cells in contrast are 'end-cells' lacking gene expression upon germination.

The embryo and endosperm are surrounded by maternal cell layers constituting the seed coat, pericarp, or integument with several important roles (Weber et al. 1995; Weber et al. 2005; Zhang et al. 2007b). Seed coat growth and differentiation proceed coordinately with those of the endosperm and embryo. The seed coat is the site of phloem unloading and metabolic conversions of phloem sap constituents (for example, cleavage of sucrose) to better adapt them for seed filling. The seed coat may also be a source of plant hormones and, in many cases, is needed for the maintenance of seed dormancy. The seed coat plays an important protective role both physical and chemical due to the accumulation of phytoalexins and antifeedants. Finally, the seed coat and other tissues of the seed may be photosynthetically active and thereby supply the embryo with part of its carbohydrate requirements (Weber et al. 1995).

20.3 Examples of Large-Scale Omics Studies of Seed Tissues: A Focus on Starch Metabolism

Seed tissues have been individually studied by the omics approaches with a view to dissect the molecular processes underlying seed quality. The most comprehensive analyses have been carried out at the transcriptome and/or proteome levels in various species, including *M. truncatula*, soybean, *Arabidopsis*, sugar beet, barley, wheat, maize, rice, and tomato (Table 20.1). The recent development of laser-capture micro dissection (LCM) enables various cell types to be isolated for RNA extraction and analysis. RNA obtained from LCM-dissected seed regions of several monocot and dicot species was hybridized to Affymetrix Genechips to generate spatial gene expression profiles during seed development (Day et al. 2009; Le et al. 2007, 2010; Tauris et al. 2009; see also http://seedgenenetwork.net/). In soybean, the results indicate that at least 22,000 diverse mRNAs are required for the formation of a globular-stage soybean embryo. Genes specifically expressed in the diverse seed compartments (e.g., endosperm, hilum, suspensor, and embryo proper) were identified, highlighting the specialization of each tissue. Comprehensive data sets arising from LCM have proved useful to help characterizing seed-expressed genes of interest (Zuber et al. 2010a), and to allow building models for seed tissue function, coordination, and regulation, as shown in barley for the translocation of zinc from the phloem to the storage sites in the grain (Tauris et al. 2009) and in *Arabidopsis* through the profiling of TF genes in seed compartments (Le et al. 2010).

During the last decade, many omics approaches were applied to isolated tissues of cereal grains (Table 20.1). Drea et al. (2005) surveyed 888 genes expressed during wheat caryopsis development using a high-throughput mRNA *in situ* hybridization method. Of these genes, 89 are expressed in specific cell types and 118 were enriched in a single cell type but expressed in more than one cell type. Other genes follow the same temporal and expression patterns in the endosperm and maternal tissues, suggesting similar physiological events in both tissues. Genes encoding CW–modifying

Table 20.1 Example of biological processes highlighted by omics studies of seed tissues

Seed Tissue	Seed Stage	Species	Nbre of Sequences [a]	Examples of Ontological Classes Related to Specific Tissues	References
Comparative Omics Studies of Maternal and Filial Tissues					
Δ Seed coat*	Onset of storage protein synthesis	*Medicago truncatula*	600 spots	Δo Sulfur metabolism (glutathione synthesis) and transport; transcriptional regulation through CCHC type zinc finger TF	Gallardo et al. 2007; Verdier et al. 2008; Zuber et al. 2009
o Endosperm			>3000 transcripts	• Arg synthesis, transcriptional regulation by B3 domain TFs	
• Embryo			41 TFs profiles by qRT-PCR		
Cell types from:	Throughout development	Arabidopsis	>10 000 transcripts from	o TF classes with possible roles in regulating endosperm differentiation/function early in seed development	Day et al. 2009; Le et al. 2007; Le et al. 2010
Δ Seed coat*			Laser Capture Microdissection (LCM) Gene Chip dataset	• Δ TF classes with possible roles in regulating events required for maturation in late seed development	http://seedgenenetwork.net/
o Endosperm		and soybean			
• Embryo					
Perisperm*	Mature stage	Sugar beet	759 spots	Sugar and polysaccharide metabolism	Catusse et al. 2008
♦Root				♦ Storage proteins; protein synthesis, folding, and turnover; components of cell structure	
• Cotyledon				• Neoglucogenesis; TCA cycle; lipid, AA and secondary metabolisms; defence processes	
o Endosperm	Mature stage	Tomato	711 spots	o Defence processes	Sheoran et al. 2005
• Embryo				o • Storage proteins	
◊ Aleurone	Mature stage	Barley	575 spots	◊ o Defence-related proteinaceous inhibitors	Finnie and Svensson 2003
o Endosperm			850 spots	• Antioxidant proteins	
• Embryo			1,000 spots		

Table 20.1 (continued)

Seed Tissue	Seed Stage	Species	Nbre of Sequences [a]	Examples of Ontological Classes Related to Specific Tissues	References
Δ Pericarp* o <u>Endosperm</u> • Embryo	Throughout development	Barley	~2,400 transcripts	Δ Sets of proteins related to PCD (includes proteases and EREBP TFs) o • ABA-mediated regulation of reserve synthesis in the endosperm and of desiccation tolerance in the embryo	Sreenivasulu et al. 2004, 2006
Δ Pericarp* • Embryo sac	Embryo-genesis	Barley	362 genes (*in silico* analysis)	Δ Lipid degradation; proteolysis • Translation; nuclear organization	Zhang et al. 2004
Δ Pericarp* o <u>Endosperm</u> • Embryo	From anthesis until 24 DAF.	Barley	~12,000 seed expressed sequences	Δ Starch synthesis and degradation o Sugar metabolism	Radchuk et al. 2009
Δ Pericarp* ♦ Nucellus* ◊ Aleurone o <u>Endosperm</u>	Caryopsis development	Wheat	888 genes (mRNA *in situ* hybridization)	Δo Defence processes (lipid transfer protein) ♦ ◊ Cell wall modifying enzymes o Storage proteins	Drea et al. 2005
Omics Studies of Particular Seed Tissue Types					
<u>Endosperm</u>	Early and late development	Wheat	250 spots	CNS metabolisms, protein synthesis and turnover (10 dap); Stress/defence and storage (36 dap)	Vensel et al. 2005
<u>Endosperm</u>	Throughout development	Maize	5326 genes (sequencing)	Carbohydrate metabolism	Lai et al. 2004
<u>Endosperm</u>	Onset of seed filling	Maize	632 spots	Metabolic processes, protein synthesis and destination, defence, cell rescue, cell death	Méchin et al. 2004
<u>Endosperm</u>	Throughout development	Maize	409 spots 1849 sequences (from EST)	Cell organization (actin, tubulin), respiration, antioxidant proteins (early stages); proteases (switch towards storage); Carbohydrate metabolism (seed filling stages); Starch/protein balance (late stages)	Méchin et al. 2007 Prioul et al. 2008a, b

Table 20.1 (continued)

Seed Tissue	Seed Stage	Species	Nbre of Sequences [a]	Examples of Ontological Classes Related to Specific Tissues	References
<u>Endosperm</u>	During development	Maize	4032 sequences (from EST)	Storage, transcription, and various metabolic processes	Verza et al. 2005
<u>Endosperm</u>	Storage	Rice	317 proteins	Antioxidants, potential thioredoxin targets	Xu et al. 2010
Embryo	Maturation (30 dap)	Maize	299 proteins	Globulin accumulation	Saleem et al. 2009
Embryo parts: • <u>Cotyledons</u>	Embryogenesis	Arabidopsis	~10,000 genes	• Protein synthesis, folding, destination, modification; energy metabolism; cell cycle and DNA processing, cellular communication, ionic homeostasis	Casson et al. 2005
♦ Root				♦ Cell growth, transport, metabolism, transcription	Spencer et al. 2007

Underlined are the main storage tissues. Stars indicate maternal tissues. [a]Number of coding sequences, transcripts or proteins (*spots in 2D gels*), depending on the technology used for identification.

Abbreviations: CCHC, Cys, Cys, His, Cys; *CNS* carbon, nitrogen, sulfur; *EREBP* ethylene-responsive element binding protein; *PCD* programmed cell death; *TCA* citric acid cycle; and *TF* transcription factor enzymes (e.g., β-expansin) are among the good example that are expressed in both the nucellar projection of maternal origin and the adjacent modified aleurone layer of the endosperm, possibly to establish complex transfer cell morphology, implying the elaboration of specialized CWs. A large proportion of seed tissue-specifically expressed genes were in the starchy endosperm and several of them encode proteins with protective roles against pathogens (e.g., α-thionin and α-amylase inhibitors). This result is reminiscent of proteomics data, showing a predominant accumulation of α-amylase/trypsin and chymotrypsin inhibitors in the endosperm of barley and maize grains (Finnie and Svensson 2003; Méchin et al. 2004), probably to protect the starch reserves against α-amylase activities of pathogens.

As outlined in the previous section, the main function of the maternal tissues is to ensure the uptake of nutrients from the phloem, then to convert them for translocation to the embryo. Accordingly, genes encoding proteins involved in the cleavage of polysaccharides (such as β-amylase and glucan endo-1,3-β-glucosidase) or in sucrose conversion (e.g., vacuolar invertase) were preferentially expressed in the pericarp of barley, where they can metabolize sucrose coming from the phloem or remobilize starch. It is assumed that the resulting mono—and disaccharides can be transported through the cells of the nucellar projection, released into the endosperm cavity to nurture the filial tissue (Sreenivasulu et al. 2002).

Recently, Radchuk et al. (2009) have comprehensively analyzed starch synthesis and degradation in distinct tissues of the developing barley caryopsis by combining gene expression profiling studies, *in situ* localization, and measurements of key metabolites and enzyme activities. They showed that starch synthesis in the pericarp is largely performed by the same genes as in leaves, and that it is mobilized following its accumulation to ensure the maintenance of sink strength in the young caryopsis and probably to regulate carbon flux to endosperm and embryo. Based on gene expression patterns and enzyme activities, they proposed two different pathways for starch degradation in maternal tissues. One pathway implies α-amylases 1 & 4 and β-amylase 1 in pericarp, nucellus, and nucellar projection tissues that undergo PCD. Another pathway is deduced for living pericarp and chlorenchyma cells, where transient starch breakdown correlates with expression of chloroplast-localized β-amylases 5, 6, and 7, glucan, water dikinase, phosphoglucan, isoamylase 3, and disproportionating enzyme. In the endosperm, starch synthesis appears much more complex and involves several endosperm-specific genes (Radchuk et al. 2009). Extensive omics studies have been carried out on this filial tissue in maize, wheat, rice, and barley with a view to understanding the metabolic control of starch accumulation (Table 20.1). For example, transcriptomics, proteomics, and metabolomics approaches were employed to provide an overview of the processes related to maize endosperm development (Méchin et al. 2007; Prioul et al. 2008a). A focus on the enzymes of carbohydrate metabolism revealed that invertases and their resulting products (hexoses) reached maximum levels at 10 DAP (end of embryogenesis) and remained high until 21 DAP (reserve synthesis), which is consistent with the need for hexoses in early endosperm development. Interestingly, pyruvate-Pi-dikinase (PPDK) accumulates in the late filing period (21 DAP onwards). This enzyme may play a central role in the establishment of endosperm reserves composition by regulating the production of PPi, a regulator of ADP-glucose synthesis. PPDK is down-regulated in the *o2* mutant seeds that accumulate more proteins and less starch, lending further support for a role of PPDK in controlling the switch from starch synthesis to SP accumulation possibly through PPi-dependent restriction of ADP-glucose synthesis (Prioul et al. 2008b).

20.4 Partitioning of Metabolic Activities Between Seed Tissues

20.4.1 Omics Analyses Allow Metabolic Network Reconstruction in Seed Tissues

Based on the identification and ontological classification of gene products present in a tissue, metabolic networks can be reconstructed, which is a crucial step in systems biology with an ultimate goal of *in silico* modelling and simulation. Using proteomics approach, Catusse et al. (2008) elegantly demonstrated the feasibility of such study by identifying 121 biochemical functions in the mature sugar beet seed based on the placement in the corresponding metabolic pathways of 561 proteins identified from the embryo and perisperm tissues. Several metabolism pathways were completely identified, such as glycolys is, FA β-oxidation, glyoxylate cycle, or starch metabolism. This study revealed that the seed proteome exhibits tissue-specific features, unveiling a compartmentalization of metabolic activities between the radicle, cotyledons, and perisperm. In the root proteome, the major functional categories represented are SPs, protein folding, turnover, and synthesis, and components of the cell structure. For the cotyledons, the main functional categories represented are the glyoxylate cycle, TCA cycle, lipid metabolism, sterols, amino acid metabolism, defense reaction, and secondary metabolism. In the case of the perisperm, the major functional category revealed is sugar and polysaccharide metabolism, which is consistent with a role of this filial tissue in storing starch. Among the very abundant proteins of the perisperm was a purple acid phosphatase (phytase), whose role in germinating seeds is to hydrolyze phytate, the main storage form of phosphorus. The detection of phytate in the embryo (a living tissue), but not in the perisperm (a dead tissue) of mature seeds, raises the possibility that a compartmentalization of phytate and phytase activity might preserve the integrity of the phytate reserves up to germination.

In *M. truncatula*, a comprehensive analysis of transcripts and proteins present in the three main tissues (seed coat, endosperm, embryo) at the onset of SP synthesis allowed reconstruction of the amino acid metabolic pathways presumably active at this stage with 49 identified biochemical steps (Gallardo et al. 2007). These data revealed a partitioning of amino acid pathways between maternal and filial tissues, reflecting a cross-talk between these tissues. The genes encoding the first enzymes of the pathway leading to the synthesis of tryptophan (chorismate synthase and anthranilate synthase) are predominantly expressed in the seed coat and endosperm, whereas the genes involved in the last steps of tryptophan synthesis (indole 3-glycerol-P synthase and tryptophan synthase) are predominantly expressed in the embryo. This raises the possibility of a transfer of the intermediary precursor forms from the seed coat to the embryo for end-product synthesis. Knowing that tryptophan acts as a feed-back inhibitor of the first enzyme of its biosynthetic pathway (anthranilate synthase; Tozawa et al. 2001), this compartmentalization between the seed coat and the embryo may favor tryptophan accumulation at the onset of seed filling. Although seed coat-located transcripts for the early steps of tryptophan synthesis were up regulated during seed filling, those for the final steps (embryo-located) fall sharply during seed filling. Stimulating the last steps of this pathway could therefore help to increase tryptophan level, which is very low in legume seeds. Other amino acid biosynthetic pathways were found to be compartmentalized between seed tissues. For example, transcripts for enzymes of the arginine biosynthetic pathway were preferentially located in the embryo, where they presumably serve to supply arginine for incorporation into SPs that contain approximately 10 % arginine residues. In contrast, transcripts for L-asparaginase, involved in the conversion of asparagine to aspartate, and for glutamate decarboxylase, which converts glutamate to γ-aminobutyrate, were preferentially expressed in the integument. This supports a role of this maternal tissue in metabolizing glutamate and asparagine, the latter being delivered by the phloem, before translocation to the embryo. This finding may represent a major control point for embryo metabolism, γ-aminobutyrate regulating the carbon flux into the TCA cycle and aspartate being the precursor of methionine, threonine, and lysine. These results, along with those from Catusse et al. (2008), are indicative of a metabolic control of seed development through the partitioning of metabolic pathways between seed tissues and raise the question of the involvement of transport systems between the different tissues (see section Transport Processes).

20.4.2 The Case of Sulfur Metabolism Compartmentalized Between Seed Tissues

To manipulate complex metabolisms in seeds, it is necessary to unravel the tissue sites of biochemical pathways and understand the regulation underlying metabolic fluxes. Examining the partitioning of sulfur biosynthetic pathways across tissue compartments and their regulation is a major goal with a view to increase the content of sulfur amino acids in seeds. Sulfur is a constituent of the amino acids methionine and cysteine, the latter controlling protein folding and/or activity through reversible disulfide bond formation. Furthermore, various compounds are derived from sulfur-containing amino acids, such as glutathione, *S*-adenosylmethionine ethylene, and biotin; all of which are important for seed germination and/or seedling growth. In plants, sulfur is usually taken up by the roots as sulfate and distributed within the tissues in this form. The first enzyme of sulfate reduction (ATP-sulfurylase) catalyzes the formation of 5'-adenylyl sulfate (APS), which can be utilized in two pathways (Fig. 20.1). In the first pathway, APS is converted by the action

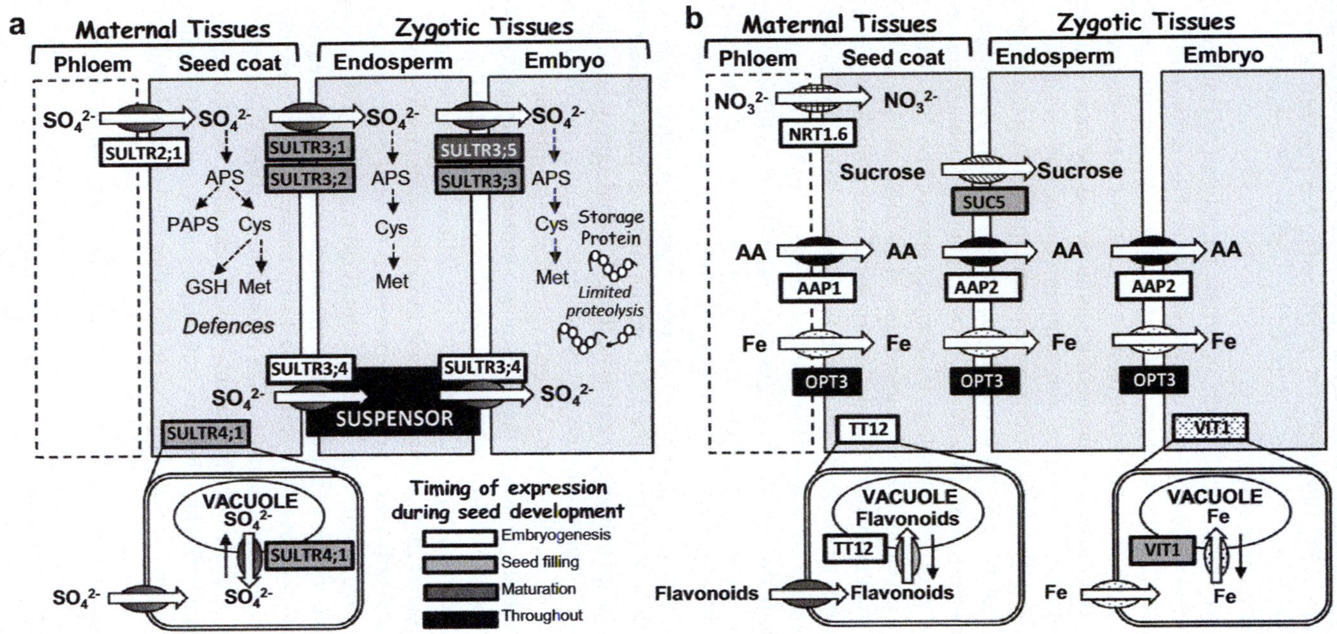

Fig. 20.1 Proposed localisation and function of identified transporters in developing dicot seeds based on omics and/or functional studies. **a** Sulfate transporters in relation with sulfur assimilation pathways. **b** Other transporters. This figure was drawn according to data from Almagro et al. 2008; Awazuhara et al. 2005; Baud et al. 2005; Debeaujon et al. 2001; Gallardo et al. 2007; Hirner et al. 1998; Kim et al. 2006; Roschzttardtz et al. 2009; Stacey et al. 2002, 2008; and Zuber et al. 2010a, b. Abbreviation: *AA* amino acids, *AAP* amino acid transporter, *APS* adenylyl sulfate, *Cys* cysteine, *Fe* iron, *GSH* glutathione, *Met* methionine, NO_3^{2-} nitrate, *NRT* nitrate transporter, *OPT* oligopeptide transporter, *PAPS* 3'-phosphoadenyl sulfate, SO_4^{2-} sulfate, *SUC* sucrose transporter, *SULTR* sulfate transporters, *TT* transparent testa transporter, and *VIT* vacuolar iron transporter

of APS kinase to 3'-phosphoadenylyl sulfate (PAPS), a sulfur donor for the sulfation of defence-related secondary metabolites (glucosinolates, phytoalexins, and sulfoflavonoids). In the second pathway, APS is reduced to provide sulfide ions for cysteine synthesis through two reactions catalyzed by adenosine 5'-phosphosulfate reductase and sulfite reductase, respectively.

In *M. truncatula*, transcripts for enzymes of sulfate reduction (ATP sulfurylase) and cysteine synthesis were detected in both the seed coat and the embryo at the onset of seed filling, illustrating the need for sulfate reduction and cysteine synthesis in the two tissues. Cysteine is utilized for protein synthesis, but is also the precursor of two pathways; one leads to the formation of glutathione and the other enables *de novo* synthesis of methionine. Based on transcriptomics data, the metabolic steps leading to the synthesis of glutathione from cysteine and that leading to the synthesis of PAPS from APS seem to be preferentially active in the embryo-surrounding tissues at the onset of seed filling (Fig. 20.1). These findings, which are consistent with expression patterns of the homolog *Arabidopsis* genes (The Bio-Array Resource database; Toufighi et al. 2005), suggest that at 14 DAP, sulfate in the tissues surrounding the dicot embryo is mainly incorporated into glutathione, a major antioxidant and defence-related secondary metabolites, whereas most of the sulfate entering the dicot embryo is utilized for the synthesis of cysteine. This compartmentalization of metabolic pathways may regulate the availability of sulfur-containing amino acids in embryo cells and protect it from intrinsic stress (i.e., desiccation, PCD), environmental changes, and pathogen ingress through the synthesis of defence molecules in the surrounding tissues. This event could also apply to monocot seeds, since a gene encoding a probable GST, which protects cells against ROS by conjugating them to glutathione, was more than five-fold upregulated in the pericarp of barley grains compared to the embryo (Sreenivasulu et al. 2002) during early caryopsis development. Such antioxidant systems are also present in the endosperm of cereals (Table 20.1), where they may play a protective role and/or allow the initiation of PCD that progresses as starch accumulates (Li et al. 2010).

The exploration of expression profiling data for the following enzymes of sulfur assimilation is indicative of methionine synthesis through distinct pathways in the seed tissues. Methionine synthase enzymes are present in the seed coat, endosperm, and the embryo during early stages of development (Gallardo et al. 2007), which parallel the observations made in barley at the transcriptome level (Sreenivasulu et al. 2002). Methionine synthase catalyses the last step of the *de novo* biosynthesis of methionine, but also regenerates methionine in the course of the *S*-methyl cycle leading to methylation of nucleic acids, proteins, lipids, and other metabolites using AdoMet as a methyl donor. The high level of transcripts for several methyl transferases in the *M. truncatula* embryo as compared to the other seed tissues is indicative of intensive AdoMet-dependent methylation reactions in the embryo at the onset of seed filling that may regenerate methionine for biosynthetic activities. In the seed coat, a gene encoding Hcy *S*-methyl transferase was preferentially expressed early during seed filling. This enzyme converts *S*-methylmethionine (SMM) to methionine, suggesting that the seed coat is metabolizing SMM, a transport and storage form of methionine unique to plants, coming from the phloem to provide methionine for biosynthetic activities in the seed coat and/or for translocation to the embryo. The abundance of the seed coat and endosperm-located methionine synthase and AdoMet synthetase decreased during seed filling. These enzymes are fundamental in controlling the transition from a quiescent to a highly active metabolic state during germination (Gallardo et al. 2002). Therefore, their disappearance at both transcript and protein levels during seed filling reflects a metabolic shift in these tissues from a highly active to a low active state as the embryo accumulates the storage compounds. In contrast, the observation that embryo-located S-adenosyl Hcy hydrolase and methionine synthase transcripts persist at the end of seed development reflects a prolonged metabolic activity in the embryo and is consistent with the finding that both the *de novo* methionine biosynthetic pathway and the activated methyl cycle operate very early during seed germination. Such a metabolic shutdown in the tissues surrounding the embryo has been inferred for other types of metabolism (Buitink et al. 2006; Hajduch et al. 2005). Further evidence for a reduction in metabolic activities in the endosperm during seed filling is the sharp decrease of glycolytic enzymes (e.g., phosphoglyce rate kinase, phosphoglyceromutase, and aldolase).

20.4.3 Proteolytic Activities in Maternal and Filial Tissues, and PCD

The nutrient contribution of maternal tissues to embryo development was illustrated by *in vitro* culture of seeds. Whereas intact seeds were able to initiate the accumulation of reserve proteins when developed *in vitro* without nitrogen, the isolated embryos were not (Gallardo et al. 2006). Consistent with this observation, the embryo-surrounding tissues contain various proteases at the switch towards seed filling that may play a role in the mobilization of nitrogen for the embryo, a process that could be related to PCD events in these tissues (Table 20.1 and Fig. 20.2). In *M. truncatula,* some proteases are specific to the endosperm (aspartic protease and oligopeptidase A) or seed coat (20 S proteasome subunit and leucine aminopeptidase), and others are present in these two tissues (serine proteases of the subtilisin type). Their accumulation profiles suggest developmentally and spatially regulated proteolytic activities (Gallardo et al. 2007). An endosperm-localized subtilisin may have a function in the switch of the developmental program to seed filling as it was specifically detected in seeds at the transition stage between embryogenesis and storage compound synthesis. Conversely, a seed coat-associated subtilisin-like protein accumulated later during the seed-filling phase and could provide nitrogen for storage activities within the embryo. In monocot seeds, subtilisin-like serine protease transcripts were identified in the degenerating endosperm (Sreenivasulu et al. 2004) along with proteinase inhibitors (e.g., subtilisin-chymotryps in inhibitor protein; Finnie and Svensson 2003) that may fine-tune the regulation of protease activity to avoid degradation of protein reserves (Fig. 20.2). A transcriptome analysis of developing barley seed tissues combined with the analysis of coexpressed

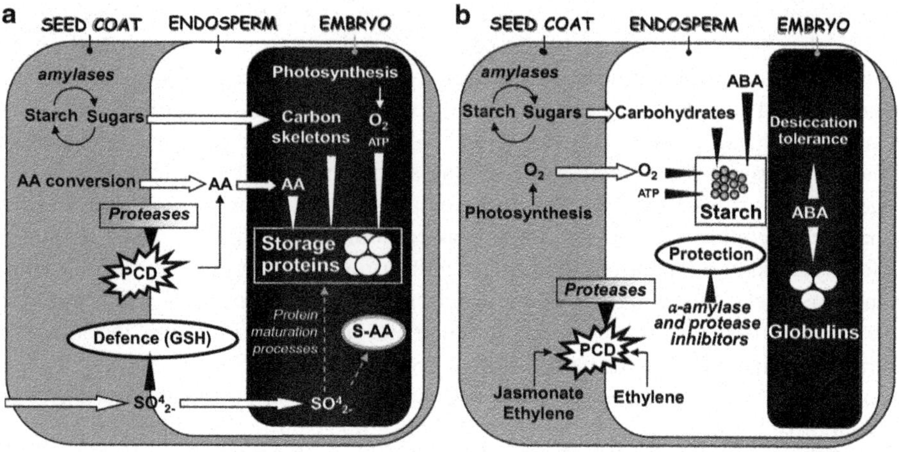

Fig. 20.2 Seed-tissue partitioning of processes related to reserve accumulation and differentiation in dicot (**a**) and monocot (**b**) seeds. These data were taken from studies listed in Table 20.1 for dicot seeds, accumulating high amount of proteins in the embryo during seed filling, and for cereal grains whose major reserves (starch) are stored in the endosperm. Abbreviations: *AA* amino acids, *ABA* abscisic acid, *ATP* adenosine triphosphate, *GSH* glutathione, O_2 oxygen, *PCD* programmed cell death, *S* sulfur and SO_4^{2-} sulfate

gene sets and the identification of *cis*-regulatory elements in ortholog rice gene promoter regions allowed building a regulatory model of PCD in the different seed tissues (Sreenivasulu et al. 2006). PCD seems to be under the hormonal control of both JA and ethylene *via* EREBP TFs in the maternal pericarp tissue that undergo a complete degeneration during late maturation. In the starchy filial tissue, whose cell structure remains intact while undergoing cell death, PCD apparently involves only the ethylene pathway, but employs distinct gene family members from those active in the pericarp, and a different set of proteases and TFs (Fig. 20.2). These studies underline the value of genome-wide transcriptome analysis of separated seed tissues for the elucidation of complex regulatory networks in specific cell types, although such regulatory models require experimental validation.

20.5 Regulation of Seed-Tissue Metabolic Specialization

20.5.1 Transport Processes

The uptake and availability of nutrients in the different seed compartments throughout development both play important roles in the spatiotemporal regulation of metabolic pathways. The identification of proteins responsible for nutrient transport within developing seeds and their fine characterization in term of localization and activity timing are necessary steps to decipher the control of tissue metabolic specialization. Affymetrix Gene Chip and LCM technologies have been combined to profile the mRNA sets present in different seed regions and compartments throughout development in soybean and *Arabidopsis* (Harada-Goldberg *Arabidopsis* LCM Gene Chip Data Set, GEO accession series GSE12404; Le et al. 2010). Transcript levels were quantified in 40 and 31 seed compartments from preglobular—to early maturation-stage seeds in soybean and *Arabidopsis*, respectively, and a transcriptomics web-based database (http://seedgenenetwork.net/) was built. Such databases constitute a precious tool for identifying candidate transporters that could play a role in specific seed tissues (without *a priori* approach). Also, the databases allow the exploration of gene expression in seed compartments for any gene of interest, thus providing seed-specific functional information.

Here, we illustrate the utility of these data in the characterization of transport processes occurring in developing seeds by providing the example of sulfur assimilation and sulfate transporters in *Arabidopsis*. As reported in a previous section, the seed tissues accommodate distinct pathways for sulfur assimilation. Sulfation and glutathione pathways are both related to defence mechanisms and preferentially occur in the seed coat, whereas cysteine is synthesized in the three seed tissues (Figs. 20.1 and 20.2). This compartmentalization implies an active exchange of sulfate between the seed compartments. Sulfate is delivered to plant tissues and cell compartments by specific sulfate transporters (SULTR), which are encoded by large gene families constituted by 12 members in *Arabidopsis*. Five of the 12 genes are strongly expressed in developing seeds (http://seedgenenetwork.net/, Le et al. 2010). *Sultr3;4* is preferentially expressed in suspensor cells and in the endosperm, whereas *Sultr2;1*, *Sultr3;1* *Sultr4;1*, and *Sultr4;2* are preferentially expressed in the seed coat. The further comparison of their sites and timing of expression with genes for sulfate assimilation pathways indicates three of these transporters may provide sulfate for specific assimilation pathways. Genes encoding SULTR4;1, SULTR4;2, and SULTR2;1 are co-expressed (correlation coefficient above 0.5 $p<0.02$) with genes encoding enzymes of glutathione synthesis, glutathione synthase 1 & 2, and with adenylyl sulfate kinase, that catalyses the synthesis of PAPS involved in the sulfation pathway. This spatiotemporal co-expression suggests a role of these transporters in providing sulfur for sulfation pathway and glutathione synthesis in the seed coat.

As transcript level does not necessarily reflect protein level and activity, further studies are needed to demonstrate the function of such transporters, generally including the analysis of knock-out and/or over-expressing mutants. Recently, such studies allowed the functional characterization of several sulfate transporters involved in seed development (Fig. 20.1a). First, the group 3 plasmalemma-predicted sulfate transporters (SULTR3) were shown to have a role in sulfate translocation between seed compartments (Zuber et al. 2010a). SULTR3;1, SULTR3;2, and SULTR3;3 may be involved in providing sulfur for cysteine synthesis, whereas SULTR3;5 may rather furnish sulfur for protein maturation processes (Fig. 20.2). Second, an RNAi mutant for the *Sultr2;1* gene was reported to have a decreased seed glutathione content, supporting a role of SULTR2;1 in providing sulfur for glutathione synthesis in the seed coat (Awazuhara et al. 2005). Finally, the tonoplast transporter SULTR4;1

may be important in maintaining redox homeostasis presumably through the delivery of sulfate into the cytosol for the synthesis of thiol-containing molecules, such as glutathione (Zuber et al. 2010b).

The combination of mutant analysis with omics approaches (e.g., transcriptomics, proteomics, metabolomics, and fluxomics) allowed the identification of various other transporters in developing seeds as depicted in Fig. 20.1b, controlling phloem loading processes (nitrate transporter NRT1.6, amino acid transporter AAP1, and oligopeptide transporter OPT3), the translocation of nutrients from maternal tissues to the endosperm and/or embryo (sucrose transporter SUC5, amino acid transporter AAP2, and oligopeptide transporter OPT3), or facilitation of their entry into the vacuole (vacuolar iron and transparent testa transporters VIT1 and TT12). A road map for zinc trafficking from the phloem to the storage sites in the developing grain was elegantly built by Tauris et al. (2009), based on expression profiling data of zinc transporters and chelating agents in four different tissue types (transfer cells, aleurone layer, endosperm, and embryo) isolated by LCM.

Although significant advances were made towards elucidating the transportome of seed tissues, several transporter classes remain to be identified or functionally studied in seeds (e.g., Ca or Mn transporters). Innovations, such as Synchrotron X-ray fluorescence (SXRF), provide powerful new tools for the investigation of seed transport processes. The SXRF technique detects the abundance of various elements, including Zn, Fe, Mn, Cu, Co and Ca, with a high spatial resolution and detection sensitivity and has the major advantage of not requiring sample pretreatment, allowing the collection of *in vivo* data (Punshon et al. 2009). SXRF has been previously used to analyze dry mature *Arabidopsis* seeds and generated crucial information on metal localization and on the function of the tonoplastic Fe transporter VIT1 in seeds (Fig. 20.1b; Kim et al. 2006).

20.5.2 Transcriptional Regulation

Transcriptional regulation is an important determinant of cell-type specificity, and not unexpectedly, a number of TFs specifically expressed in the seed or its component tissues have been characterized. We have summarized in Table 20.2 currently available transcriptomics data on TF mRNAs found exclusively or preferentially in seed tissues. In most cases, TF mRNA expression was identified using microarrays or high-throughput quantitative RT-PCR (Verdier et al. 2008). Localization of TF gene expression in specific seed tissues has been refined by new technologies, such as LCM, coupled with amplification of low amounts of RNA before hybridization (Day et al. 2008) and, in certain cases, also confirmed with reporter gene activity (Le et al. 2010).

In many cases, particularly for non-model species, the functions of the TF sequences identified by post-genomics approaches have not yet been defined by mutant analysis. However, from tissue-specific localization and seed developmental stage-specific expression a role can often be inferred. Most of the TF families

Table 20.2 List of transcription factors mainly expressed in seed tissues

	Seed Tissue-Specificity	Gene Accession Number	TF/*Nearest Homologue*	TF Family	Expression at Developmental Stages (if known)
DICOTYLEDON	*List of TF Identified in Arabidopsis Seeds (Day et al. 2008; Le et al. 2010)*				
	Endosperm	AT1G02580	MEA (MEDEA)	Polycomb group TF	MG
		AT5G50470	HAP5 A-LIKE	CCAAT-HAP5	Glob
		AT5G07160	AtbZIP72	bZIP	Cot
		AT5G54070	AT-HSFA9	heat stress TF	PMG
		At4g23750	CRF2	AP2-EREBP	
		At4g11400		ARID/BRIGHT DNA-binding domain protein	
		AT4G21080	DOF zinc finger	DOF Zn finger	Glob, Cot
		At4g21080	DOF4.5	C2C2 DOF Zn Finger	Glob
		At4g38000		C2C2 DOF Zn Finger	Cot
		At5g56200, At2g15740		C2C2 DOF Zn Finger	Glob
		At5g14960	DEL2	E2 F-DP	
		At1g49190	ARR19	GARP-ARR-B	Glob + heart stage
		At5g07210	ARR21	GARP-ARR-B	
		At4g25530	FWA/HDG6	HB	
		At4g18870	HSF14	heat shock TF	Glob, Cot
		At2g32370	HDG3	Homeobox-leucine zipper family protein	
		AT3G03260	HDG8, ATML1-like	Homeobox-leucine zipper family protein	Glob, Cot
		At2g34880	MEE27	JUMONJI	
		At2g26320, At5g26630	AGL33, AGL35	MADS	Glob
		AT5G26650	AGL36	MADS	Glob
		At1g65300	AGL38/PHE2	MADS	Glob
		At5g40430	MYB22	Myb protein	
		At5g17800 At5g11510	MYB56, MYB3R4	MYB	

Table 20.2 (continued)

Seed Tissue-Specificity	Gene Accession Number	TF/*Nearest Homologue*	TF Family	Expression at Developmental Stages (if known)
	AT3G27785	MYB118	Myb protein	Cot
	At3g56520		NAC	
	At2g01810		PHD	
Seed coat	AT3G02940	AtMYB107	Myb protein	MG, PMG
Endosperm + seed coat	AT4G21080	DOF zinc finger	C2C2-DOF Zn Finger	Glob, Cot
	AT4G18870	HSF1-like	heat shock TF	Glob, Cot
	AT5G23650	Myb family TF	Myb protein	Glob, Cot
	AT2G26320	AGL33	MADS	Glob, Cot
	AT3G05860 AT3G66656 AT5G26630	AGL45 AGL91 AGL35	MADS	Glob
	AT5G42910	AtbZIP15	bZIP	Glob, Cot
	AT3G10590	Myb-related protein	Myb protein	Glob, Cot
Endosperm + embryo	AT1G55600	MINISEED3	WRKY DNA-BINDING PROTEIN 10	Glob
	At2g35670	FIS2	C2H2	
	AT5G42640	C2H2 zinc finger	C2H2 Zn Finger	Glob
	AT5G07500	PEI1	C3H	Cot, MG, PMG
	At3g03260	HDG8	Homeobox-leucine zipper family protein	
	AT3G04100	AGL57	MADS	Glob
	AT1G77950	AGL67	MADS	MG, PMG
	At5g60440	AGL62	MADS	Glob, Cot, MG, PMG
	At4g36590	AGL40	MADS	Glob
	At1g65330	PHE1/AGL37	MADS	Glob
	At3g27785	MYB118	MYB	
Endosperm + embryo + seed coat	AT3G23060	C3 H-Zinc Finger	C3HC4-Zn Finger	Glob
	AT1G34410	ARF21	Auxin response factor	Glob
	AT5G35770	SAP	SAP	Cot

Table 20.2 (continued)

	Seed Tissue-Specificity	Gene Accession Number	TF/*Nearest Homologue*	TF Family	Expression at Developmental Stages (if known)
		AT1G28300	LEC2	ABI3—VP1	Glob, Cot
		AT3G26790	FUS3	ABI3—VP1	Glob, Cot, MG, PMG
		AT1G21970	LEC1	CCAAT—HAP3	Glob, Cot
		AT5G47670	L1L	CCAAT—HAP3	Glob, Cot, MG
		AT5G07260	Homeodomain TF	HB	Glob, Cot, MG
		AT2G32370	HDG3	Homeobox-leucine zipper family protein	Glob, Cot
		AT5G65070	AGL69/MAF4	MADS	MG, PMG
		AT3G42860	CCHC zinc knuckle	Zn knuckle CCHC-type family protein	PMG
	List of TF identified in M. truncatula (Verdier et al. 2008)				
	Embryo	AC143338	B3	B3 domain	Cot-bent
		AC149131	B3	B3 domain	Cot-bent
		AC149134	bZIP	bZIP	MG
		AC124214	bHLH	bHLH	Cot-bent
	Endosperm	AC146855		DOF Zn Finger	Cot-linear
		TC107215	MybSt1	Myb protein	PMG
		AC125474	ZF-HD	Zn Finger	Cot-linear
	Endosperm + embryo	AC144729	AP2/ERF	AP2/ERF	PMG
		AC151709	CBF	CBF	Cot-linear
	Seed coat	TC104338		AP2/ERF	PMG
		TC97994	MYB R2R3	Myb protein	Cot
	Seed coat + endosperm	AC148528		bHLH	Cot-linear
		AC143340		Zn Finger like, RING type	Cot-bent
MONOCOTYLEDON	*List of TF Identified in Wheat (Wan et al. 2008)*				
	Embryo	Ta.42.1.S1	TaVPI	ABI3B3	Seed filling
		Ta.140.1.S1	TaEmBP	bZIP	Seed filling
		Ta.7266.1.S1	TaMyb3	Myb protein	Seed filling

Table 20.2 (continued)

Seed Tissue-Specificity	Gene Accession Number	TF/*Nearest Homologue*	TF Family	Expression at Developmental Stages (if known)
Endosperm	Ta.893.1.S1	TaSPA	bZIP	Transition cell division-seed filling
	Ta.23689.1.S1	TaPBF	DOF Zn Finger	Transition cell division-seed filling
	Ta.24098.1.S1	TaGAmyb	R2R3	Cell division, seed filling
	TaAffx.37139.1.S1	TaNAC	NAM DNA domain	Seed filling
Seed coat	Ta.7431.1.A1	TaARF	ARF	Cell division
	Ta.7721.1.S1	TaYab2	C2C2-YABBY	Cell division
List of TF Identified in Maize (Wang et al. 2010)				
Endosperm	GO359981–GO359982	ZmbHLH3	bHLH	Seed filling
	GO359992	ZmbHLH6	bHLH	Transition cell division-seed filling
	GO359993	ZmbHLH7	bHLH	Early developmental stage
	GO359994	ZmbHLH8	bHLH	
List of TF identified in Rice (Tu et al. 2008)				
Endosperm	BI802782	B-box domain containing protein	Zn finger	Cell division
	BI797220	CONSTANS type domain containing protein	Zn finger	Cell division
	BI798728	GATA type domain containing protein	Zn finger	Cell division
Endosperm + embryo	BI797724	Alfin-1	Alfin	Cell division

The timing of expression relative to embryo developmental stages is given as following: *Glob* Globular stage (3–4 Days After Pollination, DAP), *Cot* Cotyledon stage (7–8 DAP), *MG* mature stage (13–14 DAP), and *PMG* post-mature stage (18–19 DAP).

described in Table 20.2 are implicated in seed tissue specialization: embryo maturation, storage protein accumulation, differentiation in cotyledon development, and seed size regulation. Some TF classes were associated to particular seed tissues, including DOF family gene members, which were predominantly expressed in the embryo surrounding tissues, mainly in the endosperm. DOF TFs participate in the regulation of many processes exclusive to plants. In cereal grains, they activate SP gene expression within the endosperm (Mena et al. 1998). Many of these factors act collectively to regulate gene expression. For example, in wheat, TaGAmyb TF is implicated in endosperm and embryo grain filling, but requires an interaction with a BPBF (prolamine-box binding factors) TF of the DOF class (Diaz et al. 2002) to activate endosperm-specific genes. As a second example, the type I MADS-box TF PHERES1 (PHE1) interacts with AGL (AGAMOUS-LIKE) 28, AGL40, and AGL62, whose genes are all co-expressed with *PHE1* in the embryo, highlighting their involvement in the same developmental process (De Folter et al. 2005).

Putative orthologs of TFs listed in Table 20.2, where better characterized in a model species, may give an insight into the putative function. An example of this type of 'translational genomics' approach would be the identification of B3 (AC149131) found in *Medicago* as a putative ortholog of FUS3, a determinant of cotyledon identity and regulator of gene expression during late embryogenesis in *Arabidopsis* (Meinke et al. 1994). Putative orthologs may also exist between monocots and dicots; for example, bHLH3 (maize) being a putative ortholog of SPATULA (SPT-*Arabidopsis*). SPT is a member of the phytochrome interacting factor (PIF) subfamily of bHLH TFs required to maintain gene repression in dormant seeds of *Arabidopsis* and mediates the germination response to temperature (Penfield et al. 2005).

It is also important to note that a TF may exert an effect in a tissue other than that in which it is expressed either by migration of the TF protein or *via* a diffusible signal produced by target genes. An example is the action of the protoderm-derived FUSCA3 TF on the *Arabidopsis* embryo (Tsuchiya et al. 2004). Although the roles of most regulatory genes identified by these global approaches are not well understood, genomics data should facilitate the identification of regulatory networks underlying seed development and compartmentalization, the hierarchy and precise roles of cell type-specific TFs being important components.

20.5.3 Oxygen Availability and Energy State

The specialization of each tissue may be controlled by the tissue-specific genetic programs of either maternal (seed coat) or zygotic (embryo) origin, but it can also be controlled by the energy and oxygen status of the different tissues. Due to its location in legume seeds, the embryo initially develops in an environment of low light and oxygen availability, which may affect ATP production and biosynthetic activities. The embryos of some seeds become photosynthetically active, providing oxygen and ATP, and thus controlling biosynthetic fluxes during development

(Rolletschek et al. 2005; Fig. 20.2). This could be a prerequisite for the initiation of seed filling, dominated by protein and/or lipid synthesis in the embryo of legume seeds. In cereal grains, Rolletschek et al. (2004) established temporal and spatial maps of oxygen and ATP distribution and, related them to the differentiation of seed tissues. This study revealed that growing lateral and peripheral regions of the filial starchy endosperm remained well supplied with oxygen due to pericarp photosynthesis. They found high ATP levels and storage activities in regions with favourable oxygen supply, leading to propose a model for the regulation of starch accumulation in barley grains. In this model, sugars are supplied from the maternal vein and flow through the nucellar projection and transfer cells towards the peripheral regions of the endosperm. Thus, the median zone of endosperm becomes favourably supplied with both sugar and oxygen, allowing high ATP generation and storage activities (Fig. 20.2). Accordingly, several ATP-synthase isoforms, which play a central role in energy production, were detected at the transcript or protein levels in the endosperm of cereals at the beginning of the storage phase (Balmer et al. 2006; Sreenivasulu et al. 2004). These data suggest a regulatory role of energy status on storage activity in seeds, which is driven by oxygen availability in the endosperm of cereal grains and by photosynthesis in the embryo of legume seeds.

20.6 Concluding Remarks

The comprehensive omics studies from separated seed tissues have significantly increased our understanding of the molecular and physiological processes taking place in specific cell types and the way they interact to make the seed. The main task for the future is the reliable and automated acquisition and assembly of such high-throughput data into plant metabolic models. Computational modelling, based on a mathematical description of a metabolic network, could be employed to exploit large-scale data sets in the emerging field of systems biology. Significant progress in the area of systems biology has been achieved for humans, where a computational method was developed that uses a variety of different tissue-specific molecular data sources, including literature-based knowledge, transcriptomics, proteomics, metabolomics, and phenotypic data to reconstruct functional metabolic network models of various tissues (Jerby et al. 2010). Such models can then be used to predict the metabolic state of a tissue under various genetic and physiological conditions without requiring additional context-specific molecular data. The first advances in this area are emerging in plants. A stoichiometric model of primary metabolism in developing endosperm of barley during starch accumulation was developed by FBA, a predictive computational approach exploiting biochemical, physiological, proteomics and genomics data (Grafahrend-Belau et al. 2009). This model can be used to investigate the metabolic ability of the endosperm to synthesize starch in response to genetic or environmental perturbations. In plants, a real challenge for the future is the integration of large-scale omics data derived from different tissues to undertake a computational reconstruction of tissue-specific metabolic models.

References

Almagro A, Hua SL, Tsay YF (2008) Characterization of the *Arabidopsis* nitrate transporter NRT1.6 reveals a role of nitrate in early embryo development. Plant Cell 20:3289–3299

Awazuhara M, Fujiwara T, Hayashi H, Watanabe-Takahashi A, Takahashi H, Saito K (2005) The function of SULTR2;1 sulfate transporter during seed development in *Arabidopsis thaliana*. Physiol Plant 125:95–105

Balmer Y, Vensel WH, DuPont FM, Buchanan BB, Hurkman WJ (2006) Proteome of amyloplasts isolated from developing wheat endosperm presents evidence of broad metabolic capability. J Exp Bot 57:1591–1602

Baud S, Wuilleme S, Lemoine R, Kronenberger J, Caboche M, Lepiniec L, Rochat C (2005) The AtSUC5 sucrose transporter specifically expressed in the endosperm is involved in early seed development in Arabidopsis. Plant J 43:824–836

Berger F, Grini PE, Schnittger A (2006) Endosperm: an integrator of seed growth and development. Curr Opin Plant Biol 9:664–670

Buitink J, Leger JJ, Guisle I, Vu BL, Wuillème S, Lamirault G, Bars A L, Meur NL, Becker A, Küster H, and Leprince O (2006) Transcriptome profiling uncovers metabolic and regulatory processes occurring during the transition from desiccation-sensitive to desiccation-tolerant stages in *Medicago truncatula* seeds. Plant J 47:735–750

Casson S, Spencer M, Walker K, Lindsey K (2005) Laser-capture microdissection for the analysis of gene expression during embryogenesis of *Arabidopsis*. Plant J 42:111–123

Catusse J, Strub JM, Job C, Van Dorsselaer JM, Job D (2008) Proteome-wide characterization of sugarbeet seed vigor and its tissue specific expression. Proc Natl Acad Sci USA 105:10262–10267

Day RC, Herridge RP, Ambrose BA, Macknight RC (2008) Transcriptome analysis of proliferating *Arabidopsis* endosperm reveals biological implications for the control of syncytial division, cytokinin signaling, and gene expression regulation. Plant Physiol 148:1964–1984

Day RC, Müller S, Macknight RC (2009) Identification of cytoskeleton-associated genes expressed during *Arabidopsis* syncytial endosperm development. Plant Signal Behav 4:883–886

De Folter S, Immink RGH, Kieffer M, Parenicova L, Henz SR, Weigel D, Busscher M, Kooiker M, Colombo L, Kater MM, Davies B, Angenent GC (2005) Comprehensive interaction map of the *Arabidopsis* MADS box transcription factors. Plant Cell 17:1424–1433

Debeaujon I, Peeters AJM, Léon-Kloosterziel KM, Koornneef M (2001) The TRANSPARENT TESTA12 gene of *Arabidopsis* encodes a multidrug secondary transporter-like protein required for flavonoid sequestration in vacuoles of the seed coat endothelium. Plant Cell 13:853–872

Diaz I, Vicente-Carbajosa J, Abraham Z, Martínez M, Isabel-La MI, Carbonero P (2002) The GAMYB protein from barley interacts with the DOF transcription factor BPBF and activates endosperm-specific genes during seed development. Plant J 29:453–464

Drea S, Leader DJ, Arnold BC, Shaw P, Dolan L, Doonan JH (2005) Systematic spatial analysis of gene expression during wheat caryopsis development. Plant Cell 17:2172–2185

Finnie C, Svensson B (2003) Feasibility study of a tissue-specific approach to barley proteome analysis: aleurone layer, endosperm, embryo and single seeds. J Cereal Sci 38:217–227

Gallardo K, Job C, Groot SPC, Puype M, Demol H, Vandekerckhove J, Job D (2002) Importance of methionine biosynthesis for *Arabidopsis* seed germination and seedling growth. Physiol Plant 116:238–247

Gallardo K, Kurt C, Thompson R, Ochatt S (2006) *In vitro* culture of immature *M. truncatula* grains under conditions permitting embryo development comparable to that observed *in vivo*. Plant Sci 170:1052–1058

Gallardo K, Firnhaber C, Zuber H, Hericher D, Belghazi M, Henry C, Kuster H, Thompson RD (2007) A combined proteome and transcriptome analysis of developing *Medicago truncatula* seeds. Mol Cell Proteomics 6:2165–2179

Goldberg RB, de Paiva G, Yadegari R (1994) Plant embryogenesis-zygote to seed. Science 266:605–614

Grafahrend-Belau E, Schreiber F, Koschützki D, Junker BH (2009) Flux balance analysis of barley seeds: a computational approach to study systemic properties of central metabolism. Plant Physiol 149:585–598

Hajduch M, Ganapathy A, Stein JW, Thelen JJ (2005) A systematic proteomic study of seed filling in soybean. Establishment of high-resolution two-dimensional reference maps, expression profiles, and an interactive proteome database. Plant Physiol 137:1397–1419

Hirner B, Fischer WN, Rentsch D, Kwart M, Frommer WB (1998) Developmental control of H+/amino acid permease gene expression during seed development of *Arabidopsis*. Plant J 14:535–544

Jerby L, Shlomi T, Ruppin E (2010) Computational reconstruction of tissue-specific metabolic models: application to human liver metabolism. Mol Syst Biol 6:401

Kawashima T, Goldberg RB (2010) The suspensor: not just suspending the embryo. Trends Plant Sci 15:23–30

Kim SA, Punshon T, Lanzirotti A, Li L, Alonso JM, Ecker JR, Kaplan J, Guerinot ML (2006) Localization of iron in *Arabidopsis* seed requires the vacuolar membrane transporter VIT1. Science 314:1295–1298

Lai J, Dey N, Kim CS, Bharti AK, Rudd S, Mayer KF, Larkins BA, Becraft P, Messing J (2004) Characterization of the maize endosperm transcriptome and its comparison to the rice genome. Genome Res 14:1932–1937

Le BH, Wagmaister JA, Kawashima T, Bui AQ, Harada JJ, Goldberg RB (2007) Using genomics to study legume seed development. Plant Physiol 144:562–574

Le BH, Cheng C, Bui AQ, Wagmaister JA, Henry KF, Pelletier J, Kwong L, Belmonte M, Kirkbride R, Horvath S, Drews GN, Fischer RL, Okamuro JK, Harada JJ, Goldberg RB (2010) Global analysis of gene activity during *Arabidopsis* seed development and identification of seed-specific transcription factors. Proc Natl Acad Sci U S A 107:8063–8070

Li CY, Li WH, Li C, Gaudet DA, Laroche A, Cao LP, Lu ZX (2010) Starch synthesis and programmed cell death during endosperm development in triticale (×*Triticosecale* Wittmack). J Integr Plant Biol 52:602–615

Lopes MA, Larkins BA (1993) Endosperm origin, development, and function. Plant Cell 5:1383–1399

Méchin V, Balliau T, Château-Joubert S, Davanture M, Langella O, Négroni L, Prioul JL, Thévenot C, Zivy M, Damerval C (2004) A two-dimensional proteome map of maize endosperm. Phytochemistry 65:1609–1618

Méchin V, Thévenot C, Le Guilloux M, Prioul JL, Damerval C (2007) Developmental analysis of maize endosperm proteome suggests a pivotal role for pyruvate orthophosphate dikinase. Plant Physiol 143:1203–1219

Meinke DW, Franzmann LH, Nickle TC, Yeung EC (1994) Leafy cotyledon mutants of *Arabidopsis*. Plant Cell 6:1049–1064

Mena M, Vicente-Carbajosa J, Schmidt RJ, Carbonero P (1998) An endosperm-specific DOF protein from barley, highly conserved in wheat, binds to and activates transcription from prolamin-box of a native B-hordein promoter in barley endosperm. Plant J 16:53–62

Molvig L, Tabe LM, Eggum BO, Moore AE, Craig S, Spencer D, Higgins TJV (1997) Enhanced methionine levels and increased nutritive value of seeds of transgenic lupins (*Lupinus angustifolius* L.) expressing a sunflower seed albumin gene. Proc Natl Acad Sci U S A 94:8393–8398

Olsen OA (2004) Nuclear endosperm development in cereals and *Arabidopsis thaliana*. Plant Cell 16:S214-S227

Opsahl-Ferstad HG, Le Deunff E, Dumas C, Rogowsky PM (1997) ZmEsr, a novel endosperm-specific gene expressed in a restricted region around the maize embryo. Plant J 12:235–246

Penfield S, Josse EM, Kannangara R, Gilday AD, Halliday KJ, Graham IA (2005) Cold and light control seed germination through the bHLH transcription factor SPATULA. Curr Biol 15:1998–2006

Prioul JL, Méchin V, Lessard P, Thévenot C, Grimmer M, Chateau-Joubert S, Coates S, Hartings H, Kloiber-Maitz M, Murigneux A, Sarda X, Damerval C, Edwards KJ (2008a) A joint transcriptomic, proteomic and metabolic analysis of maize endosperm development and starch filling. Plant Biotechnol J 6:855–869

Prioul JL, Méchin V, Damerval C (2008b) Molecular and biochemical mechanisms in maize endosperm development: the role of pyruvate-Pi-dikinase and Opaque-2 in the control of C/N ratio. C R Biol 331:772–779

Punshon T, Guerinot ML, Lanzirotti A (2009) Using synchrotron X-ray fluorescence microprobes in the study of metal homeostasis in plants. Ann Bot 103:665–672

Radchuk VV, Borisjuk L, Sreenivasulu N, Merx K, Mock HP, Rolletschek H, Wobus U, Weschke W (2009) Spatiotemporal profiling of starch biosynthesis and degradation in the developing barley grain. Plant Physiol 150:190–204

Rolletschek H, Radchuk R, Klukas C, Schreiber F, Wobus U, and Borisjuk L (2005) Evidence of a key role for photosynthetic oxygen release in oil storage in developing soybean seeds. New Phytol 167:777–786

Rolletschek H, Weschke W, Weber H, Wobus U, Borisjuk L (2004) Energy state and its control on seed development: starch accumulation is associated with high ATP and steep oxygen gradients within barley grains. J Exp Bot 55:1351–1359

Roschzttardtz H, Conéjéro G, Curie C, Mari S (2009) Identification of the endodermal vacuole as the iron storage compartment in the *Arabidopsis* embryo. Plant Physiol 151:1329–1338

Saleem M, Lamkemeyer T, Schützenmeister A, Fladerer C, Piepho HP, Nordheim A, Hochholdinger F (2009) Tissue specific control of the maize (*Zea mays* L.) embryo, cortical parenchyma, and stele proteomes by RUM1 which regulates seminal and lateral root initiation. J Proteome Res 8:2285–2297

Sheoran IS, Olson DJ, Ross AR, Sawhney VK (2005) Proteome analysis of embryo and endosperm from germinating tomato seeds. Proteomics 5:3752–3764

Spencer MWB, Casson SA, Lindsey K (2007) Transcriptional profiling of the *Arabidopsis* embryo. Plant Physiol 143:924–940 (Erratum in: Plant Physiol 143 (2007) 1982)

Sreenivasulu N, Altschmied L, Panitz R, Hähnel U, Michalek W, Weschke W, Wobus U (2002) Identification of genes specifically expressed in maternal and filial tissues of barley caryopses: a cDNA array analysis. Mol Genet Genomics 266:758–767

Sreenivasulu N, Altschmied L, Radchuk V, Gubatz S, Wobus U, Weschke W (2004) Transcript profiles and deduced changes of metabolic pathways in maternal and filial tissues of developing barley grains. Plant J 37:539–553

Sreenivasulu N, Radchuk V, Strickert M, Miersch O, Weschke W, Wobus U (2006) Gene expression patterns reveal tissue-specific signaling networks controlling programmed cell death and ABA-regulated maturation in developing barley seeds. Plant J 47:310–327 (Erratum in: Plant J 47 (2006) 987)

Stacey MG, Koh S, Becker J, Stacey G (2002) AtOPT3, a member of the oligopeptide transporter family, is essential for embryo development in Arabidopsis. Plant Cell 14:2799–2811

Stacey MG, Patel A, McClain WE, Mathieu M, Remley M, Rogers EE, Gassmann W, Blevins DG, Stacey G (2008) The Arabidopsis AtOPT3 protein functions in metal homeostasis and movement of iron to developing seeds. Plant Physiol 146:589–601

Tabe LM, Droux M (2002) Limits to sulfur accumulation in transgenic lupin seeds expressing a foreign sulfur-rich protein. Plant Physiol 128:1137–1148

Tauris B, Borg S, Gregersen PL, Holm PB (2009) A roadmap for zinc trafficking in the developing barley grain based on laser capture microdissection and gene expression profiling. J Exp Bot 60:1333–1347

Thompson RD, Hueros G, Becker H, Maitz M (2001) Development and functions of seed transfer cells. Plant Sci 160:775–783

Toufighi K, Brady SM, Austin R, Ly E, Provart NJ (2005) The botany array resource: e-northerns, expression angling, and promoter analyses. Plant J 43:153–163

Tozawa Y, Hasegawa H, Terakawa T, Wakasa K (2001) Characterization of rice anthranilate synthase alpha-subunit genes *OASA1* and *OASA2*. Tryptophan accumulation in transgenic rice expressing a feedback-insensitive mutant of OASA1. Plant Physiol 126:1493–1506

Tsuchiya Y, Nambara E, Naito S, McCourt P (2004) The FUS3 transcription factor functions through the epidermal regulator TTG1 during embryogenesis in *Arabidopsis*. Plant J 37:73–81

Tu Q, Dong H, Yao H, Fang Y, Dai Ce, Luo H, Yao J, Zhao D, Li D (2008) Global identification of significantly expressed genes in developing endosperm of rice by expression sequence tags and cDNA array approaches. J Integr Plant Biol 50:1078–1088

Ufaz S, Galili G (2008) Improving the content of essential amino acids in crop plants: goals and opportunities. Plant Physiol 147:954–961

Vensel WH, Tanaka CK, Cai N, Wong JH, Buchanan BB, Hurkman WJ (2005) Developmental changes in the metabolic protein profiles of wheat endosperm. Proteomics 5:1594–1611

Verdier J, Kakar K, Gallardo K, Le Signor C, Aubert G, Schlereth A, Town CD, Udvardi MK, Thompson RD (2008) Gene expression profiling of *M. truncatula* transcription factors identifies putative regulators of grain legume seed filling. Plant Mol Biol 67:567–580

Verza NC, E Silva TR, Neto GC, Nogueira FT, Fisch PH, de Rosa VE Jr, Rebello MM, Vettore AL, da Silva FR, Arruda P (2005) Endosperm-preferred expression of maize genes as revealed by transcriptome-wide analysis of expressed sequence tags. Plant Mol Biol 59:363–374

Wan YF, Poole RL, Huttly AK, Toscano-Underwood C, Feeney K, Welham S, Gooding MJ, Mills C, Edwards KJ, Shewry PR, Mitchell RAC (2008) Transcriptome analysis of grain development in hexaploid wheat. BMC Genomics 9:121

Wang K, Han XF, Dong K, Gao LY, Li HY, Ma WJ, Yan YM, Ye XG (2010) Characterization of seed proteome in Brachypodium distachyon. J Cereal Sci 52:177–186

Weber H, Borisjuk L, Heim U, Buchner P, Wobus U (1995) Seed coat-associated invertases of fava-bean control both unloading and storage functions—cloning of cDNAs and cell-type-specific expression. Plant Cell 7:1835–1846

Weber H, Borisjuk L, Wobus U (2005) Molecular physiology of legume seed development. Annu Rev Plant Biol 56:253–279

Xu SB, Yu HT, Yan LF, Wang T (2010) Integrated proteomic and cytological study of rice endosperms at the storage phase. J Proteome Res 9:4906–4918

Zhang H, Sreenivasulu N, Weschke W, Stein N, Rudd S, Radchuk V, Potokina E, Scholz U, Schweizer P, Zierold U, Langridge P, Varshney RK, Wobus U, Graner A (2004) Large-scale analysis of the barley transcriptome based on expressed sequence tags. Plant J 40:276–290

Zhang WH, Zhou Y, Dibley KE, Tyerman SD, Furbank RT, Patrick JW (2007b) Nutrient loading of developing seeds. Func Plant Biol 34:314–331

Zuber H, Aubert G, Davidian J-C, Thompson R, Gallardo K (2009) Sulphur metabolism and transport in developing seeds. In: Sirko A, De Kok LJ, Haneklaus S, Hawkesford MJ, Rennenber H, Saito K, Schnug E, Stulen I (eds) Sulphur metabolism in plants regulatory aspects significance of sulfur in the food chain, agriculture and the environment. Backhuys Publishers, Leiden, pp 113–117

Zuber H, Davidian JC, Aubert G, Aimé D, Belghazi M, Lugan R, Heintz D, Wirtz M, Hell R, Thompson R, Gallardo K (2010a) The seed composition of *Arabidopsis* mutants for the group 3 sulfate transporters indicates a role in sulfate translocation within developing seeds. Plant Physiol 154:913–926

Zuber H, Davidian JC, Wirtz M, Hell R, Belghazi M, Thompson R, Gallardo K (2010b) Sultr4;1 mutant seeds of *Arabidopsis* have an enhanced sulfate content and modified proteome suggesting metabolic adaptations to altered sulfate compartmentalization. BMC Plant Biol 10:78

Part VI
Discovery-Driven Seed and Yield Improvement

Chapter 21
Marker-Aided Breeding Revolutionizes Twenty-First Century Crop Improvement

Rodomiro Ortiz

Abstract The ever increasing human population always needs more healthy and nutritious food, produced in environmentally sustainable ways. Marker-aided breeding significantly contributes towards this priority goal. Molecular markers are mainly identifiable DNA sequences present in the genome and follow the Mendelian inheritance. In present time, a broad range of molecular markers are available for various crops. Advances in crop genome sequencing, high resolution genetic mapping, and precise phenotyping largely help the discovery of functional alleles and allelic variation associated with traits of interest for plant breeding. This chapter provides a brief overview on DNA markers and their use in crop breeding with examples in rice (as the model for inbreeding species) and maize (as an out-crossing species). Molecular marker-aided breeding undoubtedly speeds the conventional breeding process and makes crop improvement more precise. Availability of physical maps, genomes sequences, and high-throughput technologies will also facilitate in developing new molecular breeding approaches in this twenty-first century.

Keywords Crop improvement · DNA markers · Maize · Modern day agriculture · Molecular breeding · Rice

21.1 Introduction

Advances in molecular and biological technologies have been revolutionized crop genetic enhancement and preceded 'Green Gene Revolution'. About 2 decades ago, the potential uses of DNA markers for gene tagging and marker-aided breeding were envisaged by several researchers (Xu and Crouch 2008). DNA markers: (i) provide reliable estimates of genetic diversity; (ii) improve screening efficiency for many traits through their linkage with alleles with small and large effects; and (iii) increase our understanding of biology and architecture of quantitative traits at the genome level. Marker-aided breeding mainly depends on using specific DNA sequences (mainly physically located within or nearby genes of interest) to select desired alleles, such as host plant resistance, quality, or yield. Significant progress has

R. Ortiz (✉)
Swedish University of Agricultural Sciences, Department of Plant Breeding and Biotechnology, Sundsvagen 14, Box 101, SE 23053 Alnarp, Sweden
e-mail: rodomiro.ortiz@slu.se

G. K. Agrawal, R. Rakwal (eds.), *Seed Development: OMICS Technologies toward Improvement of Seed Quality and Crop Yield*,
DOI 10.1007/978-94-007-4749-4_21,

been made in understanding complex quantitative traits and their genetics in this genomics age (Walsh 2001). DNA marker maps are facilitating the use of marker-aided breeding in modern day agriculture, because many markers of known location are interspersed at relative short intervals throughout the genome. Subsequent association analysis allows linking marker types to traits of interest.

Genomics research continues to provide new DNA markers for aided-breeding. It can suggest candidate genes for target traits and may also help to enhance selection responses by selecting directly on genotypes. Several molecular markers are noted in individuals from different environments in population genomics (Stinchcombe and Hoekstra 2008). Population genomics helps to identify genes underlying ecologically important traits and to describe the fitness consequences of naturally occurring variation at these loci. Likewise, omics technologies are unraveling regulatory networks, which integrate pathways from gene to function (Keurentjes et al. 2008). Advances in genome sequencing and omics technologies will result in added cost effective DNA marker tools, bioinformatics, and new genetic resources.

DNA markers are used as landmarks for finding the chromosome segments bearing target gene(s) and to facilitate thereby their selection and further utilization. These markers are very useful for incorporating genes affected by the environments or to pyramid genes for host plant resistance to pathogens and pests. DNA marker-aided breeding methods are also used for improving simple inherited as well as quantitative traits in both public and private sectors (Cooper et al. 2004; Dwivedi et al. 2007). The use of DNA markers in crop breeding requires polymorphism in target population(s), markers linked to traits of interest, precise phenotyping, low cost, and high-throughput genotyping systems.

DNA marker-aided breeding approaches are less controversial than transgenic crops and are also being advocated for organic agriculture (van Bueren et al. 2010). They also provide insights about the gene pool of breeding material, means for enhancing the introgression of host plant resistance genes from wild relatives, and testing gene pyramiding, which are important targets in crop improvement. This chapter provides an overview on DNA marker-aided breeding and its applications in model cereal crops rice and maize to improve inbreeding and out-crossing crops.

21.2 DNA Markers in Plant Breeding

Molecular markers are descriptors that offer dependable and reproducible results for characterizing specific genotypes; i.e., the environment and genotype-by-environment interaction should not affect them. DNA markers have been developed rapidly over the two decades and now can be authentically used for delivering genetically enhanced populations or new genotypes with higher yield, quality, and input-use efficiency in a timely and cost-effective manner. Amplified fragment length polymorphism (AFLP), inter-simple sequence repeat (ISSR), sequence-characterized amplified region (SCAR), sequence tag site (STS), cleaved amplified polymorphic sequence (CAPS), microsatellites or simple sequence repeat (SSR), expressed sequenced tag (EST), single nucleotide polymorphism (SNP), and diversity arrays technology (DArT^{TM}) are some of the current DNA markers, which have been used for molecular breeding in crops (Semagn et al. 2006a,b). As these techniques differ in their principles, protocols, and applications, a careful consideration is required before choosing any of them for routine use in crop genetic enhancement.

SNPs are evenly distributed across the genome. Unlike gel-based markers, the SNPs assay system can be more easily automated. Recently, SNP markers have been exploited for developing a high-throughput genotyping system (Lu et al. 2009). The developed SNP chip system is a powerful tool for large-scale diversity analysis, gene mapping, or evaluation of maize germplasm, and is likely to drive the maize breeding programs (Ortiz et al. 2010, and references therein). Genome-wide association mapping is expected to become available in near future due to a rapid increase in publically available SNP markers. Such a genetic approach coupled with selective genotyping is most likely to accelerate the process discovering functional allelic variation-associated with agronomic traits of interest.

Some DNA marker systems, often used in plants (e.g., SSRs and SNPs), are mainly depend on specific sequence information, which could result in high cost per data point and may significantly limit the ability of breeding programs to obtain a high return to investment. Such a shortcoming may also preclude their routine use as marker-aided breeding. Microarray-based genotyping techniques offer the highest throughput among the other available techniques, and with their proper setup and software, should allow handling/processing hundreds to thousands of individual samples in a cost effective and short time period (Gupta et al. 2008). DArT is one of the recently developed dominant DNA marker techniques that was used in some crop and other plants, including barley (Wenzl et al. 2004), sorghum (Mace et al. 2008), rice (Jaccoud et al. 2001; Reinke 2006), wheat (Akbari et al. 2006; Crossa et al. 2007), banana (Risterucci et al. 2009; Hippolyte et al. 2010), and cassava (Xia et al. 2005). DArT is a microarray hybridization-based technique, which enables simultaneous genotyping of several hundred loci spread over the genome. This technique can detect polymorphism simultaneously at numerous genomic loci without prior sequence information, thereby making it an important marker tool for a quick, powerful (high resolution), and low-cost assessment of the large germplasm collections, breeding populations, quantitative trait loci (QTL) tagging, genome background screening, simultaneous marker-assisted selection (MAS) at several loci, or accelerated introgression of selected genomic regions.

Adequate germplasm diversity is a prerequisite for the success of a breeding program. But as advanced populations and elite stocks are bred, this diversity may suffer and reach a point of diminishing returns. Hence, new diversity needs to be incorporated into breeding materials. As the genome sequences are becoming available for increasing number of crop plants, it will be theoretically possible to test relationship between molecular polymorphism and phenotypes for every gene using forward and reverse genetics. Reverse genetics starts from the sequence information and tries to find the corresponding function. Targeting-induced local lesions in genomes (TILLING) is one of the reverse genetics tools that provides valuable information about new traits in crops (McCallum et al. 2000). TILLING searches the genomes of mutagenized organisms for mutations in a chosen gene (typically a single base-pair substitution), thereby extending the old practice of using existing variation for functional genetic discovery (Comai and Henikoff 2006).

The targeted use of landraces and wild species held in gene banks can be facilitated by integrating the whole-genome mapping approaches and DNA-marker analyses along with cloning of QTL and natural polymorphism. An extension of TILLING is EcoTILLING, which allows natural alleles at a locus to be characterized across many germplasm accessions, enabling both SNP discovery and haplotyping at these loci (Comai et al. 2004). EcoTILLING is therefore another high-throughput, low-cost reverse genetics technique for rapid discovery of natural nucleotide diversity in the natural populations. EcoTILLING screens the lines with desired allele (e.g., SNP discovery and genotyping) to select parents for crossing and further use of their offspring in MAS. Similarly, a gene-centered approach, named mutant-assisted gene identification and characterization (or MAGIC), has been advocated as an effective tool for exploring novel variation and is a valuable approach to harness natural diversity and define genetic networks in crops, such as maize (Johal et al. 2008). MAGIC uses the Mendelian mutants or other genetic variants in a trait of interest as reporters to identify novel genes and variants created and refined by nature over millions of years of evolution.

21.3 Genetic Maps, Association Genetics, and QTL

Molecular markers and genetic maps are now available for most of the important food crops. The regions within genomes that contain genes associated with a particular quantitative trait are known as QTL (Collard et al. 2005). They can be better understood as major genes having pleiotropic effects on other traits, or being different from major genes because alleles at QTL are constrained to have only small effects on the phenotypic character, to be modifiers of the expression of major loci, or have alleles with a range of effects. The alleles with large effects cause recognition of the locus as a gene with major effects, and segregation of alleles with small effects gives rise to quantitative variations. Main-effect and epistatic QTL are therefore recognized as the major types, and their effects, which affect specific phenotypes, may vary considerably (Angaji 2009). Main-effect QTL are defined as single Mendelian factors at which effects (additive or dominance) on a given phenotype arise from allelic substitution and are detected by marker-trait associations using single-factor analysis of variance or interval mapping. Epistatic QTL are those affecting traits determined by interactions between alleles of at least two loci and are detected by associations between trait values and multi-locus marker genotypes using epistatic models. QTL models considering the response of complex phenotypes under stressed environments can improve the understanding of the genetic causes underlying various stress tolerances.

DNA markers could be used to tag QTL to evaluate their contributions to the phenotypes by selecting favorable alleles at these loci in a marker-aided selection scheme, aiming to accelerate the selection of target traits and genetic advancement. Marker-trait associations have been established for a diverse array of traits in crops, and research on marker/QTL validation and refinement is increasingly common (Dwivedi et al. 2007). Simulation research suggests that the allocation of experimental resources has a crucial effect on power of QTL detection as well as on accuracy and precision of QTL estimates (Schön et al. 2004; Stich et al. 2010; Wang et al. 2007b). For example, large mapping populations led to both low probability of a false discovery rate (FDR) for a QTL and increased power, which could lead to large MAS responses (Bernardo 2004). A putative QTL, ensuing from mapping research, should be reported as a true QTL if it was found using a stringent significance level (e.g., $\alpha \cong 0.0001$). As noted by Malosetti et al. (2008), multitrait and multienvironment QTL mixed models help to identify genome regions responsible for genetic correlations, whether caused by pleiotropy or genetic linkage, and can also show how genetic correlations depend on the environmental conditions. Following their approach, they were able to find 36 QTL affecting grain yield, anthesis-silking interval, male flowering, ear number, and plant height in maize. Selective genotyping and pooled DNA analysis using materials from the two tails of the phenotypic distribution of a population also provide an effective alternative to analysis of the entire population for genetic mapping of QTL with relatively small effects, as well as linked and interacting QTL (Xu et al. 2009b). In theory, one 384-well plate that includes diverse germplasm (e.g., cultivars, landraces, recombinant inbreds, doubled haploids, introgression lines and mutants) should be able to cover almost all major genes or QTL controlling agronomic traits in a crop species. For example, Xu et al. (2009a) assembled and genotyped 1,600 maize plants and lines using a 1,536 SNP chip, which allowed testing the feasibility of this one-step simultaneous marker-trait association analysis for a large number of agronomic traits.

Candidate gene-based mapping, genome-wide linkage disequilibrium (LD), and association analyses are routinely used in addition to classical QTL mapping to identify markers, broadly applicable to breeding programs (Sorkheh et al. 2008; Sneller et al. 2009). In contrast to linkage mapping, LD mapping relies on surveys of natural variations. Association mapping involves searching for genotype-phenotype correlations in unrelated individuals and often is more rapid and cost-effective than traditional linkage mapping (Myles et al. 2009). Valid association analysis can be done with appropriate statistical methods in plant breeding populations. The most significant marker may not be however closest to the functional gene (Breseghello and Sorrels 2006). Nonetheless, LD mapping offers greater precision in QTL location than family-based linkage analysis, and therefore, it leads to more efficient MAS and facilitate new gene discovery, as well as, assist effectively to connect sequence diversity with heritable phenotypic differences (Mackay and Powell 2007). The LD is determined by their physical distance across chromosomes and has proven to be useful for dissecting complex traits because it offers a fine scale-mapping due to historical recombination. For example, Crossa et al. (2007) used DArT markers to find associations with resistance to stem rust, leaf rust, yellow rust, and powdery mildew, plus grain yield in five historical wheat international multi-environment trials. Similarly, associations between AFLP markers and complex quantitative traits were investigated in Nordic two-row spring barley cultivars by Kraakman et al. (2004). Many of the associated markers were located in regions where earlier QTL were found for yield and yield components. These results indicate that association mapping approaches can be a viable method for mapping complex QTL, which often needs costly measurements.

Many breeding populations used in association mapping studies could have unknown pedigree, which may lead to an excess of spurious results. Care must be therefore taken during association analyses to control for the increased rate of false positive results arising from population structure and various interrelationships (Mackay and Powell 2007). Statistical methods that account for population structure and familial relatedness are available. The genetic structure of an association mapping panel can be also estimated by genome-wide background markers. It can be further accounted for in association analysis. In this regard, Yu et al. (2009) indicated that kinship construction with subsets of the whole marker panel and subsequent model testing with multiple phenotypic traits provides *ad hoc* information on whether the number of markers is sufficient to quantify genetic relationships among individuals.

21.4 Marker-Aided Breeding

Conventional plant breeding (from selecting parents for crossing to releasing new cultivars) can take about 1.5–2 decades. Hence, any shortcuts, which do not penalize the science behind the plant breeding, are always welcomed by the agriculturists or more specifically by plant breeders. DNA markers, tightly linked to agronomically important genes, are used as molecular tags for MAS, which involves using the presence or absence of a marker as a substitute for, or to assist in phenotypic selection (Collard et al. 2005). Therefore, MAS approaches need to be more efficient, reliable, and cost effective than the conventional plant breeding approaches. The cost effectiveness of MAS needs to be considered on a case by case basis (Dreher et al. 2003), because its costs depend on the inheritance of the trait, the method of phenotypic screening or trait recording, testing environment(s), time used, and labor costs. DNA markers are also used for genotyping relatively few samples in foreground selection or for background selection. Foreground selection is the precise transfer of genomic regions of interest, whereas background selection accelerates the recovery of the genomic constitution of the recurrent parent (Gupta et al. 2010). The MAS has been more widely employed for simply inherited than multigenic traits, although it can be also used for quantitative traits with low heritability or difficulties to score, and to combine multiple traits. Babu et al. (2004) suggest that MAS feasibility depends on the number of target genes to be transferred, the distance between the flanking markers and the target gene(s), the number of genotypes selected in each breeding generation, the nature of germplasm, and the technical options available at the marker level. There are various examples of commercial breeding programs, which use the MAS technique in product development and as a tool in breeding schemes for cereals, legumes, oilseeds, vegetables, ornamentals, and tree crops. The levels of efficiency in selecting the trait at early generations and their characterization at later generations are the most important tangible deliverables of MAS-based techniques.

Marker-assisted backcrossing (MABC) has been frequently cited as a milestone for the use of DNA markers as indirect selection tools in cereal, legume and vegetable breeding programs (Collard and Mackill 2008; Gupta et al. 2010). MABC of single trait seems to be the most powerful approach that uses DNA markers effectively (Semagn et al. 2006a,b). Frisch and Melchinger (2001) indicated that in MABC for simultaneously introgressing two genes the least number of marker data points may be needed in breeding programs with three backcross generations. This MABC approach uses three or four selection steps on the basis of presence of the target genes and selection indices.

Advanced backcross QTL (AB-QTL) and mapping as you go (MAYG) are another two marker-aided breeding approaches in which QTL are initially detected in the segregating breeding population and thereafter used in MAS; i.e., the QTL do not require validation. Positive alleles from a donor parent, often from unadapted germplasm or wild species, are found to be tagged and rapidly transferred into elite cultivars through AB-QTL (Tanksley and Nelson 1996), as shown in tomato (Fulton et al. 2000; Bernacchi et al. 1998a,b) and rice (Moncada et al. 2001; Thomson et al. 2003) breeding. MAYG continually revises the estimates of QTL allele effects by remapping the new elite germplasm generated over selection cycles, and thereby, ensures that the QTL estimates remain relevant to the targeted breeding population (Podlich et al. 2004). In this second approach, QTL alleles for complex traits may vary in values as the breeding material changes during the selection process. Therefore, the DNA markers associated with newly detected QTL replace the previous ones that were lost during the breeding process.

There are several successful examples of marker-assisted gene pyramiding (i.e., combining into a single genotype, a series of target genes identified in distinct parents) in cotton, grain crops and vegetables (Gupta et al. 2010; Ye and Smith 2010). They include pyramiding major genes for host plant resistance to pathogens and pests, as well as QTL for host plant resistance, plant height, root traits, quality, and yield components. The best gene pyramiding method corresponds to an optimal succession of crosses over several generations (Servin et al. 2004). Individuals can be selected and mated according to their genotype using the pyramiding method. The best scheme combines eight target genes in three generations less than the reference genotype selection method with random mating, while requiring fewer genotypings. Wang et al. (2007a) used population genetic theory and QU-GENE (a breeding simulation platform) to assess various population sizes and marker assay numbers for gene pyramiding in wheat. They found that the minimum required population size could be greatly reduced by enriching the frequency of desirable alleles in the F_2 generation of single-cross and in the F_1 generation of backcross and top cross (or three-way cross)populations, whereas for a top cross of three adapted lines from an existing breeding program the population size was minimized with a three-stage selection strategy in the F_1 generation of the top cross, the F_2 generation of the top cross, and doubled haploid derived-lines. Ye and Smith (2010) provide some guidelines for designing an efficient marker-based gene pyramiding strategy. They suggest that the parents with fewer target genes should enter early in the breeding schedule, to first do crosses causing strong repulsion linkage, consider more crosses per generation, and to backcross before assembling more genes.

Breeding of out-crossing crops considers population improvement, inbred development, and hybrid releases. Marker-assisted recurrent selection (MARS) has been used for improving grain yield by the maize private seed sector (Crosbie et al. 2006). In MARS, one cycle of MAS (based on phenotypic data and marker scores) is initially applied in an F_2 population, followed by three cycles of selection based on marker scores only. Such population improvement scheme leads to increased grain yield in the performance of the derived parental lines of new hybrids. Bernardo and Charcosset (2006) indicated that it was advantageous for MARS to exploit only the QTL with large effects and ignore those with small effects, even if the locations of all QTL were known. Likewise, they concluded that known QTL in MARS are most beneficial for traits controlled by a moderately large number (e.g., 40) of QTL.

Modern day genomics research provides knowledge and tools that definitely increases the efficiency and precision of marker-aided breeding in crops (Varshney et al. 2005). Genomic selection (GES) has been recently proposed to overcome the nature of bi-parental mating designs for QTL detection as well as some of the used statistical methods that are ill-suited to the traits' multigenic nature (Heffner et al. 2009 and references therein). GES predicts the breeding values of lines in a population by analyzing their phenotypes and high-density marker scores, and incorporates all marker information in the prediction model, thereby avoiding biased marker effect estimates and capturing more of the variation due to small-effect QTL. GES, in theory, should significantly speed up the breeding cycles as well as enhance genetic gains per unit time, and will change the role of phenotyping, which, in this marker-aided breeding approach, would serve to update prediction models rather than for selecting breeding materials. Wong and Bernardo (2008), using simulation research, showed that GES is superior to MARS and phenotypic selection in terms of gain per unit cost and time for a long generation crop such as oil palm, which requires approximately 19 years per breeding cycle. They also recommend using GES for other perennial crops with long generation intervals, high costs of maintaining breeding plantations, and small population sizes for selection. Likewise, Bernardo (2009) suggests that genome-wide selection could facilitate the introgression of exotic germplasm in maize. His simulation results indicated that seven to eight cycles of genome-wide selection starting in the F_2 generation rather than in a backcross population will be a useful strategy for the rapid improvement of an adapted × exotic cross.

An accurate prediction of genetic values of new breeding materials, whose phenotypes are yet to be observed, are needed to attain rapid selection progress for reducing phenotyping costs. In this regard, Gianola and van Kaam (2008) advocate making use of phenotypic and genomic data simultaneously. Very recent research by Crossa et al. (2010) shows that that models including DNA marker information had higher predictive ability than pedigree-based models in maize and wheat breeding populations. Gains in predictive ability due to inclusion of DNA markers ranged from 7.7 to 35.7 % in wheat, whereas correlations between observed and predictive values achieved up to 0.79 in maize. A Bayesian approach to analyze complex can help plant geneticists and breeders in exploiting DNA marker and phenotypic data on pedigreed populations as available from ongoing breeding programs (Bink et al. 2008). Pérez et al. (2010) suggest using the Bayesian linear regression package in R statistical software for genomic-enabled selection, which often involves analyzing large amounts of phenotypic and molecular marker data and requires specialized computer programs.

21.5 Progress of Marker-Aided Breeding in Maize and Rice

As noted in previous sections, several DNA marker systems are available for aided-breeding in various crops. In the first decade of this millennium, there was a substantial uptake of MAS methods in the private seed sector for maize breeding (Eathington et al. 2007),whereas MAS uptake was slower for breeding barley and wheat, which could reflect the crop breeding system and cultivar types in the developed world's agriculture: out-crossing and mostly F_1 hybrids for the former and inbreeding and pure lines for the last two crops. Although rice is a self-pollinating cereal, there has been an increase use of marker-aided breeding in public and private sectors because of the use of F_1 hybrids in Asia.

Maize genomics is well advanced, but heavily focused on temperate germplasm rather than on the substantially more diverse tropical maize germplasm that includes landraces and other several cultivar types. Diversity analysis at genetic, molecular, and functional levels is therefore important to boost translational genomics from temperate to tropical maize. Advances in a number of techniques are accelerating the discovery of functional alleles- and allelic variation-associated with traits of interest in maize breeding, including linkage- and association-based mapping, genome sequencing (Schnable et al. 2009; Vielle-Calzada et al. 2009), genome-wide transcript profiling (Swanson et al. 2009), haplotype map (Gore et al. 2009), comprehensive association genetics research using methods such as nested association mapping (Buckler et al. 2009; McMullen et al. 2009), seed DNA-based genotyping systems (Gao et al. 2008), and precise phenotyping. For example, Harjes et al. (2008) showed that variation at the *Lycopene epsilon cyclase* (*lcyE*) locus alters flux down α-carotene *versus* β-carotene branches of the carotenoid pathway using association analysis, linkage mapping, expression analysis, and mutagenesis. Four natural *lcyE* polymorphisms accounted for 58 % of the variation in these two branches and a three-fold difference in pro-vitamin A compounds. MAS for favorable *lcyE* alleles will therefore assist plant breeders for developing maize germplasm with high level of pro-vitamin A in their grains.

Marker-aided breeding has been used in maize for improving grain yield and quality, abiotic stress tolerance, and host plant resistance to pathogens and pests. For example, MABC has been used for rapid conversion of maize lines to quality protein maize (QPM) germplasm in Africa (Danson et al. 2006) and South Asia (Babu et al. 2005). This DNA-aided breeding approach reduces genetic drag and time for a line conversion program, and proved valuable for the rapid development of specialty maize germplasm. Following this MABC approach, an early maturing, high-yielding QPM hybrid with 30 % higher lysine and 40 % more tryptophan than the original hybrid was released in India (Gupta et al. 2009).

Private maize breeding programs have reported larger rates of genetic gain with MARS (Crosbie et al. 2006; Eathington et al. 2007). However, reports in the literature from public maize breeding programs are inconsistent and generally less promising (Xu and Crouch 2008). For example, Ribaut and Ragot (2007) indicated that grain yield of MABC-derived hybrids was consistently higher than that of control hybrids under severe water-stress condition, but MABC-derived hybrids and controls showed similar grain yield under mild water-stress condition. These findings suggest that the genetic regulation is highly influenced by drought-prone environment and its stress intensity. Furthermore, Campos et al. (2004) concluded that most of the identified putative drought-tolerance QTLs are likely to be of limited use in applied breeding because, they depend on genetic background or to their sensitivity to the environment, coupled with a general lack of understanding of the biophysical bases of these context dependencies. Nonetheless, genomics tools remain useful in finding a new way of optimizing maize breeding for high yield under drought-prone environments. SNP markers may be used both in routine large-scale genomics-assisted marker development and in gene discovery when breeding maize under drought stress.

The availability of a high-density linkage map (Harushima et al. 1998), complete genome sequence (International Rice Genome Sequencing Project 2005), and intensive QTL mapping for several traits (Ismail et al. 2007) have been facilitating and accelerating the genetic gains in rice breeding. Analysis of the publicly available genome sequences helps to identify various markers for aided-breeding in a broad array of rice germplasm (e.g., against blast) (Fjellstrom et al. 2006). More recently, a low-cost genotyping platform based on oligonucleotide microarrays became available in rice (Edwards et al. 2008) and could be used for rapid and cost-effective gene mapping. McNally et al. (2009) assessed the distribution of 160,000 NR SNPs among 20 diverse rice cultivars and landraces, revealing the breeding history and genotypic/phenotypic diversity of the rice crop. Such an experimental approach provides new insights into the rice diversity and gene–trait relationships for their further use in the genetic enhancement of rice. Research efforts are still on to fill the gap between the genome and the phenotype (Han and Zhang 2008; Raghuvanshi et al. 2010), which may lead to routine genomics-assisted breeding for rice crop.

Much of the progress in marker-aided breeding of rice has been in gene pyramiding for host plant resistance to blast and bacterial blight (Toenniessen et al. 2003), and submergence tolerance (Xu et al. 2004). The use of DNA markers in MAS for rice was recently updated (Jena and Mackill 2008). MABC has been successfully used in introgressing major genes or QTL with large effect into widely-grown rice mega cultivars. The economic impact of MAS was also demonstrated in rice for tolerance to salinity and phosphorous deficiency (Alpuerto et al. 2009). MAS could save at least three to six years in the breeding cycle and result in incremental economic benefits over 25 years in the range of US$ 50 to US$ 900 million compared to conventional approach, depending on the country, stress, and time lags.

Screening of genebank accessions for submergence tolerance was an important activity at the International Rice Research Institute (IRRI, Los Baños, Philippines) in the1970s. Together with their national partners, they combined semi-dwarf plants with submergence tolerance in the early 1990s, mapped the submergence tolerance (*Sub1*) gene on to chromosome 9 in the late 1990s (Xu and Mackill 1996), and continued to fine mapping and marker development initiation in the early 2000s (Xu et al. 2000). Crosses were made throughout the 2000s leading to the development of *Swarna-Sub1* and other submergence-tolerant cultivars, which were later released in Bangladesh, India, Indonesia, Myanmar, and Philippines (Ram et al. 2002). MABC accelerated the breeding of submergence tolerance in popular rice mega cultivars (Septiningsih et al. 2008), which are preferred by both farmers and consumers due to its quality traits. The complex nature of the *Sub1* gene involves a set of mechanisms that basically halts or reduces shoot elongation (response to ethylene) and maintaining high carbohydrate reserves after flooding, resulting in higher survival and faster recovery (Fukao et al. 2006; Fukao and Bailey-Serres 2008; Jackson and Ram 2003; Xu et al. 2006). The IRRI breeders utilized the marker-assisted backcrossing approach and successfully achieved transfer of major QTL to a popular cultivar in about 2 or 3 years (Septiningsih et al. 2008). IRRI has bred 11 *Sub1* lines retaining the same agronomic and quality traits as of their parents. These rice cultivars showed excellent field performance, showing virtually no yield loss even after 17 days of submergence (data from Orissa, India). Over 68,000 minikits of *Sub1* cultivars were distributed in India and approximately 38,500 tons of seed was made available to other Asian countries in the year 2010 for planting about 1 million ha (Data SOURCE). It has been also estimated that about 1 million ha of flood-prone areas could benefit from the *Sub1* cultivars in the sub-Saharan Africa. The *Sub1* cultivars also showed good survival and low reduction in yield in Nigeria.

21.6 Concluding Remarks

Marker-aided breeding was restricted initially to a few economically important cereal crops. However, it has been expanding to other crops due to lowering costs, efficiency, and easiness of the enhanced DNA marker technology. Today, marker-aided breeding is applied to a broad range of crops and could facilitate domesticating entirely new crops. Furthermore, as noted by Moose and Mumm (2008), molecular-aided breeding has contributed significantly to gene discovery and functions, which led to new hypothesis for basic plant biology research, as well as to expanding useful genetic diversity for crop improvement, increasing favorable gene action, and the selection efficiency.

Genomics is the study of the entire genome of any organisms with the aid of DNA markers, which includes both DNA sequencing and fine-scale genetic mapping.

Advances in crop genomics are providing useful and informative data to identify useful DNA markers for both germplasm characterization and marker-aided breeding. Genomics-assisted breeding along with bioinformatics capability and other omics resources is becoming one of widely-accepted approaches for crop improvement programs worldwide. Genomics tools are routinely used by most private-sector breeding programs. They are also helping to dissect the genetic structure of relevant germplasm and to understand gene pools and germplasm (heterotic) groups. Genomics research also provide insights into the allelic content of genetic resources for potential use in breeding, or in screening early generation breeding populations to select segregants with the desired combinations of marker alleles associated with beneficial traits (which reduces the scale of costly phenotypic evaluations), and for establishing the genetic identity (or fingerprinting) of their products (Ortiz et al. 2010). Private sector breeding programs are also using low cost, high-throughput automation for DNA marker analysis (e.g., Dayteg et al. 2007). A primary strategic goal for international and national breeding programs should be the routine integration of molecular markers derived from new genetic resources to new elite breeding lines.

There are yet number of important issues associated with the utilization of marker-aided breeding in crop plants, which need to be addressed appropriately to exploit this molecular breeding approach in its full potential. Some of those issues are high-throughput precision phenotyping systems for QTL mapping, improved understanding of genotype by environment interaction and epistasis, and development of publicly available computational tools (Xu and Crouch 2008). Due to advances towards cheap genotyping and precise phenotyping, Bernardo (2008) envisages that marker-aided breeding of complex traits will likely focus on predictive methods for marker-based selection. Nonetheless, one of the remaining challenges in plant breeding is to identify gene combination showing significant impact on crop improvement. It has been suggested that integration of multidisciplinary research of molecular plant breeding will facilitate the acquisition of such missing knowledge (Moose and Munn 2008). As illustrated by Cattivelli et al. (2008), plant breeding progress in stressful environments needs to blend knowledge on traits sustaining yield under stress and use accurate phenoytping plus genotyping methods to accumulate the most effective genes into elite germplasm without detrimental effects on yield potential. Such a crop breeding approach will warrant the development of genetically enhanced cultivars with high yield potential and high yield stability, especially for adapting to the global changing climate.

References

Akbari M, Wenzl P, Caig V, Carling J, Xia L, Yang S, Uszynski G, Mohler V, Lehmensiek A, Kuchel H, Hayden MJ, Howes N, Sharp P, Vanghan P, Rathmell B, Huttner E, Kilian A (2006) Diversity arrays technology (DArT) for high-throughput profiling of the hexaploid wheat genome. Theor Appl Genet 113:1409–1420

Alpuerto VLEB, Norton GW, Alwang J, Ismail AM (2009) Economic impact analysis of marker-assisted breeding for tolerance to salinity and phosphorous deficiency in rice. Appl Econ Perspect Pol 31:779–792

Angaji SA (2009) QTL mapping: a few key points. Intl J Appl Res Nat Prod 2:1–3
Babu R, Nair SK, Prasanna BM, Gupta HS (2004) Integrating marker-assisted selection in crop breeding—prospects and challenges. Curr Sci 87:607–619
Babu R, Nair SK, Kumar A, Venkatesh S, Sekhar JC, Singh NN, Srinivasan G, Gupta HS (2005) Two-generation marker-aided backcrossing for rapid conversion of normal maize lines to quality protein maize (QPM). Theor Appl Genet 111:888–897
Bernacchi D, Beck-Bunn T, Emmatty D, Eshed Y, Inai S, Lopez J, Petiard V, Sayama H, Uhlig J, Zamir D, Tanksley S (1998a) Advanced backcross QTL analysis of tomato. II. Evaluation of near-isogenic lines carrying single-donor introgressions for desirable wild QTL-alleles derived from *Lycopersicom hirsutum* and *L. pimpinellifolium.* Theor Appl Genet 97:170–180
Bernacchi D, Beck-Bunn T, Eshed Y, Lopez J, Petiard V, Uhlig J, Zamir D, Tanksley S (1998b) Advanced backcross QTL analysis of tomato. I. Identification of QTLs for traits of agronomic importance from *Lycopersicom hirsutum.* Theor Appl Genet 97:381–397
Bernardo R (2004) What proportion of declared QTL in plants are false? Theor Appl Genet 109:419–424
Bernardo R (2008) Molecular markers and selection for complex traits in plants: learning from the last 20 years. Crop Sci 48:1649–1664
Bernardo R (2009) Genomewide selection for rapidintrogression of exotic germplasm in maize. Crop Sci 49:419–425
Bernardo R, Charcosset A (2004) Usefulness of gene information in marker-assisted recurrent selection: a simulation appraisal. Crop Sci 46:614–621
Bink MCAM, Boer MP, ter Braak CJF, Jansen J, Voorrips RE, van de Weg WE (2008) Bayesian analysis of complex traits in pedigreed plant populations. Euphytica 161:85–96
Breseghello F, Sorrells ME (2006) Association analysis as a strategy for improvement of quantitative traits in plants. Crop Sci 46:1323–1330
Buckler ES, Holland JB, Bradbury PJ, Acharya CB, Brown PJ, Browne C, Ersoz E, Flint-Garcia S, Garcia A, Glaubitz JC, Goodman MM, Harjes C, Guill K, Kroon DE, Larsson S, Lepak NK, Li H, Mitchell SE, Pressoir G, Peiffer JA, Oropeza Rosas M, Rocheford TR, Cinta Romay M, Romero S, Salvo S, Sanchez Villeda H, da Silva HS, Sun Q, Tian F, Upadyayula N, Ware D, Yates H, Yu J, Zhang Z, Kresovich S, McMullen MD (2009) The genetic architecture of maize flowering time. Science 325:714–718
Campos H, Cooper M, Habben JE, Edmeades GO, Schussler JR (2004) Improving drought tolerance in maize: a view from industry. Field Crops Res 90:19–34
Cattivelli C, Rizza F, Badeck FW, Mazzucotelli E, Mastrangelo AM, Francia E, Marè C, Tondelli A, Stanca AM (2008) Drought tolerance improvement in crop plants: an integrated viewfrom breeding to genomics. Field Crops Res 105:1–14
Collard BCY, Mackill DJ (2008) Marker-assisted selection: an approach for precision plant breeding in the twenty-first century. Phil Trans R Soc B 363:557–572
Collard BCY, Jahufer MZZ, Brouwer JB, Pang ECK (2005) An introduction to markers, quantitative trait loci (QTL) mapping and marker-assisted selection for crop improvement: the basic concepts. Euphytica 142:169–196
Comai L, Henikoff S (2006) TILLING: practical single-nucleotide mutation discovery. Plant J 45:684–694
Comai L, Young K, Till BJ, Reynolds SH, Greene EA, Codomo CA, Enns LC, Johnson JE, Burtner C, Odden AR, Henikoff S (2004) Efficient discovery of DNA polymorphisms in natural populations by Ecotilling. Plant J 37:778–786
Cooper M, Smith OS, Graham G, Arthur L, Feng L, Podlich DW (2004) Genomics, genetics, and plant breeding: a private sector perspective. Crop Sci 44:1907–1913
Crosbie TM, Eathington SR, Johnson GR, Edwards M, Reiter R, Stark S, Mohanty RG, Oyervides M, Buehler RE, Walker AK, Dobert R, Delannay X, Pershing JC, Hall MA, Lamkey KR (2006) Plant breeding: past, present, and future. In: Lamkey KR, Lee M (eds) Plant breeding: the Arnel R. Hallauer international symposium. Blackwell Publishing, Ames
Crossa J, Burgueño J, Dreisigacker S, Vargas M, Herrera-Foessel SA, Lillemo M, Singh RP, Trethowan R, Warburton M, Franco J, Reynolds M, Crouch JH, Ortiz R (2007) Association

analysis of historical bread wheat germplasm using additive genetic covariance of relatives and population structure. Genetics 177:1889–1913
Crossa J, de los Campos G, Pérez P, Gianola D, Burgueño J, Araus JL, Makumbi D, Singh RP, Dreisigacker S, Yan J, Arief V, Banziger M, Braun HJ (2010) Prediction of genetic values of quantitative traits in plant breeding using pedigree and molecular markers. Genetics 186:713–724
Danson JW, Mbogori M, Kimani M, Lagat M, Kuria A, Diallo A (2006) Marker assisted introgression of *opaque2* gene intoherbicide resistant elite maize inbred lines. Afr J Biotechnol 5:2417–2422
Dayteg C, Tuvesson S, Merker A, Jahoor A, Kolodinska-Brantestam A (2007) Automation of DNA marker analysis for molecular breeding in crops: practical experience of a plant breeding company. Plant Breed 126:410–415
Dreher K, Khairallah M, Ribaut JM, Morris M (2003) Money matters (I): costs of field and laboratory procedures associated with conventional and marker-assisted maize breeding at CIMMYT. Mol Breed 11:221–234
Dwivedi SL, Crouch JH, Mackill DJ, Xu Y, Blair MW, Ragot M, Upadhyaya HD, Ortiz R (2007) The molecularizationof public sector crop breeding: progress, problems and prospects. Adv Agron 95:163–319
Eathington SR, Crosbie TR, Edwards MD, Reiter RS, Bull JK (2007) Molecular markers in a commercial breeding program. Crop Sci 47:S154–S163
Edwards JD, Janda J, Sweeney MT, Gaikwad AB, Liu B, Leung H, Galbraith DW (2008) Development and evaluation of a high-throughput, low-cost genotyping platform based on oligonucleotide microarrays in rice. Plant Methods 4:13
Fjellstrom R, McClung AM, Shank AR (2006) SSR markers closely linked to the *Pi-z* locus are useful for selection of blast resistance in a broad array of rice germplasm. Mol Breed 17:149–157
Frisch M, Melchinger AE (2001) Marker-assisted backcrossing for simultaneous introgression of two genes.Crop Sci 41:1716–1725
Fukao T, Bailey-Serres J (2008) Submergence tolerance conferred by *Sub1A* is mediated by SLR1 and SLRL1 restriction of gibberellin responses in rice. Proc Natl Acad Sci U S A 105:16814–16819
Fukao T, Xu K, Ronald PC, Bailey-Serres J (2006) A variable cluster of ethylene response factor-like genes regulates metabolic and developmental acclimation responses to submergence in rice. Plant Cell 18:2021–2034
Fulton TM, Grandillo S, Beck-Bunn T, Fridman E, Frampton A, López J, Pétiard V, Uhlig J., Zamir D, Tanksley SD (2000) Advanced backcross QTL analysis of a *Lycopersicon esculentum* × *Lycopersicon parviflorum* cross. Theor Appl Genet 100:1025–1042
Gao S, Martinez C, Skinner DJ, Krivanek AF, Crouch JH, Xu Y (2008) Development of a seed DNA-based genotyping system for marker-assisted selection in maize. Mol Breed 22:477–494
Gianola D, van Kaam JBCHM (2008) Reproducing kernel hilbert spaces regression methods for genomic assisted prediction of quantitative traits. Genetics 178:2289–2303
Gore MA, Chia J-M, Elshire RJ, Sun Q, Ersoz ES, Hurwitz BL, Peiffer JA, McMullen MD, Grills GS, Ross-Ibarra J, Ware DH, Buckler ES (2009) A first-generation haplotype map of maize. Science 326:1115–1117
Gupta PK, Rustgi S, Mir RR (2008) Array-based high-throughput DNA markers for crop improvement. Heredity 101:5–18
Gupta HS, Agrawal PK, Mahajan V, Bisht GS, Kumar A, Verma P, Srivastava A, Saha S, Babu R, Pant MC, Mani VP (2009) Quality protein maize for nutritional security:rapid development of short duration hybridsthrough molecular marker assisted breeding. Curr Sci 96:230–237
Gupta PK, Kumar, J, Mir RR, Kumar A (2010) Marker-assisted selection as a component of conventional plant breeding. Plant Breed Rev 33:145–217
Han B, Zhang Q (2008) Rice genome research: current status and future perspectives. Plant Genome 1:71–76

Harjes CE, Rocheford TR, Bai L, Brutnell TP, Kandianis CB, Sowinski SG, Stapleton AE, Vallabhaneni R, Williams M, Wurtzel ET, Yan JB, Buckler ES (2008) Natural genetic variation in lycopene epsilon cyclase tapped for maize biofortification. Science 319:330–333

Harushima Y, Yano M, Shomura A, Sato M, Shimano T, Kuboki Y, Yamamoto T, Lin SY, Antonio BA, Parco A, Kajiya H, Huang N, Yamamoto K, Nagamura Y, Kurata N, Khush GS, Sasaki T (1998) A high-density rice genetic linkage map with 2275 markers using a single F_2 population. Genetics 148:479–494

Heffner EL, Sorrells ME, Jannink JL (2009) Genomic selection for crop improvement. Crop Sci 49:1–12

Hippolyte I, Bakry F, Seguin M, Gardes L, Rivallan R, Risterucci A-M, Jenny C, Perrier X, Carreel F, Argout X, Piffanelli P, Khan IA, Miller RNG, Pappas JG, Mbéguié-A-Mbéguié D, Matsumoto T, De Bernardinis V, Huttner E, Kilian A, Baurens F-C, D'Hont A, Cote F, Courtois B, Glaszmann JC (2010) Asaturated SSR/DArT linkage map of *Musa acuminata* addressing genome rearrangements among bananas. BMC Plant Biol 10:65

International Rice Genome Sequencing Project (2005) The map-based sequence of the rice genome. Nature 436:793–800

Ismail AM, Heuer S, Thomson MJ, Wissuwa M (2007) Genetic and genomic approaches to develop rice germplasm for problem soils. Plant Mol Biol 65:547–570

Jaccoud D, Peng K, Feinstein D, Kilian A (2001) Diversity arrays: a solid state technology for sequence information independent genotyping. Nucl Acids Res 29:e25

Jackson MB, Ram PC (2003) Physiological and molecular basis of susceptibility and tolerance of rice plants to complete submergence. Ann Bot 91:227–241

Jena KK, Mackill DJ (2008) Molecular markers and their use in marker-assisted selection in rice. Crop Sci 48:1266–1276

Johal GS, Balint-Kurti P, Weil CF (2008) Mining and harnessing naturalvariation: a little magic. Crop Sci 48:2066–2073

Keurentjes JJB, Koornneef M, Vreugdenhi D (2008) Quantitative genetics in the age of omics. Curr Opin Plant Biol 11:123–128

Kraakman ATW, Niks RE, Van Den Berg PMMM, StamP, Van Eeuwijk FA (2004) Linkage disequilibrium mapping of yield and yield stability in modern spring barley cultivars. Genetics 168:435–446

Lu Y, Yan J, Guimarães CT, Taba S, Hao Z, Gao S, Chen S, Li J, Zhang S, Vivek BS, Magorokosho C, Mugo S, Makumbi D, Parentoni SN, Shah T, Reng T, Crouch JH, Xu Y (2009) Molecular characterization of global maize breeding germplasm based on genome-wide single nucleotide polymorphisms. Theor Appl Genet 120:93–115

Mace ES, Xia L, Jordan DR, Halloran K, Parh DK, Huttner E, Wenzl P, Kilian A (2008) DArT markers: diversity analyses and mapping in *Sorghum bicolor.* BMC Genomics 9:26

Mackay I, Powell W (2007) Methods for linkage disequilibrium mapping in crops. Trends Plant Sci 12:57–63

Malosetti M, Ribaut JM, Vargas M, Crossa J, van Eeuwijk FA (2008) A multi-trait multi-environment QTL mixed model with an application to drought and nitrogen stress trials in maize (*Zea mays* L.). Euphytica 161:241–257

McCallum CM, Comai L, Greene EA, Henikoff S (2000) Targeting induced local lesions in genomes (TILLING) forplant functional genomics. Plant Physiol 123:439–442

McMullen MD, Kresovich S, Villeda HS, Bradbury P, Li H, Sun Q, Flint-Garcia S, Thornsberry J, Acharya C, Bottoms C, Brown P, Browne C, Eller M, Guill K, Harjes C, Kroon D, Lepak N, Mitchell SE, Peterson B, Pressoir G, Romero S, Oropeza Rosas M, Salvo S, Yates H, Hanson M, Jones E, Smith S, Glaubitz JC, Goodman M, Ware D, Holland JB, Buckler ES (2009) Genetic properties of the maize nested association mapping population. Science 325:737–740

McNally KL, Childs KL, Bohnert R, Davidson RM, Zhao K, Ulat VJ, Zeller G, Clark RM, Hoen DR, Bureau TE, Stokowski R, Ballinger DG, Frazer KA, Cox DR, Padhukasahasram B, Bustamante CD, Weigel D, Mackill DJ, Bruskiewich RM, Rätsch G, Buell CR, Leung H, Leach JE (2009) Genomewide SNP variation reveals relationships among landraces and modern varieties of rice. Proc Natl Acad Sci U S A 106:12273–12278

Moncada P, Martínez CP, Borrero J, Chatel M, Gauch Jr H, Guimaraes E, Tohme J, McCouch SR (2001) Quantitative trait loci for yield and yield components in an *Oryza sativa* × *Oriza rufipogon* BC_2F_2 population evaluated in an upland environment. Theor Appl Genet 102:41–52

Moose SP, Munn RH (2008) Molecular plant breeding as the foundation for 21st century crop improvement. Plant Physiol 147:969–977

Myles S, Peiffer J, Brown PJ, Ersoz ES, Zhang Z, Costich DE, Buckler ES (2009) Association mapping: critical considerations shift from genotyping to experimental design. Plant Cell 21:2194–2202

Ortiz R, Taba S, Chávez Tovar VH, Mezzalama M, Xu Y, Yan J, Crouch JH (2010) Conserving and enhancing maize genetic resources as global public goods—a perspective from CIMMYT. Crop Sci 50:13–28

Pérez P, de los Campos G, Crossa J, Gianola D (2010) Genomic-enabled prediction based on molecular markers and pedigree using the Bayesian linear regression package in R. Plant Genome 3:106–116

Podlich DW, Winkler CR, Cooper M (2004) Mapping as you go: an effective approach for marker-assisted selection of complex traits. Crop Sci 44:1560–1571

Raghuvanshi S, Kapoor M, Tyagi S, Kapoor S, Khurana P, Khurana J, Tyagi A (2010) Rice genomics moves ahead. Mol Breed 26:257–273

Ram PC, Singh BB, Singh AK, Ram P, Singh PN, Singh HP, Boamfa I, Harren F, Santosa E, Jackson MB, Setter TL, Reuss J, Wade LJ, Singh VP, Singh RK (2002) Submergence tolerance in rainfed lowland rice: physiological basis and prospects for cultivar improvement through marker-aided breeding. Field Crops Res 76:131–152

Reinke R (2006) Evaluating diversity array technology (DArT) for the NSW rice breeding program. Rural Industries Research and Development Corporation, Canberra

Ribaut JM, Ragot M (2007) Marker-assisted selection to improve drought adaptation inmaize: the backcross approach, perspectives, limitations, and alternatives. J Exp Bot 58:351–360

Risterucci AM, Hippolyte I, Perrier X, Xia L, Caig V, Evers M, Huttner E, Kilian A, Glaszmann JC (2009) Development and assessment of diversity arrays technology for high-throughput DNA analyses in *Musa*. Theor Appl Genet 119:1093–1103

Schnable PS, Ware D, Fulton RS, Stein JC, Wei F, Pasternak S, Liang C, Zhang J, Fulton L, Graves TA, Minx P, Reily AD, Courtney L, Kruchowski SS, Tomlinson C, Strong C, Delehaunty K, Fronick C, Courtney B, Rock SM, Belter E, Du F, Kim K, Abbott RM, Cotton M, Levy A, Marchetto P, Ochoa K, Jackson SM, Gillam B, Chen W, Yan L, Higginbotham J, Cardenas M, Waligorski J, Applebaum E, Phelps L, Falcone J, Kanchi K, Thane T, Scimone A, Thane N, Henke J, Wang T, Ruppert J, Shah N, Rotter K, Hodges J, Ingenthron E, Cordes M, Kohlberg S, Sgro J, Delgado B, Mead K, Chinwalla A, Leonard S, Crouse K, Collura K, Kudrna D, Currie J, He R, Angelova A, Rajasekar S, Mueller T, Lomeli R, Scara G, Ko A, Delaney K, Wissotski M, Lopez G, Campos D, Braidotti M, Ashley E, Golser W, Kim H, Lee S, Lin J, Dujmic Z, Kim W, Talag J, Zuccolo A, Fan C, Sebastian A, Kramer M, Spiegel L, Nascimento L, Zutavern T, Miller B, Ambroise C, Muller S, Spooner W, Narechania A, Ren L, Wei S, Kumari S, Faga B, Levy MJ, McMahan L, Van Buren P, Vaughn MW, Ying K, Yeh CT, Emrich SJ, Jia Y, Kalyanaraman A, Hsia AP, Barbazuk WB, Baucom RS, Brutnell TP, Carpita NC, Chaparro C, Chia JM, Deragon JM, Estill JC, Fu Y, Jeddeloh JA, Han Y, Lee H, Li P, Lisch DR, Liu S, Liu Z, Nagel DH, McCann MC, SanMiguel P, Myers AM, Nettleton D, Nguyen J, Penning BW, Ponnala L, Schneider KL, Schwartz DC, Sharma A, Soderlund C, Springer NM, Sun Q, Wang H, Waterman M, Westerman R, Wolfgruber TK, Yang L, Yu Y, Zhang L, Zhou S, Zhu Q, Bennetzen JL, Dawe RK, Jiang J, Jiang N, Presting GG, Wessler SR, Aluru S, Martienssen RA, Clifton SW, McCombie WR, Wing RA, Wilson RK (2009) The B73 maize genome: complexity, diversity, and dynamics. Science 326:1112–1115

Schön CC, Utz HF, Groh S, Truberg B, Openshaw S, Melchinger AE (2004) Quantitative trait locus mapping based on resampling in a vast maize testcross experiment and its relevance to quantitative genetics for complex traits. Genetics 167:485–498

Semagn K, Bjørnstad Å, Ndjiondjop MN (2006a) An overview of molecular marker methods for plants. Afr J Biotechnol 5:2540–2568

Semagn K, Bjørnstad Å, Ndjiondjop MN (2006b) Progress and prospects of marker assisted backcrossing as a tool in crop breeding programs. Afr J Biotechnol 5:2588–2603

Septiningsih EM, Pamplona AM, Sanchez DL, Neeraja CN, Vergara GV, Heuer S, Ismail AM, Mackill DJ (2008) Development of submergence tolerant rice cultivars: the *Sub1* locus and beyond. Ann Bot 103:151–160

Servin B., Martin OC, Mézard M, Hospital F (2004) Toward a theory of marker-assisted gene pyramiding. Genetics 168:513–523

Sneller CH, Mather DE, Crepieux S (2009) Analytical approaches and population types for finding and utilizing QTL in complex plant populations. Crop Sci 49:363–380

Sorkheh K, Malysheva-Otto LV, Wirthensohn MG, Tarkesh-Esfahani S, Martínez-Gómez P (2008) Linkage disequilibrium, genetic association mapping and gene localization in crop plants. Genet Mol Biol 31:805–814

Stich B, Utz HF, Piepho HP, Maurer HP, Melchinger AE (2010) Optimum allocation of resources for QTL detection using a nested association mapping strategy in maize. Theor Appl Genet 120:553–561

Stinchcombe JR, Hoekstra HE (2008) Combining population genomics and quantitative genetics: finding the genes underlying ecologically important traits. Heredity 100:158–170

Swanson-Wagner RA, DeCook R, Jia Y, Bancroft T, Ji T, Zhao X, Nettleton D, Schnable PS (2009) Paternal dominance of trans-eQTL influences gene expression patterns in maize hybrids. Science 326:1118–1119

Tanksley SD, Nelson JC (1996) Advanced backcross QTL analysis: a method for simultaneously discovery and transfer of valuable QTLs from unadapted germplasm into elite breeding lines. Theor Appl Genet 92:191–203

Thomson MJ, Tai TH, McClung AM, Lai XM, Hinga ME, Lobos KB, Xu Y, Martinez CP, McCouch SR (2003) Mapping quantitative trait loci for yield, yield components and morphological traits in an advanced backcross population between *Oryza rufipogon* and the *Oryza sativa* cultivar Jefferson. Theor Appl Genet 107:479–493

Toenniessen GH, O'Toole JC, DeVries J (2003) Advances in plant biotechnology and its adoption in developing countries. Curr Opin Plant Biol 6:191–198

van Bueren ETL, Backes G, de Vriend H, Østergård H (2010) The role of molecular markers and marker assisted selection in breeding for organic agriculture. Euphytica 175:51–64

Varshney RK, Graner A, Sorrells ME (2005) Genomics-assisted breeding for crop improvement. Trends Plant Sci 10:621–630

Vielle-Calzada JP, Martínez de la Vega O, Hernández-Guzmán G, Ibarra-Laclette E, Alvarez-Mejía C, Vega-Arreguín JC, Jiménez-Moraila B, Fernández-Cortés A, Corona-Armenta G, Herrera-Estrella L, Herrera-Estrella A (2009) The *Palomero* genome suggests metal effects on domestication. Science 326:1078

Walsh, B (2001) Quantitative genetics in the era of genomics. Theor Pop Biol 59:175–184

Wang J, Chapman SC, Bonnett DG, Rebetzke GJ, Crouch J (2007a) Application of population genetic theoryand simulation models to efficiently pyramidmultiple genes via marker-assisted selection. Crop Sci 47:582–590

Wang J, Wan X, Li H, Pfeiffer WH, Crouch J, Wan J (2007b) Application of identified QTL-marker associations in rice quality improvement through a design-breeding approach. Theor Appl Genet 115:87–100

Wenzl P, Carling J, Kudrna D, Jaccoud D, Huttner E, Kleinhofs A, Kilian A (2004) Diversity arrays technology (DArT) for whole-genome profiling of barley. Proc Natl Acad Sci U S A 101:9915–9920

Wong CK, Bernardo R (2008) Genomewide selection in oil palm: increasing selection gain per unit time and cost with small populations. Theor Appl Genet 116:815–824

Xia L, Peng K, Yang S, Wenzl P, de Vicente MC, Fregene M, Kilian A (2005) DArT for high-throughput genotyping of cassava (*Manihot esculenta*) and its wild relatives. Theor Appl Genet 110:1092–1098

Xu K, Mackill DJ (1996) A major locus for submergence tolerance mapped on rice chromosome 9. Mol Breed 2:219–224

Xu Y, Crouch JH (2008) Marker-assisted selection in plant breeding: from publications to practice. Crop Sci 48:391–407

Xu K, Xu X, Ronald PC, Mackill DJ (2000) A high-resolution linkage map in the vicinity of the rice submergence tolerance locus *Sub1*. Mol Gen Genet 263:681–689

Xu K, Deb R, Mackill DJ (2004) A microsatellite marker and a codominant PCR-based marker for marker-assisted selection of submergence tolerance in rice. Crop Sci 44:248–253

Xu K, Xu X, Fukao T, Canlas R, Maghirang-Rodriguez R, Heuer S, Ismail AM, Bailey-Serres J, Ronald PC, Mackill DJ (2006) *Sub1* A is an ethylene-response-factor-like gene that confers submergence tolerance to rice. Nature 442:705–708

Xu Y, Lu Y, Yan J, Babu R, Hao Z, Gao S, Zhang S, Li J, Vivek BS, Magorokosho C, Mugo S, Makumbi D, Taba S, Palacios P, Guimarães CT, Araus JL, Wang G, Davenport GF, Crossa J, Crouch JH (2009a) SNP chip-based genome wide scans for germplasm evaluation, marker-trait association analysis and development of a molecular breeding platform in maize. In: Proceedings of 14th Australasian plant breeding conference (APBC) & 11th Congress of the Society for the Advancement of Breeding Research in Asia and Oceania (SABRAO), Cairns Convention Centre, Cairns, 10–14 August 2009. http://open.irri.org/sabrao/images/stories/conference/site/…/apb09final00307.pdf

Xu Y, Skinner DJ, Wu H, Palacios-Rojas N, Araus JL, Yan J, Gao S, Warburton ML, Crouch JH (2009b) Advances in maize genomics and their value for enhancing genetic gains from breeding. Intl J Plant Genomics 2009:957602

Ye G, Smith KF (2010) Marker-assisted gene pyramiding for cultivar development. Plant Breed Rev 33:219–256

Yu J, Zhang Z, Zhu C, Tabanao DA, Pressoir G, Tuinstra MR, Kresovich S, Todhunter RJ, Buckler ES (2009) Simulation appraisal of the adequacy of number of background markers for relationship estimation in association mapping. Plant Genome 2:63–77

Chapter 22
Metabolomics-Assisted Crop Breeding Towards Improvement in Seed Quality and Yield

David Toubiana and Aaron Fait

Abstract Seed complex traits, such as oil, protein or starch content, but also seed size, vigor, and dormancy represent agronomic traits of value. The exact understanding of their underlying regulatory mechanisms may be central to the development of future crop cultivars and goal orientated breeding strategies, sustaining high yields, seed nutritional quality or increased oil content. Seed traits are inherently associated with seed metabolism and plant-seed carbon-nitrogen allocation, thus a metabolomics based approach can provide a comprehensive understanding of seed metabolism and more generally of seed quality. During its different developmental stages and from the maturation phase to germination, the seed is characterized by distinct metabolite signatures, which may associate to yield related traits, rendering their identification useful as metabolic markers in the development of metabolomics-assisted breeding strategies. That said, the scientific knowledge on biochemical pathways in a cell is limited by the small number of identifiable metabolites (few hundreds) as compared to the thousands present at any given moment in a cell. Moreover, the integration of different metabolomics platforms allowing the identification and quantification of known and unknown metabolites remains a non-trivial step in deciphering complete metabolomes. Last, the superimposition of metabolite data and morpho-physiological traits requires correct data handling and elaboration. Without requiring a prior knowledge of biochemical reactions, correlation based network analysis represents an attractive approach to study the mode of interaction of known metabolites, to suggest unknown candidates for pathway elucidation and to identify association between metabolites and yield related traits.

Keywords Correlation analysis · Metabolomics-assisted crop breeding · Metabolic markers · Network analysis · Seed metabolism · Seed quality

A. Fait (✉) · D. Toubiana
Jacob Blaustein Institutes for Desert Research, French Associates Institute for Agriculture and Biotechnology of Drylands Midreshet Ben-Gurion, Ben-Gurion University of the Negev French Associates, 84990 Negev, Israel
e-mail: fait@bgu.ac.il

G. K. Agrawal, R. Rakwal (eds.), *Seed Development: OMICS Technologies toward Improvement of Seed Quality and Crop Yield*,
DOI 10.1007/978-94-007-4749-4_22,

22.1 Introduction

Metabolites constitute the vocabulary for all physiological processes occurring inside the cell. The metabolome is a collection of molecules engaged in chemical reactions sustaining the survival of living organisms. An average metabolome at any given time point in the lifecycle of a plant is estimated to contain more than 3,000 compounds (Kind et al. 2009) and the variety of metabolites in the plant kingdom is overwhelming and assumed to reach the hundreds of thousands (Fiehn 2001). Metabolites differ in structure, size, and chemico-physical properties. According to their engagement into metabolic processes, they are categorized into central (primary) and specialized (secondary) classes.[1] The tissue specific qualitative and quantitative state of metabolites at a certain time point is defined as cellular metabolism at the steady state. However, cellular alterations, caused by endogenous (e.g., developmental cues) or exogenous factors (e.g., environmental cues) enforce system reorganization likely to include changes in gene expression, posttranscriptional/post-translational regulation, and enzymatic activity. These processes lead to a concerted shift in the metabolic profile to reach a new cellular steady state. The change in metabolite levels in response to the aforementioned cues can be traceable in the phenotype, e.g., changes occurring in a tomato or a strawberry during development (Carrari et al. 2006; Fait et al. 2008) or in a plant response to stress (Matthew et al. 2009). Therefore, specific phenotypes or even the absence of a phenotype (i.e., silent phenotype) can be disclosed by metabolic profiling (Meyer et al. 2007; Weckwerth et al. 2004). As such, metabolic markers are good candidates to be used as diagnostic or predictive tools in plant breeding (Steinfath et al. 2010; van Dongen and Schauer 2010). The reflection of phenotypic polymorphism onto metabolic processes has also been observed in seed morphological traits as well as seed postharvest physiology (Angelovici et al. 2011; Borisjuk et al. 2004; Focks and Benning 1998). The mechanisms at the basis of this association are to be found in concerted changes of seed metabolism during development, at which the buildup of unbound as well as bound metabolites can be observed, possibly providing a bridge between seed development and the regulation of post-harvest events (Angelovici et al. 2010; Kimura and Nambara 2010). Whilst the functional role of distinct set of molecules in the regulation of seed traits needs further elucidation, we can hypothesize that a trait can be traced back to a quantitatively and qualitatively defined group of metabolites (i.e., metabolic signatures). The understanding of the mechanisms regulating

[1] The use of the adjectives 'primary' and 'secondary' to categorize classes of metabolic processes is at times confusing. Primary or central metabolism is considered as being constitutively active. Examples of central metabolic processes are glycolysis, TCA cycle, OPPP, amino acid biosynthesis. In contrast, secondary or specialized metabolism is likely induced by exogenous and endogenous cues. Alternatively, metabolites actively integrated into biochemical pathways are referred to as *primary metabolites*, whereas the class of *secondary metabolites* does not have an immediate functionality in the pathways themselves, but often possesses important ecological functions. That said, groups as FAs as well as certain polyamines and nonprotein amino acids can correspond to both categories rendering classification artificial if not misleading. Here, we will adopt the use of central and specialized metabolism rather than primary and secondary.

seed quality traits, such as starch, oil, and protein content, is essential due to their enormous economic importance, particularly in the currently challenging scenarios characterized by the reduced availability of conventional fuels, increasing global temperatures, and limited water resources (Sen 2009). The carbohydrates, lipids, and proteins deposited in seeds are substantial sources of nutrition for humans and livestock. Thus, it is pivotal to perpetuate and constantly ameliorate seed nutritional and industrial values, respectively. Seed nutritional quality traits and other seed complex traits, such as seed size, seed dormancy, seed germination, and seedling vigor are clearly dependent on the metabolic processes during seed development. Therefore, metabolic markers may also render useful in the amelioration of seed nutritional and industrial values.

However, whilst important information from a genetic, molecular, and physiological standpoint has been added in the recent years on valuable seed complex traits (Jander et al. 2004; Koornneef et al. 2002; Martin et al. 2010; Thelen and Ohlrogge 2002), amelioration of seed quality is hampered by consistent gaps in our understanding of the regulation of nutritional associated seed traits and, generally, in plant metabolism (Kottapalli et al., 2008; Mazur et al. 1999). This is notwithstanding the fact that the development-induced metabolic changes observed in a seed are likely affected by posttranslational events, metabolic regulation (feedback, feed forward), and allosteric regulation to a greater extent than currently perceived (Angelovici et al. 2010; Helmstaedt et al. 2001; Weber et al. 2005). Henceforth, to correctly describe the structure and integration of the regulatory networks involved in the development of a metabolic trait, a systems biology approach is compelling. That being said, the integration of data from different omics platforms (e.g., central and specialized metabolite profiling, transcriptomics, and proteomics) collected on fine-time scales or by exploiting population genetics [i.e., introgression lines (ILs), recombinant-inbred lines (RILs), and near-isogenic lines (NILs) as well as mutants (e.g. ethyl methane sulphonate mutants) and transgenic collections] delineates the key approach for deciphering the topology of a highly dynamic cellular network.

Despite the complexity of inter omics regulation and crosstalk, metabolic profiling represents a relatively fast and affordable tool to monitor the effects of genetic modifications, environmental conditions, stress responses, and their combinations on the metabolism of an organ, a tissue or a group of cells (Nelson et al 2008). Moreover, the superimposition of metabolic variability (metabolic QTL, mQTL)[2] with morphological traits polymorphism can provide the basis for the development of metabolic signatures to be potentially used as markers in future breeding strategies (Keurentjes 2009; Shluepmann et al. 2003).This chapter presents a summary of recent advances in relating the metabolic phenotype with respect to crop yield parameters. As such, we are going to describe briefly: (i) the technologies and strategies utilized for metabolomics-assisted breeding; (ii) the steps necessary to successfully relate metabolic profiles to data stemming from different backgrounds; (iii) examples of the seed traits holding potential for the improvement of seed nutritional/industrial values *via* the application of metabolomics based strategies; and

[2] Metabolites are defined as quantitative traits (QT) implying the quantitative continuous change in their content across a population.

lastly (iv) the pitfalls and limitations of the current approaches and suggestions for future improvements.

22.2 State of the Art Metabolomics-Assisted Breeding Technologies: An Overview

Particularly, considering the extensive global demand for seed and grain crops and the need for increased nutritional supplement, broad-scale metabolic-assisted breeding is a must rather than 'just' a necessity (Carrari et al. 2003). Although measurements of metabolite content have been carried out for several decades now (Jellum 1977), studies have been focusing usually only on a handful of metabolites (Diers et al. 1992; Moose et al. 2004; Song and Zhang 2007; Zhao et al. 2006). Metabolomics itself is a rather novel and still evolving scientific area (Fernie and Schauer 2009; Weckwerth 2010) benefiting to a great extent from technical endeavors that have been achieved in other scientific fields and related omics platforms. With the advent of high-throughput technologies, large-scale metabolomics profiling has become a feasible tool for plant phenotyping, functional genomics studies, and a diagnostic aid to assess plant response to, and interaction with, environmental conditions or genetic perturbation (Guy et al. 2008a,b; Messerli et al. 2007). For instance, NMR, a method introduced by Rabi et al. (1938) describing molecular beams, has been greatly adapted into an array of scientific fields, amongst others into metabolomics. NMR associated metabolic profiling owns chief abilities in the unequivocal determination of atomic structures (Meiler and Will 2002)—of particular use in metabolic flux analysis, where cycling or parallel metabolic processes can be resolved (Bothwell and Griffin 2011; Sekiyama and Kikuchi 2007). NMR based applications have found a main niche of appliance in the metabolic profiling of living cells in mammalian metabolomics (Fernie et al. 2004) given to their significant advantage a noninvasive use and atomic resolution. Additional metabolomics profiling platforms are based on a DIMS, where *via* ESI-MS metabolic fingerprints can be detected. A powerful tool for DIMS is comprised of the FT-ICR MS achieving ultrahigh, accurate mass resolution by trapping and accumulating ions and thereby enabling differentiation of isomeric structures (Bedair and Summer 2008). A less costly alternative to FT-ICR, however achieving lower resolving powers, is the orbit rap mass analyzer, where by sensing the frequency of harmonic ion oscillation along a central spindle of cycling trapped ions the mass to charge ratio is ascertained (Bedair and Summer 2008). For diagnostic as well as basic science purposes plant metabolomics shall resolve in a comprehensive description of the metabolic processes taking place under distinct states, e.g., different tissues and treatments (van Dongen et al. 2009), environmental cues (Gavaghan et al. 2010; Sanchez et al. 2011), developmental stages (Angelovici et al. 2009; Carrari et al. 2006; Fait et al. 2006, 2008; Moing et al. 2011), or across extensive genotypic variance (Keurentjes 2006; Schauer et al. 2005). As such, a more comprehensive high-throughput low-cost type of analysis is required, with relatively accessible data for immediate elaboration. The currently standard technology and analytical tool that has fostered itself in a significant portion of studies on plant metabolome, is the high-throughput chromatography (gas or liquid) coupled to MS. Hyphenated platforms include single or triple quad GC, GC-TOF, HPLC, or UPLC MS. Also, due to their relatively long and extensive usage in metabolomics, a number of robust protocols have been established (Allwood et al. 2009; Fernie and Schauer 2008; Lisec et al. 2006) and derived applications for semi-automated analysis as well as constantly updated libraries (e.g.,GMD@CSB.DB) are available (Erban et al. 2007; Halket et al. 2005; Hummel et al. 2010; Kopka et al. 2005; Luedemann et al. 2008; Strehmel et al. 2008; Tohge and Fernie 2009). In particular: (i) GC-MS holds capabilities to detect hundreds of compounds during one scan (Roessner et al. 2001); (ii) GC-TOF-MS provides faster scan times and sensibility (Fernie and Schauer 2008); and (iii) the major advantage of LC-MS is its broad detection range of primary as well as specialized metabolites (Tohge et al. 2005). Usually, each technology can be applied to a preferred set of metabolites (for an overview of central mass analyzers and their preferred range se, Fig. 22.1). The hybridization of platforms, such as the use of a MS detector coupled to GC and UPLC or orbit rap machines (Lu et al. 2010), and parallel analysis of the same extract *via* combination of complementary technologies (Moing et al. 2011) can increase the resolving power and extend the range of molecular components to be determined. Generally, the ultimate target of these and next generation analyzers, such as the Field Asymmetric Waveform Ion Mobility Spectrometry (FAIMS)—MS (Fernie and Schauer 2008) is the efficient, rapid, and cost-effective unequivocal identification of as many known and unknown compounds as possible. The pace of developments in the area holds a great potential of enhancing the analysis of metabolomes, metabolones,[3] and metabolic networks. Yet, a bottleneck in the translation of the data generated to conclusions and guidelines for scientists and breeders on their study systems is represented by the nontrivial data integration and by our yet limited understanding of the regulatory mechanisms at the basis of structure of the metabolite network. Henceforth, to aid in conclusion drawing increasing attention in the recent years has been given to the development of algorithms for the reduction of data complexity, of tools for a user-friendly visualization of the metabolic data onto metabolic maps and of software for repository and integration of heterogeneous data [PageMan (Usadel et al. 2006), MapMan (Thimm et al. 2004), Paintomics (Garcia-Alcalde et al. 2011), and TMEV (Saeed et al. 2006)]. As an example from within the private sector, web-based Phenom-Network Inc. (http://phnserver.phenome-networks.com/) holds phenotypic (including metabolite profiles) data from several plant segregating populations as well as from the animal kingdom.

In the next section, we will describe correlation analysis of metabolite profiles and its potential in providing information with predictive power.

[3] A *metabolon* is a temporary structural-functional complex formed between sequential enzymes of a metabolic pathway, held together by noncovalent interactions (Srere 1985).

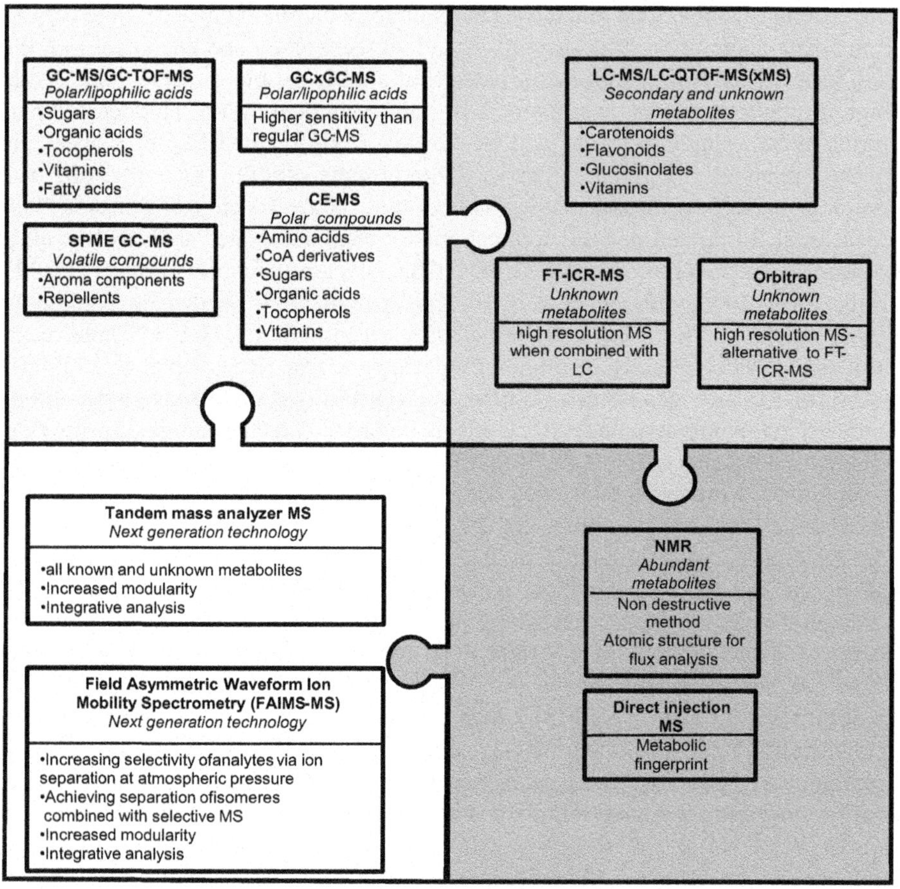

Fig. 22.1 State of the art and envisioned technologies for metabolic profiling. Different central mass analyzers and their preferred range of analytes. Future trends point to tandemly connected mass analyzers to detect the whole range of metabolites. Abbreviations: *MS* mass spectrometry, *GC* gas chromatography, *LC* liquid chromatography, *TOF* time of flight, *NMR* nuclear magnetic resonance, *CE* capillary electrophoresis, *SPME* solid phase micro extraction, and *FT-ICR* Fourier transform: ion-cyclotron resonance

22.3 Correlation (Network) Analysis: A Tool to Identify Functional Groups of Metabolites

The great advantage of metabolomics as opposed to transcriptomics is its species independent potential for broad-scale metabolic profiling. Its potential is unfolded when it is performed on a vast array of samples, such as entire mapping populations, RIL, IL, and NIL. The integration between population genetics and metabolomics allows identifying loci regulating the levels of metabolites (mQTL) (Fig. 22.2), e.g., vitamin E (Chander et al. 2008; Haiyan et al. 2010) or the colocalization of meta-

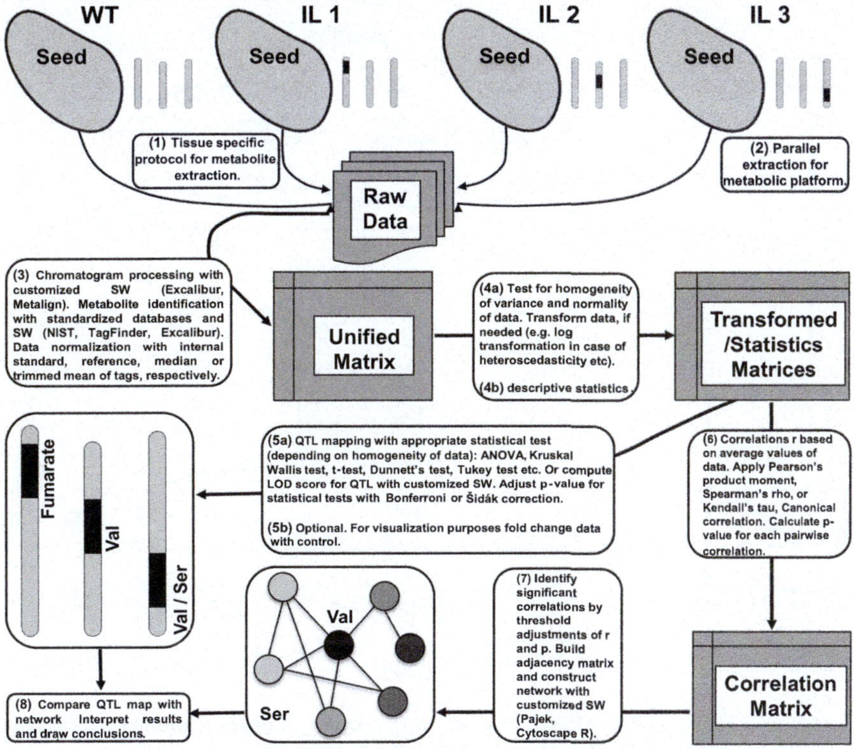

Fig. 22.2 Statistical and correlation-network-based analysis for QTL identification and metabolite coregulation in mapping and segregating populations. The figure depicts current methodologies in the analysis of mQTL, suggests the statistical tests involved in generating a mQTL map, and analyzes the coregulation of metabolites *via* correlation-network-based analysis in mapping and segregating populations. By intersecting the results, the resolving power is increased, revealing greater insights into the biological question. Abbreviation: *SW* software

bolic traits with traits related to yield or morphology. However, to date, solely a few broad-scale metabolic studies have been carried out, exploiting extensive genetic diversity in *Arabidopsis* (Keurentjes et al. 2006; Lisec et al. 2008; Meyer et al. 2007), rice (Kusano et al. 2007), maize (Harrigan et al. 2007a,b), tomatoes (Phuc et al. 2010; Schauer et al. 2006; Toubiana et al., submitted). On the other hand the strength of a correlation-based analysis (CA) is fostered by the exploitation of large datasets, including multiple variables (Steinhauser et al. 2008). Purely mathematical, based on similar or opposing behavior of metabolites, CA may suggest a relation between mass fragments, although the biochemical reactions linking the originating molecules are unknown. As such CA allows constructing networks visualizing metabolites associations based on existing metabolic pathways or regulatory mechanisms and proposing novel ones (e.g., Phuc et al. 2010; Schauer et al. 2006; Steuer et al. 2003) (Fig. 22.2). Moreover, functional associations can be queried for by incorporating data from various divergent backgrounds (Fig. 22.3). Com-

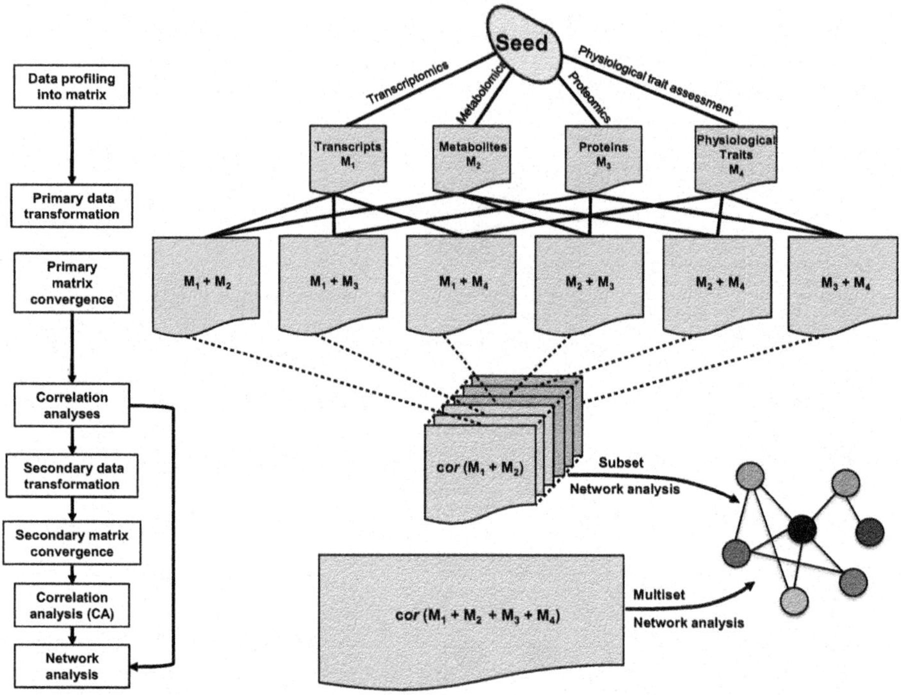

Fig. 22.3 Combining data of different origins *via* CA. Different omics platforms and phenotyping techniques lead to collections of qualitatively and quantitatively heterogeneous data, posing a challenge in finding functional relationships. Attributed to its pure mathematical nature and assessing the linearity of data vectors, CA is an aiding tool in the evaluation of data sets of divergent origins. To enable comparative analytical power between data sets of different origin, data transformation in form of Z-score transformation is needed prior to primary matrix convergence. The primary matrix convergence unites data of two independent sample assessment data sets into one respective matrix. Each independent matrix may serve as a functional platform for CA-based network analysis. Secondary matrix convergence unifies all matrices into one single matrix. After successful CA, significant pairwise or canonical correlations may be viewed as cartographic networks establishing a systems biology viewpoint

bining datasets of different origins (multiple matrices)includes the normalization of data at various levels, partitioning of data into dependent or independent sets, the application of parametric and nonparametric CA, significance testing by bootstrapping, jackknife method, or permutation test, but also the correlation of sets of variables as applied in canonical correlation analysis (CCA)(Fig. 22.2). A major disadvantage of this approach is the maze of options one may be confronted with—knowing which parameter, when, and how to choose. Also, the multidimensionality of datasets may be confounding, quickly evolving into biased results. Therefore, CA must be handled carefully and understanding of the algorithms being used may be beneficial. However, once mastering CA, the potential is enormous. For instance, aCCA was conducted by Meyer et al. (2007) and Lisec et al. (2008)on data stemming from *Arabidopsis* IL and RIL populations. The authors eventually proposed a

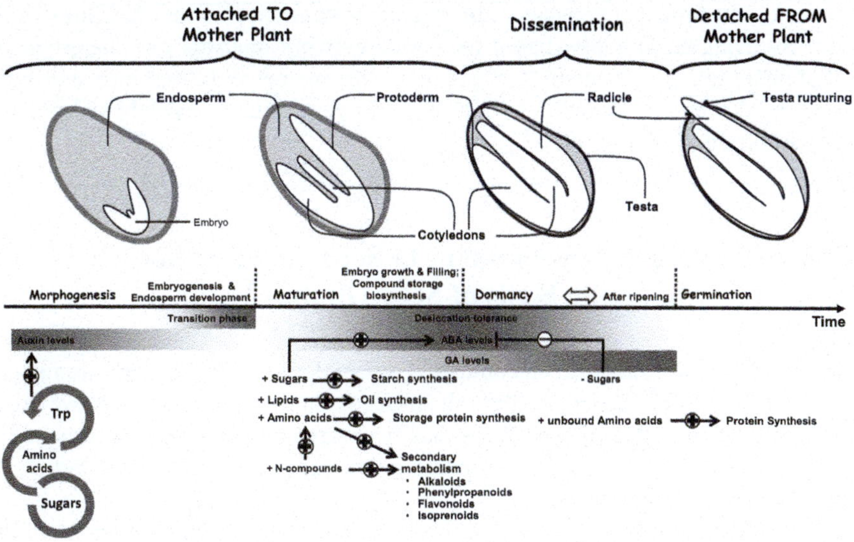

Fig. 22.4 Metabolites and their roles during seed development, maturation, and dormancy. The phases from seed development over to maturation tailed by a postharvest quiescent state are coordinated by the interplay and crosstalk of components of various cellular origin, incorporating hormones, chromatin remodeling, and metabolic signaling. It depicts functional metabolites and their identified specificities during these phases, in particular their inductive (+) or suppressing (−) effects on other cell compounds

metabolic signature associated with biomass and plant growth. Similarly, Schauer et al. (2006) have correlated fruit metabolite composition with that of yield associated morphological traits, showing that metabolites closely associated with photoassimilates imported to the fruit are tightly linked to harvest index or traits such as brix, whereas metabolites incorporated into pathways relatively distant to imported photoassimilates were described as morphology independent. Toubiana et al. (submitted) have firstly used tomato seed metabolite compositions and correlated them to fruit metabolite compositions as well as to an array of morphological traits. The comparative analysis identified inter-organ metabolite associations and demonstrated that the level of metabolites in a seed is negatively correlated to the morphological traits of the mother plant.

Via network visualization tools, correlation matrices can be depicted as networks (Kose et al. 2001), allowing to relate metabolite-metabolite associations to known (Tikunov et al. 2005) or putatively novel metabolic pathways (Weckwerth and Fiehn 2002; Figs. 22.2, 22.3, and 22.4), identifying closely related metabolites forming functional clusters, and shedding light onto the interaction between metabolic processes, such as allosteric regulations (Bartlem et al. 2000; Carrari et al. 2003; Geigenberger 2003). Cartographic visualization of correlation matrices provides additional valuable information on the topology and structure of the correlation network. Parameters, such as robustness, connectivity, and density of a network, can reflect for instance the degree of resilience of metabolism to alteration. By using cartography-based metabolic network analysis on a heterogeneous set of quantitative traits in an IL population, Lippman et al. (2007) also showed that harvest index acts as the main hub during fruit development reestablishing the sink source bottleneck paradigm.

22.4 From Seed Development to Seed Germination: The Metabolic Basis of Seed Quality Traits

A handful of studies have been conducted by employing metabolomics to foster the amelioration of nutritional quality traits in seeds. In the present section, we provide examples where metabolic profiling has been applied (or holds the potential) to develop markers or metabolic engineering strategies aimed at enhancing of yield increase and seed quality traits.

The process spanning from seed development to seed germination[4] accommodates the connection between two generations sustaining species survival. It is a complex process incorporating various defined developmental milestones and involving a concerted activity of central and specialized metabolism regulated by a developmental program integrating transcriptional and posttranscriptional elements (Keurentjes et al. 2008; Xie et al. 2008) (Fig. 22.4). *En route* to germination, seed development is initiated by morphogenesis, embodying embryogenesis, and endosperm development (De Smet et al. 2010), followed by a maturation (Weber et al. 2005) and desiccation (or maturation drying) phases in orthodox seeds,[5] which may or may not be tailed by a quiescent postharvest period of various lengths (Finch-Savage et al. 2010).

22.4.1 *Seed Maturation, Between Biofuel and Nutrition*

Maturation commences during the transition phase from maternal to filial control (Weber et al. 2005) (Fig. 22.4) with a high degree of concerted interplay of numerous signaling pathways, incorporating metabolic, and hormonal signals as well as genetic regulation (Santos-Mendoza et al. 2008; Stone et al. 2008; Wobus and Weber 1999). Seed maturation can be simplistically described as the process of storage reserves biosynthesis, which varies significantly between seeds of different backgrounds. As such, in legume seeds protein accumulation is the predominant form of storage compound often closely related to the nitrogen state in the seeds, e.g.,

[4] Defined as the emergence of the radicle from the seed testa (Baskin and Baskin 2004; Finch-Savage and Leubner-Metzger 2006).

[5] Orthodox seeds are desiccation tolerant and storable under dry states for long periods of time (Angelovici et al. 2010).

for peas (Salon et al. 2001; Weigelt et al. 2008) and soybeans (Hernández-Sebastiá et al. 2005). In oil-seed crops, such as sunflower and oilseed rape, the storage of oil in forms of TAG represents the primary form of storage compound (Graham 2008), where the TAGs remain in the seed packaged as oil bodies until seed germination (Hsieh and Huang 2004; Murphy and Vance 1999). In cereal grains, such as barley (Radchuk et al. 2009), carbohydrates are predominantly stored in the form of starch.

Storage reserves accumulation within the seed depends on central metabolic processes(i.e., particularly C-N balance), determining its development into a starchy, oil, or protein rich seed. Increased protein content is of paramount importance in regions where seeds and grains display the main constituent for the human diet, whereas oil-rich seeds may be of great interest for biofuel production. That said efforts are being dedicated in metabolic engineering to enhance accumulation of storage for nutritional and industrial purposes. One approach involves the amelioration of amino acid integration within the seed. By integrating metabolic engineering and metabolomics, Weigelt et al. (2008) have demonstrated that the upregulation of amino acid permease (VfAAP1) in the pea seed caused increased levels of nitrogen reflected in an overall elevated content of amino acids and protein. Simultaneously, levels of carbon have been depleted, which was explained with the costs necessary for each produced gram of nitrogen (Sinclair 1998). Weigelt's study also suggests that the balance of bound as well as unbound carbon and nitrogen plays a more significant role in the development of the seed than the absolute levels of carbon and nitrogen *per se*.

At the end of maturation, a shift in metabolism has been observed in different species, with increasing content of several unbound compounds (Angelovici et al. 2010). These metabolic findings have been accompanied by other, monitored shifts in the transcript programs of seed filling, identifying defined assets of mRNA accumulating in the dry seeds, as well as proteins and enzymatic machineries (Agrawal et al. 2008; Hajduch et al. 2005; Gallardo et al. 2007; Houston et al. 2009; Rajjou et al. 2007; Ruuska et al. 2002; White et al. 2000; Xu et al. 2008). In spite of the accumulating evidences suggesting for a role during the post-harvest quiescent state as well as early processes triggered at seed imbibition (Angelovici et al. 2010), the functional significance of these late molecular rearrangements remains to be fully grasped. Among the free metabolites consistently accumulating at the very end of the maturation period are amino acids (Fait et al. 2006; Weigelt et al. 2008). Beyond composing the pool of building blocks for protein synthesis, and hence from a biotechnological standpoint determining a seed nutritive and market value, amino acids in general are involved in a wide array of metabolic processes associated with specialized metabolite biosynthesis (Aharoni and Galili 2011; Drew and Demain 1977; Gonda et al. 2010), mitochondrial respiration (Weigelt et al. 2008), and stress response (Yoshiba et al. 1995). The complexity and redundancy of amino acid metabolism, however, renders difficult to estimate their potential use in developing strategies for a targeted improve of crop quality. Generally, pathways of amino acids biosynthesis are highly branched and interconnected (Less and Galili 2009), increasing the magnitude of difficulty for deciphering amino acid regulation. Metabolomics has been recently applied in concomitance with genome-wide gene expression analysis to investigate the mechanisms regulating the amino acids network in seeds *via* seed-targeted modulation of amino acid metabolism. For instance, the amino acid aspartic acid branches into the biosynthesis of lysine (an essential amino acid for the human diet found in low amount in seeds), methionine, threonine, and isoleucine, and is also involved in the conversion from threonine into glycine (Less and Galili 2009). The latter are associated with cellular processes (Rebeille et al. 2006), pathogen interactions (Thines et al. 2007), energy production (Mooney et al. 2002), and photorespiration (Joshi et al. 2006). Albeit being a good candidate for the improvement of seed nutritional quality (Rate and Greenberg 2001; Hudson et al. 2006), the regulation of lysine metabolism from a holistic viewpoint in the developing seed has failed to be established. Only recently, Angelovici et al. (2009) have used *Arabidopsis* knockout mutants performing GC-MS-based metabolic profiling and microarray analyses to comprehensively study lysine-induced changes of genome-wide patterns of gene expression and of central metabolic processes. Their findings indicate that lysine is associated to genes regulating plant vigor, while concomitantly suppressing a number of genes associated with heat shock response. Bioinformatics analyses confirmed that genes encoding for biosynthetic enzymes of the aforementioned aspartic acid family pathway (AFP), are generally coregulated with genes associated with plant growth, whereas genes encoding for catabolic enzymes of the AFP are coregulated with stress-associated genes. Additionally, in contrast with former suggestions, metabolomics-based characterization of changes during seed development showed that altered lysine content has little consistent influence on the metabolic network, with the exception of alteration in the level of TCA cycle intermediates. In a separate study, seed-targeted enhanced glutamic acid to GABA metabolism led to a dramatic alteration in FAs metabolism, increased protein content, and to the accumulation of nitrogen compounds (Fait et al. 2011). The metabolomics-based analysis suggested that manipulating the regulation of 'peripheral' pathways, specifically during seed maturation, can boost the content of amino acids and protein while maintaining intact fundamental processes in seed development, hence having applications in agriculture and medicine too (Akama et al. 2009). To a similar aim, studies in rice have attempted to enhance the level of aromatic amino acids (Ishihara et al. 2007; Tozawa et al. 2001; Yamada et al. 2008). Being the precursors of potentially toxic alkaloids, the manipulation of their content in edible parts of the plant requires a comprehensive understanding of the metabolite regulation, followed by extensive diagnostic analysis. Using nontargeted LC-MS analysis, Matsuda et al. (2005) found no marked effects on the metabolite profile when expressing of *OASA1D* (encoding a mutant anthranilate synthase α-subunit)in potato, aside moderated changes in the indole metabolism and indole amines tryptamine and serotonin in rice calli, seeds, and plants (Ishihara et al. 2007; and unpublished data). Taken together, metabolic profiling employed in parallel to metabolic engineering provides a fundamental tool for the development of effective strategies aimed at improving of seed quality as well as for powerful diagnosis to ensure the safety of engineered products.

During seed maturation, sugars exert dual functionality in the seed: (i) serve to sustain metabolism, and (ii) act as signals promoting or suppressing reactions in the cell. The ratio between sucrose and hexoses in the seed embryo has been suggested as the main component triggering the actual transition phase from morphogenesis to maturation; the significance of sugar signaling during seed development has been extensively reviewed (Weber et al. 2005). Moreover, differences in sugar ratios are associated with the composition of transfer cells, supporting the development of an epidermis-localized sucrose uptake system (Offler et al. 2003). In general, sucrose signals have regulatory attributes acting on a transcriptional and posttranscriptional level, affecting gene expression and metabolic enzyme activity, eventually controlling storage and differentiation processes (Gibson 2005; Rolland et al. 2006).

A specific relationship has been attributed to sugars and ABA, which is the primary phytohormone regulating the maturation and dormancy phase of the seed. By using *ABI* mutants, it was demonstrated that the ABA transduction chain is required for sugar signaling, inducing starch biosynthesis (Rook et al. 2006). On the other hand, glucose has been shown to be tightly linked to the biosynthesis and catabolism of ABA (Leon and Sheen 2003; Zhu et al. 2009). It is not surprising henceforth that some sugar-sensing mutants are allelic to known mutations in ABA synthesis or sensitivity (Leon and Sheen 2003; Rook et al. 2001; Smeekens 2000; Weber et al. 2005 and references therein). As such, the hormone to sugar crosstalk is a finely-concerted network involving several levels of regulation. Inducible/development-targeted transgenic systems and parallel integrative metabolite/hormone profiling approaches shall be used to acquire a comprehensive description of hormone-sugar integration during seed development to be exploited for ameliorating seed traits, including germination, dormancy, and vivipary.

22.4.2 *Predicting Dormancy versus Vivipary via Metabolite Networking?*

Seed development ends with mature seeds, entering the postharvest period and characterized by different degrees of quiescence and dormancy. Seed dormancy can be characterized as the inability of an intact vital seed to germinate, despite being exposed to suitable conditions (Bewley 1997). Diversity in dormancy regulation between and also within plant species makes it the focus of ecology and evolutionary research (Baskin and Baskin 1998), particularly in those habitats where the regulation of germination timing translates to plant survival (Finch-Savage and Leubner-Metzger 2006). The genetics of seed dormancy has been studied particularly in *Arabidopsis* where, by exploiting the ecotypic diversity, several mapping populations have been generated and characterized (Alonso-Blanco et al. 2003; Bentsink et al. 2006, 2010; Teng et al. 2008). On the contrary, viviparous seeds germinate whilst still being attached to the mother plant before dissemination (not undergoing any state of maturation drying), a common prerequisite to dormancy (Elmqvist and Cox 1996; Farnsworth 2000; van der Pijl 1983). Mostly, viviparous plant species are affiliated with regions lacking seasonal changes and sustaining high humidity throughout the year, with wet or flooded environments (Farnshworth 2000). Timber species are often associated with recalcitrant[6] behavior (Vazquez-Yanes and Arechiga 1996; Young and Young 1992). Recalcitrant species are also categorized as non-dormant. They preserve metabolic activity throughout ontogeny and differentiate themselves from viviparous species mainly by germinating immediately or shortly after dispersal (Farnsworth 2000). Aside the evolutionary significance, vivipary is a deleterious phenotype in agriculture, and forestry affecting crop yield *via* the associated trait of preharvest sprouting and consistently impairing efforts for seed bank conservation.

Dormancy and vivipary are closely associated with the levels of phytohormones ABA and its antagonist GA playing the key regulators (Farnsworth and Farrant 1998; Koornneef et al. 1998; Nonogaki et al. 2010). The understanding of antagonistic counter play of these hormones is vitally important for agronomic aspects in order to promote or suppress germination, or to ensure uniformity. The dormancy status is acquired gradually by the accumulation of ABA during seed maturation and reaches its peak in harvest-ripe seeds (Karssen et al. 1993; Ooms et al. 1993). During 'after-ripening', a highly-flexible period between species as well as within species (Donohue et al. 2005; Manz et al. 2005), dormancy is relieved accompanied by a depletion of ABA and an accumulation of GA (Nonogaki et al. 2010). After-ripening is usually dependent on factors, such as moisture, oil/starch storage balance, temperature, and seed covering structures (Manz et al. 2005). The relevance of understanding the dynamics of metabolic networks structure in respect to seed dormancy and vivipary lies in the long-known relation between ABA concentrations in seeds and soluble sugars (Finkelstein et al. 2008; Gibson 2004; Rook et al. 2006; Yuan and Wysocka-Diller 2006). Zhu et al. (2009) assessed the effect of glucose on ABA metabolism and seed germination in rice grains. They could lucidly demonstrate that delay of germination could be induced by exogenous glucose. Furthermore, by qRT-PCR analyses they showed that the amount of transcripts of ABA-catabolic genes were reduced in the presence of glucose, inferring that rather than having an inducible effect on ABA-synthesizing genes, glucose suppresses the synthesis of ABA-catabolic genes and by that sustaining higher levels of ABA delaying the process of germination. In dormant species, dormancy is imposed on the seed during late-maturation desiccation, when sugar levels augment in the embryo (Black et al. 1996; Leprince et al. 1990). For sucrose in particular, distinct functionalities have been demonstrated. As such, it was shown that sucrose is an important factor preparing the seed for maturation drying by stabilizing cell membranes *via* vitrification (Farnsworth 2000; Vertucci and Farrant 1995) or substituting water with hydroxyl groups (Crowe et al. 1992; Farnsworth 2000; Hoekstra et al. 1994; Kermode 1990). Evaluation of accumulation of disaccharides and oligosaccharides in seeds prior to desiccation has been subject to numerous studies, such as for wheat (Black et al. 1996), *Brassica campestris* (Leprince et al. 1990), soybean (Blackman et al. 1992), pea (Corbineau et al. 2000), bean (Bailly et al. 2001), and *Arabi-*

[6] Seeds that fail to go through storage-like periods, such as dry chilling; hence the term "recalcitrant" to storage (Roberts 1973).

dopsis (Fait et al. 2006). All studies have demonstrated a correlation between the acquisition of desiccation tolerance and accumulation of oligosaccharides, such as raffinose and stachyose. None of the studies, however, could ascertain the relationship for the directional cause and effect. Corbineau et al. (2000) suggested that the biosynthesis of the oligosaccharides is driven by the rate of water loss. Bailly et al. (2001) suggested that the accumulation could also have resulted from the conversion of monosaccharides to reduce respiratory substrates availability, consequently downregulating metabolic activity promoting the quiescent state of the seed. More recent findings suggest that not solely sugar regulates desiccation tolerance acquisition in seeds. It has been argued that the array of different behaviors are attributable to species-specific seed internal matrix formation (Buitink and Leprince 2008), which incorporates the interplay of metabolites of different backgrounds (sugars, organic acids, and amino acids), but also proteins and salts.

Thus, seed maturation drying and induction into dormancy are coordinated by the crosstalk of sugar signaling, hormonal balance, and chromatin remodeling (Angelovici et al. 2010; Finkelstein and Gibson 2002; Vincent-Carbajosa and Carbonero 2005). Deciphering their regulatory mechanisms remains a challenge. Nonetheless by monitoring the levels of metabolite(s) known to be highly associated with either vivipary or dormancy and their association within the metabolic and hormonal network of the developing seed, will lead to the elucidation of the genetic and physiological basis of these complex phenotypes and aid to the development of efficient strategies to ameliorate crops.

22.5 Limitations and Future Perspectives

The potential for seed nutritional/industrial quality enhancement using metabolites as targets and metabolic signatures as predictive tool are extensive exacerbated by the gap in knowledge at present on metabolism and metabolic regulation during seed development. Increasingly sensitive high-through put technologies will accelerate the process of data generation and result in ever-large complex, multidimensional datasets. In addition, the shift from model plant toward crop-based research and its implementation exploiting mapping populations and by means of field experiments will leave no alternative other than consistent data reduction and correlation-based analyses. Albeit the complexity and the multidimensionality of this approach current computational capacities and readily accessible online tools allow the evaluation of heterogeneous data quite rapidly. None the less, a major pitfall to understand metabolic regulations, interplay, and networks is the limited number of measurable annotated metabolites (100s) as compared to the vast pool of metabolites believed to exist in a given moment in a plant cell (in the thousands) (Fiehn 2001). To conclude, not annotated compounds can represent key “players” in metabolic pathways regulation, bridging biochemical reactions between known metabolites measured as correlating to one another. Thus, in order to foster the development of predictive models and metabolomics-assisted conclusion drawing for breeding purposes, non-targeted approaches based on accurate mass spectrometry and correlation network analysis (to link not annotated to known metabolites) should be employed.

References

Agrawal GK, Hajduch M, Graham K, Thelen JJ (2008) In-depth investigation of the soybean seed-filling proteome and comparison with a parallel study of rapeseed. Plant Physiol 148:504–518

Aharoni A, Galili G (2011) Metabolic engineering of the plant primary-secondary metabolism interface. Curr Opin Biotechnol 22:239–244

Akama K, Kanetou J, Shimosaki S, Kawakami K, Tsuchikura S, Takaiwa F (2009) Seed-specific expression of truncated OsGAD2 produces GABA-enriched rice grains that influence a decrease in blood pressure in spontaneously hypertensive rats. Transgenic Res 18:865–876

Allwood JW, Erband A, de Koning S, Dunn WB, Luedemann A, Lommen A, Kay L, Löscher R, Kopka J, Goodacre R (2009) Inter-laboratory reproducibility of fast chromatography-electron impact-time of flight mass spectrometry (GC-EI-TOF/MS) based plant metabolomics. Metabolomics 5:479–496

Alonso-Blanco C, Bentsink L, Hanhart CJ, Vries HBE, Koornneef M (2003) Analysis of natural allelic variation at seed dormancy loci of Arabidopsis thaliana. Genetics 164:711–729

Angelovici R, Fait A, Zhu X, Szymanski J, Feldmesser E, Fernie AR, Galili G (2009) Deciphering transcriptional and metabolic networks associated with lysine metabolism during Arabidopsis seed development. Plant Physiol 151:2058–2072

Angelovici R, Galili G, Fernie AR, Fait A (2010) Seed desiccation: a bridge between maturation and germination. Trends Plant Sci 15:211–218

Angelovici R, Fait A, Fernie AR, Galili G (2011) A seed high-lysine trait is negatively associated with the TCA cycle and slows down Arabidopsis seed germination. New Phytol 189:148–159

Bailly C, Audigier C, Ladonne F, Wagner MH, Coste F, Corbineau F, Côme D (2001) Changes in oligosaccharides content and antioxidant enzyme activities in developing bean seeds as related to acquisition do drying tolerance and seed quality. J Exp Bot 52:701–708

Bartlem D, Lambein I, Okamoto D, Itaya A, Uda Y, Kijima F, Tamaki Y, Nambara E, Naito S (2000) Mutation in the threonine synthase gene results in an over-accumulation of soluble methionine in Arabidopsis. Plant Physiol 123:101–110

Baskin CC, Baskin JM (1998) Seeds: ecology, biography, and evolution of dormancy and germination. In: Baskin CC, Baskin JM (eds). Academic, San Francisco, pp 666

Baskin JM, Baskin CC (2004) A classification system for seed dormancy. Seed Sci Res 14:1–16

Bedair M, Sumner LW (2008) Current and emerging mass-spectrometry technologies for metabolomics. Trends Anal Chem 27:238–250

Bentsink L, Hanson J, Hanhart CJ, Blankestijn-de Vries H, Coltrane C, Keizer P, El-Lithy M, Alonso-Blanco C, de Andres MT, Reymond M, van Eeuwijk F, Smeekens S, Koornneef M (2010) Natural variation for seed dormancy in Arabidopsis is regulated by additive genetic molecular pathways. Proc Natl Acad Sci U S A 107:4264–4269

Bentsink L, Jowett J, Hanhart CJ, Koornneef M (2006) Cloning of DOG1, a quantitative trait locus controlling seed dormancy in Arabidopsis. Proc Natl Acad Sci U S A 103:17042–17047

Bewley JD (1997) Seed germination and dormancy. Plant Cell 9:1055–1066

Black M, Corbineau F, Grzesik M, Guy P, Come D (1996) Carbohydrate metabolism in the developing and maturing wheat embryo in relation to its desiccation tolerance. J Exp Bot 47:161–169

Blackman SA, Obendorf RL, Leopold AC (1992) Maturation proteins and sugar in desiccation tolerance of developing soybean seeds. Plant Physiol 100:225–230

Borisjuk L, Rolletschek H, Radchuk R, Weschke W, Wobus U, Weber H (2004) Seed development and differentiation: a role for metabolic regulation. Plant Biol 6:375–386

Bothwell JHF, Griffin JL (2011) An introduction to biological nuclear magnetic resonance spectroscopy. Biol Rev 86:493–510

Buitink J, Leprince O (2008) Intracellular glasses and seed survival in the dry state. CR Biol 331:788–795

Carrari F, Urbanczyk-Wochniak E, Willmitzer L, Fernie AR (2003) Engineering central metabolism in crop species: learning the system. Metab Eng 5:191–200

Carrari F, Baxter C, Usadel B, Urbanczyk-Wochniak E, Zanor MI, Nunes-Nesi A, Nikiforova V, Centero D, Ratzka A, Pauly M, Sweetlove LJ, Fernie AR (2006) Integrated analysis of metabolite and transcript levels reveals the metabolic shifts that underlie tomato fruit development and highlight regulatory aspects of metabolic network behavior. Plant Physiol 142:1380–1396

Chander S, Guo YQ, Yang XH, Yan JB. Zhang YR, Song TM, Li JS (2008) Genetic dissection of tocopherol content and composition in maize grain using quantitative trait loci analysis and the candidate gene approach. Mol Breeding 22:353–365

Corbineau F, Picard MA, Fougereoux JA, Ladonne F, Côme D (2000) Effects of dehydration conditions on desiccation tolerance of developing pea seeds as related to oligosaccharide content and cell membrane properties. Seed Sci Res 10:329–339

Crowe JH, Hoekstra FA, Crowe LM (1992) Anhydrobiosis. Annu Rev Physiol 54:579–599

De Smet I, Lau S, Mayer U, Jürgens G (2010) Embryogenesis—the humble beginnings of plant life. Plant J 61:959–970

Diers BW, Keim P, Fehr WR, Schoemaker RC (1992) RFLP analysis of soybean seed protein and oil content. Theo Appl Genet 83:608–612

Donohue K, Dorn L, Griffith C, Kim E, Aguilera A, Polisetty CR, Schmitt J (2005) The evolutionary ecology of seed germination of Arabidopsis thaliana: variable natural selection on germination timing. Evolution 59:758–770

Drew SW, Demain AL (1977) Effect of primary metabolites on secondary metabolism. Annu Rev Microbiol 31:343–356

Elmqvist T, Cox PA (1996) The evolution of vivipary in flowering plants. Oikos 77:3–9

Erban A, Schauer N, Fernie AR, Kopka J (2007) Nonsupervised construction and application of mass spectral and retention time index libraries from time-of-flight gas chromatography-mass spectrometry metabolite profiles. Methods Mol Biol 385:19–38

Fait A, Angelovici R, Less H, Ohad I, Urbanczyk-Wochniak E, Fernie AR, Galili G (2006) Arabidopsis seed development and germination is associated with temporally distinct metabolic switches. Plant Physiol 142:839–854

Fait A, Hanhineva K, Beleggia R, Dai N, Rogachev I, Nikiforova VJ, Fernie AR, Aharoni A (2008) Reconfiguration of the achene and receptacle networks during strawberry fruit development. Plant Physiol 148:730–750

Fait A, Nesi AN, Angelovici R, Lehmann M, Pham PA, Song L, Haslam RP, Napier JA, Galili G, Fernie AR (2011) Targeted enhancement of glumate to g-aminobutyrate conversion in Arabidopsis seeds affects carbon-nitrogen balance and storage reserves in a development-dependent manner. Plant Physiol 157:1026–1042

Farnsworth EJ (2000) The ecology and physiology of viviparous and recalcitrant seeds. Annu Rev Ecol Syst 31:107–138

Farnsworth EJ, Farrant JM (1998) Reductions in abscisic acid are linked with viviparous reproduction in mangroves. Am J Bot 85:760–769

Fernie AR, Schauer N (2009) Metabolomics-assisted breeding: a viable option for crop improvement? Trends Genet 25:39–48

Fernie AR, Trethewey RN, Krotzky AJ, Willmitzer L (2004) Metabolite profiling: from diagnostics to systems biology. Nat Rev Mol Cell Biol 5:763–769

Fiehn O (2001) Combining genomics, metabolome analysis, and biochemical modelling to understand metabolic networks. Comp Funct Genom 2:155–168

Finch-Savage WE, Leubner-Metzger G (2006) Seed dormancy and the control of germination. New Phytol 171:501–523

Finch-Savage WE, Clay HA, Lynn JR, Morris K (2010) Towards a genetic understanding of seed vigour in small-seeded crops using natural variation in Brassica oleracea. Plant Sci 179:582–589

Finkelstein RR, Gibson SI (2002) ABA and sugar interactions regulating development: cross-talk or voices in a crowd? Curr Opin Plant Biol 5:26–32

Finkelstein R, Reeves W, Ariizumi T, Steber C (2008) Molecular aspects of seed dormancy. Annu Rev Plant Biol 59:387–415

Focks N, Benning C (1998) Wrinkled 1: a novel, low seed-soil-mutant Arabidopsis with a deficiency in the seed-specific regulation of carbohydrate metabolism. Plant Physiol 118:91–101

Gallardo K, Firnhaber C, Zuber H, Hericher D, Belghazi M, Henry C, Kuster H, Thompson RD (2007) A combined proteome and transcriptome analysis of developing Medicago truncatula seeds. Mol Cell Proteomics 6:2165–2179

Garcia-Alcalde F, Garcia-Lopez F, Dopazo J, Conesa A (2011) Paintomics: a web based tool for the joint visualization of transcriptomics and metabolomics data. Bioinformatics 27:137–139

Gibson SI (2004) Sugar and phytohormone response pathways: navigating a signalling network. J Exp Bot 55:253–264

Gibson SI (2005) Control of plant development and gene expression by sugar signaling. Curr Opin Plant Biol 8:93–102

Gonda I, Bar E, Portnoy V, Lev S, Burger J, Schaffer AA, Tadmor Y, Gepstein S, Giovannoni JJ, Katzir N, Lewinsohn E (2010) Branched-chain and aromatic amino acid catabolism into aroma volatules in Cucumis melo L. fruit. J Exp Bot 61:1111–1123

Graham IA (2008) Seed storage oil mobilization. Annu Rev Plant Biol 59:115–142

Guy C, Kaplan F, Kopka J, Selbig J, Hincha DK (2008a) Metabolomics of temperature stress. Physiol Plant 132:220–235

Guy C, Kopka J, Moritz T (2008b) Plant metabolomics coming of age. Physiol Plant 132:113–116

Haiyan L, Huancheng L, Yingpeng H, Xiaoxia W, Weili T, Guifeng L, Winben L (2010) Identification of QTL underlying vitamin E contents in soybean seed among multiple environments. Theor Appl Gen 120:1405–1413

Hajduch M, Ganapathy A, Stein JW, Thelen JJ (2005) A systematic proteomic study of seed filling in soybean. Establishment of high-resolution two-dimensional reference maps, expression profiles, and an interactive proteome database. Plant Physiol 137:1397–1419

Halket JM, Waterman D, Przyborowska AM, Patel RKP, Fraser PD, Bramley PM (2005) Chemical derivatization and mass spectral libraries in metabolic profiling by GC/MS and LC/MS/MS. J Exp Bot 56:219–243

Harrigan GG, Martino-Catt S, Glenn KC (2007a) Metabolomics, metabolic diversity and genetic variation in crops. Metabolomics 3:259–272

Harrigan GG, Stork LG, Riordan SG, Reynolds TL, Ridley WP, Masucci JD, MacIsaac S, Halls SC, Orth R, Smith RG, Wen L, Brown WE, Welsch M, Riley R, McFarland D, Pandravada A, Glenn KC (2007b) Impact of genetics and environment on nutritional and metabolite components of maize grain. J Agric Food Chem 55:6177–6185

Helmstaedt K, Krappmann S, Braus GH (2001) Allosteric regulation of catalytic activity: Escherichia coli aspartate transcarbamoylase versus yeast chorismatemutase. Microbiol Mol Biol R 65:404–421

Hernández-Sebastià C, Marsolais F, Saravitz C, Israel D, Dewey RE, Huber SC (2005) Free amino acid profiles suggest a possible role for asparagine in the control of storage-product accumulation in developing seeds of low- and high-protein soybean lines. J Exp Bot 56:1951–1963

Hoekstra FA, Haigh AM, Tetteroo FAA, van Roekel T (1994) Changes in soluble sugars in relation to desiccation tolerance in cauliflower seeds. Seed Sci Res 4:143–147

Houston NL, Hajduch M, Thelen JJ (2009) Quantitative proteomics of seed filling in castor: comparison with soybean and rapeseed reveals differences between photosynthetic and nonphotosynthetic seed metabolism. Plant Physiol 151:857–868

Hsieh K, Huang AHC (2004) Endoplasmatic reticulum, oleosins, and oils in seeds and tapetum cells. Plant Physiol 136:3427–3434

Hudson AO, Singh BK, Leustek T, Gilvarg C (2006) An LL-diaminopimelate aminotransferase defines a novel variant of the lysine biosynthesis pathway in plants. Plant Physiol 140:292–301

Hummel J, Strehmel N, Selbig J, Walther D, Kopka J (2010) Decision tree supported substructure prediction of metabolite from GC-MS profiles. Metabolomics 6:322–333

Ishihara A, Matsuda F, Miyagawa H, Wakasa K (2007) Metabolomics for metabolically manipulated plants: effects of tryptophan overproduction. Metabolomics 3:319–334

Jander G, Norris SR, Joshi V, Fraga M, Rugg A, Yu S, Li L, Last RL (2004) Application of a high-throughput HPLC-MS/MS assay to Arabidopsis mutant screening; evidence that threonine aldolase plays a role in seed nutritional quality. Plant J 39:465–475

Jellum E (1977) Profiling of human body fluids in healthy and diseased states of using gas chromatography and mass spectrometry, with special reference to organic acids. J Chromatogr 143:427–462

Joshi V, Laubengayer KM, Schauer N, Fernie AR, Jander G (2006) Two Arabidopsis threonine aldoases are nonredundant and compete with threonine deaminase for a common substrate pool. Plant Cell 18:3564–3575

Kermode AR (1990) Regulatory mechanisms involved in the transition from seed development to germination. Crit Rev Plant Sci 9:155–195

Keurentjes JJB (2009) Genetical metabolomics: closing in on phenotypes. Curr Opin Plant Biol 12:223–230

Keurentjes JJB, Fu J, de Vos CHR, Lommen A, Hall RD, Bino RJ, Van Der Plas LHW, Jansen RC, Vreugdenhil D, Koornneef M (2006) The genetics of plant metabolism. Nat Genet 38:842–849

Keurentjes JJB, Sulpice R, Gibon Y, Steinhauser MC, Fu J, Koornneef M, Stitt M, Vreugdenhil D (2008) Integrative analysis of genetic variation in enzyme activities of primary carbohydrate metabolism reveal distinct modes of regulation in Arabidopsis thaliana. Genome Biol 9:r129

Kimura M, Nambara E (2010) Stored and noesynthesized mRNA in Arabidopsis seeds: effect of cycloheximide and controlled deterioration treatment on the resumption of transcription during imbibition. Plant Mol Biol 73:119–129

Kind T, Scholz M, Fiehn O (2009) How large is the metabolome? A critical analysis of data exchange practices in chemistry. PLoS One 4:e5440

Koornneef M, Leon-Kloosterziel KM, Schwartz SH, Zeevart JAD (1998) The genetic and molecular dissection of abscisic biosynthesis and signal transduction in Arabidopsis. Plant Physiol Biochem 36:83–89

Koornneef M, Bentsink L, Hilhorst H (2002) Seed dormancy and germination. Curr Opin Plant Biol 5:33–36

Kottapalli KR, Payton P, Rakwal R, Agrawal GK, Shibato J, Burow M, Puppala N (2008) Proteomics analysis of mature seed of four peanut cultıvars using two-dimensional gel electrophoresis reveals distinct differential expression of storage, anti-nutritional, and allergenic proteins. Plant Sci 175:321–329

Kusano M, Fukushima A, Kobayashi M, Hayashi N, Jonsson P, Moritz T, Ebana K, Saito K (2007) Application of a metabolomic method combining one-dimensional and two-dimensional gas chromatography-time-of-flight/mass spectrometry to metabolic phenotyping of natural variants in rice. J Chromatogr B 855:71–79

Leon P, Sheen J (2003) Sugar and hormone connections. Trends Plant Sci 8:110–116

Leprince O, Bronchart R, Deltour R (1990) Changes in starch and soluble sugars in relation to the acquisition of desiccation tolerance during maturation of Brassica campestris seed. Plant Cell Environ 13:539–546

Less H, Galili G (2009) Coordinations between gene modules control the operation plant amino acid metabolic networks. BMC Sys Biol 3:14

Lippman ZB, Semel Y, Zamir D (2007) An integrated view of quantitative trait variation using tomato interspecific introgression lines. Curr Opin Genet Dev 17:545–552

Lisec J, Schauer N, Kopka J, Willmitzer L, Fernie AR (2006) Gas chromatography mass spectrometry-based metabolite profiling in plants. Nat Prot 1:387–396

Lisec J, Meyer RC, Steinfath M, Redestig H, Becher M, Witucka-Wall H, Fiehn O, Törjek O, Selbig J, Altmann T, Willmitzer L (2008) Identification of metabolic and biomass QTL in Arabidopsis thaliana in a parallel analysis of RIL and IL populations. Plant J 53:960–972

Lu W, Clasquin MF, Melamud E, Amador-Noguez D, Caudy AA, Rabinowitz JD (2010) Metabolomic analysis via reversed-phase ion-pairing liquid chromatography coupled to a stand alone orbitrap mass spectrometer. Anal Chem 82:3212–3221

Luedemann A, Strassburg K, Erban A, Kopka J (2008) TagFinder for the quantitative analysis of gas chromatography-mass spectrometry (GC-MS)-based metabolite profiling experiments. Bioinformatics 24:732–737

Manz B, Muller K, Kuzera B, Volke F, Leubner-Metzger G (2005) Water uptake and distribution in germinating tobacco seeds investigated in vivo by nuclear magnetic resonance imaging. Plant Physiol 138:1538–1551

Martin RC, Pluskota WE, Nonogaki H (2010) Seed germination. In: Pua EC, Davey MR (eds) Plant developmental biology—biotechnological perspectives, vol 1. Springer, Berlin, pp 383–404

Matsuda F, Miyazawa H, Wakasa K, Miyagawa H (2005) Quantification of indole-3-acetic acid and amino acid conjugates in rice by liquid chromatography-electrospray ionization-tandem mass spectrometry. Biosci Biotech Bioch 69:778–783

Matthew T, Zhou W, Rupprecht J, Lim L, Thomas-Hall SR, Doebbe A, Kurse O, Hankamer B, Marx UC, Smith SM, Schenk PM (2009) The metabolome of chlamydomonas reinhardtii following induction of anaerobic H-2 production by sulfur depletion. J Biol Chem 284:23415–23425

Mazur B, Krebbers E, Tingey S (1999) Gene discovery and product development for grain quality traits. Science 285:372–375

Meiler J, Will M (2002) Genius: a genetic algorithm for automated structure elucidation from 13C NMR spectra. J Am Chem Soc 124:1868–1870

Messerli G, Nia VP, Trevisian M, Kolbe A, Schauer N, Geigenberger P, Chen J, Davison AC, Fernie AR, Zeeman SC (2007) Rapid classification of phenotypic mutants of Arabidopsisvia metabolite fingerprinting. Plant Physiol 143:1484–1492

Meyer RC, Steinfath M, Lisec J, Becher M, Witucka-Wall H, Törjek O, Fiehn O, Eckardt Ä, Willmitzer L, Selbig J, Altmann T (2007) The metabolic signature related to high plant growth rate in Arabidopsis thaliana. Proc Natl Acad Sci U S A 104:4759–4764

Moing A, Aharoni A, Biais B, Rogachev I, Meir S, Brodsky L, Allwood JW, Erban A, Dunn WB, Kay L, de Koning S, de Vos RCH, Jonker H, Mumm R, Deborde C, Maucourt M, Bernillon S, Gibon Y, Hansen TH, Husted S, Goodacre R, Kopka J, Schjoerring JK, Rolin D, Hall RD (2011) Extensive metabolic cross-talk in melon fruit revealed by spatial and developmental combinatorial metabolomics. New Phytol 190:683–696

Mooney BP, Miernyk JA, Randall DD (2002) The complex fate of alpha-ketoacids. Annu Rev Plant Biol 53:357–375

Moose SP, Dudley JW, Rocheford TR (2004) Maize selection passes the century mark: a unique resource for 21st century genomics. Trends Plant Sci 9:358–364

Murphy DJ, Vance J (1999) Mechanisms of lipid-body formation. Trends Biochem Sci 24:109–115

Nelson T, Gandotra N, Tausta SL (2008) Plant cell types: reporting and sampling with new technologies. Curr Opin Plant Biol 11:567–573

Nonogaki H, Bassel GW, Bewley JD (2010) Germination – Still a mystery. Plant Sci 179:574–581

Offler CE, McCurdy DW, Patrick JW, Talbot MJ (2003) Transfer cells: cells specialized for a special purpose. Annu Rev Plant Biol 54:431–454

Ooms JJJ, Léon-Kloosterziel KM, Bartels D, Koornneef M, Karssen CM (1993) Acquisition of desiccation tolerance and longevity in seeds of Arabidopsis thaliana (A comparative study using abscisic acid-intensive abi3 mutants). Plant Physiol 102:1185–1191

Phuc TD, Prudent M, Sulpice R, Causse M, Fernie AR (2010) The influence of fruit load on the tomato pericarp metabolome in Solanum chmielewskii introgression line population. Plant Physiol 154:1128–1142

Rabi II, Zacharias JR, Millman S, Kusch P (1938) A new method of measuring nuclear magnetic moment. Phys Rev 53:318–318

Radchuk VV, Borisjuk L, Sreenivasulu N, Merx K, Mock HP, Rolletschek H, Wobus U, Weschke W (2009) Spatiotemporal profiling of starch biosynthesis and degradation in the developing barley grain. Plant Physiol. 150:190–204

Rajjou L, Miche L, Huguet R, Job C, Job D (2007) The use of proteome and transcriptome profiling in the understanding of seed germination and identification of intrinsic markers determining seed quality, germination efficiency and early seedling vigour. In: Adkins SW, Ashmore

SE, Navie SC (eds) Seeds: biology, development and ecology. CABI Publishing, Wallingford, pp 149–158
Rate DN, Greenberg JT (2001) The Arabidopsis aberrant growth and death2 mutant shows resistance to Pseudomonas syringae and reveals a role for NPR1 in suppressing hypersensitive cell death. Plant J 27:203–211
Rebeille F, Jabrin S, Bligny R, Loizeau K, Gambonnet B, Van Wilder V, Douce R, Ravanel S (2006) Methionine catabolism in Arabidopsis cells is initiated by a gamma-cleavage process and leads to S-methylcysteine and isoleucine syntheses. Proc Natl Acad Sci U S A 103:15687–15692
Roberts EH (1973) Predicting the storage of life seeds. Seed Sci Technol 1:499–514
Roessner U, Luedemann A, Brust D, Fiehn O, Linke T, Willmitzer L, Fernie AR (2001) Metabolic profiling allows comprehensive phenotyping of genetically or environmentally modified plant systems. Plant Cell 13:11–29
Rolland F, Baena-Gonzalez E, Sheen J (2006) Sugar sensing and signaling in plants: conserved and novel mechanisms. Annu Rev Plant Biol 57:675–709
Rook F, Corke F, Card R, Munz G, Smith C, Bevan MW (2001) Impaired sucrose induction mutants reveal the modulation of sugar-induced starch biosynthetic gene expression by abscisic acid signalling. Plant J 26:421–433
Rook F, Hadingham SA, Li Y, Bevan WM (2006) Sugar and ABA response pathways and the control of gene expression. Plant Cell Environ 29:426–434
Ruuska SA, Girke T, Benning C, Ohlrogge JB (2002) Contrapuntal networks of gene expression during Arabidopsis seed filling. Plant Cell 14:1191–1206
Saeed AI, Bhagabati NK, Braisted JC, Liang W, Sharov V, Howe EA, Li J, Thiagarajan M, White JA, Quackenbush J (2006) TM4 microarray software suite. Methods Enzymol 411:134–193
Salon C, Munier-Jolain NG, Duc G, Voisin AS, Grandgirard D, Larmure A, Emery RJN, Ney B (2001) Grain legume seed filling in relation to nitrogen acquisition: a review and prospects with particular reference to pea. Agronomie 21:539–552
Sanchez DH, Pieckenstain FL, Escaray F, Erban A, Kraemer U, Udvardi MK, Kopka J (2011) Comparative ionomics and metabolomics in extremophile and glycophytic Lotus species under salt stress challenge the metabolic pre-adaptation hypothesis. Plant Cell Environ 34:605–617
Santos-Mendoza M, Dubreucq B, Baud S, Parcy F, Caboche M, Lepiniec L (2008) Deciphering gene regulatory networks that control seed development and maturation in Arabidopsis. Plant J 54:608–620
Schauer N, Zamir D, Fernie AR (2005) Metabolic profiling of leaves and fruit of wild species tomato: a survey of the Solanum lycopersicum complex. J Exp Bot 56:297–307
Schauer N, Semel Y, Roessner U, Gur A, Balbo I, Carrari F, Pleban T, Perez-Melis A, Bruedigam C, Kopka J, Willmitzer L, Zamir D, Fernie AR (2006) Comprehensive metabolic profiling and phenotyping of interspecific introgression lines for tomato improvement. Nat Biotechnol 24:447–454
Sekiyama Y, Kikuchi J (2007) Towards dynamic metabolic network measurements by multi-dimensional NMR-based fluxomics. Phytochem 68:2320–2329
Sen Z (2009) Global warming threat on water resources and environment: a review. Env Geol 57:321–329
Sinclair TR (1998) Historical changes in harvest index and crop nitrogen accumulation. Crop Sci 38:683–643
Smeekens S (2000) Sugar-induced signal transduction in plants. Annu Rev Plant Physiol Plant Mol Biol 51:49–81
Song XL, Zhang TZ (2007) Identification of quantitative trait loci controlling seeds physical and nutrient traits in cotton. Seed Sci Res 17:243–251
Srere PA (1985) The metabolon. Trends Biochem Sci 10:109–110.
Steinfath M, Strehmel N, Peters R, Schauer N, Groth D, Hummel J, Steup M, Selbig J, Kopka J, Geigenberger P, van Dongen JT (2010) Discovering plant metabolic biomarkers for phenotype prediction using an untargeted approach. Plant Biotechnol J 8:900–911
Steuer R, Kurths J, Fiehn O, Weckwerth W (2003) Interpreting correlations in metabolic networks. Biochem Soc Trans 31:1476–1478

Stone SL, Braybrook SA, Paula SL, Kwong LW, Meuser J, Pelletier J, Hsieh TF, Fischer RL, Goldberg RB, Harada JJ (2008) Arabidopsis LEAFY COTYLEDON2 induces maturation traits and auxin activity: implications for somatic embryogenesis. Proc Natl Acad Sci U S A 105:3151–3156

Strehmel N, Hummel J, Erban A, Strassburg K, Kopka J (2008) Retention index thresholds for compound matching in GC-MS metabolite profiling. J Chromatogr B 871:182–190

Teng S, Rognoni S, Bentsink L, Smeekens S (2008) The Arabidopsis GSQ5/DOG1 Cvi allele is induced by the ABA-mediated sugar signalling pathway, and enhances sugar sensitivity by stimulating ABI4 expression. Plant J 55:372–381

Thelen JJ, Ohlrogge JB (2002) Metabolic engineering of fatty acid biosynthesis in plants. Metab Eng 4:12–21

Thimm O, Bläsing O, Gibon Y, Nagel A, Meyer S, Krüger P, Selbig J, Müller LA, Rhee SY, Stitt M (2004) MAPMAN: a user-driven tool to display genomics data sets onto diagrams of metabolic pathways and other biological processes. Plant J 37:914–939

Thines B, Katsir L, Melotto M, Niu Y, Mandaokar A, Liu G, Nomura K, He SY, Howe GA, Browse J (2007) JAZ repressor proteins are targets of the SCF(COI1) complex during jasmonate signalling. Nature 448:661–665

Tohge T, Fernie AR (2009) Web-based resources for mass-spectrometry-based metabolomics: a user's guide. Phytochemistry 70:450–456

Tohge T, Nishiyama Y, Hirai MY, Yano M, Nakajima J, Awazuhara M, Inoue E, Takahashi H, Goodenowe DB, Kitayama M, Noji M, Yamazaki M, Saito K (2005) Functional genomics by integrated analysis of metabolome and transcriptome of Arabidopsis plants over-expressing an MYB transcription factor. Plant J 42:218–235

Tozawa Y, Hasegawa H, Terakawa T, Wakasa K (2001) Characterization of rice anthranilate synthase alpha-subunit genes OASA1 and OASA2. Tryptophan accumulation in transgenic rice expressing a feedback-insensitive mutant of OASA1. Plant Physiol 126:1493–1506

Usadel B, Nagel A, Steinhauser D, Gibon Y, Blasing OE, Redestig H. Sreenivasulu N, Krall L, Hanna MA, Poree F, Fernie AR, Stitt M (2006) PageMan: an interactive ontology tool to generate, display, and annotate overview graphs for profiling experiments. BMC Bioinformatics 7:535

Van Der Pijl L (1983) Principles of dispersal in higher plants. Springer, Berlin, pp 214

van Dongen JT, Fröhlich A, Ramirez-Aguilar SJ, Schauer N, Fernie AR, Erban A, Kopka J, Clark J, Langer A, Geigenberger P (2009) Transcript and metabolite profiling of the adaptive response to mild decreases in oxygen concentration in the roots of Arabidopsis plants. Ann Bot 103:269–280

van Dongen JT, Schauer N (2010) Metabolic marker as a selection tool in plant breeding. ISB News Report. Information Systems for Biotechnology, Virginia Tech., Blacksburg

Vazquez-Yanes C, Arechiga MR (1996) Ex situ conservation of tropical rain forest seed: problems and perspectives. Interciencia 21:293–298

Vertucci CW, Farrant JM (1995) Acquisition and loss of desiccation tolerance. In: Kigel J, Galili G (eds) Seed development and germination. Marcel Dekker, New York, pp 237–271

Vincent-Carbajosa J, Carbonero P (2005) Seed maturation: developing intrusive phase to accomplish a quiescent state. Int J Dev Biol 49:645–651

Weber H, Borisjuk L, Wobus U (2005) Molecular physiology of legume seed development. Annu Rev Plant Biol 56:253–279

Weckwerth W (2010) Metabolomics: an integral technique in systems biology. Bioanalysis 2:829–836

Weckwerth W, Fiehn O. (2002) Can we discover novel pathways using metabolomics analysis? Curr Opin Biotechnol 13:156–160

Weckwerth W, Loureiro ME, Wenzel K, Fiehn O (2004) Differential metabolic networks unravel the effects of silent plant phenotypes. Proc Natl Acad Sci U S A 101:7809–7814

Weigelt K, Küster H, Radchuk R, Müller M, Weichert H, Fait A, Fernie AR, Saalbach I, Weber H (2008) Increasing amino acid supply in pea embryos reveals specific interactions of N and C metabolism, and highlights the importance of mitochondrial metabolism. Plant J 55:909–926

White JA, Todd J, Newman T, Focks N, Girke T, de Ilarduya OM, Jaworski JG, Ohlrogge JB, Benning C (2000) A new set of Arabidopsis expressed sequence tags form developing seeds. The metabolic pathway from carbohydrates to seed oil. Plant Physiol 124:1582–1594

Xie Z, Kapteyn J, Gang DR (2008) A systems biology investigation of the MEP/terpenoid and shikimate/phenylpropanoid pathways points to multiple levels of metabolic control in sweet basil glandular trichomes. Plant J 54:349–361

Xu SB, Ti L, Deng ZY, Chong K, Xue Y, Wang T (2008) Dynamic proteomic analysis reveals a switch between central carbon metabolism and alcoholic fermentation in rice filling grains. Plant Physiol 148:980–925

Yamada T, Matsuda F, Kasai K, Fukuoka S, Kitamura K, Tozawa Y, Miygawa H, Wakasa K (2008) Mutation of a rice gene encoding a phenylalanine biosynthetic enzyme results in accumulation of phenylalanine and tryptophan. Plant Cell 20:1316–1229

Yoshiba Y, Kiyosue T, Katagiri T, Ueda H, Mizoguchi T, Yamaguchi-Shinozaki K, Wada K, Harada Y, Shinozaki K (1995) Correlation between the induction of a gene for delta I-pyrroline-5-carboxylate synthetase and the accumulation of proline in Arabidopsis thaliana under osmotic stress. Plant J 7:751–760

Young JA, Young CG (1992) Seeds of woody plants of North America. Dioscorides Press, Portland, pp 407

Yuan K, Wysocka-Diller J. (2006) Phytohormone signalling pathways interact with sugars during seed germination and seedling development. J Exp Bot 57:3359–3367

Zhao JY, Becker HC, Zhang DQ, Zhang YF, Ecke W (2006) Conditional QTL mapping of oil content in rapeseed with respect to protein content and traits related to plant development and grain yield. Theo Appl Genet 113:33–38

Zhu G, Ye N, Zhang J (2009) Glucose-induced delay of seed germination in rice is mediated by the suppression of ABA catabolism rather than an enhancement of ABA biosynthesis. Plant Cell Physiol 50:644–651

Chapter 23
A Role for "Omics" Technologies in Exploration of the Seed Nutritional Quality

Marc Galland, Imen Lounifi, Gwendal Cueff, Aurélie Baldy, Halima Morin, Dominique Job and Loïc Rajjou

Abstract In the next 20 years, the world population is expected to reach a total of 8.3 billion people. Agriculture will have to produce sufficient food knowing that presently and globally agricultural production nearly matches world consumption. Yet, malnutrition is not only a matter of food amount but also concerns food nutritional quality. In particular, the nutritional value and quality of seeds should be improved to alleviate malnutrition and provide a well-balance diet. Progress in seed biology has considerably benefited from the rise in the last decade of the two model plants *Arabidopsis* and rice. Along with their genome sequences obtained respectively in 2000 for *Arabidopsis* and in 2005 for rice, functional genomics became possible because of the rapid development of their mutant libraries, full-length cDNA libraries, stock centers, web-accessible databases, and information portals, such as TAIR (The Arabidopsis Information Resource). Then, the development of high-throughput technological breakthroughs (e.g. DNA and protein array, mass spectrometry) helped to survey the omics state (transcriptome, proteome, and metabolome) of seeds at different developmental and environmental conditions. These approaches fuel candidate genes for seed quality (composition, germination vigor and capacity, good resistance to stress, etc.) that can be confirmed using functional genomics resources. Finally, exploitation of the confirmed candidate genes by plant breeders should improve seed nutritional quality and yield. In this chapter, we discuss how global "omics" technologies can help to find new candidate genes relevant for improvement of seed nutritional quality. Examples of omics application in unraveling the rice seed biology are particularly discussed.

L. Rajjou (✉) · M. Galland · I. Lounifi · G. Cueff · A. Baldy · H. Morin
Institut Jean-Pierre Bourgin, Unité Mixte de Recherche 1318, INRA-AgroParisTech, team "Physiology of Seed Germination", INRA Centre de Versailles-Grignon, Route de St-Cyr (RD10), 78026, Versailles Cedex, France
e-mail: loic.rajjou@agroparistech.fr

M. Galland (✉) · L. Rajjou · I. Lounifi · G. Cueff · D. Job
AgroParisTech, Unité de Formation et de Recherche de Physiologie Végétale, 16 rue Claude Bernard, 75231, Paris Cedex 05, France
e-mail: marc.galland@versailles.inra.fr

D. Job
Unité Mixte de Recherche 5240, CNRS-Université Claude Bernard Lyon-Institut National des Sciences Appliquées-Bayer Crop Science, Bayer Crop Science, Centre de Recherche de La Dargoire, 14-20 rue Pierre Baizet, 69263, Lyon Cedex 9, France

G. K. Agrawal, R. Rakwal (eds.), *Seed Development: OMICS Technologies toward Improvement of Seed Quality and Crop Yield*,
DOI 10.1007/978-94-007-4749-4_23,

Keywords Metabolomics · Nutrition · Proteomics · Rice · Seed · Transcriptomics

23.1 Introduction

In flowering plants, seeds develop from ovules (Baroux et al. 2002; Finch-Savage and Leubner-Metzger 2006; Linkies et al. 2010). The ovule is the structure providing the female reproductive cells. Flowering plant ovules contain the two outer layers called the integument(s), the nucellus, and the embryo sac in its center. After pollination, a pollen tube is formed, through which two sperm nuclei are released into the ovule. One sperm nucleus combines with the egg cell that leads to development of the diploid embryo. A nutritive tissue called endosperm surrounds this embryo. The endosperm is derived from the fusion of the second sperm nucleus with the two polar nuclei of the central cell to give rise to this triploid tissue. This double fertilization is unique to flowering plants (Frohlich and Chase 2007). Finally, a maternal tissue of ovular origin, the integument(s), gives rise to the seed coat (testa). This testa is surrounding the embryo and the endosperm to form the mature seed. Depending on the species, there is a considerable heterogeneity of seeds in terms of morphology and physiology as deeply discussed in the first introductory chapter (Muller-Landau 2003). During the seed-maturation program, storage compounds (such as carbohydrates, lipids, proteins, vitamins, or minerals) are accumulated in the embryo, the endosperm, or in the seed coat. These compounds protect the embryo during seed storage and are used as nutrients for germination and seedling growth.

Plants are the elementary element for livestock and human food. Seed represents the major plant organ consumed. It is worth noting that cereal grains alone provide more than 50 % of the protein and 75 % of caloric intake in human diets (Cordain 1999). However, some essential amino acids, such as lysine, are in low content in most cereal grains. In contrast, legume seed proteins are comparatively deficient in both the sulfur and aspartate-derived amino acids (Azevedo et al. 2006; Molvig et al. 1997; Waddell 1958). Therefore, crop seeds lack balanced essential amino acids, with important consequences on both animal and human health. This feature is also true for balanced essential fatty acids, vitamins, cofactors, minerals, and others nutritional quality factors. Given this, a major challenge for the future is to enhance the value of major crop seeds for animal feed and human nutrition use through both conventional breeding approaches and novel biotechnologies. During the last decade in crop sciences, seed biology has been highly benefited from the model plants *Arabidopsis* (The Arabidopsis Genome Initiative 2000; Somerville and Koornneef 2002; Koornneef and Meinke 2010; North et al. 2010) and rice (Goff et al. 2002; Shimamoto and Kyozuka 2002). Omics-based technologies (transcriptomics, proteomics, and metabolomics) have allowed seed profiling of gene transcription, protein expression, and/or small molecule synthesis and metabolism at genome level, as evidence by other chapters in this book. These investigations represent novel opportunities for utilizing omics technologies in rapid identification of markers with nutritional importance that might help to create modern quality crops (Agrawal et al. 2011).

23.2 From Arabidopsis to Rice to Understand Crop Species

23.2.1 Arabidopsis: The Undisputed Model of Plant Biologists

Arabidopsis was proposed as a model plant by Laibach (1943) as early as the year 1943. Today, *Arabidopsis* has become perhaps the most widely-studied plant. This small-sized flowering plant has several advantages including: (i) simple growth requirements and a short-lived life cycle (i.e., two months from seed to seed); (ii) selection of homozygote mutants due to its small-genome diploid species; (iii) self-fertilization but also cross with other individuals to give rise in both cases to a numerous seed progeny; and finally (iv) easy to transform with *Agrobacterium tumefaciens* using different reproducible transformation techniques (Valentine 2003; Zupan and Zambryski 1995).

During the past 25 years, classical genetic studies of collected *Arabidopsis* mutants have produced a wealth of information in identifying mutants of nutrient uptake, hormonal response, and flowering, serving as proof-of-concept as to how easy and powerful such mutants are in characterizing key physiological pathways (Somerville and Koornneef 2002). In the last decade alone, research in *Arabidopsis* has made tremendous progress towards understanding it biology. First of all, the sequencing of the *Arabidopsis* Columbia ecotype was completed in 2000 (The Arabidopsis Genome Initiative 2000). Then, functional genomics were considerably accelerated by the generation of several mutant collections (e.g., T-DNA insertion, gene trap and enhancer trap, EMS *mutagenesis*, and RNAi knock-out) to disrupt gene function and monitor gene expression. Most of these mutant collections are publicly accessible from stock centers. An overview of forward and reverse genetics projects of *Arabidopsis* and seed resources is presented in Table 23.1. Finally, high-throughput omics technologies are being implemented in *Arabidopsis* to uncover transcriptomes, proteomes, and metabolomes of this model weed plant (McCourt and Benning 2010). We can reasonably argue that *Arabidopsis* has unified the efforts of plant biologists to reveal the secrets of its genome.

However, generated datasets *via* omics or other large-scale approaches are becoming a big concern as to how to manage, integrate, and public dissemination. This necessity has given birth to the so-called systems biology to understand and predict the whole *Arabidopsis* system based on the integration of different layers of information (Baerenfaller et al. 2008). Research in *Arabidopsis* and the outcomes have been valuable to other plants, including crops. Indeed, as part of the *Brassicaceae* family, *Arabidopsis* is a close relative of *Brassica napus* (rapeseed), which is an important crop for edible oil production and biofuel research. *B. oleracea* (cabbage) is also an important contributor to human nutrition. Hence, *Arabidopsis*

Table 23.1 Worldwide *Arabidopsis* forward and reverse genetics resources, (*From* http://www.arabidopsis.org/portals/mutants/worldwide.jsp, accessed May 4th 2011)

Resource name	Type	Background line	Available from
AGRIKOLA	RNAi knockout (individual lines)	Col	ABRC NASC
Alonso, Crosby and Ecker	Simple insert (sets of pools and individual lines)	Col-0	ABRC NASC
Biological Research Center, Hungary	Promoter trap and activation tag (individual lines)	Col	Contact authors
Bressan, Yokoi and Koiwa	Activation tag (sets of pools)	C24	ABRC NASC
Characterized mutants	Misc. individual lines	Misc.	ABRC NASC
CSHL	Gene trap and enhancer trap (individual lines)	Ler	ABRC NASC
EXOTIC/JIC Gene Trap	Gene trap transposon (individual lines)	Ler	ABRC NASC
Feldmann	Simple insert (sets of pools and individual lines)	Ws-2	ABRC NASC
GABI-Kat *	T-DNA insertion (individual lines)	Col	ABRC NASC
GARNet—JIC SM	Single insert Ds-Spm (insertion lines)	Col-0	ABRC NASC
Haseloff/U. Cambridge	Enhancer trap (individual lines)	C24	ABRC NASC
IMA*	Ds insertion (individual lines)	Ler	ABRC NASC
INRA-Versailles FLAG_FST*	Promoter trap (individual lines)	Ws	INRA
INRA	Promoter trap (set of pools)	Ws-4	ABRC NASC
JIC Activate	Activation trap (individual lines)	Ler	ABRC NASC
Jack	Enhancer trap (set of pools)	Col-6 (gl1)	ABRC NASC
Misc. Insertion Collections	Misc.	Misc.	ABRC NASC
Misc. T-DNA pools for forward screening	Misc.	Misc.	ABRC NASC
Poethig/U. Penn	Enhancer trap (individual lines)	Col	ABRC NASC
RIKEN	Ds transposon insertion (individual lines))	No-O	RIKEN BRC
RIKEN	Activation trap (individual lines)	Col	RIKEN BRC
SAIL (formerly GARLIC/TMRI)	T-DNA insertion (individual lines)	Col	ABRC NASC
SALK	T-DNA insertion (individual lines)	Col	ABRC NASC
Confirmed SALK insertion lines	T-DNA insertion (PCR-validated homozygous or heterozygous individual lines)	Col	ABRC NASC
Scheible and Somerville	Activation tag (set of pools)	Col-2	ABRC NASC
Sussman and Amasino	Simple insert (set of pools)	Ws-2	ABRC NASC
TAMARA	Tn mediated activation tag (individual lines)	Col-0	NASC
TILLING	EMS mutagenized (individual lines)	Col er-105	ABRC NASC
Weigel	Activation tag (set of pools)	Col-7	ABRC NASC
Wisconsin Ds-Lox	Ds-Lox insertion (individual lines)	Col	ABRC NASC
Wisconsin KO	T-DNA insertion (set of pools)	Ws	ABRC NASC

seed biology investigations remarkably accelerate the acquisition of knowledge of gene functions involved in seed quality (Hsieh et al. 2011; Le et al. 2010; North et al. 2010).

23.2.2 The Rise of Rice: The Need for a Monocot and Cereal Model Plant

Despite numerous advantages of *Arabidopsis* as a model plant and research achievements in that, *Arabidopsis*-driven discoveries cannot always be transferred to other plant species, including crops. Importantly, many crop species, essential to mankind, belong to monocots and more specifically to the cereal plant family. Historically, maize studies provided numerous valuable results in cereal genetics, including identification of genes involved in flavonoid and starch metabolisms (Candela and Hake 2008). Other agricultural crop species (such as barley, wheat, or rice) were also studied for both basic and applied research. Yet, at the dawn of the genomic era, it became utterly clear that cereal scientists could greatly benefit from a model species with its genome sequenced. Compared to other crop species, rice has a relatively small genome of 389 Mb, which is about 10- and 42-folds smaller than maize (2,500 Mb (Messing and Dooner 2006; Rabinowicz and Bennetzen 2006)) and wheat (16,500 Mb (Jaiswal et al. 2006; Ware et al. 2002)), respectively. Moreover, as a diploid species, rice gives access to classical Mendelian inheritance, and therefore, mutant production is possible. Finally, one of the most valuable perspectives is perhaps the synteny of the rice genome sequence with the genomes of other cereals. Having this, comparative genomics between grasses can allow retrieving gene functions in other economically important cereals (Bennetzen and Ma 2003; Bolot et al. 2009; Moore et al. 1995).

These reasons had driven the build-up of the International Rice Genome Sequencing Project (IRGSP) consortium in the year 1998 that gathered sequencing centers scattered in ten countries (Sasaki and Burr 2000). This joint effort led to the establishment of a complete and mapped sequence of the rice genome originating from a single inbred line from *Oryza sativa* ssp. *japonica* cv. Nipponbare (International Rice Genome Sequencing Project 2005).

23.2.3 Arabidopsis and Rice Structural Genomics

The *Arabidopsis* Information Resource (TAIR) recently reannotated the entire *Arabidopsis* genome (TAIR10, www.Arabidopsis.org/news; November 2010), revealing the presence of a total 33,602 genes (27,416 protein-coding genes, 4,827 pseudogenes or transposable element genes, and 1,359 ncRNAs). This result is strikingly close to the original number of predicted genes when the *Arabidopsis* genome sequence was published in 2000 (The Arabidopsis Genome Initiative 2000). In case of rice, the rice genome was independently annotated by two different groups, namely The Institute for Genome Research (TIGR) and the Rice Annotation Project (RAP) (Ohyanagi et al. 2006; Ouyang et al. 2007; Tanaka et al. 2008; Yuan et al. 2003). These independent annotations have led to contrasting levels of predicted protein-coding genes, the total number of which is estimated to range between 30,000 and 43,000 genes (Itoh et al. 2007; Ohyanagi et al. 2006; Ouyang et al. 2007; Tanaka et al. 2008).

Although protein-coding genes are of key interest for seed biologists and plant breeders, one interesting feature of a plant genome is the presence of DNA transposable elements (TEs) (Feschotte 2008; Feschotte et al. 2002). These mobile TEs can insert themselves at another genome position either by self-copying or by excision/insertion. In *Arabidopsis*, around 10 % of the total genome is estimated to be composed of TEs, a relatively small percentage as compared to rice (about 30 % of the total genome) (Itoh et al. 2007; Ouyang et al. 2007). In general, the cereal genomes are known to be rich in TEs, which are perhaps responsible for their large genome size. Recent studies illustrate the multiple mechanisms through which TEs regulate gene expression and function, and ultimately seed phenotypes (Bennetzen 2005; Chantret et al. 2005; Kashkush et al. 2003).

Finally, one interesting common feature of the *Arabidopsis* and rice genomes relies in their highly duplicated structure. The alignment of chromosome sections against the rest of the genomic sequences showed that as much as 60 % of the *Arabidopsis* genome was duplicated (Ermolaeva et al. 2003). Similarly, the rice genome underwent whole-genome duplication, as 65 % of the genome shows homology with another chromosomal region (Yu et al. 2005). Moreover, individual gene duplications in rice seem to occur at a high rate that could explain the loss in microcolinearity compared to other cereal genomes (Yu et al. 2005).

Taken together, despite evident divergent evolution at the phenotype level, intriguing similarities can be found at the genome level particularly when looking at the close number of protein-coding genes in *Arabidopsis* and rice (even if the rice gene number must be refined). Moreover, the greater size of the rice genome relatively to *Arabidopsis* is mostly due to TEs. Finally, the genomes of these two model species underwent duplication, which is fundamental concerning functional biology where one particular gene often has family relatives elsewhere in the genome.

23.2.4 Functional Genomics of Arabidopsis and Rice

One fundamental goal of plant functional genomics is to assign a precise biological role to each gene in order to facilitate crops improvement. Following computational genome annotation, the *Arabidopsis* and rice predicted genes must be experimentally confirmed. Several methods exist that can be either characterized as 'top-down' (phenotype to genotype) or 'bottom-up' (genotype to phenotype) approaches, and are called forward and reverse genetics, respectively. Historically, forward genetics consisted in the mutagenization of large sets of seeds or plantlets in the hope that it would deliver clear phenotypes. Good examples come from the hormonal field, where collections of mutants were screened for ABA (Koornneef et al. 1989; Meurs et al. 1992; Ooms et al. 1993) and GAs (Nambara et al. 1991; Talon et al. 1990; Yamaguchi et al. 2001) responses that disturb seed development, seed storability, or seed germination. Then, positional cloning assisted by DNA markers allowed to map and characterize the genes of interest involved in seed quality. This method yielded considerable results particularly in *Arabidopsis*, where milestone discoveries were made in (but not restricted to) photosynthesis, signaling pathways, response to change in environmental conditions, and genetic features. Yet, the need to fine map the mutations responsible for the phenotypes slowed considerably the pace of scientific progress because one had to produce numerous progenies before being able to localize the gene of interest beneath the 5–6 kb level.

The progressive refinement of protocols for transformation of *Arabidopsis* by *A. tumefaciens* plant led to a bonanza of insertional mutants, where a known sequence (T-DNA) is introduced randomly in the genomes (Bechtold and Pelletier 1998; Bent 2000). This together with *ad hoc* stock center facilities and publicly web-accessible portals, providing information on position of T-DNA insertion genome wide by systematically sequencing the flanking T-DNA sequences tag (Alonso and Stepanova 2003). While this is mostly true for *Arabidopsis*, the picture is a bit more complicated for rice. *Agrobacterium*-mediated rice transformation is still technically challenging. Nonetheless, considerable resources are available because 172,000 flanking sequences have been generated that spans about half of rice total genes (27,551 from a total of 57,142 rice genes) (Jung et al. 2008; Krishnan et al. 2009; Zhang et al. 2006). These rice mutant lines can be ordered to and obtained from the producing institutions *via* two centralized web interfaces, RiceGE and OryGenesDB (Droc et al. 2006). Complementary to the knock-outstrategy, gain-of-function transgenic lines were generated for both *Arabidopsis* and rice species (Kondou et al. 2009). All of these genetic resources are valuable tools for the functional characterization of genes involved in controlling seed quality.

23.2.5 Model Seeds as Useful Tools

Since the dawn of agriculture some 10,000 years ago, mankind relies on plant seeds for its nutrition through feed and food production (Tresset and Vigne 2011). A better understanding of seed biology could yield important tools and knowledge for plant breeders to select key genes involved in, for example, accumulation of storage compounds or the control of seed germination. Seeds of *Arabidopsis* and rice have been the support for basic and applied research for some 25 years for *Arabidopsis*, and some studies in rice dated back to the early 1970s (Rensink and Buell 2004). Since then, great progress in seed biology has been made and our comprehension of gene expression regulation and seed metabolism control has been enlightened from virtually nowhere (Fitzgerald et al. 2009; Gutierrez et al. 2007; Holdsworth et al. 2008a,b; North et al. 2010). Further, the accumulated knowledge on these model species has allowed to find applications for the improvement of seed quality in a wide range of crops as exemplified by the TF GLABRA2 (GL2) that has been found to regulate seed oil accumulation in *Arabidopsis* (Shen et al. 2006). Four ortholog genes were characterized in *B. napus*, *B. rapa*, and *B. oleracea*, and are considered as valuable molecular markers for oilseed rape breeding focusing on improvement of seed oil content and quality (Chai et al. 2010). Thus, opportunities for improving the nutritional value of seeds have become increasingly evident.

23.3 Contribution of Omics Technologies to Decipher Biology and Improve Nutritional Value of Seeds

Humans require a varied and well-balanced food containing a complex mixture of carbohydrates, lipids, amino acids (proteins), and micronutrients (e.g., minerals and vitamins) in order to maintain optimal health. However, macronutrient and micronutrient levels of seed-based food vary widely. How genomic and postgenomic approaches can help to improve nutritional value of seeds?

23.3.1 Oleaginous Seeds

Arabidopsis is a commonly-used model plant for studying *oilseed* biology and metabolism. The pattern of FA accumulation in *Arabidopsis* seeds is closely paralleled with the increase of seed weight (Baud and Lepiniec 2009). The oil content expressed on a DW basis has been determined for 360 *Arabidopsis thaliana* accessions (O'Neill et al. 2003). As per report (O'Neill et al. 2003), the model oil content is around 38 % of DW, with most accessions lying within the range of 33–43 %. Predominant FA species are polyunsaturated FAs (PUFAs; 16:0, 18:0, 18:1, 18:2 and 18:3) and VLCFAs ($C \geq 20$). Interestingly, the seed oil fatty acyl compositions are remarkably conserved within the natural genetic diversity of *Arabidopsis* (O'Neill et al. 2003). Transcriptomics approaches used to decipher changes in gene expression during seed development were resulted in identification of many genes encoding core FA synthesis enzymes (Baud et al. 2009; Tranbarger et al. 2011). These genes could be used as molecular markers and to draw comparative genome alignment of *Arabidopsis* and other oleaginous crops (e.g., *Brassica* spp.). It is worth noting that the seed-specific overexpression of genes coding respectively for cytosolic glycerol-3-phosphate dehydrogenase (Vigeolas et al. 2007), glycerol-3-phosphate acyltransferase (Jain et al. 2000), lysophosphatidic acid acyltransferase (Zou et al. 1997), and DAGAT (Jako et al. 2001) has led to increase in oil content of transgenic rape seeds. The *WRI-1* TF, regarded as a central regulator of seed FA synthesis, also constitutes a promising candidate for manipulating the amount of FAs produced in transgenic plants. Studies of the *wri-1* mutant of *Arabidopsis* have shown that the glycolytic pathway plays a fundamental role in lipid biosynthesis, since this mutant has reduced lipid content due to the limited activity of several glycolytic enzymes (Focks and Benning 1998). In line with this finding, proteomics study of sunflower lines differing in seed oil contents has shown differences in protein abundance of some glycolytic enzymes (Hajduch et al. 2007). The characterization of glycolytic pathway in sunflower seeds indicates that key enzymes are responsible for establishing the levels of soluble carbohydrates in seeds, perhaps in a similar way to *Arabidopsis*, and hence highlight the importance of these activities in regulating substrate entry into the pathway (Troncoso-Ponce et al. 2009).

Oil seeds constitute a source of vitamin E (tocopherols and tocotrienols). Vitamin E was first discovered as an essential dietary nutrient for mammals (Schneider 2005). Tocopherols and tocotrienols are lipophilic antioxidants that are synthesized by all plants and are particularly abundant in seeds. While genes involved in the tocopherol pathway are well characterized in *Arabidopsis*, their biological function in seeds is unclear. Transcriptomics analyses of tocopherol-deficient mutant loci in *Arabidopsis* were used to investigate the functions of tocopherols (Sattler et al. 2006). Induction of six HSP genes in tocopherol-deficient mutant seeds was observed, suggesting that the tocopherol deficiency has the biological consequences, including protection of a group of proteins. Interestingly, seed-specific overexpression of the heat stress TF HSF in both sunflower and tobacco (*Nicotiana tabacum*) enhances the accumulation of HSP101 and small HSPs and improves resistance to seed deterioration (Prieto-Dapena et al. 2006). These findings point out the importance of the control of expression of the stress proteins in seed storability and viability. Seed viability is a major parameter that reflects their nutritional quality (Rajjou and Debeaujon 2008).

23.3.2 Leguminous Seeds

Leguminous seeds (e.g., lentils, beans, peas, etc.) are suitable sources of protein in feed and food. Legumes tend to be poor in the sulfur amino acids (cysteine and methionine) but are rich in lysine by contrast with cereal grains. Leguminous and cereals seeds, therefore, allow a well-balanced protein supply. By combining proteomics and QTL mapping approaches, the genetic architecture of seed proteome variability was recently highlighted in pea (Bourgeois et al. 2011). It was proposed that the accumulation of the major SP families was under the control of a limited number of loci. This idea is in accordance with the fact that a relatively small number of genes encode SSPs regardless of the plant species (Miernyk and Hajduch 2011). Precursor forms of the SSPs (e.g., prolamins, 2S albumins, 7S globulins, and 11/12S globulins) are synthesized during seed maturation in the ER, and the processed forms are accumulated in protein storage vacuoles to form the protein bodies. Seed proteomics analyses showed that these SSPs are subjected to a wide range of PTMs, which may affect the structure of protein bodies, digestibility, and allergenic properties of these proteins (Arc et al. 2011). Indeed, seed protein allergy is becoming increasingly prevalent, notably in peanut, sesame, and soybean. However, allergenic proteins are poorly described. Proteomics allows identifying allergenic protein isoforms. The knowledge of conserved IgE binding epitopes in common food allergens represents a useful tool for predicting cross-reactivity to certain foods and to orientate plant breeding to produce novel crops more appropriate to food.

23.3.3 Cereals Seed Biology: Seed Quality

The cultivated cereals, members of the Poaceae family of the angiosperms, represent relevant economical, scientific, and social issues. These plants have been studied for decades to elucidate their biology in order to improve their quality. Rice, wheat, and barley have extensive genetic, molecular, and genomic resources (Bruskiewich et al. 2006; Feuillet et al. 2008; Sreenivasulu et al. 2008a). Transcriptomics and proteomics have been used to relate changes in transcript and protein abundances in developing cereal grains (Agrawal and Rakwal 2006; Finnie and Svensson 2009; Martínez-Barajas et al. 2011; Sreenivasulu et al. 2010; Venu et al. 2011; Vensel et al. 2005; Wan et al. 2009). These works revealed key biological function and provided a way to improve cereal grain nutritional quality by manipulating key regulatory genes. For instance, transcriptomics and proteomics have unveiled the high complexity of the regulation of genes involved in starch synthesis. The functions of individual enzyme isoforms have provided new insights into how linear polymer chains and branch linkages are synthesized in cereals (Jeon et al. 2010). In particular, genetic analyses of characteristic mutants have formed the basis of a new model outlining the role of various enzyme isoforms in cereal starch production (Jeon et al. 2010). Omics technologies are instrumental toward a better comprehension of seed biology and in consequence for seed improvement. The difficulty in studies undertaken on cereals, generally monocots, lies in the large diversity among species. A plant model is primordial as a basis for study. Rice, a monocot species, was chosen to be a model plant because of its small genome size (389 Mb), extensive genetic map, relative easiness of transformation, and its close relatedness to major cereals (Izawa and Shimamoto 1996; Shimamoto and Kyozuka 2002; Agrawal and Rakwal 2006). Despite the fact that a great deal of progress has been made in the field of functional genomics, it is still hard to make links between results obtained from different species even if the studied specie is similar to the cereal plant model.

23.3.4 Influence of Germination Process on Nutritional Compounds

Germination is the period starting upon imbibition of a nondormant dry mature seed until the radicle protrudes from the seed coat (Bewley 1997). During seed germination *sensus tricto* (before radicle emergence), tremendous biochemical changes occur. The quiescent seed state is characterized by a reduced respiration, low-water content (10–15 %), and low metabolism. Upon seed water imbibition, seed respiration and its water content increase rapidly (Weitbrecht et al. 2011). The coincident increases in oxygen and water uptake reinduce the energetic metabolism (Fait et al. 2006; Howell et al. 2006; Sreenivasulu et al. 2008b). More precisely, there is continuity between the seed maturation and germination programs (Fait et al. 2006; Gallardo et al. 2007). In *Arabidopsis* seeds, aromatic amino acids, glucose-6-P and fructose-6-P strongly accumulate during the first 24 h of germination (Fait et al. 2006).

Moreover, *Arabidopsis* germinated seeds and seedlings mainly use TAG as a source of energy by their beta-oxidation in dedicated organelles (i.e., peroxisomes; Germain et al. 2001). Firstly, proteomics studies demonstrated that lipid and FA catabolism related-enzymes accumulate during the first 24h of germinating *Arabidopsis* seeds (Gallardo et al. 2001; Rajjou et al. 2006). Also, the deficiency in two acyl-CoA-oxidases genes seriously impairs the germination of *Arabidopsis* seeds as they fail to remobilize storage TAGs (Pinfield-Wells et al. 2005). In rice seeds, starch is the main storage compound from which energy is produced for germination and seedling growth (Fincher 1989; Yamakawa et al. 2007). The expression of genes encoding the starch-metabolizing enzymes is strongly regulated during rice grain filling and germination. As in *Arabidopsis* seeds, glucose-6-P and fructose-6-P strongly accumulate during the first 24h of rice germination (Howell et al. 2009). Free sugars probably fuel the glycolysis and TCA cycle because almost all transcripts encoding genes related to the glycolysis and TCA cycle are upregulated within the first 24 h of imbibition (Howell et al. 2009).

Building up a new plant requires energy but also elementary building blocks, such as amino acids. In the mature dry seed, proteins are stored in specialized structures that derive from the ER (Müntz 1998). More precisely, one can distinguish the protein bodies that originate from the ER and the protein storage vacuoles derived from the endomembrane trafficking system (Jolliffe et al. 2005). SSPs form a source of carbon, nitrogen, and last but not least, sulfur (Herman and Larkins 1999). In germinating *Arabidopsis* seeds, the amount of 12S globulin precursor forms decreases, while the amount of its α- and β-subunits increases (Gallardo et al. 2001). Yet, it is also noteworthy that some SPs are resynthesized during germination (Rajjou et al. 2006). Comparative proteomics profiling of dry and germinated rice seeds showed imbibition-associated protein changes (Kim et al. 2008; Yang et al. 2007). It was found that the content of soluble proteins, including some allergens, is lowered in germinated rice seeds as compared to dry seeds. Efforts and trials have been made to produce hypoallergenic rice using an enzymatic treatment (Yamada et al. 2005; Yang et al. 2007). In addition to allergens, several studies were undertaken on phytic acid (myo-inositol 1,2,3,4,5,6-hexakisphosphate), an antinutritional and antiecological factor. Phytic acid, the main storage compound of phosphorus, acts as a scavenger of metallic cations (namely K^+, Ca^{2+}, Mg^{2+}, Fe^{2+}, Zn^{2+}, and Mn^{2+}), decreasing their bioavailability and nutritional benefits (Bohn et al. 2008). Besides this nutrition issue, undigested phytic acid is associated with a negative ecological impact by contributing to the increase of phosphate release in manure, which is an environmental matter (Dao 2003). Transgenic approaches have been implemented to decrease phytic acid bioavailability in rice, corn, barley, and soybean using enzymatic degradation by means of barley phytases and production of low phytic acid lines (Sung el al. 2005). A set of low phytic acid rice mutant lines has been generated towards a goal to increase the nutritional quality. The rice ABC transporter gene *OsMRP5* has been proposed to be potentially involved in the process of phytic acid compartmentalization and/or associated transport (Xu et al. 2009). Proteomics and bioinformatics analyses of this low phytic acid rice line (mutated for the *OsMRP5* gene) revealed that triose phosphate isomerase and fructose bisphosphate aldolase, two major differentially expressed proteins, are involved in *myo*-inositol metabolism. Accumulation of SPs was also considerably decreased in the low phytic acid line (Emami et al. 2010). This work displayed the potential of proteomics and bioinformatics profiling methods for safety assessment of novel foods.

Altogether, these considerations have significant consequences from a nutritional point of view because TAG, minerals, starch, and SP breakdown could improve seed digestibility through a simple and economic process, such as germination.

23.3.5 Aging Influence on Seed Nutritional Value

In the emerging concerns about global changes, the attention is progressively focusing on the conservation of plant biodiversity. In particular, the Key Millennium Seed Bank and the Svalbard Global Seed Vault have been funded in order to save plant and agricultural relevant seed germplasms (Charles 2006). Seed viability and aging have therefore emerged as key issues that must be understood and managed.

Reductions in germination capability and seedling establishment constitute two hallmarks of seed viability (Rajjou and Debeaujon 2008). In particular, seed aging is characterized by an increase in ROS and protein damages (Bailly et al. 2008; Rajjou et al. 2008) and a lowering of DNA repair (Waterworth et al. 2010) and translational capacities (Rajjou et al. 2008). These altered cell metabolisms could impair seed nutritional value as the lipidic and proteic seed composition dramatically changes. For instance, rice seed proteins are oxidized during a controlled deterioration treatment (Fig. 23.1). Interestingly, the sensitivity of the endosperm and embryo proteins to the controlled deterioration treatment is very different (Fig. 23.1).

Upon artificial or natural seed aging, the polyunsaturated FA seed content declines steadily and lipid peroxidation by H_2O_2 can be easily detected (Devaiah et al. 2007). These lipid peroxidations can be prevented by tocochromanols that collectively form the so-called vitamin E (Mène-Saffrané and DellaPenna 2010). Tocochromanols encompass both tocotrienols and tocopherols, which are characterized by the presence of a polar group together with a long aliphatic carbon chain saturated (tocopherols) or polyunsaturated (tocotrienols) (DellaPenna and Pogson, 2006). Moreover, each tocopherol and tocotrienol possesses four isoforms, termed α- to δ-, which are differentially accumulated in seeds with contrasting antioxidant activity (DellaPenna and Pogson 2006). The α-tocopherol isoform, which has the most powerful vitamin E activity of all tocochromanols, could quench up to 120

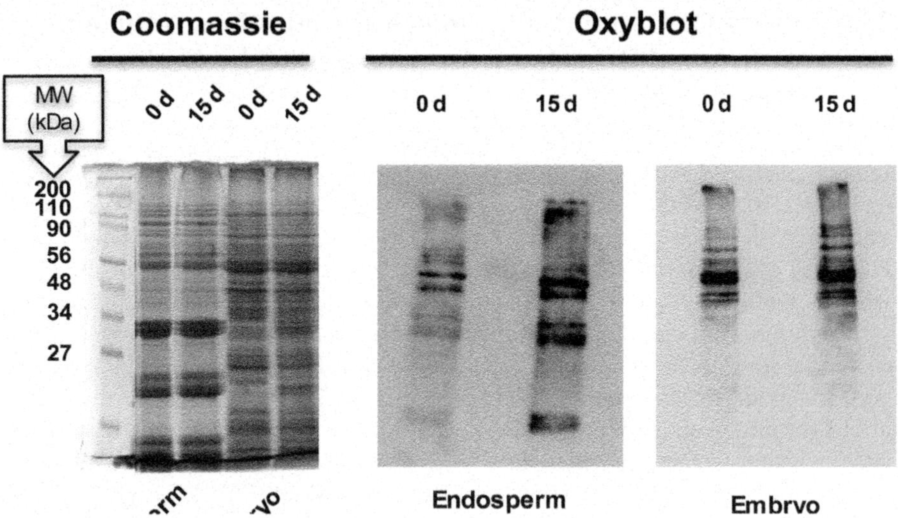

Fig. 23.1 Detection of oxidized proteins in the embryo and endosperm of control or rice seeds that underwent a controlled deterioration treatment (CDT). Rice seeds were equilibrated at 85 % relative humidity (RH) and 20 °C for 3 days. Then, control seeds (0d) were directly dried back at 32 % RH at 20 °C while deteriorated seeds were placed during 15 days (15d) at 85 % RH and 40 °C. **a** Protein stain with Coomassie brilliant blue and **b** western blot related to anti-DNP immunoassay (Oxyblot) are shown. Proteins were then extracted and detection of oxidized proteins by carbonylation was performed by derivatization of protein extracts with 2,4-dinitrophenylhydrazine and immunological detection of the DNP adducts with monoclonal anti-DNP antibody (OxyBlot Oxidized Protein Detection Kit; Chemicon) as described previously (Job et al. 2005). Abbreviation: *d* day

singlet oxygen radicals before being degraded (Fukuzawa et al. 1982). In seeds, tocopherols have been shown to act as key determinants of seed resistance to aging because a mutant impaired in all forms of tocopherol biosynthesis exhibits a considerable vigor loss during accelerated aging (Sattler et al. 2004).

Recently, genetic modulation of the protein L-isoaspartylmethyltransferase 1 (PIMT1) using T-DNA insertion mutant and transgenic lines allowed to demonstrate that protein repair of age-related protein damage contributes importantly to seed longevity (Ogé et al. 2008). The PIMT enzyme activity reduces the level of deamidated and isomerized aspartyl and asparaginyl residues that represent a significant part of the spontaneous damage to proteins (Reissner and Aswad 2003). Alteration of protein conformation by isoaspartyl accumulation has been described as a root case of seed aging. Although oxidative stress is known to enhance these protein alterations, isoaspartyl formation is not chemically an oxidation. This suggests that in addition to oxidative processes other mechanisms contribute to aging and vigor loss. How this metabolism of protein modification control seed survival remains to be elucidated. The identification of altered proteins that are repaired by PIMT will be promising to discover important actors of seed longevity (Chen et al. 2010b).

23.4 Proof-Of-Concept: Physiological Conditions Influence Nutritional Rice Quality as Highlighted by Omics Experiments

In Asia, populations rely heavily on rice for their food intake as rice calorie supply represents 30 % of the total calorie uptake in Asia (IRRI 2009). The rice grain is available either as a polished grain (white rice) or as a whole grain (brown rice). Brown rice is made up of the endosperm, the embryo, and the bran of the grain, whereas white rice is restricted to the endosperm. Unfortunately, most of the consumed rice is white rice that is less interesting from a nutritional point of view. Most of the rice nutritional compounds beneficial to human health are present in the embryo and bran, which are removed by polishing (Fig. 23.2). For instance, the removal of the bran layers eliminates γ-oryzanol (a group of esters of trans-ferulic acid and phytosterols) that has beneficial effects on cholesterol rates and also possesses anti-inflammatory properties (Miller and Engel 2006). Still, the brown rice seed in the dry state is hard to cook and contains phytates, a storage form of phosphate that cannot be assimilated in the human tractus (Hurrell 2003). Phytates are also known to significantly lower the assimilation of several important trace elements, in particular, iron and zinc (Hurrell 2003). As mentioned above, one proposed way to lower the rice seed phytate content is germination (Hurrell 2003).

The seed protein quality is also a critical determinant in the nutritional value of rice. Proteomics constitute an appropriate tool to investigate rice features (see for reviews, Rakwal and Agrawal 2003; Agrawal and Rakwal 2006, 2011; Agrawal et al. 2009). Proteomics approaches have allowed deciphering endosperm protein variations among various rice varieties. The granule-bound starch synthase (Waxy locus) has been proposed as a useful marker to identify differences between *japonica* and *indica* cultivars and associated to the organoleptic properties of cooked rice (Abe et al. 1996). Similarly, a seed bran proteomics study has suggested several proteins as effective biomarkers for aromatic rice (Trisiriroj et al. 2004).

Gene expressed in the rice seed embryo during germination has been documented since a long time (Bhat and Padayatty 1975). During germination, a dramatic gene reprogramming occurs as 67 % of the transcripts change in abundance as compared to transcripts in the dry seed (Howell et al. 2009). More precisely, the major transcript variations occurred between 3 h and 12 h and to a lesser extent between 1 h and 3 h of seed imbibition (Howell et al. 2009). For example, transcripts related to amino acid biosynthesis are strongly accumulated between 3 h and 12 h of germination (Fig. 23.3b). In contrast, the transcript profiles related to amino acid biosynthesis between 0 h and 3h of germination hardly show any change (Fig. 23.3a). Therefore, at the transcriptome level, a major shift occurs at 3 h of germination for genes related to amino acid biosynthesis. In accordance with this hypothesis, metabolome analysis demonstrated that several amino acids strongly accumulate in the embryo of germinating rice seeds during the first 24 h of germination (Fig. 23.3b; Howell et al. 2009). Valine, leucine, and phenylalanine, three indispensable amino acids for human nutrition were among the most accumulated (Fig. 23.3b). It is noteworthy that amino acid accumulation is delayed by a few hours as compared to

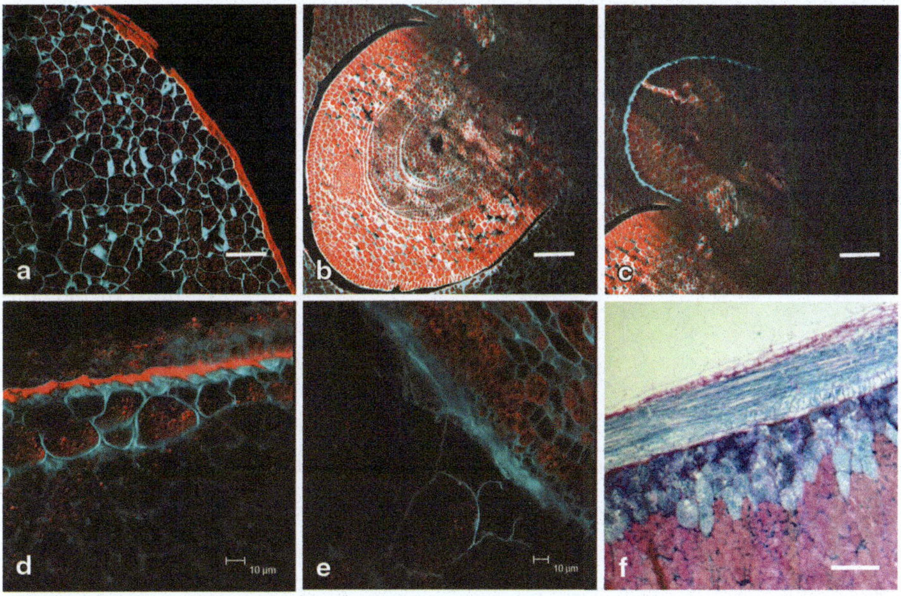

Fig. 23.2 High compartmentalization of some nutritional compounds within the rice seed. Flavonoids were localized by confocal microscopy in the albumen (**a**), and in the rice embryo (**b**, **c**) using DPBA (*red* channel). Scale bars represent 100 µm. (**d**) Lipids were localized in the albumen (**d**) and in the scutellum/albumen region (**e**) by confocal microscopy using Nile Red (*red* channel). In (**a**, **b**, **c**, **d**, **e**), cell walls were visualized (*green* channel) using calcofluor fluorescence. Scale bars represent 10 µm. (**f**) Protein bodies (*blue*) and complex sugars (including starch) were visualized using an APS-Napthol Blue Black coloration on 5 µm thick-sections visualized by optical microscopy. Scale bar represents 100 µm

the amino acid biosynthesis related transcript accumulation (Fig. 23.3). In countries where rice is the main staple food, it is then conceivable that the germination of a small part of that rice could improve the amino acid nutritional value of rice meals.

Iron deficiency is an important health issue in countries where white rice composes most of the daily meals. Indeed, white rice has a poor Fe content in contrast to brown rice. During germination, it was shown that almost half of the Fe-deficiency related genes are expressed at high levels in the rice seed (Nozoye et al. 2007). It is also interesting to observe that strong modifications in Fe metabolism occur in the germinating rice seed. In the rice genome, three genes (*OsNAS1, OsNAS2,* and *OsNAS3*) code for the *nicotianamide synthase*, an enzyme responsible for nicotianamide biosynthesis from AdoMet (Inoue et al. 2003). Nicotianamide plays an important role in metallic ion transport within developing tobacco seeds (Takahashi et al. 2003). The three *OsNAS* transcripts are either present in high-abundance or strongly accumulated during rice seed germination (Nozoye et al. 2007). More precisely, gene-reporter experiments also showed that expression of the three *OsNAS* genes is strongly induced in the embryo of germinating rice seeds (Nozoye et al. 2007). Strikingly, during germination, the *OsNAS1* gene expression is also highly induced in the endosperm tissue. If nicotianamide is accumulated in the endosperm

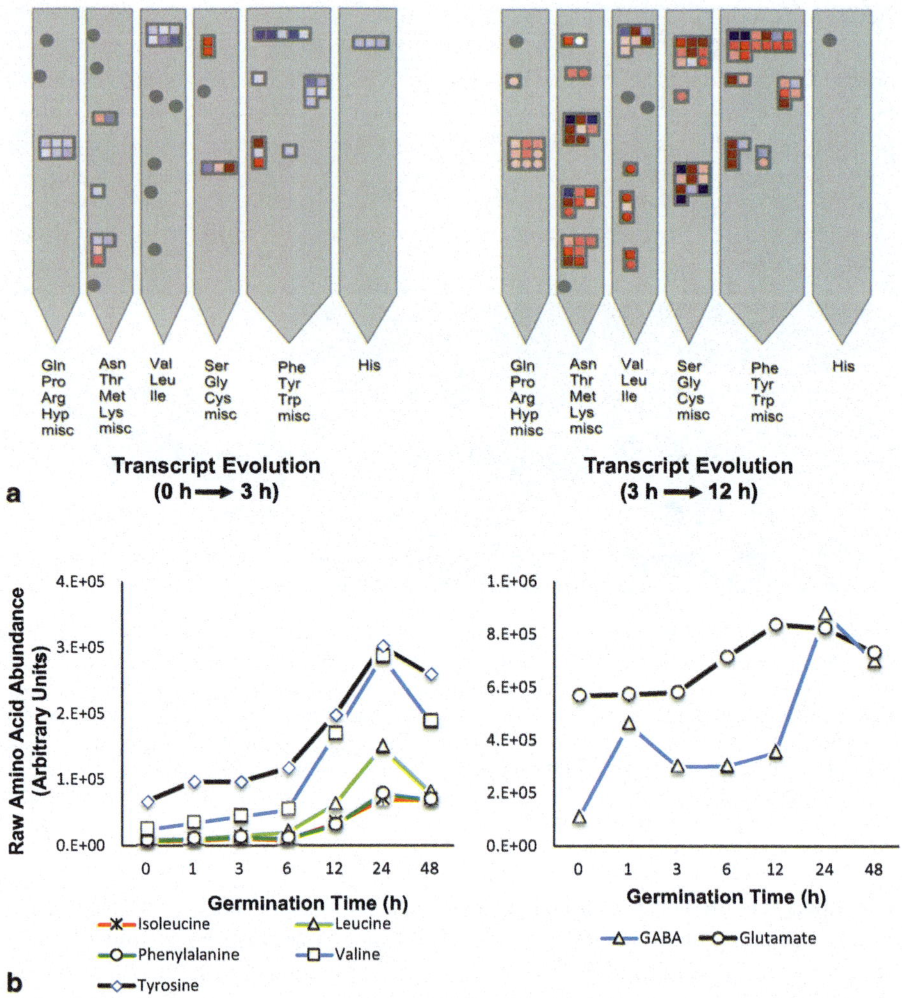

Fig. 23.3 Transcriptome and metabolome survey of rice seed germination highlights major changes in amino acid metabolism. **a** MapMan metabolism overview showing fold-change comparisons of rice embryo amino acid-related transcriptome at different time points. Red color indicates transcripts that are upaccumulated (i.e. more abundant at 3 h than 0 h) while blue color indicates downaccumulated transcripts. **b** Selected amino acid raw abundance that shows accumulation as a function of the rice embryo germination time. Presented data come from the work of Howell et al. (2009)

during germination, it could help to Fe transport and other metallic ions. As stated above, phytic acid is the main storage form of cations (including Fe) and the germination process lowers phytic acid content. Therefore, nicotianamide could subsequently transport the Fe and other cations originating from phytic acid degradation. Thereby, it is reasonable to propose that germinating rice seeds could help to attenuate trace elements malnutrition due to favored white rice consumption.

Induction of rice seed germination seems a promising perspective to improve rice nutritional value. Nevertheless, seed germination vigor and capacity is highly dependent on storage conditions (Rajjou and Debeaujon 2008). In developing countries, rice seeds are often stored in ambient air temperature and humidity, which can be high in main-rice producing countries, such as Vietnam and Thailand. High temperature and humidity are known to cause seed aging that severely impairs seed germination because of the loss of function of protein translation, DNA repair, and detoxifying apparatus (Rajjou et al. 2008; Waterworth et al. 2010). Therefore, in aged rice seeds, the nutritional benefits of germination will be reduced. Concerning seed quality, bad storage conditions increase the seed oxidative status that in turn causes the accumulation of lipid peroxides (Rajjou and Debeaujon 2008). As a consequence, reactive aldehydes can be produced, altering seed nutritional valor (Murthy and Sun 2000; Shin et al. 2009). A comprehensive study of rice seed aging by omics technologies will certainly help to assess the impact of aging on rice seed nutritional value.

23.5 Concluding Remarks

The improvement of seed-based nutrition can be seen as equally important as the increase of the raw agricultural production. To do so, plant biologists need to fuel plant breeders with candidate genes for several key traits, such as seed protein, oil, and vitamin, low phytate, and high germination vigor. To this end, omics technologies can help plant biologists to select the important genes for a trait of interest. Then, functional genomics will be needed to demonstrate the importance of the selected candidate genes. Concerning *Arabidopsis*, the wealth of genetic and technical resources will continue to serve rapeseed and other *Brassicaceae* knowledge transfer. Concerning cereals, several emerging plant models in addition to rice are barley and *Brachypodium distachyon* (Opanowicz et al. 2008; The International Brachypodium Initiative 2010). Thus, the cereal plant biologist community seems far less unified under a single plant model. While this can be seen as a limiting factor for cereal science, comparative genomics will also certainly bring functional information on each single cereal species.

Acknowledgements This work was supported by the French Ministry for the Economy, Industry, and Employment (Ph.D. thesis support to ImenLounifi and post-doctoral position support to Dr. Marc Galland) in the frame of the Nutrice project.

References

Abe T, Gusti RS, Ono M, Sasahara T (1996) Variations in glutelin and high molecular weight endosperm proteins among subspecies of rice (*Oryza sativa* L.) detected by two-dimensional electrophoresis. Genes Genet Syst 71:63–68

Agrawal GK, Rakwal R (2006) Rice proteomics: a cornerstone for cereal food crop proteomes. Mass Spectrom Rev 25:1–53

Agrawal GK, Rakwal R (2011) Rice proteomics: a move toward expanded proteome coverage to comparative and functional proteomics uncovers the mysteries of rice and plant biology. Proteomics 11:1630–1649

Agrawal GK, Jwa NS, Rakwal R (2009) Rice proteomics: ending phase I and the beginning of phase II. Proteomics 9:935–963

Agrawal GK, Job D, Zivy M, Agrawal VP, Bradshaw R, Dunn MJ, Haynes PA, van Wijk KJ, Kikuchi S, Renaut J, Weckwerth W, Rakwal R (2011) Time to articulate a vision for the future of plant proteomics—a global perspective. An initiative for establishing the international plant proteomics organization (INPPO). Proteomics 11:1559–1568

Alonso JM, Stepanova AN (2003) T-DNA mutagenesis in *Arabidopsis*. Methods Mol Biol 236:177–187

Arc E, Galland M, Cueff G, Godin B, Lounifi I, Job D, Rajjou L(2011) Reboot the system thanks to protein post-translational modifications and proteome diversity: how quiescent seeds restart their metabolism to prepare seedling establishment. Proteomics 11:1606–1618

Azevedo RA, Lancien M, Lea PJ (2006) The aspartic acid metabolic pathway, an exciting and essential pathway in plants. Amino acids 30:143–162

Baerenfaller K, Grossmann J, Grobei MA, Hull R, Hirsch-Hoffmann M, Yalovsky S, Zimmermann P, Grossniklaus U, Gruissem W, Baginsky S (2008) Genome-scale proteomics reveals *Arabidopsis thaliana* gene models and proteome dynamics. Science 320:938–941

Bailly C, El-Maarouf-Bouteau H, Corbineau F (2008) From intracellular signaling networks to cell death: the dual role of reactive oxygen species in seed physiology. C R Biol 331:806–814

Baroux C, Spillane C, Grossniklaus U (2002) Evolutionary origins of the endosperm in flowering plants. Genome Biol 3:10261–10265

Baud S, Lepiniec L (2009) Regulation of *de novo* fatty acid synthesis in maturing oilseeds of *Arabidopsis*. Plant Physiol Biochem 47:448–455

Baud S, Wuilleme S, To A, Rochat C, Lepiniec L (2009) Role of WRINKLED1 in the transcriptional regulation of glycolytic and fatty acid biosynthetic genes in *Arabidopsis*. Plant J 60:933–947

Bechtold N, Pelletier G (1998) In planta Agrobacterium-mediated transformation of adult *Arabidopsis thaliana* plants by vacuum infiltration. Methods Mol Biol 82:259–266

Bennetzen JL (2005) Transposable elements, gene creation and genome rearrangement in flowering plants. Curr Opin Genet Dev 15:621–627

Bennetzen JL, Ma J (2003) The genetic colinearity of rice and other cereals on the basis of genomic sequence analysis. Curr Opin Plant Biol 6:128–133

Bent AF (2000) *Arabidopsisin planta* transformation. Uses, mechanisms, and prospects for transformation of other species. Plant Physiol 124:1540–1547

Bewley JD (1997) Seed germination and dormancy. Plant Cell 9:1055–1066

Bohn L, Meyer AS, Rasmussen SK (2008) Phytate: impact on environment and human nutrition. A challenge for molecular breeding. J Zhejiang Univ Sci B 9:165–191

Bhat SP, Padayatty JD (1975) Transcriptional events during early phase of germination of rice embryo. Nature 256:227–228

Bolot S, Abrouk M, Masood-Quraishi U, Stein N, Messing J, Feuillet C, Salse J (2009) The 'inner circle' of the cereal genomes. Curr Opin Plant Biol 12:119–125

Bourgeois M, Jacquin F, Cassecuelle F, Savois V, Belghazi M, Aubert G, Quillien L, Huart M, Marget P, Burstin J (2011) A PQL (protein quantity loci) analysis of mature pea seed proteins identifies loci determining seed protein composition. Proteomics 11:1581–1594

Bruskiewich R, Metz T, McLaren G (2006) Bioinformatics and crop information systems in rice research. Int Rice Res Notes 31:5–12

Candela H, Hake S (2008) The art and design of genetic screens: maize. Nat Rev Genet 9:192–203

Chai G, Bai Z, Wei F, King GJ, Wang C, Shi L, Dong C, Chen H, Liu S (2010) *BrassicaGLABRA2* genes: analysis of function related to seed oil content and development of functional markers. Theor Appl Genet 120:1597–1610

Chantret N, Salse J, Sabot F, Rahman S, Bellec A, Laubin B, Dubois I, Dossat C, Sourdille P, Joudrier P, Gautier MF, Cattolico L, Beckert M, Aubourg S, Weissenbach J, Caboche M, Bernard M, Leroy P, Chalhoub B (2005) Molecular basis of evolutionary events that shaped the hardness locus in diploid and polyploid wheat species (*Triticum* and *Aegilops*). Plant Cell 17:1033–1045

Charles D (2006) Species conservation. A 'forever' seed bank takes root in the Arctic. Science 312:1730–1731

Chen T, Nayak N Majee SM, Lowenson J, Schäfermeyer KR, Eliopoulos AC, Lloyd TD, Dinkins R, Perry SE, Forsthoefel NR, Clarke SG, Vernon DM, Zhou ZS, Rejtar T, Downie AB (2010b) Substrates of the *Arabidopsis thaliana* protein isoaspartyl methyltransferase 1 identified using phage display and biopanning. J Biol Chem 285:37281–37292

Cordain L (1999) Cereal grains: humanity's double edged sword. In: Artemis S (ed) Evolutionary aspects of nutrition and health diet, exercise, genetics and chronic disease, vol 84. Karger, Basel, pp 19–73

Dao TH (2003) Polyvalent cation effects on myo-inositol hexakis dihydrogenphosphate enzymatic dephosphorylation in dairy wastewater. J Environ Qual 32:694–701

DellaPenna D, Pogson BJ (2006) Vitamin synthesis in plants: tocopherols and carotenoids. Annu Rev Plant Biol 57:711–738

Devaiah SP, Pan X, Hong Y, Roth M, Welti R, Wang X (2007) Enhancing seed quality and viability by suppressing phospholipase D in *Arabidopsis*. Plant J 50:950–957

Droc G, Ruiz M, Larmande P, Pereira A, Piffanelli P, Morel JB, Dievart A, Courtois B, Guiderdoni E, Périn C (2006) OryGenesDB: a database for rice reverse genetics. Nucl Acids Res 34:D736–D740

Emami K, Morris NJ, Cockell SJ, Golebiowska G, Shu QY, Gatehouse AM (2010) Changes in protein expression profiles between a low phytic acid rice (*Oryza sativa* L. Ssp. japonica) line and its parental line: a proteomic and bioinformatic approach. J Agric Food Chem 58:6912–6922

Ermolaeva MD, Wu M, Eisen JA, Salzberg SL (2003) The age of the *Arabidopsis thaliana* genome duplication. Plant Mol Biol 51:859–866

Fait A, Angelovici R, Less H, Ohad I, Urbanczyk-Wochniak E, Fernie AR, Galili G (2006) *Arabidopsis* seed development and germination is associated with temporally distinct metabolic switches. Plant Physiol 142:839–854

Feschotte C (2008) Transposable elements and the evolution of regulatory networks. Nat Rev Genet 9:397–405

Feschotte C, Jiang N, Wessler SR (2002) Plant transposable elements: where genetics meets genomics. Nat Rev Genet 3:329–341

Feuillet C, Langridge P, Waugh R (2008) Cereal breeding takes a walk on the wild side. Trends Genet 24:24–32

Fincher GB (1989) Molecular and cellular biology associated with endosperm mobilization in geminating cereal grain. Annu Rev Plant Physiol Plant Mol Biol 40:305–346

Finch-Savage WE, Leubner-Metzger G (2006) Seed dormancy and the control of germination. New Phytol 171:501–523

Finnie C, Svensson B (2009) Barley seed proteomics from spots to structures. J Proteomics 72:315–324

Fitzgerald MA, McCouch SR, Hall RD (2009) Not just a grain of rice: the quest for quality. Trends Plant Sci 14:133–139

Focks N, Benning C (1998) Wrinkled 1: a novel, low seed-soil-mutant *Arabidopsis* with a deficiency in the seed-specific regulation of carbohydrate metabolism. Plant Physiol 118:91–101

Frohlich MW, Chase MW (2007) After a dozen years of progress the origin of angiosperms is still a great mystery. Nature 450:1184–1189

Fukuzawa K, Tokumura A, Ouchi S, Tsukatani H (1982) Antioxidant activities of tocopherols on Fe^{2+}-ascorbate-induced lipid peroxidatoin in lecithin liposomes. Lipids 17:511–513

Gallardo K, Job C, Groot SPC, Puype M, Demol H, Vandekerckhove J, Job D (2001) Proteomic analysis of *Arabidopsis* seed germination and priming. Plant Physiol 126:835–848

Gallardo K, Firnhaber C, Zuber H, Hericher D, Belghazi M, Henry C, Kuster H, Thompson RD (2007) A combined proteome and transcriptome analysis of developing *Medicago truncatula* seeds. Mol Cell Proteomics 6:2165–2179

Germain V, Rylott EL, Larson TR, Sherson SM, Bechtold N, Carde JP, Bryce JH, Graham IA, Smith SM (2001) Requirement for 3-ketoacyl-CoA thiolase-2 in peroxisome development, fatty acid beta-oxidation and breakdown of triacylglycerol in lipid bodies of *Arabidopsis* seedlings. Plant J 28:1–12

Goff SA, Ricke D, Lan TH, Presting G, Wang R, Dunn M, Glazebrook J, Sessions A, Oeller P, Varma H, Hadley D, Hutchison D, Martin C, Katagiri F, Lange BM, Moughamer T, Xia Y, Budworth P, Zhong J, Miguel T, Paszkowski U, Zhang S, Colbert M, Sun WL, Chen L, Cooper B, Park S, Wood TC, Mao L, Quail P, Wing R, Dean R, Yu Y, Zharkikh A, Shen R, Sahasrabudhe S, Thomas A, Cannings R, Gutin A, Pruss D, Reid J, Tavtigian S, Mitchell J, Eldredge G, Scholl T, Miller RM, Bhatnagar S, Adey N, Rubano T, Tusneem N, Robinson R, Feldhaus J, Macalma T, Oliphant A, Briggs S (2002) A draft sequence of the rice genome (*Oryza sativa* L. ssp. japonica). Science 296:92–100

Gutierrez L, Van Wuytswinkel O, Castelain M, Bellini C (2007) Combined networks regulating seed maturation. Trends Plant Sci 12:294–300

Hajduch M, Casteel JE, Tang S, Hearne LB, Knapp S, Thelen JJ (2007) Proteomic analysis of near-isogenic sunflower varieties differing in seed oil traits. J Proteome Res 6:3232–3241

Herman EM, Larkins BA (1999) Protein storage bodies and vacuoles. Plant Cell 11: 601–614

Holdsworth MJ, Bentsink L, Soppe WJJ (2008a) Molecular networks regulating *Arabidopsis* seed maturation, after-ripening, dormancy and germination. New Phytol 179:33–54

Holdsworth MJ, Finch-Savage WE, Grappin P, Job D (2008b) Postgenomics dissection of seed dormancy and germination. Trends Plant Sci 13:7–13

Howell KA, Narsai R, Carroll A, Ivanova A, Lohse M, Usadel B, Millar AH, Whelan J (2009) Mapping metabolic and transcript temporal switches during germination in rice highlights specific transcription factors and the role of RNA instability in the germination process. Plant Physiol 149:961–980

Hsieh TF, Shin J, Uzawa R, Silva P, Cohen S, Bauer MJ, Hashimoto M, Kirkbride RC, Harada JJ, Zilberman D, Fischer RL (2011) Regulation of imprinted gene expression in *Arabidopsis* endosperm. Proc Natl Acad Sci U S A 108:1755–1762

Hurrell RF (2003) Influence of vegetable protein sources on trace element and mineral bioavailability. J Nutr 133:2973S–2977S

International Rice Genome Sequencing Project (2005) The map-based sequence of the rice genome. Nature 436:793–800

International Rice Research Institute, World Rice Statistics (2009) http://ricestat.irri.org:8080/wrs/. Accessed 1 Oct 2012

Inoue H, Higuchi K, Takahashi M, Nakanishi H, Mori S, Nishizawa NK (2003) Three rice nicotianamine synthase genes, OsNAS1, OsNAS2, and OsNAS3 are expressed in cells involved in long-distance transport of iron and differentially regulated by iron. Plant J 36:366–381

Itoh T, Tanaka T, Barrero RA, Yamasaki C, Fujii Y, Hilton PB, Antonio BA, Aono H, Apweiler R, Bruskiewich R, Bureau T, Burr F, Costa de Oliveira A, Fuks G, Habara T, Haberer G, Han B, Harada E, Hiraki AT, Hirochika H, Hoen D, Hokari H, Hosokawa S, Hsing Y, Ikawa H, Ikeo K, Imanishi T, Ito Y, Jaiswal P, Kanno M, Kawahara Y, Kawamura T, Kawashima H, Khurana JP, Kikuchi S, Komatsu S, Koyanagi KO, Kubooka H, Lieberherr D, Lin YC, Lonsdale D, Matsumoto T, Matsuya A, McCombie WR, Messing J, Miyao A, Mulder N, Nagamura Y, Nam J, Namiki N, Numa H, Nurimoto S, O'Donovan C, Ohyanagi H, Okido T, Oota S, Osato N, Palmer LE, Quetier F, Raghuvanshi S, Saichi N, Sakai H, Sakai Y, Sakata K, Sakurai T, Sato F, Sato Y, Schoof H, Seki M, Shibata M, Shimizu Y, Shinozaki K, Shinso Y, Singh NK, Smith-White B, Takeda JI, Tanino M, Tatusova T, Thongjuea S, Todokoro F, Tsugane M, Tyagi AK, Vanavichit A, Wang A, Wing RA, Yamaguchi K, Yamamoto M, Yamamoto N, Yu Y, Zhang H, Zhao Q, Higo K, Burr B, Gojobori T, Sasaki T (2007) Curated genome annotation of *Oryza sativa* ssp. japonica and comparative genome analysis with *Arabidopsis thaliana*. Genome Res 17:175–183

Izawa T, Shimamoto K (1996) Becoming a model plant: the importance of rice to plant science. Trends Plant Sci 1:95–99

Jain RK, Coffey M, Lai K, Kumar A, MacKenzie SL (2000) Enhancement of seed oil content by expression of glycerol-3-phosphate acyltransferase genes. Biochem Soc Trans 28:958–961

Jaiswal P, Ni J, Yap I, Ware D, Spooner W, Youens-Clark K, Ren L, Liang C, Zhao W, Ratnapu K, Faga B, Canaran P, Fogleman M, Hebbard C, Avraham S, Schmidt S, Casstevens TM, Buckler ES, Stein L, McCouch S (2006) Gramene: a bird's eye view of cereal genomes. Nucl Acids Res 34:D717–D723

Jako C, Kumar A, Wei Y, Zou J, Barton DL, Giblin EM, Covello PS, Taylor DC (2001) Seed-specific over-expression of an *Arabidopsis* cDNA encoding a diacylglycerol acyltransferase enhances seed oil content and seed weight. Plant Physiol 126:861–874

Jeon JS, Ryoo N, Hahn TR, Walia H, Nakamura Y (2010) Starch biosynthesis in cereal endosperm. Plant Physiol Biochem 48:383–392

Job C, Rajjou L, Lovigny Y, Belghazi M, Job D (2005) Patterns of protein oxidation in *Arabidopsis* seeds and during germination. Plant Physiol 138:790–802

Jolliffe NA, Craddock CP, Frigerio L (2005) Pathways for protein transport to seed storage vacuoles. Biochem Soc Trans 33:1016–1018

Jung KH, An G, Ronald PC (2008) Towards a better bowl of rice: assigning function to tens of thousands of rice genes. Nat Rev Genet 9:91–101

Kashkush K, Feldman M, Levy AA (2003) Transcriptional activation of retrotransposons alters the expression of adjacent genes in wheat. Nat Genet 33(1):102–106

Kim ST, Kang SY, Wang Y, Kim SG, Hwang du H, Kang KY (2008) Analysis of embryonic proteome modulation by GA and ABA from germinating rice seeds. Proteomics 8:3577–3587

Kondou Y, Higuchi M, Takahashi S, Sakurai T, Ichikawa T, Kuroda H, Yoshizumi T, Tsumoto Y, Horii Y, Kawashima M, Hasegawa Y, Kuriyama T, Matsui K, Kusano M, Albinsky D, Takahashi H, Nakamura Y, Suzuki M, Sakakibara H, Kojima M, Akiyama K, Kurotani A, Seki M, Fujita M, Enju A, Yokotani N, Saitou T, Ashidate K, Fujimoto N, Ishikawa Y, Mori Y, Nanba R, Takata K, Uno K, Sugano S, Natsuki J, Dubouzet JG, Maeda S, Ohtake M, Mori M, Oda K, Takatsuji H, Hirochika H, Matsui M (2009) Systematic approaches to using the FOX hunting system to identify useful rice genes. Plant J 57:883–894

Koornneef M, Meinke D (2010) The development of *Arabidopsis* as a model plant. Plant J 61:909–291

Koornneef M, Hanhart CJ, Hilhorst HWM, Karssen CM (1989) *In vivo* inhibition of seed development and reserve protein accumulation in recombinants of abscisic acid biosynthesis and responsiveness mutants in *Arabidopsis thaliana*. Plant Physiol 90:463–469

Krishnan HB, Natarajan SS (2009) A rapid method for depletion of rubisco from soybean (*Glycine max*) leaf for proteomic analysis of lower abundance proteins.Phytochemistry 70:1958–1964

Laibach F. (1943) *Arabidopsis thaliana* (L.) Heynh. als object fur genetische und entwicklungsphysiologische untersuchungen. Bot Archiv 44:439–455

Le BH, Cheng C, Bui AQ, Wagmaister JA, Henry KF, Pelletier J, Kwong L, Belmonte M, Kirkbride R, Horvath S, Drews GN, Fischer RL, Okamuro JK, Harada JJ, Goldberg RB (2010) Global analysis of gene activity during *Arabidopsis* seed development and identification of seed-specific transcription factors. Proc Natl Acad Sci U S A 107:8063–8070

Linkies A, Graeber K, Knight C, Leubner-Metzger G (2010) The evolution of seeds. New Phytol 186: 817–831

Martínez-Barajas E, Delatte T, Schluepmann H, de Jong GJ, Somsen GW, Nunes C, Primavesi LF, Coello P, Mitchell RA, Paul MJ (2011) Wheat grain development is characterized by remarkable trehalose 6-phosphate accumulation pregrain filling: tissue distribution and relationship to SNF1-related protein kinase1 activity. Plant Physiol 156: 373–381

McCourt P, Benning C (2010) *Arabidopsis*: A rich harvest 10 years after completion of the genome sequence. Plant J 61: 905–908

Mène-Saffrané L, DellaPenna D (2010) Biosynthesis, regulation and functions of tocochromanols in plants. Plant Physiol Biochem 48:301–309

Messing J, Dooner HK (2006) Organization and variability of the maize genome. Curr Opin Plant Biol 9:157–163

Meurs C, Basra AS, Karssen CM, van Loon LC (1992) Role of abscisic acid in the induction of desiccation tolerance in developing seeds of *Arabidopsis thaliana*. Plant Physiol 98:1484–1493

Miernyk JA, Hajduch M (2011) Seed proteomics. J Proteomics 74:389–400

Miller A., Engel KH (2006) Content of gamma-oryzanol and composition of steryl ferulates in brown rice (*Oryza sativa* L.) of European origin. J Agric Food Chem 54:8127–8133

Molvig L, Tabe LM, Eggum BO, Moore AE, Craig S, Spencer D, Higgins TJV (1997) Enhanced methionine levels and increased nutritive value of seeds of transgenic lupins (*Lupinus angustifolius* L.) expressing a sunflower seed albumin gene. Proc Natl Acad Sci U S A 94:8393–8398

Moore G, Devos KM, Wang Z, Gale MD (1995) Cereal genome evolution. Grasses, line up and form a circle. Curr Biol 5:737–739

Muller-Landau HC (2003) Seeds of understanding of plant diversity. Proc Natl Acad Sci U S A 100:1469–1471

Murthy UM, Sun WQ (2000) Protein modification by Amadori and Maillard reactions during seed storage: roles of sugar hydrolysis and lipid peroxidation. J Exp Bot 51(348):1221–1228

Müntz K (1998) Deposition of storage proteins. Plant Mol Biol 38:77–99

Nambara E, Akazawa T, McCourt P (1991) Effects of the gibberellin biosynthetic inhibitor uniconazol on mutants of *Arabidopsis*. Plant Physiol 97:736–738

North H, Baud S, Debeaujon I, Dubos C, Dubreucq B, Grappin P, Jullien M, Lepiniec L, Marion-Poll A, Miquel M, Rajjou L, Routaboul JM, Caboche M (2010) *Arabidopsis* seed secrets unravelled after a decade of genetic and omics-driven research. Plant J 61:971–981

Nozoye T, Inoue H, Takahashi M, Ishimaru Y, Nakanishi H, Mori S, Nishizawa NK (2007) The expression of iron homeostasis-related genes during rice germination. Plant Mol Biol 64:35–47

Ogé L, Bourdais G, Bove J, Collet B, Godin B, Granier F, Boutin JP, Job D, Jullien M, Grappin P (2008) Protein repair L-isoaspartyl methyltransferase 1 is involved in both seed longevity and germination vigor in *Arabidopsis*. Plant Cell 20:3022–3037

Ohyanagi H, Tanaka T, Sakai H, Shigemoto Y, Yamaguchi K, Habara T, Fujii Y, Antonio BA, Nagamura Y, Imanishi T, Ikeo K, Itoh T, Gojobori T, Sasaki T (2006) The Rice Annotation Project Database (RAP-DB): hub for *Oryza sativa* ssp. japonica genome information. Nucl Acids Res 34:D741–D744

O'Neill CM, Gill S, Hobbs D, Morgan C, Bancroft I (2003) Natural variation for seed oil composition in *Arabidopsis thaliana*. Phytochemistry 64:1077–1090

Ooms JJJ, Léon-Kloosterziel KM, Bartels D, Koornneef M, Karssen CM (1993) Acquisition of desiccation tolerance and longevity in seeds of *Arabidopsis thaliana* (A comparative study using abscisic acid-intensive abi3 mutants). Plant Physiol 102:1185–1191

Opanowicz M, Vain P, Draper J, Parker D, Doonan JH (2008) *Brachypodium distachyon*: making hay with a wild grass. Trends Plant Sci 13:172–177

Ouyang S, Zhu W, Hamilton J, Lin H, Campbell M, Childs K, Thibaud-Nissen F, Malek RL, Lee Y, Zheng L, Orvis J, Haas B, Wortman J, Buell CR (2007) The TIGR rice genome annotation resource: improvements and new features. Nucl Acids Res 35:D883–D887

Pinfield-Wells H, Rylott EL, Gilday AD, Graham S, Job K, Larson TR, Graham IA (2005) Sucrose rescues seedling establishment but not germination of *Arabidopsis* mutants disrupted in peroxisomal fatty acid catabolism. Plant J 43:861–872

Prieto-Dapena P, Castaño R, Almoguera C, Jordano J (2006) Improved resistance to controlled deterioration in transgenic seeds. Plant Physiol 142:1102–1112

Rabinowicz PD, Bennetzen JL (2006) The maize genome as a model for efficient sequence analysis of large plant genomes. Curr Opin Plant Biol 9:149–156

Rajjou L, Debeaujon I (2008) Seed longevity: survival and maintenance of high germination ability of dry seeds. C R Biologies 331:796–805

Rajjou L, Belghazi M, Huguet R, Robin C, Moreau A, Job C, Job D (2006) Proteomic investigation of the effect of salicylic acid on *Arabidopsis* seed germination and establishment of early defense mechanisms. Plant Physiol 141:910–923

Rajjou L, Lovigny Y, Groot SPC, Belghazi M, Job C, Job D (2008) Proteome-wide characterization of seed aging in *Arabidopsis*: a comparison between artificial and natural aging protocols. Plant Physiol 148:620–641

Rakwal R, Agrawal GK (2003) Rice proteomics: current status and future perspectives. Electrophoresis 24:3378–3389

Reissner KJ, Aswad DW (2003) Deamidation and isoaspartate formation in proteins: unwanted alterations or surreptitious signals? Cell Mol Life Sci 60:1281–1295

Rensink WA, Buell CR (2004) *Arabidopsis* to rice. Applying knowledge from a weed to enhance our understanding of a crop species. Plant Physiol 135:622–629

Sasaki T, Burr B (2000) International Rice Genome Sequencing Project: the effort to completely sequence the rice genome. Curr Opin Plant Biol 3:138–141

Sattler SE, Gilliland LU, Magallanes-Lundback M, Pollard M, DellaPenna D (2004) Vitamin E is essential for seed longevity and for preventing lipid peroxidation during germination. Plant Cell 16:1419–1432

Sattler SE, Mène-Saffrané L, Farmer EE, Krischke M, Mueller MJ, DellaPenna D (2006) Nonenzymatic lipid peroxidation reprograms gene expression and activates defense markers in *Arabidopsis* tocopherol-deficient mutants. Plant Cell 18:3706–3720

Schneider C (2005) Chemistry and biology of vitamin E. Mol Nutr Food Res 49:7–30

Shen B, Sinkevicius KW, Selinger DA, Tarczynski MC (2006) The homeobox gene *GLABRA2* affects seed oil content in *Arabidopsis*. Plant Mol Biol 60:377–387

Shimamoto K, Kyozuka J (2002) Rice as a model for comparative genomics of plants. Annu Rev Plant Biol 53:399–419

Shin JH, Kim SR, An G (2009) Rice aldehyde dehydrogenase 7 is needed for seed maturation and viability. Plant Physiol 149:905–915

Somerville C, Koornneef M (2002) A fortunate choice: the history of *Arabidopsis* as a model plant. Nat Rev Genet 3:883–889

Sreenivasulu N, Graner A, Wobus U (2008a) Barley genomics: an overview. Int J Plant Genomics 2008: 486258

Sreenivasulu N, Usadel B, Winter A, Radchuk V, Scholz U, Stein N, Weschke W, Strickert M, Close TJ, Stitt M, Graner A, Wobus U (2008b) Barley grain maturation and germination: metabolic pathway and regulatory network commonalities and differences highlighted by new MapMan/PageMan profiling tools. Plant Physiol 146:1738–1758

Sreenivasulu N, Borisjuk L, Junker BH, Mock HP, Rolletschek H, Seiffert U, Weschke W, Wobus U (2010) Barley grain development: toward an integrative view. Int Rev Cell Mol Biol 281:49–89

Sung HG, Shin HT, Ha JK, Lai HL, Cheng KJ, Lee JH (2005) Effect of germination temperature on characteristics of phytase production from barley. Bioresour Technol 96:1297–1303

Takahashi M, Terada Y, Nakai I, Nakanishi H, Yoshimura E, Mori S, Nishizawa NK (2003) Role of nicotianamine in the intracellular delivery of metals and plant reproductive development. Plant Cell 15:1263–1280

Talon M, Koornneef M, Zeevaart JA (1990) Endogenous gibberellins in *Arabidopsis thaliana* and possible steps blocked in the biosynthetic pathways of the semidwarf ga4 and ga5 mutants. Proc Natl Acad Sci U S A 87:7983–7987

Tanaka T, Antonio BA, Kikuchi S, Matsumoto T, Nagamura Y, Numa H, Sakai H, Wu J, Itoh T, Sasaki T, Aono R, Fujii Y, Habara T, Harada E, Kanno M, Kawahara Y, Kawashima H, Kubooka H, Matsuya A, Nakaoka H, Saichi N, Sanbonmatsu R, Sato Y, Shinso Y, Suzuki M, Takeda JI, Tanino M, Todokoro F, Yamaguchi K, Yamamoto N, Yamasaki C, Imanishi T, Okido T, Tada M, Ikeo K, Tateno Y, Gojobori T, Lin YC, Wei FJ, Hsing YI, Zhao Q, Han B, Kramer MR, McCombie RW, Lonsdale D, O'Donovan CC, Whitfield EJ, Apweiler R, Koyanagi KO, Khurana JP, Raghuvanshi S, Singh NK, Tyagi AK, Haberer G, Fujisawa M, Hosokawa S, Ito Y, Ikawa H, Shibata M, Yamamoto M, Bruskiewich RM, Hoen DR, Bureau TE, Namiki N, Ohyanagi H, Sakai Y, Nobushima S, Sakata K, Barrero RA, Sato Y, Souvorov A, Smith-White B, Tatusova T, An S, An G, OOta S, Fuks G, Messing J, Christie KR, Lieberherr D, Kim H, Zuccolo A, Wing RA, Nobuta K, Green PJ, Lu C, Meyers BC, Chaparro C, Piegu B, Panaud

O, Echeverria M (2008) The rice annotation project database (RAP-DB): 2008 update. Nucl Acids Res. 36:D1028–D1033

The *Arabidopsis* Genome Initiative (2000) Analysis of the genome sequence of the flowering plant *Arabidopsis thaliana.* Nature 408:796–815

The International Brachypodium Initiative (2010) Genome sequencing and analysis of the model grass *Brachypodium distachyon*. Nature 463:763–768

Tranbarger TJ, Dussert S, Joët T, Argout X, Summo M, Champion A, Cros D, Omore A, Nouy B, Morcillo F (2011) Regulatory mechanisms underlying oil palm fruit mesocarp maturation, ripening, and functional specialization in lipid and carotenoid metabolism. Plant Physiol 156:564–584

Tresset A, Vigne JD (2011) Last hunter-gatherers and first farmers of Europe. C R Biol 334:182–189

Trisiriroj A, Jeyachok N, Chen ST (2004) Proteomics characterization of different bran proteins between aromatic and nonaromatic rice (*Oryza sativa* L. ssp. indica). Proteomics 4:2047–2057

Troncoso-Ponce MA, Kruger NJ, Ratcliffe G, Garcés R, Martínez-Force E (2009) Characterization of glycolytic initial metabolites and enzyme activities in developing sunflower (*Helianthus annuus* L.) seeds. Phytochemistry 70:1117–1122

Valentine L (2003) Agrobacterium tumefaciens and the plant: the David and Goliath of modern genetics. Plant Physiol 133:948–955

Vensel WH, Tanaka CK, Cai N, Wong JH, Buchanan BB, Hurkman WJ (2005) Developmental changes in the metabolic protein profiles of wheat endosperm. Proteomics 5:1594–1611

Venu RC, Sreerekha MV, Nobuta K, Belo A, Ning Y, An G, Meyers BC, Wang GL (2011) Deep sequencing reveals the complex and coordinated transcriptional regulation of genes related to grain quality in rice cultivars. BMC Genomics 12:190

Vigeolas H, Waldeck P, Zank T, Geigenberger P (2007) Increasing seed oil content in oil-seed rape (*Brassica napus* L.) by over-expression of a yeast glycerol-3-phosphate dehydrogenase under the control of a seed-specific promoter. Plant Biotechnol J 5:431–441

Waddell J (1958) Supplementation of plant proteins with amino acids. In: Altschul AM (ed) Processed plant protein foodstuffs. Academic, New York, pp 307–351

Wan Y, Underwood C, Toole G, Skeggs P, Zhu T, Leverington M, Griffiths S, Wheeler T, Gooding M, Poole R, Edwards KJ, Gezan S, Welham S, Snape J, Mills EN, Mitchell RA, Shewry PR (2009) A novel transcriptomic approach to identify candidate genes for grain quality traits in wheat. Plant Biotechnol J 7:401–410

Ware DH, Jaiswal P, Ni J, Yap IV, Pan X, Clark KY, Teytelman L, Schmidt SC, Zhao W, Chang K, Cartinhour S, Stein LD, McCouch SR (2002) Gramene, a tool for grass genomics. Plant Physiol 130:1606–1613

Waterworth WM, Masnavi G, Bhardwaj RM, Jiang Q, Bray CM, West CE (2010) A plant DNA ligase is an important determinant of seed longevity. Plant J 63:848–860

Weitbrecht K, Müller K, Leubner-Metzger G (2011) First off the mark: early seed germination. J Exp Bot 62:3289–3309

Xu XH, Zhao HJ, Liu QL, Frank T, Engel KH, An G, Shu QY (2009) Mutations of the multi-drug resistance-associated protein ABC transporter gene 5 result in reduction of phytic acid in rice seeds. Theor Appl Genet 119:75–83

Yamada C, Izumi H, Hirano J, Mizukuchi A, Kise M, Matsuda T, Kato Y (2005) Degradation of soluble proteins including some allergens in brown rice grains by endogenous proteolytic activity during germination and heat-processing. Biosci Biotechnol Biochem 69:1877–1883

Yamaguchi S, Kamiya Y, Sun T (2001) Distinct cell-specific expression patterns of early and late gibberellin biosynthetic genes during *Arabidopsis* seed germination. Plant J 28:443–453

Yamakawa H, Hirose T, Kuroda M, Yamaguchi T (2007) Comprehensive expression profiling of rice grain filling-related genes under high temperature using DNA microarray. Plant Physiol 144:258–277

Yang P, Li X, Wang X, Chen H, Chen F, Shen S (2007) Proteomic analysis of rice (*Oryza sativa*) seeds during germination. Proteomics 7:3358–3368

Yu J, Wang J, Lin W, Li S, Li H, Zhou J, Ni P, Dong W, Hu S, Zeng C, Zhang J, Zhang Y, Li R, Xu Z, Li S, Li X, Zheng H, Cong L, Lin L, Yin J, Geng J, Li G, Shi J, Liu J, Lv H, Li J, Wang J, Deng Y, Ran L, Shi X, Wang X, Wu Q, Li C, Ren X, Wang J, Wang X, Li D, Liu D, Zhang X, Ji Z, Zhao W, Sun Y, Zhang Z, Bao J, Han Y, Dong L, Ji J, Chen P, Wu S, Liu J, Xiao Y, Bu D, Tan J, Yang L, Ye C, Zhang J, Xu J, Zhou Y, Yu Y, Zhang B, Zhuang S, Wei H, Liu B, Lei M, Yu H, Li Y, Xu H, Wei S, He X, Fang L, Zhang Z, Zhang Y, Huang X, Su Z, Tong W, Li J, Tong Z, Li S, Ye J, Wang L, Fang L, Lei T, Chen C, Chen H, Xu Z, Li H, Huang H, Zhang F, Xu H, Li N, Zhao C, Li S, Dong L, Huang Y, Li L, Xi Y, Qi Q, Li W, Zhang B, Hu W, Zhang Y, Tian X, Jiao Y, Liang X, Jin J, Gao L, Zheng W, Hao B, Liu S, Wang W, Yuan L, Cao M, McDermott J, Samudrala R, Wang J, Wong GKS, Yang H (2005) The Genomes of *Oryza sativa*: a history of duplications. PLoS Biol 3:e38

Yuan Q, Ouyang S, Liu J, Suh B, Cheung F, Sultana R, Lee D, Quackenbush J, Buell CR (2003) The TIGR rice genome annotation resource: annotating the rice genome and creating resources for plant biologists. Nucl Acids Res 31:229–233

Zhang J, Li C, Wu C, Xiong L, Chen G, Zhang Q, Wang S (2006) RMD: a rice mutant database for functional analysis of the rice genome. Nucl Acids Res 34:D745–D748

Zou J, Katavic V, Giblin EM, Barton DL, MacKenzi SL, Keller WA, Hu X, Taylor DC (1997) Modification of seed oil content and acyl composition in the brassicaceae by expression of a yeast sn-2 acyltransferase gene. Plant Cell 9:909–923

Zupan JR, Zambryski P (1995) Transfer of T-DNA from Agrobacterium to the plant cell. Plant Physiol 107:1041–1047

Chapter 24
Using Genome-Enabled Technologies to Address Allergens in Seeds of Crop Plants: Legumes as a Case Study

Lena Y. C. Soo, Nicole E. Walczyk and Penelope M. C. Smith

Abstract In this chapter, we discuss how genome-enabled technologies have improved identification of legume seed allergens and are contributing to the development of hypoallergenic legume crops. We show how proteomic identification of allergens will be driven by the availability of plant genomes. The expression of the major allergens can be reduced or eliminated by gene silencing, mutation, or genome editing. The limitation of these technologies to agricultural performance and food quality, and the clinical consequences of modified food products are described.

Keywords Legumes · Seed allergens · Food allergy · Hypoallergenic · Gene silencing · Mutagenesis

24.1 Introduction

24.1.1 Legume Seeds as Food

Legume seeds are an important source of protein in many regions of the world (Duranti 2006). They are becoming an increasingly popular food, particularly in developed countries, due to the benefits associated with low meat consumption (Duranti and Scarafoni 1999). Legume seeds, common as human foods, are peas, soybean, peanuts, lentils, beans, and chickpeas, while other legumes (such as lupin) are gradually being introduced to the consumer (Magni et al. 2004). Legume seeds are known to be beneficial to health when incorporated into human diets (Chassaigne et al. 2007; Trinidad et al. 2010; Venn and Mann 2004) both for their protein and other components. Besides protein, legume seeds contain trace nutrients that have been linked to disease prevention. Isoflavones, a plant defence compound found prevalently in legumes, has estrogenic, antiangiogenic, antioxidant, and anticancer

P. M. C. Smith (✉) · L. Y. C. Soo · N. E. Walczyk
School of Biological Sciences, University of Sydney, Sydney, Australia
e-mail: penny.smith@sydney.edu.au

N. E. Walczyk
Department of Chemistry and Biomolecular Sciences, Macquarie University, Macquarie, Australia

G. K. Agrawal, R. Rakwal (eds.), *Seed Development: OMICS Technologies toward Improvement of Seed Quality and Crop Yield*,
DOI 10.1007/978-94-007-4749-4_24, © Springer Science+Business Media Dordrecht 2012

properties (Dixon 1999; Dixon and Ferreira 2002). A link has been proposed between consumption of soy isoflavones and the reduced risk of breast and prostate cancer (Adlercreutz 1998; Lamartiniere 2000; Setchell and Cassidy 1999). Tocopherols, which are phytochemicals found in grain legumes, are involved in the prevention of cardiovascular and eye pathologies (Boschin and Arnoldi 2011).

24.1.2 Allergy to Legume Seeds

Despite their benefits, some components of legume seeds cause allergenic reactions in susceptible people. Peanut and soybean are included in the 'Big 8' food allergens, which are responsible for 90 % of all food allergies in USA (Krska et al. 2004). Peanut, together with tree nut, affect more than 1 % of the US population, i.e., approximately 3 million people (Sicherer et al. 2010). A study conducted in 2010 has reported the prevalence for peanut allergy in US to be rising, particularly among children (Sicherer et al. 2010). Peanut allergy has significantly increased with a prevalence of 2.1 % reported for subjects under 18 years old in 2010, compared to 1.2 % in 2002 and 0.6 % in 1997.

Regional dietary habits also play an important role in the prevalence of legume allergy or any type of food allergy (Dalal et al. 2002; San Ireneo et al. 2008). For example, in Spain, allergy to lentils is more common than allergy to peanut (Pascual et al. 1999), and in France, lupin allergy is becoming more common since the introduction of lupin seed flour used in a mix with wheat flour (Campbell and Yates 2010). A study comparing Israeli and UK Jewish children showed that peanut allergy affected only 0.17 % of the children raised in Israel, while it is ten-fold higher (1.85 %) in the UK (Du Toit et al. 2008). In India, lentils and chickpeas are the major legumes that cause allergy (San Ireneo et al. 2008; Sánchez-Monge et al. 2000b).

Mechanism of Allergy Reaction An allergy is an adverse reaction by the immune system to an allergen. Allergens can be either chemicals or proteins, with the latter being more common as food allergens. Food allergy includes several immune mechanisms but the most common is IgE-mediated food allergy, also classified as a type I hypersensitivity reaction. An IgE-mediated food reaction involves a three-step process. The first step is known as a sensitization. When an atopic individual is exposed to an allergen for the first time, the helper T cells present in the immune system direct B cells to produce an antibody, called IgE antibodies. The allergen-specific antibodies then bind to high-affinity receptors present on the surface of mast cells and these IgE primed mast cells are circulated and remain in the immune system. At this stage, the subject will not display any visible allergic reaction. However, when this individual is subsequently exposed to the same allergen, elicitation will occur. In this process, the allergen binding to the IgE-produced from previous exposure will stimulate the attached mast cells to release histamine and cytokines. These substances cause clinical symptoms including swelling, hives and rashes, and rhinitis, but also more severe clinical conditions such as asthma and anaphylaxis that can be life threatening. The severity of the reaction depends largely on the amount of IgE produced in response to the allergen. If the IgE being produced does not exceed the threshold point, no clinical reaction is observed. However, the individual will be regarded as sensitized to the food allergen. For some people with peanut allergy, the threshold value to elicit a reaction can be as low as 100 μg (Wichers et al. 2004).

Allergies to legumes are common in atopic patients, who are sensitive to foods. Severe reactions, such as asthma and anaphylaxis, have been reported for peanut (Sampson 1990) and soybean allergy (Sicherer et al. 2000), with less aggressive reactions reported for other legumes.

For peanut allergy, once sensitivity is acquired, it may continue into adulthood. However, approximately 20 % of young infants, who are allergic to peanuts, are likely to outgrow their allergy, particularly those with lower peanut-IgE level (Skolnick et al. 2001). Factors contributing to legume allergy are not clear. One study has linked mutations in the *Filaggrin* gene, which causes epithelial barrier dysfunction as a significant risk factor for peanut allergy (Brown et al. 2011).

24.1.3 Legume Seed Allergens

Legume allergens identified to date are represented by 13 protein families (Riascos et al. 2010). Among them, four protein families and superfamilies include the majority of legume allergens: cupins (including the 7S and 11S globulins); the prolamin superfamily (including the nonspecific (ns)-LTPs and 2S storage albumins); profilins; and the larger group of PR proteins (mostly composed of homologues of the major birch pollen allergen, Bet v1) (Riascos et al. 2010). Other protein families, such as the oleosins, cysteine proteases, Kunitz trypsin inhibitors, calcium-binding proteins, and the seed biotinylated proteins, account for the remaining allergenic proteins in leguminous crops. Table 24.1 shows the legume allergens identified and their protein families' classification.

At present, the most effective way to manage food allergy is by avoidance. However, biotechnological engineering enhanced by genome-enabled technologies may give a better understanding of allergy. Furthermore biotechnological engineering can be used in attempts to reduce the allergenicity of plants.

24.2 Identification of Allergens

24.2.1 IgE, the Key to Food Allergen Identification

The identification of allergens of type I hypersensitivity (IgE-mediated) is based on the ability of these allergens to bind IgE. The use of IgE from serum as a diag-

Table 24.1 Legume allergens classified according to their families

Protein family		Legume	Allergen	Size (kDa)	Epidemiology (reference)	Isoform
Cupin superfamily	7S vicilins	Peanut	Ara h 1	63.5	>90 (Burks et al. 1995), 65 (Kleber-Janke et al. 1999)	✓
		Soybean	Gly m 5 (Gly m Bd 60 K)	50–71	43 (Holzhauser et al. 2009)	✓
		Soybean	Gly m Bd 28 K	26	No data	No data
		Lentil	Len c 1	50	77 (López-Torrejón et al. 2003)	✓
		Peas	Pis s 1	44	>50 (Sánchez-Monge et al. 2004)	✓
		Peas	Pis s 2	63	>50 (Sánchez-Monge et al. 2004)	✓
		Beans	Pha v phaseolin	~30	One anaphylatic patient (Rougé et al. 2011)	No data
		Lupins	Lup an 1	40–70	75 (Goggin et al. 2008)	✓
	11S legumins	Peanut	Ara h 3	60 (precursor)	45 (Rabjohn et al. 1998)	✓
		Peanut	Ara h 4	36	53 (Kleber-Janke et al. 1999)	✓
		Soybean	Gly m 6	52.4–61.2	37.6 (Holzhauser et al. 2009)	✓
		Lupins	Conglutin α	43	83 (Dooper et al. 2009)	✓
Prolamins	2S albumins	Peanut	Ara h 2	17.3	>90 (Burks et al. 1992), 85 (Kleber-Janke et al. 1999)	✓
		Peanut	Ara h 6	14.5	38 (Kleber-Janke et al. 1999), >90 (Codreanu et al. 2011)	✓
		Peanut	Ara h 7	14–17	43 (Kleber-Janke et al. 1999)	✓
		Soybean	Gly m 2S albumin	12	55 (Vissers et al. 2011)	✓
		Cowpeas	Vig u	41, 55	No data	No data
		Lupins	Conglutin δ	13	66 (Dooper et al. 2009)	✓
	nsLTPs (nonspecific lipid transfer proteins)	Peanut	Ara h 9	9	7 (Vereda et al. 2011), 64 (Krause et al. 2009), 90 (Lauer et al. 2009)	✓
		Soybean	Gly m 1	7	>90 (González et al. 1991)	✓
		Lentil	Len c 3	9	No data	✓
		Beans	Pha v 3	10	80 (Zoccatelli et al. 2010)	✓

Table 24.1 (continued)

Protein family		Legume	Allergen	Size (kDa)	Epidemiology (reference)	Isoform
PR proteins	PR-10	Peanut	Ara h 8	17	85 (Mittag et al. 2004a), 39.7 (Codreanu et al. 2011)	✓
		Soybean	Gly m 4	17	25 (Lin et al. 2006), 70 (Ballmer-Weber et al. 2007)	✓
		Mung beans	Vig r 1	16.2	80 (Mittag et al. 2005)	✓
	PR-3	Beans	Pha v chitinase	32	87.5 (Sánchez-Monge et al. 2000a)	No data
Profilins		Peanut	Ara h 5	14	13 (Kleber-Janke et al. 1999)	✓
		Soybean	Gly m 3	14	25 (Mittag et al. 2004b), 69 (Rihs et al. 1999)	✓
		Mung beans	Vig r 5	15	70 (Mittag et al. 2005)	No data
Other protein families	Oleosin	Peanut	Ara h 10	16	No data	✓
		Peanut	Ara h 11	14	No data	✓
	Hull	Soybean	Gly m 2	8	No data	✓
		Soybean	Gly m 50 K	50	51.7 (Codina et al. 2002)	No data
	Plant specific protein	Soybean	Gly m 39 K	39	33 (Gu et al. 2001)	No data
	Kunitz soybean trypsin inhibitor	Soybean	Gly m TI	20	6 (Lin et al. 2006)	No data
	Cysteine protease	Soybean	P34 (Gly m Bd 30 K)	30–34	65 (Babiker et al. 2000)	No data
	Lectin	Soybean	Gly m agglutinin	9	31.2 (Lin et al. 2006)	No data
		Beans	Pha V Pha	~70, ~60, ~50	One anaphylatic patient (Rougé et al. 2011)	No data
	Lectin-like glycoprotein	Lupins	Conglutin γ	16, 49	No data	✓
	Seed biotinylated protein	Soybean	Seed biotinylated protein	70	50 (Riascos et al. 2009)	No data
		Lentil	Len c 2	66	41 (Sánchez-Monge et al. 2000b)	✓

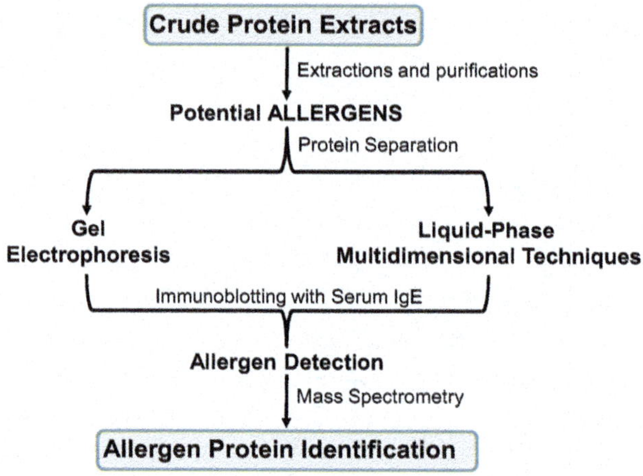

Fig. 24.1 General approach in allergen identification. (Modified from González-Buitrago et al. 2007)

nostic indicator can be complicated by the fact that not all individuals having IgE against a particular protein in their sera react clinically to that protein. Thus, before allergen identification can be attempted, sera from clinically reactive patients must be collected. Defining clinical reaction can be problematic. The best method is to test all patients with a double-blind placebo-controlled food challenge (DBPCFC). However, for seriously allergic individuals this may risk anaphylaxis, and hence, the case history of allergic reactions and skin prick tests are often as an alternative. Skin prick tests involve the pricking of the skin with an allergen extract (or purified allergen) and assessment of the size of the wheal that develops around the point of contact. It is possible to have a skin prick reaction but not react clinically to the allergen; therefore, results must be treated with caution. However, a large wheal after a skin prick test is often indicative of a clinical reaction to the protein. In addition, as each atopic individual has a unique spectrum of proteins that they react to and there are groups of individuals that react to different sets of allergens from the same source, serum should be collected from as many individuals as possible if the full spectrum of allergens in a seed is to be determined. It is important that sera from as many individuals as possible is tested with an allergen source in order to determine a clear picture of the allergenicity of different seed components. Identification of allergens can utilize different kinds of techniques, but the principal behind the techniques used to identify an allergen is generally the same (Fig. 24.1). In most cases, the protein (allergen) is extracted, purified, and separated, and its allergenic potential is tested by inducing specific binding to IgE and finally identification. For less complicated proteins, extraction and purification may be done with conventional protein isolation methods, such as anion-exchange chromatography. However, in the case of legumes in which the protein contents are far more complex, recombinant protein techniques are sometimes utilized to enable isolation, particularly in the case where a protein presents isoforms. After it is clear that a protein binds IgE, it is further dissected into smaller fragments to identify the particular region on the protein that is the active binding site of IgE (epitope).

The efficiency in allergen identification is parallel with the amount of sequences (protein or gene) available. Unless the respective sequences are in database, the identification process can be long and laborious, especially with SSPs where it is further complicated by the presence of PTM and multigene families. However, advanced progress in genome-based techniques has enabled whole genome of a species to be sequenced rapidly. Soybean is the only legume plant important in terms of allergy that has had its genome sequenced. This could provide a platform to study other legume allergens.

24.2.2 Genome-Enabled Technologies in Allergen Identification

Initial studies trying to identify allergens were done long before the advent of genome sequencing. Progress was relatively slow and studies tended to concentrate on a single allergen. With the increase in genome sequencing in recent years, and the availability of genomes for nonmodel plants, the ability to identify allergens quickly and consistently has been enhanced. Proteomics analysis based on the IgE-binding properties of allergens, combined with the availability of a complete genome allows fast analysis of the allergen profile of a seed. This is particularly important where a particular allergen has a number of different isoforms. Using traditional methods, one isoform may be identified, but the presence of others with similar characteristics may have taken much longer to establish. With the availability of genome sequence, it is possible to predict proteins that are likely to be isoforms of an allergen and target these for investigation. It is important to characterize all isoforms and determine their capacity to bind IgE as any approach to reducing allergenicity needs to take the isoforms into account (see subsection, Gene Silencing for Allergen Reduction).

Identification of the peanut allergen Ara h2 is a good example of the traditional approach to allergen identification. Ara h2 is a 2S albumin protein (SSP) and is one of the major and earliest characterized legume allergens (Burks et al. 1992). Crude seed protein was first fractionated with anion-exchange chromatography and the fractions screened with serum of peanut allergic patients to identify proteins that bound IgE. The amino acid sequence of a doublet of approximately 17 kDa that bound IgE was determined and later identified as Ara h2. Oligonucleotides based on the amino acid sequence were used to identify cDNA clones from a peanut cDNA library (Stanley et al. 1997). Once the amino acid sequence was deduced from the cDNA, it was possible to show that the other band of the original doublet was a second isoform of Ara h2 (Chatel et al. 2003), which was later reported to have a different IgE-binding capacity (Hales et al. 2004).

Identification of *Lupinus angustifolius* allergens, although not relying on genome information, illustrates the value of DNA sequence resources for the plant under study. Proteomics analysis of a whole protein extract directed the identification of a particular class of protein as allergenic. A whole protein extract of lupin seed was separated *via* 2-DGE, followed by immunoblotting to determine which

Table 24.2 BLAST search results for proteins similar to each soybean allergen. Genes that are likely to code for isoforms of the allergen for identical copies of gene (isoform) based on soybean genome database (Phytozome). Identical copies of gene are mostly obtained for soy allergen that are of seed storage proteins origin (e.g., the Kunitz protease inhibitor allergen, Gly m TI, returned with 0 hit), indicating the complexity of seed storage proteins

Allergen	Number of genes	Ontology
Gly m 1	4	Prolamin nsLTPs
Gly m 3	7	Profilins
Gly m 4	>10	PR protein PR-10
Gly m 5 (Beta subunit)	3	Cupin 7S globulins
Gly m Bd 28 k	2	Cupin 7S globulins
Gly m 2S albumin	2	Prolamin 2S albumins
Gly m 39 k	2	Soybean plant specific protein

proteins bound IgE from the sera of lupin allergic individuals (Goggin et al. 2008). Proteins that bound IgE were identified by MS with peptides matched to the available cDNA sequences from lupin seeds. Due to the lack of a genome sequence, many IgE-binding spots could not be identified. The identity of β-conglutin as a potential allergen was confirmed after purification of this protein, and it was named Lup an1 (Goggin et al. 2008). However, although it was clear that the analyzed spots represented more than one isoform of β-conglutin, it was not possible to determine how many forms were present in the seed. The study used serum from individuals allergic only to lupin (not peanut), and hence it is possible that with other sera (e.g., of individuals allergic to both lupin and peanut) other proteins would be identified as allergens as suggested by other studies of lupin allergy (Holden et al. 2008). Subsequent sequencing of large numbers of lupin seed ESTs followed by the isolation of full-length clones for all SSPs provided the data to allow the initial proteomics data (Goggin et al. 2008) to be reassessed (Foley et al. 2011). Seven genes encoding β-conglutin were identified and many of the spots of the original 2-D gel could be defined as being expressed from one of these genes. At least six of the seven β-conglutin proteins bound IgE from individuals allergic to lupin but not peanut (Foley et al. 2011). Additionally, cDNAs encoding multiple forms of each of the other lupin SSPs were identified and used to pinpoint some of these proteins on the 2-D gel of *L. angustifolius* seed proteins, forming the basis of a reference map for lupin proteome. This illustrates how the availability of DNA sequence analysis for an allergenic plant can drive allergen identification. Without a complete genome sequence of *L. angustifolius*, it is not possible to be sure there are not other genes encoding β-conglutin and that there are no allergens that are not SSPs. The genome sequence will also assist in identifying isoforms of allergenic proteins that may share common peptides after digestion with trypsin. At present, the only sequenced genome of an allergenic legume is soybean. Table 24.2 shows the number of genes likely to encode isoform of particular soybean allergens based on the soybean genome sequence.

With the availability of the gene sequence encoding allergens, it is possible to generate recombinant proteins. Recombinant proteins may be useful for confirmation of allergenicity, although the possibility that recombinant allergens do not fold in the same manner as native allergens should be taken into account. Recombinant allergens are especially useful when the purification of the allergen is challenging or where many protein isoforms of the allergen exist. It is difficult or impossible to obtain the isoforms by standard purification methods due to the high degree of homogeneity, but the allergenicity of these isoforms can be tested *via* its recombinant proteins. Recombinant proteins can also be used for identification of IgE epitopes by subjecting the gene to mutagenesis to induce amino acid changes in the recombinant protein and analyzing the difference of IgE binding before and after the alteration.

A more apparent use of recombinant allergen is in allergy diagnosis and immunotherapy. Natural allergen extracts have problems, such as the difficulty of standardization, poor representation of allergens, and presence of contaminants (Valenta et al. 2011), which can be overcome by using recombinant proteins. However, the efficacy of recombinant allergens in diagnosis and treatment of allergy remains questionable, as the recombinant protein must exactly mimic the native allergen structure, which may be important for IgE-binding assays used in diagnosis. Nonetheless, these recombinant proteins are useful in providing a predictive platform for allergy diagnosis as well as studies.

24.2.3 Detection of Allergens in Food

Besides identifying allergen, a paper published recently has successfully utilized genomic tools for detecting the presence of legume allergens in a food product (Galan et al. 2011). This work has described the development of a real time-PCR method to allow the simultaneous detection of traces of lupin and soya in processed food by using mitochondrial DNA (mtDNA) as markers. It was reported that the sensitivity of the method was suitable to detect allergenic ingredients in the low milligram per kilogram range of food prepared at high temperature. The method was successfully applied to bakery (e.g., bread) and vegetarian (e.g., nonmeat sausages) food products that contain or may contain soya and/or lupin as ingredient or contaminant (Galan et al. 2011).

24.3 Genome-Enabled Technologies to Create Low Allergen Seeds

The availability of a complete genome gives advantages for identifying allergens through proteomics analysis, but also opens the way for approaches to lower allergenicity and create hypoallergenic crops. These approaches rely on the availability of DNA sequence for the allergens and the efficacy is enhanced, if the sequences of all genes encoding similar allergens (isoforms) are known. Various approaches have been used to suppress or eliminate the expression of allergens in crops and new methods for genome editing offer much promise for directly targeting the allergens genes. Ideally the aim would be to change the IgE epitopes of the protein without affecting the expression and overall physiological function of the allergens. However, because there are often multiple epitopes on the same protein and different patients react to different group of epitopes on the same protein it is not currently practically or economically feasible to modify all allergenic epitopes in an allergen to create a hypoallergenic plant.

24.3.1 Gene Silencing for Allergen Reduction

Posttranscriptional gene silencing is a naturally occurring pathway in plants, used for defence against viral infection and for developmental gene regulation (Baulcombe 2004; Chapman and Carrington 2007). It is initiated by the introduction of a transgene, viral infection, or the formation of dsRNA. An efficient approach is to induce production of small interfering RNAs (siRNA), which target the allergen transcript of interest for cleavage (Dodo et al. 2005; Ozias-Akins et al. 2010; Song et al. 2004). Using this pathway, the expression of specific allergens can be selectively suppressed. This approach has been used in a number of plants through expression of antisense constructs or cosuppression. These methods have produced reductions in allergenic proteins (see below), but the methods for inducing production of siRNAs have now been refined and are termed RNAi. RNAi can be induced through introduction of hairpin constructs, producing a transcript that will fold to form a dsRNA (Smith et al. 2000). Inclusion of an intron between the two complementary regions enhances the efficiency of silencing (Smith et al. 2000; Waterhouse et al. 1998). The transgenes can be introduced into the plant through biolistic or *Agrobacterium tumefaciens*-mediated genetic transformation. After incorporation into the plant genome, the transgene dsRNA is processed into a series of small 21–24-nucleotides siRNAs by an endogenous RNase III enzyme called 'DICER' (Bernstein et al. 2001; Hamilton and Baulcombe 1999). The siRNA associates with the RNA-induced silencing complex (RISC) in the cytoplasm directing it to the target gene. Argonaute proteins within the RISC cleave the target mRNAs within the sequence complementary to the siRNA or translation of the target gene is inhibited (Song et al. 2004).

RNAi is based on the homology of sequences, and therefore, has the potential to silence simultaneously all genes in a gene-family. This is particularly important for allergens, as they are commonly encoded by members of multigene families (Table 24.1). All isoforms can consequently be targeted to ensure an efficient suppression of the allergens. Additionally, allergenic crops can contain multiple allergenic proteins, which might exhibit high sequence homology and often belong to the same protein families. However, suppressing all allergens of one protein family might strongly influence the viability of the transgenic plant, due to the loss of their biological function. In general, only the allergens to which the majority of patients react and that do not fulfil an essential role in the seed can be potential candidates for this type of suppression. Furthermore, it is essential to use sequence fragments having least homology to other genes in plants to lower the allergenicity of a plant in order to reduce the collateral effect (Ozias-Akins et al. 2010).

An alternative approach uses a second type of small RNA, called miRNA, to induce silencing. The miRNAs are a similar size RNA molecule to siRNAs, but are produced from a specific transcript, and that once transcribed, forms a hairpin, which is cleaved by DICER to produce a single miRNA product. The miRNA acts in a similar manner to a siRNA, binding to the target gene transcript and marking it for cleavage or inhibition of translation. Methods to design artificial miRNAs have been developed based on the characteristics of binding of native miRNAs to their targets (Schwab et al. 2006). A sequence complementary to the gene of interest is used to replace the natural miRNA in a miRNA gene. This method has not been used to reduce allergen expression, but is useful in that a miRNA that targets a particular member of a gene family can be designed. The sequence used in the artificial miRNA construct is only 21-nucleotide long compared to the 100–300 bp sequences used in hairpin RNAi constructs (Smith et al. 2000), making it a more specific method for silencing. Conversely, if all members of a gene family share an identical 21-nucleotide sequence, they can all be targeted by a single artificial miRNA.

The RNAi approach has been used to reduce the allergenicity of soybean (Herman et al. 2003) and peanut (Chu et al. 2008; Dodo et al. 2008), but also rice (Tada et al. 1996), apple (Gilissen et al. 2005), tomato (Lien et al. 2006), and carrot (Peters et al. 2011). In peanut, two independent studies (Chu et al. 2008; Dodo et al. 2008) have selected the major allergen *Ara h2* for RNAi silencing experiments. As a type of trypsin inhibitor (Maleki et al. 2003), Ara h2 is thought act as a defence protein against pathogens in the seeds and in contrast to SSPs not essential for plant development making *Ara h2* a good target gene for silencing. Silencing was achieved using a hairpin construct based on the coding region of Ara h2 gene driven by an enhanced cauliflower mosaic virus (CaMV) 35S constitutive promoter. There was no obvious difference in plant performance of the transgenic plants compared to wild type. A specific sandwich enzyme-linked immunosorbent assay (ELISA) with monoclonal Ara h2-mAbs and western blotting revealed that the protein abundance of Ara h 2 in crude protein extract was decreased by 21–25 % compared to the wild type in several seeds. Furthermore, ELISA using the patient's sera showed reduced IgE binding when compared to IgE binding to a nontransgenic seed (Dodo et al. 2008). However, none of the patient sera revealed a complete elimination of IgE binding, which illustrates the limitations of this method. Analysis of later generations was not completed; hence, the stability of reduction in allergen could not be determined.

By selecting a sequence region that is highly conserved between peanut allergens Ara h2 and 6, Chu et al. (2008) produced an RNAi construct that effectively silenced both allergens simultaneously. Ara h6 has an overall homology of 59 % to Ara h2 and is a major allergen (Koppelman et al. 2005), which makes this approach very useful. The plants did not show any significant phenotypic differences compared to wild type. Even though the defence protein Ara h2 is suppressed, the transgenic lines did not show an increase in its susceptibility to the common *Aspergillus flavus* fungal infection. It is important to note that, the susceptibility was high in both transformed and wild type plants. This may suggest that Ara h2 is usually involved in the defence for pathogens other than *A. flavus*. By using immunoblotting, Chu et al. (2008) demonstrated decreased IgE recognition to both allergens in the transgenic seeds with sera from three different patients with peanut allergy. Other major allergens, such as Ara h1 and 3, did not show any notable differences. Nevertheless, a separate analysis by Stevenson et al. (2009) observed besides other changes, an increase in minor allergen Ara h10 (oleosin 1) in one of the three mentioned transformed lines. Independently generated Ara h2-silenced lines showed surprisingly large variations in protein profiles in addition to reduced Ara h2 expression (Dodo et al. 2008; Ozias-Akins et al. 2010).

In soybean, cosuppression was used to silence the major allergen Gly m Bd 30K, also known as P34 (Herman et al. 2003). Although it is a relatively minor seed constituent (less than 1 % of total seed protein), P34 is regarded as the major or immunodominant soybean allergen (see, Table 24.1). Soybean embryos were transformed using a vector containing a specific promoter and the entire ORF of P34 in sense direction. It was demonstrated that P34 was absent *via* comparison by 2-DGE and MS. Furthermore, the IgE immunoreactivity for P34 polypeptide was absent in P34-suppressed seed protein extracts, when using a serum pool from six soy-allergic individuals. Proteomics analysis of extracts from transgenic plants could not detect any other significant changes in polypeptide pattern. Seeds of the transgenic lines did not show any compositional, developmental, structural, or ultrastructural phenotypic differences when compared with control plants (Herman et al. 2003). The P34 elimination was maintained over three generations (Gallo and Sayre 2009). The commercial success of this approach will depend on the long-term stability of the gene suppressing expression of P34 construct.

24.3.2 Mutation of Allergen Genes

A second approach to creating a hypoallergenic crop is the use of mutagenesis to modify the genome of an allergenic plant. It can then be screen for expression of the eliminated gene. Theoretically, some mutation may modify the amino acid sequence of the protein, and hence that protein is no longer an allergen. However, this process is likely to be a much rare event. Screening methods (such as TILLING) relying on the availability of the genome sequence are important in screening of the mutated allergen genes.

There are different approaches to mutation of an allergen gene. These included induced mutations (such as those caused by EMS or radiation) or insertional mutagenesis *via* T-DNA or transposable elements.

Mutagenesis Using Radiation Radiation induces moderate to large-scale chromosomal changes, often resulting in deletion or rearrangement of large portions of the genome. This can be an advantage, since this treatment can have an effect on many allergen genes at once. Conversely, it may result in lethal modifications or have harmful effects on the overall phenotype of the plant.

Several soybean lines have been created that lack allergens by employing γ-ray mutagenesis. The breeding line EnB1 lacks all five subunits of the SSP glycinin, also known as Gly m6 (Odanaka and Kaizuma 1989; Teraishi et al. 2001). The mutant breeding line must therefore either contain several independent glycinin gene mutations (glycinin genes are known to be located on multiple chromosomes) or more likely a single mutation in a transacting modifier gene that controls expression of the whole gene family (Ozias-Akins et al. 2010).

The breeding line 'Kari-kei 434', which lacks the α′- subunit of β-conglycinin, was treated with γ-radiation. In this experiment, a line known as 'Tohoku 124' or 'Yumeminori' was obtained, which further lacks the α- and α′-subunits of the allergen β-conglycinin, and has reduced levels of the β-subunit (collectively Gly m5, a vicilin) (Takahashi et al. 1994). Furthermore, this line does not contain Gly m Bd28K, another vicilin allergen (Samoto et al. 1997). Samoto et al. (1997) was also able to remove 99.8 % of another allergen Gly m Bd30K (P34) from soymilk through processing (Ozias-Akins et al. 2010).

Upon elimination of both β-conglutin subunits in soybean after γ-radiation, lethal chlorosis occurred in the plants, which might have resulted from a large chromosomal deletion (Phan et al. 1996). This finding shows that a more targeted approach for the elimination of allergens is useful.

The γ-radiation was used to develop three mutants, which lacked the A3 subunit of allergenic glycinin (Manjaya et al. 2007). Among the three, two mutants were lacking α and α′-subunits of β-conglycinin (vicilin). Plant growth and physiological abnormalities were not detected. However, the allergenicity of the mutant lines was not determined.

A fast neutron mutant population has recently been described for soybean. The population consists of 23,000 independent M2 lines (Bolon et al. 2011) and promises to be an important resource for screening for low-allergen soybeans.

Chemical Mutagenesis One of the most common chemicals used for mutagenesis is EMS. Chemical mutagens tend to induce very small changes to DNA sequences: SNPs or very small insertions (Ozias-Akins et al. 2010). In some cases, this process can have an effect on the gene function and protein expression. If the result is a premature stop codon, this may lead to elimination of expression of the affected allergen. Nevertheless, it can leave the functional allergens intact, but disrupt key epitopes (Ramos et al. 2009; Riascos et al. 2010). Single amino acid substitutions in characterized allergen epitopes can have a dramatic effect on IgE recognition and binding (Helm et al. 2000; Rabjohn et al. 2002; Shin et al. 1998; Stanley et al. 1997). Since these mutations tend to be distributed randomly throughout the genome, and the resulting phenotype is not easily observed, it is important to be able to screen for the presence of mutations in the gene of interest. This is even more relevant for polyploidy crops (such as tetraploid peanuts), where duplicate genes frequently mask mutant phenotypes.

TILLING for Identification of Mutations A common screening technique for the small sequence differences induced by chemical mutagens is TILLING (McCallum et al. 2000)—a technique that is simplified by the availability of a genome sequence. TILLING is a high-throughput screening technique based on the PCR technique (Henikoff and Comai 2003; McCallum et al. 2000). With this technique, individuals carrying mutations in the genes of interest can be identified by recognizing SNPs in an induced mutant population (McCallum et al. 2000). The target gene sequence is thus required for this type of mutant screen. The process of determining the pool size and evaluation of the rate of mutation throughout the genome can be facilitated, if the genomic sequence of the allergenic plant is known. The developed TILLING populations serve as a valuable genetic resource for identifying changes to allergen genes. Nevertheless, this technique presents challenges for discriminating between homologous genes in an allotetraploid, such as peanut. To overcome this, genomic sequence of the targeted crop would give very valuable information.

A peanut TILLING population has been developed and screened for variations in the major allergens Ara h1 and Ara h2 (Knoll et al. 2011). Using EMS as a mutagen, a number of mutations in the target genes were generated. Some of the mutants for Ara h2 resulted in amino acid substitutions within allergenic epitopes. As homozygous mutant lines are not yet available, they have not been tested to see if this reduces the IgE-binding to the protein. A mutant Ara h2 gene with a modified start codon has resulted in elimination of expression of one of the two Ara h2 genes, but again the reduction in IgE binding as a result of this change has not been assessed. An Ara h1 gene with a premature stop codon was also identified and appears to result in a reduction in the protein level for one Ara h1 spot when assessed by 2-DGE.

In soybean, several TILLING populations have already been developed after mutagenesis and can be accessed easily *via* online platforms, such as http://www.soybeantilling.org/. These platforms, however, concentrate on plant performance and functional genomics. To date, no work was published on soy allergen gene mutations.

EcoTILLING for Identifying Natural Variation in Allergenic Proteins A variation of TILLING, EcoTILLING enables the determination of polymorphisms of target genes in untreated populations and has been adapted to find natural cultivars containing allergen-isoforms with lowered allergenicity (Comai et al. 2004; Henikoff and Comai 2003). In peanut, EcoTILLING has been used to find polymorphisms of Ara h2.01 in 30 *Arachis duranensis* accessions (a wild relative of *Arachis hypogaea* with a richer source of variation), available from the US germplasm collection. None of the identified polymorphisms induced major conformational changes, and were therefore thought to be neutral for their function in the plant. Nevertheless, in terms of allergenicity, one allele showed a 56–99 % reduction in IgE-binding activity. Furthermore, none of the mutations occurred within the T-cell epitopes characterized for Ara h2.01. Therefore, the potential use of this natural hypoallergenic variant for the purpose of immunotherapy was suggested (Ramos et al. 2009).

24.3.3 Genome Editing to Modify Allergenicity

The ability to target a site in specific genes for cleavage is a huge step forward for plant breeding, allowing precise engineering of the plant genome. Techniques, using zinc-finger nucleases (ZFN) (reviewed in Urnov et al. 2010) or transcription activator-like effector nucleases (TALEN) (Miller et al. 2011), have been developed specifically for this purpose. The techniques rely on the induction of double-stranded breaks at specific sites in the target gene inducing repair mechanisms that often introduce modifications to the gene. DNA-binding proteins, such as zinc-finger proteins and transcription activator-like effectors, are used to target specific sites in the genome. By fusing these proteins to a nuclease, such as the cleavage domain of the FokI restriction enzyme, the nuclease can be targeted to the region of the genome at which the protein binds. The binding proteins have been engineered to modify their DNA-binding specificity and increase the range of sites they target (Miller et al. 2011). On-line resources are available to aid in the development of ZFN or TALEN that target particular genes. Once engineered, the constructs can be introduced using standard plant transformation methods and mutants in the gene of interest selected from the transformants.

Genome editing with ZFN has been demonstrated for some plant species (Sander et al. 2011) including soybean (Curtin et al. 2011), and it is likely to provide an efficient method to target mutations to allergen genes.

24.3.4 Limitations and Challenges When Creating Low Allergen Seeds

Most crops contain various allergens with essential physiological functions, whose elimination may have effects on the physiology and agricultural performance of the plant. The SSPs are generally degraded to provide energy for seed germination, hence care would need to be taken that this was not affected in reduced allergen seeds. Also since the allergenic SSPs are often major components of the seed, it is possible that properties like taste may be affected in plants with reduced allergen (Krause et al. 2010).

With RNAi, it is rare to have complete suppression of expression, and therefore, it is unlikely the allergen will be completely eliminated. For peanuts, even trace amount of residual allergens can elicit reactions in a very atopic patient (Hourihane et al. 1997), which is underlined by the fact that the IgE in patients' sera still detected the allergens in Ara h2- and 6-silenced peanuts (Chu et al. 2008; Dodo et al. 2008). The transformed peanuts therefore still present a risk for patients allergic to the allergens. However, for some allergenic proteins, a reduction of allergen may provide some benefit.

It is also possible that reduction in the expression of one SSP may lead to an increase in another allergenic SSP to compensate. An example is in lines with naturally low Ara h1 composition, where mediator release assay of rat basophilic leukaemia sensitized with IgE from peanut allergic individuals showed no reduction in allergenicity for these lines. It was proposed that other proteins had compensated for the lack of Ara h1 and the implication was that these were also allergens (Krause et al. 2010).

What these limitations mean is that each plant that is produced where an allergen has been suppressed or eliminated must be carefully assessed, taking into account the range of allergens different patients react to and the possibility of unexpected compensatory effects on the seed metabolism or allergenicity.

To date, it is not sufficiently practicable to determine the overall allergenicity of a hypoallergenic crop the product. The presence of multiple allergens in one food source makes the collection of allergenicity data in a clinical setting *via* the gold standard DBPCFC (Peeters et al. 2007) test risky for allergic patients. Therefore, it has not been used for evaluation of any allergen reduced transgenic lines (Ozias-Akins et al. 2010). Additionally, animal models have not been satisfactorily correlated with diverse human allergic sensitization (Goodman et al. 2008). Therefore, the analysis of the allergenicity of crops is mostly done on protein level, utilizing proteomics techniques (Ozias-Akins et al. 2010; Stevenson et al. 2009; Thelen 2009), and tests involving IgE binding using sera from allergic patients, *in-vitro* histamine release from sensitized human or humanized basophils (Goodman et al. 2008). Although individuals who react to an allergen produce IgE antibodies against the allergen, the presence of IgE to a particular allergen is not an absolute predictor of a clinical reaction. It has been suggested that where allergic patients generate IgE to glycans on an allergen, these IgE molecules are less likely to elicit a clinical reaction to the allergen (Altmann 2007). Together, the findings and discussion illustrate the need for careful characterization of sera (IgE) used to assess allergenicity of proteins.

Banks of sera have IgE against a range of protein epitopes (Altmann 2007; Goodman et al. 2008), and that will be a useful tool to assess changes of allergenicity as a result of manipulation of allergen gene expression and modification. This information can be obtained using protein array assays, such as immuno solid-phase allergen chip (ISAC). Furthermore many of the methods mentioned are not standardized, although efforts have been made to standardize commercial assays, such as ELISA (van Ree et al. 2008). Importantly, it must be recognized that there are no absolute thresholds of serum IgE binding that provide absolute measurement of safety or risk (Goodman et al. 2008).

Additionally, a range of publications on peanuts have shown that the allergenic potential of a crop can change during processing, e.g., by roasting or boiling (Maleki et al. 2000, 2003; Mondoulet et al. 2003, 2005; Pomes et al. 2006). For example, the IgE-binding capacity of purified Ara h1 and Ara h2 was observed to be increased in roasted peanuts and decreased in boiled peanuts, compared to raw samples. Although considering these secondary effects can be difficult, its needs to be taken in account to ensure the safety of the consumers, since there is no standardization on the manufacturing process such as on roasting temperatures.

Despite strict safety assessment and control of GM crops, there is a great public resistance against GM food and it is not known wether the acceptance by consumers will rise by providing possible recovery from allergy.

24.4 Concluding Remarks

As the genomes of more crop plants become available, it will become easier and faster to identify allergens and characterize their allergenic epitopes. Genome-enabled technologies will also allow more efficient modification of plants to reduce allergenicity. However, many of the current techniques do not allow complete elimination of allergenic proteins and the foods produced will still remain a risk to allergic individuals. Modifications could create new allergenic epitopes or changes in expression of other allergens. Hence, screening of the resulting plants must take into account all allergens, not just those modified. This will require development of more sophisticated strategies for clinically evaluating allergenicity. Nevertheless, these studies while illustrating the challenges for eliminating allergens show that at least reduction of allergenicity is attainable. For some foods, this will make a major difference for allergenic individuals. The utilization of genome-enabled methods is a basis for advancing research in this area, especially identification and characterization of allergenic molecules. Genomic tools that improve our knowledge of allergens and their structure will help us address the central question: why some protein are allergens while others are not?

References

Adlercreutz H (1998) Epidemiology of phytoestrogens. Baillieres Clin Endocrinol Metab 12:605–623

Altmann F (2007) The Role of protein glycosylation in allergy. Int Arch Allergy Immunol 142:99–115

Babiker EE, Azakami H, Ogawa T, Kato A (2000) Immunological characterization of recombinant soy protein allergen produced by *Escherichia coli* expression system. J Agric Food Chem 48:571–575

Ballmer-Weber BK, Holzhauser T, Scibilia J, Mittag D, Zisa G, Ortolani C, Oesterballe M, Poulsen LK, Vieths S, Bindslev-Jensen C (2007) Clinical characteristics of soybean allergy in Europe: a double-blind, placebo-controlled food challenge study. J Allergy Clin Immunol 119:1489–1496

Baulcombe D (2004) RNA silencing in plants. Nature 431:356–363

Bernstein E, Caudy AA, Hammond SM, Hannon GJ (2001) Role for a bidentate ribonuclease in the initiation step of RNA interference. Nature 409:363–366

Bolon YT, Haun WJ, Xu WW, Grant D, Stacey MG, Nelson RT, Gerhardt DJ, Jeddeloh JA, Stacey G, Muehlbauer GJ, Orf JH, Naeve SL, Stupar RM, Vance CP (2011) Phenotypic and genomic analyses of a fast neutron mutant population resource in soybean. Plant Physiol 156:240–253

Boschin G, Arnoldi A (2011) Legumes are valuable sources of tocopherols. Food Chem 127: 1199–1203

Brown SJ, Asai Y, Cordell HJ, Campbell LE, Zhao Y, Liao H, Northstone K, Henderson J, Alizadehfar R, Ben-Shoshan M, Morgan K, Roberts G, Masthoff LJN, Pasmans SGMA, van den Akker PC, Wijmenga C, Hourihane JO'B, Palmer CNA, Lack G, Clarke A, Hull PR, Irvine AD, McLean WHI (2011) Loss-of-function variants in the filaggrin gene are a significant risk factor for peanut allergy. J Allergy Clin Immunol 127(3):661–667 (ISSN 0091-6749, 10.1016/j.jaci.2011.01.031)

Burks AW, Williams LW, Connaughton C, Cockrell G, O'Brien TJ, Helm RM (1992) Identification and characterization of a second major peanut allergen, Ara h II, with use of the sera of patients with atopic dermatitis and positive peanut challenge. J Allergy Clin Immunol 90:962–969

Burks AW, Cockrell G, Stanley JS, Helm RM, Bannon GA (1995) Recombinant peanut allergen *Ara hI* expression and IgE binding in patients with peanut hypersensitivity. J Clin Invest 96:1715–1721

Campbell CP, Yates DH (2010) Lupin allergy: a hidden killer at home, a menace at work; occupational disease due to lupin allergy. Clin Exp Allergy 40:1467–1472

Chapman EJ, Carrington JC (2007) Specialization and evolution of endogenous small RNA pathways. Nat Rev Genet 8:884–896

Chassaigne H, Norgaard JV, van Hengel AJ (2007) Proteomics-based approach to detect and identify major allergens in processed peanuts by capillary LC-Q-TOF (MS/MS). J Agri Food Chem 55:4461–4473

Chatel JM, Bernard H, Orson FM (2003) Isolation and characterization of two complete *Ara h2* Isoforms cDNA. Int Arch Allergy Immunol 131:14–18

Chu Y, Faustinelli P, Ramos ML, Hajduch M, Stevenson S, Thelen JJ, Maleki SJ, Cheng H, Ozias-Akins P (2008) Reduction of IgE binding and nonpromotion of *Aspergillus flavus* fungal growth by simultaneously silencing *Ara h2* and *Ara h6* in peanut. J Agric Food Chem 56:11225–11233

Codina R, Ardusso L, Lockey RF, Crisci CD, Jaén C, Bertoya NH (2002) Identification of the soybean hull allergens involved in sensitization to soybean dust in a rural population from Argentina and N-terminal sequence of a major 50 KD allergen. Clin Exp Allergy 32:1059–1063

Codreanu F, Collignon O, Roitel O, Thouvenot B, Sauvage C, Vilain A-C, Cousin MO, Decoster A, Renaudin JM, Astier C, Monnez JM, Vallois P, Morisset M, Moneret-Vautrin DA, Brulliard M, Ogier V, Castelain MC, Kanny G, Bihain BE, Jacquenet S (2011) A novel immunoassay using recombinant allergens simplifies peanut allergy diagnosis. Int Arch Allergy Immunol 154:216–226

Comai L, Young K, Till BJ, Reynolds SH, Greene EA, Codomo CA, Enns LC, Johnson JE, Burtner C, Odden AR, Henikoff S (2004) Efficient discovery of DNA polymorphisms in natural populations by Ecotilling. Plant J 37:778–786

Curtin SJ, Zhang F, Sander JD, Haun WJ, Starker C, Baltes NJ, Reyon D, Dahlborg EJ, Goodwin MJ, Coffman AP, Dobbs D, Joung JK, Voytas DF, Stupar RM (2011) Targeted mutagenesis of duplicated genes in soybean with zinc finger nucleases. Plant Physiol 156:466–473

Dalal I, Binson I, Reifen R, Amitai Z, Shohat T, Rahmani S, Levine A, Ballin A, Somekh E (2002) Food allergy is a matter of geography after all: sesame as a major cause of severe IgE-mediated food allergic reactions among infants and young children in Israel. Allergy 57:362–365

Dixon RA (1999) Isoflavonoids: biochemistry, molecular biology and biological functions. In: Sankawa U (ed) Comprehensive natural products, vol 1. Elsevier, Oxford, pp 773–823

Dixon RA, Ferreira D (2002) Genistein. Phytochemistry 60:205–211

Dodo H, Konan K, Viquez O (2005) A genetic engineering strategy to eliminate peanut allergy. Curr Allergy Asthma Rep 5:67–73

Dodo HW, Konan KN, Chen FC, Egnin M, Viquez OM (2008) Alleviating peanut allergy using genetic engineering: the silencing of the immunodominant allergen *Ara h 2* leads to its significant reduction and a decrease in peanut allergenicity. Plant Biotechnol J 6:135–145

Dooper MM, Plassen C, Holden L, Lindvik H, Faeste CK (2009) Immunoglobulin E cross-reactivity between lupine conglutins and peanut allergens in serum of lupine-allergic individuals. J Investig Allergol Clin Immunol 19:283–291

Du Toit G, Katz Y, Sasieni P, Mesher D, Maleki SJ, Fisher HR, Fox AT, Turcanu V, Amir T, Zadik-Mnuhin G, Cohen A, Livne I, Lack G (2008) Early consumption of peanuts in infancy is associated with a low prevalence of peanut allergy. J Allergy Clin Immunol 122:984–991

Duranti M (2006) Grain legume proteins and nutraceutical properties. Fitoterapia 77:67–82

Duranti M, Scarafoni A (1999) Modification of storage protein content and quality in legume seeds. J New Seeds 1:17–35

Foley RC, Gao LL, Spriggs A, Soo LYC, Goggin DE, Smith PMC, Atkins CA, Singh KB (2011) Identification and characterisation of seed storage protein transcripts from *Lupinus angustifolius*. BMC Plant Biol 11:59

Galan AMG, Brohée M, de Andrade Silva E, van Hengel AJ, Chassaigne H (2011) Development of a real-time PCR method for the simultaneous detection of soya and lupin mitochondrial DNA as markers for the presence of allergens in processed food. Food Chem 127:834–841

Gallo M, Sayre R (2009) Removing allergens and reducing toxins from food crops. Curr Opin Biotechnol 20:191–196

Gilissen LJWJ, Bolhaar STHP, Matos CI, Rouwendal GJA, Boone MJ, Krens FA, Zuidmeer L, van Leeuwen A, Akkerdaas J, Hoffmann-Sommergruber K, Knulst AC, Bosch D, van de Weg WE, van Ree R (2005) Silencing the major apple allergen *Mal d 1* by using the RNA interference approach. J Allergy Clin Immunol 115:364–369

Goggin DE, Mir G, Smith WB, Stuckey M, Smith PMC (2008) Proteomic analysis of lupin seed proteins to identify conglutin β as an allergen, Lup an 1. J Agric Food Chem 56:6370–6377

González R, Zapatero L, Caravaca F, Carreira J (1991) Identification of soybean proteins responsible for respiratory allergies. Int Arch Allergy Appl Immunol 95:53–57

González-Buitrago JM, Ferreira L, Isidoro-García M, Sanz C, Lorente F, Dávila I (2007) Proteomic approaches for identifying new allergens and diagnosing allergic diseases. Clinica Chimica Acta 385(1–2):21–27 (ISSN 0009-8981, 10.1016/j.cca.2007.07.010)

Goodman RE, Vieths S, Sampson HA, Hill D, Ebisawa M, Taylor SL, van Ree R (2008) Allergenicity assessment of genetically modified crops—What makes sense? Nat Biotechnol 26:73–81

Gu X, Beardslee T, Zeece M, Sarath G, Markwell J (2001) Identification of IgE-binding proteins in soy lecithin. Int Arch Allergy Immunol 126:218–225

Hales BJ, Bosco A, Mills KL, Hazell LA, Loh R, Holt PG, Thomas WR (2004) Isoforms of the major peanut allergen Ara h2: IgE binding in children with peanut allergy. Int Arch Allergy Immunol 135:101–107

Hamilton AJ, Baulcombe DC (1999) A species of small antisense RNA in posttranscriptional gene silencing in plants. Science 286:950–952

Helm RM, Cockrell G, Connaughton C, West CM, Herman E, Sampson HA, Bannon GA, Burks AW (2000) Mutational analysis of the IgE-binding epitopes of P34/Gly m Bd 30K. J Allergy Clin Immunol 105:378–384

Henikoff S, Comai L (2003) Single-nucleotide mutations for plant functional genomics. Annu Rev Plant Biol 54:375–401

Herman EM, Helm RM, Jung R, Kinney AJ (2003) Genetic modification removes an immunodominant allergen from soybean. Plant Physiol 132:36–43

Holden L, Sletten GBG, Lindvik H, Fæste CK, Dooper MMBW (2008) Characterization of IgE binding to lupin, peanut and almond with sera from lupin-allergic patients. Int Arch Allergy Immunol 146:267–276

Holzhauser T, Wackermann O, Ballmer-Weber BK, Bindslev-Jensen C, Scibilia J, Perono-Garoffo L, Utsumi S, Poulsen LK, Vieths S (2009) Soybean (*Glycine max*) allergy in Europe: *Gly m5* (beta-conglycinin) and *Gly m6* (glycinin) are potential diagnostic markers for severe allergic reactions to soy. J Allergy Clin Immunol 123:452–458

Hourihane JOB, Kilburn SA, Nordlee JA, Hefle SL, Taylor SL, Warner JO (1997) An evaluation of the sensitivity of subjects with peanut allergy to very low doses of peanut protein: a randomized, double-blind, placebo-controlled food challenge study. J Allergy Clin Immunol 100:596–600

Kleber-Janke T, Crameri R, Appenzeller U, Schlaak M, Becker WM (1999) Selective cloning of peanut allergens, including profilin and 2S albumins, by phage display technology. Int Arch Allergy Immunol 119:265–274

Knoll J, Ramos ML, Zeng Y, Holbrook CC, Chow M, Chen S, Maleki S, Bhattacharya A, Ozias-Akins P (2011) TILLING for allergen reduction and improvement of quality traits in peanut (*Arachis hypogaea* L.). BMC Plant Biol 11:81

Koppelman SJ, De Jong GAH, Laaper-Ertmann M, Peeters KABM, Knulst AC, Hefle SL, Knol EF (2005) Purification and immunoglobulin E-binding properties of peanut allergen Ara h 6: evidence for cross-reactivity with Ara h 2. Clin Exp Allergy 35:490–497

Krause S, Reese G, Randow S, Zennaro D, Quaratino D, Palazzo P, Ciardiello MA, Petersen A, Becker WM, Mari A (2009) Lipid transfer protein (Ara h9) as a new peanut allergen relevant for a Mediterranean allergic population. J Allergy Clin Immunol 124:771–778

Krause S, Latendorf T, Schmidt H, Darcan-Nicolaisen Y, Reese G, Petersen A, Janssen O, Becker WM (2010) Peanut varieties with reduced *Ara h1* content indicating no reduced allergenicity. Mol Nutr Food Res 54:381–387

Krska R, Welzig E, Baumgartner S (2004) Immunoanalytical detection of allergenic proteins in food. Anal Bioanal Chem 378:63–65

Lamartiniere CA (2000) Protection against breast cancer with genistein: a component of soy. Am J Clin Nutr 71:1705S–1707S

Lauer I, Dueringer N, Pokoj S, Rehm S, Zoccatelli G, Reese G, Miguel-Moncin MS, Cistero-Bahima A, Enrique E, Lidholm J, Vieths S, Scheurer S (2009) The non-specific lipid transfer protein, Ara h9, is an important allergen in peanut. Clin Exp Allergy 39:1427–1437

Lien Quynh L, Vera M, Yvonne L, Stephan S, Sophia B, Stefan V, Uwe S (2006) Reduced allergenicity of tomato fruits harvested from *Lyc e 1*, silenced transgenic tomato plants. J Allergy Clin Immunol 118:1176–1183

Lin J, Shewry PR, Archer DB, Beyer K, Niggemann B, Haas H, Wilson P, Alcocer MJ (2006) The potential allergenicity of two 2S albumins from soybean (*Glycine max*): a protein microarray approach. Int Arch Allergy Immunol 141:91–102

López-Torrejón G, Salcedo G, Martín-Esteban M, Díaz-Perales A, Pascual CY, Sánchez-Monge RJ (2003) Len c1, a major allergen and vicilin from lentil seeds: protein isolation and cDNA cloning. J Allergy Clin Immunol 112:1208–1215

Magni C, Sessa F, Accardo E, Vanoni M, Morazzoni P, Scarafoni A, Duranti M (2004) Conglutin gamma, a lupin seed protein, binds insulin *in vitro* and reduces plasma glucose levels of hyperglycemic rats. J Nutr Biochem 15:646–650

Maleki SJ, Chung SY, Champagne ET, Raufman JP (2000) The effects of roasting on the allergenic properties of peanut proteins. J Allergy Clin Immunol 106:763–768

Maleki SJ, Viquez O, Jacks T, Dodo H, Champagne ET, Chung SY, Landry SJ (2003) The major peanut allergen, *Ara h2*, functions as a trypsin inhibitor, and roasting enhances this function. J Allergy Clin Immunol 112:190–195

Manjaya JG, Suseelan KN, Gopalakrishna T, Pawar SE, Bapat VA (2007) Radiation induced variability of seed storage proteins in soybean [*Glycine max* (L.) Merrill]. Food Chem 100:1324–1327

McCallum CM, Comai L, Greene EA, Henikoff S (2000) Targeting induced local lesions in genomes (TILLING) forplant functional genomics. Plant Physiol 123:439–442

Miller JC, Tan S, Qiao G, Barlow KA, Wang J, Xia DF, Meng X, Paschon DE, Leung E, Hinkley SJ, Dulay GP, Hua KL, Ankoudinova I, Cost GJ, Urnov FD, Zhang HS, Holmes MC, Zhang L, Gregory PD, Rebar EJ (2011) A TALE nuclease architecture for efficient genome editing. Nat Biotechnol 29:143–148

Mittag D, Akkerdaas J, Ballmer-Weber BK, Vogel L, Wensing M, Becker WM, Koppelman SJ, Knulst AC, Helbling A, Hefle SL, Van Ree R, Vieths S (2004a) Ara h8, a Bet v 1-homologous allergen from peanut, is a major allergen in patients with combined birch pollen and peanut allergy. J Allergy Clin Immunol 114:1410–1417

Mittag D, Vieths S, Vogel L, Becker WM, Rihs HP, Helbling A, Wüthrich B, Ballmer-Weber BK (2004b) Soybean allergy in patients allergic to birch pollen: clinical investigation and molecular characterization of allergens. J Allergy Clin Immunol 113:148–154

Mittag D, Vieths S, Vogel L, Wagner-Loew D, Starke A, Hunziker P, Becker WM, Ballmer-Weber BK (2005) Birch pollen-related food allergy to legumes: identification and characterization of the Bet v1 homologue in mungbean (Vigna radiata), Vig r1. Clin Exp Allergy 35:1049–1055

Mondoulet L, Drumare MF, Ah-Leung S, Paty E, Scheinmann P, Wal JM, Bernard H (2003) Influence of thermal processing on the IgE binding capacity of peanut allergens. Revue Francaise D Allergologie Et D Immunologie Clinique 43:486–491

Odanaka H, Kaizuma N (1989) Mutants on soybean storage proteins induced with gamma-ray irradiation. Jpn J Breed 39:430–431

Ozias-Akins P, Chu Y, Knoll J, Bhattacharya A (2010) Mitigating allergenicity of crops. Adv Agron 107:93–121

Pascual CY, Fernandez-Crespo J, Sanchez-Pastor S, Padial MA, Diaz-Pena JM, Martin-Muñoz F, Martin-Esteban M (1999) Allergy to lentils in Mediterranean pediatric patients. J Allergy Clin Immunol 103:154–158

Peeters K, Koppelman SJ, van Hoffen E, Van Der Tas CWH, Jager CFD, Penninks AH, Hefle SL, Bruijnzeel-Koomen C, Knol EF, Knulst AC (2007) Does skin prick test reactivity to purified allergens correlate with clinical severity of peanut allergy? Clin Exp Allergy 37:108–115

Peters S, Imani J, Mahler V, Foetisch K, Kaul S, Paulus KE, Scheurer S, Vieths S, Kogel KH (2011) Dau c 1.01 and Dau c 1.02-silenced transgenic carrot plants show reduced allergenicity to patients with carrot allergy. Transgenic Res 20:547–556

Phan TH, Kaizuma N, Odanaka H, Takahata Y (1996) Specific inheritance of a mutant gene controlling alpha, beta subunits-null of beta-conglycinin in soybean (*Glycine max* (L)Merrill*)* and observation of chloroplast ultrastructure of the mutant. Breed Sci 46:53–59

Pomes A, Butts CL, Chapman MD (2006) Quantification of Ara h1 in peanuts: why roasting makes a difference. Clin Exp Allergy 36:824–830

Rabjohn P, West CM, Connaughton C, Sampson HA, Helm RM, Burks AW, Bannon GA (2002) Modification of peanut allergen Ara h3: effects on IgE binding and T cell stimulation. Int Arch Allergy Immunol 128:15–23

Ramos M, Huntley J, Maleki S, Ozias-Akins P (2009) Identification and characterization of a hypoallergenic ortholog of *Ara h2.01*. Plant Mol Biol 69:325–335

Riascos JJ, Burks WA, Pons LA, Weissinger AK, Weissinger SM (2009) Identification of a soybean seed biotinylated protein as a novel allergen. J Allergy Clin Immunol 123:S24

Riascos JJ, Weissinger AK, Weissinger SM, Burks AW (2010) Hypoallergenic legume crops and food allergy: factors affecting feasibility and risk. J Agric Food Chem 58:20–27

Rihs HP, Chen Z, Ruëff F, Petersen A, Rozynek P, Heimann H, Baur X (1999) IgE binding of the recombinant allergen soybean profilin (rGly m3) is mediated by conformational epitopes. J Allergy Clin Immunol 104:1293–1301

Rougé P, Culerrier R, Thibau F, Didier A, Barre A (2011) A case of severe anaphylaxis to kidney bean: Phaseolin (vicilin) and PHA (lectin) identified as putative allergens. Allergy 66:301–302

Samoto M, Fukuda Y, Takahashi K, Tabuchi K, Hiemori M, Tsuji H, Ogawa T, Kawamura Y (1997) Substantially complete removal of three major allergenic soybean proteins (Gly m Bd 30K, Gly m Bd 28K, and the alpha-subunit of conglycinin) from soy protein by using a mutant soybean, Tohoku 124. Biosci Biotechnol Biochem 61:2148–2150

Sampson HA (1990) Peanut anaphylaxis. J Allergy Clin Immunol 86:1–3

San Ireneo MM, Ibáñez MD, Sánchez JJ, Carnés J, Fernández-Caldas E (2008) Clinical features of legume allergy in children from a Mediterranean area. Ann Allergy Asthma Immunol 101:179–184

Sánchez-Monge R, Blanco C, Perales AD, Collada C, Carrillo T, Aragoncillo C, Salcedo G (2000a) Class I chitinases, the panallergens responsible for the latex-fruit syndrome, are induced by ethylene treatment and inactivated by heating. J Allergy Clin Immunol 106:190–195

Sánchez-Monge R, Pascual CY, Díaz-Perales A, Fernández-Crespo J, Martín-Esteban M, Salcedo G (2000b) Isolation and characterization of relevant allergens from boiled lentils. J Allergy Clin Immunol 106:955–961

Sánchez-Monge R, Lopez-Torrejón G, Pascual CY, Varela J, Martin-Esteban M, Salcedo G (2004) Vicilin and convicilin are potential major allergens from pea. Clin Exp Allergy 34:1747–1753

Sander JD, Dahlborg EJ, Goodwin MJ, Cade L, Zhang F, Cifuentes D, Curtin SJ, Blackburn JS, Thibodeau-Beganny S, Qi Y, Pierick CJ, Hoffman E, Maeder ML, Khayter C, Reyon D, Dobbs D, Langenau DM, Stupar RM, Giraldez AJ, Voytas DF, Peterson RT, Yeh JR, Joung JK (2011)

Selection-free zinc-finger-nuclease engineering by context-dependent assembly (CoDA). Nat Methods 8:67–69

Schwab R, Ossowski S, Riester M, Warthmann N, Weigel D (2006) Highly specific gene silencing by artificial microRNAs in *Arabidopsis*. Plant Cell 18:1121–1133

Setchell KDR, Cassidy A (1999) Dietary isoflavones: biological effects and relevance to human health. J Nutr 129:758S–767S

Shin DS, Compadre CM, Maleki SJ, Kopper RA, Sampson H, Huang SK, Burks AW, Bannon GA (1998) Biochemical and structural analysis of the IgE binding sites on Ara h1, an abundant and highly allergenic peanut protein. J Biol Chem 273:13753–13759

Sicherer SH, Sampson HA, Burks AW (2000) Peanut and soy allergy: a clinical and therapeutic dilemma. Allergy 55:515–521

Sicherer SH, Muñoz-Furlong A, Godbold JH, Sampson HA (2010) US prevalence of self-reported peanut, tree nut, and sesame allergy: 11-year follow-up. J Allergy Clin Immunol 125:1322–1326

Skolnick HS, Conover-Walker MK, Koerner CB, Sampson HA, Burks W, Wood RA (2001) The natural history of peanut allergy. J Allergy Clin Immunol 107:367–374

Smith CR, Knowles VL, Plaxton WC (2000) Purification and characterization of cytosolic pyruvate kinase from *Brassica napus* (rapeseed) suspension cell cultures. Implications for the integration of glycolysis with nitrogen assimilation. Eur J Biochem 267:4477–4485

Song JJ, Smith SK, Hannon GJ, Joshua-Tor L (2004) Crystal structure of argonaute and its implications for RISC slicer activity. Science 305:1434–1437

Stanley JS, King N, Burks AW, Huang SK, Sampson H, Cockrell G, Helm RM, West CM, Bannon GA (1997) Identification and mutational analysis of the immunodominant IgE binding epitopes of the major peanut allergen *Ara h2*. Arch Biochem Biophys 342:244–253

Stevenson SE, Chu Y, Ozias-Akins P, Thelen JJ (2009) Validation of gel-free, label-free quantitative proteomics approaches: applications for seed allergen profiling. J Proteomics 72:555–566

Tada Y, Nakase M, Adachi T, Nakamura R, Shimada H, Takahashi M, Fujimura T, Matsuda T (1996) Reduction of 14–16 kDa allergenic proteins in transgenic rice plants by antisense gene. FEBS Lett 391:341–345

Takahashi K, Banba H, Kikuchi A, Ito M, Nakamura S (1994) An induced mutant line lacking the alpha-subunit of beta-conglycinin in soybean [*Glycine max* (L) Merrill]. Breed Sci 44:65–66

Teraishi M, Takahashi M, Hajika M, Matsunaga R, Uematsu Y, Ishimoto M (2001) Suppression of soybean β-conglycinin genes by a dominant gene, *Scg-1*. Theor App Genet 103:1266–1272

Thelen JJ (2009) Proteomics tools and resources for investigating protein allergens in oilseeds. Regul Toxicol Pharmacol 54:S41–45

Trinidad TP, Mallillin AC, Loyola AS, Sagum RS, Encabo RR (2010) The potential health benefits of legumes as a good source of dietary fibre. Br J Nutr 103:569–574

Urnov FD, Rebar EJ, Holmes MC, Zhang HS, Gregory PD (2010) Genome editing with engineered zinc finger nucleases. Nat Rev Genet 11:636–646

Valenta R, Linhart B, Swoboda I, Niederberger V (2011) Recombinant allergens for allergen-specific immunotherapy: 10 years anniversary of immunotherapy with recombinant allergens. Allergy 66:775–783

van Ree R, Chapman MD, Ferreira F, Vieths S, Bryan D, Cromwell O, Villalba M, Durham SR, Becker WM, Aalbers M, André C, Barber D, Cistero Bahima A, Custovic A, Didierlaurent A, Dolman C, Dorpema JW, Di Felice G, Eberhardt F, Fernandez Caldas E, Fernandez Rivas M, Fiebig H, Focke M, Fötisch K, Gadermaier G, Das RG, Gonzalez Mancebo E, Himly M, Kinaciyan T, Knulst AC, Kroon AM, Lepp U, Marco FM, Mari A, Moingeon P, Monsalve R, Neubauer A, Notten S, Ooievaar-de Heer P, Pauli G, Pini C, Purohit A, Quiralte J, Rak S, Raulf-Heimsoth M, San Miguel Moncin MM, Simpson B, Tsay A, Vailes L, Wallner M, Weber B (2008) The CREATE project: development of certified reference materials for allergenic products and validation of methods for their quantification. Allergy 63:310–326

Venn BJ, Mann JI (2004) Cereal grains, legumes and diabetes. Eur J Clin Nutr 58:1443–1461

Vereda A, van Hage M, Ahlstedt S, Ibañez MD, Cuesta-Herranz J, van Odijk J, Wickman M, Sampson HA (2011) Peanut allergy: clinical and immunologic differences among patients from 3 different geographic regions. J Allergy Clin Immunol 127:603–607

Vissers YM, Jansen AP, Ruinemans-Koerts J, Wichers HJ, Savelkoul HF (2011) IgE component-resolved allergen profile and clinical symptoms in soy and peanut allergic patients. Allergy 66:1125–1127

Waterhouse PM, Graham MW, Wang MB (1998) Virus resistance and gene silencing in plants can be induced by simultaneous expression of sense and antisense RNA. Proc Natl Acad Sci U S A 95:13959–13964

Wichers HJ, de Beijer T, Savelkoul HFJ, van Amerongen A (2004) The major peanut allergen Ara h 1 and its cleaved-off *N*-terminal peptide; possible implications for peanut allergen detection. J Agric Food Chem 52:4903–4907

Zoccatelli G, Pokoj S, Foetisch K, Bartra J, Valero A, Del Mar San Miguel-Moncin M, Vieths S, Scheurer S (2010) Identification and characterization of the major allergen of green bean (*Phaseolus vulgaris*) as a non-specific lipid transfer protein (Pha v3). Mol Immunol 47:1561–1568

Chapter 25
Improving Quality and Content of Oils in Seeds: Strategies, Approaches, and Applications Towards Engineering New Oilseed Crop Plants

Thomas Roscoe, Frédéric Domergue, Martine Devic and René Lessire

Abstract Plant oils are the major sources for human nutrition. There is increasing interest in the use of plant oils as renewable sources of industrial feedstocks. In order to alleviate the increasing demand for plant oils, omics approaches have been adopted to facilitate modification of the fatty acid composition in order to improve the nutritional properties and to generate specific physiochemical properties for industrial uses. An overview of the application of omics to aid progress in the engineering of oil quality and seed yield is presented in this chapter.

Keywords Lipid metabolic pathways · Seed and yield improvement · Transcription factors · Industrial applications · Human nutrition

25.1 Introduction

Seed oils have always been used for nutrition and are now increasingly used as industrial feedstocks. Today, increasing the seed oil content and oil yield per unit area of land are of paramount importance in supplying oil for food and non food applications. Most oleaginous plants characteristically accumulate oil in seeds (rapeseed, sunflower) or in fruit (palm, olive). Most of these oilseed crops are growing in temperate climate, need only limited amounts of fertilizers and present a high yield. Moreover, harvesting conditions and oil extraction process are well controlled along with readily availability of their processing facilities.

More than 85 % of the vegetable oil production is used for human nutrition (Drexler et al. 2003). The nutritional quality of vegetable oils is related to its FA composition, which influences the preservation (oxidation of the unsaturated FAs should be avoided), and the ratio between monounsaturated/saturated FAs should

R. Lessire (✉) · F. Domergue
Laboratoire de Biogenèse Membranaire, CNRS UMR 5200, Université Bordeaux Segalen, INRA Aquitaine BP 81, Villenave d'Ornon, France
e-mail: rene.lessire@biomemb.u-bordeaux2.fr

T. Roscoe · M. Devic
Seed Development Group, Laboratoire Genome et Developement des Plantes, IRD-UPVD-CNRS UMR5096, IRD Centre de Montpellier, 911 avenue Agropolis BP5045, 34032, Montpellier cedex 1, France

G. K. Agrawal, R. Rakwal (eds.), *Seed Development: OMICS Technologies toward Improvement of Seed Quality and Crop Yield*,
DOI 10.1007/978-94-007-4749-4_25, © Springer Science+Business Media Dordrecht 2012

be high to prevent cardiac diseases (Damude and Kinney 2008b; Kinney 1996). For example, cooking oils generally contain a higher proportion of monounsaturated FAs (such as oleic acid), which are more stable under high temperature, while margarines and spreads are often rich in saturated FAs (e.g., palmitic and stearic acids). Other oils, such as salad oils, contain more polyunsaturated FAs (e.g., linoleic and α-linolenic acids). Issues of oil quality and stability have been solved by breeding and recombining of desired traits. However, one important challenge is to increase the yield of oilseeds in order to anticipate the increasing demand due to the ever increasing world population (Damude and Kinney 2007, 2008a; Domergue et al. 2005a; Kinney 1996; Napier 2007; Napier et al. 2004; Napier and Graham 2010).

Vegetable oils mainly consist of TAGs, resulting from the esterification of three FAs to a glycerol molecule. These FAs, characterized by the generic formula $CH_3(CH_2)_nCOOH$, are chemically close to fossil hydrocarbons (gas and petroleum) as described by a $CH_3(CH_2)_nCH_3$ formula. For this reason, oilseed crops, which are generally rich in FAs, are considered as an alternative source to fossil energy, which after minor modifications could be submitted to a cracking process using the same equipment as petroleum chemistry (Abbadi et al. 2004; Broun et al. 1999; Dyer et al. 2008; Nikolau et al. 2008; Ohlrogge 1994; Ohlrogge et al. 2009).

On other hand, in the plant kingdom more than 200 different FAs structures exist, which differ by the C chain length, linear, and cyclic configurations. Diverse functional groups are introduced by hydroxylation, unsaturation, and epoxidation, which could be used as synthons for the chemical industry. In recent years, plant omics biology has advanced rapidly, allowing the development of new projects in agricultural production including metabolic engineering (Drexler et al. 2003; Dyer et al. 2008; Ohlrogge 1994). Among these, the modification of FA composition and oil content of oilseeds aims to improve oil quality for nutrition and to design new oils for industrial uses (Cahoon et al. 2007; Ohlrogge 1994; Somerville and Bonetta 2001).

25.2 Seed Lipid Composition and Metabolic Pathways

The largest part of plant oil production from the major oilseed crops is consumed as food. The oils used in these markets contain various proportions of the five common and nutritionally important FAs: palmitic (16:0); stearic (18:0); oleic (18:1Δ9); linoleic (18:2Δ9,12); and linolenic (18:3Δ9,12,15) acids (Table 25.1). The properties and value of oils depend on their FA composition, and certain compositions are desirable for specific end uses (Ohlrogge 1994). Besides the usual FA composition, exotic plants may accumulate unusual FAs with short C chain length (C8:0 to C14:0, Cuphea), long C chain length (C22:1, rape, Crambe), or hydroxyl FA (C18:0-OH, castor bean) (Table 25.2), which could be used as feedstocks for industrial applications (bio lubricants, plasticizers, inks, and paints) (Dyer et al. 2008). These exotic plants have a particular interest since they possess alternative pathways for the synthesis of diverse FAs in addition to the pathways controlling the

Table 25.1 Fatty acid composition of oils from major and minor oil plants

Major oil plants	Fatty acids (%)									
	8:00	10:00	12:00	14:00	16:00	18:00	18:01	18:02	18:03	20:01
Palm				5	36	2	50	8		
Soybean					11	4	23	54	8	
Canola					4	2	60	21	10	1
Sunflower					7	5	19	68		
Linseed					6	2	19	24	47	
Coconut	7	7	48	18	9	3	6	2		
Palm kernel	3	4	48	16	8	2	15	2		

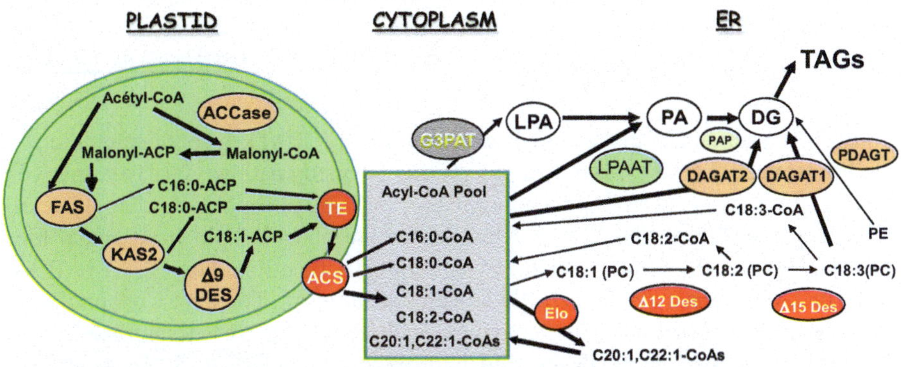

Fig. 25.1 Lipid synthesis pathway in Crucifer oilseeds. *ACCase* acetyl-CoA carboxylase, *ACS* acyl-CoA synthetase, *DAGAT* Diacylglycerol acyltransferase, *DES* Desaturase, *DG* Diacylglycerol, *Elo* Elongase, *ER* Endoplasmic reticulum, *FAS* Fatty acid synthase, *G3PAT* Glycerol-3-phosphate acyltransferase, *KAS* Ketoacyl-ACP synthase, *LPA* Lysophosphatitic acid, *LPAAT* Lyso phosphatidic acid acyltransferase, *PA* Phosophatidic acid, *PDAGT* Phospholipd diacykglyceride acyltransferase, *PC* Phosphatidylcholine, *PE* Phosphatidylethanolamine, *TAGs* triacylglycerides, and *TE* Acyl-ACP thiolase

ubiquitous FAs. As such, they represent a potential source of genes available for the modification of storage lipids in cultivated oilseeds (Cahoon et al. 2007; Somerville and Bonetta 2001).

The biosynthesis of lipids containing the five major FAs involves three subcellular compartments (Fig. 25.1). In plastids, palmitic (16:0) and stearic (18:0) acids are *de novo* synthesized from acetyl-CoA and malonyl-ACP with the nascent FAs esterified to ACP. Stearoyl-ACP is subsequently desaturated to oleic acid (18:1Δ9). A fraction of these three FAs is then exported to the cytosol to enter into the acyl-CoA and acyl-lipid pools. In the ER, oleic acid from these pools is incorporated into phosphatidylcholine and may be converted to linoleic (18:2Δ9,12) and to α-linolenic (18:3Δ9,12,15) acids by the sequential action of substrate-specific desaturases. A part of these modified fatty acyl products is incorporated as structural components of cellular membranes and in seeds, the remaining proportion may: (i) enter the acyl-CoA pool to be esterified to all three positions of TAG by the

Table 25.2 Unusual fatty acids in minor crops and transgenic plants

Unusual FA	Wild Plants		Transgenic plants	
	Content (%)		Content (%)	
γ Linolenic	*Borago officinalis*	23	*Brassica napus*	43
Caprylic (C8:0)	*Cuphea hookeriana*	50	*Brassica napus*	11
Capric (C10:0)	*Cuphea lanceolata*	83	*Brassica napus*	27
Ricinoleic (Δ12-OH, C18:1 Δ9)	*Ricinus communis*	89	*Arabidopsis thaliana*	42
			Brassica napus	16
ω7 fatty acids (C16:1, Δ9)	*Asclepias syriaca*	25	*Arabidopsis thaliana*	
and cis C:18:1,Δ11)	*Doxantha unguis-cati*	72		71
Vernolic (epoxy 12,13 C18:1Δ9)	*Vernolia galamensis*	67	*Arabidopsis thaliana*	15
Eleostearic (C18:3, trans Δ9, Δ11, Δ13)	*Momordica charantia*	57	*Glycine max*	18
Erucic (C22:1 Δ13)	*Crambe abyssinica*	56	*Brassica napus*	60
Waxes	*Simmondsia chinensis*	60	*Arabidopsis thaliana*	70

consecutive actions of glycerol-3-P acyltransferase (GPAT), lysophosphatidic acid acyltransferase (LPAAT), and DAGAT *via* the 'Kennedy pathway'; (ii) alternatively be directly transferred from phospholipids to TAGs by phospholipid:diacylglycerol (DG) acyltransferase (PDAT); or (iii) remove the phospholipid head group form DG (Dahlqvist et al. 2000), which can subsequently be utilized by DGAT or PDAT to form TAGs. Seeds synthesize TAGs with unusual FAs. The acyl-CoA pool composition is modified by diverse enzymes, such as Δ12 or Δ15 desaturases, acetylenases, conjugases, epoxydases, and hydroxylases, leading to TAGs containing polyunsaturated or hydroxyl FAs. In *Cuphea*, the plastidial acyl-ACP thioesterase selectively hydrolyzes the short chain acyl-ACP before their export to the cytosol (Dehesh et al. 1996a). Alternatively, for example in *B. napus*, the ER-located acyl-CoA elongase catalyzes to the formation of VLCFA from a oleoyl-CoA primer by the condensation of C2 units from malonyl-CoA, which may then be esterified to glycerol-3-P (Bao et al. 1998). Most of the genes coding for the different enzymes involved in the TAG biosynthesis are now well characterized and metabolic engineering approach has been undertaken to produce improved vegetable oils for industrial usage or human nutrition (Broun et al. 1999).

25.3 Manipulation of Oil Seed Composition Using Omics

Since the entire genome of *Arabidopsis thaliana* (a model for oilseeds) has been sequenced, an enormous quantity of DNA sequence data-derived from plants can be found in public databases. These databases and additional resources, such as transcriptomics data and mutant collections, constitute major tools for studying the different aspects of lipid metabolism. This abundance of sequence data has been rapidly exploited to identify genes and their probable function using homology searches and coexpression studies and, to reveal the presence of candidate genes encoding new proteins involved in lipid biosynthesis and regulation. It has been demonstrated that most of the genes encoding enzymes of lipid biosynthesis are represented by gene families frequently reflecting a diversification of isoform functions. Moreover, the development of microarrays, allowing the simultaneous expression profiling of thousands of genes, have led to better understanding of the mechanisms controlling lipid metabolism in seeds. With such omics tools in place, plant lipid metabolism has become increasingly accessible to investigation and subsequent manipulation of FA composition through genetic modification.

25.3.1 Seed Oil Composition Modifications for Industrial Applications

Medium chain FAs C8-C14 (MCF) are an important source for the industry, since they are used as cosmetics, soaps, and detergents (Dyer et al. 2008; Somerville and Bonetta 2001; Thelen and Ohlrogge 2002b). They are produced from coconut and palm kernel oils, and are also found in high proportion in the different species of *Cuphea* (Voelker et al. 1992; Voelker 1996), coconut, and California bay (*Umbellularia californica*). Lauric acid is the main product of the seed oil of California bay and it has been shown that a plastidial C12-specific acyl-ACP thioesterase was responsible for the enrichment of this FA in TAGs (Pollard et al. 1991). The acyl-ACP thioesterase was purified and a partial amino acid sequence was obtained, allowing the cloning the corresponding gene. The seed-specific expression of the cDNA of lauroyl-ACP thioesterase from California bay was resulted in 50–63 % C12:0 in transgenic canola (Voelker 1996; Voelker et al. 1992). Similar approaches have been made using genes coding from different *Cuphea* species with specificities for C8:0, C10:0, or C14:0, but no transgenic canola plants expressing one of these genes with a high content of MCFhave been generated to date (Dehesh et al. 1996a, b).

The identification of enzymes, which catalyze diverse modifications of the acyl moeties as hydroxylation, epoxydation, and double-bound conjugation (Cahoon et al. 1999; Nikolau et al. 2008) in seeds of cultivated plants, has formed the basis of the production of unusual FAs for novel industrial oils by engineering of oilseed crops (Cahoon et al. 2007; Voelker and Kinney 2001). It has been shown that all these enzymes are divergent forms of the Δ12 oleic desaturases, which catalyze the introduction of a *cis*Δ12 double bond in oleic acid to produce linoleic acid (van de Loo et al. 1995). By example, transgenic *Arabidopsis* and soybean mutants expressing the castor bean hydroxylase (FAH12) produce oil containing 20 % of ricinoleic acid (Δ12-OH oleic acid) (Broun and Somerville 1997; Lu et al. 2006; Smith et al. 2003). Coexpression the *R. communis* (castor) type-2 acyl-coenzyme A:DG acyltransferase (RcDGAT2) in FAH12 transgenic plants revealed that ricinoleic acid could increase nearly 30 % (Burgal et al. 2008). This oil enriched in ricinoleic acid has properties desirable in nylon production and for soaps and resins (Cahoon et al. 2007).

The enrichment in the content of calendulic acid (C18:3 Δ8 *trans*, 10 trans, 13 *cis*), which improves drying properties of the oil for paint, ink, and other coating applications, has been achieved by transferring the gene from *Calendula offiinalis* in *Arabidopsis* (Cahoon et al. 2006). Oilseeds with 5–18 % of α-eleostearic and α-parinaric acids were obtained by transferring the genes of two different Δ12 conjugases from *Momordica charantia* and *Impatiens balsamine* (Cahoon et al. 1999, 2001).

Plant oil with a high content of ω-7 FAs would present a new and sustainable feedstock as 1-octene could be obtained by ethenolytic metathesis. Heterologous expression of Δ9-C16:0-ACP desaturases genes responsible for the production of ω-7 FAs from milkweed (*Asclepias syriaca*) and cat's claw (*Doxantha unguis-cati*) failed (Cahoon et al. 1997, 1998). Recently, the expression of engineered plastidial C16:0-ACP desaturase in *Arabidopsis* seeds increased accumulation of ω-7 FAs up to 71 %, which is equivalent to levels found in *Doxiantha* seeds (Nguyen et al. 2010).

Another strategic goal in oilseed modification is to genetically increase the content of erucic acid ($C_{22:1}$, Δ12) in rapeseed. Erucic acid and its derivatives are feedstocks in manufacturing slip-promoting agents, surfactants, plasticizers nylon 1313, and wax coatings. Erucic acid results from the elongation of oleoyl-CoA by the acyl-CoA elongase (KCS) (Domergue et al. 2000; Puyaubert et al. 2001). A *B. napus* able to synthesize an oil containing 80 % of erucic acid would comprise an economically viable and renewable feedstock, which would significantly reduce the production cost and could meet the increasing demand. The first attempt to increase erucic acid content was genetic manipulations with the LPAAT, since in *B. napus* it had been demonstrated that this enzyme does not allow esterification of the sn-2 position of the glycerol backbone by erucoyl-CoA (Berneth and Frentzen 1990). Contrary to expectation, the content of erucic acid of transgenic rapeseed expressing the LPAAT from *Limnanthes alba*, which is able to esterify erucic acid at the sn-2 position, enabled the synthesis of trierucin but the total C22:1 content remained unchanged (Lassner et al. 1995). This result was interpreted as reflecting a low erucoyl-CoA pool available for TAG biosynthesis (Lassner et al. 1995) such that enhanced elongase activity (via KCS) was required. The condensing enzymes from the elongase complex from *B. napus* and LPAAT from *Limnanthes douglasii* have been coexpressed in high erucic acid rapeseed (HEAR), however the content of trierucin in the oil of transformants was modest (Lassner et al. 1995). In contrast, the seed-specific expression of KCS from garden nasturtium (*Tropaeolum majus*) resulted in up to an eight-fold increase in erucic acid proportions in *Arabidopsis* seed oil (Mietkiewska et al. 2004). Surprisingly, the SLC1 gene coding for a putative yeast sn-2 acyltransferase gene (SLC1-1) was introduced into *B. napus* leading to substantial increase of 8–48 % in seed oil content and increasing amounts of VLCFA in seed TAGs. Furthermore, the proportion of VLCFA found at the sn-2 position of TAGs was increased (Zou et al. 1997). Recently, increased levels of erucic acid in *B. carinata* and *B. napus* were observed by cosuppression and antisense repression of the endogenous *FAD2* gene (Jadhav et al. 2005; Nath et al. 2009).

Certain seeds, such as jojoba (*Simmondsia chinensis*), store wax esters instead of TAGs led to attempts to develop cultivated oilseeds able to synthesize these compounds. The jojoba oil is used as lubricant and cosmetics, a mixture of esters of long-chain FAs and alcohols, which result from the reduction of FA to alcohol by an acyl-CoA reductase (Cases et al. 2001) and by the esterification of the resulting alcohol to a FA by a wax synthase (WS) (Metz et al. 2000). Both enzymes have been cloned from jojoba and coexpressed with a KCS from *Lunaria annua* in *B. napus* (Metz et al. 2000) and in *Arabidopsis*. In the seeds of some *Arabidopsis* transformants, 70 % of the oil was comprised of wax esters (Lardizabal et al. 2000).

In conclusion, omics technologies have been very useful in identifying and characterizing new genes responsible for the synthesis of unusual FAs. Moreover, growing evidence demonstrate that gene transfer from wild plants to usual crops is possible and leads to the modification of FA composition of oilseeds. Nevertheless, in most of cases the production of the desire FA remains lower than observed in the wild plant, suggesting possible differences in regulation and channelling mechanisms involved in lipid metabolism in plants.

25.3.2 Improvement of Seed Oil Composition for Human Nutrition

Since higher animals lack the genes coding for the Δ12 and Δ15 desaturases, responsible for linoleic (18:2ω6) and α-linolenic (18:3ω3) acids biosynthesis, these polyunsaturated FAs (PUFAs) are obtained from their diet. This absolute dietary requirement explains the importance of plant oils in human nutrition. Nowadays, about 85 % of worldwide vegetable oils are used for food applications. For decades, breeding programs have been dedicated to improve the composition of plant oils in order to fulfil human nutrition requirements. Such a strategy led to the generation of HEAR varieties from which canola oil is produced. Canola oil contains very low proportions of the toxic compounds erucic acid (22:1) and glucosinolates as well as about 20 % linoleic (18:2ω6) and 10 % α-linolenic (18: 3ω3) acids, which results in an ω6 to ω3 ratio close to 2:1. Such a ratio is recommended by nutritionists, because numerous epidemiological studies have pointed out that inappropriate ω6 to ω3 balances result in various metabolic disorders, such as cardiovascular diseases, cancer, and autoimmune diseases (reviewed in Simopoulos 2002). Nevertheless, the human diets of most Western societies are characterized by their deficit in ω3 FAs and their consequent elevated ω6 to ω3 ratios (close to 10:1). In addition, the capacity of mammals to convert linoleic (18:2ω6) and α-linolenic (18:3ω3) acids into the corresponding very long-chain polyunsaturated FAs (ω3-VLCPUFAs) is limited. Since numerous medical studies have in particular pointed to the importance of ω3-VLCPUFAs, such as eicosapentaenoic acid (EPA; 20:5) and docosahexaenoic acid (DHA; 22:6) in human metabolism (Tapiero et al. 2002), a regular intake of these w3-VLCPUFA by consuming oil fish two to three times a week is strongly recommended. Neither natural marine fish resources nor global fisheries will be able on the long term to fulfil the ever increasing demand of the human population. For these reasons, the production of 'fish oil' FAs in oilseed crops by genetic engineering has been undertaken (Domergue et al. 2005a, b; Venegas-Calerón et al. 2010).

As marine microorganisms are the primary producers of ω3-VLCPUFAs, they have been used as gene sources in order to implement ω3-VLCPUFAs biosynthesis in plants. With the exception of an anaerobic polyketide-based pathway operating in few organisms, like *Shewanella* (Takeyama et al. 1997) or *Shizochytrium* (Metz et al. 2001), VLCPUFAs are usually made by successive aerobic desaturation and elongation steps using 18:2ω6 or 18:3ω3 acids as precursors (reviewed in Damude and Kinney 2007). Typically, in the so called ω6-pathway, 18:2ω6 is converted into 20:4ω6 by the successive action of a Δ6-desaturase, a Δ6-elongase, and a Δ5-desaturase. In the ω3-pathway, the same sequence of enzymes convert 18:3ω3 into 20:5ω3 EPA, while additional Δ5-elongase and Δ4-desaturase then convert EPA (20:5ω3) into DHA (22:6ω3). An alternative Δ8-pathway relaying on a Δ9-elongase followed by a Δ8-desaturase (instead of the conventional Δ6-desaturase – Δ6-elongase sequence) was first described in *Euglena gracilis* (Wallis and Browse 1999). In addition, ω3-desaturases were shown to interconnect the ω6-pathway together with the ω3-pathway by converting 18- and 20-C atoms ω6-PUFAs into the corresponding ω3. Finally, bifunctional proteins catalyzing two steps in VL-CPUFAs biosynthesis, like for example Δ12/ω3-desaturase (Damude et al. 2006) or Δ5/Δ6-elongase (Meyer et al. 2004), were also identified. Over the last 15 years, using genomic databases and protein homologies, a vast catalog of sequences coding for these various activities has been created by gene-cloning from various sources (algae, fungi, mosses as well as plants and animals). Functional characterization of the encoded proteins was then usually made by heterologous expression in yeast (Meyer et al. 2004). Subsequently the genes of interest were used by several groups to reverse-engineer the ω3-VLCPUFAs trait in transgenic plants, the first three reports appearing together in 2004 (Abbadi et al. 2004; Kinney et al. 2004; Qi et al. 2004).

The alternative Δ8-pathway was reconstituted in *Arabidopsis* leaves (Qi et al. 2004), yielding about 6.6 % of arachidonic acid (20:4ω6; ARA) and 3.0 % EPA (Table 25.3). Using seed-specific promoters and the classical ω6-pathway, Abbadi et al. (2004) reported the production of ARA in tobacco (1.5 %) as well as the production of both ARA and EPA in linseed (1 and 0.8 %, respectively; Table 25.3). Finally, (Kinney et al. 2004) generated soybean seeds with about 2 % ARA but more than 19 % EPA by reconstituting the classical ω6-pathway and expressing simultaneously two additional ω3-desaturases (Table 25.3). Although all these strategies relied on different combinations of genes and plant host systems, they all successfully led to the synthesis of the desired ω3-VLCPUFAs ('proof of concept'). Nevertheless, the final levels of these FAs were in most cases relatively low and in-depth lipid analyses revealed the highly complex biochemistry encountered when attempting to synthesize ω3-VLCPUFAs and to accumulate these non-native FAs in oilseed crops. First, several enzymes have shown *in planta* substrate specificities that were different from those observed when they have been functionally characterized in yeast, especially when expressed in multiple combinations. Competitive and promiscuous activities resulted in the production of non-methylene-interrupted PUFAs (Qi et al. 2004). Second, the same gene (or combination of genes) expressed in different host plants led to different FA compositions, underlining the necessity

Table 25.3 Seed-specific production of VLCPUFAs obtained by different transgenic approaches. This table summarizes the different achievements to produce VLCPUFAs in seeds using transgenic approaches. For simplicity, the exact genes used have not been indicated but the reference to find such information and additional details are included

Year	Host plant	Pathway	Additional genes	ω3-VLCPUFAs yields	Reference
2004	*Linum usitatissimum* (Linseed)	Classical Δ6-pathway	none	ARA (2,7 %), EPA (1,6 %)	Abbadi et al. (2004)
2004	*Glycine max* (soybean)	Classical Δ6-pathway	ω3-desaturases	ARA (2,2 %), EPA (19,6 %)	Kinney et al. (2004)
2005	*Arabidopsis thaliana*	Classical Δ6-pathway[1]	None	ARA (1,6 %), EPA (3,2 %)	Robert et al. (2005)
2005	*Arabidopsis thaliana*	Classical Δ6-pathway	Δ4-desaturase, Δ5/Δ6-elongase	DHA (0,5 %)	Robert et al. (2005)
2005	*Brassica juncea*	Classical Δ6-pathway	ω3- and Δ12-desaturase, Δ6-elongase	ARA (5,4 %), EPA (8,1 %)	Wu et al. (2005)
2005	*Brassica juncea*	Classical Δ6-pathway	Δ4-desaturase, Δ5/Δ6-elongase, LPAT[2,3]	DHA (0,2 %)	Wu et al. (2005)
2006	*Glycine max* (soybean)	Classical Δ6-pathway	RNAi for Δ15-desaturase[4]	ARA (2,1 %)	Chen et al. (2006)
2008	*Arabidopsis thaliana*	Acyl-CoA Δ6-pathway	None	EPA (0,045 %)	Hoffmann et al. (2008)
2008	*Glycine max* (soybean)	Acyl-CoA Δ6-pathway[5]	None	ARA (2,5 %), EPA (0,03 %)	Kajikawa et al. (2008)
2010	*Brassica carinata*	Classical Δ6-pathway	ω3- and Δ12-desaturase	ARA (5,7 %), EPA (20,4 %)	Cheng et al. (2010)

[1] A bifunctional Δ5/Δ6-desaturase was used in this study
[2] LPAT stands for LysoPhosphatidyl Acyl Transferase
[3] For DHA biosynthesis, these 3 genes were added to the classical Δ6-pathway already supplemented with ω3-, Δ12-desaturases and Δ6-elongase (9 genes construct)
[4] An inverted repeat structure was used to down-regulate the endogenous Δ15-desaturase
[5] According to Kajikawa et al. (2004)

to test many different strategies and to perform iterative approaches together with lipidomic analyses in order to find the best combination for producing high levels of the desired FAs (trial and error approach) (Abbadi et al. 2004). Third, kinetic studies together with acyl-CoAs profiling analyses in yeast and plants demonstrated that the elongases used acyl-CoAs as substrates whereas most of the desaturases were active on glycerolipid-linked acyl-chains (Abbadi et al. 2004; Domergue et al. 2003). This 'substrate dichotomy' led to inefficient channelling of intermediates between the different activities, the products of Δ6-desaturase being for example poorly transferred from glycerolipids to the acyl-CoA pool for further elongation. Fourth, the different strategies to implement ω3-VLCPUFAs in plants showed that the TAG assembling machineries of the different host plants were not optimal for the accumulation of the desired nonnative FAs, leading to the presence in the oil of relatively high amounts of metabolic intermediates, in contrast to the situation found in fish oils. Finally, coexpression of ω3-desaturases in addition to the classical enzymes required for ω3-VLCPUFAs biosynthesis (Δ6-desaturase+Δ6-elongase+Δ5-desaturase) appeared crucial in order to boost the production of ω3-VLCPUFAs (Kinney et al. 2004).

Since these pioneering works, many studies on the production of VLCPUFAs in plants have been reported (Chen 2010; Chen et al. 2006; Hoffmann et al. 2008; Kajikawa et al. 2008; Petrie et al. 2010b; Robert et al. 2005; Wu et al. 2005). These attempts differed in the targeted FA (ARA, EPA, or DHA), the metabolic route chosen (conventional or acyl-CoA pathway), the set of genes used as well as in the host organism, and the promoters selected. If the use of Δ6-acyl-CoA desaturase allowed to circumvent the bottleneck resulting from 'substrate dichotomy' (Domergue et al. 2005a, b), the reconstitution of an exclusively acyl-CoA dependant pathway in plants did not significantly improve the yields of ARA, EPA, or DHA for reasons which remain obscure (Hoffmann et al. 2008; Kajikawa et al. 2008; Petrie et al. 2010b). In contrast, using the conventional Δ6-pathway together with Δ12- and ω3-desaturases, (Cheng et al. 2010) succeeded very recently in producing up to 20 % EPA in zero-erucic *B. caritana*, confirming that the choice of the host organism as well as the use of additional ω3-desaturases are key factors for reaching high yields of ω3-VLCPUFAs in oleaginous crops. Finally, if the reported DHA levels in seeds are rather low [< 1.5 %: (Robert et al. 2005; Wu et al. 2005)], the recent development of a rapid assay to test seed-specific multistep recombinant pathways in leaf tissue (Petrie et al. 2010a) should enable to characterize genes allowing efficient conversion of EPA to DHA. In addition, since PKS-dependent DHA synthesis was shown to function in plants, this alternative pathway may represent another solution to produce ω3-VLCPUFAs in oleaginous crops. Finally, it should be added that increasing the antioxidant seed oil content by enhancing Vitamin E biosynthesis, which prevents the oxidation of the stored oil (in particular that of ω3-PUFAs), is currently undertaken in parallel as a complementary strategy (Hunter and Cahoon 2007).

In conclusion, genomic projects have been very useful for the identification of genes coding for the different activities involved in ω3-VLCPUFAs biosynthesis.

In a second step, iterative transgenic approaches together with lipidomic analyses of the engineered oils resulting from the different strategies allowed the identification of the best gene combinations and host plants to produce these beneficial FAs in oilseed crops. Finally, the recent report on the production of around 20 % EPA in zero-erucic *B. caritana* (Cheng et al. 2010) suggests that the production of ω3-VLCPUFAs in oilseed crops by transgenic metabolic engineering will very soon reach the production phase.

25.3.3 Bottlenecks of Oil Seed Composition Using Gene Transfer

Despite the success in modifying cultivated oilseeds to produce novel industrial oils in engineered plants, the amounts of unusual FAs are considerably lower than those found in seeds from wild plants (Table 25.2), implying the existence of 'bottlenecks' in FA flux (Bach et al. 2008; Cahoon et al. 2007; Drexler et al. 2003). In the case of transgenic crop plants modified to synthesize unusual unsaturated or hydroxyl FAs, it has been shown that unusual FAs accumulate bound to PC (Cahoon et al. 2006) or in the acyl-CoA pool (Larson et al. 2002). This was not the case in the exotic plant of origin where the unusual FAs accumulate in large quantities in the oil, suggesting that efficient membrane editing and channelling of unusual FAs are apparently lacking in the engineered plants. Similarly, the introduction of the sole gene of the Δ6 desaturase was not sufficient for the production of petroselenic FA in tobacco (Cahoon et al. 1992). It has been necessary to decrease the oleic acid biosynthesis and to transfer the gene of the specific enzymes (Δ4 desaturase, Δ6 Acyl-ACP thiosterase, KAS 1) responsible for the synthesis of this unusual FA (Cahoon and Ohlrogge 1994; Cahoon et al. 1992). These results strongly indicate the existence of a selective channel for the synthesis of unusual FAs. An additional degree of complexity has become evident recently, since it has been reported that key enzymes in TAG biosynthesis, such as DAGAT 1 and 2 on one hand and GPAT 8 and 9 in other hand, are located in different ER subdomains (Shockey et al. 2006). In view of these results, it could be hypothesized that the enzymes belonging to distinct pathways are organized in a large metabolon complex and locate in different domains of the ER membrane. In order to improve the yield of the production of unusual FAs in seeds of transgenic plants: (i) the introduction of the genes controlling the complete biosynthesis pathway seems necessary; (ii) the encoded proteins should be also targeted in the appropriate membrane domain; and (iii) the competing activities of the endogenous enzymes involved in the synthesis of endogenous TAGs should be reduced. Another limitation of the overproduction of unusual FAs in engineering plants is the activation of futile cycling of unusual FAs through degradative pathways (i.e, β-oxidation) in peroxisomes, and that may lead to a decrease in the accumulation of the unusual FAs, which are not efficiently incorporated into TAGs (Eccleston and Ohlrogge 1998; Moire et al. 2004).

25.4 Discovery-Driven Seed and Yield Improvement

25.4.1 Enhancing Seed Oil Content with Omics Technologies

Yield improvement in oilseeds can be addressed by increasing the quantity of oil per seed, by increasing the size of the seed, or by increasing the number of seeds per plant. Increases in seed oil content have been achieved principally through conventional breeding aided by the identification of QTL and facilitated by marker-assisted selection of favourable alleles. Recent attempts to improve seed oil content have been made by transgenic approaches aimed at stimulating the provision of precursors for lipid synthesis, such as glycerol-3-P (Vigeolas et al. 2007) and acetyl-CoA (Abbadi et al. 2004; Marillia et al. 2003), by increasing the rate of FA synthesis (Roesler et al. 1997) and by enhancing TAG assembly by stimulating glycerolipid acyltransferase activity (Jain et al. 2000; Maisonneuve et al. 2010; Taylor et al. 2001; Weselake et al. 2008). Attempts have also been made to alter the flux of carbon to favor TAG synthesis by manipulating those TFs that exert developmental control over reserve accumulation (Mu et al. 2008) or factors that control sugar metabolism and FA biosynthesis (Shen et al. 2010). The application of transcriptomics, proteomics, and metabolomics to this transgenic material is providing valuable insight into the gene regulatory networks and is revealing the importance of posttranscriptional and metabolic regulation.

25.4.2 Engineering Yield by Altering the Expression of Genes Related to the Control of Lipid Synthesis and Carbon Partitioning

Fatty acid (FA) Synthesis

ACCase The multisubunit form of ACCase, considered to be important in the regulation of FA synthesis in plants (Ohlrogge and Jaworski 1997), uses acetyl-CoA as a substrate for the synthesis of malonyl-CoA, the precursor for chain elongation during *de novo* FA synthesis in the plastid (Fig. 25.1). Overexpression of the cytosolic multifunctional form of ACCase in the plastids of rapeseed resulted in a modest increase in seed lipid content (Roesler et al. 1997). In contrast, overexpression of the gene encoding the biotin carboxyl carrier protein 2 (BCCP2) subunit of the plastidial ACCase resulted in a lower seed oil content, elevated sugar, and protein contents (Thelen and Ohlrogge 2002a). This phenotype was associated with an incomplete biotinylation of BCCP2, resulting in the formation of nonfunctional ACCase complexes and, in consequence, a severely reduced ACCase activity. Chen et al. (2009) conducted a combined transcriptome and proteome analysis of what constitutes a unique dominant-negative mutant in *de novo* FA synthesis. These analyses revealed the complexity of interactions among components of the glycolytic pathway, pyruvate dehydrogenase, FA synthase, and lipid biosynthetic enzymes, highlighting the importance of both transcriptional and posttranscriptional control mediated by feedback and feed-forward regulation. The study provides insight into the regulation of carbon partitioning between oil and protein, since the analyses revealed an increased synthesis of certain amino acids and an enhanced SP content in seeds deficient in *de novo* FA synthesis. This finding confirms that oil and protein synthesis pathways share common carbon precursors, and that reducing malonyl-CoA flow into FAs increases the availability of C for amino acid and protein synthesis.

Fatty acid (FA) Utilization

LPAAT and DAGAT TAGs are assembled in the ER *via* the sequential acylation of glycerol-3-P under the control of membrane bound glycerolipid acyltransferases. LPAAT controls the conversion of 1-acyl-glycerol-3-P to phosphatidic acid, which will be dephosphorylated to form DG. The transformation of *Arabidopsis* and *B. napus* with a variant yeast LPAAT altered the stereochemical composition at *sn-2* of the TAG and resulted in an unexpected increase in seed oil content and size (Taylor et al. 2001; Zou et al. 1997). Similarly, the expression of *B. napus* genes encoding microsomal LPAAT isoforms in seeds of *Arabidopsis* resulted in an unanticipated enhancement of seed total FA content and an increase in seed weight (Maisonneuve et al. 2010). In parallel studies, overexpression of the gene encoding DAGAT1, thought to represent a limiting step in acyl-CoA dependent TAG synthesis, led to an enhanced DAGAT activity and resulted in an increase in seed oil content and size (Jako et al. 2001). Each of these reports argued that an increased flux through the Kennedy pathway enhanced the sink size for storage lipid synthesis and depleted pathway intermediates, which induced feedback stimulation of FA synthesis. This hypothesis was validated in part by the work of Sharma et al. (2008), who conducted transcript and hormone metabolite profiling of microspore derived embryos isolated from rapeseed lines over-expressing DAGAT1 and differing in seed oil content. The objective was to obtain insight into the regulation and synthesis of TAG based on the prediction that a greater DAGAT1 activity would induce secondary metabolic or regulatory effects which contribute to the phenotype. Despite a partially representative seed array, this expectation was justified since transcriptional changes in seed-expressed genes were observed in the transgenic embryos, including those encoding other Kennedy pathway enzymes which were upregulated. Relatively few genes were altered in expression. These included genes encoding enzymes controlling FA and lipid synthesis, transport, storage, and metabolism but also included the genes associated with acetyl-CoA production, mitochondrial PDC, and cytosolic isocitrate lyase which were upregulated. Of the lipid metabolism related genes present on the microarray, approximately one-fifth were regulated in the *DAGAT1* overexpressing lines. At least one quarter of the regulated genes were of unknown identity or function and these constitute genes whose potential to influence or optimise TAG accumulation merits further investigation.

25.4.3 Transcription Factors (TFs) Related to the Seed Maturation Program

Regulation of Lipid Synthesis in Seeds

TFs Transcriptional regulation is an important component in the control of storage lipid synthesis in seeds as evidenced by the coordinate expression of groups of genes and proteins involved in carbon mobilization, FA synthesis, and TAG assembly during seed development as revealed using transcriptomics analyses (Ruuska et al. 2002). It has been argued that TFs may control several or all of the genes in a pathway and thereby constitute the most effective means to manipulate carbon flow towards TAG synthesis. Transcriptomics analyses have provided insight as to the nature of the gene regulatory networks that coordinate changes in metabolism during seed development and have led to the identification of regulatory factors whose expressions are correlated with changes in gene expression and metabolism. Such factors represent candidates for altering carbon flow into lipid synthesis in seeds. TFs regulating storage lipid synthesis have also been identified: (i) as genetic loci controlling diverse aspects of seed maturation; (ii) by screens for altered lipid composition or content in mutagenesis programs; or (iii) by extrapolation from inferred functional homologies.

LEAFY COTYLEDON1 (LEC1) The *LEC1* encodes a homolog of the HAP3 subunit of the CCAAT box binding factor. LEC1 is a transcriptional regulator that is involved in determination of cotyledon identity and is required for embryo maturation. Ectopic expression of *LEC1* induces the expression of embryo-specific genes, including those controlling the synthesis of seed reserves in vegetative cells (Lotan et al. 1998). Transcriptomic analysis of germinating seeds overexpressing *LEC1*, and characterized by elevated FA and lipid contents, revealed that genes related to FA metabolism represented a significant proportion of the upregulated genes, including approximately two-thirds of the genes encoding plastidial FA synthesis enzymes, as were genes encoding seed SP and oleosin (Mu et al. 2008). The study also showed that LEC1 enhancement of lipid synthesis was partially dependent on ABI3 and WRI1 and that, FUSCA3 may be essential for the stimulation of FA synthesis by LEC1. The maize *ZmLEC1* gene was shown to enhance maize seed oil content primarily by enhancing accumulation in the embryo and accompanied by an increase in embryo size, however negative effects on seed germination and leaf growth compromise the use of LEC1 to enhance seed yield (Shen et al. 2010).

B3 Domain TFs LEC2, FUSCA3,and ABI3 contain a B3 DNA-binding domain and are recognized as major regulators of seed maturation. Transcriptome analyses of seedlings ectopically expressing *LEC2* revealed the expression of transcripts characteristic of the maturation phase of seed development, including seed SPs and OB proteins, indicating that *LEC2* controls a transcriptional network of genes expressed during seed development (Braybrook et al. 2006). Furthermore, the promoters of these maturation genes contain RY motifs that are directly bound by LEC2 (Braybrook et al. 2006). FA- and lipid synthesis-related genes (with the exception of OB proteins) were not identified as regulated genes in the *LEC2* transcriptome, yet ectopic expression of *LEC2* induced TAG accumulation in leaves, which was argued to be a consequence of the regulation of *WRI1* by *LEC2* (Baud et al. 2007).

A delay (Wang et al. 2007) or reduction (Chiron et al. 2007) in the *FUSCA3* expression is associated with lower seed total FA contents, implying that FUSCA3 is a transcriptional activator of FA or lipid biosynthetic genes. Mining of *Arabidopsis* seed transcriptome databases revealed that the expression of most of the genes controlling plastidial FA biosynthesis was coincident with that of FUSCA3 (Wang et al. 2007). Induction of FA synthesis genes after stable and transient expression of *FUSCA3* in plants and protoplasts confirmed FUSCA3 as a major regulator of seed lipid synthesis.

The total FA content in seeds of the *abi3* mutant was reduced as was the quantity of VLCFAs (Chiron et al. 2007). The constitutive expression of *VP1*, a homolog of *ABI3*, in an *Arabidopsisabi3* mutant followed by microarray analyses led to the identification of a large number of *VP1/ABI3*-regulated genes (Suzuki et al. 2003). Results showed the ectopic expression of many seed-specific genes, including SSPs, oleosins, and LEAs, confirming that these factors are regulators of seed-specific gene expression (Suzuki et al. 2003), although *ABI3* may act as a coregulator of certain FA- and storage lipid-related genes.

WRINKLED1 (WRI1) Seeds of the *wri1* mutant are characterized by a severe reduction in TAG, an unchanged protein content, a greater content of starch and sugars, a consequence of lower activities of several glycolytic pathway enzymes (Focks and Benning 1998). An early transcriptomics analysis of the effect of *wri1* mutation during seed development revealed regulation of relatively few genes, the majority of which were expressed at a lower level in *wri1* compared to wild type. Few of the regulated genes encoding enzymes of plastidial glycolysis and *de novo* FA biosynthesis were downregulated, implying the importance of metabolic regulation of these pathways (Ruuska et al. 2002). *WRI1* was subsequently identified as an AP2/ERBP class TF and constitutive expression led to an increase in seed oil and the accumulation of TAG in leaves (Cernac and Benning 2004). Transcriptomics analyses of *WRI1* overexpression plants during seedling growth confirm that the regulated genes include those controlling *de novo* FA synthesis in plastids, uptake of glycolytic intermediates conversion to acetyl-CoA and malonyl-CoA, and synthesis of cofactors lipoic acid and biotin required respectively for ACCase and the pyruvate dehydrogenase complex (Maeo et al. 2009). Overexpression of *B. napus* genes encoding WRI1 isoforms in *Arabidopsis* seeds resulted in increased oil content, seed weight, and seed size, a consequence of increased cell size. Polymorphisms in *WRI1* correlated with rapeseed lines that vary in oil content (Liu et al. 2010). Similarly, overexpression of maize *ZmWRI1* greatly increased the oil content in the embryo without affecting protein content, whereas starch content was reduced by 60 % (Shen et al. 2010). This finding suggests that WRI1 may constitute a way in which to regulate C flux between starch and oil in maize embryos and, in combination with overexpression of genes (such as DAGAT1) may lead to enhanced yields at the crop level (Shen et al. 2010).

GLABRA2 GLABRA2 (*GL2*) is a homeobox gene necessary for trichome and root hair development. *GL2* was identified as a regulator of seed oil accumulation in a screen of mutants affected in seed oil content. Loss of *GL2* enhances seed oil content up to 8 % but does not affect seed size (Shen et al. 2006). The overexpression of *Brassica GL2* homolog in *Arabidopsis* seeds resulted in enhanced oil content in seeds exhibiting a GL2 suppression phenotype (Chai et al. 2010). The manner in which GL2 exerts over seed oil content remains obscure, however, it is not mediated through the control of seed expressed phospholipase D zeta isoforms by GL2 (Shi et al. 2010).

DOF-type factors A DOF TF has been implicated in the regulation of maize endosperm SPs by binding to the prolamin box and by interaction with the *O2* transcription regulator (Vicente-Carbajosa et al. 1997). Expression of the soybean DOF4 and DOF11 factors were reported to increase total seed FA and lipid content, accompanied by increase in seed size and weight in *Arabidopsis* (Wang et al. 2007). Transcriptomic analyses revealed that a large proportion of the regulated genes are of unknown function. The *DOF4* and *DOF11* were shown to transactivate the β-subunit of ACCase and a long chain acyl-CoA synthetase gene and to repress a seed SP gene *CRA1*. Furthermore, whereas DOF11 increased all classes of FAs, DOF4 increased selectively C18:3 and C20:1 FAs.

In conclusion, DNA microarrays have been used to examine changes in gene expression associated with transgenesis or to correlate changes in phenotype after mutagenesis. Such analyses provide information as to the effect of altered expression of a gene on its direct targets and indirect insight into the existence of an extended regulatory network. Individual and combined omics analyses have been performed on seed coat, endosperm, and embryo during seed development, revealing coordinated changes in lipid and SP biosynthesis and the activation of specific pathways. Comparative omics now allows access into the nature and description of gene regulatory networks that underlie the metabolic pathways in the different tissues of seeds and fruits. Omics technologies will constitute an increasingly important platform to facilitate the application of systems biology to the processes governing seed development. Together, omics technologies will lead to an understanding of how the seeds of diverse species synthesize their characteristic components and will allow new tools to be developed to facilitate the development and selection of new crops.

25.4.4 Enhancing Seed Oil Content by Increasing Seed Size

Above, we have illustrated how omics approaches are leading to an understanding of the regulation of lipid biosynthesis and to the partitioning of carbon to oil reserves. The manipulation of the expression of genes encoding enzymes of FA synthesis or utilization or altering the expression of regulatory genes may result in increases in seed oil content and that, this is frequently accompanied by increases in seed size. Thus, the final seed size can be determined by the amount of reserves accumulated. Alternatively, the size of the embryo sac, which is established during early development and is genetically controlled, can also influence the final seed size and may constitute a determinant of seed filling, leading to an increase of seed reserves by providing a larger 'container' for storage.

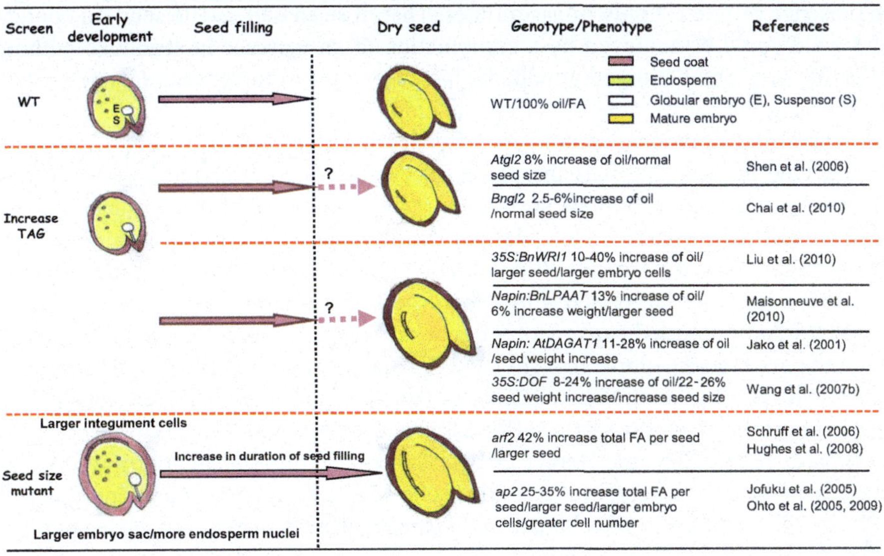

Fig. 25.2 Determinants of seed size and seed mass

25.4.5 *Determinants of Seed Size and Seed Mass*

The seed of flowering plants is composed of three main components: the **embryo** embedded in the nutritive **endosperm** and protected by the **seed coat** (Fig. 25.2). The formation of a viable seed requires the coordinate development of these three compartments from distinct genetic origins. Although embryo and endosperm both result from fertilization, the embryo is diploid with one maternal and one paternal copy of the genome, whereas the endosperm is triploid with an additional maternal copy. The integuments of the ovule will give rise to the seed coat after fertilization, and therefore, they are essentially composed of diploid maternal tissues. Genetic approaches have identified several loci involved in the control of seed size and their recent molecular characterization has helped to unravel some of the developmental relationships between the different seed compartments (Nowack et al. 2010). Since increasing seed size, and possibly as a consequence seed weight, may influence the yield of the crop, the results and perspective arising from the genetic manipulation of seed size in *Arabidopsis* has gained interest among plant breeders. However, if seed filling is limited by the amount of assimilate available for the seeds, an increase of seed weight could only be achieved by a decrease in seed numbers. Such a relation is not strict, since reduced fertility (resulting in a reduced seed number) does not dramatically influence seed size and seed weight in *Arabidopsis* (Alonso-Blanco et al. 1999). Additionally, efforts to improve agronomic traits like seed protein or seed oil content in *Brassica* and soybean have shown that changes in total seed protein levels are often inversely proportional to changes in seed oil levels in presence of fixed supply of assimilates. Therefore an increase in seed oil content and yield could be achieved by a combination of an increase in seed size without affecting seed number together with the increase of the ratio between oil and protein by favoring the partitioning of C towards lipids.

Several developmental processes influence the final size of a seed (Sun et al. 2010). One process involves the parent-of-origin effects. Balanced maternal and paternal genome contributions are a requirement for successful seed development (Jullien and Berger 2010). This parental conflict takes place principally in the endosperm involves DNA methylation (Xiao et al. 2006) and results in the formation of either smaller or larger seeds in *Arabidopsis* (Scott et al. 1998). Indeed, it was proposed that imprinted expression of genes of the polycomb complex (PcG) is an evolutionary conserved mechanism used to balance parental genome contributions in the embryo nourishing tissues (Erilova et al. 2009). However, the relationship between imprinting (for glossary see Box 25.1) and dosage imbalance is more complex than predicted (Jullien and Berger 2010). The control of parental contribution constitutes the basis of heterosis or hybrid vigour. Maize F_1 hybrids and allopolyploids, such as oil seed rape, grow bigger and stronger than their parents. Heterosis and hybrid vigour result from genome-wide changes and interactions between paternal and maternal genomes involving epigenetic regulation established during early seed development and during further vegetative growth, such as the control of circadian clock (Chen 2010). A greater understanding of genes and regulatory mechanisms for polyploids and hybrid vigour will aid the selection of the best combination of parents for the increased production of seeds, fruits, biomass, and metabolites, such as lipids and carbohydrates for the growing demand of food and biofuels.

Box 25.1 Glossary

Genomic imprinting: Epigenetic process that involves methylation and histone modifications in order to achieve a selective monoallelic (monoparental) gene expression without altering the DNA sequence.

Parental conflict: This hypothesis states that the inequality between parental genomes due to imprinting is a result of the differing interests of each parent in terms of the evolutionary fitness of their genes.

Epigenetics: Alteration of gene expression without changes in the DNA sequence which is maintained during cell division.

Polycomb complex (PcG): A group of repressive chromatin proteins that maintain states of gene expression throughout development.

Heterosis or hybrid vigour: The occurrence of a genetically superior offspring from mixing the genes of its parents.

Harvest Index (HI): The ratio of grain yield to biological yield or biomass.

A second mechanism of controlling endosperm development is independent of parent-of-origin. It involves HAIKU (IKU1 and IKU2) and MINISEED3, which are TFs expressed in the endosperm (Garcia et al. 2003; Luo et al. 2005; Wang et al. 2010). These mutations are sporophytic recessive and the mutated plants produce seeds smaller than wild type. A fourth gene *SHORT HYPOCOTYL UNDER BLUE 1 (SHB1)*, initially characterized as a gain-of-function allele increasing seed size (Zhou et al. 2009), has been shown to bind to the promoter of *IKU2* and *MINI3* in order to regulate endosperm development during early seed development and subsequently embryo cell division and elongation. These two processes demonstrate the importance of the development of the nourishing tissues in the final seed size and that, the development of the integuments is controlled by signaling originating in the endosperm.

A third pathway involves genes that act principally in the maternal sporophyte (integuments, nucellus), such as *APETALA2 (AP2)*, *TRANSPARANT TESTA GLABROUS 2 (TTG2)*, and *ARF2* and the cytochrome P450 *KLUH/CYP78A5*. Loss-of-function of *AP2* and *ARF2* increase seed size, whereas mutation in *TTG2 and KLUH/CYP78A5* produces a reverse effect (Adamski et al. 2009; Garcia et al. 2005; Jofuku et al. 2005; Ohto et al. 2005, 2009; Schruff et al. 2006). This third pathway indicates that signals originating in the seed coat can influence endosperm nuclei proliferation and embryo size, and illustrates the existence of reciprocal cross-talk between these three seed compartments.

Whereas heterosis and hybrid vigour are well utilized by plant breeders for seed yield improvement, only a limited number of studies has addressed the potential of storage lipid and protein content in seed size mutants. Comprehensive studies of the quantity and quality of seed reserves have been performed essentially with *ap2* and *arf2* mutant lines, since loss-of-function of these genes results in larger seeds than those of wild type plants (Hughes et al. 2008; Jofuku et al. 2005; Ohto et al. 2005, 2009).

AP2 encodes a TF and was initially characterized as an homeotic gene involved in the specification of flower organ identity (Jofuku et al. 1994). The seeds of the *ap2* mutant are larger than wild type seeds as a result of an increase in the elongation of integument cells and of the division rate of endosperm nuclei and embryo cells (Ohto et al. 2009). Remarkably, the large seed phenotype was accompanied by an increase in seed weight without a compensatory decrease in total seed yield. Furthermore, the increase in seed weight is due in part to an increase in both total protein and oil content without significant qualitative changes in FA composition (Jofuku et al. 2005). These results indicate that AP2 activity affects source-sink relations. Although *AP2* transcripts have been detected at all stages of development in every organ and tissue examined, reciprocal crosses have demonstrated that seed size is predominantly maternally controlled by *AP2*. To explain how *AP2* can maternally influence embryo size and endosperm development, it was proposed that AP2 could activate an integument CW bound invertase, which can regulate sugar levels in the syncytial endosperm (Ohto et al. 2005, 2009). The altered H/S ratio measured in *ap2* seeds during seed development may promote an extended period of cell division, which could account for the increase in cell number of *ap2* embryos. Similarly, in *WRI1* (another member of the AP2/EREP family), a modified sugar metabolism may influence seed size although by acting through different mechanisms (Liu et al. 2010).

ARF2 is a member of a family of TFs, mediating gene expression in response to auxin. *ARF2* is a repressor of cell division and elongation and organ growth. Mutation of *ARF2* results in extra cell divisions in the integuments before fertilization leading to the formation of an enlarged seed cavity and a dramatic increase of both seed size and seed weight (Schruff et al. 2006). As in the case of *ap2* seeds, *arf2* seeds contain more oil than the seeds of wild type plants without major differences in FA composition. The conserved FA composition in a seed size mutant is important for technology transfer to crop plants, such as *B. napus*. The main obstacle for the direct utilization of these seed-size mutants in agronomy is that *ap2* and *arf2* plants display undesirable pleiotropic effects, in particular a reduced fertility, which counteracts the advantage of the enlarged seed phenotype. Although *AP2* and *ARF2* act maternally from the integuments, they are expressed in most plant organs and tissues. To circumvent this problem, Hughes et al. (2008) have supplemented *arf2* plants with a wild type copy only in sepals and petals to restore a fairly normal flower morphology and hence, fertility. The analysis of this transgenic material allows estimating contribution of increased seed size due to a reduced fertility and the contribution of the *arf2* mutation alone. Their results unambiguously demonstrate that *arf2* seeds were larger and contained more oil than those of wild type when seed set was constant. Furthermore, to fully assess the performance of the partially complemented *arf2* plants, the authors have applied the concept of harvest index (HI) to Arabidopsis. HI is defined as the ratio of grain yield to biomass (dry mass of aerial organs). The HI value of the partially complemented *arf2* plants was dramatically increased compared to that of the *arf2* mutant since more seeds were produced; the HI, however, remained lower than that of wild type. One reason was that the restoration of fertility was greatly improved but was not complete. The second reason is that *arf2* mutation causes the over growth of many organs in addition to seeds and as a consequence *arf2* plants partially complemented or not, produce a higher biomass than wild type. However, with the increased demand for biofuels production, the partially complemented *arf2* plants may represent a good compromise for increasing both oil yield and biomass (Hughes et al. 2008).

In conclusion, the seed is complex sink structure. The most important factor that influences seed size in dicots is the initial volume of the seed cavity at early seed development determined by (i) the growth of the integuments, and (ii) the endosperm nuclei proliferation and timing of its cellularization. Once established, the embryo will grow to fill the space and may modify its final size by the control of cell division and elongation as well as the level of storage compounds. The manipulation of the development of this structure by genetic and epigenetic mechanisms combined with the control of the metabolism of storage compounds may constitute a powerful means to increase TAG accumulation in oilseed crops.

25.5 Concluding Remarks

Genetic engineering using omics has contributed substantial progress to the knowledge of plant lipid metabolism. New genes, encoding enzymes responsible for the synthesis of unusual FAs and regulatory factors controlling the maturation as well the size and the mass of the seed, have been characterized. All these data have allowed substantial progress in the modification of the FA composition of oilseeds using genomic engineering. Although transgenic plants producing FAs for industrial purposes have been obtained, they can be considered as renewable sources of feedstocks since in most cases the contents of valuable FAs are not sufficient to compete economically with mineral oil derivatives. Additional progress for improving storage product accumulation and yield should be made. As a first approach, a better knowledge of the biosynthesis pathway regulation should be undertaken. For example, the substrate specificity of the different acyltransacylases could be determined more precisely using new tools such as lipidomics based on MS/MS. As it has been recently demonstrated that a single base mutation in the encoding gene could lead to enhancement of the activity of DAGAT 1 and DAGAT 2 (Köhler et al. 2010; Lardizabal et al. 2008; Snyder et al. 2009; Weselake et al. 2009), a strategy of TILLING seems to be promising and could be also useful for increasing oil content by a modification of promoters regulating the expression of genes encoding for these enzymes. The carbon flow into the seed oil is also a major factor and studies are undertaken to understand the mechanisms, which are responsible for the orientation of this flow to the storage lipid pathway over other reserves. The aim of these studies will be to identify genes controlling the balance between carbohydrate and lipid pathways in order by genetic engineering to enhance the carbon flow towards seed oil synthesis.

The combination of these approaches and manipulation of the seed development by genetic and epigenetic mechanisms may be expected to lead in the coming years to tailor-made oilseeds in which very high amounts of oil may constitute alternative source of energy or industrial precursors or which could also constitute a benefit for human nutrition.

References

Abbadi A, Domergue F, Bauer J, Napier JA, Welti R, Zahringer U, Cirpus P, Heinz E (2004) Biosynthesis of very-long-chain polyunsaturated fatty acids in transgenic oilseeds: constraints on their accumulation. Plant Cell 16:2734–2748

Adamski NM, Anastasiou E, Eriksson S, O'neill CM, Lenhard M (2009) Local maternal control of seed size by KLUH/CYP78A5-dependent growth signaling. Proc Natl Acad Sci U S A 106:20115–20120

Alonso-Blanco C, Blankestijn-de Vries H, Hanhart CJ, Koornneef M (1999) Natural allelic variation at seed size loci in relation to other life history traits of *Arabidopsis thaliana*. Proc Natl Acad Sci U S A 96:4710–4717

Bach L, Michaelson LV, Haslam R, Bellec Y, Gissot L, Marion J, Da Costa M, Boutin JP, Miquel M, Tellier F, Domergue F, Markham JE, Beaudoin F, Napier JA, Faure JD (2008) The very-long-chain hydroxy fatty acyl-CoA dehydratase PASTICCINO2 is essential and limiting for plant development. Proc Natl Acad Sci U S A 105:14727–14731

Bao X, Pollard M, Ohlrogge J (1998) The biosynthesis of erucic acid in developing embryos of *Brassica rapa*. Plant Physiol 118:183–190

Baud S, Mendoza MS, To A, Harscoet E, Lepiniec L, Dubreucq B (2007) WRINKLED1 specifies the regulatory action of LEAFY COTYLEDON2 towards fatty acid metabolism during seed maturation in *Arabidopsis*. Plant J 50:825–838

Berneth R, Frentzen M (1990) Utilization of erucoyl-CoA by acyltransferases from developing seeds of *Brassica napus* (L.) involved in triacylglycerol biosynthesis. Plant Sci 67:21–28

Braybrook SA, Stone SL, Park S, Bui AQ, Le BH, Fischer RL, Goldberg RB, Harada JJ (2006) Genes directly regulated by LEAFY COTYLEDON2 provide insight into the control of embryo maturation and somatic embryogenesis. Proc Natl Acad Sci U S A 103:3468–3473

Broun P, Somerville C (1997) Accumulation of ricinoleic, lesquerolic, and densipolic acids in seeds of transgenic *Arabidopsis* plants that express a fatty acyl hydroxylase cDNA from castor bean. Plant Physiol 113:933–942

Broun P, Gettner S, Somerville C (1999) Genetics engineering of plant lipids. Annu Rev Nutrition 19:197–216

Burgal J, Shockey J, Lu C, Dyer J, Larson T, Graham I, Browse J (2008) Metabolic engineering of hydroxy fatty acid production in plants: RcDGAT2 drives dramatic increases in ricinoleate levels in seed oil. Plant Biotechnol J 6:819–831

Cahoon EB, Ohlrogge JB (1994) Metabolic Evidence for the involvement of a Δ4-palmitoyl-acyl carrier protein desaturase in petroselinic acid synthesis in coriander endosperm and transgenic tobacco cells. Plant Physiol 104:827–837

Cahoon EB, Shanklin J, Ohlrogge JB (1992) Expression of a coriander desaturase results in petroselinic acid production in transgenic tobacco. Proc Natl Acad Sci U S A 89:11184–11188

Cahoon EB, Coughlan SJ, Shanklin J (1997) Characterization of a structurally and functionally diverged acyl-acyl carrier protein desaturase from milkweed seed. Plant Mol Biol 33:1105–1110

Cahoon EB, Shah S, Shanklin J, Browse J (1998) A determinant of substrate specificity predicted from the acyl-acyl carrier protein desaturase of developing cat's claw's seed. Plant Physiol 117:593–598

Cahoon EB, Carlson TJ, Ripp KG, Schweiger BJ, Cook GA, Hall SE, Kinney AJ (1999) Biosynthetic origin of conjugated double bonds: production of fatty acid components of high-value drying oils in transgenic soybean embryos. Proc Natl Acad Sci U S A 96:12935–12940

Cahoon EB, Ripp KG, Hall SE, Kinney AJ (2001) Formation of conjugated Δ8,Δ10-double bonds by Δ12-oleic-acid desaturase-related enzymes. J Biol Chem 276:2637–2643

Cahoon EB, Dietrich CR, Meyer K, Damude HG, Dyer JM, Kinney AJ (2006) Conjugated fatty acids accumulate to high levels in phospholipids of metabolically engineered soybean and Arabidopsis seeds. Phytochemistry 67(12):1166–1176

Cahoon EB, Shockey JM, Dietrich CR, Gidda SK, Mullen RT, Dyer JM (2007) Engineering oilseeds for suitable production of industrial and nutritional feedstocks: solving bottlenecks in fatty acid flux. Curr Opin Plant Biol 10:236–244

Cases S, Stone SJ, Zhou P, Yen E, Tow B, Lardizabal KD, Voelker T, Farese RV Jr (2001) Cloning of DGAT2, a second mammalian diacylglycerol acyltransferase, and related family members. J Biol Chem 276:38870–38876

Cernac A, Benning C (2004) WRINKLED1 encodes an AP2/EREB domain protein involved in the control of storage compound biosynthesis in *Arabidopsis*. Plant J 40:575–585

Chai G, Bai Z, Wei F, King GJ, Wang C, Shi L, Dong C, Chen H, Liu S (2010) *Brassica GLABRA2* genes: analysis of function related to seed oil content and development of functional markers. Theor Appl Genet 120:1597–1610

Chen ZJ (2010) Molecular mechanisms of polyploidy and hybrid vigor. Trends Plant Sci 15:57–71

Chen RM, Ogawa MK, Oe M, Ochia M, Kawashima H (2006) Expression of Δ6,Δ5 desaturase and GLEO elongase genes from *Mortierella alpina* for production of arachidonic acid in soybean (*Glycine max* (L.) *Merrill*) seeds. Plant Sci 170:399–406

Chen M, Mooney BP, Hajduch M, Joshi T, Zhou M, Xu D, Thelen JJ (2009) System analysis of an *Arabidopsis* mutant altered in de novo fatty acid synthesis reveals diverse changes in seed composition and metabolism. Plant Physiol 150:27–41

Cheng B, Wu G, Vrinten P, Falk K, Bauer J, Qiu X (2010) Towards the production of high levels of eicosapentaenoic acid in transgenic plants: the effects of different host species, genes and promoters. Transgenic Res 19:221–229

Chiron H, Delseny M, Roscoe T (2007) Transcriptional regulation of lipid biosynthesis in crucifer seeds. In: Benning C, Ohlrogge J (eds) Current advances in the biochemistry and cell biology of plant lipids. Aardvark Global, Salt Lake City, pp 253–260

Dahlqvist A, Ståhl U, Lenman M, Banas A, Lee M, Sandager L, Ronne H, Stymne S (2000) Phospholipid:diacylglycerol acyltransferase: an enzyme that catalyzes the acyl-CoA-independent formation of triacylglycerol in yeast and plants. Proc Natl Acad Sci U S A 97:6487–6492

Damude HG, Kinney AJ (2007) Engineering oilseed plants for a sustainable, land-based source of long chain polyunsaturated fatty acids. Lipids 42:179–185

Damude HG, Kinney AJ (2008a) Engineering oilseeds to produce nutritional fatty acids. Physiol Plant 132:1–10

Damude HG, Kinney AJ (2008b) Enhancing plant seed oils for human nutrition. Plant Physiol 147:962–968

Damude HG, Zhang H, Farrall L, Ripp KG, Tomb JF, Hollerbach D, Yadav NS (2006) Identification of bifunctional delta12/omega3 fatty acid desaturases for improving the ratio of omega3 to omega6 fatty acids in microbes and plants. Proc Natl Acad Sci U S A 103:9446–9451

Dehesh K, Edwards P, Hayes T, Cranmer AM, Fillatti J (1996a) Two novel thioesterases are key determinants of the bimodal distribution of acyl chain length of *Cuphea palustris* seed oil. Plant Physiol 110:203–210

Dehesh K, Jones A, Knutzon DS, Voelker TA (1996b) Production of high levels of 8:0 and 10:0 fatty acids in transgenic canola by overexpression of Ch FatB2, a thioesterase cDNA from *Cuphea hookeriana*. Plant J 9:167–172

Domergue F, Chevalier S, Créach A, Cassagne C, Lessire R (2000) Purification of the acyl-CoA elongase complex from developing rapeseed and characterization of the 3-ketoacyl-CoA synthase and the 3-hydroxyacyl-CoA dehydratase. Lipids 35:487–494

Domergue F, Abbadi A, Ott C, Zank TK, Zähringer U, Heinz E (2003) Acyl carriers used as substrates by the desaturases and elongases involved in very long-chain polyunsaturated fatty acids biosynthesis reconstituted in yeast. J Biol Chem 278:35115–35126

Domergue F, Abbadi A, Heinz E (2005a) Relief for fish stocks: oceanic fatty acids in transgenic oilseeds. Trends Plant Sci 10:112–116

Domergue F, Abbadi A, Zähringer U, Moreau H, Heinz E (2005b) *In vivo* characterization of the first acyl-CoA Δ6-desaturase from a member of the plant kingdom, the microalga *Ostreococcus tauri*. Biochem J 389:483–490

Drexler H, Spiekermann P, Meyer A, Domergue F, Zank T, Sperling P, Abbadi A, Heinz E (2003) Metabolic engineering of fatty acids for breeding of new oilseed crops: strategies, problems and first results. J Plant Physiol 160:779–802

Dyer JM, Stymne S, Green AG, Carlsson AS (2008) High-value oils from plants. Plant J 54:640–655

Eccleston VS, Ohlrogge JB (1998) Expression of lauroyl-acyl carrier protein thioesterase in *Brassica napus* seeds induces pathways for both fatty acid oxidation and biosynthesis and Implies a set point for triacylglycerol accumulation. Plant Cell 10:613–622

Erilova A, Brownfield L, Exner V, Rosa M, Twell D, Mittelsten Scheid O, Hennig L, Köhler C (2009) Imprinting of the polycomb group gene MEDEA serves as a ploidy sensor in *Arabidopsis*. PLoS Genet 5:e1000663

Focks N, Benning C (1998) Wrinkled 1: a novel, low seed-soil-mutant *Arabidopsis* with a deficiency in the seed-specific regulation of carbohydrate metabolism. Plant Physiol 118:91–101

Garcia D, Saingery V, Chambrier P, Mayer U, Jürgens G, Berger F (2003) *Arabidopsis* haiku mutants reveal new controls of seed size by endosperm. Plant Physiol 131:1661–1670

Garcia D, Fitz Gerald JN, Berger F (2005) Maternal control of integument cell elongation and zygotic control of endosperm growth are coordinated to determine seed size in *Arabidopsis*. Plant Cell 17:52–60

Hoffmann M, WagnerM, Abbadi A, Fulda M, Feussner I (2008) Metabolic engineering of3-very long chain polyunsaturated fatty acid production by an exclusively acyl-CoA-dependent pathway. J Biol Chem 283:22352–22362

Hughes R, Spielman M, Schruff MC, Larson TR, Graham IA, Scott RJ (2008) Yield assessment of integument-led seed growth following targeted repair of auxin response factor 2. Plant Biotechnol J 6:758–769

Hunter SC, Cahoon EB (2007) Enhancing vitamin E in oilseeds: unraveling tocopherol and tocotrienol biosynthesis. Lipids 42:97–108

Jadhav A, Katavic V, Marillia EF, Michael Giblin E, Barton DL, Kumar A, Sonntag C, Babic V, Keller WA, Taylor DC (2005) Increased levels of erucic acid in *Brassica carinata* by co-suppression and antisense repression of the endogenous *FAD2* gene. Metab Eng 7:215–220

Jain RK, Coffey M, Lai K, Kumar A, MacKenzie SL (2000) Enhancement of seed oil content by expression of glycerol-3-phosphate acyltransferase genes. Biochem Soc Trans 28:958–961

Jako C, Kumar A, Wei Y, Zou J, Barton DL, Giblin EM, Covello PS, Taylor DC (2001) Seed-specific over-expression of an *Arabidopsis* cDNA encoding a diacylglycerol acyltransferase enhances seed oil content and seed weight. Plant Physiol 126:861–874

Jofuku KD, den Boer BG, Van Montagu M, Okamuro JK (1994) Control of *Arabidopsis* flower and seed development by the homeotic gene APETALA2. Plant Cell 6:1211–1225

Jofuku KD, Omidyar PK, Gee Z, Okamuro JK (2005) Control of seed mass and seed yield by the floral homeotic gene APETALA2. Proc Natl Acad Sci U S A 102:3117–3122

Jullien PE, Berger F (2010) Parental genome dosage imbalance deregulates imprinting in *Arabidopsis*. PLoS Genet 6:e1000885

Kajikawa M, Yamato KT, Kohzu Y, Nojiri M, Sakuradani E, Shimizu S, Sakai Y, Fukuzawa H, Ohyama K (2004) Isolation and characterization of delta(6)-desaturase, an ELO-like enzyme and delta(5)-desaturase from the liverwort Marchantia polymorpha and production of arachidonic and eicosapentaenoic acids in the methylotrophic yeast Pichia pastoris. Plant Mol Biol 54(3):335-352

Kajikawa M, Matsui K, Ochiai M, Tanaka Y, Kita Y, Ishimoto M, Kohzu Y, Shoji S, Yamato KT, Ohyama K, Fukuzawa H, Kohchi T (2008) Production of arachidonic and eicosapentaenoic acids in plants using bryophyte fatty acid 6-desaturase, 6-elongase, and 5-desaturase genes. Biosci Biotechnol Biochem 72:435–444

Kinney AJ (1996) Designer oils for better nutrition. Nat Biotechnol 14:946–946

Kinney AJ, Cahoon EB, Damude HG, Hitz WD, Kolar CW, Liu ZB (2004) Production of very long chain polyunsaturated fatty acids in oilseed plants. E.I. Dupont De Nemours and Company, WO 2004/071467 A2. International Patent Application published under the Patent Cooperation Treaty, WIPO, Geneva, Switzerland

Köhler C, Mittelsten Scheid O, Erilova A (2010) The impact of the triploid block on the origin and evolution of polyploid plants. Trends Genet 26:142–148

Lardizabal KD, Metz JG, Sakamoto T, Hutton WC, Pollard MR, Lassner MW (2000) Purification of a jojoba embryo wax synthase, cloning of its cDNA, and production of high levels of wax in seeds of transgenic *Arabidopsis*. Plant Physiol 122:645–656

Lardizabal K, Effertz R, Levering C, Mai J, Pedroso MC, Jury T, Aasen E, Gruys K, Bennett K (2008) Expression of *Umbelopsis ramanniana* DGAT2A in seed increases oil in soybean. Plant Physiol 148:89–96

Larson TR, Edgell T, Byrne J, Dehesh K, Graham IA (2002) Acyl CoA profiles of transgenic plants that accumulate medium-chain fatty acids indicate inefficient storage lipid synthesis in developing oilseeds. Plant J 32:519–527

Lassner MW, Levering CK, Davies HM, Knutzon DS (1995) Lysophosphatidic acid acyltransferase from Meadowfoam mediates insertion of erucic acid at the sn-2 position of triacylglycerol in transgenic rapeseed oil. Plant Physiol 109:1389–1394

Liu J, Hua W, Zhan G, Wei F, Wang X, Liu G, Wang H (2010) Increasing seed mass and oil content in transgenic *Arabidopsis* by the overexpression of *wri1*-like gene from *Brassica napus*. Plant Physiol Biochem 48:9–15

Lotan T, Ohto M, Yee KM, West MA, Lo R, Kwong RW, Yamagishi K, Fischer RL, Goldberg RB, Harada JJ (1998) Arabidopsis LEAFY COTYLEDON1 is sufficient to induce embryo development in vegetative cells. Cell 93(7):1195-1205

Lu C, Fulda M, Wallis JG, Browse J (2006) A high-throughput screen for genes from castor bean that boost hydroxy fatty acid accumulation in seed oils of transgenic *Arabidopsis*. Plant J 45:847–856

Luo M, Dennis ES, Berger F, Peacock WJ, Chaudhury A (2005) *MINISEED3* (*MINI3*), a *WRKY* family gene, and *HAIKU2* (*IKU2*), a leucine-rich repeat (*LRR*) *KINASE* gene, are regulators of seed size in *Arabidopsis*. Proc Natl Acad Sci U S A 102:17531–17536

Maeo K, Tokuda T, Ayame A, Mitsui N, Kawai T, Tsukagoshi H, Ishiguro S, Nakamura K (2009) An AP2-type transcription factor, WRINKLED1, of *Arabidopsis thaliana* binds to the AW-box sequence conserved among proximal upstream regions of genes involved in fatty acid synthesis. Plant J 60:476–487

Maisonneuve S, Bessoule JJ, Lessire R, Delseny M, Roscoe TJ (2010) Expression of rapeseed microsomal lysophosphatidic acid acyltransferase isozymes enhances seed oil content in *Arabidopsis*. Plant Physiol 152:670–684

Marillia EF, Micallef BJ, Micallef M, Weninger A, Pedersen KK, Zou J, Taylor DC (2003) Biochemical and physiological studies of *Arabidopsis thaliana* transgenic lines with repressed expression of the mitochondrial pyruvate dehydrogenase kinase. J Exp Bot 54:259–270

Metz JG, Pollard MR, Anderson L, Hayes TR, Lassner MW (2000) Purification of a Jojoba embryo fatty acyl-coenzyme a reductase and expression of Its cDNA in high erucic acid rapeseed. Plant Physiol 122:635–644

Metz JG, Roessler P, Facciotti D, Levering C, Dittrich F, Lassner M, Valentine R, Lardizabal K, Domergue F, Yamada A, Yazawa K, Knauf V, Browse J (2001) Production of polyunsaturated fatty acids by polyketide synthases in both prokaryotes and eukaryotes. Science 293:290–293

Meyer A, Kirsch H, Domergue F, Abbadi A, Sperling P, Bauer J, Cirpus P, Zank TK, Moreau H, Roscoe TJ, Zähringer U, Heinz E (2004) Novel fatty acid elongases and their use for the reconstitution of docosahexaenoic acid biosynthesis. J Lipid Res 45:1899–1909

Mietkiewska E, Giblin EM, Wang S, Barton DL, Dirpaul J, Brost JM, Katavic V, Taylor DC (2004) Seed-Specific heterologous expression of a nasturtium *FAE* gene in *Arabidopsis* results in a dramatic Increase in the proportion of erucic acid. Plant Physiol 136:2665–2675

Moire LE, Rezzonico E, Goepfert S, Poirier Y (2004) Impact of unusual fatty acid synthesis on futile cycling through beta-oxydation and on gene expression in transgenic plants. Plant Physiol 134:432–442

Mu J, Tan H, Zheng Q, Fu F, Liang Y, Zhang J, Yang X, Wang T, Chong K, Wang XJ, Zuo J (2008) *LEAFY COTYLEDON1* is a key regulator of fatty acid biosynthesis in *Arabidopsis*. Plant Physiol 148:1042–1054

Napier JA (2007) The Production of unusual fatty acids in transgenic plants. Annu Rev Plant Biol 58:295–319

Napier JA, Graham IA (2010) Tailoring plant lipid composition: designer oilseeds come of age. Curr Opin Plant Biol 13:330–337

Napier JA, Beaudoin F, Michaelson LV, Sayanova O (2004) The production of long chain polyunsaturated fatty acids in transgenic plants by reverse-engineering. Biochimie 86:785–792

Nath UK, Wilmer JA, Wallington EJ, Becker HC, Möllers C (2009) Increasing erucic acid content through combination of endogenous low polyunsaturated fatty acids alleles with Ld-LPAAT+Bn-fae1 transgenes in rapeseed (*Brassica napus* L.). Theo Appl Genet 118:765–773

Nguyen HT, Mishra G, Whittle E, Bevan SA, Merlo AO, Walsh TA, Shanklin J (2010) Metabolic engineering of seeds can achieve levels of omega-7 fatty acids comparable with the highest levels found in natural plant sources. Plant Physiol 154:1897–1904

Nikolau BJ, Perera MA, Brachova L, Shanks B (2008) Platform biochemicals for a biorenewable chemical industry. Plant J 54:536–545

Nowack MK, Ungru A, Bjerkan KN, Grini PE, Schnittger A (2010) Reproductive cross-talk: seed development in flowering plants. Biochem Soc Trans 38:604–612

Ohlrogge JB (1994) Design of new plant products: engineering of fatty acid metabolism. Plant Physiol 104:821–826

Ohlrogge JB, Jaworski JG (1997) Regulation of fatty acid synthesis. Annu Rev Plant Physiol Plant Mol Biol 48:109–136

Ohlrogge J, Allen D, Berguson B, DellaPenna D, Shachar-Hill Y, Stymne S (2009) Driving on biomass. Science 324:1019–1020

Ohto MA, Fischer RL, Goldberg RB, Nakamura K, Harada JJ (2005) Control of seed mass by *APETALA2*. Proc Natl Acad Sci U S A 102:3123–3128

Ohto MA, Floyd SK, Fischer RL, Goldberg RB, Harada JJ (2009) Effects of APETALA2 on embryo, endosperm, and seed coat development determine seed size in *Arabidopsis*. Sex Plant Reprod 22:277–289

Petrie JR, Shrestha P, Liu Q, Mansour M, Wood CC, Zhou XR, Nichols PD, Green AG, Singh SP (2010a) Rapid expression of transgenes driven by seed-specific constructs in leaf tissue: DHA production. Plant Methods 6:8

Petrie JR, Shrestha P, Mansour MP, Nichols PD, Liu Q, Singh SP (2010b) Metabolic engineering of omega-3 long-chain polyunsaturated fatty acids in plants using an acyl-CoA delta6-desaturase with omega3-preference from the marine microalga *Micromonas pusilla*. Metab Eng 12:233–240

Pollard MR, Anderson L, Fan C, Hawkins DJ, Davies HM (1991) A specific acyl-ACP thioesterase implicated in medium-chain fatty acid production in immature cotyledons of *Umbellularia californica*. Arch Biochem Biophys 284:306–312

Puyaubert J, Garbay B, Costaglioli P, Dieryck W, Roscoe TJ, Renard M, Cassagne C, Lessire R (2001) Acyl-CoA elongase expression during seed development in *Brassica napus*. Biochim Biophys Acta-Mol Cell Biol Lipids 1533:141–152

Qi B, Fraser T, Mugford S, Dobson G, Sayanova O, Butler J, Napier JA, Stobart AK, Lazarus CM (2004) Production of very long chain polyunsaturated omega-3 and omega-6 fatty acids in plants. Nat Biotechnol 22:739–745

Robert SS, Singh SP, Zhou XR, Petrie JR, Blackburn SI, Mansour PM, Nichols PD, Liu Q, Green AG (2005) Metabolic engineering of *Arabidopsis* to produce nutritionally important DHA in seed oil. Funct Plant Biol 32:473–479

Roesler K, Shintani D, Savage L, Boddupalli S, Ohlrogge J (1997) Targeting of the Arabidopsis homomeric acetyl-coenzyme A carboxylase to plastids of rapeseeds. Plant Physiol 11375–81

Ruuska SA, Girke T, Benning C, Ohlrogge JB (2002) Contrapuntal networks of gene expression during *Arabidopsis* seed filling. Plant Cell 14:1191–1206

Schruff MC, Spielman M, Tiwari S, Adams S, Fenby N, Scott RJ (2006) The AUXIN RESPONSE FACTOR 2 gene of *Arabidopsis* links auxin signalling, cell division, and the size of seeds and other organs. Development 133:251–261

Scott RJ, Spielman M, Bailey J, Dickinson HG (1998) Parent-of-origin effects on seed development in *Arabidopsis thaliana*. Development 125:3329–3341

Sharma N, Anderson M, Kumar A, Zhang Y, Giblin EM, Abrams SR, Zaharia LI, Taylor DC, Fobert PR (2008) Transgenic increases in seed oil content are associated with the differential expression of novel *Brassica*-specific transcripts. BMC Genomics 9:619–636

Shen B, Sinkevicius KW, Selinger DA, Tarczynski MC (2006) The homeobox gene *GLABRA2* affects seed oil content in *Arabidopsis*. Plant Mol Biol 60:377–387

Shen B, Allen WB, Zheng P, Li C, Glassman K, Ranch J, Nubel D, Tarczynski MC (2010) Expression of *ZmLEC1* and *ZmWRI1* increases seed oil production in maize. Plant Physiol 153:980–987

Shi L, Yu Y, Katavic V, Haughn GW, Kunst L (2010) Investigation of potential roles of phospholipase D zêta in Arabidopsis thaliana seed oil accumulation. 19th International Symposium on Plant Lipids, Cairns, Poster 18, 11–16 July 2010

Shockey JM, Gidda SK, Chapital DC, Kuan JC, Dhanoa PK, Bland JM, Rothstein SJ, Mullen RT, Dyer JM (2006) Tung tree DGAT1 and DGAT2 have nonredundant functions in triacylglycerol biosynthesis and are localized to different subdomains of the endoplasmic reticulum. Plant Cell 18:2294–2313

Simopoulos AP (2002) Omega-3 fatty acids in inflammation and autoimmune diseases. J Am Coll Nutr 21:495–505

Smith MA, Moon H, Chowira G, Kunst L (2003) Heterologous expression of a fatty acid hydroxylase gene in developing seeds of *Arabidopsis thaliana*. Planta 217:507–516

Snyder CL, Yurchenko OP, Siloto RM, Chen X, Liu Q, Mietkiewska E, Weselake RJ (2009) Acyltransferase action in the modification of seed oil biosynthesis. New Biotechnol 26:11–16

Somerville CR, Bonetta D (2001) Plants as factories for technical materials. Plant Physiol 125:168–171

Sun X, Shantharaj D, Kang X, Ni M (2010) Transcriptional and hormonal signaling control of *Arabidopsis* seed development. Curr Opin Plant Biol 13:611–620

Suzuki M, Ketterling MG, Li QB, McCarty DR (2003) Viviparous1 alters global gene expression patterns through regulation of abscisic acid signaling. Plant Physiol 132:1664–1677

Takeyama H, Takeda D, Yazawa K, Yamada A, Matsunaga T (1997) Expression of the eicosapentaenoic acid synthesis gene cluster from *Shewanella* sp. in a transgenic marine cyanobacterium, *Synechococcus* sp. Microbiology 143:2725–2731

Tapiero H, Ba GN, Couvreur P, Tew KD (2002) Polyunsaturated fatty acids (PUFA) and eicosanoids in human health and pathologies. Biomed Pharmacoth 56:215–222

Taylor DC, Katavic V, Zou JT, MacKenzie SL, Keller WA, An J, Friesen W, Barton DL, Pedersen K, Giblin EM, Ge Y, Dauk M, Sonntag C, Luciw T, Males D (2001) Field testing of transgenic rapeseed cv.Hero transformed with a yeast *sn-2* acyltransferase results in increased oil content, erucic acid content and seed yield. Mol Breed 8:317–322

Thelen JJ, Ohlrogge JB (2002a) Both antisense and sense expression of biotin carboxyl carrier protein isoform 2 inactivates the plastid acetyl-coenzyme A carboxylase in *Arabidopsis thaliana*. Plant J 32:419–431

Thelen JJ, Ohlrogge JB (2002b) Metabolic engineering of fatty acid biosynthesis in plants. Metab Eng 4:12–21

van de Loo FJ, Broun P, Turner S, Somerville C (1995) An oleate 12-hydroxylase from *Ricinus communis* L. is a fatty acyl desaturase homolog. Proc Natl Acad Sci U S A 92:6743–6747

Venegas-Calerón M, Sayanova O, Napier JA (2010) An alternative to fish oils: Metabolic engineering of oil-seed crops to produce omega-3 long chain polyunsaturated fatty acids. Prog Lipid Res 49:108–119

Vicente-Carbajosa J, Moose SP, Parsons RL, Schmidt RJ (1997) A maize zinc-finger protein binds the prolamin box in zein gene promoters interacts with the basic leucine zipper transcriptional activator Opaque2. Proc Natl Acad Sci U S A 94:7685–7690

Vigeolas H, Waldeck P, Zank T, Geigenberger P (2007) Increasing seed oil content in oil-seed rape (*Brassica napus* L.) by over-expression of a yeast glycerol-3-phosphate dehydrogenase under the control of a seed-specific promoter. Plant Biotechnol J 5:431–441

Voelker T (1996) Plant acyl-ACP thioesterases: chain-length determining enzymes in plant fatty acid biosynthesis. Genet Eng 18:111–133

Voelker T, Kinney AJ (2001) Variations in the biosynthesis of seed-storage lipids. Annu Rev Plant Physiol Plant Mol Biol 52:335–361

Voelker TA, Worrell AC, Anderson L, Bleibaum J, Fan C, Hawkins DJ, Radke SE, Davies HM (1992) Fatty acid biosynthesis redirected to medium chains in transgenic oilseed plants. Science 257:72–74

Wallis JG, Browse J (1999) The Delta8-Desaturase of *Euglena gracilis*: an alternate pathway for synthesis of 20-carbon polyunsaturated fatty acids. Arch Biochem Biophys 365:307–316

Wang H, Guo J, Lambert KN, Lin Y (2007) Developmental control of *Arabidopsis* seed oil biosynthesis. Planta 226:773–783

Wang A, Garcia D, Zhang H, Feng K, Chaudhury A, Berger F, Peacock WJ, Dennis ES, Luo M (2010) The VQ motif protein IKU1 regulates endosperm growth and seed size in Arabidopsis. Plant J 63:670–679

Weselake RJ, Shah S, Tang M, Quant PA, Snyder CL, Furukawa-Stoffer TL, Zhu W, Taylor DC, Zou J, Kumar A, Hall L, Laroche A, Rakow G, Raney P, Moloney MM, Harwood JL (2008) Metabolic control analysis is helpful for informed genetic manipulation of oilseed rape (*Brassica napus*) to increase seed oil content. J Exp Bot 59:3543–3549

Weselake RJ, Taylor DC, Rahman MH, Shah S, Laroche A, McVetty PB, Harwood JL (2009) Increasing the flow of carbon into seed oil. Biotechnol Adv 27:866–878

Wu G, Truksa M, Datla N, Vrinten P, Bauer J, Zank T, Cirpus P, Heinz E, Qiu X (2005) Stepwise engineering to produce high yields of very long-chain polyunsaturated fatty acids in plants. Nat Biotechnol 23:1013–1017

Xiao W, Brown RC, Lemmon BE, Harada JJ, Goldberg RB, Fischer RL (2006) Regulation of seed size by hypomethylation of maternal and paternal genomes. Plant Physiol 142:1160–1168

Zou J, Katavic V, Giblin EM, Barton DL, MacKenzi SL, Keller WA, Hu X, Taylor DC (1997) Modification of seed oil content and acyl composition in the brassicaceae by expression of a yeast sn-2 acyltransferase gene. Plant Cell 9:909–923

Zhou Y, Zhang X, Kang X, Zhao X, Zhang X, Ni M (2009) SHORT HYPOCOTYL UNDER BLUE1 associates with MINISEED3 and HAIKU2 promoters *in vivo* to regulate *Arabidopsis* seed development. Plant Cell 21:106–117

Chapter 26
Integrating Omics in Food Quality and Safety Assessment

Howard Davies and Louise Shepherd

Abstract A concise overview is provided on the use, and potential for further development of omics approaches in crop improvement, food safety assessment, and in nutrigenomics research with specific emphasis on crop plant-based dietary components.

Keywords Food · Human health · Nutrigenomics · Omics technology · Safety · Seed

26.1 Introduction

26.1.1 The Rationale for Safe, High Quality, and Nutritious Products from Seeds

Agricultural production and food delivery chains are facing difficult and challenging times as evidenced by major fluctuations in the prices of major world seed commodities, such as wheat, rice, and maize, and the resulting political and social ramifications. Between the years 2007 and 2008, the price of rice grain increased 3 fold, and wheat and maize more than doubled (FAO 2010a). FAO's (2011) latest cereal production forecast stands at 2302 million tonnes. At this revised level, world cereal production would now be slightly below overall utilization, leading to a further decline in world stocks. Whilst international cereal prices subsequently declined after 2008, reflecting ample global cereal supplies, in 2009/10 food prices in most low-income food-deficit countries remained above the precrisis level of early 2008, negatively affecting access to food by vulnerable populations. More recently, the international market prices of wheat and maize have again doubled from the low experienced in 2009.It is not surprising therefore that food security has moved up the political agenda as we face burgeoning population growth coupled with increases in fuel and fertilizer prices and crop losses in certain parts of the world due to extreme climatic conditions (FAO 2010b). This, together with pressures to farm

H. Davies (✉) · L. Shepherd
James Hutton Institute, Invergowrie, DD2 5DA, Dundee, UK
e-mail: howard.davies@hutton.ac.uk

G. K. Agrawal, R. Rakwal (eds.), *Seed Development: OMICS Technologies toward Improvement of Seed Quality and Crop Yield,*
DOI 10.1007/978-94-007-4749-4_26,

sustainably whilst protecting biodiversity and nonrenewable natural resources will require a significant rethink of how we value and maintain food supplies and how we reduce food wastage. Research and development initiatives to improve agricultural productivity and the delivery of next generation crops and production systems fit for purpose are important drivers of a successful food security platform, but must complement a broader range of scientific as well as political issues. A sobering thought is that the United Nations has predicted that farms will need to produce 70 % more food over the next 40 years.

Improving yields, and even maintaining yield, under conditions where climate change and extreme environmental events will have major adverse impacts through biotic and abiotic stresses, are now major crop improvement targets. Water and nutrient efficiencies are high on the agenda but the continued demand for high quality, safe, and more nutritious plant-based foods means that plant breeders face a particularly challenging time. However, due to on-going advances in molecular genetics and associated technologies we are living through paradigm shifts in the potential for advanced, knowledge-based plant breeding. Discovery and isolation of important genes, and analysis of their functions underpins this potential, where all omics approaches have a role to play in the meeting this challenge (Langridge and Fleury 2011).

26.2 Linking Omics to the Improvement of Seed Quality and Nutritional Value

There are many articles, which cover the principles of omics technologies, their strengths and weaknesses (see for example, Davies 2009; HESI 2010, and references therein). Omics technologies provide coverage of gene, protein, and metabolite analysis that is unsurpassed compared with traditional targeted approaches. Indeed, a growing number of examples indicate that profiling approaches can be used to expose significant sources of variation, e.g., in the composition of plants caused by genetic background, breeding method, growing environment (site, season), genotype-environment interactions, and crop cultural practices. The variability in composition of harvested seed and grain is unsurprising given the plasticity of the plant genome and diverse environmental influences on development (Harrigan et al. 2010). Whilst breeders have long been aware of such variation from tried and tested targeted analytical approaches, the broad-scale, so called 'unbiased' analysis offered by omics provides new insights into our understanding of the true extent of variation in biological expression parameters relevant to human and animal health and nutrition.

Metabolomics, as a case study, is helping to provide targets for plant breeding by linking gene expression and allelic variation to variation in metabolite complement (functional genomics), and is also being deployed to better assess the potential impacts of climate change and reduced input agricultural systems on crop composition. Skogerson et al. (2010) used metabolomics to reveal genotype-related diversity in maize with mean values of identified metabolites covering the range of 1.5–93 folds. Similarly, Anttonen et al. (2010) revealed significant effects of genotype, location, and season on the maize proteome, independent of genotype. The use of omics in evaluating biological variation in crops is well covered by articles published from an ILSI (HESI) workshop (HESI 2010).

Major advancements in crop improvement programs will derive from a systems-based understanding of biological processes driving yield, seed quality, and nutritional value. The components of a successful strategy include:

1. Available genome sequences, EST, and full-length enriched cDNA libraries (for glossary see Box 26.1). Genome sequences are available for a growing number of crop species (Mochida and Shinozaki 2010).
2. High-throughput genotyping, cloning, transformation, and gene tagging approaches (forward and reverse genetics).
3. High-throughput/broad scale omics (transcriptomic, proteomic, and metabolomic) platforms to integrate gene structure with function.
4. Extensive biological resources (e.g., mapping and mutation populations) required to associate gene and allelic diversity with omics and trait data.
5. High-throughput phenotyping platforms.
6. Strong bioinformatics and other computational biology capabilities.

Box 26.1 Glossary

AFLP (Amplified Fragment Length Polymorphism): A PCR based tool used in molecular biology research to detect polymorphisms in DNA. Genomic DNA is digested using restriction enzymes, and a subset of DNA fragments amplified using PCR.

cDNA Library: A pool of cloned cDNA (complementary DNA) fragments cloned into host cells.

EST (Expressed Sequence Tag): A short sub-sequence of a cDNA sequence used in the identification of genes being expressed in a particular tissue.

DNA Marker: A gene or DNA sequence with a known physical location on the chromosome, and whose pattern of inheritance can be followed. Used to trace or identify specific regions of a gene on a chromosome.

Biomarker: A biological measurement that facilitates the prediction or reality of specific parameters, e.g. resistance or susceptibility to disease, agronomic performance, system resilience.

Interactome: The whole set of molecular interactions in cells.

QTL (Quantitative Trait Locus): A region of DNA, associated with a particular phenotypic trait, that can be located on one or more chromosomes.

SNP (Single Nucleotide Polymorphism): Variations that occur as a result of a single nucleotide base (A, T, C or G) change in a DNA sequence, and result in different alleles.

SSR (Simple Sequence Repeat): SSRs, also known as microsatellites or short-tandem repeats (STRs), are repeating sequences of 1–6 base pairs of DNA, and can be used as markers.

The review of Mochida and Shinozaki (2010) provides comprehensive and contemporary information on the inter-relationships between omics and their potential deployment in crop and seed improvement.

26.3 Omics and Food Safety Assessment

A fuller evaluation of the compositional variation of raw agricultural commodities and downstream products will emerge through the development of comparative metabolomics databases that can be expanded and modified by the international community. This information can be used to benchmark any measured differences between a particular crop and the extent of 'acceptable' variation within the framework of a history of safe use of the crop species in question. There is an on-going debate over the potential value of much broader scale, more unbiased approaches in evaluating the safety of food. In recent years, the major controversy over the safety of plant-derived foods has undoubtedly focused on GM crops, leading to scientific and political tensions (Davies 2009). This has stimulated the development of risk assessment procedures for GM crops at European and international levels. Central to these procedures are comparative analyses of the GM seeds and crops with appropriate non GM comparators that have a history of safe use. Whilst there is currently no indication of a breach of safety concerning commercially grown GM crops, arguments have been developed for the use of omics approaches within the risk assessment process. The objective would be to broaden the comparative analysis and thereby reduce the level of any uncertainties. This has stimulated a growing number of publications on the subject many of which are summarised in Table 26.1 on a plant/crop specific basis. One of the most comprehensive uses of the omics technologies in the analysis of GM crops comes from Barros et al. (2010), who used a transcriptomics, proteomics, and metabolomics to analyze three maize lines (a control and two transgenic lines modified with two different genes). All plants were grown at a single location over three years, allowing assessment of the variation caused by year of harvest and genotype. All three omics analyses clearly differentiated seed grown over the three growing seasons for all samples, indicating once again that environment has a strong effect on gene, protein, and metabolite expression. Indeed, the data generated showed that the growing season had a stronger overall effect on the transcriptome, proteome, and metabolome of the three

Table 26.1 Omics technologies in addressing food quality and safety

Seed	Technology	Trait	Gene/GM Event	Reference
Arabidopsis	Transcriptomics	Abiotic stress tolerance	*ABF3* (transcription factor)	Abdeen et al. (2010)
	Proteomics	Over-expression of β-glucuronidase & kanamycin	*gus* & *nptII*	Reubelt et al. (2006)
		Increased homogentistic acid	*hppd* (*p*-hydroxyphenylpyruvate dioxygenase)	
		Increased tocopherol	*gtmt* (γ-tocopherolmethyltransferase)	
	Transcriptomics	Selectable marker protein phosphinothricin acetyl transferase (PAT)	*Bar*	Ren et al. (2009b)
	Metabolomics		*Bar*	Ren et al. (2009a)
	Metabolomics	Increased dhurrin production	*CYP79A1*, *CYP71E1* & *UGT85B1*	Kristensen et al. (2005)
Maize	Transcriptomics Proteomics Metabolomics	Insect resistant (Bt), Glyphosate tolerance (Roundup Ready)	MON810: *Cry1Ab & NK603: cp4 epsps (5-enolpyruvylshikimate-3-phosphate)*	Barros et al. (2010); van Dijk et al. (2010)
	Proteomics	Insect resistant (Bt)	MON810: *Cry1Ab*	Coll et al. (2010)
	Proteomics	Insect resistant (Bt)	MON810: *Cry1Ab*	Batista and Oliveira (2010)
	Proteomics	Insect resistant (Bt)	MON810: *Cry1Ab*	Zolla et al. (2008)
	Metabolomics	Insect resistant	MON810: *Cry1Ab*	Piccioni et al. (2009)
	Metabolomics	Insect resistant	MON810: *Cry1Ab*	Leon et al. (2009)
	Metabolomics	Insect resistant	MON810: *Cry1Ab*	Levandi et al. (2008)
	Metabolomics	Rpd3-type histone deacetylase	*ZmRpd3-101*	Castro and Manetti (2007)
	Metabonomics	Rpd3-type histone deacetylase	*ZmRpd3-101* & MON810: *Cry1Ab*	Manetti et al. (2004)
	Metabolomics	Insect resistant	*Cry34Ab12 + Cry35Ab1*	Herman et al. (2007)
		Insect resistant	*Cry 1F*	Herman et al. (2004)
	Metabolomics	Insect resistant	MON810: *Cry1Ab*	Rodríguez-Nogales et al. (2007)
	Metabolomics	Insect resistant	MON810: *Cry1Ab*	Manetti et al. (2006)
Wheat	Transcriptomics Proteomics	High glutenin	LMW-GS	Scossa et al. (2008)
	Transcriptomics	High phytase	*phy*-A	Gregersen et al. (2005)
		High glutenin	*Glu-1Ax*	Baudo et al. (2006)

Table 26.1 (continued)

Seed	Technology	Trait	Gene/GM Event	Reference
		High glutenin	HMW-*Dx5* & *Glu-1Ax*	Baker et al. (2006)
		High glutenin	HMW-*Dx5* & *Glu-1Ax*	Shewry et al. (2007)
	Metabolomics	High glutenin	HMW-*Dx5*	Stamova et al. (2009)
Barley	Transcriptomics	Fungal resistance	(1,3-1,4)-β-glucanase (gluB)	Kogel et al. (2010)
	Metabolomics	Improved nutritional value	Endochitinase (ChGP)	
Rice	Proteomics	Production of recombinant human protein	hGM-CSF	Luo et al. (2009)
	Transcriptomics	Antibody production	ScFVT84.66	Batista et al. (2008)
	Transcriptomics Metabolomics	Increased tryptophan production	OASA1D (Anthranilate synthase α-subunit)	Dubouzet et al. (2007)
	Metabolomics	Rice blast resistance	RCH10, RAC22, β-1,3-Glu & B-RIP	Jiao et al. (2010)
		Sheath blight resistance	RC24 & β-1,3-glu	
	Metabolomics	Insect resistant	*Cry1Ac*	Zhou et al. (2009)
		Lepidoptera resistance	*sck*	
	Metabolomics	Increased tryptophan production	OASA1D (Anthranilate synthase α-subunit)	Ishihara et al. (2007)
	Metabolomics	Increased tryptophan production	OASA1D (Anthranilate synthase α-subunit)	Wakasa et al. (2006)
	Metabolomics	Herbicide resistance	*bar*	Oberdoerfer et al. (2005)
Oil Seed Rape	Transcriptomics	Increased oil content	Diacylglycerol acyltransferase (DGAT)	Sharma et al. (2008)
	Transcriptomics	Herbicide tolerant	Oxy235, MS8xRF3, MS1XRF1, MS1XRF2, HCN92 & T45	Schmidt et al. (2008)
		Glyphosate tolerant (Roundup Ready)	GT73	
Soybean	Metabolomics	Glyphosate tolerance (Roundup Ready)	*cp4 epsps* (5-enolpyruvylshikimate-3-phosphate)	García et al. (2009)
		Insect resistance	*cry1Ac*	Berman et al. (2009)
		Glyphosate tolerance (Roundup Ready)	*cp4 epsps* (5-enolpyruvylshikimate-3-phosphate)	García-Villalba et al. (2008)
Pea	Proteomics	Insect resistance	*α*AI1 (*α-amylase* inhibitor)	Chen et al. (2009a)
	Proteomics	Insect resistance	*α*AI1 (*α-amylase* inhibitor)	Islam et al. (2009)

maize genotypes than the genetic modification. Similar results have been reported by Baudo et al. (2006) and Batista et al. (2008) for wheat and rice, respectively.

To date, omics technologies have not been formally applied to the food safety assessment of transgenic crops but this does not preclude their inclusion in the future, despite some strong arguments as to why they should not be used (Chassy 2010). The requirements and current status of omics applied to food safety assessment have been previously discussed (Davies 2009; HESI 2010; ILSI 2004, 2008).

26.4 Omics and Biomarkers

26.4.1 Plant Breeding

Biomarkers are essentially qualitative or quantitative agents measurements made on biological systems that can predict the properties of such systems before specific features become evident. Thus, in molecular plant breeding DNA markers (such as SSRs, AFLPs, and SNPs; for glossary see Box 26.1) can be used to predict the inheritance of traits without the need for phenotyping. However, such technologies are successful in cases where monogenetic traits or dominant genes predominate. For polygenetic traits, polyploidy, and epistatically or environmentally influenced traits, such approaches are far more limited in their success. It has already been stated that omics approaches can be used to assess the sources and extents of compositional variability in crop plants and that a combination of omics approaches can add value to plant improvement programs by identifying the genes and genetic control mechanisms regulating phytochemical composition. The use of metabolomics as a biomarker platform in plant breeding is not yet a reality but indications are promising. For example, Meyer et al. (2007) assessed the effect of metabolic composition on the biomass of recombinant inbred lines of *Arabidopsis* under controlled conditions. The data reveal a strong relationship between the expression of plant phenotypes and metabolic composition. Schauer et al. (2006) correlated, in a set of introgression lines, metabolite QTLs from tomato fruit derived from a commercial × exotic cross to phenotypic traits such as yield, harvest index, seed number, and total soluble solids content of tomato. The study confirmed known correlations, such as those between sugars, organic acids, and total soluble solids, establishing the applicability of this approach and its utility for identifying novel relationships.

Steinfath et al. (2010) have taken this approach one step further in demonstrating that established metabolomics techniques can be used to predict agronomically important quality phenotypes independent of environmental variables. This may be particularly useful where genome sequences are not available for the crop in question or when marker-assisted breeding becomes limited. Initial trials indicate that a prediction model could be used for a wide variety of genotypes. Ideally, a specific biomarker would relate to a single, highly predictive metabolite, but metabolite fingerprinting (an overall profile/pattern of metabolites/proteins such as that derived *via* MS or NMR spectroscopy) would also be valuable even if the underpinning components could not be identified. Steinfath et al. (2010) consider that a successful biomarker will most likely comprise a range of compounds between two and 100, but would most likely have a predictability of less than 100 %. When databases are developed using metabolite profiles from breeding populations and cultivars, it would be also possible to screen for novel phenotypes in existing datasets once those relationships are identified. Therefore, biomarker discovery for new and improved traits in plant breeding has real promise and a combination of metabolic and genetic markers may prove to be extremely powerful combination in cases of more complex traits (Gartner et al. 2009).

26.4.2 Human Health

It is well accepted that specific plant metabolites (e.g., certain polyphenols, vitamins, isoflavones, carotenoids, and terpenoids) have significant impacts on human health (Hall et al. 2008). This has stimulated the science of nutrigenomics, which aims to comprehensively understand the response of the body to dietary and food factors through various omics technologies. Ultimately, through the development of appropriate suites of human gene, protein, and metabolite biomarkers linked to food intake, it may be possible to tailor the diet of individuals with known or predicted genetic susceptibility to certain diseases (Davis and Hord 2005; Ovensá et al. 2008; Rezzi et al. 2007).

The Human Metabolome Database (HMDB; http://www.hmdb.ca/; accessed on July 13th 2011) is a freely available electronic database housing detailed information on small molecule metabolites found in the human body. The database currently contains over 7,900 metabolite entries, including both water-soluble and lipid-soluble metabolites, and ~7,200 proteins (and DNA) sequences are linked to the metabolite entries. Four additional databases contain equivalent information on ~1,500 drugs, 2,900 common toxins and environmental pollutants, pathway diagrams for 350 human metabolic and disease pathways. Importantly, the FooDB website (http://foodbs.org/; accessed on July 13th 2011) contains equivalent information on ~2,000 food components and food additives. This will be an increasingly important tool for applications in biomarker development linking food and health as food contains not only staple ingredients, but also a multitude of less prevalent components representing potentially physiologically relevant bioactives (Long et al. 2006). Examples of the application of omics technologies in nutrition and phytomedical research are provided by Zhang et al. (2007), Shyur and Yang (2008), Kato (2008), Kussmann et al. (2008). However, studies are not confined to humans with Ametaj et al. (2010) using metabolomics to show that feeding large amounts of barley grain to dairy cattle resulted in major increases in a number of potentially harmful rumen metabolites. Their study suggests that metabolomics could be used as a reliable tool to study the interactions between nutrition and health in dairy cows.

Omics approaches can also be used to assess adverse as well as positive effects as evidenced by links between conventional toxicological analysis and gene, protein, and metabolite expression. The expectations raised by toxicogenomics approaches are earlier and more sensitive detection of toxicity. Furthermore, toxicogenomics will provide a better understanding of the mechanism of toxicity and may facilitate the prediction of toxicity of unknown compounds found in foods (Stierum et al. 2005). This might include, for example, safety issues surrounding the plant material used in botanical dietary supplements, which could arise from misidentification of a botanical, substitution of a different species, and use of incorrect parts of a plant.

26.5 Concluding Remarks

Omics approaches offer unprecedented opportunities to unravel the complexities of the genetic regulation of plant form, compositional and function, bridging the boundaries between DNA sequence variation, gene expression, translation into functional proteins, and metabolite fluxes. The technologies continue to advance allowing deeper and more rapid coverage, which in turn requires smarter computational and biological approaches to establish the interactive networks, giving rise to the phenotype and phenotypic plasticity. In the area of crop improvement, and given the investment to date, the challenge now is to exploit the continuing advances in genomics with 'post genomics' approaches to deliver a stream of crop improvement stories based on the most advanced plant breeding approaches, which can encompass many types of omics data. As this article has illustrated, the uses of omics approaches are multifactorial and not only related to plant genetic improvement, but to many issues where knowledge can be extracted from large datasets encompassing gene, protein, and metabolite networks. Successful biomarkers, capable of associating plant components with agronomic performance, specific health benefits or safety risks may well be delivered by only one of the omics technologies. However, a more systems based understanding of the "interactome" is the bigger challenge with arguably the biggest rewards at the end.

References

Abdeen A, Schnell J, Miki B (2010) Transcriptome analysis reveals absence of unintended effects in drought-tolerant transgenic plants overexpressing the transcription factor *ABF3*. BMC Genomics 11:69–80

Ametaj BN, Zebeli Q, Saleem F, Psychogios N, Lewis MJ, Dunn SM, Xia J, Wishart DS (2010) Metabolomics reveals unhealthy alterations in rumen metabolism with increased proportion of cereal grain in the diet of dairy cows. Metabolomics 6:583–594

Anttonen MJ, Lehesranta S, Auriola S, Röhlig RM, Engel KH, Kärenlampi SO (2010) Genetic and environmental influence on maize kernel proteome. J Proteome Res 9:6160–6168

Baker JM, Hawkins ND, Ward JL, Lovegrove A, Napier JA, Shewry PR, Beale MH (2006) A metabolomic study of substantial equivalence of field-grown genetically modified wheat. Plant Biotech J 4:381–392

Barros E, Lezar S, Anttonen MJ, Van Dijk JP, Röhlig RM, Kok EJ, Engel KH (2010) Comparison of two GM maize varieties with a near-isogenic non-GM variety using transcriptomics, proteomics and metabolomics. Plant Biotechnol J 8:436–451

Batista R, Oliveira M (2010) Plant natural variability may affect safety assessment data. Regul Toxicol Pharmacol 58:S8–S12

Batista R, Saibo N, Lourenço T, Oliveira MM (2008) Microarray analyses reveal that plant mutagenesis may induce more transcriptomic changes than transgene insertion. Proc Natl Acad Sci U S A 105:3640–3645

Baudo MM, Lyons R, Powers S, Pastori GM, Edwards KJ, Holdsworth MJ, Shewry PR (2006) Trangenesis has less impact on the transcriptome of wheat grain than conventional breeding. Plant Biotechnol J 4:369–380

Berman KH, Harrigan GG, Riordan SG, Nemeth MA, Hanson C, Smith M, Sorbet R, Zhu E, Ridley WP (2009) Compositions of seed, forage, and processed fractions of insect-protected soybean MON 87701 are equivalent to those of conventional soybean. J Agric Food Chem 57:11360–11369

Castro C, Manetti M (2007) A multiway approach to analyze metabonomic data: a study of maize seeds development. Anal Biochem 371:194–200

Chassy BM (2010) Can omics inform a food safety assessment? Regul Toxicol Pharmacol 58:S62–S70

Chen H, Bodulovic G, Hall PJ, Moore A, Higgins TJV, Djordjevis MA, Rolfe BG (2009a) Unintended changes in protein expression revealed by proteomic analysis of seeds from transgenic pea expressing a bean α-amylase inhibitor gene. Proteomics 9:4406–4415

Coll A, Nadal A, Collado R, Capellades G, Kubista M, Messeguer J, Pla M (2010) Natural variation explains most transcriptional changes among maize plants of MON810 and comparable non-GM varieties subjected to two N-fertilization farming practices. Plant Mol Biol 73:349–362

Davies HV (2009) A role for "omics" technologies in food safety assessment. Food Control 21:1601–1611

Davis CD, Hord NG (2005) Nutritional 'omics' technologies for elucidating the role(s) of bioactive food components in colon cancer prevention. J Nutr 135:2694–2697

Dubouzet JG, Ishihara A, Matsuda F, Miyagawa H, Iwata H, Wakasa K (2007) Integrated metabolomic and transcriptomic analyses of high-tryptophan rice expressing a mutant anthranilate synthase alpha subunit. J Exp Bot 58:3309–3321

FAO (2010a) Crop prospects and food situation. No. 4, December 2010

FAO (2010b) The state of Food Insecurity in the world. ISBN 978–92-5–106610-2

FAO (2011) Crop prospects and food situation. No. 2, June 2011

García MC, García B, García-Ruiz C, Gómez A, Cifuentes A, Marina ML (2009) Rapid characterisation of (glyphosate tolerant) transgenic and non-transgenic soybeans using chromatographic protein profiles. Food Chem 113:1212–1217

García-Villalba R, León C, Dinelli G, Segura-Carretero A, Fernández-Gutiérrez A, Garcia-Cañas V, Cifuentes A (2008) Comparative metabolomic study of transgenic versus conventional soybean using capillary electrophoresis-time-of-flight spectrometry. J Chromatogr A 1195:164–173

Gartner T, Steinfath M, Andorf S, Lisec J, Meyer RC, Altmann T, Willmitzer L, Selbig J (2009) Improved heterosis prediction by combining information on DNA and metabolic markers. PLoS ONE 4:e5220

Gregersen PL, Brich-Pedersen H, Holm PB (2005) A microarray-based comparative analysis of gene expression profiles during grain development in transgenic and wild type wheat. Transgenic Res 14:887–905

Hall RD, Brouwer ID, Fitzgerald MA (2008) Plant metabolomics and its potential application for human nutrition. Physiol Plant 132:162–175

Harrigan GG, Glenn KC, Ridley WP (2010) Assessing the natural variability in crop composition. Regul Toxicol Pharmacol 58:S13–S20

Herman RA, Phillips AM, Collins RA, Tagliani LA, Clauseen FA, Graham CD, Bickers BL, Harris TA, Prochaska LM (2004) Compositional equivalency of Cry1F corn event TC6275 and conventional corn (*Zea mays* L.). J Agric Food Chem 52:2726–2734

Herman RA, Storer NP, Phillips AM, Prochaska LM, Windles P (2007) Compositional assessment of event DAS-59122-7 maize using substantial equivalence. Regul Toxicol Pharmacol 47:37–47

HESI (2010) HESI workshop on evaluating biological variation in non-transgenic crops. Regul Toxicol Pharmacol 58 (special supplement)

ILSI (2004) Nutritional and safety assessments of foods and feeds nutritionally improved through biotechnology. Compr Rev Food Sci Food Saf 3:35–104

ILSI (2008) Recent developments in the safety and nutritional assessment of nutritionally improved foods and feeds. Comp Rev in Food Sci Food Saf 7:50–113

Ishihara A, Matsuda F, Miyagawa H, Wakasa K (2007) Metabolomics for metabolically manipulated plants: effects of tryptophan overproduction. Metabolomics 3:319–334

Islam N, Campbell PM, Higgins TJV, Hirano H, Akhurst RJ (2009) Transgenic peas expressing an alpha-amylase inhibitor gene from beans show altered expression and modification of endogenous proteins. Electrophoresis 30:1863–1868

Jiao Z, Si XX, Li GK, Zhang ZM, Xu XP (2010) Unintended compositional changes in transgenic rice seeds (*Oryza sativa* L.) studied by spectral and chromatographic analysis coupled with chemometrics methods. J Agric Food Chem 58:1746–1754

Kato Kauffman KJ, Prakash P, Edwards JS (2003) Advances in flux balance analysis. Curr Opin Biotechnol 14:491–496

Kato H (2008) Nutrigenomics: the cutting edge and Asian perspectives. Asia Pac J Clin Nutr 17:S12–S15.

Kogel KH, Voll LM, Schäfer P, Jansen C, Wu Y, Langen G, Imani J, Hofmann J, Schmiedl A, Sonnewald S, von Wettstein D, Cook RJ, Sonnewald U (2010) Transcriptome and metabolome profiling of field-grown transgenic barley lack induced differences but show cultivar-specific variances. Proc Nat Acad Sci U S A 107:6198–6203

Kristensen C, Morant M, Olsen CE, Ekstrøm CT, Galbraith DW, Møller BL, Bak S (2005) Metabolic engineering of dhurrin in transgenic *Arabidopsis* plants with marginal inadvertent effects of the metabolome and transcriptome. Proc Natl Acad Sci U S A 102:1779–1784

Kussmann M, Rezzi S, Daniel H (2008) Profiling techniques in nutrition and health research. Curr Opin Biotech 19:83–99

Langridge P, Fleury D (2011) Making the most of 'omics' for crop breeding. Trends Biotechnol 29:33–40

Leon C, Rodriguez-Meizoso I, Lucio M, Garcia-Cañas V, Ibañez E, Schmitt-Kopplin P, Cifuentes A (2009) Metabolomics of transgenic maize combining Fourier transform-ion cyclotron resonance-mass spectrometry, capillary electrophoresis-mass spectrometry and pressurized liquid extraction. J Chromatogr A 1216:7314–7323

Levandi T, Leon C, Kaljurand M, Garcia-Canas V, Cifuentes A (2008) Capillary electrophoresis time-of-flight mass spectrometry for comparative metabolomics of transgenic versus conventional maize. Anal Chem 80:6329–6335

Long M, Millar DJ, Kimura Y, Donovan G, Rees J, Fraser PD, Bramley PM, Bolwell GP (2006) Metabolite profiling of carotenoid and phenolic pathways in mutant and transgenic lines of tomato: identification of a high antioxidant fruit line. Phytochemistry 67:1750–1757

Luo J, Ning T, Sun Y, Zhu J, Zhu Y, Lin Q, Yang D (2009) Proteomic analysis of rice endosperm cells in response to expression of hGM-CSF. J Proteome Res 8:829–837

Manetti C, Bianchetti C, Bizzarri M, Casciani L, Castro C, D'Ascenzo G, Delfini M, Di Cocco ME, Laganà A, Miccheli A, Motto M, Conti F (2004) NMR-based metabonomic study of transgenic maize. Phytochemistry 65:3187–3198

Manetti C, Bianchetti C, Casciani L, Castro C, Di Cocco ME, Miccheli A, Motto M, Conti F (2006) A metabonomic study of transgenic maize (*Zea mays*) seeds revealed variations is osmolytes and branched amino acids. J Exp Bot 57:2613–2625

Meyer RC, Steinfath M, Lisec J, Becher M, Witucka-Wall H, Törjek O, Fiehn O, Eckardt Ä, Willmitzer L, Selbig J, Altmann T (2007) The metabolic signature related to high plant growth rate in *Arabidopsis thaliana*. Proc Natl Acad Sci U S A 104:4759–4764

Mochida K, Shinozaki K (2010) Genomics and bioinformatics resources for crop improvement. Plant Cell Physiol 51:497–523

Oberdoerfer RB, Shillito RD, De Beuckeleer M, Mitten DH (2005) Rice (*Oryza sativa* L.) containing the *bar* gene is compositionally equivalent to the nontransgenic counterpart. J Agric Food Chem 53:1457–1465

Ovesná J, Slabý O, Toussaint O, Kodícek M, Marsík P, Pouchová V, Vaněk T (2008) High throughput 'omics' approaches to assess the effects of phytochemicals in human health studies. Brit J Nutr 99:ES127–ES134

Piccioni F, Capitani D, Zolla L, Mannina L (2009) NMR metabolic profiling of transgenic maize with the *Cry1A(b)* gene. J Agric Food Chem 57:6041–6049

Ren Y, Lv J, Wang T, Li L, Peng Y, Qu LJ (2009a) A comparative proteomics approach to detect unintended effects in transgenic *Arabidopsis*. J Genet Genomics 36:629–639

Ren Y, Wang T, Peng Y, Xia B, Qu LJ (2009b) Distinguishing transgenic from non-transgenic *Arabidopsis* plants by 1H NMR-based metabolic fingerprinting. J Genet Genomics 36:621–628

Reubelt MC, Lipp M, Reynolds TL, Schmuke JJ, Astwood JD, DellaPenna D, Engel KH, Jany KD (2006) Application of two-dimensional gel electrophoresis to interrogate alterations in the proteome of genetically modified crops. 3. Assessing unintended effects. J Agric Food Chem 54:2169–2177

Rezzi S, Ramadan Z, Fay LB, Kochhar S (2007) Nutritional metabonomics: applications and perspectives. J Proteome Res 6:513–525

Rodríguez-Nogales JM, Cifuentes A, García MC, Marina, ML (2007) Characterization of protein fractions from Bt-transgenic and non-transgenic maize varieties using perfusion and monolithis RP-HPLC. Maize differentiation by multivariate analysis. J Agric Food Chem 55:3835–3842

Schauer N, Semel Y, Roessner U, Gur A, Balbo I, Carrari F, Pleban T, Perez-Melis A, Bruedigam C, Kopka J, Willmitzer L, Zamir D, Fernie AR (2006) Comprehensive metabolic profiling and phenotyping of interspecific introgression lines for tomato improvement. Nat Biotechnol 24:447–454

Schmidt AM, Sahota R, Pope DS, Lawrence TS, Belton MP, Rott ME (2008) Detection of genetically modified canola using multiplex PCR coupled with oligonucleotide microarray hybridization. J Agric Food Chem 56:6791–6800

Scossa F, Laudencia-Chinqcuanco D, Anderson OD, Vensel WH, Lafiandra D, D'Ovidio R, Masci S (2008) Comparative proteomic and transcriptomic profiling of a bread wheat cultivar and its derived transgenic lines overexpressing a low molecular weight glutenin subunit gene in the endosperm. Proteomics 8:2948–2966

Sharma N, Anderson M, Kumar A, Zhang Y, Giblin EM, Abrams SR, Zaharia LI, Taylor DC, Fobert PR (2008) Transgenic increases in seed oil content are associated with the differential expression of novel *Brassica*-specific transcripts. BMC Genomics 9:619–636

Shewry PR, Baudo M, Lovegrove A, Powers S, Napier JA, Ward JL, Baker JM, Beale MH (2007) Are GM and conventionally bred cereals really different? Trends Food Sci Technol 18:201–209

Shyur LF, Yang NS (2008) Metabolomics for phytomedicine research and drug development. Curr Opin Chem Biol 12:66–71

Skogerson K, Harrigan GG, Reynolds TL, Halls SC, Ruebelt M, Iandolino A, Pandravada A, Glenn KC, Fiehn O (2010) Impact of genetics and environment on the metabolite composition of maize grain. J Agric Food Chem 58:3600–3610

Stamova BS, Roessner U, Suren S, Laudencia-Chingcuanco D, Bacic A, Beckles DM (2009) Metabolic profiling of transgenic wheat over-expressing the high molecular-weight Dx5 glutenin subunit. Metabolomics 5:239–252

Steinfath M, Strehmel N, Peters R, Schauer N, Groth D, Hummel J, Steup M, Selbig J, Kopka J, Geigenberger P, van Dongen JT (2010) Discovering plant metabolic biomarkers for phenotype prediction using an untargeted approach. Plant Biotechnol J 8:900–911

Stierum R, Heijne W, Kienhuis A, van Ommen B, Groten J (2005) Toxicogenomics concepts and applications to study hepatic effects of food additives and chemicals. Toxicol Appl Pharmacol 207:S179–S188

van Dijk JP, Leifert C, Barros E, Kok EJ (2010) Gene expression profiling for food safety assessment: examples in potato and maize. Regul Toxicol Pharmacol 58:S21–S25

Wakasa K, Hasegawa H, Nemoto H, Matsuda F, Miyazawa H, Tozawa Y, Morino K, Komatsu A, Yamada T, Terakawa T, Miyagawa H (2006) High-level tryptophan accumulation in seeds of transgenic rice and its limited effects on agronomic traits and seed metabolite profile. J Exp Bot 57:3069–3078

Zhang X, Yap Y, Wei D, Chen G, Chen F (2007) Novel omics technologies in nutrition research. Biotechnol Adv 26:169–176

Zhou J, Ma C, Xu H, Yuan K, Lu X, Zhu Z, Wu Y, Xu G (2009) Metabolic profiling of transgenic rice with *cryIAc* and *sck* genes: an evaluation of unintended effects at metabolic level by using GC-FID and GC-MS. J Chromatogr B 877:725–732

Zolla L, Rinalducci S, Antonioli P, Righetti PG (2008) Proteomics as a complementary tool for identifying unintended side effects occurring in transgenic maize seeds as a result of genetic modifications. J Proteome Res 7:1850–1861

Index

G. K. Agrawal, R. Rakwal (eds.), *Seed Development: OMICS Technologies toward Improvement of Seed Quality and Crop Yield*,
DOI 10.1007/978-94-007-4749-4, © Springer Science+Business Media Dordrecht 2012

T

V

SPRINGER NATURE

GPSR Compliance

The European Union's (EU) General Product Safety Regulation (GPSR) is a set of rules that requires consumer products to be safe and our obligations to ensure this.

If you have any concerns about our products, you can contact us on ProductSafety@springernature.com

In case Publisher is established outside the EU, the EU authorized representative is:

Springer Nature Customer Service Center GmbH
Europaplatz 3
69115 Heidelberg, Germany

The manufacturer's authorised representative in the EU is Springer Nature Customer Service Centre GmbH, Europaplatz 3, 69115 Heidelberg, Germany. If you have any concerns regarding our products, please contact ProductSafety@springernature.com

Printed and bound by CPI Group (UK) Ltd, Croydon, CR0 4YY
30/09/2026
02225501-0019